A Physicist's Guide to Mathematica®

SECOND EDITION

Patrick T. Tam
Department of Physics and Astronomy
Humboldt State University
Arcata, California

AMSTERDAM • BOSTON • HEIDELBERG • LONDON
NEW YORK • OXFORD • PARIS • SAN DIEGO
SAN FRANCISCO • SINGAPORE • SYDNEY • TOKYO

Academic Press is an imprint of Elsevier

Academic Press is an imprint of Elsevier
30 Corporate Drive, Suite 400, Burlington, MA 01803, USA
525 B Street, Suite 1900, San Diego, California 92101-4495, USA
84 Theobald's Road, London WC1X 8RR, UK

Library of Congress Cataloging-in-Publication Data
Tam, Patrick.
A physicist's guide to Mathematica/Patrick T. Tam.
p. cm.
Includes bibliographical references and index.
ISBN 978-0-12-683192-4 (pbk. : alk. paper) 1. Physics–Data processing. 2. Mathematica (Computer file) I. Title.
QC20.7.E4T36 2008
530.150285–dc22

2008044787

British Library Cataloguing-in-Publication Data
A catalogue record for this book is available from the British Library.

ISBN: 978-0-12-683192-4

For information on all Academic Press publications visit our Web site at *www.elsevierdirect.com*

Printed in the United States of America
09 10 11 9 8 7 6 5 4 3 2 1

A Physicist's Guide to Mathematica®

SECOND EDITION

To

P.T.N.H. Jiyu-Kennett, Shunryu Suzuki

He Tin and May Yin Tam

Sandra, Teresa

Harriette, Frances

Contents

Preface to the Second Edition

Eleven years have elapsed since the publication of the first edition of this book in 1997. Then *Mathematica* 3.0 had less than 1200 built-in functions and other objects; now *Mathematica* 6.0, a major upgrade, has over 2200 of them. Also, *Mathematica* 6.0 features innovations such as real-time update of dynamic output, interface for interactive parameter manipulation, interactive graphics drawing and editing, load-on-demand curated data, and syntax coloring. Eleven years ago, *Mathematica* was well-known for its steep learning curve; the curve is no longer steep as we can now learn *Mathematica* from established courses and reader-friendly books rather than from only the definitive but formidable and encyclopedic reference, The *Mathematica* Book [Wol03].

The second edition of this book is compatible with *Mathematica* 6.0 and introduces a number of its new and best features. This new edition expands the material covered in many sections of the first edition; it includes new sections on data analysis, interactive graphics drawing, and interactive graphics manipulation; and it has a 146% increase in the number of end-of-section exercises and end-of-chapter problems. A compact disc accompanies the book and contains all of its *Mathematica* input and output. An online Instructor's Solutions Manual is available to qualified adopters of the text.

I am deeply grateful to Mervin Hanson (Humboldt State University) for being my friend, partner, and mentor from the beginning of our *Mathematica* journey. Even in his retirement, he labored over my manuscript. Without him, this book would not exist. I am much indebted to Zenaida Uy (Millersville University) whose friendship, advice, encouragement, and help sustained me during the preparation of the manuscript over eight years. Bill Titus (Carleton College) and Anthony Behof (DePaul University) deserve my heartfelt gratitude as their constructive criticisms and insightful suggestions for the manuscript were invaluable. Appreciation is due to my students for their thoughtful and helpful testing of the manuscript in class and to many readers of the first edition of the book for their valuable feedback. I wish to thank Leroy Perkins (Shasta College) for editing this preface even when there were numerous demands on his time and attention.

My special appreciation goes to William Golden (Humboldt State University). Teaching *Mathematica* with him has been a joy and an enriching experience. I am thankful to Robert Zoellner (Humboldt State University) and members of the chemistry and physics departments for their support of my *Mathematica* ventures. For their guidance, assistance, and patience in the development, production and marketing of this book, I wish to express my gratitude to Lauren Schultz Yuhasz, Gavin Becker, and Philip Bugeau at Elsevier. I would like to acknowledge Wolfram Research, Inc. for granting me permission to include *Mathematica* usage statements and help messages in this book. I am most grateful to Rev. Masters Haryo Young and Eko Little as well as the community of the Order of Buddhist Contemplatives for being my sangha refuge and to Drs. Leo Leer, Timothy Pentecost, and Nathan Shishido for maintaining and improving my health.

My deepest gratitude belongs to Sandra, my wife, for her collaboration and understanding during the writing and production of this book. Remaining calm and nurturing while living with the author is a testimony of her love and fortitude.

For corrections and updates, please visit the author's webpage at www.humboldt.edu/~ptt1/APGTM_Updates.html, or locate the book's webpage at http://elsevierdirect.com/companions/9780126831924 and then click the the update link. If you encounter difficulties with or have questions about any inputs and outputs in the book, inspect them—with *Mathematica* 6—in the notebooks on the accompanying compact disc. If the issues are not resolved, send the inputs to the kernel and examine the outputs. Offerings of comments, suggestions, and bug reports are gratefully accepted at Patrick.Tam@humboldt.edu.

Patrick T. Tam

Preface to the First Edition

Traditionally, the upper-division theoretical physics courses teach the formalisms of the theories, the analytical technique of problem-solving, and the physical interpretation of the mathematical solutions. Problems of historical significance, pedagogical value, or if possible, recent research interest are chosen as examples. The analytical methods consist mainly of working with models, making approximations, and considering special or limiting cases. The student must master the analytical skills, because they can be used to solve many problems in physics and, even in cases where solutions cannot be found, can be used to extract a great deal of information about the problems. As the computer has become readily available, these courses should also emphasize computational skills, since they are necessary for solving many important, real, or "fun" problems in physics. The student ought to use the computer to complement and reinforce the analytical skills with the computational skills in problem-solving and, whenever possible, use the computer to visualize the results and observe the effects of varying the parameters of the problem in order to develop a greater intuitive understanding of the underlying physics.

The pendulum in classical mechanics serves as an example to elucidate these ideas. The plane pendulum is used as a model. It consists of a particle under the action of gravity and constrained to move in a vertical circle by a massless rigid rod. For small angular deviations, the equation of motion can be linearized and solved easily. For finite angular oscillations, the motion is nonlinear. Yet it can still be studied analytically in terms of the energy integral and the phase diagram. The period of motion is expressed in terms of an elliptic integral. The integral can be expanded in a power series, and for small angular oscillations the expansion converges rapidly. However, numerical methods and computer programming are necessary for determining the motion of a damped, driven pendulum. The student can use the computer to explore and simulate the motion of the pendulum with different sets of values for the parameters in order to gain a deeper intuitive understanding of the chaotic dynamics of the pendulum.

Normally, physics juniors and seniors have taken a course in a low-level language such as FORTRAN or Pascal and possibly also a course in numerical analysis. Nevertheless, attempts to introduce numerical methods and computer programming into the upper-division theoretical physics courses have been largely unsuccessful. Mastering the symbols and syntactic rules of these low-level languages is straightforward; but programming with them requires too many lines of complicated and convoluted code in order to solve interesting problems. Consequently, rather than enhancing the student's problem-solving skills and physical intuition, it merely adds a frustrating and ultimately nonproductive burden to the student already struggling in a crowded curriculum.

Mathematica, a system developed recently for doing mathematics by computer, promises to empower the student to solve a wide range of problems including those that are important, real, or "fun," and to provide an environment for the student to develop intuition and a deeper understanding of physics. In addition to numerical calculations, *Mathematica* performs symbolic as well as graphical calculations and animates two- and three-dimensional graphics. The numerical capabilities broaden the problem-solving skills of the student; the symbolic capabilities relieve the student from the tedium and errors of "busy" or long-winded derivations; the graphical capabilities and the capabilities for "instant replay" with various parameter values for the problem enable the student to deepen his or her intuitive understanding of physics. These astounding interactive capabilities are sufficiently powerful for handling most problems and are surprisingly easy to learn and use. For complex and demanding problems, *Mathematica* also features a high-level programming language that can make use of more than a thousand built-in functions and that embraces many programming styles such as functional, rule-based, and procedural programming. Furthermore, to provide an integrated technical computing environment, the Macintosh and Windows versions for *Mathematica* support documents called "notebooks." A notebook is a "live textbook." It is a file containing ordinary text, *Mathematica* input and output, and graphics. *Mathematica*, together with the user-friendly Macintosh and Windows interfaces, is likely to revolutionize not only *how* but also *what* we teach in the upper-division theoretical physics courses.

Purpose

The primary purpose of this book is to teach upper-division and graduate physics students as well as professional physicists how to master *Mathematica*, using examples and approaches that are motivating to them. This book does not replace Stephen Wolfram's ***Mathematica*: A System for Doing Mathematics by Computer** [Wol91] for *Mathematica* version 2 or **The *Mathematica* Book** [Wol96] for version 3. The encyclopedic nature of these excellent references is formidable, indeed overwhelming, for novices. My guidebook prepares the reader for easy access to Wolfram's indispensable references. My book also shows that *Mathematica* can be a powerful and wonderful tool for learning, teaching, and doing physics.

Uses

This book can serve as the text for an upper-division course on *Mathematica* for physics majors. Augmented with chemistry examples, it can also be the text for a course on *Mathematica* for chemistry majors. (For the last several years, a colleague in the chemistry department and I have team-taught a *Mathematica* course for both chemistry and physics majors.) Part I, "*Mathematica* with Physics," provides sufficient material for a two-unit, one-semester course. A three-unit, one-semester course can cover Part I, sample Part II, "Physics with *Mathematica*," require a polished *Mathematica* notebook from each student reporting a project, and include supplementary material on introductory numerical analysis discussed in many texts (see [KM90], [DeV94], [Gar94], and [Pat94]). Exposure to numerical analysis allows the student to appreciate the limitations (i.e., the accuracy and stability) of numerical algorithms and understand the differences between numerical and symbolic functions, for example, between `NSolve` and `Solve`, `NIntegrate` and `Integrate`, as well as `NDSolve` and `DSolve`. Experience suggests that a three-hour-per-week laboratory is essential to the success of both the two- and three-unit courses. For the degree requirement, either course is an appropriate addition to, if not replacement for, the existing course in a low-level language such as C, Pascal, or FORTRAN.

If a course on *Mathematica* is not an option, a workshop merits consideration. A two-day workshop can cover Chapter 1, "The First Encounter," and Chapter 2, "Interactive Use of *Mathematica*," and a one-week workshop can also include Chapter 3, "Programming in *Mathematica*." Of course, further digestion of the material may be necessary after one of these accelerated workshops.

For students who are *Mathematica* neophytes, this book can also be a supplemental text for upper-division theoretical physics courses on mechanics, electricity and magnetism, and quantum physics. For *Mathematica* to enrich rather than encroach upon the curriculum, it must be introduced and integrated into these courses gradually and patiently throughout the junior and senior years, beginning with the interactive capabilities. While the interactive capabilities of *Mathematica* are quite impressive, in order to realize its full power the student must grasp its structure and master it as a programming language. Be forewarned that learning these advanced features as part of the regular courses, while possible, is difficult. A dedicated *Mathematica* course is usually a more gentle, efficient, and effective way to learn this computer algebra system.

Finally, the book can be used as a self-paced tutorial for advanced physics students and professional physicists who would like to learn *Mathematica* on their own. While the sections in Part I should be studied consecutively, those in Part II, each focusing on a particular physics problem, are independent of each other and can be read in any order. The reader may find the solutions to exercises and problems in Appendices D and E helpful.

Organization

Part I gives a practical, physics-oriented, and self-contained introduction to *Mathematica*. Chapter 1 shows the beginner how to get started with *Mathematica* and discusses the notebook front end. Chapter 2 introduces the numerical, symbolic, and graphical capabilities of *Mathematica*. Although these features of *Mathematica* are dazzling, *Mathematica*'s real power rests on its programming capabilities. While Chapter 2 considers many elements of *Mathematica*'s programming language, Chapter 3 treats in depth five key programming elements: expressions, patterns, functions, procedures, and graphics. It also examines three programming styles: procedural, functional, and rule-based. It shows how a proper choice of algorithm and style for a problem results in a correct, clear, efficient, and elegant program. This chapter concludes with a discussion of writing packages. Examples and practice problems, many from physics, are included in Chapters 2 and 3.

Part II considers the application of *Mathematica* to physics. Chapters 4 through 6 illustrate the solution with *Mathematica* of physics problems in mechanics, electricity and magnetism, and quantum physics. Each chapter presents several examples of varying difficulty and sophistication within a subject area. Each example contains three sections: The Problem, Physics of the Problem, and Solution with *Mathematica*. Experience has taught that the Physics of the Problem section is essential because the mesmerizing power of *Mathematica* can distract the student from the central focus, which is, of course, physics. Additional problems are included as exercises in each chapter.

Appendix A relates the latest news on *Mathematica* version 3.0 before this book goes to press. Appendix B tabulates many of *Mathematica*'s operator input forms together with the corresponding full forms and examples. Appendix C provides information about the books, journals, conferences, and electronic archives and forums on *Mathematica*. Appendices D and E give solutions to selected exercises and problems.

Suggestions

The reader should study this book at a computer with a *Mathematica* notebook opened, key in the commands, and try out the examples on the computer. Although all of the code in this book is included on an accompanying diskette, directly keying in the code greatly enhances the learning process. The reader should also try to work out as many as possible of the exercises at the end of the sections and the practice problems at the end of the chapters. The more challenging ones are marked with an asterisk, and those requiring considerable effort are marked with two asterisks.

Prerequisites

The prerequisites for this book are calculus through elementary differential equations, introductory linear algebra, and calculus-based physics with modern physics. Some of the physics in Chapters 5 and 6 may be accessible only to seniors. Basic Macintosh or Windows skills are assumed.

Computer Systems

This book, compatible with *Mathematica* versions 3.0 and 2.2, is to be used with Macintosh and Microsoft-Windows-based IBM-compatible computers. While the front end or the user interface is optimized for each kind of computer system, the kernel, which is the computational engine of *Mathematica*, is the same across all platforms. As over 95% of this book is about the kernel, the book can also be used, with the omission of the obviously Macintosh- or Windows-specific comments, for all computer systems supporting *Mathematica*, such as NeXT computers and UNIX workstations.

Acknowledgments

I wish to express my deepest gratitude to Mervin Hanson (Humboldt State University), who is my partner, friend, mentor, and benefactor. Saying that I wrote this book with him is not an exaggeration. Bill Titus (Carleton College), to whom this book owes its title, deserves my heartfelt gratitude. His involvement, guidance, support, and inspiration in the writing of this book is beyond the obligation of a colleague and a friend. I am indebted to Zenaida Uy (Millersville University), whose great enthusiasm and considerable labor for my project invigorated me when I was weary and feeling low, and to her students for testing my manuscript in their *Mathematica* class. I am most grateful to Jim Feagin (California State University, Fullerton) for his careful reading of my manuscript, for being my friend and stern master, and for sharing his amazing insight into physics and *Mathematica*. Special recognition is due to my students who put up with the numerous errors in my innumerable editions of the manuscript, submitted to being the subjects of my experiments, and gave me their valuable feedback. I am eternally grateful to my wife, Sandra, of more than 30 exciting years for her labor of love in editing and proofreading the evolving manuscript and for keeping faith in me during those dark nights of writer's blues. I am thankful to my friend, David Cowsky, for revealing to me some of the subtleties of the English language.

I would like to acknowledge and thank the following reviewers for their constructive criticisms, invaluable suggestions, and much needed encouragement:

Anthony Behof, DePaul University
Wolfgang Christian, Davidson College
Robert Dickau, Wolfram Research, Inc.
Richard Gaylord, University of Illinois
Jerry Keiper, Wolfram Research, Inc.
Peter Loly, University of Manitoba
David Withoff, Wolfram Research, Inc.
Amy Young, Wolfram Research, Inc.

To Nancy Blachman (Variable Symbols, Inc., and Stanford University) and Vithal Patel (Humboldt State University), I am grateful for their interest, advice, and friendship. Special appreciation is due to my colleagues who covered my classes while I was away on many *Mathematica*-related trips. I am most appreciative of my department chair, Richard Stepp, for his support, and my dean, James Smith, for cheering me onto my *Mathematica* ventures. For their assistance, guidance, and patience in the production and marketing of this book, I would like to thank Abby Heim, Kenneth Metzner, and Zvi Ruder at Academic Press, Inc, and Joanna Hatzopoulos and her associates at Publication Services. I am much indebted to Prem Chawla, Chief Operating Officer of Wolfram Research, Inc., for granting me permission to include *Mathematica* usage statements and help messages in this book.

A special commendation to my daughter, Teresa, is in order for her patience with the sparse social calendar of our family during the development of this book. Finally, I am grateful to my physicians, David O'Brien and John Biteman, for improving and maintaining my health.

Patrick T. Tam

Part I: *Mathematica* with Physics

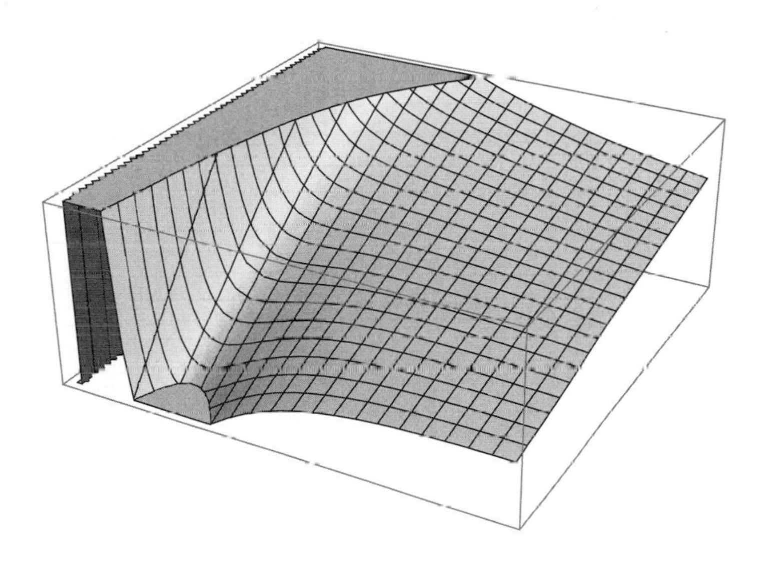

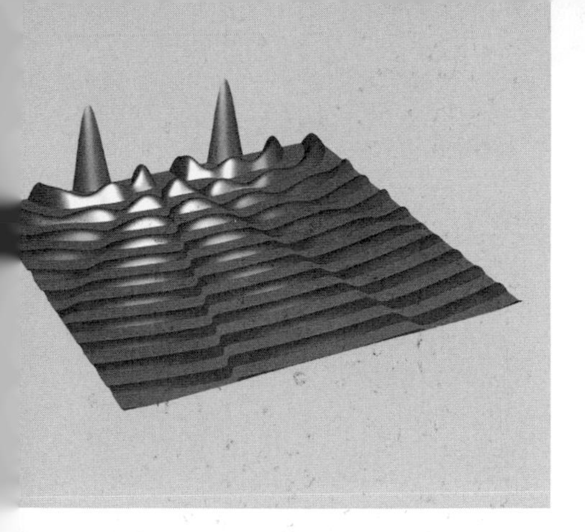

Chapter 1
The First Encounter

Mathematica consists of two parts: the kernel and the front end. The kernel conducts the computations, and the front end provides the interface between the user and the kernel. Whereas the kernel remains the same, the front end is optimized for each kind of computer system.

This chapter shows the neophyte how to get started with *Mathematica*. After introducing some basic features and capabilities of *Mathematica*, it examines the online help and warning messages provided by the kernel and then brings up the notion of "packages" that extend the built-in capabilities of *Mathematica*. It concludes with a discussion of notebook front ends.

1.1 THE FIRST TEN MINUTES

Let us begin by opening a new *Mathematica* document, called a "notebook." On Mac OS X, double-click the *Mathematica* icon ; on Windows, click **Start**, point to **All Programs**, and choose **Mathematica 6** from the **Wolfram Mathematica** program group. A *Mathematica* window appears.

To reproduce *Mathematica* input and output resembling those in this book, click **Mathematica ▸ Preferences ▸ Evaluation** for Mac OS X or **Edit ▸ Preferences ▸ Evaluation** for Windows (i.e., click the Evaluation tab in the Preferences window of the Application (Mathematica) menu for Mac OS X or the Edit menu for Windows) and then verify that **StandardForm** is selected, by default, in the drop-down menus of both "Format type of new input cells:" and "Format type of new output cells:". (Section 2.5 discusses **InputForm**, **OutputForm**, **StandardForm**, and **TraditionalForm**.) The appearance of *Mathematica* input and output may vary from version to version and platform to platform.

We are ready to do several computations. Type **2+3** and without moving the cursor, evaluate the input. To evaluate an input, press *shift+return* (hold down the *shift* key and press *return*) for Mac OS X or Shift+Enter (hold down the Shift key and press Enter) for Windows. (For further discussion of evaluating input, see Section 1.6.5.) The following appears in the notebook:

```
In[1]:= 2 + 3
Out[1]= 5
```

Note that *Mathematica* generates the labels "In[1]:=" and "Out[1]=" automatically. If *Mathematica* beeps, use the mouse to pull down the Help menu and select **Why the Beep?** to see what is happening. Otherwise, *Mathematica* performs the calculation and returns the result below the input.

Type **100!** and without moving the cursor, evaluate the input

```
In[2]:= 100!
```

Mathematica computes 100 factorial and returns the output

```
Out[2]= 93 326 215 443 944 152 681 699 238 856 266 700 490 715 968 264 381 621 468 592 ⋱
         963 895 217 599 993 229 915 608 941 463 976 156 518 286 253 697 920 827 223 ⋱
         758 251 185 210 916 864 000 000 000 000 000 000 000 000
```

In the output, the character ⋱ indicates the continuation of an expression onto the next line.

Type and evaluate the input

```
In[3]:= Expand[(x + y)^30]
```

Mathematica expands $(x+y)^{30}$. The output is

Out[3]= $x^{30} + 30\,x^{29}\,y + 435\,x^{28}\,y^2 + 4060\,x^{27}y^3 + 27\,405\,x^{26}\,y^4 + 142\,506\,x^{25}\,y^5 +$
$593\,775\,x^{24}\,y^6 + 2\,035\,800\,x^{23}\,y^7 + 5\,852\,925\,x^{22}\,y^8 + 14\,307\,150\,x^{21}\,y^9 +$
$30\,045\,015\,x^{20}\,y^{10} + 54\,627\,300\,x^{19}\,y^{11} + 86\,493\,225\,x^{18}y^{12} + 119\,759\,850\,x^{17}\,y^{13} +$
$145\,422\,675\,x^{16}\,y^{14} + 155\,117\,520\,x^{15}\,y^{15} + 145\,422\,675\,x^{14}\,y^{16} +$
$119\,759\,850\,x^{13}\,y^{17} + 86\,493\,225\,x^{12}\,y^{18} + 54\,627\,300\,x^{11}\,y^{19} + 30\,045\,015\,x^{10}\,y^{20} +$
$14\,307\,150\,x^9\,y^{21} + 5\,852\,925\,x^8\,y^{22} + 2\,035\,800\,x^7\,y^{23} + 593\,775\,x^6\,y^{24} +$
$142\,506\,x^5\,y^{25} + 27\,405\,x^4\,y^{26} + 4060\,x^3\,y^{27} + 435\,x^2\,y^{28} + 30\,x\,y^{29} + y^{30}$

Mathematica is case sensitive. That is, *Mathematica* distinguishes between uppercase and lowercase letters. For example, **Expand** and **expand** are different.

Enter and evaluate

```
In[4]:= Integrate[1/(Sin[x]^2 Cos[x]^2), x]
```

Mathematica performs the integration and returns the result

```
Out[4]= -2 Cot[2 x]
```

Besides distinguishing between uppercase and lowercase letters, *Mathematica* also insists that parentheses, curly brackets, and square brackets are different.

Enter and evaluate

```
In[5]:= Plot3D[Sin[x y], {x, -Pi, Pi}, {y, -Pi, Pi}, PlotPoints→25]
```

Mathematica displays the following graphic:

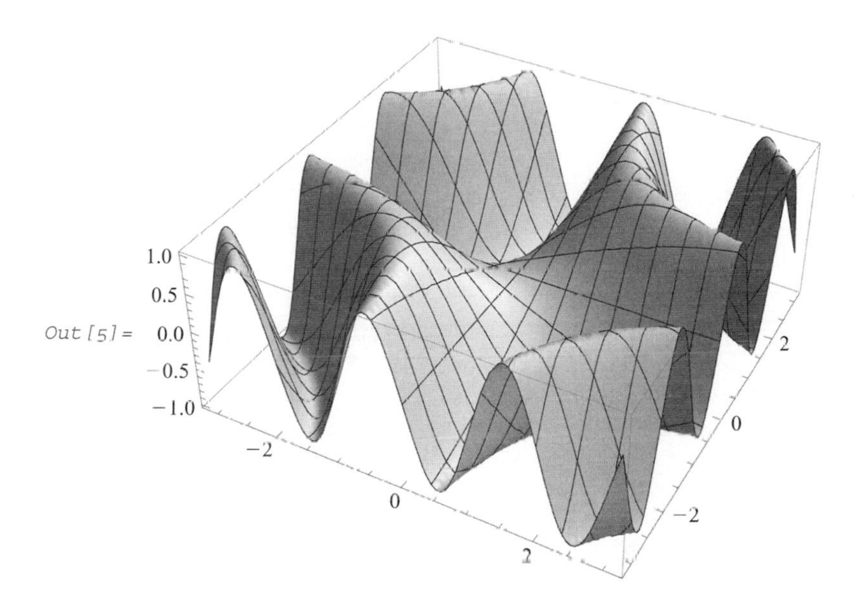

In the preceding input, the character → can be entered as **->**. Also, be sure to leave a space between the letters **x** and **y**.

Normally, having enough computer memory for *Mathematica* calculations is not a problem. Yet there are calculations that require an enormous amount of memory. If there are reasons to suspect that the computer is running out of memory, save the notebook and quit *Mathematica*; otherwise, *Mathematica* crashes and all the work is lost! To quit *Mathematica*, choose **Quit Mathematica** in the Application (Mathematica) menu for Mac OS X or **Exit** in the File menu for Windows. If available computer memory is a problem, a remedy is to let *Mathematica* perform the calculations specified in the notebook in several sessions, if possible. (Upon evaluation, **MemoryInUse[]** gives the number of bytes of memory currently being used to store all data in the current *Mathematica* kernel session, and **MaxMemoryUsed[]** gives the maximum number of bytes of memory used to store all data for the current *Mathematica* kernel session.

`MemoryInUse[$FrontEnd]` gives the number of bytes of memory used in the *Mathematica* front end.)

1.2 A TOUCH OF PHYSICS

1.2.1 Numerical Calculations

Example 1.2.1 Find the eigenvalues and eigenvectors of the Pauli matrix

$$\sigma_x = \begin{pmatrix} 0 & 1 \\ 1 & 0 \end{pmatrix}$$

```
In[1]:= pauliMatrix = {{0, 1}, {1, 0}};

In[2]:= Eigensystem[pauliMatrix]
Out[2]= {{-1, 1}, {{-1, 1}, {1, 1}}}
```

The eigenvalues of the Pauli matrix σ_x are -1 and 1, and the corresponding eigenvectors are $(-1, 1)$ and $(1, 1)$. ■

1.2.2 Symbolic Calculations

Example 1.2.2 Consider an object moving with constant acceleration a in one dimension. The initial displacement and velocity are x_0 and v_0, respectively. Determine the displacement x as a function of time t.

```
In[3]:= DSolve[{x''[t] == a, x'[0] == v0, x[0] == x0}, x[t], t]
```

Out[3]= $\left\{\left\{\text{x[t]} \to \frac{1}{2}\left(\text{a t}^2 + 2\,\text{t v0} + 2\,\text{x0}\right)\right\}\right\}$

In the input, the operator `''` in the second derivative `x''[t]` consists of two single quotation marks. ■

1.2.3 Graphics

Example 1.2.3 For an acoustic membrane clamped at radius a, the $n = 2$ normal mode of vibration is

$$z(r, \theta, t) = J_2(\omega r/v)\sin(2\,\theta)\cos(\omega\, t)$$

where z denotes the membrane displacement at polar coordinates (r, θ) and time t. J_2 is the Bessel function of order 2, v is the acoustic speed, and ω is the frequency. The boundary condition at $r = a$ requires that $J_2(\omega a/v) = 0$, which is satisfied if $\omega a/v = 5.13562$. Let $\omega = 1$ and $v = 1$. Display graphically the vibration at time $t = \pi$.

```
In[4]:= ParametricPlot3D[
          {r Cos[theta], r Sin[theta], BesselJ[2, r] Sin[2theta] Cos[Pi]},
          {r, 0, 5.13562}, {theta, 0, 2 Pi},
          PlotPoints→25, BoxRatios→{1, 1, 0.4},
          ViewPoint→{2.340, -1.795, 1.659}]
```

Out[4]=

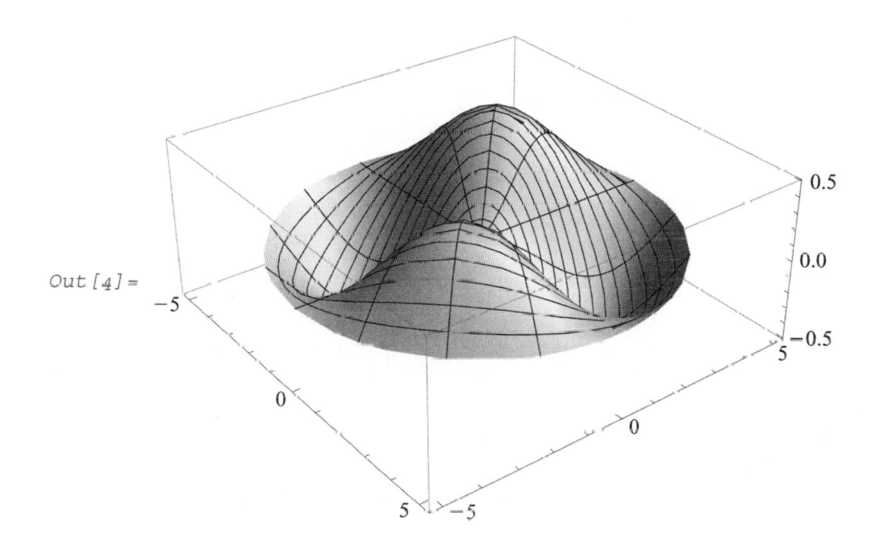

(For a brief discussion of the vibration of circular acoustic planar membrane, refer to [Cra91].) ■

1.3 ONLINE HELP

We can access information about any kernel object such as function and constant by typing a question mark followed by the name of the object and then pressing *shift* + *return* for Mac OS X or Shift + Enter for Windows. The question mark must be the first character of the input line. To obtain information on **Pi**, for example, enter

```
In[1]:= ?Pi
```

Pi is π, with numerical value $\simeq$ 3.14159. ≫

Clicking the button ≫ to the right of the usage information displays the symbol reference page, showing more information together with examples and relevant links. (For more information on the reference pages, see Section 1.6.3.) For the **DSolve** function used in Example 1.2.2, enter

```
In[2]:= ?DSolve
```

DSolve $[eqn, y, x]$ solves a differential equation for the function y, with independent variable x.
DSolve$[\{eqn_1, eqn_2, \ldots\}, \{y_1, y_2, \ldots\}, x]$ solves a list of differential equations.
DSolve$[eqn, y, \{x_1, x_2, \ldots\}]$ solves a partial differential equation. ≫

To get additional information about an object, use **??** instead of **?**:

```
In[3]:= ??DSolve
```

DSolve $[eqn, y, x]$ solves a differential equation for the function y, with independent variable x.
DSolve$[\{eqn_1, eqn_2, \ldots\}, \{y_1, y_2, \ldots\}, x]$ solves a list of differential equations.
DSolve$[eqn, y, \{x_1, x_2, \ldots\}]$ solves a partial differential equation. ≫

```
Attributes[DSolve]={Protected}

Options[DSolve]={GeneratedParameters→C}
```

When used in conjunction with **?**, the "metacharacter" ***** is a wild card that matches any sequence of ordinary characters:

```
In[4]:= ?ND*
```

NDSolve $[eqns, y, \{x, x_{min}, x_{max}\}]$ finds a numerical solution to the ordinary differential equations $eqns$ for the function y with the independent variable x in the range x_{min} to x_{max}.
NDSolve$[eqns, y, \{x, x_{min}, x_{max}\}, \{t, t_{min}, t_{max}\}]$ find a numerical solution to the partial differential equations $eqns$.
NDSolve$[eqns, \{y_1, y_2, \ldots\}, \{x, x_{min}, x_{max}\}]$ finds numerical solutions for the functions y_i. ≫

In this example, the metacharacter ***** matches the characters "`Solve`". If the specification matches more than one name, *Mathematica* returns a list of the names:

```
In[5]:= ?*Find*
```

■ System`

Find	FindList	FindSettings
FindClusters	FindMaximum	FindShortestTour
FindFit	FindMinimum	NotebookFind
FindInstance	FindRoot	NotebookFindReturnObject

■ PacletManager`

PacletFind	PacletFindAll

Here the names all include the characters "`Find`". (This list includes only names of built-in *Mathematica* objects in the **`System``** context and *Mathematica* package objects in the **`PacletManager``** context. Section 1.6 introduces *Mathematica* packages for extending the functionality of *Mathematica*; Section 3.7.1 introduces *Mathematica* contexts for organizing names.) To obtain information on an object, click the object's name in the list:

FindFit [*data*, *expr*, *pars*, *vars*] finds numerical values of the parameters *pars* that make *expr* give a best fit to *data* as a function of *vars*. The data can have the form $\{\{x_1, y_1, \ldots, f_1\}, \{x_2, y_2, \ldots, f_2\}, \ldots\}$, where the number of coordinates *x*, *y*, ... is equal to the number of variables in the list *vars*. The data can also be of the form $\{f_1, f_2, \ldots\}$, with a single coordinate assumed to take values 1, 2, ...
FindFit [*data*, {*expr*, *cons*}, *pars*, *vars*] finds a best fit subject to the parameter constraints *cons*. ≫

We can use **?** to ask for information about many operator input forms. (Appendix B lists some common operator input forms.) To access information concerning the **->** operator, for example, enter

```
In[6]:= ? ->
```

lhs −>*rhs* or *lhs* → *rhs* represents a rule that transforms *lhs* to *rhs*. ≫

1.4 WARNING MESSAGES

When *Mathematica* finds an input questionable upon evaluation, it prints one or more warning messages each consisting of a *symbol*, a message *tag*, and a brief message, in the form *symbol*: : *tag*: *message text*. For instance, entering y instead of t in the second argument of Example 1.2.2 triggers a warning message:

```
In[1]:= DSolve[{x''[t] == a, x'[0] == v0, x[0] == x0}, x[y], t]

        DSolve::deqx: Supplied equations are not
           differential equations of the given functions. >>

Out[1]= DSolve[{x''[t] == a, x'[0] == v0, x[0] == x0}, x[y], t]
```

Clicking the button ≫ to the right of the warning message displays the message reference page, showing more information. Warning messages can be helpful and educational if we regard them as liberating communications rather than dreadful indictments.

1.5 PACKAGES

Mathematica has more than 2000 built-in functions. Yet we often need a function that is not already built into *Mathematica*. In that case, we can define the function in the notebook or use one contained in a package, which is one or more files consisting of functional definitions written in the *Mathematica* language. Many standard packages come with *Mathematica*. The standard-package folders are in the Packages folder of the AddOns folder of the Mathematica folder. (For Windows, the default directory is C:\Program Files\Wolfram Research\Mathematica\6.0\AddOns\Packages. For Mac OS X, click the Mathematica icon while holding down the *control* key and then select Show Package Contents to reveal the AddOns folder.)

To use a function in a package, we must first load the package. The command for loading a package is

<< *context*`

or equivalently, **Get["***context*`**"]**, where *context*` is the context name of the package. The command **<<***context*` loads the file **init.m** in the Kernel folder of the package folder named *context*. The initialization file **init.m** then reads in the necessary files for the package. Note that the backquote ` rather than the single quotation mark ' is used in context names. The backquote, or grave accent character, is called a "context mark" in *Mathematica*. (For a discussion of *Mathematica* contexts, see Section 3.7.1.) Another command for loading a package is

Needs["*context*`**"]**

The "**<<**" command requires fewer key strokes to enter than the **Needs** command does. On the other hand, **Needs** has the advantage over "**<<**" in that it reads in a package only if the package is not already loaded, whereas "**<<**" reads in a package even if it has been loaded. This book will use **Needs** to read in a *Mathematica* package. (For further comparison of the commands "**<<**" and **Needs**, see Problem 6 in Section 3.7.5.)

After loading a package, we can obtain information about the functions defined in the package with the **?** and **??** operators. For example, we can access information about the function **VectorFieldPlot3D** defined in the package **VectorFieldPlots`**:

```
In[1]:= Needs["VectorFieldPlots`"]
```

```
In[2]:= ?VectorFieldPlot3D
```

VectorFieldPlot3D[{f_x,f_y,f_z}, {x, x_{min}, x_{max}}, {y, y_{min}, y_{max}}, {z, z_{min}, z_{max}}] generates a three–dimensional plot of the vector field given by the vector–valued function {f_x,f_y,f_z} as a function of x and y and z.
VectorFieldPlot3D[{ f_x,f_y,f_z}, {x, x_{min}, x_{max}, dx}, {y, y_{min}, y_{max}, dy}, {z, z_{min}, z_{max}, dz }] uses steps dx, dy and dz for variables x, y and z respectively. »

As mentioned in Section 1.3, clicking the button » to the right of the usage information displays the symbol reference page, showing more information together with examples and relevant links.

An important point to remember is that definitions in a package may shadow or be shadowed by other definitions. For example, let us define

```
In[3]:= PeakWavelength[T_] := (0.201405 c h)/(k T)
```

Then, load the package **BlackBodyRadiation`**:

```
In[4]:= Needs["BlackBodyRadiation`"]
```

PeakWavelength::shdw: Symbol PeakWavelength appears in multiple contexts {BlackBodyRadiation`, Global`}; definitions in context BlackBodyRadiation` may shadow or be shadowed by other definitions.

Mathematica warns that the definition of **PeakWavelength** in the package may shadow or be shadowed by our earlier definition. To illustrate the idea of shadowing, let us evaluate

```
In[5]:= PeakWavelength[5700 Kelvin]
Out[5]= 5.0838 × 10^-7 Meter
```

Mathematica returns the result in accordance with the definition of **PeakWavelength** in the package. Our earlier definition of **PeakWavelength** is shadowed or ignored. To use our definition of **PeakWavelength**, we must first execute the command **Remove[*name*]**:

```
In[6]:= Remove[PeakWavelength]
```

Let us evaluate again

```
In[7]:= PeakWavelength[5700 Kelvin]
```

Out[7]= $\frac{0.0000353342\ \text{c h}}{\text{k Kelvin}}$

Mathematica now returns the result according to our definition of **`PeakWavelength`**.

1.6 NOTEBOOK INTERFACES

This section discusses notebook interfaces, or notebook front ends, for Mac OS X and Microsoft Windows. Though notebook front ends share many standard features, a front end is customized for each kind of computer system.

1.6.1 Notebooks

For notebook interfaces, *Mathematica* documents are called notebooks. A *Mathematica* notebook contains ordinary text, *Mathematica* input and output, as well as graphics. Within a notebook, existing or modified inputs can be sent to the kernel for actual computations, animations can be generated, dynamic outputs can be updated in real time, and interfaces for interactive parameter manipulation can be created.

The basic unit of organization in a notebook is a cell. The bracket to the right of a cell marks its extent. For a new notebook, a new cell is created when we start typing, for example,

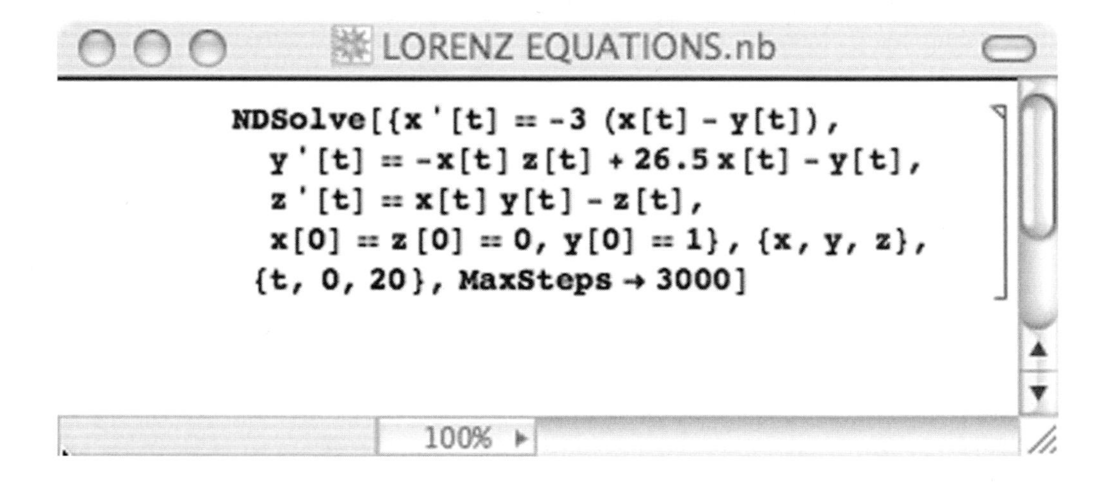

To produce another cell above or below this one or between any two cells, click when the pointer turns into a horizontal I-beam at the desired location and then type.

1.6.2 Entering Greek Letters

Mathematica recognizes a large number of special characters, in addition to the ordinary keyboard characters. Section 2.5.1 discusses the special characters in detail; this section introduces only the Greek letters. We can use Greek letters just like the ordinary keyboard letters.

To enter, for example, the letter β in a notebook:

1. When entering the letter in a cell created earlier, place the cursor at the location where the letter is to be inserted; when entering the letter in a new cell among other cells, move

the pointer to the desired location and click as it turns into a horizontal I-beam to create a horizontal line, called the cell insertion bar; when entering the letter in a new notebook, omit this step.

2. Choose **Palettes ▸ SpecialCharacters** (i.e., choose **SpecialCharacters** in the Palettes menu).
3. For *Mathematica* 6.0.0 and 6.0.1, select **Greek Letters** in the drop-down menu. For *Mathematica* 6.0.2, click the Letters button (or tab) and then the α (i.e., Greek Letters) button/tab in the row of five buttons/tabs.
4. Click the β button.
5. For *Mathematica* 6.0.0 and 6.0.1, click the Insert button to insert β in the notebook. For *Mathematica* 6.0.2, omit this step.

(Section 2.5.1.1 describes other ways to enter Greek letters.)

Note that Greek letters are special characters of *Mathematica* rather than the similar-looking ordinary keyboard characters displayed in the Symbol font. For example, "β" is the keyboard letter "b" in the Symbol font, whereas "β" is the *Mathematica* letter β. Greek letters do not have special meanings in *Mathematica*, with the exception of the letter π, which stands for the mathematical constant pi. With Greek letters, we can, for example, enter the input of Example 1.2.3 as

```
ParametricPlot3D[
 {r Cos[θ], r Sin[θ], BesselJ[2, r] Sin[2θ] Cos[π]},
 {r, 0, 5.13562}, {θ, 0, 2π},
 PlotPoints→25, BoxRatios→{1, 1, 0.4},
 ViewPoint→{2.340, -1.795, 1.659}]
```

1.6.3 Getting Help

Section 1.3 discussed getting help directly from the kernel; this section considers the help provided by the notebook front end.

The Wolfram *Mathematica* Documentation Center provides an enormous amount of useful information about the *Mathematica* system. Choosing **Help ▸ Documentation Center** displays its home page showing links to guide pages for many topics as well as the index of functions. The topics are organized under seven headings: Core Language, Mathematics and Algorithms, Data Handling & Data Sources, Systems Interfaces & Deployment, Dynamic Interactivity, Visualization and Graphics, and Notebooks and Documents.

The symbol reference pages provide information on built-in and standard-package objects such as functions and constants. Highlighting the object name in a notebook and choosing **Help ▸ Find Selected Function** display the reference page for the object. The page often comprises seven sections: Usage Information, More Information, Examples, See Also, Tutorials, Related Links, and More About. The first three sections need no elaboration. The See Also section lists the links to reference pages for related objects; Tutorials, to tutorial pages; Related Links, to Wolfram websites; and More About, to guide pages for topics in the Wolfram *Mathematica* Documentation Center mentioned earlier.

Mathematica 6.0.2 includes two more elements: Virtual Book and Function Navigator. Choosing **Help ▸ Virtual Book** opens the Virtual Book that comprises the tutorial pages. It is a revised, updated, expanded, and online edition of the *Mathematica* Book [Wol03], which is the linearly organized and encyclopedic reference of *Mathematica*. Choosing **Help ▸ Function Navigator** opens the Function Navigator, which is a hierarchical tool for navigating built-in objects such as functions and constants. It classifies the objects according to their functionalities and provides links to their reference pages.

1.6.4 Preparing Input

Input lines can be edited with standard Mac OS X or Windows techniques. Like most programming languages, *Mathematica* is very strict with input spelling. Fortunately, the **Complete Selection** command can help.

Given the initial characters of the name of a kernel object, the **Complete Selection** command returns the full name. If, for example, we type **`Fib`**, leave the cursor immediately after the letter "**b**", and choose **Complete Selection** in the Edit menu, the full name appears as

```
Fibonacci
```

which is the name of the *Mathematica* function for the Fibonacci numbers and polynomials. If there are several possible completions, a list of the names is displayed. For example, if we type **`Plot`** and choose **Complete Selection**, the following menu pops up:

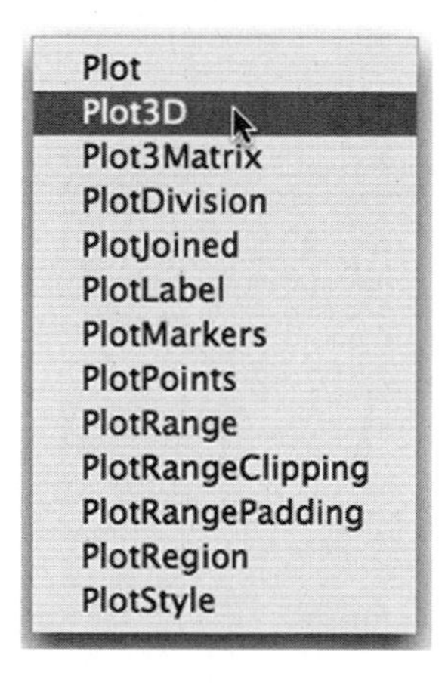

Click a name to have it pasted in the cell:

```
Plot3D
```

A function's template specifies the number, type, and location of its arguments. A template of the **Do** function is

Do[*expr*, {$i_{\max}$}]

The first argument is an expression; the second argument is an iterator.

The **Make Template** command returns a function's template. To obtain, for example, the template of the **DensityPlot** function, type the name of the function, leave the cursor at the end of the name, and choose **Make Template** in the Edit menu. The following is pasted in the cell:

DensityPlot[f, {x, x_{min}, x_{max}}, {y, y_{min}, y_{max}}]

We can now replace the arguments with our function and values. **Make Template** usually returns the simplest of several possible forms for the function. To see the other forms, use the **?** operator discussed in Section 1.3 or the "front end help" considered in Section 1.6.3.

(Section 2.5.2 introduces two-dimensional forms and explains how to enter them. For information on syntax coloring of *Mathematica* input, click **Mathematica ▸ Preferences ▸ Appearance ▸ Syntax Coloring** for Mac OS X or **Edit ▸ Preferences ▸ Appearance ▸ Syntax Coloring** for Windows. Also, see Problem 7 of Section 1.7.)

1.6.5 Starting and Aborting Calculations

To send *Mathematica* input in an input cell to the kernel, put the cursor anywhere in the cell or highlight the cell bracket, and press *shift+return* or *enter* for Mac OS X, or Shift+Enter or Enter on the numeric keypad for Windows. The kernel evaluates the input and returns the result in one or more cells below the input cell.

To abort a calculation, choose **Abort Evaluation** in the Evaluation menu. *Mathematica* aborts the current calculation, sometimes after some delay, and we may continue with further calculations. If the kernel does not respond to this command, we can abort the calculation by choosing **Local** or another appropriate item in the Quit Kernel submenu of the Evaluation menu and then by clicking **Quit** in the dialog box to disconnect the kernel from the front end and terminate the current *Mathematica* session. The notebook, however, is unaffected. Sending another input to the kernel restarts the kernel and begins a new *Mathematica* session.

1.7 PROBLEMS

In this book, straightforward, intermediate-level, and challenging problems are unmarked, marked with one asterisk, and marked with two asterisks, respectively. For solutions to selected problems, see Appendix D.

1. Type and evaluate all the inputs in Sections 1.1 through 1.5.
2. Using **?**, access information on the built-in function **RSolve**.

Answer:

> RSolve[$eqn, a[n], n$] solves a recurrence equation for $a[n]$.
> RSolve[$\{eqn_1, eqn_2, \ldots\}, \{a_1[n], a_2[n], \ldots\}, n$]
> solves a system of recurrence equations.
> RSolve[$eqn, a[n_1, n_2, \ldots], \{n_1, n_2, \ldots\}$] solves a partial recurrence equation. ≫

3. (a) To access information on cells and cell styles, choose
   Help ▸ Documentation Center ▸ Notebooks and Documents/Notebook Basics ▸ Tutorials/Notebooks as Documents
   Help ▸ Documentation Center ▸ Notebooks and Documents/Notebook Basics ▸ More About/Menu Items ▸ Format/Style
   (b) Create a notebook with the following specifications: The first cell is a title cell containing the title of the notebook, the second cell is a text cell giving your name and affiliation, and the third cell is an input cell with the input

```
Factor[45 + 63 x + 32 x^2 + 16 x^3 + 3 x^4 + x^5]
```

   (c) Evaluate the input in Part (b).

Answer:

$$(1 + x)\ (9 + x^2)\ (5 + 2x + x^2)$$

4. Enter the *Mathematica* expression $\alpha+\beta$ in an input cell of a notebook. (See Problem 3 of this section about the various kinds of *Mathematica* cells.)

Answer:

```
α + β
```

5. (a) Choose **Help ▸ Documentation Center ▸ Notebooks and Documents/ Notebook Basics ▸ More About/Menu Items ▸ Evaluation**, and access (in the Evaluation Menu) information on the **Evaluate in Place** command.
   (b) "Evaluate in place" the input

```
(1 + 2 + 3) + 4 + 5
```

   (c) Evaluate in place in the preceding input only what is enclosed by the parentheses. *Hint:* Select (i.e., highlight) the piece to be evaluated.

Answers:

```
15
(6) + 4 + 5
```

6. (a) To access information on the function **GradientFieldPlot** in the package **Vector-FieldPlots`**, display its reference page: Type its name in a notebook, highlight the name, and choose **Help ▸ Find Selected Function**.

   (b) Using the function **Needs**, load the package **VectorFieldPlots`**.

   (c) Using the function **GradientFieldPlot**, plot the electric field of an electric dipole. That is, evaluate

```
GradientFieldPlot[1/Sqrt[(x - 1)^2 + y^2] - 1/Sqrt[{x + 1}^2 + y^2],
 {x, -2, 2}, {y, -2, 2}]
```

(For a discussion of plotting electric field lines, see Section 5.1.)

Answer:

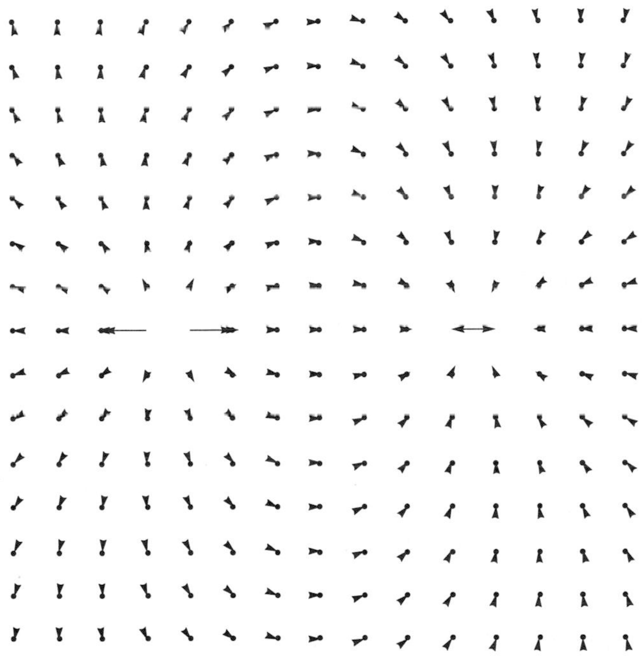

7. Choose **Mathematica ▸ Preferences ▸ Appearance ▸ Syntax Coloring** for Mac OS X or **Edit ▸ Preferences ▸ Appearance ▸ Syntax Coloring** for Windows. After

making sure that the Enable automatic syntax coloring button is checked, obtain information on syntax coloring by clicking one at a time the three buttons: Local Variables, Errors and Warnings, and Other. Using syntax coloring, identify the errors in the input

```
ParametricPlot3D[
  {r Cos[theta], r Sin[theta], BesselJ[2, r, J] Sin[2 theta] Cos[pi]},
  {r, 0, 5.13562}, {theta, 0, 2 Pi},
  PlotPoints → 25, BoxRatio → {1, 1, 0.4},
  ViewPoint → {{2.340, -1.795, 1.659}]]
```

Correct the errors and evaluate the input. *Hint:* See Example 1.2.3 in Section 1.2.3.

Answer:

The letter **J** together with the comma to its left (red), symbol **pi** (blue), **BoxRatio** (red), first left curly bracket (purple) to the right of **ViewPoint**, and last right square bracket (purple) of the input.

*8. Choose **Help ▸ Documentation Center ▸ Wolfram *Mathematica* DOCUMENTATION CENTER/Data Handling & Data Sources/Integrated Data Sources ▸ Integrated Data Sources/Physical & Chemical Data/ ParticleData**, and obtain information on the function **ParticleData**. Using the function, (a) verify that **SigmaPlus** is the *Mathematica* name for Σ^+; (b) list all properties available for the particle; (c) find the antiparticle, mass, spin, baryon number, strangeness, lifetime, and quark composition of the particle; and (d) determine the units in which mass and lifetime are given in the output. *Hint:* Internet connection may be necessary.

Answers:

```
{Antiparticle,BaryonNumber,Bottomness,Charge,ChargeStates,Charm,
 CParity,DecayModes,DecayType,Excitations,FullDecayModes,
 FullSymbol,GenericFullSymbol,GenericSymbol,GFactor,GParity,
 HalfLife,Hypercharge,Isospin,IsospinMultiplet,IsospinProjection,
 LeptonNumber,Lifetime,Mass,MeanSquareChargeRadius,
 Memberships,Parity,PDGNumber,QuarkContent,Spin,
 Strangeness,Symbol,Topness,UnobservedDecayModes,Width}
```

$\bar{\Sigma}^-$

1189.37

$\frac{1}{2}$

1

-1

8.018×10^{-11}

```
{{StrangeQuark, UpQuark, UpQuark}}
MegaelectronVoltsPerSpeedOfLightSquared
Seconds
```

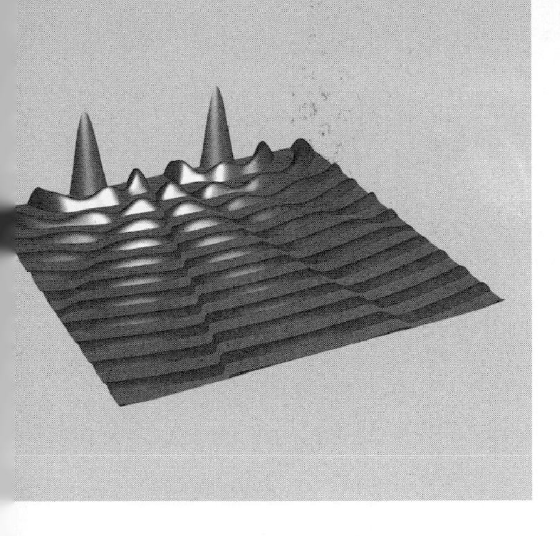

Chapter 2
Interactive Use of *Mathematica*

This chapter covers the use of *Mathematica* as a supercalculator. It does what an electronic calculator can do, and it does a lot more. We enter input and *Mathematica* returns the output. As seen in Chapter 1, the nth input is labeled "`In[`n`]:=`" and the corresponding output, "`Out[`n`]=`".

2.1 NUMERICAL CAPABILITIES

2.1.1 Arithmetic Operations

Table 2.1. Arithmetic Operations in *Mathematica*

Mathematica Operation	Symbol
Addition	+
Subtraction	−
Multiplication	*
Division	/
Exponentiation	^

Table 2.1 lists the arithmetic symbols in *Mathematica*. Here are some examples of their use:

```
In[1]:= 2.1 + 3.72
Out[1]= 5.82
```

```
In[2]:= 6.882/2
Out[2]= 3.441
```

```
In[3]:= 2^3
Out[3]= 8
```

2.1.2 Spaces and Parentheses

Any number of spaces can replace the symbol $*$ for multiplication:

```
In[1]:= 2×5
Out[1]= 10
```

```
In[2]:= 2  5
Out[2]= 10
```

(When there is only a single space between two multiplied numbers in the input, *Mathematica* inserts, by default, the multiplication sign "×" in the space.) *Mathematica* ignores any spaces put before or after arithmetic symbols:

```
In[3]:= (3   + 4)^2
Out[3]= 49
```

Parentheses are used for grouping. Although *Mathematica* observes the standard mathematical rules for precedence of arithmetic operators, parentheses should be used generously to avoid ambiguity about the order of operations. For example, it is not obvious whether **2^3^4** means **(2^3)^4** or **2^(3^4)**. Parentheses are needed for clarity:

```
In[4]:= (2^3)^4
Out[4]= 4096
```

```
In[5]:= 2^(3^4)
Out[5]= 2417851639229258349412352
```

```
In[6]:= 2^3^4
Out[6]= 2417851639229258349412352
```

As it turns out, **2^3^4** stands for **2^(3^4)**.

2.1.3 Common Mathematical Constants

Table 2.2 lists some built-in mathematical constants. The characters **π, ⅇ, °, ⅈ,** and **∞** are special characters of *Mathematica*. To enter **ⅇ**, for example, place the cursor at the desired location in the notebook, choose **Palettes ▸ BasicMathInput**, and click the **ⅇ** button.

Table 2.2. Some Mathematical Constants Known to *Mathematica*

Mathematica Name	Constant
Pi or π	π
E or **ⅇ**	e
Degree or °	$\pi/180$
I or **ⅈ**	$\sqrt{-1}$
Infinity or ∞	∞
GoldenRatio	$(1+\sqrt{5})/2$

2.1.4 Some Mathematical Functions

Some common mathematical functions built into *Mathematica* are

Sqrt[x]	square root
Exp[x]	exponential
Log[x]	natural logarithm
Log[b,x]	logarithm to base b
Factorial[n] or $n!$	factorial
Round[x]	closest integer
Floor[x]	greatest integer not larger than x
Ceiling[x]	least integer not smaller than x
Rationalize[x]	rational number approximation
Sign[x]	-1, 0 or 1 depending on whether x is negative, zero, or positive
Abs[x]	absolute value

Sin[x]	sine	**ArcSin**[x]	inverse sine
Cos[x]	cosine	**ArcCos**[x]	inverse cosine
Tan[x]	tangent	**ArcTan**[x]	inverse tangent
Csc[x]	cosecant	**ArcCsc**[x]	inverse cosecant
Sec[x]	secant	**ArcSec**[x]	inverse secant
Cot[x]	cotangent	**ArcCot**[x]	inverse cotangent

Sinh[x]	hyperbolic sine	**ArcSinh**[x]	inverse hyperbolic sine
Cosh[x]	hyperbolic cosine	**ArcCosh**[x]	inverse hyperbolic cosine
Tanh[x]	hyperbolic tangent	**ArcTanh**[x]	inverse hyperbolic tangent
Csch[x]	hyperbolic cosecant	**ArcCsch**[x]	inverse hyperbolic cosecant
Sech[x]	hyperbolic secant	**ArcSech**[x]	inverse hyperbolic secant
Coth[x]	hyperbolic cotangent	**ArcCoth**[x]	inverse hyperbolic cotangent

2.1.5 Cases and Brackets

Mathematica is case sensitive. The names of built-in *Mathematica* objects all begin with capital letters. For example, **Sqrt** and **sqrt** are different and the former is a built-in *Mathematica* object:

```
In[1]:= Sqrt[5.0]
Out[1]= 2.23607
```

```
In[2]:= sqrt[5.0]
Out[2]= sqrt[5.]
```

Since **sqrt** has not been defined, *Mathematica* returns the input unevaluated.

There are five different kinds of brackets in *Mathematica*:

(*term*)	parentheses for grouping
f[*expr*]	square brackets for functions
{*a*, *b*, *c*}	curly brackets for lists
v[[*i*]]	double square brackets for indexing list elements
(* *comment* *)	commenting brackets for comments to be ignored by the kernel

2.1.6 Ways to Refer to Previous Results

One way to refer to previous results is by their assigned names. We can assign values to variables with the operator "=":

variable = *value*

For example, we can assign a value to the variable **t**:

```
In[1]:= t = 3 + 4
Out[1]= 7
```

The variable **t** now has the value 7:

```
In[2]:= t + 2
Out[2]= 9
```

To avoid confusion, names of user-created variables should normally start with lowercase letters because built-in *Mathematica* objects have names beginning with capital letters. In the previous example, we chose the name **t** rather than **T** for the variable.

In each *Mathematica* session, the assignments of values to variables remain in effect until the values are removed from or new values are assigned to these variables. It is a good practice to remove the values as soon as they are no longer needed. We can use **t =.** or **Clear[t]** to remove the value assigned to **t**:

```
In[3]:= t =.
```

Another way to reference previous results is by using one or more percent signs:

`%`	the previous result
`%%`	the second previous result
`%%%`	the third previous result
`%n`	the result on output line `Out[n]`

Let us illustrate their use:

```
In[4]:= 2^3
Out[4]= 8
```

The symbol `%` refers to the previous result:

```
In[5]:= % + 5
Out[5]= 13
```

The symbol `%4` stands for the result on the fourth output line:

```
In[6]:= %4^2
Out[6]= 64
```

2.1.7 Standard Computations

Mathematica can do standard computations just like an electronic calculator. For example, it can evaluate the expression

$$\sqrt{\frac{(1.4 \times 10^{-25})(6.9 \times 10^{-2})}{1.1 \times 10^{-22}}}$$

```
In[1]:= expr = Sqrt[((1.4×10^-25) (6.9×10^-2))/(1.1×10^-22)]
Out[1]= 0.00937114
```

(When there is only a single space between the number and the power of ten of a real number written in scientific notation in the input, *Mathematica* inserts, by default, the multiplication sign "×" in the space.) We can use the function **NumberForm** to write the number showing two significant figures:

```
In[2]:= NumberForm[expr, 2]
Out[2]//NumberForm=
        0.0094
```

NumberForm[*expr*, *n*] prints *expr* with all approximate real numbers showing at most *n* significant figures—that is, at most *n*-digit precision. We can express the result in scientific notation:

```
In[3]:= ScientificForm[expr]
Out[3]//ScientificForm=
        9.37114 × 10^-3
```

ScientificForm[*expr*] prints *expr* with all approximate real numbers expressed in scientific notation. We can also write the number in scientific notation showing two significant figures:

```
In[4]:= ScientificForm[expr,2]
Out[4]//ScientificForm=
        9.4 × 10^-3
```

ScientificForm[*expr*, *n*] prints *expr* with all approximate real numbers expressed in scientific notation showing at most *n* significant figures—that is, at most *n*-digit precision. As mentioned in Section 2.1.6, it is a good practice to remove the values assigned to variables as soon as the values are no longer needed:

```
In[5]:= Clear[expr]
```

Example 2.1.1 In vacuum systems, pressure as low as 1.00×10^{-9} Pa is attainable. Calculate the number of molecules in a volume of $1.50\,\mathrm{m}^3$ at this pressure, if the temperature is 350 K.

For the number of molecules N, the ideal gas law gives

$$N = \frac{PV}{kT}$$

where P, V, k, and T are pressure, volume, Boltzmann's constant, and temperature, respectively. Thus, N is

```
In[6]:= ((1.00 × 10^-9)1.50)/((1.38 × 10^-23)350)
Out[6]= 3.10559 × 10^11
```

By default, *Mathematica* prints all approximate real numbers, which have exponents outside the range from −5 to 5, in scientific notation. We can write the number showing three significant figures:

```
In[7]:= ScientificForm[%,3]
Out[7]//ScientificForm=
        3.11 × 10^11
```

■

2.1.8 Exact versus Approximate Values

Mathematica treats integers and rational numbers as exact numbers. **Precision**[*x*] gives the effective number of digits of precision (i.e., significant figures) in the number *x*. For exact numbers, **Precision** returns ∞ or **Infinity**. For example, 5 and 345/678 are exact:

```
In[1]:= Precision[5]
Out[1]= ∞
```

```
In[2]:= Precision[345/678]
Out[2]= ∞
```

When we give *Mathematica* exact values as input, it tries to return exact results:

```
In[3]:= 123/456 + 456/789 + 2 Sqrt[5]
```

Out[3]= $\frac{33887}{39976} + 2\sqrt{5}$

Note that *Mathematica* considers $\sqrt{5}$ to be exact:

```
In[4]:= Precision[Sqrt[5]]
Out[4]= ∞
```

We can obtain an approximate numerical result with the function **N**:

```
In[5]:= N[%%]
Out[5]= 5.31982
```

We can also use the function **N** in its postfix form:

```
In[6]:= %%%//N
Out[6]= 5.31982
```

In general, the postfix form *expr*`//`*func* is equivalent to *func*`[`*expr*`]`.

The arguments of trigonometric functions must be in radians. The constant **Degree** is used to convert degrees to radians. Consider, for example, the evaluation of **Cos[20Degree]**:

```
In[7]:= Cos[20 Degree]
Out[7]= Cos[20°]
```

Why didn't *Mathematica* return a numerical value? Built-in mathematical constants such as **E**, **Pi**, and **Degree** are exact, and *Mathematica* does not automatically convert them to approximate numbers. We can use the function **N** to obtain an approximate numerical result:

```
In[8]:= Cos[20 Degree] //N
Out[8]= 0.939693
```

2.1.9 Machine Precision versus Arbitrary Precision

Mathematica treats approximate real numbers as either machine-precision numbers or arbitrary-precision numbers. Whereas machine-precision numbers have a fixed number of digits of precision (i.e., significant figures), arbitrary-precision numbers can have any larger number of digits of precision. For machine-precision numbers, **Precision[**x**]** returns the symbol **MachinePrecision** whose numerical value is **$MachinePrecision**. For the Macintosh and

Windows computer systems, **$MachinePrecision** equals $53 \log_{10} 2$, which is approximately 16. For arbitrary-precision numbers, **Precision** returns numbers that are greater than **$MachinePrecision**.

Unless specified otherwise, *Mathematica* considers approximate real numbers entered with fewer than **$MachinePrecision** digits to be machine-precision numbers:

```
In[1]:= Precision[3.14159]
Out[1]= MachinePrecision
```

If machine-precision numbers appear in a calculation, *Mathematica* returns a machine-precision result:

```
In[2]:= Sqrt[1.0 + 40 + 2/3]
Out[2]= 6.45497
```

```
In[3]:= Precision[%]
Out[3]= MachinePrecision
```

By default, *Mathematica* prints only six digits for the result. To see the other digits, use the function **InputForm**:

```
In[4]:= InputForm[%%]
Out[4]//InputForm=
        6.454972243679028
```

As seen in Section 2.1.8, we can use the function **N** to obtain approximate values for expressions containing only exact numbers:

```
In[5]:= N[123/456 + 456/789 + 2 Sqrt[5]]
Out[5]= 5.31982
```

N[*expr*] evaluates *expr* numerically to give a machine-precision result:

```
In[6]:= Precision[%]
Out[6]= MachinePrecision
```

The function **N** also allows us to do arbitrary-precision calculations. **N[*expr*, *n*]** evaluates *expr* numerically to give a result with *n* digits of precision. For example, we can determine the volume of a sphere of radius 2 m to 200 digits:

```
In[7]:= N[(4Pi/3)(2m)^3, 200]
Out[7]= 33.51032163829112787693486275498136409810314026000110\
          2875706607565128337500038622931869903813698258205\
          8476246256141677937569001009168539543805070500353\
          3704652528030304288206775584593028759781221859240
          74 m^3
```

```
In[8]:= Precision[%]
Out[8]= 200.
```

2.1.10 Special Functions

All the familiar special functions of mathematical physics are built into *Mathematica*. Here are some of them:

LegendreP[n, x]	Legendre polynomials $P_n(x)$
LegendreP[n, m, x]	associated Legendre polynomials $P_n^m(x)$
SphericalHarmonicY[l, m, θ, ϕ]	spherical harmonics $Y_l^m(\theta, \phi)$
HermiteH[n, x]	Hermite polynomials $H_n(x)$
LaguerreL[n, x]	Laguerre polynomials $L_n(x)$
LaguerreL[n, a, x]	generalized Laguerre polynomials $L_n^a(x)$
ClebschGordan[{j_1, m_1}, {j_2, m_2}, {j,m}]	Clebsch–Gordan coefficient
BesselJ[n, z] and **BesselY** [n, z]	Bessel functions $J_n(z)$ and $Y_n(z)$
Hypergeometric1F1[a,b,z]	confluent hypergeometric function ${}_1F_1(a;b;z)$

(There are several notations for the associated Laguerre polynomials in quantum mechanics texts. For the relations between the generalized Laguerre polynomials in *Mathematica* and these associated Laguerre polynomials, see Example 2.2.13 of this book and p. 451 of [Lib03].)

Consider, for example, the evaluation of the Clebsch–Gordan coefficient $< 1\,0\,1\,0\,|\,2\,0 >$:

```
In[1]:= ClebschGordan[{1, 0}, {1, 0}, {2, 0}]
```

Out[1]= $\sqrt{\frac{2}{3}}$

2.1.11 Matrices

Mathematica represents vectors and matrices by lists and nested lists, respectively:

$$\text{vector}: \begin{pmatrix} a \\ b \\ c \end{pmatrix} \qquad \text{list}: \quad \{a, b, c\}$$

$$\text{matrix}: \begin{pmatrix} a & b & c \\ d & e & f \\ g & h & i \end{pmatrix} \qquad \text{nested list}: \quad \{\{a, b, c\}, \{d, e, f\}, \{g, h, i\}\}$$

Some functions for vectors are

$c\,v$	product of scalar c and vector v
$u.v$	dot product of vectors u and v
Cross[u, v]	cross product of vectors u and v
Norm[v]	norm of vector v

Here are some functions for matrices:

c m	product of scalar *c* and matrix *m*
m.*n*	product of matrices *m* and *n*
Inverse[*m*]	inverse of matrix *m*
MatrixPower[*m*, *k*]	*k*th power of matrix *m*
Det[*m*]	determinant of matrix *m*
Tr[*m*]	trace of matrix *m*
Transpose[*m*]	transpose of matrix *m*
Eigenvalues[*m*]	eigenvalues of matrix *m*
Eigenvectors[*m*]	eigenvectors of matrix *m*
IdentityMatrix[*n*]	$n \times n$ identity matrix
DiagonalMatrix[*list*]	square matrix with the elements in *list* on the diagonal
MatrixForm[*list*]	*list* displayed in matrix form

Example 2.1.2 Find the inverse of the matrix

$$\begin{pmatrix} 16 & 0 & 0 \\ 0 & 14 & -6 \\ 0 & -6 & -2 \end{pmatrix}$$

In terms of a nested list, the matrix takes the form

```
In[1]:= m = {{16, 0, 0}, {0, 14, -6}, {0, -6, -2}}
Out[1]= {{16, 0, 0}, {0, 14, -6}, {0, -6, -2}}
```

where we have assigned the matrix to the variable **m** so that we can refer to it later. The function **MatrixForm** displays the matrix in familiar two-dimensional form:

```
In[2]:= MatrixForm[m]
Out[2]//MatrixForm=
```

$$\begin{pmatrix} 16 & 0 & 0 \\ 0 & 14 & -6 \\ 0 & -6 & -2 \end{pmatrix}$$

The function **Inverse** gives the inverse of the matrix:

```
In[3]:= mInv = Inverse[m]
```

Out[3]= $\left\{\left\{\frac{1}{16}, 0, 0\right\}, \left\{0, \frac{1}{32}, -\frac{3}{32}\right\}, \left\{0, -\frac{3}{32}, -\frac{7}{32}\right\}\right\}$

where we have assigned the inverse matrix to the variable **mInv**. To verify that **mInv** is the inverse of **m**, we multiply the matrices together:

```
In[4]:= m.mInv
Out[4]= {{1, 0, 0}, {0, 1, 0}, {0, 0, 1}}
```

In standard two-dimensional matrix form, the result can be displayed as

```
In[5]:= MatrixForm[%]
Out[5]//MatrixForm=
```

$$\begin{pmatrix} 1 & 0 & 0 \\ 0 & 1 & 0 \\ 0 & 0 & 1 \end{pmatrix}$$

which is the identity matrix. Again, it is a good practice to remove unneeded values assigned to variables:

```
In[6]:= Clear[m, mInv]
```

■

2.1.12 Double Square Brackets

Double square brackets allow us to pick out elements of lists and nested lists:

list[[*i*]]	the *i*th element of *list*
list[[{*i*, *j*, *k*, ...}]]	a list of the *i*th, *j*th, *k*th, ... elements of *list*
list[[*i*, *j*]]	the *j*th element in the *i*th sublist of a nested list

Example 2.1.3 Assign the name **ourlist** to the list {1, 3, 5, 7, 9, 11, 13, 15}, extract the second element, and create a list of the first, fourth, and fifth elements.

We begin by naming the list:

```
In[1]:= ourlist = {1, 3, 5, 7, 9, 11, 13, 15}
Out[1]= {1, 3, 5, 7, 9, 11, 13, 15}
```

We then pick out the second element:

```
In[2]:= ourlist[[2]]
Out[2]= 3
```

Finally, we generate a list of the first, fourth, and fifth elements:

```
In[3]:= ourlist[[{1, 4, 5}]]
Out[3]= {1, 7, 9}
```

■

Example 2.1.4 Give the name **mymatrix** to the matrix

$$\begin{pmatrix} 1 & 2 & 3 \\ 4 & 5 & 6 \\ 7 & 8 & 9 \end{pmatrix}$$

Pick out (a) the second row and (b) the second element in the third row.

In *Mathematica*, a matrix is represented as a nested list. We begin by assigning the nested list to the variable **myMatrix**:

```
In[4]:= myMatrix={{1, 2, 3}, {4, 5, 6}, {7, 8, 9}}
Out[4]= {{1, 2, 3}, {4, 5, 6}, {7, 8, 9}}
```

Here is the second row of the matrix:

```
In[5]:= myMatrix[[2]]
Out[5]= {4, 5, 6}
```

Here is the second element in the third row:

```
In[6]:= myMatrix[[3,2]]
Out[6]= 8

In[7]:= Clear[ourlist, myMatrix]
```

■

2.1.13 Linear Least-Squares Fit

Fit[*data*, *funs*, *vars***]** finds a least-squares fit to a list of data in terms of a linear combination of the functions in list *funs* of the variables in list *vars*. The argument *funs* can be any list of functions that depend only on the variables in list *vars*. When list *vars* has only one variable, **Fit[***data*, *funs*, *vars***]** takes the form

$$\texttt{Fit}\left[\{\{x_1, y_1\}, \{x_2, y_2\}, \ldots\}, \{f_1, f_2, \ldots\}, \{x\}\right],$$

where the curly brackets in the third argument $\{x\}$ are optional. If $x_1 = 1, x_2 = 2, \ldots$—that is, $x_i = i$—this can be written as

$$\texttt{Fit}\left[\{y_1, y_2, \ldots\}, \{f_1, f_2, \ldots\}, x\right]$$

Example 2.1.5 Given here are the data of distance versus time for a hot Volkswagen, where d is in meters and t in seconds. Find the equation that gives d as a function of t.

t	d	t	d
0	0	5	37.5
1	1.5	6	54.0
2	6.0	7	73.5
3	13.5	8	96.0
4	24.0	9	121.5

We begin by defining **time** and **distance**:

```
In[1]:= time = Table[i, {i, 0, 9}]
Out[1]= {0, 1, 2, 3, 4, 5, 6, 7, 8, 9}
```

```
In[2]:= distance = {0, 1.5, 6.0, 13.5, 24.0, 37.5, 54.0, 73.5, 96.0, 121.5}
Out[2]= {0, 1.5, 6., 13.5, 24., 37.5, 54., 73.5, 96., 121.5}
```

where the function **Table** will be discussed in Section 2.1.19. We then generate the nested list for the data and name it **vwdata**:

```
In[3]:= vwdata = Transpose[{time, distance}]
Out[3]= {{0, 0}, {1, 1.5}, {2, 6.}, {3, 13.5}, {4, 24.},
         {5, 37.5}, {6, 54.}, {7, 73.5}, {8, 96.}, {9, 121.5}}
```

The function **TableForm** allows us to see the data in familiar two-dimensional form:

```
In[4]:= TableForm[vwdata]
Out[4]//TableForm=
        0    0
        1    1.5
        2    6.
        3    13.5
        4    24.
        5    37.5
        6    54.
        7    73.5
        8    96.
        9    121.5
```

TableForm[*list*] prints with the elements of *list* arranged in an array of rectangular cells. Let us fit the data with a linear combination of functions, 1, t, t^2, and t^3:

```
In[5]:= Fit[vwdata, {1, t, t^2, t^3}, t]
```

Out[5]= $-1.49605\times10^{-14}+6.79999\times10^{-15}t+1.5t^2-1.86769\times10^{-16}t^3$

We can use the function **Chop** to remove terms that are close to zero. **Chop**[*expr*] replaces in *expr* all approximate real numbers with magnitude less than 10^{-10} by the exact integer 0:

```
In[6]:= Chop[%]
Out[6]= 1.5t^2
```

Thus, the formula for distance versus time is $d = 1.5t^2$. Again, it is a good idea to clear unneeded values assigned to variables as soon as possible:

```
In[7]:= Clear[time, distance, vwdata]
```

■

2.1.14 Complex Numbers

Mathematica works with complex numbers as well as real numbers. The following are some complex number operations:

Abs[*z*]	absolute value
Arg[*z*]	the argument
Re[*z*]	real part
Im[*z*]	imaginary part
Conjugate[*z*]	complex conjugate

Consider, for example, putting the expression

$$\sqrt{-7} + \ln(2 + 8i)$$

in the form $a + ib$, where a and b are approximate real numbers, and finding the complex conjugate:

```
In[1]:= N[Sqrt[-7] + Log[2 + 8 I]]
Out[1]= 2.10975 + 3.97157 i

In[2]:= Conjugate[%]
Out[2]= 2.10975 - 3.97157 i
```

2.1.15 Random Numbers

Mathematica has a built-in random number generator that generates uniformly distributed pseudorandom numbers:

RandomReal[ ]	a random real number in the range 0 to 1
RandomReal[{*min*, *max*}]	a random real number in the range *min* to *max*
RandomReal[{*min*, *max*}, *n*]	a list of *n* random real numbers in the range *min* to *max*
RandomInteger[ ]	0 or 1 with equal probability
RandomInteger[{*min*, *max*}]	a random integer in the range *min* to *max*
RandomInteger[{*min*, *max*}, *n*]	a list of *n* random integers in the range *min* to *max*

`RandomComplex[]`	a random complex number in the square defined by 0 and $1+i$
`RandomComplex[{`*min*`,` *max*`}]`	a random complex number in the rectangle defined by *min* and *max*
`RandomComplex[{`*min*`,` *max*`},` *n*`]`	a list of n random complex numbers in the rectangle defined by *min* and *max*
`RandomChoice[{`e_1`,` e_2`, ...}]`	a random choice of one of the e_i
`RandomChoice[{`e_1`,` e_2`, ...},` *n*`]`	a list of n random choices of the e_i

where a range specification of *max* instead of `{`*min*`,`*max*`}` is equivalent to `{0,`*max*`}`. For example, we can obtain a dozen random integers in the range 1 to 10:

```
In[1]:= RandomInteger[{1, 10}, 12]
Out[1]= {8, 7, 10, 4, 4, 7, 6, 1, 5, 3, 9, 5}
```

Evaluating the preceding input again gives a different dozen random integers in the range 1 to 10:

```
In[2]:= RandomInteger[{1, 10}, 12]
Out[2]= {1, 3, 6, 7, 9, 1, 8, 7, 9, 6, 6, 3}
```

Mathematica gives pseudorandom numbers rather than truly random numbers. Using **SeedRandom**, we can instruct *Mathematica* to give a particular sequence of pseudorandom numbers. `SeedRandom[`*n*`]` resets the random number generator, using the integer n as a seed. Choosing the integer 5 as the seed for example, we have

```
In[3]:= SeedRandom[5]

In[4]:= RandomReal[{0, 1}, 12]
Out[4]= {0.000790584, 0.0650192, 0.989555, 0.968768, 0.200866, 0.819521,
         0.0897634, 0.970701, 0.22991, 0.612503, 0.096816, 0.548855}
```

Resetting the random number generator with the same seed, *Mathematica* gives exactly the same sequence:

```
In[5]:= SeedRandom[5]

In[6]:= RandomReal[{0, 1}, 12]
Out[6]= {0.000790584, 0.0650192, 0.989555, 0.968768, 0.200866, 0.819521,
         0.0897634, 0.970701, 0.22991, 0.612503, 0.096816, 0.548855}
```

2.1.16 Numerical Solution of Polynomial Equations

Because there are no general analytical methods for solving polynomial equations of degree higher than four, often numerical methods are the only recourse. `NSolve[`*eqns*`,` *vars*`]` finds the numerical approximations to the roots of a polynomial equation or a system of polynomial equations. In *Mathematica*, an equation is written as *lhs* == *rhs*. It is important to distinguish between the two *Mathematica* operators " = " and " == ". The operator " = " is for assignments, whereas the operator " == " is for specifying equations.

For example, we can determine the roots of the polynomial equation

$$x^5 + 4x^3 + 3x^2 + 2x = 10$$

```
In[1]:= NSolve[x^5 + 4 x^3 + 3 x^2 + 2 x == 10, x]
Out[1]= {{x→-1. - 1.i}, {x→-1. + 1.i},
         {x→0.5 - 2.17945 i}, {x→0.5 + 2.17945 i}, {x→1.}}
```

NSolve gives solutions as rules of the form $x \to sol$. Section 2.2.7 will introduce *Mathematica* transformation rules.

We can also find the roots of the set of simultaneous polynomial equations

$$x^2 - xy - y^2 = 1$$
$$x^3 + 3xy - y^3 = 9$$

With a system of equations, the arguments of **NSolve** are lists:

```
In[2]:= NSolve[{x^2 - x y - y^2 == 1, x^3 + 3 x y - y^3 == 9}, {x, y}]
Out[2]= {{x→1.76268, y→ - 2.57952},
         {x→-0.43195 - 0.816763 i, y→0.552813 + 1.71762 i},
         {x→-0.43195 + 0.816763 i, y→0.552813 - 1.71762 i},
         {x→-2.07194 - 1.87447 i, y→-1.16444 - 1.26905 i},
         {x→-2.07194 + 1.87447 i, y→-1.16444 + 1.26905 i},
         {x→1.74511, y→0.802781}}
```

2.1.17 Numerical Integration

Not all integrations can be done analytically, and we often have to evaluate integrals numerically. **NIntegrate**[*f*, {*x*, *xmin*, *xmax*}] gives a numerical approximation to the definite integral

$$\int_{xmin}^{xmax} f(x)dx$$

Here is an example of integrals that cannot be evaluated analytically:

$$\int_0^2 \sin\left(\sqrt{1 + x^4 + \sin\left(x^3\right)}\right) dx$$

NIntegrate can find a numerical approximation to this integral:

```
In[1]:= NIntegrate[Sin[Sqrt[1 + x^4 + Sin[x^3]]], {x, 0, 2}]
Out[1]= 1.31859
```

`NIntegrate` can find numerical approximations to improper integrals provided that they converge. A definite integral is an improper integral if one or both of the limits of integration are infinite or if the integrand is infinite at some isolated points in the interval of integration. Let us illustrate with several examples the use of `NIntegrate` for evaluating improper integrals:

- Integral with an infinite limit of integration

$$\int_1^\infty \sin\left(\frac{1}{1+x^2}\right) dx$$

```
In[2]:= NIntegrate[Sin[1/(1+x^2)], {x, 1, ∞}]
Out[2]= 0.778031
```

- Integral whose limits of integration are both infinite

$$\int_{-\infty}^\infty \frac{1}{\sqrt{x^8+x^3+2}} dx$$

```
In[3]:= NIntegrate[1/Sqrt[x^8+x^3+2], {x, -∞, ∞}]
Out[3]= 1.96609
```

- Integral whose integrand is infinite at the upper limit

$$\int_{-1}^1 \sqrt{\frac{1+x^5}{1-x}}\, dx$$

```
In[4]:= NIntegrate[Sqrt[(1+x^5)/(1-x)], {x, -1, 1}] //Chop
Out[4]= 3.07224
```

As mentioned in Example 2.1.5, `Chop[`*expr*`]` replaces approximate real numbers in *expr* that are close to zero by the exact integer 0.

- Integral whose integrand is infinite at the lower limit

$$\int_0^1 \frac{e^{-t}}{\sqrt{t}} dt$$

```
In[5]:= NIntegrate[Exp[-t]/Sqrt[t], {t, 0, 1}]
Out[5]= 1.49365
```

- Integral whose integrand is infinite at an interior point of the interval of integration

$$\int_0^3 \frac{2x}{\left(\left|x^2-1/3\right|\right)^{2/3}} dx$$

```
In[6]:= NIntegrate[2 x/(Abs[x^2 - 1/3]^(2/3)), {x, 0, 3}]
        NIntegrate::ncvb:
          NIntegrate failed to converge to prescribed accuracy after
             9 recursive bisections in x near {x} =
             {0.5773502691896258420487562968763193085624642055068 8\
                 2548315259205516}. NIntegrate obtained
             8.2423 - 8.64224 × 10^-6 i and 0.000015175156814913454`
             for the integral and error estimates. >>
Out[6]= 8.2423 - 8.64224 × 10^-6 i
```

Mathematica generated a warning message indicating that the result may be incorrect. Comments on the singularity checking feature of **NIntegrate** are in order. **NIntegrate**[f, {x, $xmin$, $xmax$}] only checks for singularities at the end points of the integration interval. For checking singularities at interior points, use **NIntegrate**[f, {x, $xmin$, x_1, x_2, . . . , $xmax$}], which also checks for possible singularities at each of the interior points x_i. Because our integrand is singular at an interior point $x = \frac{1}{\sqrt{3}}$, we must enter

```
In[7]:= NIntegrate[2 x/(Abs[x^2 - 1/3[^(2/3)), {x, 0, 1/Sqrt[3], 3}]
        NIntegrate::ncvb:
         NIntegrate failed to converge to prescribed accuracy after
            9 recursive bisections in x near {x} =
            {0.5773502691896258420811705036351961738475222492545 0\
                5491835278111585}. NIntegrate obtained
            8.242308997008463` and 0.000010431937927988286`
            for the integral and error estimates. >>
Out[7]= 8.24231
```

Mathematica generated another warning message. Increasing the maximum number of recursive subdivisions, we can verify that the result is correct:

```
In[8]:= NIntegrate[2 x/(Abs[x^2 - 1/3]^(2/3)),
          {x, 0, 1/Sqrt[3], 3}, MaxRecursion -> 12]
Out[8]= 8.24231
```

- Integral that diverges

$$\int_1^\infty \frac{\ln x}{x}\,dx$$

```
In[9]:= NIntegrate[Log[x]/x, {x, 1, ∞}]
        NIntegrate::slwcon:
         Numerical integration converging too slowly; suspect one of the
           following: singularity, value of the integration is 0, highly
           oscillatory integrand, or WorkingPrecision too small. >>
```

```
NIntegrate::ncvb:
  NIntegrate failed to converge to prescribed accuracy after
    9 recursive bisections in x near {x} = {8.16907 × 10^224}.
    NIntegrate obtained 1.1241219061963112`*^10 and
    1.093432414716436`*^10 for the integral and error estimates. >>
Out[9]= 1.12412×10^10
```

Mathematica issued two warning messages to alert us to possible errors. Increasing the maximum number of recursive subdivisions does not lead to convergence:

```
In[10]:= NIntegrate[Log[x]/x, {x, 1, ∞}, MaxRecursion -> 100]
         NIntegrate::slwcon:
           Numerical integration converging too slowly; suspect one of the
              following: singularity, value of the integration is 0, highly
              oscillatory integrand, or WorkingPrecision too small. >>
         NIntegrate::inumri:
                           Log[x]
           The integrand  ------  has evaluated to Overflow, Indeterminate,
                             x
              or Infinity for all sampling points in the
              region with boundaries {{0., Overflow[]}}. >>
                          Log[x]
Out[10]= NIntegrate[ ------ , {x, 1, ∞}, MaxRecursion→100]
                            x
```

Indeed, the integral diverges and the result is wrong.

- Another integral that diverges

$$\int_0^4 \frac{1}{(4-x)^{3/2}}\,dx$$

```
In[11]:= NIntegrate[1/((4 - x)^(3/2)), {x, 0, 4}]//Chop
         NIntegrate::ncvb:
          NIntegrate failed to converge to prescribed accuracy
            after 9 recursive bisections in x near {x} =
            {3.99999999999999999999999999999999999999999999999.
               999999999373894218}. NIntegrate obtained
            9.250010462772341`*^32 and 9.220487164197568`*^32
            for the integral and error estimates. >>
Out[11]= 9.25001 × 10^32
```

The warning message alerts us to potential problems. Again, the integral diverges and the result is wrong. We must be vigilant against using **NIntegrate** to evaluate divergent integrals.

NIntegrate[f**, {**x**,**$xmin$**,**$xmax$**}, {**y**,**$ymin$**,**$ymax$**},...]** finds a numerical approximation to the multidimensional integral

$$\int_{xmin}^{xmax} dx \int_{ymin}^{ymax} dy \ldots f(x, y, \ldots)$$

Example 2.1.6 A uniformly charged disk of radius a and charge density σ is placed on the xy plane with its center at the origin. At the point (x, y, z), the electric potential due to the disk is

$$V(x, y, z) = \frac{\sigma}{4\pi\varepsilon_0} \int_0^a dr \int_0^{2\pi} d\varphi \frac{r}{\sqrt{(x - r\cos\varphi)^2 + (y - r\sin\varphi)^2 + z^2}}$$

where ε_0 is the permittivity constant. Determine V at $(a, 2a, 5a)$.

In units of $\sigma a/4\pi\varepsilon_0$, the electric potential can be expressed as

$$V(x, y, z) = \int_0^1 dr \int_0^{2\pi} d\varphi \frac{r}{\sqrt{(x - r\cos\varphi)^2 + (y - r\sin\varphi)^2 + z^2}}$$

where the unit of x, y, and z is a. For $x = 1$, $y = 2$, and $z = 5$, **NIntegrate** evaluates numerically the double integral:

```
In[12]:= NIntegrate[r/Sqrt[(1 - r Cos[φ])^2 + (2 - r Sin[φ])^2 + 5^2],
           {r, 0, 1}, {φ, 0, 2 π}]
Out[12]= 0.570012
```

Thus, the electric potential at $(a, 2a, 5a)$ equals $0.570012\,\sigma a/4\pi\varepsilon_0$. ■

Example 2.1.7 The volume of a solid is bounded by the graphs of $4x^2 - y^2 + z^2 = 0$ and $y = 3a$. If the density ρ at the point (x, y, z) is proportional to the distance between the origin and the point—that is,

$$\rho(x, y, z) = k\sqrt{x^2 + y^2 + z^2}$$

the moment of inertia of the solid about the y-axis can be expressed as

$$I_y = \int_0^{3a} \int_{-y/2}^{y/2} \int_{-\sqrt{y^2-4x^2}}^{\sqrt{y^2-4x^2}} \left(x^2 + z^2\right) k\sqrt{x^2 + y^2 + z^2}\, dz\, dx\, dy$$

Determine I_y.

Note that I_y can be written as

$$I_y = 4 \int_0^{3a} \int_0^{y/2} \int_0^{\sqrt{y^2-4x^2}} (x^2+z^2)\, k\sqrt{x^2+y^2+z^2}\, dz\, dx\, dy$$

With a change of variables ($x \to ax, y \to ay,$ and $z \to az$), I_y becomes

$$I_y = 4k\,a^6 \int_0^{3} \int_0^{y/2} \int_0^{\sqrt{y^2-4x^2}} (x^2+z^2)\, \sqrt{x^2+y^2+z^2}\, dz\, dx\, dy$$

Using **NIntegrate**, we can find a numerical approximation to the triple integral:

```
In[13]:= (4 k a^6) NIntegrate[(x^2 + z^2) Sqrt[x^2 + y^2 + z^2],
             {y, 0, 3}, {x, 0, y/2}, {z, 0, Sqrt[y^2 - 4 x^2]}]
Out[13]= 72.5991 a^6 k
```

Thus, $I_y = 72.5991\, k\,a^6$. ■

2.1.18 Numerical Solution of Differential Equations

Not all differential equations are amiable to analytical solution. Often, we must resort to numerical methods. **NDSolve**[*eqns*, *y*, {*x*, *xmin*, *xmax*}] numerically solves *eqns*, which is a differential equation together with the necessary initial conditions, for the function y with independent variable x in the range *xmin* to *xmax*.

Example 2.1.8 The equation of motion of the van der Pol oscillator can be written as

$$\frac{d^2x}{dt^2} - \varepsilon\left(1-x^2\right)\frac{dx}{dt} + x = 0$$

where ε is a positive parameter. Consider the case in which $\varepsilon = 0.5$. Solve the differential equation numerically for the initial conditions $v(0) = 0$ and $x(0) = 1.5$ and for t ranging from 0 to 7π.

Here is an example of differential equations that cannot be solved analytically. Let us numerically determine $x(t)$ for t in the range 0 to 7π and assign the result to the variable **sol**:

```
In[1]:= sol = NDSolve[{x''[t] - 0.5 (1 - x[t]^2) x'[t] + x[t] == 0,
            x'[0] == 0, x[0] == 1.5}, x, {t, 0, 7 π}]
Out[1]= {{x → InterpolatingFunction[{{0., 21.9911}}, <>]}}
```

The numerical solution to the differential equation is a list of pairs of numbers (t_i, x_i). *Mathematica* returns the solution as an interpolating function, which is an approximate function interpolating these pairs.

We can find the value of x at any time from 0 to 7π. For $t = 6$,

```
In[2]:= x[6] /. sol[[1]]
Out[2]= 1.94862
```

We can also plot $x(t)$ with t in the range 0 to 7π:

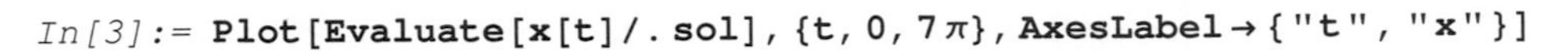

```
In[3]:= Plot[Evaluate[x[t]/. sol], {t, 0, 7π}, AxesLabel→{"t", "x"}]
```

Out[3]=

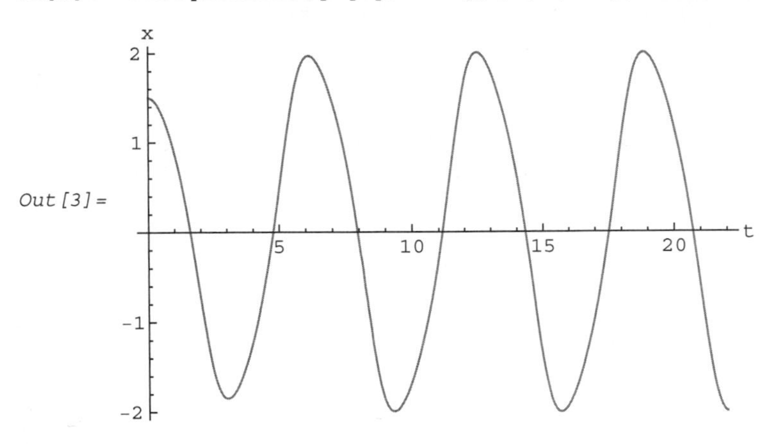

The operator **/.** and the functions **Evaluate** and **Plot** will be discussed in Sections 2.2.7 and 2.3.1.1, respectively.

To see how well the numerical solution approximates the exact solution, we substitute the numerical solution into the *lhs* (i.e., left-hand side) of the differential equation, plot the result, and observe the deviation of the curve from zero:

```
In[4]:= Plot[Evaluate[(x''[t] - 0.5(1 - x[t]^2)x'[t] + x[t]) /. sol], {t, 0, 7π},
          PlotRange→{-0.00015, 0.00015}, AxesLabel→{"t", "lhs"}]
```

Out[4]=

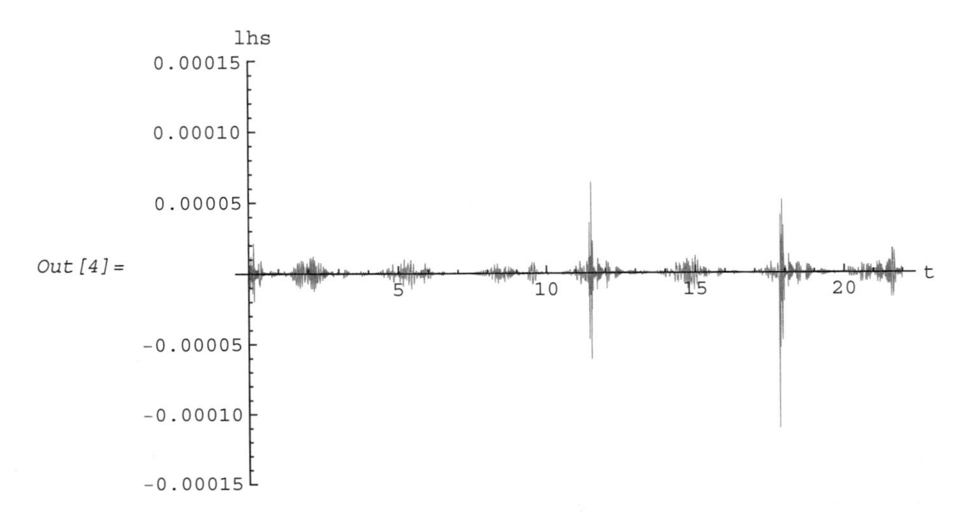

With the numerical solution, the *lhs* of the differential equation is zero to within ±0.00012.

```
In[5]:= Clear[sol]
```

(For further exploration of the van der Pol oscillator, see Exercises 36 and 37 of Section 2.3.5. [TM04] provides an introduction to nonlinear oscillations and chaos.) ■

NDSolve[*eqns*, {y_1, y_2, ...}, {*x*, *xmin*, *xmax*}] numerically solves *eqns*, which is a set of differential equations together with the necessary initial conditions, for the functions y_1, $y_2, \ldots$ with independent variable x in the range *xmin* to *xmax*.

Example 2.1.9 The magnitude of the force on a planet due to the Sun is

$$F = \frac{GMm}{r^2}$$

where G, r, M, and m are the gravitational constant, the distance of the planet from the Sun, the mass of the Sun, and the mass of the planet, respectively. The direction of the force is indicated in Figure 2.1.1. Newton's second law gives

$$\frac{d^2x}{dt^2} + \frac{GMx}{(x^2+y^2)^{3/2}} = 0$$

$$\frac{d^2y}{dt^2} + \frac{GMy}{(x^2+y^2)^{3/2}} = 0$$

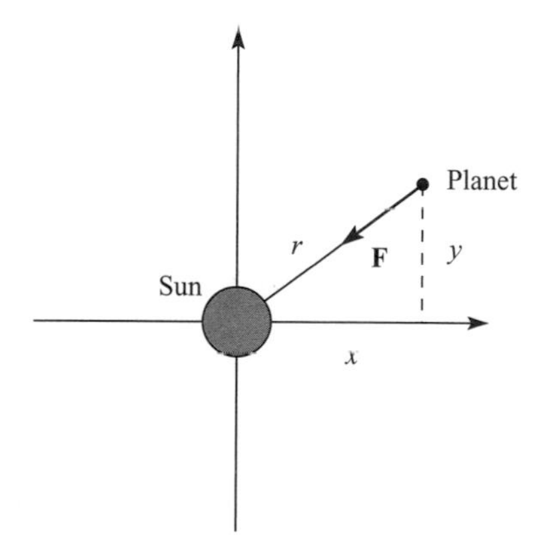

Figure 2.1.1. A planet under the gravitational attraction of the Sun.

where we have assumed that the Sun is stationary at the origin of the Cartesian coordinate system. In a system of units where the unit of length is the astronomical unit (AU)—that is, the length of the semimajor axis of Earth's orbit—and the unit of time is 1 year, $GM = 4\pi^2$ (see [GTC07]). Solve these equations of motion for $x(t)$ and $y(t)$ with t ranging from 0 to 1.6 and plot the trajectory of the planet. Let $x(0) = 1$, $y(0) = 0$, $x'(0) = -\pi$, and $y'(0) = 2\pi$.

NDSolve finds a numerical solution to the equations of motion with the specified initial conditions:

```
In[6]:= orbit = NDSolve[{x''[t] + (4 π^2) x[t]/((x[t]^2 + y[t]^2)^(3/2)) == 0,
          y''[t] + (4 π^2) y[t]/((x[t]^2 + y[t]^2)^(3/2)) == 0, x[0] == 1,
          y[0] == 0, x'[0] == -π, y'[0] == 2 π}, {x, y}, {t, 0, 1.6}]
Out[6]= {{x → InterpolatingFunction[{{0., 1.6}}, <>],
          y → InterpolatingFunction[{{0., 1.6}}, <>]}}
```

Let us plot the path of the planet:

```
In[7]:= ParametricPlot[Evaluate[{x[t], y[t]} /. orbit], {t, 0, 1.6},
          AspectRatio → Automatic, AxesLabel → {"x(AU)", "y(AU)"}]
```

Out[7]=

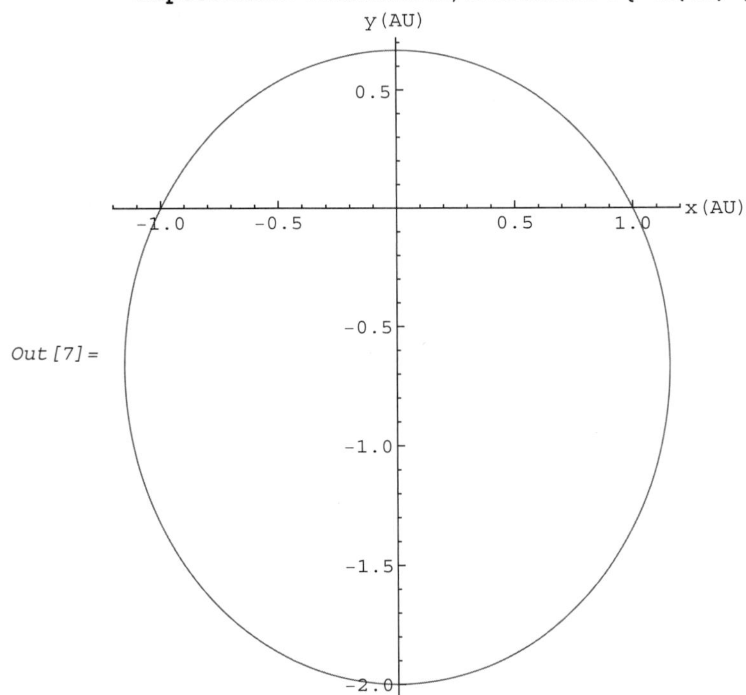

The trajectory is an ellipse with the Sun at one focus in agreement with Kepler's first law. The function **ParametricPlot** will be discussed in Section 2.3.1.7.

```
In[8]:= Clear[orbit]
```

■

2.1.19 Iterators

Iterators instruct *Mathematica* to perform certain tasks a number of times. The iterator notation is

{*imax*}	repeat *imax* times
{*i*, *imax*}	repeat with *i* running from 1 to *imax* in steps of 1
{*i*, *imin*, *imax*}	repeat with *i* running from *imin* to *imax* in steps of 1
{*i*, *imin*, *imax*, *di*}	repeat with *i* running from *imin* to *imax* in steps of *di*
{*i*, *imin*, *imax*}, {*j*, *jmin*, *jmax*}, ...	repeat with *i* running from *imin* to *imax* in steps of 1; for each *i*, let *j* go from *jmin* to *jmax* in steps of 1; and so forth

Table and **Do** are examples of functions that use iterators. **Table**[*expr*, *iterator*] generates a list of *expr*. **Do**[*expr*, *iterator*] evaluates *expr* as many times as indicated by the *iterator*.

For example, we can generate the list {1, 1/2, 1/4, 1/8, 1/16, 1/32, ..., 1/4096} with the function **Table**:

```
In[1]:= Table[1/2^(i-1), {i, 13}]
```

$$\text{Out[1]=}\ \left\{1, \frac{1}{2}, \frac{1}{4}, \frac{1}{8}, \frac{1}{16}, \frac{1}{32}, \frac{1}{64}, \frac{1}{128}, \frac{1}{256}, \frac{1}{512}, \frac{1}{1024}, \frac{1}{2048}, \frac{1}{4096}\right\}$$

Here, **i** goes from 1 to 13 in steps of 1.

We can plot $\sin(2x)$, $\sin(3x)$, and $\sin(4x)$ with the function **Do**:

```
In[2]:= Do[Print[Plot[Sin[k x], {x, -2Pi, 2Pi}, AxesLabel→{x, Sin[k x]}]],
          {k, 2, 4}]
```

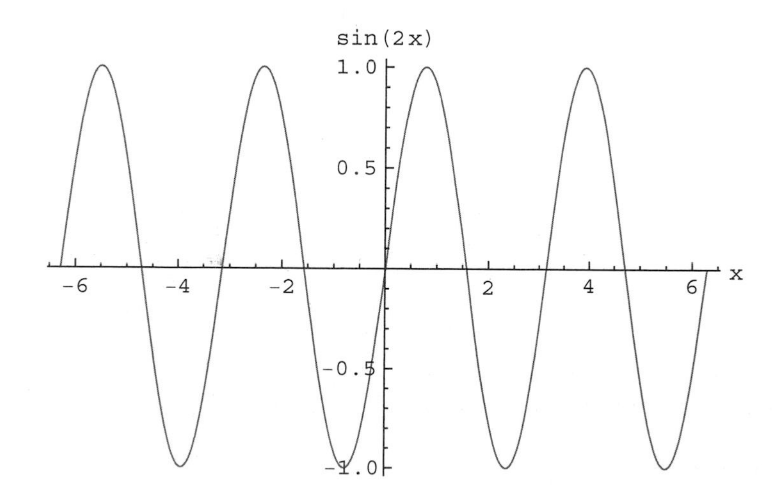

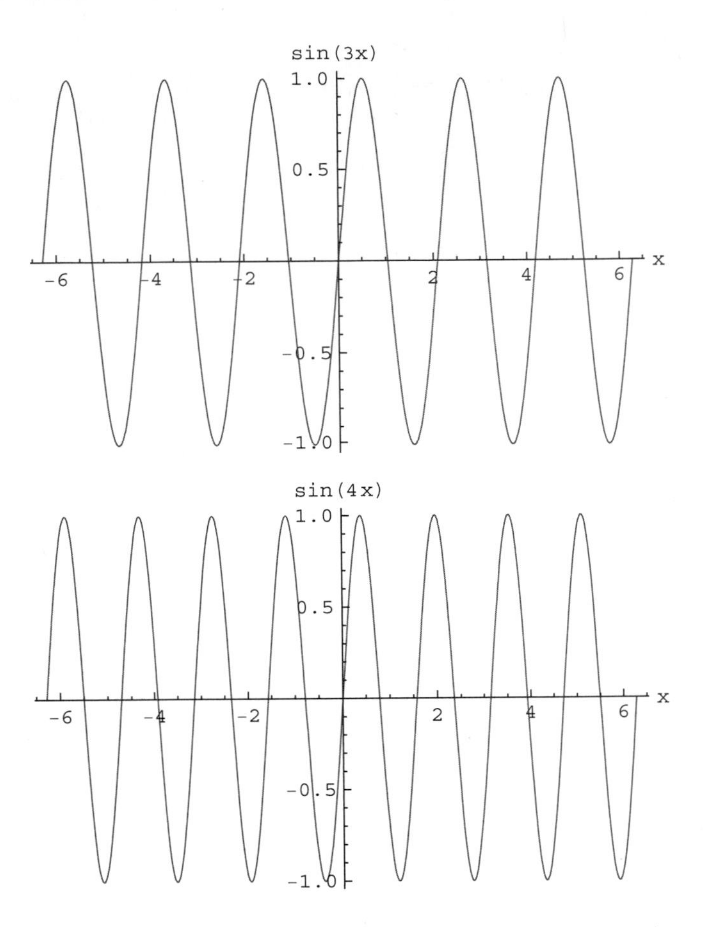

In this example, **k** runs from 2 to 4 in steps of 1. Also, **Print**[*expr*] prints *expr* that can be any expression, including graphics.

2.1.20 Exercises

In this book, straightforward, intermediate-level, and challenging exercises are unmarked, marked with one asterisk, and marked with two asterisks, respectively. For solutions to most odd-numbered exercises, see Appendix C.

1. Evaluate

$$\left(10 \times \sqrt{\frac{10.8 \times 10^3}{300}} + 4\right)^{1/3}$$

Answer:

```
4.
```

2. (a) Evaluate

$$\frac{\left(3.9122 \times 10^{2}\right)^{1/3}\left(\sqrt{2.017 \times 10^{-5}}\right)}{3.661 \times 10^{-4}}$$

(b) Using the function **NumberForm**, write the result showing four significant figures.

Answers:

```
89.7209
89.72
```

3. Evaluate the following and express the result in scientific notation showing three significant figures:

$$\frac{2.54^{3/5}\sqrt{1.15 \times 10^{-2}} + 5.11^{2/5}}{\sqrt{2.32 \times 10^{-5}}}$$

Answer:

4.38×10^{2}

4. Evaluate the following and express the result in scientific notation showing two significant figures:

$$\frac{\left(3.00 \times 10^{2}\right)^{3}\sqrt{2.7 \times 10^{7}}}{3.6 \times 10^{-8}}$$

Answer:

3.9×10^{18}

5. Evaluate the following and express the result in scientific notation showing three significant figures:

$$\frac{\left(3 \times 10^{2}\right)^{3}\left(2 \times 10^{-5}\right)^{1/3}}{\sqrt{3.63 \times 10^{-8}}}$$

Answer:

3.85×10^{9}

6. Evaluate the following and express the result in scientific notation showing five significant figures:

$$\frac{(10^{-24} \times 10^{12})}{10^{-14}} \times \sqrt{\frac{32000}{2^3}}$$

Hint: Use the functions **`N`** and **`ScientificForm`**.
Answer:

`6.3246` $\times$ `10`^{3}

7. What is the distance between two points having coordinates $(-2, 5)$ and $(5, -3)$? Express the result in the scientific notation with three significant figures.
Answer:

`1.06` $\times$ `10`^{1}

8. Determine the length a of the triangle shown.

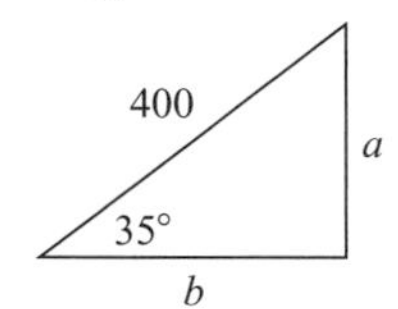

Answer:

`2.29` $\times$ `10`^{2}

9. Determine the length b of the triangle shown.

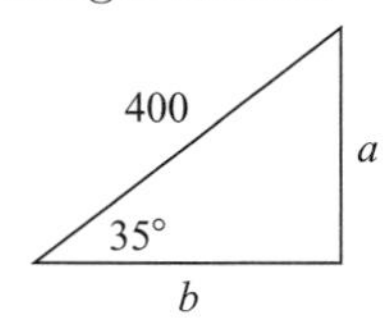

Answer:

`3.28` $\times$ `10`^{2}

10. Determine the angle θ, in degrees, of the triangle shown.

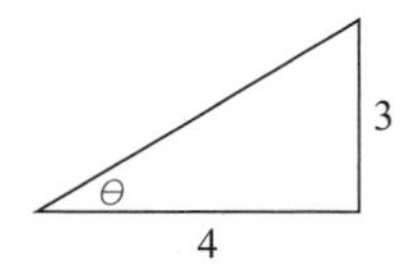

Answer:

`36.9`

11. If $\cos\theta = 2/3$, what is θ in degrees, and what is $\tan^2\theta$?
Answers:

```
48.1897
1.25
```

12. Determine the numerical value of π to 20 significant figures.
Answer:

```
3.1415926535897932385
```

13. Determine the numerical value of Euler's constant γ, whose name is **EulerGamma** in *Mathematica*, to 25 significant figures.
Answer:

```
0.5772156649015328606065121
```

14. Given the matrices

$$A = \begin{pmatrix} 1 & 0 & 1 \\ 0 & 2 & 3 \end{pmatrix}$$

$$B = \begin{pmatrix} 2 & -1 & 4 \\ 1 & 0 & -2 \\ 0 & 3 & 1 \end{pmatrix}$$

express each matrix as a nested list, compute the product AB, and express the result in standard two-dimensional matrix form.
Answer:

$$\begin{pmatrix} 2 & -4 & 3 \\ 2 & 9 & -1 \end{pmatrix}$$

15. Given the matrices A and B, determine AB and display the result in standard matrix form:

$$A = \begin{pmatrix} 1 & 0 & -1 \\ 2 & 4 & 7 \\ 5 & 3 & 0 \end{pmatrix}$$

$$B = \begin{pmatrix} 6 & 1 \\ 0 & 4 \\ -2 & 3 \end{pmatrix}$$

Answer:

$$\begin{pmatrix} 8 & -2 \\ -2 & 39 \\ 30 & 17 \end{pmatrix}$$

16. Given the matrix

$$\begin{pmatrix} 1 & 0 & -1 \\ -2 & 1 & 0 \\ 1 & -1 & 2 \end{pmatrix}$$

express it as a nested list, compute its inverse, and display the inverse in standard matrix form.
Answers:

```
{{1, 0, -1}, {-2, 1, 0}, {1, -1, 2}}
{{2, 1, 1}, {4, 3, 2}, {1, 1, 1}}
```

$$\begin{pmatrix} 2 & 1 & 1 \\ 4 & 3 & 2 \\ 1 & 1 & 1 \end{pmatrix}$$

17. Given the matrix

$$\begin{pmatrix} -2 & 1 & 3 \\ 0 & -1 & 1 \\ 1 & 2 & 0 \end{pmatrix}$$

display its inverse in standard matrix form.
Answer:

$$\begin{pmatrix} -\frac{1}{4} & \frac{3}{4} & \frac{1}{2} \\ \frac{1}{8} & -\frac{3}{8} & \frac{1}{4} \\ \frac{1}{8} & \frac{5}{8} & \frac{1}{4} \end{pmatrix}$$

18. Find the eigenvalues of the matrix

$$\begin{pmatrix} 1 & 0 & -2 \\ 0 & 0 & 0 \\ -2 & 0 & 4 \end{pmatrix}$$

Answer:

```
{5, 0, 0}
```

19. Luap Yllek, a Martian physicist, has discovered a very peculiar gas. He measured the volume and pressure of this gas and obtained the following data (V, P), where V is in Martian liters and P in Martian atmospheres: (0.608, 0.05), (0.430, 0.10), (0.304, 0.20), (0.248, 0.30), (0.215, 0.40), (0.192, 0.50). Determine the equation of the curve that best fits this set of data. Use a combination of 1, $1/V$, and $1/V^2$.

Answer:

$$-0.00120577+\frac{0.0182999}{V^2}+\frac{0.000988148}{V}$$

20. Use the method of least squares to find the best straight line for the four points (4, 5), (6, 8), (8, 10), and (9, 12).

Answer:

```
-0.288136 + 1.33898 x
```

21. The gravitational constant g can be determined by $d = (1/2)gt^2$, where d is the distance and t is the time. An experiment provides the data:

t(s)	d(ft)
0.5	4.2
1.0	16.1
1.5	35.9
2.0	64.2

Using the least-squares method, determine g.

Answer:

$$\frac{32.0656 \text{ ft}}{\text{s}^2}$$

*22. On a distant planet where air resistance is negligible and the free-fall acceleration does not vary with altitude over short vertical distances, astronauts toss a rock into the air. With the aid of a camera that takes pictures at a steady rate, they record the height of the rock as a function of time as given in the table below. (a) With the function **`Table`**, make a list of the average velocities of the rock in each time interval between a measurement and the next. Because the acceleration is constant, these average velocities equal the instantaneous velocities at the midpoints of the time intervals. (b) With the function **`Table`**, make a list of the times at the midpoints of the time intervals. (c) With the function **`Transpose`**, make a nested list of $\{t_i, v_i\}$ for i varying from 1 to 20, from the lists created in parts (a) and (b). (d) With the function **`Fit`**, find the equation of the curve that best fits the data generated in part (c). What is the acceleration?

Height of a Rock versus Time

Time (s)	Height (m)	Time (s)	Height (m)
0	5.	2.75	7.62
0.25	5.75	3.	7.25
0.5	6.4	3.25	6.77
0.75	6.94	3.5	6.2
1.	7.38	3.75	5.52
1.25	7.72	4.	4.73
1.5	7.96	4.25	3.85
1.75	8.1	4.5	2.86
2.	8.13	4.75	1.77
2.25	8.07	5.	0.58
2.5	7.9		

Answers:

```
{5., 5.75, 6.4, 6.94, 7.38, 7.72, 7.96, 8.1, 8.13, 8.07, 7.9,
 7.62, 7.25, 6.77, 6.2, 5.52, 4.73, 3.85, 2.86, 1.77, 0.58}

{3., 2.6, 2.16, 1.76, 1.36, 0.96, 0.56, 0.12, -0.24, -0.68, -1.12,
 -1.48, -1.92, -2.28, -2.72, -3.16, -3.52, -3.96, -4.36, -4.76}

{0.125, 0.375, 0.625, 0.875, 1.125, 1.375, 1.625, 1.875, 2.125, 2.375,
 2.625, 2.875, 3.125, 3.375, 3.625, 3.875, 4.125, 4.375, 4.625, 4.875}

{{0.125, 3.}, {0.375, 2.6}, {0.625, 2.16}, {0.875, 1.76},
 {1.125, 1.36}, {1.375, 0.96}, {1.625, 0.56}, {1.875, 0.12},
 {2.125, -0.24}, {2.375, -0.68}, {2.625, -1.12}, {2.875, -1.48},
 {3.125, -1.92}, {3.375, -2.28}, {3.625, -2.72}, {3.875, -3.16},
 {4.125, -3.52}, {4.375, -3.96}, {4.625, -4.36}, {4.875, -4.76}}

3.20081 - 1.63392 t
```

$$-\frac{1.63392\ \mathrm{m}}{\mathrm{s}^2}$$

23. Hooke's law states that when a force is applied to a spring of uniform material, the increase in length of the spring is proportional to the applied force; in other words, $F = kx$, where F is the force (in pounds), k is the spring constant, and x is the increase in length of the spring (in inches). An experiment provides the following data:

x	F
0.1	1.1
0.3	3.2
0.5	4.9
0.7	6.8
0.9	9.1

Using the least-squares method, determine k.

Answer:

$$\frac{9.98182\ \text{lb}}{\text{in}}$$

24. Using the functions **Re** and **Im**, determine the real and imaginary parts of the expression

$$\frac{4+i}{2+3i}$$

Answers:

$$\frac{11}{13}$$
$$-\frac{10}{13}$$

25. Express the following in the form $p + qi$, where p and q are approximate real numbers:

$$(3\pi + 7i)\cos 37° + (2 + 8i)e^{-3i+2}$$

Answer:

```
1.23869 - 55.0159 i
```

26. Express the following in the form $p + qi$, where p and q are approximate real numbers with 25 digits of precision:

$$(4\pi + 5i)\sin 25° + (3 + 5i)e^{-1+2i}$$

Answer:

```
3.178942962210888816241724 + 2.351167468050152469909314 i
```

27. (a) Using **RandomInteger**, generate a list of 20 random integers in the range 1 to 100. (b) Using the function **Max**, pick out the largest integer. *Hint:* With "front end help," obtain information about the function **Max**.

28. (a) Using **RandomReal**, generate a list of 10 random real numbers in the range 10 to 100. (b) Using the function **Min**, pick out the smallest number in the list. (c) Using the function **Floor**, determine the greatest integer that is not larger than the number picked out in part (b). *Hint:* With "front end help," obtain information about the function **Min**.

*29. The function **RandomInteger** can be used to generate the integers 0 or 1 with equal probability. Using **RandomInteger**, we can simulate the tossing of a coin N times by letting the integer 0 represent "head" and the integer 1 "tail." (a) With **RandomInteger**, generate a list of $N = 10$ zeros or ones. (b) With "front end help," obtain information about the function **Count**. Using **Count**, count the number of zeros in the list generated in part (a). (c) Repeat parts (a) and (b) for $N = 100$. (d) Repeat parts (a) and (b) for $N = 10000$, **without explicitly displaying or printing the list of zeros and ones.** (e) What can you conclude about probability and actual fractional outcome from your results?

30. Solve the equation

$$x^7 + x^5 + 2x^3 + x^2 + 1 = 0$$

Answers:

```
{{x→-0.812432},{x→-0.640787-1.07931 ⅈ},{x→-0.640787+1.07931 ⅈ},
 {x→0.254825-0.700968 ⅈ}, {x→0.254825+0.700968 ⅈ},
 {x→0.792178-0.881387 ⅈ}, {x→0.792178+0.881387 ⅈ}}
```

31. Solve the equation

$$4\beta^7 - 16\beta^4 + 17\beta^3 + 6\beta^2 - 21\beta + 10 = 0$$

Answers:

```
{{β→-1.05209-1.54511 ⅈ}, {β→-1.05209+1.54511 ⅈ},
{β→-1.}, {β→0.648515-0.68252 ⅈ},
{β→0.648515+0.68252 ⅈ}, {β→0.807146}, {β→1.}}
```

32. Solve the system of equations

$$x^2 + xy + y^2 = 1$$

$$x^3 + x^2y + xy^2 + y^3 = 4$$

Answers:

```
{{x→-1.-1.41421 ⅈ, y→-1.+1.41421 ⅈ},
{x→-1.+1.41421 ⅈ, y→-1.-1.41421 ⅈ},
{x→1.6838-0.133552 ⅈ, y→-0.683802+1.13355 ⅈ},
{x→1.6838+0.133552 ⅈ, y→-0.683802-1.13355 ⅈ},
```

```
{x→-0.683802-1.13355 i, y→1.6838+0.133552 i},
{x→-0.683802+1.13355 i, y→1.6838-0.133552 i}}
```

33. Using **NIntegrate**, evaluate the integral

$$\int_0^1 \frac{x^3}{e^x - 1}\, dx$$

Answer:

```
0.224805
```

34. Using **NIntegrate**, evaluate the integral

$$\int_{-\infty}^{\infty} H_4(x)^2 e^{-x^2}\, dx$$

where $H_n(x)$ are the Hermite polynomials.

Answer:

```
680.622
```

35. Using **NIntegrate**, evaluate the integral

$$\int_2^{\infty} \frac{x^2}{\sqrt{x^7+1}}\, dx$$

Answer:

```
1.41385
```

36. Using **NIntegrate**, evaluate the integral

$$\int_{-\infty}^{\infty} \frac{1}{x^2+2x+2}\, dx$$

Answer:

```
3.14159
```

37. Using **NIntegrate**, evaluate the integral

$$\int_0^1 \frac{\ln x}{\sqrt{1-x^2}}\, dx$$

Answer:

```
-1.08879
```

38. Using **NIntegrate**, evaluate the integral

$$\int_2^\infty \frac{1}{x-1}\,dx$$

Answer:

The integral diverges.

39. Using **NIntegrate**, evaluate the integral

$$\int_0^1 \frac{1}{(x-1)^2}\,dx$$

Answer:

The integral diverges.

40. Using **NIntegrate**, evaluate the integral

$$\int_0^3 \frac{1}{(|x-1|)^{2/3}}\,dx$$

Answer:

```
6.77976
```

41. Evaluate the double integral

$$\int_{-a}^0 \int_{-\sqrt{a^2-x^2}}^0 \sqrt{x^2+y^2}\,dy\,dx$$

Hint: Make a change of variables: $x \to ax$ and $y \to ay$.
Answer:

$0.523599\,a^3$

42. Using **NIntegrate**, evaluate the triple integral

$$\int_0^{8/5} \int_{y/4}^{\sqrt{4-y^2}/3} \int_0^{4-9x^2-y^2} x\left(x^2+y^2+z^2\right)^{1/2}\,dz\,dx\,dy$$

Answer:

```
0.54469
```

*43. The volume of a solid is bounded by the graphs $x^2 - y^2 + z^2 = a^2$, $y = 0$, and $y = 4a$. If the density at the point (x, y, z) is directly proportional to the distance from the y-axis to the point—that is,

$$\rho(x, y, z) = k\sqrt{x^2 + z^2}$$

the moment of inertia of the solid about the y-axis can be expressed as

$$I_y = \int_0^{4a} \int_{-\sqrt{a^2+y^2}}^{\sqrt{a^2+y^2}} \int_{-\sqrt{a^2+y^2-x^2}}^{\sqrt{a^2+y^2-x^2}} k\left(x^2 + z^2\right)^{3/2} dz\, dx\, dy$$

Using **NIntegrate**, evaluate I_y. *Hint:* Make a change of variables: $x \to ax$, $y \to ay$, and $z \to az$.

Answer:

```
1078.95 a^6 k
```

44. Solve the differential equation

$$\frac{d^2y}{dx^2} = 2x + y + 3\frac{dy}{dx}$$

with $y(2) = 1$ and $y'(2) = -1$. Find y (2.2) and also plot $y(x)$ with x ranging from 2.0 to 2.3.

Answers:

```
{{y→InterpolatingFunction[{{2., 2.3}}, <>]}}
0.851094
```

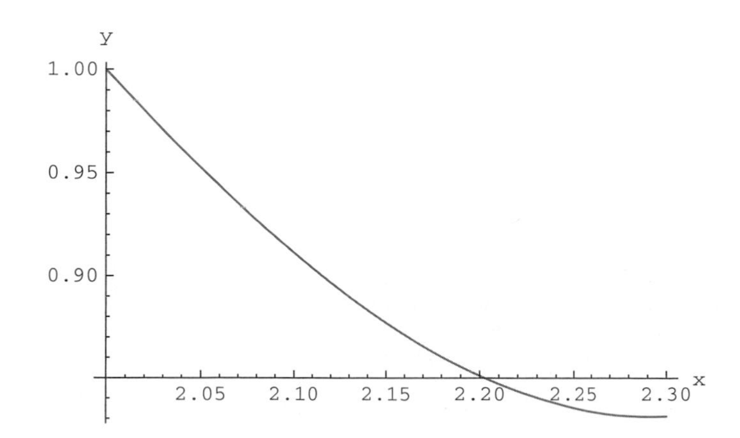

45. Solve the differential equation

$$\frac{d^2y}{dx^2} + \sin^2 x \frac{dy}{dx} + 3y^2 = e^{-x^2}$$

with $y(0) = 1$ and $y'(0) = 0$. Determine $y(1)$, $y(2)$, and $y(3)$ and also plot $y(x)$ with x in the range 0 to 3.

Answers:

```
{{y→InterpolatingFunction[{{0., 3.}}, < >]}}
{0.345837, -0.155485, -0.465162}
```

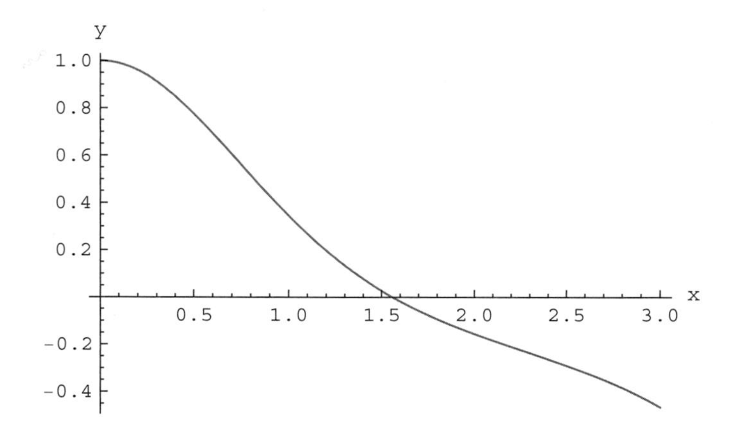

46. Solve the differential equation

$$\frac{d^2y}{dx^2} + 5x\frac{dy}{dx} - \left(1 - x^3\right) y = 0$$

with $y(0) = 0$ and $y'(0) = 1$. Determine $y(1.8)$ and also plot $y(x)$ with x ranging from 0 to 4.

Answers:

```
{{y→InterpolatingFunction[{{0., 4.}}, < >]}}
0.544786
```

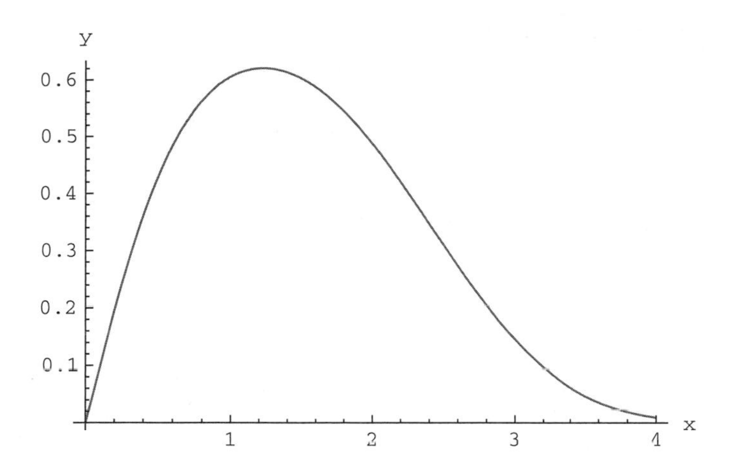

47. Solve the differential equation

$$y'' = \sin(t)y + t$$

with $y(0) = 0$ and $y'(0) = 1$. Plot $y(t)$ with t in the range 0 to 6.4.

Answers:

```
{{y → InterpolatingFunction[{{0., 6.4}}, <>]}}
```

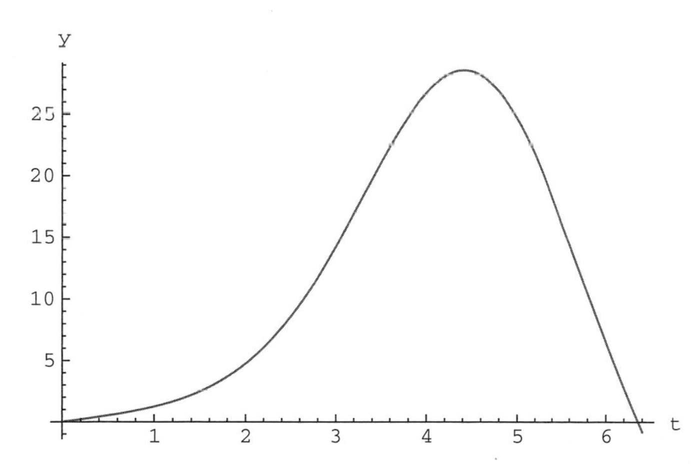

*48. A uniform ring of radius a and total charge q is placed on the xy plane with its center at the origin. Determine the electric potential at the point $(2a, 3a, 5a)$ in units of $q/(4\pi\varepsilon_0 a)$, where ε_0 is the permittivity constant. *Hint:* Use **NIntegrate**.

Answer:

```
0.161174
```

*49. A uniform disk of radius a and total charge q is placed on the xy plane with its center at the origin. Determine the electric potential at the point $(2.5a, 1.2a, -3.7a)$. *Hint:* Choose a convenient system of units and use **`NIntegrate`**.

Answer:

```
0.215092
```

in units of $q/(4\pi\varepsilon_0 a)$, where ε_0 is the permittivity constant.

2.2 SYMBOLIC CAPABILITIES

Section 2.1 considered the numerical capabilities of *Mathematica*. *Mathematica* can work with variables or symbols without assigned values as well as with numbers. This section discusses its symbolic capabilities. Most concepts and techniques developed for numerical computations can be easily generalized to symbolic computations.

2.2.1 Transforming Algebraic Expressions

Mathematica has many functions for transforming algebraic expressions. Some of them are

`Apart`[*expr*, *var*]	rewrite a rational expression as a sum of terms with minimal denominators, treating all variables other than *var* as constants
`Cancel`[*expr*]	reduce a fraction to lowest terms
`Collect`[*expr*, *x*]	collect together terms involving the same power of x
`Expand`[*expr*]	expand out products and positive integer powers
`ExpandAll`[*expr*]	apply **`Expand`** to all subexpressions
`ExpandDenominator`[*expr*]	apply **`Expand`** to the denominator
`ExpandNumerator`[*expr*]	apply **`Expand`** to the numerator
`Factor`[*expr*]	factor *expr* completely—that is, until no factor can be factored further
`FactorTerms`[*expr*, *x*]	pull out factors that do not depend on x
`Together`[*expr*]	reduce a sum of fractions to a single fraction with the least common denominator as denominator
`ComplexExpand`[*expr*]	expand *expr* assuming that all variables are real
`PowerExpand`[*expr*]	expand nested powers, powers of products, logarithms of powers, and logarithms of products; that is, expand out $(ab)^c, (a^b)^c$, etc. (Use **`PowerExpand`** with caution because it is oblivious to issues of branches of multi-valued functions.)

`Simplify[expr]`	return the simplest form by applying various standard algebraic transformations
`FullSimplify[expr]`	return the simplest form by applying a wide range of transformations

Let us illustrate the use of these functions with several examples.

Using the function **Apart**, we can find the partial fraction decomposition of the rational expression

$$\frac{5x^2 - 4x + 16}{(x^2 - x + 1)^2 (x - 3)}$$

```
In[1]:= Apart[(5x^2 - 4x + 16)/(((x^2 - x + 1)^2)(x - 3)), x]
```

$$\text{Out[1]=}\ \frac{1}{-3+x} + \frac{-3-2x}{(1-x+x^2)^2} + \frac{-2-x}{1-x+x^2}$$

Applying the function **Cancel** to the fraction

$$\frac{y^2 - 5y + 4}{y - 1}$$

reduces it to lowest terms:

```
In[2]:= Cancel[(y^2 - 5y + 4)/(y - 1)]
Out[2]= -4 + y
```

Let us apply **Expand**, **ExpandAll**, **ExpandDenominator**, and **ExpandNumerator** to

$$\frac{(x + 3)(x - 1)^2}{(x^2 + 1)(x + 5)^2}$$

and observe the differences among the results:

```
In[3]:= myexpr = ((x + 3)(x - 1)^2)/((x^2 + 1)(x + 5)^2)
```

$$\text{Out[3]=}\ \frac{(-1+x)^2(3+x)}{(5+x)^2(1+x^2)}$$

```
In[4]:= Expand[myexpr]
```

$$\text{Out[4]=}\ \frac{3}{(5+x)^2(1+x^2)} - \frac{5x}{(5+x)^2(1+x^2)} + \frac{x^2}{(5+x)^2(1+x^2)} + \frac{x^3}{(5+x)^2(1+x^2)}$$

```
In[5]:= ExpandAll[myexpr]
```

$$\text{Out[5]=}\ \frac{3}{25+10x+26x^2+10x^3+x^4} - \frac{5x}{25+10x+26x^2+10x^3+x^4} + \frac{x^2}{25+10x+26x^2+10x^3+x^4} + \frac{x^3}{25+10x+26x^2+10x^3+x^4}$$

```
In[6]:= ExpandDenominator[myexpr]
```

$$Out[6]= \frac{(-1+x)^2 (3+x)}{25+10x+26x^2+10x^3+x^4}$$

```
In[7]:= ExpandNumerator[myexpr]
```

$$Out[7]= \frac{3-5x+x^2+x^3}{(5+x)^2 (1+x^2)}$$

The results, which are equivalent algebraic expressions, have different forms; that is, the four expanding functions produce different effects.

The function **Factor** factors

$$16y^2 - 25y^4$$

completely:

```
In[8]:= Factor[16 y^2 - 25 y^4]
Out[8]= -y^2(-4 + 5 y) (4 + 5 y)
```

Applying **Expand** to this result gives back the original expression:

```
In[9]:= Expand[%]
Out[9]= 16 y^2 - 25 y^4
```

Consider simplifying

$$\frac{1}{x^2-16} - \frac{x+4}{x^2-3x-4}$$

The obvious function to use is **Simplify**:

```
In[10]:= Simplify[(1/(x^2 - 16)) - ((x + 4)/(x^2 - 3 x - 4))]
```

$$Out[10]= \frac{4+x}{4+3x-x^2} + \frac{1}{-16+x^2}$$

Contrary to expectation, **Simplify** does not always return the "simplest" form. Let us try the function **Together**:

```
In[11]:= Together[(1/(x^2 - 16)) - ((x + 4)/(x^2 - 3 x - 4))]
```

$$Out[11]= \frac{-15-7x-x^2}{(-4+x)(1+x)(4+x)}$$

One can argue that the result is still not in the "simplest" form because the denominator is factored. That is just a matter of preference. We can apply **ExpandDenominator** to the output:

```
In[12]:= ExpandDenominator[%]
```

Out[12]= $\dfrac{-15-7x-x^2}{-16-16x+x^2+x^3}$

In general, to get an expression into a desired form, experiment with different combinations of transforming functions.

Because **FullSimplify** knows a wide range of transformations, it can simplify, slowly at times, some expressions that elude **Simplify**:

```
In[13]:= Simplify[
           (Sqrt[(2 + x) / (7 + x)]Sqrt[(3 - x^2) / (25 - x^2)]Sqrt[(5 + x) (2 - x)]) /
            (Sqrt[4 - x^2]Sqrt[3 - x^2])]
```

Out[13]= $\dfrac{\sqrt{\frac{2+x}{7+x}}\sqrt{10-3x-x^2}\sqrt{\frac{-3+x^2}{-25+x^2}}}{\sqrt{3-x^2}\sqrt{4-x^2}}$

```
In[14]:= FullSimplify[
           (Sqrt[(2 + x) / (7 + x)] Sqrt[(3 - x^2) / (25 - x^2)] Sqrt[(5 + x) (2 - x)]) /
            (Sqrt[4 - x^2]Sqrt[3 - x^2])]
```

Out[14]= $\dfrac{1}{\sqrt{-(-5+x)(7+x)}}$

Whereas **Simplify** modified the expression slightly, **FullSimplify** simplified it.

Mathematica does not automatically convert $(a^b)^c$ to a^{bc} nor $(ab)^c$ to $a^c b^c$ because these conversions are certain to be correct only if c is an integer or a and b are real and nonnegative. These conversions can be done, at our discretion, with the function **PowerExpand**:

```
In[15]:= √(x^2) (y^5) //PowerExpand
```

Out[15]= $x\,y^{5/2}$

As mentioned in Section 2.1.8, *expr*// **PowerExpand** is the postfix form of **PowerExpand** [*expr*].

```
In[16]:= Clear[myexpr]
```

2.2.2 Transforming Trigonometric Expressions

With the exception of **ComplexExpand**, **Simplify**, and **FullSimplify**, the algebraic transformation functions introduced in Section 2.2.1 treat trigonometric functions as indivisible objects and leave them unchanged in algebraic manipulations. For instance, **Expand** does not rewrite the trigonometric expression $\sin(x)\cos(x)$ as $\sin(2x)/2$:

```
In[1]:= Expand[Sin[x] Cos[x]]
Out[1]= Cos[x] Sin[x]
```

Expand left the trigonometric functions unaltered. However, **ComplexExpand** expands, for instance, $\cos(x + iy)$, assuming that x and y are real:

```
In[2]:= ComplexExpand[Cos[x + I y]]
Out[2]= Cos[x] Cosh[y] - i Sin[x] Sinh[y]
```

Also, **Simplify** transforms, for example, $\sin^2(x) + \cos^2(x)$ into 1:

```
In[3]:= Simplify[Sin[x]^2 + Cos[x]^2]
Out[3]= 1
```

As with algebraic expressions, **FullSimplify** can simplify, sometimes slowly, some trigonometric expressions that are opaque to **Simplify**, because **FullSimplify** tries a wide range of transformations:

```
In[4]:= Simplify[Tan[x]^2 (3 + 3 Tan[x]^2 + Tan[x]^4)]
Out[4]= Tan[x]^2 (3 + 3 Tan[x]^2 + Tan[x]^4)

In[5]:= FullSimplify[Tan[x]^2 (3 + 3 Tan[x]^2 + Tan[x]^4)]
Out[5]= -1 + Sec[x]^6
```

Whereas **Simplify** returned the expression unchanged, **FullSimplify** made a remarkable simplification.

Mathematica provides several other functions for manipulating trigonometric expressions:

TrigExpand [*expr*]	expand trigonometric expressions out into a sum of terms
TrigFactor [*expr*]	factor trigonometric expressions into products of terms
TrigReduce [*expr*]	apply trigonometric multiple angle identities
TrigToExp [*expr*]	write trigonometric functions in terms of exponentials
ExpToTrig [*expr*]	write exponentials in terms of trigonometric functions
FunctionExpand [*expr*]	expand out certain trigonometric functions

Let us apply **TrigReduce** to the trigonometric expression $\sin^2(x)\cos(y)\sin(z)$:

```
In[6]:= TrigReduce[Sin[x]^2 Cos[y] Sin[z]]
Out[6]= 1/8 (Sin[2 x - y - z] - 2 Sin[y - z] +
            Sin[2 x + y - z] - Sin[2 x - y + z] + 2 Sin[y + z] - Sin[2 x + y + z])
```

TrigReduce turns products and powers of trigonometric functions into sums of trigonometric functions with combined arguments. Now apply **TrigFactor** to the result generated by **TrigReduce**:

```
In[7]:= TrigFactor[%]
Out[7]= Cos[y] Sin[x]^2 Sin[z]
```

We get back the original expression! In this case, **TrigFactor** is the inverse of **TrigReduce**.

Applying **TrigExpand**, for instance, to $\cos(x + y + z)$ gives a sum of products of trigonometric functions:

```
In[8]:= TrigExpand[Cos[x + y + z]]
Out[8]= Cos[x]Cos[y]Cos[z] - Cos[z]Sin[x]Sin[y] -
          Cos[y]Sin[x]Sin[z] - Cos[x]Sin[y]Sin[z]
```

Applying the function **ExpToTrig**, for example, to e^{iz} verifies Euler's formula:

```
In[9]:= ExpToTrig[Exp[I z]]
Out[9]= Cos[z] + i Sin[z]
```

FunctionExpand can expand out trigonometric functions whose arguments are rational multiples of π, integer or half-integer multiples of inverse trigonometric functions, or expressions of the form **I** *expr*, where **I** represents the imaginary unit $\sqrt{-1}$:

```
In[10]:= FunctionExpand[Tan[I (x + I y)] ((Sec[I x]/Sin[5π/3])^2)/
           (Sin[I y]/(Cos[π/5] Tan[ArcCos[z]/2]))^3]
```

$$\text{Out[10]=} -\frac{(1+\sqrt{5})^3 (1-z)^{3/2}\, \text{Csch}[y]^3\, \text{Sech}[x]^2\, \text{Tanh}[x + i\, y]}{48 (1+z)^{3/2}}$$

There is a simple method for proving trigonometric identities: To prove the identity *lhs* == *rhs*, apply **Simplify** to it and show that *Mathematica* returns **True**. Let us prove several trigonometric identities:

(a) $(1 - \cot\theta)^2 + (1 - \tan\theta)^2 = (\sec\theta - \csc\theta)^2$

(b) $2\csc 4\theta + 2\cot 4\theta = \cot\theta - \tan\theta$

(c) $\tan\dfrac{\alpha}{2}\cot\dfrac{\beta}{2} - \cot\dfrac{\alpha}{2}\tan\dfrac{\beta}{2} = \dfrac{2(\cos\beta - \cos\alpha)}{\sin\alpha\sin\beta}$

```
In[11]:= Simplify[(1 - Cot[θ])^2 + (1 - Tan[θ])^2 == (Sec]θ] - Csc[θ])^2]
Out[11]= True

In[12]:= Simplify[2 Csc[4 θ] + 2 Cot[4 θ] == Cot[θ] - Tan[θ]]
Out[12]= True

In[13]:= Simplify[Tan[α/2]Cot[β/2] - Cot[α/2]Tan[β/2] ==
           2 (Cos[β] - Cos[α]) / (Sin[α]Sin[β])]
Out[13]= True
```

2.2.3 Transforming Expressions Involving Special Functions

In addition to expanding out certain trigonometric functions as shown in Section 2.2.2, **FunctionExpand** also tries to expand out special functions, when possible, reducing compound arguments to simpler ones:

```
In[1]:= FunctionExpand[Zeta[1 - z]]
```

$$\text{Out[1]= } 2^{1-z}\pi^{-z}\,\text{Gamma[z] Sin}\left[\frac{1}{2}\pi\,(1-z)\right]\text{Zeta[z]}$$

where **Zeta[z]** is the Riemann zeta function $\zeta(z)$ and **Gamma[z]** is the Euler gamma function $\Gamma(z)$. **FunctionExpand** can establish relations involving special functions:

```
In[2]:= FunctionExpand[Beta[a, b] == Gamma[a]Gamma[b]/Gamma[a + b]]
Out[2]= True
```

where **Beta[a,b]** is the Euler beta function $B(a, b)$. **FunctionExpand** can also transform derivatives of special functions:

```
In[3]:= FunctionExpand[(1 - x^2)D[LegendreP[n, x], x]] // Simplify
Out[3]= (1 + n) (x LegendreP[n, x] - LegendreP[1 + n, x])
```

where **LegendreP[n,x]** gives the Legendre polynomial $P_n(x)$ and **D[LegendreP[n, x], x]** gives the derivative of $P_n(x)$ with respect to x. (Section 2.2.9 will discuss differentiation.)

In addition to simplifying algebraic and trigonometric expressions as discussed in Sections 2.2.1 and 2.2.2, **FullSimplify** also tries to simplify, slowly at times, expressions involving special functions using a wide range of transformations:

```
In[4]:= FullSimplify[
          BesselJ[-n, z] BesselJ[n - 1, z] + BesselJ[-n + 1, z]BesselJ[n, z]]
```

$$\text{Out[4]= } \frac{2\,\text{Sin}[n\,\pi]}{\pi\,z}$$

where **BesselJ[n, z]** gives the Bessel function of the first kind $J_n(z)$. We can use **FullSimplify** to verify relations that escape **Simplify**:

```
In[5]:= Simplify[Beta[a, b + 1] == (b/(a + b))Beta[a, b]]
```

$$\text{Out[5]= Beta[a, 1 + b]} == \frac{\text{b Beta[a, b]}}{\text{a + b}}$$

```
In[6]:= FullSimplify[Beta[a, b + 1] == (b/(a + b))Beta[a, b]]
Out[6]= True
```

2.2.4 Using Assumptions

In manipulating expressions, *Mathematica* makes as few assumptions about the variables and symbols without assigned values as possible so that the results are as general as possible.

Sometimes, we do want *Mathematica* to have more specific information about certain variables and symbols in transforming expressions. We can give the information by including assumptions as additional arguments for several transforming functions:

Refine[*expr*, *assum*]	refine *expr* using assumptions *assum*
Simplify[*expr*, *assum*]	simplify *expr* with assumptions *assum*
FullSimplify[*expr*, *assum*]	full simplify *expr* with assumptions *assum*
FunctionExpand[*expr*, *assum*]	function expand *expr* with assumptions *assum*

Assumptions can be entered as equations, inequalities, domain specifications, and logical combinations of these. Quantities appearing algebraically in inequalities are always assumed to be real. Domains can be specified as

Element[*x*, *dom*] or $x \in dom$	assert that x is an element of the domain *dom*
Element[{x_1, x_2, ...}, *dom*] or {x_1, x_2, ...}, $\in$ *dom*	assert that all the x_i are elements of the domain *dom*

Some of the domains supported by *Mathematica* are

Complexes	the domain of complex numbers
Reals	the domain of real numbers
Rationals	the domain of rational numbers
Integers	the domain of integers

To enter the operator ∈, position the cursor at the desired location in the notebook, choose **Palettes ▸ BasicMathInput**, and click the ∈ button. Let us illustrate the use of assumptions with several examples.

Simplify does nothing to $\ln(x^r)$:

```
In[1]:= Simplify[Log[x^r]]
Out[1]= Log[x^r]
```

Assuming x is a positive number and r is a real number, it simplifies the logarithmic expression:

```
In[2]:= Simplify[Log[x^r], x>0 && r ∈ Reals]
Out[2]= r Log[x]
```

where *assum*$_1$ **&&** *assum*$_2$ assumes both *assum*$_1$ and *assum*$_2$. (Section 2.2.17 will elaborate on logical operators such as **&&**.)

Simplify leaves $\cos\left(\frac{n\pi}{2} - x\right)$ unaltered:

```
In[3]:= Simplify[Cos[n π/2 - x]]
Out[3]= Cos[n π/2 - x]
```

Assuming $(n-1)/4$ is an integer (i.e., $n = \ldots, -11, -7, -3, 1, 5, 9, \ldots$), it simplifies the trigonometric expression:

```
In[4]:= Simplify[Cos[n π/2 - x], (n - 1)/4 ∈ Integers]
Out[4]= Sin[x]
```

FullSimplify leaves unchanged the expression $I_{-n}(z) - I_n(z)$:

```
In[5]:= FullSimplify[BesselI[-n, z] - BesselI[n, z]]
Out[5]= BesselI[-n, z] - BesselI[n, z]
```

where **BesselI[n, z]** gives the modified Bessel function of the first kind $I_n(z)$. With the assumption that n is an integer, it reduces the expression to 0:

```
In[6]:= FullSimplify[BesselI[-n, z] - BesselI[n, z], n ∈ Integers]
Out[6]= 0
```

FullSimplify validates the inequality

$$|\alpha + i\beta| \leq |\alpha| + |\beta|$$

only if both α and β are real numbers:

```
In[7]:= FullSimplify[Abs[α + i β] ≤ Abs[α] + Abs[β]]
Out[7]= Abs[α + i β] ≤ Abs[α] + Abs[β]

In[8]:= FullSimplify[Abs[α + i β] ≤ Abs[α] + Abs[β], {α, β} ∈ Reals]
Out[8]= True
```

De Moivre's Theorem states that if

$$z = r(\cos\theta + i\sin\theta)$$

and n is a positive integer, then

$$z^n = r^n(\cos n\theta + i\sin n\theta)$$

FullSimplify confirms the theorem:

```
In[9]:= z = r (Cos[θ] + i Sin[θ])
Out[9]= r (Cos[θ] + i Sin[θ])

In[10]:= FullSimplify[z^n == (r^n) (Cos[n θ] + i Sin[n θ]),
          r ≥ 0 && n > 0 && n ∈ Integers]
Out[10]= True
```

Whereas **Simplify** and **FullSimplify** try to simplify expressions, **Refine**[*expr*, *assum*] shows what *Mathematica* would return if the symbols in *expr* were replaced by explicit numerical expressions satisfying the assumptions *assum*. Consider $\sqrt{-b}$ for any positive real number b:

```
In[11]:= Refine[Sqrt[-b], b > 0]
Out[11]= i √b
```

For example, let $b = 3$ and *Mathematica* returns

```
In[12]:= Sqrt[-3]
Out[12]= i √3

In[13]:= Clear[z]
```

The product law for logarithms states that if x and y are positive numbers, then $\ln(xy) = \ln x + \ln y$. **FunctionExpand** abides by the product law:

```
In[14]:= FunctionExpand[Log[x y], x > 0 && y > 0]
Out[14]= Log[x] + Log[y]
```

The function **Assuming** provides another way to pose assumptions for the transforming functions discussed in this section. **Assuming**[*assum*, *transfunc*[*expr*]] evaluates *transfunc*[*expr*] with the assumptions *assum*, and *transfunc* is one of the transforming functions. For example, **Refine** transforms $\cos(\theta + n\pi)$ into $-\cos(\theta)$ if n is an odd integer:

```
In[15]:= Assuming[(n + 1)/2 ∈ Integers, Refine[Cos[θ + n π]]]
Out[15]= -Cos[θ]
```

For another example, **Simplify** transforms $\sqrt{\cosh^2 x - 1}$ into $-\sinh x$, if x is negative:

```
In[16]:= Assuming[x < 0, Simplify[√(Cosh[x]^2 - 1)]]
Out[16]= -Sinh[x]
```

(In programming to be discussed in Chapter 3, we can use **Assuming** to specify assumptions for a collection of operations. **Assuming** only affects functions that accept the **Assumptions** option.)

2.2.5 Obtaining Parts of Algebraic Expressions

Several of the functions for obtaining parts of algebraic expressions are

Coefficient[*poly*, x^i]	coefficient of x^i in the expanded form of polynomial *poly*
Coefficient[*poly*, x, 0]	terms independent of x in the expanded form of *poly*
Coefficient[*poly*, $x^i y^j z^k \ldots$]	coefficient of $x^i y^j z^k \ldots$ in the expanded form of *poly*

Numerator[*expr*]	numerator of the fractional expression *expr*
Denominator[*expr*]	denominator of the fractional expression *expr*

For example, consider the polynomial in two variables:

$$(a + 2bx^2)\,(3 + 4x + 7y^2)^2$$

Let us name it **ourpoly**

```
In[1]:= ourpoly = (a + 2 b x^2) (3 + 4 x + 7 y^2)^2
```
Out[1]= $(a + 2\,b\,x^2)\,(3 + 4\,x + 7\,y^2)^2$

Expand shows its expanded form:

```
In[2]:= ourpoly // Expand
```
Out[2]= $9\,a + 24\,a\,x + 16\,a\,x^2 + 18\,b\,x^2 + 48\,b\,x^3 + 32\,b\,x^4 + 42\,a\,y^2 + 56\,a\,x\,y^2 + 84\,b\,x^2\,y^2 + 112\,b\,x^3\,y^2 + 49\,a\,y^4 + 98\,b\,x^2\,y^4$

Coefficient gives the coefficient of $x^2\,y^4$:

```
In[3]:= Coefficient[ourpoly, x^2 y^4]
```
Out[3]= $98\,b$

Using **Collect** introduced in Section 2.2.1, we can collect together terms involving the same powers of x:

```
In[4]:= Collect[ourpoly,x]
```
Out[4]= $9\,a + 32\,b\,x^4 + 42\,a\,y^2 + 49\,a\,y^4 + x\,(24\,a + 56\,a\,y^2) + x^3\,(48\,b + 112\,b\,y^2) + x^2\,(16\,a + 18\,b + 84\,b\,y^2 + 98\,b\,y^4)$

We can pick out the coefficient of x^2:

```
In[5]:= Coefficient[ourpoly, x^2]
```
Out[5]= $16\,a + 18\,b + 84\,b\,y^2 + 98\,b\,y^4$

Here is the part independent of x:

```
In[6]:= Coefficient[ourpoly, x, 0]
```
Out[6]= $9\,a + 42\,a\,y^2 + 49\,a\,y^4$

Here is the constant term independent of both x and y:

```
In[7]:= Coefficient[%, y, 0]
```
Out[7]= $9\,a$

Consider the fractional expression

$$\frac{\sqrt{2x^4 - x^2 + 1}}{x^2 - 4}$$

Let us name it **ourexpr**:

```
In[8]:= ourexpr = Sqrt[(2x^4 - x^2 + 1)]/(x^2 - 4)
```

$$\text{Out[8]=}\ \frac{\sqrt{1 - x^2 + 2x^4}}{-4 + x^2}$$

Numerator gives the numerator:

```
In[9]:= Numerator[ourexpr]
```

$$\text{Out[9]=}\ \sqrt{1 - x^2 + 2x^4}$$

Denominator extracts the denominator:

```
In[10]:= Denominator[ourexpr]
```

$$\text{Out[10]=}\ -4 + x^2$$

(Section 3.1.3.1 will discuss general ways for obtaining parts of expressions with the functions **Part**, **Extract**, and **Select**.)

```
In[11]:= Clear[ourpoly, ourexpr]
```

2.2.6 Units, Conversion of Units, and Physical Constants

Units can be tagged to numbers in calculations.

Example 2.2.1 A race car accelerates along a straight line from 6.50 m/s at a rate of 5.00 m/s^2. What is the speed of the car after it has traveled 29.5 m?

An equation of kinematics is

$$v^2 = v_0^2 + 2ax$$

In this problem, $v_0 = 6.50\,\text{m/s}$, $a = 5.00\,\text{m/s}^2$, and $x = 29.5\,\text{m}$. Thus, v equals

```
In[1]:= Sqrt[(6.50 m/s)^2 + 2(5.00 m/s^2) (29.5 m)]
```

$$\text{Out[1]=}\ 18.3644\sqrt{\frac{\text{m}^2}{\text{s}^2}}$$

Mathematica does not automatically convert **Sqrt[z^2]** to **z** nor **Sqrt[a b]** to **Sqrt[a] Sqrt[b]** because these conversions are certain to be correct only if **z**, **a**, and **b** are real and nonnegative. The function **PowerExpand**, introduced in Section 2.2.1, can simplify the answer to the problem:

```
In[2]:= NumberForm[PowerExpand[%], 3]
```

```
Out[2]//NumberForm=
        18.4 m
        ------
          s
```

where **NumberForm**[*expr*, *n*] prints *expr* with all approximate real numbers showing at most *n* significant figures. ■

The package **Units`** contains the functions **Convert**, **ConvertTemperature**, and **SI** for conversion of units. To use these functions, we must first load the package:

```
In[3]:= Needs["Units`"]
```

Convert[*old*, *new*] converts *old* to a form involving the combination of units *new*. The names of units must be spelled out in full. What follow are some unit names.

- Electrical units
  Amp, **Coulomb**, **Farad**, **Gilbert**, **Henry**, **Mho**, **Ohm**, **Siemens**, **Statampere**, **Statcoulomb**, **Statfarad**, **Stathenry**, **Statohm**, **Statvolt**, **Volt**
- Units of length
  AU, **Centimeter**, **Fathom**, **Feet**, **Fermi**, **Foot**, **Inch**, **LightYear**, **Meter**, **Micron**, **Mil**, **Mile**, **NauticalMile**, **Parsec**, **Rod**, **Yard**
- Units of time
  Day, **Hour**, **Minute**, **Month**, **Second**, **Year**
- Units of mass
  AtomicMassUnit, **Gram**, **Kilogram**, **MetricTon**, **Slug**, **SolarMass**
- Units of force
  Dyne, **Newton**, **Poundal**, **PoundForce**
- Units of radiation
  Becquerel, **Curie**, **Gray**, **Rad**, **Roentgen**, **Rontgen**, **Rutherford**
- Units of power
  HorsePower, **Watt**
- Magnetic units
  BohrMagneton, **Gauss**, **Maxwell**, **NuclearMagneton**, **Oersted**, **Tesla**, **Weber**
- Units of pressure
  Atmosphere, **Bar**, **InchMercury**, **MillimeterMercury**, **Pascal**, **Torr**
- Units of energy
  BTU, **Calorie**, **ElectronVolt**, **Erg**, **Joule**, **Rydberg**, **Therm**
- Others
  Hertz, **Knot**, **Liter**

For instance, we can convert the density of water $1.00 \times 10^3\,\mathrm{kg/m^3}$ at STP to $\mathrm{g/cm^3}$:

```
In[4]:= Convert[1.00×10^3 Kilogram/Meter^3, Gram/Centimeter^3]
Out[4]= 1.Gram
        -----------
        Centimeter^3
```

Prefixes can be given for units. They must be entered as separate words. Some prefixes used in the SI system are **Centi**, **Deca**, **Deci**, **Femto**, **Giga**, **Hecto**, **Kilo**, **Mega**, **Micro**, **Milli**, **Nano**, **Peta**, **Pico**, and **Tera**. As an example, convert 55.0 mi/h to km/h:

```
In[5]:= NumberForm[Convert[55.0 Mile/Hour, Kilo Meter/Hour], 3]
Out[5]//NumberForm=
        88.5 Kilo Meter
        ---------------
             Hour
```

Temperature conversions require **ConvertTemperature**[*temp*, *old*, *new*] that converts *temp* from the *old* scale to the *new* scale. The function **Convert** cannot be used because temperature conversions are not multiplicative. Some temperature scales are **Celsius**, **Fahrenheit**, and **Kelvin**. Let us, for example, convert 50°F to Celsius:

```
In[6]:= ConvertTemperature[50, Fahrenheit, Celsius]
Out[6]= 10
```

SI[*expr*] converts *expr* to the SI system. As an example, a superconducting magnet generates a magnetic field of 3×10^5 G. What is the field in the SI system?

```
In[7]:= SI[3×10^5 Gauss]
Out[7]= 30 Tesla
```

The package **PhysicalConstants`** contains the values of many physical constants. Again, full names of the constants must be invoked. Here are some available constants:

AccelerationDueToGravity	**ElectronComptonWavelength**	**PlanckConstant**
AgeOfUniverse	**ElectronMagneticMoment**	**PlanckConstantReduced**
AvogadroConstant	**ElectronMass**	**ProtonMagneticMoment**
BohrRadius	**FineStructureConstant**	**ProtonMass**
BoltzmannConstant	**GravitationalConstant**	**RydbergConstant**
DeuteronMass	**HubbleConstant**	**SpeedOfLight**
EarthMass	**MagneticFluxQuantum**	**StefanConstant**
EarthRadius	**MolarGasConstant**	**VacuumPermeability**
ElectronCharge	**NeutronMass**	**VacuumPermittivity**

After loading the package

```
In[8]:= Needs["PhysicalConstants`"]
```

we can determine, for example, the rest energy of an electron:

```
In[9]:= Convert[ElectronMass SpeedOfLight^2, Joule]
Out[9]= 8.1871 × 10^-14 Joule
```

2.2.7 Assignments and Transformation Rules

We can assign symbolic as well as numerical values to variables. To make an assignment, enter

variable = *value*

For example, let us expand $(a+b)(a-b)$ and assign the value to the variable **t**:

```
In[1]:= t = (a + b) (a - b) // Expand
Out[1]= a^2 - b^2
```

Now, **t** has the value $a^2 - b^2$. Thus, adding b^2 to **t** gives a^2:

```
In[2]:= t + b^2
Out[2]= a^2
```

Names of variables can consist of any number of alphanumeric characters. In addition to permitting the use of Greek letters, *Mathematica* also permits the use of subscripts and other letterlike forms to be discussed in Section 2.5. The only restriction is that a name cannot begin with a number. For instance, **3x** is not an acceptable name, whereas **x3** is (**3x** stands for **3** times **x**). By convention, user-defined variables should begin with lowercase letters because names of built-in *Mathematica* objects start with uppercase letters. Yet we may forgo this convention if it collides with the common symbol usage in physics. For example, we use for volume the uppercase letter **V** rather than the lowercase letter **v**. In any case, for user-defined variables, be sure not to use names such as **C**, **D**, **E**, **I**, **N**, and **O** that have built-in meanings in *Mathematica*.

When new values are assigned to variables, old values are discarded. For example, assign the value d^2 to **t** in the previous example and then add b^2 to it:

```
In[3]:= t = d^2
Out[3]= d^2

In[4]:= t + b^2
Out[4]= b^2 + d^2
```

The answer for $\texttt{t} + b^2$ is no longer a^2 since **t** now has the value d^2.

Assignments remain in a *Mathematica* session until the values are changed or removed. Forgetting that certain variables were assigned values earlier in a session and proceeding to use them as formal variables is a common mistake. Let us, for example, solve numerically the differential equation

$$\frac{d^2x}{dt^2} + \frac{dx}{dt} + x = 0$$

with $x = 1$ and $dx/dt = 0$ at $t = 0$:

```
In[5]:= NDSolve[{x''[t] + x'[t] + x[t] == 0, x'[0] == 0, x[0] == 1}, x, {t, 0, 5}]
        NDSolve::dsvar: d^2 cannot be used as a variable. >>
Out[5]= NDSolve[{x[d^2] + x'[d^2] + x''[d^2] == 0, x'[0] == 0, x[0] == 1}, x, {d^2, 0, 5}]
```

We get an error message. What happened is that **t** was assigned the value d^2 in the previous example.

To avoid the mistake of forgetting previous assignments, we should develop the habit of removing unneeded values as soon as possible with either the deassignment operator **=.** or the function **Clear**. Let us apply **Clear** to **t** and solve the differential equation again:

```
In[6]:= Clear[t]

In[7]:= NDSolve[{x''[t] + x'[t] + x[t] == 0, x'[0] == 0, x[0] == 1}, x, {t, 0, 5}]
Out[7]:= {{x -> InterpolatingFunction[{{0., 5.}}, < >]}}
```

The equation is solved.

Example 2.2.2 In the free electron model of metal, the number per unit volume of electrons with an energy between ε_1 and ε_2 is given as

$$n = \frac{8\sqrt{2}\,\pi\, m^{3/2}}{h^3} \int_{\varepsilon_1}^{\varepsilon_2} \frac{\varepsilon^{1/2}}{e^{(\varepsilon - \varepsilon_F)/kT} + 1}\, d\varepsilon$$

where m is the electron mass, h is Planck's constant, k is Boltzmann's constant, and T is the temperature. (In the integral, we have replaced the chemical potential μ by the Fermi energy ε_F, as $|\mu - \varepsilon_F|/\varepsilon_F \ll 1$.) The Fermi energy ε_F can be determined from

$$\varepsilon_F = \left(\frac{3}{8\pi}\right)^{2/3} \frac{h^2}{2m} \left(\frac{N}{V}\right)^{2/3}$$

where N/V is the electron number density. For sodium at room temperature (293 K), what is the number per unit volume of electrons with an energy between the Fermi energy and 0.15 eV above the Fermi energy?

Since each sodium atom contributes one valence electron to become a conduction electron, the number density of free electrons is the same as that of the sodium atoms and therefore

$$N/V = \frac{\rho N_A}{M}$$

where N_A is Avogadro's number, and ρ and M are, respectively, the density and mass per mole of sodium. The formula for the Fermi energy becomes

$$\varepsilon_F = \left(\frac{3}{8\pi}\right)^{2/3} \frac{h^2}{2m} \left(\frac{\rho N_A}{M}\right)^{2/3}$$

To obtain numerical values for the physical constants and to make unit conversions, we load two packages (if they are not already loaded):

```
In[8]:= Needs["PhysicalConstants`"]

In[9]:= Needs["Units`"]
```

Let us make several assignments:

```
In[10]:= m = ElectronMass
Out[10]= 9.10938 × 10^-31 Kilogram

In[11]:= N = AvogadroConstant
Out[11]= 6.02214 × 10^23 / Mole

In[12]:= h = PlanckConstant
Out[12]= 6.62607 × 10^-34 Joule Second
```

where the Greek letter **N** is the alias of N_A. (For a reminder on using Greek letters, see Section 1.6.2.)

ElementData["*name***", "***property***"]** gives the value of the specified property for the chemical element "*name*". Let us make a couple more assignments:

```
In[13]:= M = ElementData["Sodium", "AtomicWeight"] (10^(-3)) Kilogram/Mole
Out[13]= 0.022989770 Kilogram / Mole

In[14]:= ρ = ElementData["Sodium", "Density"] Kilogram/Meter^3
Out[14]= 968. Kilogram / Meter^3
```

The Fermi energy is

```
In[15]:= ϵ = Convert[((3/(8 π))^(2/3)) ((h^2)/(2 m)) ((ρN/M)^(2/3)) //
            PowerExpand, ElectronVolt]
Out[15]= 3.14725 ElectronVolt
```

where **ϵ** is the alias of ε_F.

In order to evaluate the integral numerically for the number per unit volume of electrons with an energy between the Fermi energy and 0.15 eV above the Fermi energy, we must express the integral in terms of dimensionless quantities. Let us define

$$x = \frac{\varepsilon - \varepsilon_F}{\text{eV}}, \quad e = \varepsilon_F/\text{eV}, \quad b = \frac{\text{eV}}{kT}$$

and write

$$n = \frac{8\sqrt{2}\pi m^{3/2}}{h^3} \int_0^{0.15} \frac{(x+e)^{1/2}(\text{eV})^{3/2}}{e^{bx}+1}\, dx$$

Since

```
In[16]:= e = ϵ /ElectronVolt
Out[16]= 3.14725

In[17]:= b = (ElectronVolt/Convert[BoltzmannConstant 293 Kelvin, ElectronVolt])
Out[17]= 39.6058
```

we have

```
In[18]:= n = Convert[(8 Sqrt[2] π (m^(3/2)) (ElectronVolt^(3/2)) / (h^3))
           NIntegrate[((x + e)^(1/2)) / (Exp[b x] + 1), {x, 0, 0.15}],
          Meter^-3] // PowerExpand

         2.11683 × 10^26
Out[18]= ---------------
            Meter^3

In[19]:= Clear[M, ρ, m, N, ϵ, e, h, b, n]
```

■

Clearing the declared values as they are no longer needed is definitely a good habit. Should we also routinely apply at the beginning of a calculation the function **Clear** to all the variables to be used later just in case they possess preassigned values? That is a good practice if we know ahead of time the names of the variables. Unlike other languages such as C and Pascal, *Mathematica* does not require commencing calculations with variable declarations. Therefore, we often do not know all the variable names until the end of the computation. Beginning a new *Mathematica* session, like starting with a clean slate, for each problem is perhaps a better, albeit time-consuming practice. An alternative, almost as good as starting with a clean slate, is to execute the command **Clear["Global`*"]**, which clears the values for all symbols in the **Global`** context, except those that are protected. (Section 3.3.5 will discuss symbol protection, and Section 3.7.1 will introduce *Mathematica* contexts.) Automatic syntax coloring also lets us know whether a symbol in the **Global`** context has an assigned value. Without assigned values, global symbols in *Mathematica* input are, by default, blue. As soon as values are assigned to them, they become black.

A simple way to avoid forgetting earlier assignments is to refrain from making assignments whenever possible. Replacements of variables with values can be made with transformation rules. A transformation rule is written as *variable* **->** *value*, with no space between "**-**" and "**>**".

The transformation rule is applied to an expression with the replacement operator `/.`, with no space between the slash "`/`" and the dot "`.`". To make a replacement, write *expr* `/.` variable

`->` value

We can read "`/.`" as "given that" and "`->`" as "goes to." For example, "**`2 x + 10 /. x -> 10`**" is interpreted as **`2 x + 10`** given that **`x`** goes to **`10`**:

```
In[20]:= 2 x + 10 /. x → 10
Out[20]= 30
```

Note that *Mathematica* automatically replaces "`->`" with "`→`", which is a *Mathematica* special character having the same built-in meaning as that of "`->`". (Section 2.5.1 will discuss the special characters of *Mathematica*.) The advantage of a replacement over an assignment is that it only affects the specific expression. The variable **`x`** in the preceding example can still be used as a formal variable. Let us ask for **`x`**:

```
In[21]:= x
Out[21]= x
```

Indeed, **`x`** has no global value.

We can also replace a subpart of an expression with a value by writing

expr `/.` *lhs* `->` *rhs*

where the value *rhs* is substituted for the subpart *lhs* in the expression *expr*. For example, we can determine the rest energy of an electron in MeV. Since the packages **`Units``** and **`PhysicalConstants``** are already loaded, we have, for the rest energy of an electron,

```
In[22]:= Convert[ElectronMass SpeedOfLight^2, ElectronVolt] /.
           ElectronVolt → 10^-6 MeV
Out[22]= 0.510999 MeV
```

2.2.8 Equation Solving

Section 2.1.16 discussed solving equations numerically. This section considers exact solution of equations. `Solve[`*lhs* `==` *rhs*`,` *var*`]` attempts to solve exactly an equation for the variable *var*. For example, consider solving the equation

$$x^2 - 3x - 10 = 0$$

```
In[1]:= Solve[x^2 - 3 x - 10 == 0, x]
Out[1]= {{x → -2}, {x → 5}}
```

`Solve` returns a nested list of transformation rules stating that the equation is satisfied if x is replaced by -2 or 5. We can verify the solutions with the replacement operator `/.` (again, with no space between the slash and the dot):

```
In[2]:= x^2 - 3x - 10 == 0 /. %
Out[2]= {True, True}
```

Mathematica reports that the equation is true if x is replaced by -2 or 5. To obtain a list of the solutions, apply the transformation rules to **x** in the form

```
In[3]:= x /. %%
Out[3]= {-2, 5}
```

Although **Solve** is intended primarily for solving polynomial equations, it can solve some other equations. As an example, consider a one-dimensional relativistic particle of mass m moving at a velocity v. Its momentum is

$$p = \frac{mv}{\sqrt{1 - v^2/c^2}}$$

where c is the speed of light. Let us solve for v:

```
In[4]:= Solve[p == (m v)/Sqrt[1 - v^2/c^2], v]
```

$$\text{Out[4]= } \left\{\left\{v \to -\frac{c\,p}{\sqrt{c^2\,m^2 + p^2}}\right\}, \left\{v \to \frac{c\,p}{\sqrt{c^2\,m^2 + p^2}}\right\}\right\}$$

The first solution is, of course, physically unacceptable because it implies that the velocity and momentum have opposite directions. For another example, solve the equation

$$9\cos^2 x + 8\cos x = 1$$

```
In[5]:= Solve[9 Cos[x]^2 + 8 Cos[x] == 1, x]
      Solve::ifun :
      Inverse functions are being used by Solve, so some solutions may not
        be found; use Reduce for complete solution information. >>
```

$$\text{Out[5]= } \left\{\{x \to -\pi\}, \{x \to \pi\}, \left\{x \to -\text{ArcCos}\left[\frac{1}{9}\right]\right\}, \left\{x \to \text{ArcCos}\left[\frac{1}{9}\right]\right\}\right\}$$

The warning message alerts us to the fact that for every solution to an equation of the form $\cos(x) = a$, there are an infinite number of solutions differing by multiples of 2π. (For information on the function **Reduce** mentioned in the warning message, see the last paragraph of this section, Section 2.2.8.)

Mathematically, complicated equations such as transcendental equations and polynomial equations with degrees higher than 4 do not always have exact solutions. Consider, for example, the transcendental equation

$$-5 + x^2 e^{-x} + x = 0$$

```
In[6]:= Solve[-5 + (x^2) Exp[-x] + x == 0, x]
        Solve::tdep : The equations appear to involve the
            variables to be solved for in an essentially non-algebraic way. >>
Out[6]= Solve[-5 + x + e^-x x^2 == 0, x]
```

Mathematica cannot find exact solutions for the equation and returns the input slightly rearranged. For another example, consider solving the polynomial equation

$$x^5 - 3x^4 + 2x^3 + x = 4$$

```
In[7]:= Solve[x^5 - 3 x^4 + 2 x^3 + x == 4, x]
Out[7]= {{x → Root[-4 + #1 + 2 #1^3 - 3 #1^4 + #1^5 &, 1]},
         {x → Root[-4 + #1 + 2 #1^3 - 3 #1^4 + #1^5 &, 2]},
         {x → Root[-4 + #1 + 2 #1^3 - 3 #1^4 + #1^5 &, 3]},
         {x → Root[-4 + #1 + 2 #1^3 - 3 #1^4 + #1^5 &, 4]},
         {x → Root[-4 + #1 + 2 #1^3 - 3 #1^4 + #1^5 &, 5]}}
```

The result indicates that **Solve** fails to produce explicit solutions. (**Root**[*f*, *k*] represents the *k*th root of the polynomial equation *f*[*x*] **== 0**, where *f* must be a pure function, to be introduced in Section 3.3.1.) We can, of course, determine approximate solutions to the equation numerically:

```
In[8]:= %//N
Out[8]= {{x → 2.15819}, {x → -0.637669 - 0.748825 i}, {x → -0.637669 + 0.748825 i},
         {x → 1.05857 - 0.89183 i}, {x → 1.05857 + 0.89183 i}}
```

Solve[{*lhs*$_1$ == *rhs*$_1$, *lhs*$_2$ == *rhs*$_2$, ...}, {*var*$_1$, *var*$_2$, ...}] attempts to solve a set of equations exactly for the variables *var*$_1$, *var*$_2$, etc. For instance, we can solve the system of equations

$$x + y = z$$

$$10 - 6x - 2z = 0$$

$$6x - 24 - 4y = 0$$

```
In[9]:= Solve[{x + y == z, 10 - 6 x - 2 z == 0, 6 x - 24 - 4 y == 0}, {x, y, z}]
Out[9]= {{x → 2, y → -3, z → -1}}
```

Here is a problem from direct current circuits.

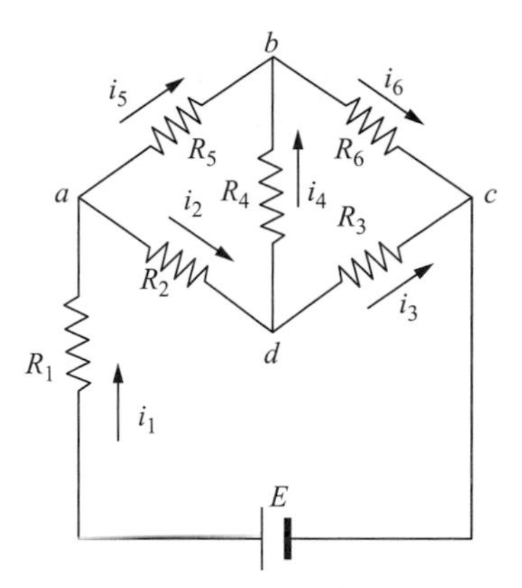

Figure 2.2.1. A multiloop circuit.

Example 2.2.3 For the circuit in Figure 2.2.1, find all currents in terms of the resistances and the emf. Assume that the internal resistance of the battery is negligible.

We have arbitrarily assigned directions to the currents. If we assumed a wrong direction for a current, its resulting value will be negative but the magnitude will be correct. Kirchhoff's rules applied to the circuit give

junction *a*	$i_1 - i_2 - i_5 = 0$
junction *b*	$i_6 - i_4 - i_5 = 0$
junction *d*	$i_2 - i_3 - i_4 = 0$
loop *abda*	$-i_5 R_5 + i_4 R_4 + i_2 R_2 = 0$
loop *dbcd*	$-i_4 R_4 - i_6 R_6 + i_3 R_3 = 0$
loop *adca*	$-i_2 R_2 - i_3 R_3 + E - i_1 R_1 = 0$

Since **E** is a built-in *Mathematica* object representing the exponential constant e, let **emf** be the alias of E. **Solve** solves the system of linear equations:

```
In[10]:= sol = Solve[{i1 - i2 - i5 == 0, i6 - i4 - i5 == 0, i2 - i3 - i4 == 0,
          -i5 R5 + i4 R4 + i2 R2 == 0, -i4 R4 - i6 R6 + i3 R3 == 0,
          -i2 R2 - i3 R3 + emf - i1 R1 == 0}, {i1, i2, i3, i4, i5, i6}] // Simplify
Out[10]= {{i1 -> (emf (R4 R5 + R3 (R4 + R5) + R4 R6 + R5 R6 + R2 (R3 + R4 + R6)))/
             (R1 (R4 R5 + R3 (R4 + R5) + R4 R6 + R5 R6 + R2 (R3 + R4 + R6)) +
               R3 (R5 R6 + R4 (R5 + R6)) + R2 (R5 R6 + R3 (R5 + R6) + R4 (R5 + R6))),
          i6 -> (emf (R2 (R3 + R4) + R3 (R4 + R5)))/
             (R1 (R4 R5 + R3 (R4 + R5) + R4 R6 + R5 R6 + R2 (R3 + R4 + R6)) +
               R3 (R5 R6 + R4 (R5 + R6)) + R2 (R5 R6 + R3 (R5 + R6) + R4 (R5 + R6))),
          i3 -> (emf ((R2 + R5) R6 + R4 (R5 + R6)))/
             (R1 (R4 R5 + R3 (R4 + R5) + R4 R6 + R5 R6 + R2 (R3 + R4 + R6)) +
               R3 (R5 R6 + R4 (R5 + R6)) + R2 (R5 R6 + R3 (R5 + R6) + R4 (R5 + R6))),
```

```
i2 → (emf (R3 R5 + R5 R6 + R4 (R5 + R6)))/
   (R1 (R4 R5 + R3 (R4 + R5) + R4 R6 + R5 R6 + R2 (R3 + R4 + R6)) +
     R3 (R5 R6 + R4 (R5 + R6)) + R2 (R5 R6 + R3 (R5 + R6) + R4 (R5 + R6))),
i4 → (emf (R3 R5 - R2 R6))/
   (R1 (R4 R5 + R3 (R4 + R5) + R4 R6 + R5 R6 + R2 (R3 + R4 + R6)) +
     R3 (R5 R6 + R4 (R5 + R6)) + R2 (R5 R6 + R3 (R5 + R6) + R4 (R5 + R6))),
i5 → (emf (R3 R4 + R2 (R3 + R4 + R6)))/
   (R1 (R4 R5 + R3 (R4 + R5) + R4 R6 + R5 R6 + R2 (R3 + R4 + R6)) +
     R3 (R5 R6 + R4 (R5 + R6)) + R2 (R5 R6 + R3 (R5 + R6) + R4 (R5 + R6)))}}
```

Checking answers with special cases is a good practice because computer algebra systems are known to have bugs and because we can gain insight into the problems. Let us consider the case in which $R_2 = R_3 = R_5 = R_6 = R$. The symmetry of the circuit requires that $i_4 = 0$ and $i_2 = i_3 = i_5 = i_6 = i_1/2$:

```
In[11]:= sol /. {R2 → R, R3 → R, R5 → R, R6 → R} // Simplify
```

$$Out[11]= \left\{\left\{i1 \to \frac{emf}{R + R1},\ i6 \to \frac{emf}{2\,(R + R1)},\ i3 \to \frac{emf}{2\,(R + R1)},\ i2 \to \frac{emf}{2\,(R + R1)},\ i4 \to 0,\ i5 \to \frac{emf}{2\,(R + R1)}\right\}\right\}$$

Indeed, it is so.

```
In[12]:= Clear[sol]
```

■

In addition to **Solve**, *Mathematica* provides two other functions for solving equations: **Reduce** gives all the possible solutions to a set of equations and **RSolve** solves recurrence equations. For information on **Reduce**, see Example 2.2.15 in Section 2.2.17 and Exercises 95 and 106 in Section 2.2.20; for information on **RSolve**, see Exercises 102–104 in Section 2.2.20.

2.2.9 Differentiation

D[f, x] gives

$$\frac{\partial f}{\partial x}$$

the partial derivative of a function f of several variables with respect to x. It simply gives the ordinary derivative of f with respect to x if f is a function of x only. For instance, we can differentiate

$$\frac{\cos 4x}{1 - \sin 4x}$$

simplify the result, and evaluate it at, for example, $x = 0$:

```
In[1]:= D[Cos[4 x]/(1 - Sin[4 x]), x]
```

$$\text{Out[1]= } \frac{4\,\text{Cos}[4\,x]^2}{(1-\text{Sin}[4\,x])^2} - \frac{4\,\text{Sin}[4\,x]}{1-\text{Sin}[4\,x]}$$

```
In[2]:= FullSimplify[%]
```

$$\text{Out[2]= } -\frac{4}{-1+\text{Sin}[4\,x]}$$

```
In[3]:= % /. x → 0
Out[3]= 4
```

`D[`f`, {`x`, `n`}]` gives

$$\frac{\partial^n f}{\partial x^n}$$

the nth partial derivative of f with respect to x. For example, we can determine the fourth derivative of

$$4x^2 - 5x + 8 - \frac{3}{x}$$

```
In[4]:= D[4 x^2 - 5 x + 8 - 3/x, {x, 4}]
```

$$\text{Out[4]= } -\frac{72}{x^5}$$

`D[`f`, `x_1`, `x_2`, ...]` gives

$$\frac{\partial}{\partial x_1}\frac{\partial}{\partial x_2}\cdots f$$

the multiple partial derivative of f with respect to $x_1, x_2, \ldots$. Let us find the second partial derivative

$$\frac{\partial^2 f}{\partial x\,\partial y}$$

for

$$f(x, y) = x^3y^2 - 2x^2y + 3x$$

```
In[5]:= D[(x^3)(y^2) - 2(x^2)y + 3 x, x, y]
```

$$\text{Out[5]= } -4\,x + 6\,x^2\,y$$

`Dt[`f`]` gives the differential (also called the total differential) df. The following are two examples illustrating the use of the function `Dt` for differentials. (For a discussion of differentials, see [Stewart03].)

Example 2.2.4 A rectangular box measures 75 cm × 60 cm × 40 cm, and each dimension is correct to within 0.2 cm. Estimate the largest possible fractional error in the volume.

If x, y, and z denote the dimensions of the box, its volume V is

```
In[6]:= V = x y z
Out[6]= x y z
```

and its differential dV is

```
In[7]:= Dt[V]
Out[7]= y z Dt[x] + x z Dt[y] + x y Dt[z]
```

That is,

$$dV = y\,z\,dx + x\,z\,dy + x\,y\,dz$$

With the given data, the largest possible error in the volume is

```
In[8]:= % /. {x → 75 cm, y → 60 cm, z → 40 cm,
           Dt[x] → 0.2 cm, Dt[y] → 0.2 cm, Dt[z] → 0.2 cm}
Out[8]= 1980. cm^3
```

and the largest possible fractional error is

```
In[9]:= %/(x y z) /. {x → 75 cm, y → 60 cm, z → 40 cm}
Out[9]= 0.011
```

which is approximately 1%.

```
In[10]:= Clear[V]
```

■

Example 2.2.5 The internal energy U of a system is a function of its entropy S, volume V, and number of particles n:

$$U = U(S, V, n)$$

The partial derivatives of U with respect to the independent variables can be expressed in terms of the temperature T, pressure P, and chemical potential μ:

$$\left(\frac{\partial U}{\partial S}\right)_{V,n} = T$$

$$\left(\frac{\partial U}{\partial V}\right)_{S,n} = -P$$

$$\left(\frac{\partial U}{\partial n}\right)_{S,V} = \mu$$

Find the differential dU.

The differential dU is

```
In[11]:= dU == Dt[U[S, V, n]]
Out[11]= dU == Dt[n] U^(0,0,1)[S, V, n] + Dt[V] U^(0,1,0)[S, V, n] + Dt[S]U^(1,0,0)[S, V, n]
```

where `U`$^{(i,j,k)}$`[S,V,n]` is the partial derivative of U taken i times with respect to S, j times with respect to V, and k times with respect to n. We can write the result in more familiar notation:

```
In[12]:= %/. {Dt[S] → dS, Dt[V] → dV, Dt[n] → dn,
              U^(1,0,0)[S, V, n] → T, U^(0,1,0)[S, V, n] → -P, U^(0,0,1)[S, V, n] → μ}
Out[12]= dU == -dV P + dS T + dn μ
```

which is a well-known relation in thermodynamics. ■

`Dt[`f`, `x`]` gives

$$\frac{df}{dx}$$

the total derivative of a function f of several variables with respect to x. (It simply gives the ordinary derivative of f with respect to x if f is a function of x only.) Whereas all other variables are held constant in partial differentiation of f with respect to x, here all variables are allowed to vary with x in total differentiation of f with respect to x. `Dt[`f`,`x`, Constants -> {`c_1`, `c_2`,...}]` gives the total derivative of f with respect to x with the specification that c_1, $c_2, \ldots$ are constants. Let us illustrate the use of the function `Dt` for total derivatives with three examples: two from calculus and one from classical mechanics.

Example 2.2.6 If

$$z = x^2\, y + 3\, x\, y^4$$

find the total derivative $\frac{dz}{dt}$, evaluate it when $x = \sin 2t$ and $y = \cos t$, and determine its value at $t = 0$.

With the assignment

```
In[13]:= z = (x^2) y + 3 x (y^4)
Out[13]= x^2 y + 3 x y^4
```

the total derivative $\frac{dz}{dt}$ is

```
In[14]:= Dt[z, t]
Out[14]= 2 x y Dt[x, t] + 3 y^4 Dt[x, t] + x^2 Dt[y, t] + 12 x y^3 Dt[y, t]
```

If $x = \sin 2t$ and $y = \cos t$, it becomes

```
In[15]:= %/.{x→Sin[2t],y→Cos[t]}
Out[15]= 6Cos[t]^4 Cos[2t]+4Cos[t] Cos[2t] Sin[2t]-
          12Cos[t]^3 Sin[t]Sin[2t]-Sin[t] Sin[2t]^2
```

At $t = 0$, its value is

```
In[16]:= %/.t→0
Out[16]= 6
```

■

Example 2.2.7 Given the equation of a circle:

$$x^2 + y^2 = 25$$

find

$$\frac{dy}{dx}$$

and determine an equation of the tangent to the circle at the point (3, 4).

Let us differentiate both sides of the equation of the circle and solve for $\frac{dy}{dx}$:

```
In[17]:= Dt[x^2+y^2==25, x]
Out[17]= 2x+2yDt[y, x]==0

In[18]:= Solve[%, Dt[y, x]]
Out[18]= {{Dt[y, x] → -x/y}}
```

That is, $\frac{dy}{dx} = -x/y$. Thus, the slope of the tangent to the circle at the point (3, 4) is

```
In[19]:= m=Dt[y, x]/.%[[1]]/.{x→3, y→4}
Out[19]= -3/4
```

In the point-slope form of the equation of a line, the equation of the tangent is therefore

```
In[20]:= y-4==m(x-3)
Out[20]= -4+y==-3/4(-3+x)
```

This equation can be written as

```
In[21]:= %//Simplify
Out[21]= 3x+4y==25
```

■

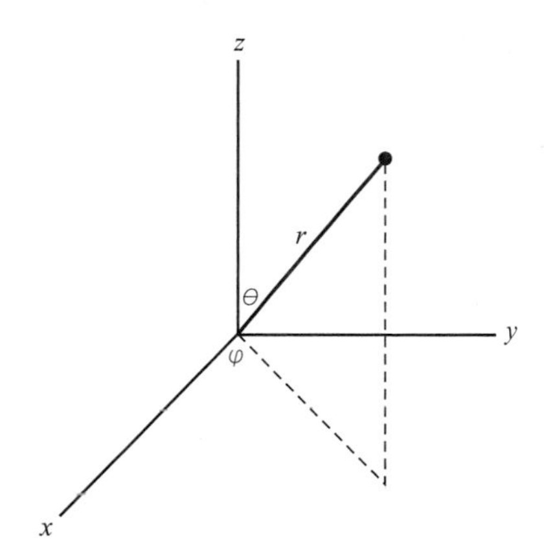

Figure 2.2.2. Spherical coordinates (r, θ, φ).

Example 2.2.8 The spherical coordinates (r, θ, φ) of a point are shown in Figure 2.2.2. Let $\mathbf{e}_1, \mathbf{e}_2$, and $\mathbf{e}_3$ be the unit vectors along the x, y, and z directions, respectively. The unit vectors $\mathbf{e}_r, \mathbf{e}_\theta$, and $\mathbf{e}_\varphi$ along the r, θ, and φ directions, respectively, can be expressed as

$$\mathbf{e}_r = \mathbf{e}_1 \sin\theta\cos\varphi + \mathbf{e}_2 \sin\theta\sin\varphi + \mathbf{e}_3 \cos\theta$$

$$\mathbf{e}_\theta = \mathbf{e}_1 \cos\theta\cos\varphi + \mathbf{e}_2 \cos\theta\sin\varphi - \mathbf{e}_3 \sin\theta$$

$$\mathbf{e}_\varphi = -\mathbf{e}_1 \sin\varphi + \mathbf{e}_2 \cos\varphi$$

Whereas the unit vectors $\mathbf{e}_1$, $\mathbf{e}_2$, and $\mathbf{e}_3$ are always fixed, the unit vectors $\mathbf{e}_r$, $\mathbf{e}_\theta$, and $\mathbf{e}_\varphi$ vary with time t when r, θ, and φ are functions of t. Any vector $\mathbf{A}$ can be represented in terms of its spherical components:

$$\mathbf{A} = A_r\mathbf{e}_r + A_\theta\mathbf{e}_\theta + A_\varphi\mathbf{e}_\varphi$$

(a) Show

$$\begin{aligned}\frac{d\mathbf{A}}{dt} &= \left(\frac{dA_r}{dt} - A_\theta\frac{d\theta}{dt} - A_\varphi \sin\theta\frac{d\varphi}{dt}\right)\mathbf{e}_r \\ &+ \left(\frac{dA_\theta}{dt} + A_r\frac{d\theta}{dt} - A_\varphi \cos\theta\frac{d\varphi}{dt}\right)\mathbf{e}_\theta \\ &+ \left(\frac{dA_\varphi}{dt} + A_r \sin\theta\frac{d\varphi}{dt} + A_\theta \cos\theta\frac{d\varphi}{dt}\right)\mathbf{e}_\varphi\end{aligned}$$

(b) In spherical coordinates the position vector is

$$\mathbf{R} = r\mathbf{e}_r$$

Show that the velocity **v** can be expressed as

$$\mathbf{v} = \frac{d\mathbf{R}}{dt} = \left(\frac{dr}{dt}\right)\mathbf{e}_r + \left(r\frac{d\theta}{dt}\right)\mathbf{e}_\theta + \left(r\sin\theta\frac{d\varphi}{dt}\right)\mathbf{e}_\varphi$$

```
In[22]:= Clear["Global`*"]
```

(a) To prove the equation, we first enter it:

```
In[23]:= Dt[A, t, Constants→{e1, e2, e3}] ==
          (Dt[Ar, t, Constants→{e1, e2, e3}] -
             Aθ Dt[θ, t, Constants→{e1, e2, e3}] -
             Aφ Sin[θ] Dt [φ, t, Constants→{e1, e2, e3}])er +
           (Dt[Aθ, t, Constants→{e1, e2, e3}] + Ar Dt[θ, t, Constants→
                {e1, e2, e3}] - Aφ Cos[θ] Dt [φ, t, Constants→{e1, e2, e3}])eθ +
           (Dt[Aφ, t, Constants→{e1, e2, e3}] +
             Ar Sin[θ] Dt [φ, t, Constants→{e1, e2, e3}] +
             Aθ Cos[θ] Dt [φ, t, Constants→{e1, e2, e3}])eφ;
```

Then we show that it evaluates to **`True`** after the appropriate replacements have been made:

```
In[24]:= %/. {A->Ar er + Aθ eθ + Aφ eφ} /.
          {er->e1 Sin[θ] Cos[φ] + e2 Sin[θ]Sin[φ] + e3 Cos[θ],
           eθ->e1 Cos[θ] Cos[φ] + e2 Cos[θ]Sin[φ] - e3 Sin[θ],
           eφ->-e1 Sin[φ] + e2 Cos[φ]} //Simplify
Out[24]= True
```

(b) The well-known equation for the velocity in spherical coordinates is a special case of the equation established in part (a):

```
In[25]:= %%/. {A→R, Ar→r, Aθ→0, Aφ→0}
Out[25]= Dt[R, t, Constants→{e1, e2, e3}] ==
           er Dt[r, t, Constants→{e1, e2, e3}] +
            eθ r Dt[θ, t, Constants→{e1, e2, e3}] +
            eφ r Dt[φ, t, Constants→{e1, e2, e3}] Sin[θ]
```

■

2.2.10 Integration

`Integrate``[f, x]` gives the indefinite integral

$$\int f dx$$

For example, we can find

$$\int \frac{x^4}{a^2 + x^2}\, dx$$

```
In[1]:= Integrate[(x^4)/(a^2+x^2), x]
```

Out[1]= $-a^2 x + \frac{x^3}{3} + a^3 \text{ArcTan}\left[\frac{x}{a}\right]$

Note that *Mathematica* does not include an arbitrary constant of integration in the result. Differentiating the result should, of course, recover the integrand:

```
In[2]:= D[%, x]
```

Out[2]= $-a^2 + x^2 + \frac{a^2}{1 + \frac{x^2}{a^2}}$

To show that this expression is indeed the integrand, use the function **Simplify**:

```
In[3]:= Simplify[%]
```

Out[3]= $\frac{x^4}{a^2 + x^2}$

Integrate[*f*, {*x*, *xmin*, *xmax*}] gives the definite integral

$$\int_{xmin}^{xmax} f\,dx$$

For example, we can evaluate

$$\int_0^1 \frac{1}{1+x+x^2}\,dx$$

```
In[4]:= Integrate[1/(1+x+x^2), {x, 0, 1}]
```

Out[4]= $\frac{\pi}{3\sqrt{3}}$

For another example, let us find

$$\int_0^{\pi} \sin^p x\,dx$$

```
In[5]:= Integrate[Sin[x]^p, {x, 0, π}]
```

Out[5]= $\text{If}\left[\text{Re}[p] > -1, \frac{\pi^{3/2}\,\text{Sec}\left[\frac{p\pi}{2}\right]}{\text{Gamma}\left[\frac{1}{2} - \frac{p}{2}\right]\text{Gamma}\left[1 + \frac{p}{2}\right]},\right.$
$\left.\text{Integrate}\left[\text{Sin}[x]^p, \{x, 0, \pi\}, \text{Assumptions} \to \text{Re}[p] \le -1\right]\right]$

where the output indicates that the result is

$$\frac{\pi^{3/2}\,\text{Sec}\left[\frac{p\pi}{2}\right]}{\text{Gamma}\left[\frac{1}{2}-\frac{p}{2}\right]\text{Gamma}\left[1+\frac{p}{2}\right]}$$

if $\text{Re}(p) > -1$ is true and that *Mathematica* cannot do the integral if it is false. For a definite integral, we can include a trailing argument, called an option, in the form **Assumptions ->** *assumptions* to declare assumptions about the parameters; *assumptions* can be equations, inequalities, domain specifications, or lists or logical combinations of these. Quantities appearing algebraically in inequalities are always assumed to be real. Domains can be specified as explained in Section 2.2.4. (Section 2.3.1.2 will discuss *Mathematica* options.) With the assumption that $\text{Re}(p) > -1$, the integral becomes

```
In[6]:= Integrate[Sin[x]^p, {x, 0, π}, Assumptions→Re[p] > -1]
```

$$\text{Out[6]=}\quad \frac{\pi^{3/2}\,\text{Sec}\left[\frac{p\pi}{2}\right]}{\text{Gamma}\left[\frac{1}{2}-\frac{p}{2}\right]\text{Gamma}\left[1+\frac{p}{2}\right]}$$

which can be expressed as $\sqrt{\pi}$ **Gamma[** $\frac{1+p}{2}$ **]/Gamma[1 +** $\frac{p}{2}$ **]**:

$$\text{In[7]:= FullSimplify}\left[\% == \frac{\sqrt{\pi}\ \text{Gamma}\left[\frac{1+p}{2}\right]}{\text{Gamma}\left[1+\frac{p}{2}\right]}\right]$$

```
Out[7]= True
```

As shown in Section 2.2.4, we can also use the function **Assuming** to declare assumptions. For example, consider the integral

$$\int_0^{\pi} \sin m x \sin n x \, dx$$

where m and n are integers and $m = n$. Together with **Assuming**, **Integrate** finds the value of the integral:

```
In[8]:= Assuming[{m ∈ Integers, n ∈ Integers, n == m},
          Integrate[Sin[m x] Sin[n x], {x, 0, π}]]
```

$$\text{Out[8]=}\quad \frac{\pi}{2}$$

There are integrals that *Mathematica* cannot do. For instance, it does not know how to evaluate the integral

$$\int_0^1 \tan(\cos x) dx$$

```
In[9]:= Integrate[Tan[Cos[x]], {x, 0, 1}]
Out[9]= ∫₀¹ Tan[Cos[x]] dx
```

Mathematica returns the input unevaluated. We can, of course, obtain an approximate value for the integral:

```
In[10]:= N[%]
Out[10]= 1.16499
```

Integrate[*f*, {*x*, *xmin*, *xmax*}, {*y*, *ymin*, *ymax*}] gives the multiple integral

$$\int_{xmin}^{xmax} dx \int_{ymin}^{ymax} dy\, f$$

For an example, find

$$\int_0^1 dx \int_0^{2-2x} dy\left(x^2 + y^2 + 1\right)$$

```
In[11]:= Integrate[x^2 + y^2 + 1, {x, 0, 1}, {y, 0, 2 - 2x}]
Out[11]= 11/6
```

We end this section with an example from quantum physics.

Example 2.2.9 Planck's blackbody radiation (also known as cavity radiation) formula is

$$R(\lambda, T) = \frac{2\pi hc^2}{\lambda^5} \frac{1}{e^{hc/\lambda kT} - 1}$$

where $R(\lambda, T)$ is the spectral radiancy as a function of wavelength λ and temperature T, k is Boltzmann's constant, c is the speed of light, and h is Planck's constant. Derive the Stefan–Boltzmann law that relates the radiant intensity $R(T)$ and temperature T by

$$R(T) = \sigma T^4$$

in which

$$\sigma = \frac{2\pi^5 k^4}{15c^2h^3}$$

is called the Stefan–Boltzmann constant.

$R(\lambda, T)d\lambda$ is defined as the energy radiated with wavelengths in the interval λ to $\lambda + d\lambda$ per unit time from unit area of the surface. Thus, the radiant intensity $R(T)$, which is the total energy radiated at all wavelengths per unit time from unit area, is given by

$$R(T) = \int_0^\infty R(\lambda, T)\, d\lambda$$

Integrate evaluates the integral and validates the Stefan–Boltzmann law:

```
In[12]:= Integrate[((2πhc^2)/(λ^5))(1/(Exp[(hc)/(λkT)]-1)),
           {λ, 0, ∞}, Assumptions→((ch)/(kT))>0]/.
          ((2(π^5)(k^4))/(15(c^2)(h^3)))→σ
Out[12]= T^4 σ
```

■

2.2.11 Sums

Sum[*f*, {*i*, *imin*, *imax*}] evaluates the sum

$$\sum_{i=imin}^{imax} f$$

The second argument of the function **Sum** is an example of iterators introduced in Section 2.1.19. As noted there, *imin* has a default value of 1—that is, if *imin* is omitted, it assumes the value of 1. Let us, for example, generate the sum

$$1 + \frac{1}{x} + \frac{1}{x^2} + \frac{1}{x^3} + \frac{1}{x^4} + \frac{1}{x^5} + \frac{1}{x^6}$$

```
In[1]:= Sum[1/x^i, {i, 0, 6}]
```

$$\text{Out[1]=}\ 1 + \frac{1}{x^6} + \frac{1}{x^5} + \frac{1}{x^4} + \frac{1}{x^3} + \frac{1}{x^2} + \frac{1}{x}$$

We can also evaluate the sum

$$1 + \frac{1}{2^4} + \frac{1}{3^4} + \frac{1}{4^4} + \frac{1}{5^4} + \frac{1}{6^4} + \frac{1}{7^4}$$

```
In[2]:= Sum[1/i^4, {i, 7}]
```

$$\text{Out[2]=}\ \frac{33\,654\,237\,761}{31\,116\,960\,000}$$

The result is exact. To determine the approximate numerical value, use the function **N**:

```
In[3]:= N[%]
Out[3]= 1.08154
```

Sum[*f*, {*i*, *imin*, *imax*, *di*}] evaluates the sum with i increasing in steps of di. To generate the sum

$$1+\frac{1}{x^2}+\frac{1}{x^4}+\frac{1}{x^6}+\frac{1}{x^8}+\frac{1}{x^{10}}$$

we enter

```
In[4]:= Sum[1/x^i, {i, 0, 10, 2}]
```

Out[4]= $1+\frac{1}{x^{10}}+\frac{1}{x^8}+\frac{1}{x^6}+\frac{1}{x^4}+\frac{1}{x^2}$

Here, **i** increases in steps of 2.

Sum[*f*, {*i*, *imin*, *imax*}, {*j*, *jmin*, *jmax*}, ...] evaluates the multiple sum

$$\sum_{i=imin}^{imax}\sum_{j=jmin}^{jmax}\cdots f$$

For instance, we can generate the sum

$$\sum_{i=1}^{3}\sum_{j=1}^{i}\frac{1}{x^i+y^j}$$

```
In[5]:= Sum[1/(x^i + y^j), {i, 1, 3}, {j, 1, i}]
```

Out[5]= $\frac{1}{x+y}+\frac{1}{x^2+y}+\frac{1}{x^3+y}+\frac{1}{x^2+y^2}+\frac{1}{x^3+y^2}+\frac{1}{x^3+y^3}$

We can also evaluate the sum

$$\sum_{i=1}^{2}\sum_{j=0}^{i+1}\frac{1}{(i^2+j^3)^2}$$

```
In[6]:= Sum[1/(i^2 + j^3)^2, {i, 2}, {j, 0, i + 1}]
```

Out[6]= $\frac{21372503}{15568200}$

Sum can find the sum of an infinite series, if it is convergent. To determine the sum of the series

$$\sum_{n=1}^{\infty}\frac{1}{(4n+1)(4n-3)}$$

enter

```
In[7]:= Sum[1/((4n+1)(4n-3)), {n,∞}]
        1
Out[7]= -
        4
```

To find the sum of the series

$$\sum_{n=0}^{\infty} \frac{(-1)^n x^{2n}}{2^{2n}(n!)^2}$$

enter

```
In[8]:= Sum[(((-1)^n)x^(2n))/((2^(2n))(n!^2)), {n, 0, ∞}]
Out[8]= BesselJ[0, x]
```

which is the Bessel function of the first kind $J_0(x)$.

Sum can find symbolic or indefinite sums. For example, consider the infinite series

$$\sum_{i=1}^{\infty} \frac{1}{j(j+1)}$$

Its nth partial sum is

```
In[9]:= Sum[1/(j(j+1)), {j, n}]
          n
Out[9]= -----
        1 + n
```

What follows is an example from quantum physics.

Example 2.2.10 Determine the degeneracy of the energy levels of (a) the hydrogen atom and (b) the three-dimensional isotropic harmonic oscillator.

(a) For labeling the energy eigenstates of the hydrogen atom, the four quantum numbers are the principal quantum number n, azimuthal or orbital quantum number l, magnetic quantum number m_l, and spin magnetic quantum number m_s. The allowed values of these quantum numbers are

$$n = 1, 2, 3, \ldots$$

$$l = 0, 1, 2, \ldots, (n-1)$$

$$m_l = -l, -l+1, \ldots, 0, 1, 2, \ldots, +l$$

$$m_s = -\frac{1}{2}, +\frac{1}{2}$$

For the hydrogen atom, the energy eigenvalues are

$$E_n = -\frac{R}{n^2}$$

where R is the Rydberg constant. Because the energy E_n depends only on n, there are two or more eigenstates that correspond to the same energy. These eigenstates are said to be degenerate. For each n, l ranges from 0 to $(n-1)$; for each l, m_l can take on $(2l+1)$ values; for each m_l, m_s can assume two values. Thus, the total number of energy eigenstates corresponding to E_n—that is, the degeneracy of E_n—is

$$\sum_{l=0}^{n-1} 2(2l+1)$$

Mathematica finds this symbolic or indefinite sum:

```
In[10]:= Sum[2 (2 l + 1), {l, 0, n - 1}] // Simplify
Out[10]= 2 n^2
```

(b) For labeling the energy eigenstates of the three-dimensional isotropic harmonic oscillator, the three quantum numbers are n_1, n_2, and n_3, which are nonnegative integers—that is, $0, 1, 2, 3 \ldots$. The energy eigenvalues are

$$E_n = \left(n + \frac{3}{2}\right)\hbar\omega, \qquad n = n_1 + n_2 + n_3$$

where $\hbar$ is Planck's constant divided by 2π and the oscillator frequency $\omega = \sqrt{k/m}$, in which m is the mass and k is the spring constant. Eigenstates with quantum numbers n_1, n_2, and n_3 such that $n_1 + n_2 + n_3 = n$ have the same energy E_n and are degenerate. For each n, n_1 ranges from 0 to n; for each n_1, n_2 can take on $(n - n_1 + 1)$ values; for each n_2, the only value for n_3 is $(n - n_1 - n_2)$. Therefore, the degeneracy of E_n is

$$\sum_{n_1=0}^{n} (n - n_1 + 1)$$

Mathematica finds this sum:

```
In[11]:= Sum[(n - n1 + 1), {n1, 0, n}] // Factor
Out[11]= 1/2 (1 + n) (2 + n)
```

■

There are sums that the function **Sum** cannot find. In these cases, *Mathematica* returns the results in symbolic forms. For definite sums, we can obtain numerical approximations. For example, *Mathematica* fails to find the sum

$$\sum_{n=1}^{\infty} \frac{n^2}{(2\,e^n + 1)}$$

and returns the result in a symbolic form:

```
In[12]:= Sum[(n^2)/(2 Exp[n] + 1), {n, 1, ∞}]
```

$$\text{Out[12]}= \sum_{n=1}^{\infty} \frac{n^2}{1+2\,e^n}$$

The function **N** gives a numerical approximation for this sum:

```
In[13]:= N[%]
Out[13]= 0.943293
```

N[Sum[...]] calls **NSum**. **NSum[***f*, {*i*, *imin*, *imax*}**]** gives a numerical approximation to the sum

$$\sum_{i=imin}^{imax} f$$

As mentioned previously, the default value for *imin* is 1 for the iterator. If we only wish to obtain numerical approximations, calling **NSum** rather than **N[Sum[...]]** saves time because *Mathematica* does not have to waste time trying to find exact results first before resorting to calling **NSum**:

```
In[14]:= Timing[N[Sum[(n^2)/(2 Exp[n] + 1), {n, ∞}]]]
Out[14]= {0.155069,0.943293}

In[15]:= Timing[NSum[(n^2)/(2 Exp[n] + 1), {n, ∞}]]
Out[15]= {0.014337,0.943293}
```

where **Timing[***expr***]** evaluates *expr* and returns a list of the time in seconds used, together with the result obtained. **NSum** takes less than 1/10 of the time that **N[Sum[...]]** does in evaluating the sum.

2.2.12 Power Series

Series[*f*, {*x*, x_0, *n*}**]** finds a power series representation for f in $(x - x_0)$ to order $(x - x_0)^n$. For example, we can determine the power series expansion about $u = 0$ for

$$e^{\sin u}$$

to order u^7:

```
In[1]:= Series[Exp[Sin[u]], {u, 0, 7}]
```

$$\text{Out[1]}= 1 + u + \frac{u^2}{2} - \frac{u^4}{8} - \frac{u^5}{15} - \frac{u^6}{240} + \frac{u^7}{90} + O[u]^8$$

O[*u***]**^{n} stands for a term of order u^n.

Normal[*series*] converts a power series to an ordinary expression—that is, removes the term O[$x-x_0$]^n. For example, **Normal** removes the term **O[u]**^{8} from the series in the preceding output:

```
In[2]:= Normal[%]
```

$$\text{Out[2]}= 1+u+\frac{u^2}{2}-\frac{u^4}{8}-\frac{u^5}{15}-\frac{u^6}{240}+\frac{u^7}{90}$$

For another example, let us approximate the integral

$$\int_0^{0.3} \left(1+x^4\right)^{1/3}\, dx$$

First, expand the integrand in a series about $x=0$ up to order x^8:

```
In[3]:= Series[(1+x^4)^(1/3), {x, 0, 8}]
```

$$\text{Out[3]}= 1+\frac{x^4}{3}-\frac{x^8}{9}+O[x]^9$$

Second, remove the term **O[x]**^{9}.

```
In[4]:= Normal[%]
```

$$\text{Out[4]}= 1+\frac{x^4}{3}-\frac{x^8}{9}$$

Finally, integrate the terms of the series:

```
In[5]:= Integrate[%, {x, 0, 0.3}]
Out[5]= 0.300162
```

Let us compare the result with that from numerical integration:

```
In[6]:= NIntegrate[(1+x^4)^(1/3), {x, 0, 0.3}]
Out[6]= 0.300162
```

The results agree to six decimal places.

We conclude this section with an example from introductory physics.

Example 2.2.11 The electric potential V on the axis of a uniformly charged disk of radius a and charge density σ is given as

$$V = k\sigma\pi \int_0^a \frac{2r}{\sqrt{r^2+x^2}} dr$$

where x is the distance from the disk along the axis and k is the Coulomb constant (see [SJ04]). Determine V, and verify that it approaches the electric potential produced by a point charge in the limit $x >> a$.

The electric potential V is

```
In[7]:= k σ π Integrate[(2 r)/Sqrt[r^2 + x^2],
          {r, 0, a}, Assumptions → {a > 0, x > 0}]
```

Out[7]= $2\,k\,\pi\left(-x+\sqrt{a^2+x^2}\right)\sigma$

To show that V approaches the electric potential produced by a point charge in the limit $x >> a$, make the substitution $x = a/y$ and then find a power series representation for V in y:

```
In[8]:= Series[% /. x → a/y, {y, 0, 2}] // PowerExpand
```

Out[8]= $a\,k\,\pi\,\sigma\,y + O[y]^3$

Since $y << 1$ in the limit $x >> a$, we need to keep only the first term of the series:

```
In[9]:= Normal[%]
```

Out[9]= $a\,k\,\pi\,y\,\sigma$

Now, replace y with a/x:

```
In[10]:= % /. y → a/x
```

Out[10]= $\frac{a^2\,k\,\pi\,\sigma}{x}$

The factor $\sigma\pi a^2$ is just the total charge q of the disk. Thus, V equals

```
In[11]:= % /. σ π a^2 → q
```

Out[11]= $\frac{k\,q}{x}$

That is the electric potential produced by a point charge of charge q. ∎

2.2.13 Limits

Limit[$f(x)$, x -> x_0] determines the limit of $f(x)$ as x approaches x_0. For example, find

$$\lim_{x\to\infty}\frac{\sqrt{9x^2+2}}{3-4x}$$

```
In[1]:= Limit[Sqrt[9 x^2 + 2]/(3 - 4 x), x → ∞]
```

Out[1]= $-\frac{3}{4}$

`Limit`[$f(x)$, x `->` x_0, `Direction -> -1`] determines the limit of $f(x)$ as x approaches x_0 from larger values. We can find

$$\lim_{x\to 3^+} \frac{2x^2}{9-x^2}$$

```
In[2]:= Limit[(2 x^2)/(9 - x^2), x → 3, Direction → -1]
Out[2]= -∞
```

`Limit`[$f(x)$, x `->` x_0, `Direction -> 1`] determines the limit of $f(x)$ as x approaches x_0 from smaller values. We can also find

$$\lim_{x\to 3^-} \frac{2x^2}{9-x^2}$$

```
In[3]:= Limit[(2 x^2)/(9 - x^2), x → 3, Direction → 1]
Out[3]= ∞
```

The last two results show that the limits as x approaches x_0 from larger and smaller values may be different for a discontinuous function.

2.2.14 Solving Differential Equations

`DSolve`[*eqn*, y`[`x`]`, x] solves the differential equation *eqn* for the function y`[`x`]`, with independent variable x. For example, solve

$$\frac{d^2y}{dx^2} - 3\frac{dy}{dx} - 18y = x\,e^{4x}$$

```
In[1]:= DSolve[y''[x] - 3 y'[x] - 18 y[x] == x Exp[4 x], y[x], x]
```

$$\text{Out[1]} = \left\{\left\{y[x] \to -\frac{1}{196} e^{4x} (5 + 14x) + e^{-3x} C[1] + e^{6x} C[2]\right\}\right\}$$

`C[1]` and `C[2]` are the arbitrary constants.

For another example, determine $x(t)$, the position at time t, for a damped harmonic oscillator with the equation of motion

$$\frac{d^2x}{dt^2} = -\omega_0^2 x - \gamma\frac{dx}{dt}$$

```
In[2]:= DSolve[x''[t] + γ x'[t] + (ω0^2) x[t] == 0, x[t], t]
```

$$\text{Out[2]} = \left\{\left\{x[t] \to e^{\frac{1}{2}t\left(-\gamma - \sqrt{\gamma^2 - 4\,\omega0^2}\right)} C[1] + e^{\frac{1}{2}t\left(-\gamma + \sqrt{\gamma^2 - 4\,\omega0^2}\right)} C[2]\right\}\right\}$$

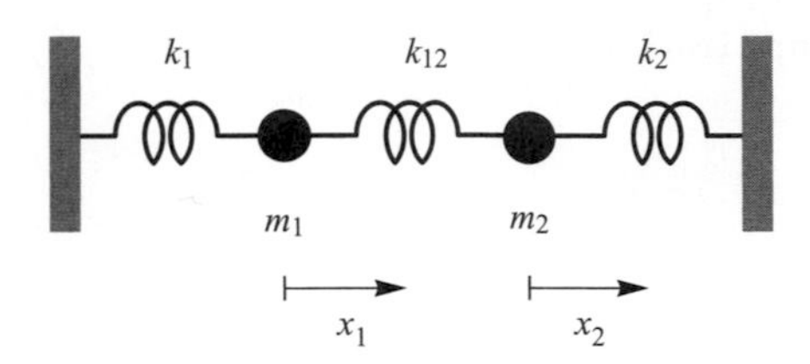

Figure 2.2.3. Two coupled harmonic oscillators.

Again, `C[1]` and `C[2]` are the arbitrary constants. (For more information on the damped harmonic oscillator, refer to [TM04].)

`DSolve[{`eqn_1`, `eqn_2`, ...}, {`y_1`[`x_1`, ...], ...}, {`x_1`, ...}]` solves the list of differential equations `{`eqn_1`, `eqn_2`, ...}` for the functions y_1`[`x_1`, ...]`, ..., with independent variables $x_1, \ldots$. Boundary conditions may be included in the list of equations. Let us solve

$$\frac{d^2y}{dx^2} + 8\frac{dy}{dx} + 16y = 0$$

with $y = 2$ and $dy/dx = 1$ when $x = 0$:

```
In[3]:= DSolve[{y''[x] + 8 y'[x] + 16 y[x] == 0, y[0] == 2, y'[0] == 1}, y[x], x]
Out[3]= {{y[x] → e^(-4 x) (2 + 9 x)}}
```

What follows is an example from classical mechanics.

Example 2.2.12 Two simple harmonic oscillators are coupled together by a spring, as in Figure 2.2.3. While m_1 is held fixed, m_2 is displaced to the right by a distance a from its equilibrium position. At time $t = 0$, both masses are released. Determine the subsequent motion. Consider only the case in which $k_1 = k_2 = k_{12} = k$ and $m_1 = m_2 = m$.

Let x_1 and x_2 be the displacements of m_1 and m_2, respectively. Each displacement is measured from the equilibrium position, and the direction to the right is positive. The force on m_1 is $-kx_1 - k(x_1 - x_2)$, and the force on m_2 is $-kx_2 - k(x_2 - x_1)$. Thus, the equations of motion are

$$m\frac{d^2x_1}{dt^2} + 2kx_1 - kx_2 = 0$$

$$m\frac{d^2x_2}{dt^2} + 2kx_2 - kx_1 = 0$$

At $t = 0$,

$$\frac{dx_1}{dt} = 0$$

$$\frac{dx_2}{dt} = 0$$

$$x_1 = 0$$

$$x_2 = a$$

Let $\omega^2 = k/m$. **DSolve** solves the equations of motion subject to the initial conditions:

```
In[4]:= DSolve[{x1''[t] + 2(ω^2)x1[t] - (ω^2)x2[t] == 0,
          x2''[t] + 2(ω^2)x2[t] - (ω^2)x1[t] == 0, x1'[0] == 0,
          x2'[0] == 0, x1[0] == 0, x2[0] == a}, {x1[t], x2[t]}, t]
```

$$Out[4] = \left\{\left\{x1[t] \to \frac{1}{2} a \left(Cos[t\,\omega] - Cos\left[\sqrt{3}\, t\,\omega\right]\right),\right.\right.$$
$$\left.\left. x2[t] \to \frac{1}{2} a \left(Cos[t\,\omega] + Cos\left[\sqrt{3}\, t\,\omega\right]\right)\right\}\right\}$$

There are two normal or characteristic frequencies: $\sqrt{k/m}$ and $\sqrt{3k/m}$. ■

2.2.15 Immediate versus Delayed Assignments and Transformation Rules

As discussed in Section 2.2.7, we make an immediate assignment by writing *variable* = *value*. We can also make a delayed assignment by typing *variable* := *value* (with no space between : and =). What is the difference between these assignments? With an immediate assignment, *value* is evaluated as soon as the assignment is made; for a delayed assignment, *value* is evaluated later when it is requested. Let us illustrate the difference between the two assignments with the **RandomReal** function that generates, upon request, a random real number in the range 0 to 1. To make an immediate assignment to the variable **imasm**, we enter

```
In[1]:= imasm = RandomReal[]
Out[1]= 0.388517
```

The **RandomReal** function is evaluated immediately, and the variable **imasm** is assigned a permanent value until it is changed or removed. To make a delayed assignment to the variable **dlasm**, we write

```
In[2]:= dlasm := RandomReal[]
```

In this case, the **RandomReal** function is not evaluated, and no value is assigned to the variable **dlasm** until it is called. Now call these variables several times with the **Table** function introduced in Section 2.1.19:

```
In[3]:= Table[imasm, {4}]
Out[3]= {0.388517, 0.388517, 0.388517, 0.388517}
```

```
In[4]:= Table[dlasm, {4}]
Out[4]= {0.778027, 0.839587, 0.450894, 0.0963999}
```

Whereas the value for **imasm** remains the same, that for **dlasm** varies from call to call.

As shown in Section 2.2.7, we can specify an immediate transformation rule in the form *expr* /. *lhs* -> *rhs*. We can also specify a delayed transformation rule in the form *expr* /. *lhs* :> *rhs* (with no space between : and >). The difference between these rules rests on when *rhs* is evaluated. Whereas *rhs* of an immediate rule is evaluated upon specification of the rule, *rhs* of a delayed rule is evaluated afresh when the rule is invoked. Let us use the **RandomReal** function to illustrate the difference between these rules:

```
In[5]:= {f[a], f[a], f[a], f[a]} /. f[a] → RandomReal[]
Out[5]= {0.742471, 0.742471, 0.742471, 0.742471}

In[6]:= {f[a], f[a], f[a], f[a]} /. f[a] :> RandomReal[]
Out[6]= {0.137166, 0.319445, 0.882133, 0.114903}
```

In the first case, the same transformation rule is used for all replacements because **RandomReal[]** is evaluated right away. In the second case, the delayed rule varies from replacement to replacement since **RandomReal[]** is evaluated afresh each time.

2.2.16 Defining Functions

An obvious definition for a function is *name*[arg_1, arg_2,...] := *body*. To illustrate the flaw of this definition, we define the function $f(x) = x^3 - 1$:

```
In[1]:= f[x] := x^3 - 1
```

Let us call **f[x]**:

```
In[2]:= f[x]
Out[2]= -1 + x^3
```

As expected, it gives $x^3 - 1$. Now, call **f[y]** and **f[2]**:

```
In[3]:= f[y]
Out[3]= f[y]

In[4]:= f[2]
Out[4]= f[2]
```

Contrary to expectation, *Mathematica* does not return $y^3 - 1$ and 7 for **f[y]** and **f[2]**, respectively. The problem is that although *Mathematica* knows **f[x]**, it does not recognize **f[y]** and **f[2]**.

To define a function, we need to write

name[arg_1_, arg_2_, ...] := *body*

where "_", called a blank, stands for "any expression." The symbol *arg*_ means any expression to be named *arg* in the *body* of the definition. To show that this definition works, we return to the previous example:

```
In[5]:= f[x_] := x^3 - 1

In[6]:= f[y]
Out[6]= -1 + y^3

In[7]:= f[2]
Out[7]= 7
```

The results for `f[y]` and `f[2]` are indeed correct.

We can use the immediate assignment operator "=" instead of the delayed assignment operator ":=" in a definition if we prefer immediate evaluation of the body of the function upon definition. Yet, in most cases, we would rather have the bodies of the functions evaluated when they are called. When in doubt, use the delayed assignment operator ":=".

In Section 2.2.7, we discussed the advisability of clearing as often as possible unneeded values assigned to variables. It is also a good practice to remove unneeded definitions for functions as soon as possible.

```
In[8]:= Clear[f]
```

Let us illustrate the writing of functions with a physics problem.

Example 2.2.13 The normalized hydrogenic eigenfunctions are given as

$$\psi_{nlm}(r, \theta, \phi) = R_{nl}(r) Y_{lm}(\theta, \phi)$$

where R_{nl} and Y_{lm} are the radial functions and spherical harmonics, respectively. In terms of the associated Laguerre polynomials L_p^q and the Bohr radius a, R_{nl} can be expressed as

$$a^{-3/2} \frac{2}{n^2} \sqrt{\frac{(n-l-1)!}{[(n+l)!]^3}} F_{nl}\left(\frac{2r}{na}\right)$$

with

$$F_{nl}(x) = x^l e^{-x/2} L_{n-l-1}^{2l+1}(x)$$

Find the radial functions for the $1s, 2s, 2p$, and $3s$ states.

In terms of the generalized Laguerre polynomials `LaguerreL[p, q, x]` in *Mathematica*, the associated Laguerre polynomials $L_p^q(x)$ are given by

$$L_p^q(x) = (p+q)!\,\texttt{LaguerreL[p, q, x]}$$

([Lib03] compares three common notations for the associated Laguerre polynomials. The notation adopted here is that used in [Lib03] and [Mes00].) Thus, $F_{nl}(x)$ can be defined as

```
In[9]:= F[n_, l_, x_] :=
          ((x^l) Exp[-x/2] ((n+l)!) LaguerreL[n-l-1, 2l+1, x])
```

$R_{nl}(r)$ can be written as

```
In[10]:= R[n_, l_, r_] := ((a^(-3/2)) (2/(n^2))
           Sqrt[((n-l-1)!)/(((n+l)!)^3)]F[n, l, 2r/(na)])
```

For $R_{10}(r)$, $R_{20}(r)$, $R_{21}(r)$, and $R_{30}(r)$, we have

```
In[11]:= R[1, 0, r]
```

Out[11]= $\frac{2\,e^{-\frac{r}{a}}}{a^{3/2}}$

```
In[12]:= R[2, 0, r]
```

Out[12]= $\frac{e^{-\frac{r}{2a}}\left(2-\frac{r}{a}\right)}{2\sqrt{2}\,a^{3/2}}$

```
In[13]:= R[2, 1, r]
```

Out[13]= $\frac{e^{-\frac{r}{2a}}\,r}{2\sqrt{6}\,a^{5/2}}$

```
In[14]:= R[3, 0, r]
```

Out[14]= $\frac{2e^{-\frac{r}{3a}}\left(27\,a^2-18\,a\,r+2\,r^2\right)}{81\sqrt{3}\,a^{7/2}}$

```
In[15]:= Clear[F, R]
```

■

Section 2.1.8 pointed out that `N[`*expr*`]` has a special input form, namely the postfix form *expr*`//N`. For example, `Sin[50Degree]//N` and `N[Sin[50Degree]]` are identical:

```
In[16]:= N[Sin[50Degree]] == Sin[50Degree]//N
Out[16]= True
```

In general, functions with a single argument have two special input forms:

func `@` *arg*	prefix form of *func*`[`*arg*`]`
arg `//` *func*	postfix form of *func*`[`*arg*`]`

We can, for example, determine `Sqrt[3.2]` with the prefix form `Sqrt@3.2` or the postfix form `3.2//Sqrt`:

```
In[17]:= Sqrt@3.2
Out[17]= 1.78885
```

```
In[18]:= 3.2 // Sqrt
Out[18]= 1.78885
```

If the argument contains *Mathematica* operators, we must be careful in using the prefix form. For instance, **f@x y + z** equals **f[x] y + z** rather than **f[x y + z]**. To get **f[x y + z]**, we must write **f@(x y + z)**. On the other hand, the postfix form is usually forgiving; for example, **x y + z // f** gives **f[x y + z]**. When in doubt, parenthesize the argument.

To illustrate the use of prefix form, consider a problem from quantum mechanics.

Example 2.2.14 The time-independent Schrödinger equation is

$$H\psi = E\psi$$

where H is the Hamiltonian operator, E is the energy eigenvalue, and ψ is the corresponding eigenfunction. For a particle with mass m moving in a one-dimensional potential $V(x)$, the Hamiltonian operator is given by

$$H = -\frac{\hbar^2}{2m}\frac{d^2}{dx^2} + V(x)$$

Solve the time-independent Schrödinger equation for a free particle in one dimension.

The prefix special input form is most suited for working with operators. If the Hamiltonian is defined in *Mathematica* by

```
In[19]:= H[V_]@ψ_ := ((-(ℏ^2)/(2m))D[ψ, {x, 2}] + V ψ)
```

the time-independent Schrödinger equation in one dimension is

```
In[20]:= H[V[x]]@ψ[x] == Ε ψ[x]

                        ℏ² ψ''[x]
Out[20]= V[x] ψ[x] - ----------- == Ε ψ[x]
                           2 m
```

with the alias the Greek letter **Ε** for E because the standard keyboard **E** is the exponential constant e in *Mathematica*. (For a reminder on using Greek letters, see Section 1.6.2.) With a free particle, $V = 0$. Let us show that e^{ikx} is a free-particle eigenfunction by substituting

```
In[21]:= ψ[x] = Exp[I k x]
Out[21]= ⅇ^(i k x)
```

into

```
In[22]:= H[0]@ψ[x] == Ε ψ[x]

          ⅇ^(i k x) k² ℏ²
Out[22]= ---------------- == ⅇ^(i k x) Ε
               2 m
```

We see that e^{ikx} is indeed an eigenfunction, and the energy eigenvalue is given by

```
In[23]:= Solve[%, E]
```

Out[23]= $\{\{E \to \frac{k^2 \hbar^2}{2m}\}\}$

(For an excellent discourse on using *Mathematica* to do quantum mechanics, see [Fea94].)

```
In[24]:= Clear[H, ψ]
```

■

Section 1.3 showed how to access information about built-in *Mathematica* objects with the operators **?** and **??**. We can also obtain information about user-defined functions. For example, let us define a function **f**:

```
In[25]:= f[x_]:=x^3
```

Using the operator **?** on **f** yields its definition together with the fact that it is a symbol in the context **Global`**:

```
In[26]:= ?f
```

```
Global`f
```

f[x_] := x^3

(Section 3.7.1 will discuss the notion of contexts.) We can specify usage messages for user-defined functions in the form

f::**usage** = "*usage message text*"

For example, let us give **f** a usage statement:

```
In[27]:= f::usage = "f[x] returns the cube of x."
Out[27]= f[x] returns the cube of x.
```

Using **?** on **f** now retrieves the usage statement:

```
In[28]:= ?f
```

```
f [x] returns the cube of x.
```

To obtain additional information about `f`, use `??`:

```
In[29]:= ??f
```

f [x] returns the cube of x.

```
f[x_] := x^3
In[30]:= Clear[f]
```

2.2.17 Relational and Logical Operators

Relational operators compare two expressions and return **True** or **False** whenever comparisons are possible. *Mathematica* has several relational operators or functions:

lhs == *rhs* or **Equal**[*lhs*, *rhs*]	return **True** if *lhs* and *rhs* are identical
lhs != *rhs* or **Unequal**[*lhs*, *rhs*]	return **False** if *lhs* and *rhs* are identical
lhs > *rhs* or **Greater**[*lhs*, *rhs*]	yield **True** if *lhs* is determined to be greater than *rhs*
lhs >= *rhs* or **GreaterEqual**[*lhs*, *rhs*]	yield **True** if *lhs* is determined to be greater than or equal to *rhs*
lhs < *rhs* or **Less**[*lhs*, *rhs*]	yield **True** if *lhs* is determined to be less than *rhs*
lhs <= *rhs* or **LessEqual**[*lhs*, *rhs*]	yield **True** if *lhs* is determined to be less than or equal to *rhs*

For example, consider

```
In[1]:= f[x_] := x^2

In[2]:= g[x_] := x^3

In[3]:= g[2] > f[3] + 1
Out[3]= False
```

We can combine relational operators:

```
In[4]:= 8 - 5 == 3 == 6/2
Out[4]= True

In[5]:= 4 > 3 <= 2
Out[5]= False
```

When *Mathematica* is unable to compare the expressions, it returns the input:

```
In[6]:= 1 + x > 2 y
Out[6]= 1 + x > 2 y
```

Mathematica supports several logical operators or functions:

! *expr* or **Not**[*expr*]	give **False** if *expr* is **True**, and **True** if it is **False**
$expr_1$ **&&** $expr_2$ **&&**... or **And**[$expr_1$, $expr_2$, ...]	evaluate the expressions in order from left to right, giving **False** immediately if any of them are **False**, and **True** if they are all **True**
$expr_1$ **||** $expr_2$ **||**... or **Or** [$expr_1$, $expr_2$, ...]	evaluate the expressions from left to right, giving **True** immediately if any of them are **True**, and **False** if they are all **False**
Xor [$expr_1$, $expr_2$, ...]	give **True** if an odd number of the $expr_i$ are **True** and the rest are **False**; give **False** if an even number of the $expr_i$ are **True** and the rest are **False**

For example, consider

```
In[7]:= 4 > 3 || 5 <= 10
Out[7]= True

In[8]:= Xor[4 > 3, 5 <= 10, 7 < 2]
Out[8]= False
```

Whereas **||** is inclusive, **Xor** is exclusive. In other words, **||** returns **True** when one or both expressions are **True**, but **Xor** gives **True** only when one expression is **True** and the other is **False**.

With $expr_1$ **&&** $expr_2$ **&&** ..., *Mathematica* evaluates the expressions from left to right, stopping at the first expression that is **False**:

```
In[9]:= EvenQ[g[2] + 1] && 1/0 < 10
Out[9]= False
```

EvenQ[*expr*] gives **True** if *expr* is an even integer, and **False** otherwise. The second expression was not evaluated; otherwise, *Mathematica* would have given a warning message. This can be seen by reversing the expressions:

```
In[10]:= 1/0 < 10 && EvenQ[g[2] + 1]
```

Power::infy : Infinite expression $\frac{1}{0}$ encountered. >>

Less::nord : Invalid comparison with ComplexInfinity attempted. >>

```
Out[10]= False
```

When *Mathematica* cannot ascertain the truth of the expressions, it returns the input:

```
In[11]:= a || b || c
Out[11]= a || b || c
```

```
In[12]:= Clear[f, g]
```

We conclude this section with a physics example illustrating the use of relational and logical operators.

Example 2.2.15 Consider a particle of mass m moving in a one-dimensional box with walls at $x = 0$ and $x = a$. Inside the box, the energy eigenvalue and the corresponding eigenfunction can be expressed as

$$E = \frac{\hbar^2 k^2}{2m}$$

$$\phi(x) = A \cos kx + B \sin kx$$

where $\hbar$ is Planck's constant divided by 2π. The boundary conditions are

$$\phi(0) = 0$$

$$\phi(a) = 0$$

Determine A and k.

Let us begin with a definition for the eigenfunction:

```
In[13]:= ϕ[x_] := A Cos[k x] + B Sin[k x]
```

Solve gives

```
In[14]:= Solve[{ϕ[0] == 0, ϕ[a] == 0}, {A, k}]
```

```
Solve::ifun :
  Inverse functions are being used by Solve, so some solutions may
     not be found; use Reduce for complete solution information. >>
```

```
Out[14]= {{A→0, k→0}, {A→0, k→-π/a}, {A→0, k→π/a}}
```

Heeding the warning message, we use the function **Reduce** to obtain complete solution information:

```
In[15]:= Reduce[{ϕ[0] == 0, ϕ[a] == 0, a≠0}, {A, k}]
Out[15]= (C[1]∈Integers && ((B≠0 && a≠0 && A == 0 && k == 2πC[1]/a) ||
            (a≠0 && A == 0 && k == (π + 2πC[1])/a))) ||
         ((a k - π)/(2π) ∈ Integers && a≠0 && B == 0&&A == 0)
```

where **Reduce**[*eqns*, *vars*] reduces the equations *eqns* while maintaining all possible solutions for the variables *vars*. Arranged in order of decreasing precedence, the relational and logical operators in the result are **==**, **≠** (or **! =**), **∈**, **∉**, **&&**, and **||**. The result shows that the nontrivial solutions are $A = 0$ and

$$k_n = n\pi/a$$

where n is any integer. In fact, n must be positive because $n = 0$ implies that the particle cannot be found anywhere and because the inclusion of negative integers for n yields no new eigenfunctions. (For more on why n must be positive, see Section 4.1 of [Lib03]; for further exploration of the particle in a box problem, see Exercise 95 of Section 2.2.20.)

```
In[16]:= clear[ϕ]
```

■

2.2.18 Fourier Transforms

The Fourier transform, if it exists, of the function $f(x)$ is

$$g(u) = \left(\frac{a}{2\pi}\right)^{\frac{1}{2}} \int_{-\infty}^{\infty} f(x)e^{-iaux}dx \tag{2.2.1}$$

and the inverse Fourier transform of $g(u)$ is

$$f(x) = \left(\frac{a}{2\pi}\right)^{\frac{1}{2}} \int_{-\infty}^{\infty} g(u)e^{iaux}du \tag{2.2.2}$$

where a is a positive real constant. With $a = 1$, the variables x and u usually represent, respectively, position and wave number or time and angular frequency. With $a = 1/\hbar$, where

$\hbar = h/2\pi$ and h is Planck's constant, x and u often represent position and momentum or time and energy. (For a discussion of the Fourier transform and an introduction to the theory of generalized functions, see [DK67] or [Lea04].)

Mathematica provides several functions that are useful for evaluating Fourier transforms:

FourierTransform[$f(x)$, x, u, **FourierParameters** -> {α, β}]	give $\sqrt{\frac{\lvert\beta\rvert}{(2\pi)^{1-\alpha}}}\int_{-\infty}^{\infty} f(x)e^{i\beta ux}dx$
InverseFourierTransform[$g(u)$, u, x, **FourierParameters** -> {α, β}]	give $\sqrt{\frac{\lvert\beta\rvert}{(2\pi)^{1+\alpha}}}\int_{-\infty}^{\infty} g(u)e^{-i\beta ux}du$
DiracDelta [$x - x_0$]	Dirac delta function with singularity located at $x = x_0$
UnitStep[x]	unit step function, equal to 1 for $x \geq 0$ and 0 for $x < 0$
HeavisideTheta[x]	Heaviside theta function, equal to 1 for $x > 0$ and 0 for $x < 0$
Piecewise[{{val_1, $cond_1$}, {val_2, $cond_2$}...}, val]	piecewise function with values val_i in the regions defined by the conditions $cond_i$; if none of the $cond_i$ apply, the function assumes the value val, which has the default 0

The last argument for both **FourierTransform** and **InverseFourierTransform** is a named optional argument or, simply, an option that is specified in terms of a transformation rule introduced in Section 2.2.7. The default values for α and β are 0 and 1, respectively. That is, if the rule **FourierParameters** -> $\{\alpha, \beta\}$ is omitted, α assumes the value of 0 and β takes on the value of 1. For both **FourierTransform** and **InverseFourierTransform**, we can also include, as a trailing argument, the option **Assumptions** -> *assumptions* discussed in Section 2.2.10 to declare assumptions about the parameters in the functions $f(x)$ and $g(u)$; *assumptions* can be equations, inequalities, domain specifications, or lists or logical combinations of these.

Comparing the definition of **FourierTransform** and Equation 2.2.1 reveals that the Fourier transform of $f(x)$ is given by

```
FourierTransform[f[x], x, -u, FourierParameters → {0, a}]
```

Similarly, the inverse Fourier transform of $g(u)$ is given by

```
InverseFourierTransform[g[u], u, -x, FourierParameters → {0, a}]
```

Let us set $a = 1$ and $u = k$ and determine the Fourier transforms of several functions. Consider

$$f(x) = \frac{1}{1 + x^2}$$

The Fourier transform of $f(x)$ is

```
In[1]:= FourierTransform[1/(1+x^2), x, -k]
```

$$\text{Out[1]}= e^{-\text{Abs}[k]}\sqrt{\frac{\pi}{2}}$$

The inverse Fourier transform is

```
In[2]:= InverseFourierTransform[%, k, -x]
```

$$\text{Out[2]}= \frac{1}{1 + x^2}$$

which is, of course, $f(x)$. Now find the Fourier transform of the Dirac delta function:

```
In[3]:= FourierTransform[DiracDelta[x - x0], x, -k]
```

$$\text{Out[3]}= \frac{e^{-i k x0}}{\sqrt{2\pi}}$$

Finally, determine the Fourier transform of

$$f(x) = \begin{cases} i\sqrt{2b}e^{-bx} & x > 0 \\ 0 & x < 0 \end{cases}$$

with $b > 0$:

```
In[4]:= FourierTransform[I Sqrt[2 b] Exp[-b x] HeavisideTheta[x], x, -k]
```

$$\text{Out[4]}= \frac{i\sqrt{b}}{(b + i k)\sqrt{\pi}}$$

Consider a particle in one dimension. If $\psi(x)$ and $\varphi(p)$ are, respectively, its wave functions in configuration space and momentum space, then $\psi^*(x)\psi(x)dx$ is the probability of finding the particle's position between x and $x + dx$ and $\varphi^*(p)\varphi(p)dp$ is the probability of finding the particle's momentum between p and $p + dp$. Furthermore, the wave functions are transforms of each other:

$$\varphi(p) = \left(\frac{1}{2\pi\hbar}\right)^{\frac{1}{2}} \int_{-\infty}^{\infty} \psi(x)e^{\frac{-ipx}{\hbar}}dx \tag{2.2.3}$$

$$\psi(x) = \left(\frac{1}{2\pi\hbar}\right)^{\frac{1}{2}} \int_{-\infty}^{\infty} \varphi(p)e^{\frac{ipx}{\hbar}}dp \tag{2.2.4}$$

Comparing Equations 2.2.3 and 2.2.4 with Equations 2.2.1 and 2.2.2 establishes that these transforms are just Fourier and inverse transforms if we identify u as momentum p and a as $1/\hbar$. For example, a wave function in configuration space has the form

$$\psi(x) = \begin{cases} Ae^{ip_0 x/\hbar} & |x| < b \\ 0 & |x| > b \end{cases}$$

where A and b are positive real constants. We can plot the probability density $P(x) = \psi(x)^* \psi(x)$:

```
In[5]:= Plot[Piecewise[{{1, Abs[x] < 1}}], {x, -2, 2},
         AxesLabel → {"x (b)", "P (x) (A^2)"}, PlotStyle -> Thickness[0.01]]
```

Out[5]=

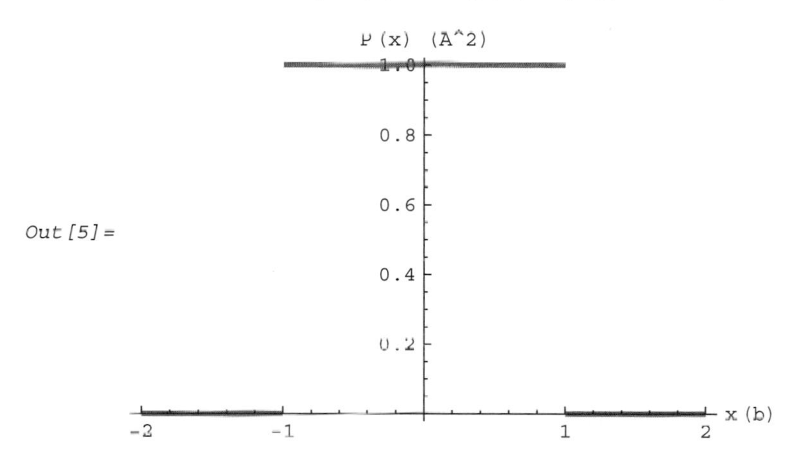

with $P(x)$ and x expressed in units of A^2 and b, respectively. (The function **Plot** will be discussed in Section 2.3.1.1.) The uncertainty Δx in the particle's position is approximately $2b$. The wave function in momentum space is the Fourier transform of $\psi(x)$:

```
In[6]:= Simplify[
         FourierTransform[Piecewise[{{A Exp[(I p0 x)/ℏ], Abs[x] < b}}], x,
           -p, FourierParameters → {0, 1/ℏ}, Assumptions → b > 0] //
          ExpToTrig // TrigReduce, ℏ > 0]
```

$$\text{Out[6]=} \quad \frac{A\sqrt{\frac{2}{\pi}}\sqrt{\hbar}\,\text{Sin}\left[\frac{b\,(p - p0)}{\hbar}\right]}{p - p0}$$

Let us plot the probability density $P(p) = \varphi^*(p)\varphi(p)$:

```
In[7]:= Plot[(Sin[p - 10]/(p - 10))^2, {p, -5, 20}, PlotRange → All,
         AxesLabel → {"p (ℏ/b)", "P(p) ((2/πℏ) (Ab)^2)"}]
```

Out[7]=

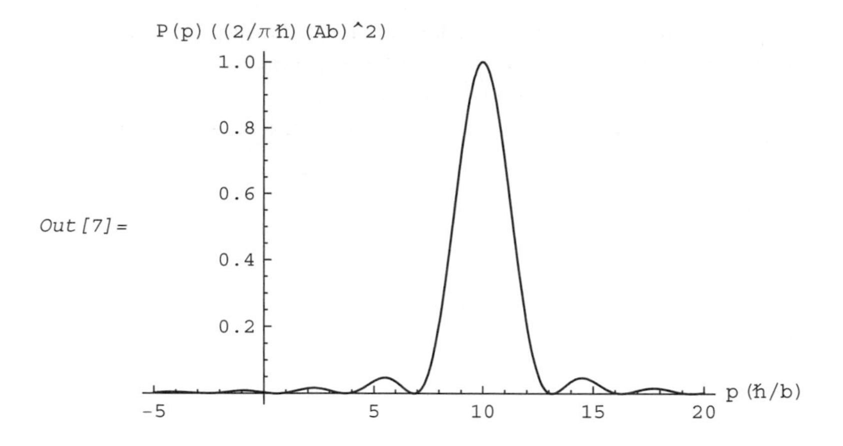

with $P(p)$ and p expressed in units of $2A^2b^2/\pi\hbar$ and $\hbar/b$, respectively. We have also set $p_0 = 10$ in the graph. The probability density has a sharp peak at p_0 and falls off rapidly on both sides to zero at $p = p_0 \pm \pi\hbar/b$. Therefore, a reasonable estimate of the uncertainty Δp in the particle's momentum is $2\pi\hbar/b$. The product of the uncertainties in position and momentum is

$$\Delta x \Delta p \approx 4\pi\hbar$$

This is consistent with Heisenberg's uncertainty relations.

2.2.19 Evaluating Subexpressions

Hitherto we have always asked *Mathematica* to evaluate the whole current cell. We can also ask *Mathematica* to evaluate only a selected subexpression in the cell. To evaluate a subexpression, select or highlight it and press ⌘+*return* for Mac OS X or Shift+Ctrl+Enter for Windows.

For example, enter the integral

$$\int \sqrt{a + b\sin^2\theta}\cos\theta\, d\theta$$

```
In[1]:= Integrate[Sqrt[a + b Sin[θ]^2] Cos[θ], θ]
```

$$\text{Out[1]=}\ \frac{a\,\text{Log}\left[b\,\text{Sin}[\theta] + \sqrt{b}\sqrt{a + b\,\text{Sin}[\theta]^2}\right]}{2\sqrt{b}} + \frac{1}{2}\text{Sin}[\theta]\sqrt{a + b\,\text{Sin}[\theta]^2}$$

Differentiating the output and then simplifying the result should recover the integrand:

```
In[2]:= D[%, θ] // Simplify
```

$$\text{Out[2]=}\ \frac{\text{Cos}[\theta]\sqrt{2\,a + b - b\,\text{Cos}[2\,\theta]}}{\sqrt{2}}$$

Yet this is not exactly the integrand. Applying the trigonometric manipulating functions **TrigExpand**, **TrigFactor**, or **TrigReduce** to it produces no further changes. Manipulating its subexpressions may be the only way to retrieve the integrand.

Let us copy and paste the previous output in an Input cell and wrap the function **TrigExpand** around the radicand:

$$\frac{\text{Cos}[\theta]\sqrt{\text{TrigExpand}[2\,a + b - b\,\text{Cos}[2\,\theta]]}}{\sqrt{2}}$$

Select the edited radicand, and press ⌘+*return* for Mac OS X or Shift+Ctrl+Enter for Windows. *Mathematica* returns in place

$$\frac{\text{Cos}[\theta]\sqrt{(2\,a + b - b\,\text{Cos}[\theta]^2 + b\,\text{Sin}[\theta]^2)}}{\sqrt{2}}$$

Wrap the function **Simplify** around $b - b\,\text{Cos}[\theta]^2$:

$$\frac{\text{Cos}[\theta]\sqrt{\left(2\,a + \text{Simplify}\left[b - b\,\text{Cos}[\theta]^2\right] + b\,\text{Sin}[\theta]^2\right)}}{\sqrt{2}}$$

Select $\text{Simplify}\left[b - b\,\text{Cos}[\theta]^2\right]$, and press ⌘+*return* for MacOSX or Shift+Ctrl+Enter for Windows. *Mathematica* generates in place

$$\frac{\text{Cos}[\theta]\sqrt{(2\,a + b\,\text{Sin}[\theta]^2 + b\,\text{Sin}[\theta]^2)}}{\sqrt{2}}$$

Finally, apply the function **Simplify** to the entire expression:

$$In[5]:= \frac{\text{Cos}[\theta]\sqrt{(2\,a + b\,\text{Sin}[\theta]^2 + b\,\text{Sin}[\theta]^2)}}{\sqrt{2}} \;//\; \text{Simplify}$$

$$Out[5]= \text{Cos}[\theta]\sqrt{a + b\,\text{Sin}[\theta]^2}$$

This is the integrand!

For manipulating subexpressions and expressions, the AlgebraicManipulation palette of the Palettes menu provides an alternative to entering transforming functions such as **Simplify** and **TrigExpand** directly from the keyboard. Let us copy and paste the output "Out[2]" in an Input cell:

$$\frac{\text{Cos}[\theta]\sqrt{2\,a + b - b\,\text{Cos}[2\,\theta]}}{\sqrt{2}}$$

Select the radicand in the numerator and click **TrigExpand[■]** in the palette. *Mathematica* gives

$$\frac{\text{Cos}[\theta]\sqrt{(2\,\text{a}+\text{b}-\text{b}\,\text{Cos}[\theta]^2+\text{b}\,\text{Sin}[\theta]^2)}}{\sqrt{2}}$$

In this expression, highlight **b-bCos[θ]²** and click **Simplify[■]** in the palette. *Mathematica* generates in place

$$\frac{\text{Cos}[\theta]\sqrt{(2\,\text{a}+\text{b}\,\text{Sin}[\theta]^2+\text{b}\,\text{Sin}[\theta]^2)}}{\sqrt{2}}$$

Finally, select the entire expression and click **Simplify[■]**. *Mathematica* returns in place

$$\text{Cos}[\theta]\sqrt{\text{a}+\text{b}\,\text{Sin}[\theta]^2}$$

which is the integrand.

After an in-place evaluation of a subexpression, **%** refers only to the result of the subexpression evaluation rather than the content of the entire cell. For example, type

$$\sqrt{2\,\text{a}+\text{b}-\text{b}\,\text{Cos}[\theta]^2+\text{b}\,\text{Sin}[\theta]^2}$$

Highlight **b-bCos[θ]²**, choose AlgebraicManipulation in the Palettes menu, and click **Simplify[■]**. *Mathematica* returns in place

$$\sqrt{2\,\text{a}+\text{b}\,\text{Sin}[\theta]^2+\text{b}\,\text{Sin}[\theta]^2}$$

Now, **%** refers to the result of the previous in-place evaluation:

```
In[10]:= %
Out[10]= b Sin[θ]^2
```

2.2.20 Exercises

In this book, straightforward, intermediate-level, and challenging exercises are unmarked, marked with one asterisk, and marked with two asterisks, respectively. For solutions to most odd-numbered exercises, see Appendix C.

1. Find the partial fraction decomposition of

$$\frac{x^2+2x-1}{2x^3+3x^2-2x}$$

   Answer:

$$\frac{1}{2\,\text{x}}-\frac{1}{10\,(2+\text{x})}+\frac{1}{5\,(-1+2\,\text{x})}$$

2. Find the partial fraction decomposition of

$$\frac{4x^3 - 27x^2 + 5x - 32}{30x^5 - 13x^4 + 50x^3 - 286x^2 - 299x - 70}$$

Answer:

$$-\frac{668}{323(1+2x)} - \frac{9438}{80155(-7+3x)} + \frac{24110}{4879(2+5x)} + \frac{48935+22098x}{260015(5+x+x^2)}$$

3. Factor

$$x^4 + 2x^3 - 3x - 6$$

Answer:

$(2+x)(-3+x^3)$

4. Factor

$$6t^3 + 9t^2 - 15t$$

Answer:

$3(-1+t)t(5+2t)$

5. Factor

$$ax^2 + ay + bx^2 + by$$

Answer:

$(a+b)(x^2+y)$

6. Factor

$$2x^4y^6 + 6x^2y^3 - 20$$

Answer:

$2(-2+x^2y^3)(5+x^2y^3)$

7. Simplify

$$\frac{1}{x^2 - 16} - \frac{x+4}{x^2 - 3x - 4}$$

Answer:

$$\frac{-15 - 7\,x - x^2}{(-4 + x)\,(1 + x)\,(4 + x)}$$

8. Simplify

$$\frac{\sqrt{4 - x^2} + \dfrac{x^2}{\sqrt{4 - x^2}}}{4 - x^2}$$

Answer:

$$\frac{4}{(4 - x^2)^{3/2}}$$

9. Simplify

$$\frac{\dfrac{1}{x^2} - \dfrac{1}{y^2}}{\dfrac{1}{x} + \dfrac{1}{y}}$$

Answer:

$$\frac{-x + y}{x\,y}$$

10. Add and simplify

$$\frac{3}{x^3 - x} + \frac{4}{x^2 + 2x + 1}$$

Answer:

$$\frac{3 - x + 4\,x^2}{(-1 + x)\,x\,(1 + x)^2}$$

11. Add and simplify

$$\frac{3}{x^3 - x} + \frac{4}{x^2 + 4x + 4}$$

Answer:

$$\frac{12 + 8\,x + 3\,x^2 + 4\,x^3}{-4\,x - 4\,x^2 + 3\,x^3 + 4\,x^4 + x^5}$$

12. Transform

$$\frac{x^3 + x}{x - 1}$$

into

$$x^2 + x + 2 + \frac{2}{x - 1}$$

Answer:

$$2 + \frac{2}{-1 + x} + x + x^2$$

13. Transform

$$\frac{x^4 - 2x^2 + 4x + 1}{x^3 - x^2 - x + 1}$$

into

$$x + 1 + \frac{1}{x - 1} + \frac{2}{(x - 1)^2} - \frac{1}{x + 1}$$

Answer:

$$1 + \frac{2}{(-1 + x)^2} + \frac{1}{-1 + x} + x - \frac{1}{1 + x}$$

14. Simplify

$$\frac{\sqrt{\frac{2+x}{7+x}}\,\sqrt[3]{\frac{3-x^2}{25-x^2}}\,\sqrt[3]{5 + x}}{\sqrt{4 - x^2}\,\sqrt[3]{3 - x^2}}$$

Answer:

$$\frac{1}{(5 - x)^{1/3}\sqrt{-(-2 + x)(7 + x)}}$$

15. Using **`PowerExpand`**, expand

$$\sqrt[3]{54x^3y^6z}$$

Assume x, y, and z are positive and real.

Answer:

$3\,2^{1/3}\,x\,y^2\,z^{1/3}$

16. Using **PowerExpand**, simplify

$$\frac{\sqrt{6xy^2}}{\sqrt{3x^5}}$$

Assume x and y are positive and real.

Answer:

$\frac{\sqrt{2}\,y}{x^2}$

17. Using **PowerExpand**, simplify

$$\ln\left(\frac{\sqrt{x}}{x}\right) + \ln\sqrt[4]{ex^2}$$

Assume x is positive and real.

Answer:

$\frac{1}{4}$

18. Simplify

$$\sec^2 x\,(1 + \sec 2x)$$

Answer:

2 sec [2 x]

19. Simplify

$$(\sec\alpha - 2\sin\alpha)(\csc\alpha + 2\cos\alpha)\sin\alpha\cos\alpha$$

Answer:

$\cos[2\,\alpha]^2$

20. Simplify

$$\sec\theta\tan\theta(\sec\theta+\tan\theta)+(\sec\theta-\tan\theta)$$

Answer:

```
sec[θ]^3 + Tan[θ]^3
```

21. Expand $\sin 7\theta$.

Answer:

```
7 cos[θ]^6 sin[θ] - 35 cos[θ]^4 sin[θ]^3 + 21 cos[θ]^2 sin[θ]^5 - sin[θ]^7
```

22. Transform

$$4\sin x\cos^2 x\,(1-8\sin^2 x)$$

into

$$-3\sin x-\sin 3x+2\sin 5x$$

Answer:

```
-3 sin[x] - sin[3 x] + 2 sin[5 x]
```

23. Transform

$$\sin^3 x\sin 3x+\cos^3 x\cos 3x$$

into

$$\cos^3 2x$$

Answer:

```
cos[2 x]^3
```

24. Transform

$$2\sin\left(\frac{x+y}{2}\right)\cos\left(\frac{x-y}{2}\right)$$

into

$$\sin x+\sin y$$

Answer:

```
sin[x] + sin[y]
```

25. Factor

$$-\sin(\alpha-\beta-\gamma)+\sin(\alpha+\beta-\gamma)+\sin(\alpha-\beta+\gamma)-\sin(\alpha+\beta+\gamma)$$

Answer:

```
4 sin[α] sin[β] sin[γ]
```

26. Expand $\sin(4\tan^{-1}x)$.

Answer:

$$\frac{4\,x}{(1+x^2)^2}-\frac{4\,x^3}{(1+x^2)^2}$$

27. Simplify

$$\sec x\tan x\,(\sec x+\tan x)+(\sec x-\tan x)$$

Answer:

```
sec[x]^3 + Tan[x]^3
```

28. Prove that

$$\frac{1-\cos\theta}{1-\csc\theta}-\frac{1+\cos\theta}{1+\csc\theta}=2\tan\theta\,(\sin\theta-\sec\theta)$$

Answer:

```
True
```

29. Prove that

$$\frac{\sec x-\csc x}{\tan x+\cot x}=\frac{\tan x-\cot x}{\sec x+\csc x}$$

Answer:

```
True
```

30. Prove that

$$\frac{2\sin 2x-\sin 3x}{\cos x}=\sin x\left(8\sin^2\frac{x}{2}+\sec x\right)$$

Answer:

```
True
```

31. Simplify

$$\Gamma(z)\Gamma(-z)$$

where $\Gamma(z)$ is the Euler gamma function.

Answer:

```
  π Csc [π z]
- ----------
      z
```

32. Simplify

$$\Gamma(z)\Gamma(1-z)$$

where $\Gamma(z)$ is the Euler gamma function.

Answer:

```
π Csc [π z]
```

33. Expand out

$$\frac{dH_n(x)}{dx}$$

where $H_n(x)$ are the Hermite polynomials.

Answer:

```
2 n HermiteH[-1 + n, x]
```

34. Simplify

$$-2x^{2n}K_n(x) + \pi I_n(x)[-x^{2n} + (ix)^{2n}\cos(n\pi)]\csc(n\pi)$$

where $I_n(x)$ are the modified Bessel functions of the first kind and $K_n(x)$ are the modified Bessel functions of the second kind.

Answer:

```
π((ⅈ x)^(2 n) BesselI[n, x] Cot[n π] - x^(2 n) BesselI[-n, x] Csc[n π])
```

35. Show that

$$\frac{d\Gamma(z+1)}{dz} = \Gamma(z) + z\frac{d\Gamma(z)}{dz}$$

where $\Gamma(z)$ is the Euler gamma function.

Answer:

```
True
```

36. Show that

$$\Gamma(z) = (z-1)\Gamma(z-1)$$

where $\Gamma(z)$ is the Euler gamma function.

Answer:

```
True
```

37. Expand out

$$\Gamma(z)\Gamma\left(z+\frac{1}{2}\right)$$

where $\Gamma(z)$ is the Euler gamma function.

Answer:

$2^{1-2z}\sqrt{\pi}$ Gamma[2z]

38. Show that

$$\mathrm{B}(p,q+1) = \left(\frac{q}{p}\right)\mathrm{B}(p+1,q)$$

where $\mathrm{B}(p,q)$ is the Euler beta function.

Answer:

```
True
```

39. Assuming n is a half-integer such as $\ldots, -11/2, -7/2, -3/2, 1/2, 5/2, 9/2, \ldots$, simplify $\sin(n\pi - \theta)$.

Answer:

Cos[θ]

40. Assuming n is an even integer, simplify $\cos(\theta - n\pi)$.

Answer:

```
cos [θ]
```

41. If $-\pi/2 \leq u \leq \pi/2$, simplify $\sin^{-1}(\sin u)$.

Answer:

```
u
```

42. Find the coefficient of x^3 in the expanded form of the polynomial

$$(1 + x - 2y)^6$$

Answer:

$20 - 120\,y + 240\,y^2 - 160\,y^3$

43. Find the term independent of x in the expanded form of the polynomial

$$(4a + x)^2(7 + bx^2)^4$$

Answer:

$38\,416\,a^2$

44. Find the coefficient of x^4y^3 in the expanded form of the polynomial

$$(4x + 5y)^7$$

Answer:

1 120 000

45. Find the numerator of the rational expression

$$\frac{x^2 + 3x + 2}{x^2 - 1}$$

Answer:

$2 + 3\,x + x^2$

46. Find the denominator of the fractional expression

$$\frac{3+\sqrt{4x+y}}{(x+2y)^2}$$

Answer:

```
(x + 2 y)^2
```

47. Using the standard packages **Units`** and **PhysicalConstants`**, calculate the de Broglie wavelength of (a) an electron with kinetic energy of 10 eV and (b) a proton with kinetic energy of 10 MeV. The de Broglie wavelength λ is given by

$$\lambda = \frac{h}{\sqrt{2mE_k}}$$

Where h is Planck's constant, m is the mass, and E_k is the kinetic energy.

Answers:

```
3.8783×10^-10 Meter
9.0508×10^-15 Meter
```

*48. The rotational energy levels of HCl are given by

$$\mathcal{E}_l = \left(\frac{\hbar^2}{2I}\right) l(l+1)$$

where

$$l = 0, 1, 2, 3, \ldots$$

$\hbar$ is Planck's constant divided by 2π, and the moment of intertia

$$I = \frac{m_{\mathrm{H}} m_{\mathrm{Cl}}}{m_{\mathrm{H}} + m_{\mathrm{Cl}}} r^2$$

in which m_{H} is the mass of the hydrogen atom, m_{Cl} is the mass of the chlorine atom, and r is the interatomic separation. Let $r = 1.27 \times 10^{-10}$ m. Using the standard packages **Units`** and **PhysicalConstants`** and the function **ElementData**, (a) determine m_{H} and m_{Cl} in kg, (b) calculate the moment of inertia I, (c) determine the rotational energy level of HCl for $l = 4$, and (d) express in electron volts the rotational energy obtained in part (c).

Answers:

```
1.67372 × 10^-27 kilogram
5.88711 × 10^-26 kilogram
2.62492 × 10^-47 kilogram Meter^2
4.23678 × 10^-21 Joule
0.0264439 ElectronVolt
```

49. Solve

$$4x + 5y = 5$$
$$6x + 7y = 7$$

Answer:

```
{{x→0, y→1}}
```

50. Solve

$$t^2 + 4t - 8 = 0$$

Answers:

$$\{\{t \to 2(-1-\sqrt{3})\}, \{t \to 2(-1+\sqrt{3})\}\}$$

51. Solve

$$\sqrt{x+2} + 4 = x$$

Answer:

```
{{x→7}}
```

52. Solve

$$\sqrt{y-5} - \sqrt{y} = 1$$

Answer:

```
{}
```

53. Solve

$$\sqrt{2x-3} - \sqrt{x+7} = 2$$

Answer:

```
{{x→42}}
```

54. (a) Solve the system of equations

$$x = y + z$$
$$by + ax = f$$
$$-g + cz - by = 0$$

for x, y, and z in terms of a, b, c, f, and g. (b) In part (a), *Mathematica* returned the solution in terms of transformation rules. Apply these rules to the system of equations with the replacement operator "`/.`" in order to verify the solution.

Answers:

$$\left\{\left\{x \to -\frac{-bf - cf - bg}{ab + ac + bc},\ y \to -\frac{-cf + ag}{ab + ac + bc},\ z \to -\frac{-bf - ag - bg}{ab + ac + bc}\right\}\right\}$$

```
{{True, True, True}}
```

55. For the circuit shown, (a) find all currents in terms of the emf and the resistances and (b) obtain numerical values for the currents with $E = 12\,\text{V}$, $R_1 = 2.0\,\Omega$, $R_2 = 4.0\,\Omega$, $R_3 = 6.0\,\Omega$, $R_4 = 2.0\,\Omega$, and $R_5 = 3.0\,\Omega$. Assume that the internal resistance of the battery is negligible.

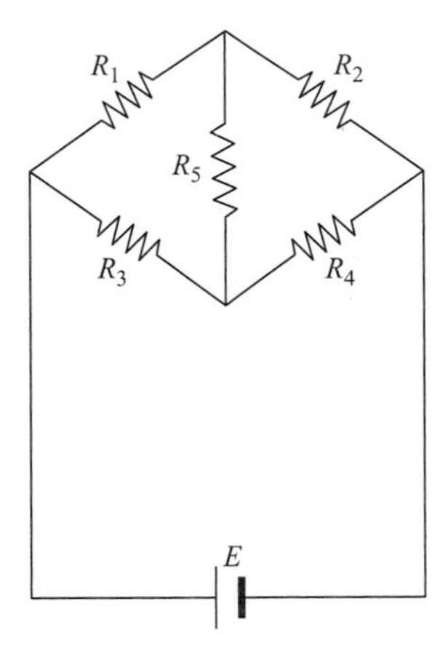

Answers:

$$\left\{\left\{i3 \to \frac{\text{emf}\,(R2\,R5 + R1\,(R2 + R4 + R5))}{R2\,(R4\,R5 + R3\,(R4 + R5)) + R1\,(R2\,(R3 + R4) + R4\,R5 + R3\,(R4 + R5))},\right.\right.$$

$$i6 \to \frac{\text{emf}\,(R3\,R4 + R3\,R5 + R4\,R5 + R2\,(R3 + R5) + R1\,(R2 + R4 + R5))}{R2\,(R4\,R5 + R3\,(R4 + R5)) + R1\,(R2\,(R3 + R4) + R4\,R5 + R3\,(R4 + R5))},$$

$$i4 \to \frac{\text{emf}\,(R1\,(R2 + R5) + R2\,(R3 + R5))}{R2\,(R4\,R5 + R3\,(R4 + R5)) + R1\,(R2\,(R3 + R4) + R4\,R5 + R3\,(R4 + R5))},$$

```
i5 → emf (R2 R3 - R1 R4) / (R2 (R4 R5 + R3 (R4 + R5)) + R1 (R2 (R3 + R4) + R4 R5 + R3 (R4 + R5))),

i1 → emf (R2 R3 + R4 R5 + R3 (R4 + R5)) / (R2 (R4 R5 + R3 (R4 + R5)) + R1 (R2 (R3 + R4) + R4 R5 + R3 (R4 + R5))),

i2 → emf (R1 R4 + R4 R5 + R3 (R4 + R5)) / (R2 (R4 R5 + R3 (R4 + R5)) + R1 (R2 (R3 + R4) + R4 R5 + R3 (R4 + R5)))}}

{{i3 → 1.28571 A, i6 → 3.85714 A, i4 → 2.14286 A,
  i5 → 0.857143 A, i1 → 2.57143 A, i2 → 1.71429 A}}
```

56. For the circuit shown, (a) find all currents in terms of the emfs and the resistances and (b) obtain numerical values for the currents with $E_1 = 12\,\text{V}$, $E_2 = 8\,\text{V}, R_1 = 20.0\,\Omega, R_2 = 15.0\,\Omega, R_3 = 3.0\,\Omega, R_4 = 6.0\,\Omega, R_5 = 12.0\,\Omega$, and $R_6 = 2.0\,\Omega$. Assume that the internal resistances of the batteries are negligible.

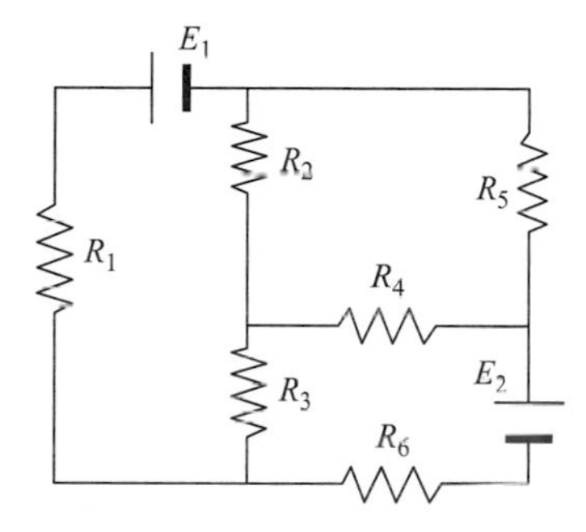

Answers:

```
{{i5 → (E2 ((R1 + R3) R4 + R2 (R3 + R4)) + E1 (R3 R4 + R2 (R3 + R4 + R6))) /
    (R1 (R4 R5 + R3 (R4 + R5) + R4 R6 + R5 R6 + R2 (R3 + R4 + R6)) +
     R3 (R5 R6 + R4 (R5 + R6)) + R2 (R5 R6 + R3 (R5 + R6) + R4 (R5 + R6))),
  i6 → (E1 (R2 (R3 + R4) + R3 (R4 + R5)) +
      E2 (R3 (R4 + R5) + R1 (R2 + R4 + R5) + R2 (R3 + R4 + R5))) /
    (R1 (R4 R5 + R3 (R4 + R5) + R4 R6 + R5 R6 + R2 (R3 + R4 + R6)) +
      R3 (R5 R6 + R4 (R5 + R6)) + R2 (R5 R6 + R3 (R5 + R6) + R4 (R5 + R6))),
  i1 → (E2 (R2 (R3 + R4) + R3 (R4 + R5)) +
      E1 (R4 R5 + R3 (R4 + R5) + R4 R6 + R5 R6 + R2 (R3 + R4 + R6))) /
    (R1 (R4 R5 + R3 (R4 + R5) + R4 R6 + R5 R6 + R2 (R3 + R4 + R6)) +
      R3 (R5 R6 + R4 (R5 + R6)) + R2 (R5 R6 + R3 (R5 + R6) + R4 (R5 + R6))),
  i2 → (E2 (R1 R4 - R3 R5) - E1 (R3 R5 + R5 R6 + R4 (R5 + R6))) /
    (R1 (R4 R5 + R3 (R4 + R5) + R4 R6 + R5 R6 + R2 (R3 + R4 + R6)) +
      R3 (R5 R6 + R4 (R5 + R6)) + R2 (R5 R6 + R3 (R5 + R6) + R4 (R5 + R6))),
  i3 → (E2 (R2 R5 + R1 (R2 + R4 + R5)) - E1 ((R2 + R5) R6 + R4 (R5 + R6))) /
```

```
    (R1 (R4 R5 + R3 (R4 + R5) + R4 R6 + R5 R6 + R2 (R3 + R4 + R6)) +
      R3 (R5 R6 + R4 (R5 + R6)) + R2 (R5 R6 + R3 (R5 + R6) + R4 (R5 + R6))),
  i4 → (-E2 ((R2 + R3) R5 + R1 (R2 + R5)) + E1 (-R3 R5 + R2 R6)) /
    (R1 (R4 R5 + R3 (R4 + R5) + R4 R6 + R5 R6 + R2 (R3 + R4 + R6)) +
      R3 (R5 R6 + R4 (R5 + R6)) + R2 (R5 R6 + R3 (R5 + R6) + R4 (R5 + R6)))}}
```

```
{{i5 → 0.481 A, i6 → 1.15 A, i1 → 0.596 A,
  i2 → -0.116 A, i3 → 0.556 A, i4 → -0.671 A}}
```

57. Calculate the derivative of

$$y(x) = \sin\left(e^{x^2}\right)$$

Answer:

$2\, e^{x^2}\, x \,\mathrm{Cos}\left[e^{x^2}\right]$

58. Differentiate

$$y(x) = \frac{x^{3/4}\sqrt{x^2+1}}{(3x+2)^5}$$

Answer:

$$\frac{6 - 51\,x + 14\,x^2 - 39\,x^3}{4\,x^{1/4}\,(2+3\,x)^6\,\sqrt{1+x^2}}$$

59. Find the derivative of

$$y(x) = x\sinh^{-1}(x/3) - \sqrt{9+x^2}$$

Answer:

$\mathrm{ArcSinh}\left[\frac{x}{3}\right]$

60. Find the second derivative of

$$y(x) = e^x \sin x$$

Answer:

$2\, e^x \cos[x]$

61. Given

$$f = \frac{\sin xy}{\cos(x+y)}$$

find

$$\frac{\partial f}{\partial x}$$

Answer:

```
y Cos[x y] Sec[x + y] + Sec[x + y] Sin[x y] Tan[x + y]
```

62. Given

$$u = \ln\left(x^2 + y\right)$$

find

$$\frac{\partial^3 u}{\partial y^2 \partial x}$$

Answer:

$$\frac{4\,x}{\left(x^2 + y\right)^3}$$

63. (a) Given

$$z = x^2 + 3xy - y^2$$

Find the differential dz. (b) If x changes from 2 to 2.05 and y changes from 3 to 2.96, calculate dz.

Answers:

```
(2 x + 3 y) Dt[x] + (3 x - 2 y) Dt[y]
0.65
```

64. The base radius and height of a right circular cone measure 10 cm and 25 cm, respectively, to within 0.1 cm. Estimate the largest possible fractional error in the calculated volume.

Answer:

```
0.024
```

65. The ideal gas law can be written as

$$P = \frac{nRT}{V}$$

where P, n, R, T, and V are the pressure, number of moles, gas constant, temperature, and volume, respectively. If the rate of increase of the temperature is α and that of the volume is β, find the rate of increase of the pressure. *Hint:* n and R are constants.

Answer:

$$\frac{\text{n R (V}\,\alpha - \text{T}\,\beta)}{\text{V}^2}$$

66. Given the equation of a curve:

$$x^3 + y^3 = 6xy$$

Find $\frac{dy}{dx}$ and determine the equation of the tanget at the point $(3, 3)$.

Answers:

$$\left\{\left\{\text{Dt[y, x]} \to \frac{\text{x}^2 - 2\,\text{y}}{2\,\text{x} - \text{y}^2}\right\}\right\}$$

$$\text{x} + \text{y} == 6$$

67. Let the functions $f(t)$ and $r(t)$ be related by the equation

$$f(t) = \frac{dr(t)}{dt}$$

To within an additive constant, $r(t)$ is given in terms of $f(t)$ by

$$r(t) = \int f(t)dt$$

Determine $r(t)$ for

$$f(t) = 5\ln t - e^{-3t}$$

Answer:

$$\frac{e^{-3\,t}}{3} - 5\,t + 5\,t\,\text{Log[t]}$$

68. Let the functions $f(t)$ and $r(t)$ be related by the equation

$$f(t) = \frac{dr(t)}{dt}$$

To within an additive constant, $r(t)$ is given in terms of $f(t)$ by

$$r(t) = \int f(t)dt$$

Determine $r(t)$ for

$$f(t) = \frac{t}{\sqrt{d^2 + t^2}}$$

where d is a constant.

Answer:

$\sqrt{d^2 + t^2}$

69. Evalute

$$\int_0^1 \frac{4}{1 + x^2} dx$$

What is special about this integral?

Answer:

π

The integral evaluates to π exactly.

70. Consider the integral

$$\int \frac{x}{a^3 + x^3} dx$$

(a) Evaluate the integral. (b) Differentiate the result and recover the integrand.

Answers:

$$\frac{2\sqrt{3}\,\text{ArcTan}\left[\frac{-a + 2x}{\sqrt{3}\,a}\right] - 2\,\text{Log}[a + x] + \text{Log}[a^2 - a x + x^2]}{6a}$$

$$\frac{x}{a^3 + x^3}$$

71. Consider the integral

$$\int \frac{2x^2 - x + 4}{x^3 + 4x} dx$$

(a) Evaluate the integral. (b) Differentiate the result of part (a) and recover the integrand.

Answers:

$$-\frac{1}{2}\text{ArcTan}\left[\frac{x}{2}\right] + \text{Log}[x] + \frac{1}{2}\text{Log}[4 + x^2]$$

$$\frac{4 - x + 2x^2}{4x + x^3}$$

72. Evalute

$$\int \sin^2 x \cos x \ln(\sin x) dx$$

Answer:

$$\frac{1}{9}(-1 + 3\,\text{Log}[\text{Sin}[x]])\,\text{Sin}[x]^3$$

73. Assuming that the constant a is both real and positive, evaluate the integral

$$\int_{-\infty}^{\infty} x^2 e^{-2ax^2} dx$$

Answer:

$$\frac{\sqrt{\frac{\pi}{2}}}{4\,a^{3/2}}$$

74. Assuming that $a > 0$ and $m \geq 0$, evaluate

$$\int_0^{\infty} \frac{\cos^2 mx}{a^2 + x^2} dx$$

Answer:

$$\frac{\left(1 + e^{-2\,a\,m}\right)\pi}{4\,a}$$

75. Assuming that $p > -1$, find

$$\int_0^{\pi} x \sin^p x\, dx$$

Answer:

$$\frac{\pi^{3/2}\,\text{Gamma}\left[\frac{1+p}{2}\right]}{2\,\text{Gamma}\left[1+\frac{p}{2}\right]}$$

76. Evaluate

$$\int_{\pi/6}^{\pi/4}\int_{\tan x}^{\sec x}(y+\sin x)dydx$$

Answer:

$$\frac{1}{24}\left(\pi+12\left(-1+\sqrt{2}+4\,\text{ArcTanh}\left[2-\sqrt{3}\right]-4\,\text{ArcTanh}\left[\text{Tan}\left[\frac{\pi}{8}\right]\right]+\text{Log}\left[\frac{3}{2}\right]\right)\right)$$

77. Consider the wave function

$$\psi(x) = \frac{A}{b^2+x^2}$$

where b is a positive constant. (a) Normalize the wave function. That is, find a constant A such that

$$\int_{-\infty}^{\infty}|\psi(x)|^2dx = 1$$

(b) Find the expectation value of x. Is *Mathematica* necessary for evaluating the integral? (c) Determine the expectation value of x^2.

Answers:

$$\left\{A\to b^{3/2}\sqrt{\frac{2}{\pi}}\right\}$$

0

b^2

78. Find the sum of the series

$$\sum_{n=0}^{\infty}\frac{(-1)^n}{n!}$$

Answer:

$$\frac{1}{e}$$

79. Find the sum of the series

$$\sum_{n=0}^{\infty} \frac{\cos^n x}{2^n}$$

Answer:

$$-\frac{2}{-2 + \text{Cos}[x]}$$

80. Find the sum of the series

$$\sum_{n=1}^{\infty} \frac{n^2 + 3n + 1}{(n^2 + n)^2}$$

Answer:

2

81. Find the indefinite sum

$$1^3 + 2^3 + 3^3 + 4^3 + \cdots + n^3$$

Answer:

$$\frac{1}{4} n^2 (1 + n)^2$$

82. Find a power series approximation, up to terms of order x^8, for

$$\sqrt[3]{1 + x^4}$$

Answer:

$$1 + \frac{x^4}{3} - \frac{x^8}{9} + O[x]^9$$

83. Find a power series approximation for $\sec(x)$ about $x = \pi/3$, accurate to order $(x - \pi/3)^3$.

Answer:

$$2 + 2\sqrt{3}\left(x - \frac{\pi}{3}\right) + 7\left(x - \frac{\pi}{3}\right)^2 + \frac{23\left(x - \frac{\pi}{3}\right)^3}{\sqrt{3}} + O\left[x - \frac{\pi}{3}\right]^4$$

84. Find a power series approximation, up to terms of order x^{11}, for

$$\ln\left(\frac{1+x}{1-x}\right)$$

Answer:

$$2\,x + \frac{2\,x^3}{3} + \frac{2\,x^5}{5} + \frac{2\,x^7}{7} + \frac{2\,x^9}{9} + \frac{2\,x^{11}}{11} + O[x]^{12}$$

85. Compute the limit as x goes to infinity of

$$\frac{\log(x-a)}{(a-b)(a-c)} + \frac{\log 2(x-b)}{(b-c)(b-a)} + \frac{\log(x-c)}{(c-a)(c-b)}$$

where log is the logarithmic function to the base 10.

Answer:

$$-\frac{\text{Log}[2]}{(a-b)\,(b-c)\,\text{Log}[10]}$$

*86. Consider the definite integral

$$\int_0^{1/3} x^2 \tan^{-1}(x^4)dx$$

(a)Find a power series approximation, up to terms of order x^{54}, for the integrand. (b) Integrate the result of part (a) to obtain an approximate value for the definite integral. (c) In the power series approximation for the integrand, determine the minimum number of terms needed to obtain for the definite integral an approximate value that matches, to six significant figures, the one given by **NIntegrate**. (d) Using the "front end help" discussed in Section 1.6.3, ascertain whether the approximate value given by **NIntegrate** is correct to six significant figures. (e) Find the radius of convergence of the series in part (a) and then use the Alternating Series Estimation Theorem to determine an approximate value for the integral correct to six significant figures. Does it agree with the one given by **NIntegrate**?

Answers:

$$x^6 - \frac{x^{14}}{3} + \frac{x^{22}}{5} - \frac{x^{30}}{7} + \frac{x^{38}}{9} - \frac{x^{46}}{11} + \frac{x^{54}}{13}$$

0.0000653195

Two terms are needed.

The approximate value given by **`NIntegrate`** is normally correct to at least six significant figures.
The radius of convergence is $R = 1$.

```
0.0000653195
```

87. Compute the limit as x goes to infinity of

$$\frac{1}{e^x - e^{(x-x^{-2})}}$$

Answer:

```
0
```

88. Find

$$\lim_{x\to\pi^-} \frac{\sin x}{1-\cos x}$$

Answer:

```
0
```

89. Find

$$\lim_{x\to 0^+} (1+\sin 4x)^{\cot x}$$

Answer:

e^4

90. Find

$$\lim_{x\to 1} \frac{\ln x}{x-1}$$

Answer:

```
1
```

91. Find

$$\lim_{t\to 0} \frac{\sqrt{t^2+9}-3}{t^2}$$

Answer:

$$\frac{1}{6}$$

92. Solve the differential equation

$$\frac{d^2y}{dx^2} - 6\frac{dy}{dx} + 13y = e^x \cos x$$

Answer:

$$\left\{\left\{y[x] \to \frac{1}{65} e^x \left(65 e^{2x} C[2] \cos[2x] - 4 \sin[x] + \cos[x]\left(7 + 130 e^{2x} C[1] \sin[x]\right)\right)\right\}\right\}$$

93. Solve the differential equation

$$y'' - 2y' = \sin 4x$$

Answer:

$$\left\{\left\{y[x] \to \frac{1}{2} e^{2x} C[1] + C[2] + \frac{1}{40} \cos[4x] - \frac{1}{20} \sin[4x]\right\}\right\}$$

94. Solve the differential equation

$$y'' + y = e^x + x^3$$

with the initial conditions $y(0) = 2$ and $y'(0) = 0$.

Answer:

$$\left\{\left\{y[x] \to \frac{1}{2}\left(e^x - 12x + 2x^3 + 3\cos[x] + 11\sin[x]\right)\right\}\right\}$$

*95. A particle of mass m moves in a one-dimensional box with walls at $x = 0$ and $x = a$. The potential can be expressed as

$$V(x) = \begin{cases} 0 & 0 < x < a \\ \infty & \text{elsewhere} \end{cases}$$

Inside the box, the time-independent Schrödinger equation is

$$-\frac{\hbar^2}{2m}\frac{d^2\phi}{dx^2} = E\phi$$

where $\hbar$ is Planck's constant divided by 2π, E is the energy eigenvalue, and ϕ is the corresponding eigenfunction. Letting

$$E = \frac{\hbar^2 k^2}{2m}$$

we can rewrite the time-independent Schrödinger equation as

$$\frac{d^2\phi}{dx^2} + k^2\phi = 0$$

(For the reason why E cannot be negative, see [Gas 96].) (a) Show that inside the box the eigenfunction has the form

$$\phi(x) = C_1 \cos kx + C_2 \sin kx$$

(b) As the eigenfunction vanishes outside the box, we must impose the boundary conditions

$$\phi(0) = 0$$

$$\phi(a) = 0$$

Show that the condition $\phi(0) = 0$ implies that $C_1 = 0$.
(c) For nontrivial solutions, $C_2 \neq 0$. Thus, the condition $\phi(a) = 0$ implies

$$\sin ka = 0$$

Use **Simplify** to show that

$$k_n = n\pi/a$$

are solutions to this equation, where n is any integer. In fact, n must be positive because $n = 0$ implies that the particle cannot be found anywhere and because the inclusion of negative integers for n yields no new eigenfunctions. (For more on why n must be positive, see [SMM97].)
(d) Using the "front end help" discussed in Section 1.6.3, obtain information on the function **Reduce**. Use **Reduce** to confirm that the solutions of

$$\sin ka = 0$$

are indeed

$$k_n = n\pi/a$$

where n is any integer.

(e) Use the normalization condition

$$\int_0^a \phi^2 dx = 1$$

to show that

$$C_2 = \sqrt{\frac{2}{a}}$$

Thus, the eigenenergies and normalized eigenfunctions are

$$E_n = \frac{n^2\pi^2\hbar^2}{2ma^2}$$

$$\phi_n(x) = \sqrt{\frac{2}{a}}\sin\left(\frac{n\pi x}{a}\right)$$

$$n = 1, 2, 3, 4, \ldots$$

(f) Show that eigenfunctions corresponding to different eigenenergies are orthogonal. That is, show

$$\int_0^a \phi_n\phi_m dx = 0$$

when $n \neq m$.

Answers:

```
{{φ[x] → C[1] Cos[k x] + C[2] Sin[k x]}}
```

```
C[1] == 0
```

```
True
```

$$\texttt{n[1]} \in \texttt{Integers \&\& a} \neq \texttt{0 \&\&} \left(\texttt{k} == \frac{2\pi\,\texttt{n[1]}}{\texttt{a}} \;||\; \texttt{k} == \frac{\pi + 2\pi\,\texttt{n[1]}}{\texttt{a}}\right)$$

$$\texttt{C2} \to \frac{\sqrt{2}}{\sqrt{\texttt{a}}}$$

$$\texttt{En} == \frac{\texttt{n}^2\,\pi^2\,\hbar^2}{2\,\texttt{a}^2\,\texttt{m}}$$

$$\phi\texttt{n} == \sqrt{2}\,\sqrt{\frac{1}{\texttt{a}}}\;\texttt{Sin}\left[\frac{\texttt{n}\,\pi\,\texttt{x}}{\texttt{a}}\right]$$

```
0
```

96. For a particle of mass m moving in a one-dimensional box with walls at $x = 0$ and $x = a$, the energy eigenvalues and normalized eigenfunctions are

$$E_n = \frac{n^2\pi^2\hbar^2}{2ma^2}$$

$$\phi_n(x) = \sqrt{\frac{2}{a}}\sin\left(\frac{n\pi x}{a}\right)$$

$$n = 1, 2, 3, 4, \ldots$$

(a) For the energy eigenvalues, define a function with the quantum number n as the argument. Evaluate the four lowest energies—that is, for $n = 1, 2, 3$, and 4. (b) For the eigenfunctions, define a function with two arguments: the quantum number n and the position x. Determine the eigenfunctions for $n = 1, 2, 3$, and 4.

Answers:

$$\left\{\frac{\pi^2\hbar^2}{2\,a^2\,m}, \frac{2\,\pi^2\hbar^2}{a^2\,m}, \frac{9\,\pi^2\hbar^2}{2\,a^2\,m}, \frac{8\,\pi^2\hbar^2}{a^2\,m}\right\}$$

$$\left\{\sqrt{2}\sqrt{\frac{1}{a}}\,\mathrm{Sin}\left[\frac{\pi\,x}{a}\right], \sqrt{2}\sqrt{\frac{1}{a}}\,\mathrm{Sin}\left[\frac{2\,\pi\,x}{a}\right], \sqrt{2}\sqrt{\frac{1}{a}}\,\mathrm{Sin}\left[\frac{3\,\pi\,x}{a}\right], \sqrt{2}\sqrt{\frac{1}{a}}\,\mathrm{Sin}\left[\frac{4\,\pi\,x}{a}\right]\right\}$$

97. For the one-dimensional harmonic oscillator, the energy eigenvalues are

$$E_n = \hbar\omega\left(n + \frac{1}{2}\right), \qquad n = 0, 1, 2, \ldots$$

where ω is the oscillator frequency. Also, the normalized eigenfunctions are

$$\phi_n(x) = 2^{-n/2}(n!)^{-1/2}\left(\frac{m\omega}{\hbar\pi}\right)^{1/4} H_n\left(\sqrt{m\omega/\hbar}\,x\right)\exp\left(-\frac{m\omega}{2\hbar}x^2\right), \qquad n = 0, 1, 2, \ldots$$

where m is the mass, x is the displacement, and $H_n(\xi)$ are the Hermite polynomials. (a) For the energy eigenvalues, define a function with the quantum number n as the argument. Evaluate the first four lowest energies—that is, for $n = 0, 1, 2, 3$. (b) For the eigenfunctions, define a function with two arguments: the quantum number n and displacement x. Determine the eigenfunctions for $n = 0, 1, 2$, and 3. *Hint*: In *Mathematica*, **`HermiteH[n, x]`** gives the Hermite polynomial $H_n(x)$.

Answers:

$$\left\{\frac{\omega\hbar}{2}, \frac{3\,\omega\hbar}{2}, \frac{5\,\omega\hbar}{2}, \frac{7\,\omega\hbar}{2}\right\}$$

$$\left\{\frac{e^{-\frac{m x^2 \omega}{2 \hbar}} \left(\frac{m \omega}{\hbar}\right)^{1/4}}{\pi^{1/4}}, \frac{\sqrt{2}\, e^{-\frac{m x^2 \omega}{2 \hbar}}\, x \left(\frac{m \omega}{\hbar}\right)^{3/4}}{\pi^{1/4}},\right.$$

$$\left.\frac{e^{-\frac{m x^2 \omega}{2 \hbar}} \left(-2 + \frac{4 m x^2 \omega}{\hbar}\right) \left(\frac{m \omega}{\hbar}\right)^{1/4}}{2 \sqrt{2}\, \pi^{1/4}}, \frac{e^{-\frac{m x^2 \omega}{2 \hbar}} \left(-12 x \sqrt{\frac{m \omega}{\hbar}} + 8 x^3 \left(\frac{m \omega}{\hbar}\right)^{3/2}\right) \left(\frac{m \omega}{\hbar}\right)^{1/4}}{4 \sqrt{3}\, \pi^{1/4}}\right\}$$

98. Find the Fourier transform of the function

$$\psi(x) = \begin{cases} 1 - |x| & |x| < 1 \\ 0 & |x| \geq 1 \end{cases}$$

In Equation 2.2.1, let $a = 1/\hbar$ and $u = p$.

Answer:

$$-\frac{\sqrt{\frac{2}{\pi}}\, \hbar^{3/2} \left(-1 + \text{Cos}\left[\frac{p}{\hbar}\right]\right)}{p^2}$$

99. Determine the Fourier transform of the function

$$\psi(x) = \begin{cases} 0 & |x| \geq b \\ A & 0 < x < b \\ -A & 0 > x > -b \end{cases}$$

where A and b are positive real constants. In Equation 2.2.1, let $a = 1$ and $u = k$.

Answer:

$$-\frac{i A \sqrt{\frac{2}{\pi}} \left(-1 + \text{Cos}[b k]\right)}{k}$$

100. Find the Fourier transform of the function

$$f(x) = \sqrt{b} e^{-b|x|}$$

with $b > 0$. In Equation 2.2.1, let $a = 1$ and $u = k$.

Answer:

$$\frac{b^{3/2} \sqrt{\frac{2}{\pi}}}{b^2 + k^2}$$

101. Find the inverse Fourier transform of the function

$$g(k) = \frac{1}{1 + ik^3}$$

In Equation 2.2.2, let $a = 1$ and $u = k$. Verify the result by evaluating its Fourier transform.

Answer:

$$\frac{1}{3}\sqrt{2\pi}\left(e^{x}\,\text{HeavisideTheta}[-x] + (-1)^{1/3}\left(e^{-\frac{1}{2}\left(1+i\sqrt{3}\right)x} + \frac{2\,e^{-\frac{\left(-i+\sqrt{3}\right)x}{i+\sqrt{3}}}}{-1+i\sqrt{3}}\right)\text{HeavisideTheta}[x]\right)$$

102. **RSolve**[*eqn*, *a*[*n*], *n*] solves a recurrence equation for $a[n]$. Using **RSolve**, solve the recurrence equation

$$a_{n+1} + a_{n-1} = \frac{2n}{x}a_n$$

Answer:

```
{{a[n] → BesselJ[n, x]C[1] + BesselY[n, x]C[2]}}
```

103. After n years, the balance A_n in an account with principal P and annual percentage rate r (expressed as a decimal) is given by the equation

$$A_n = A_{n-1} + rA_{n-1}$$

together with the end condition

$$A_0 = P$$

(a) **RSolve**[{*eqn*, *endcond*$_1$, ...}, *a*[*n*], *n*] solves a recurrence equation with end conditions. Using **RSolve**, solve for A_n and thus determine a formula for compound interest. (b) If \$1000 is invested at an annual percentage rate of 8.0%, determine the balance in the account after 30 years.

Answers:

```
{{A[n] → P(1 + r)^n}}
10062.7
```

104. `RSolve[{`*eqn*`,` *endcond*$_1$`, ...},` *a*`[`*n*`],` *n*`]` solves a recurrence equation with end conditions. Using **RSolve**, solve the recurrence equation

$$f_n = n f_{n-1}$$

together with the end condition

$$f_0 = 1$$

Answer:

```
{{f[n] → Gamma[1 + n]}}
```

105. Using the AlgebraicManipulation palette in the Palettes menu, transform

$$\frac{1-\cos 2x}{(\csc x+\cot x)(1-\cos x)(\tan 2x+\cot 2x)}$$

into

$$\frac{1-\cos 2x}{2\sin x\csc 4x}$$

Answer:

$$\frac{1 - \text{Cos}[2\,x]}{(2\,\text{Csc}[4\,x]\,\text{Sin}[x])}$$

*106. A boat that has mass m and initial velocity v_0 moves along the x-direction. It is slowed by a frictional force

$$F = -\alpha e^{\beta v}$$

where α and β are positive constants. Determine (a) the motion of the boat (i.e., $x(t)$) and (b) the stopping distance. *Hint:* Use **Reduce**.

Answers:

$$\frac{t\,\beta - t\,\beta\,\text{Log}\left[e^{-v0\,\beta} + \frac{t\,\alpha\,\beta}{m}\right] - \frac{e^{-v0\,\beta}\,m\,\text{Log}\left[1 + \frac{e^{v0\,\beta}\,t\,\alpha\,\beta}{m}\right]}{\alpha}}{\beta^2}$$

$$\frac{e^{-v0\,\beta}\,m\,(-1 + e^{v0\,\beta} - v0\,\beta)}{\alpha\,\beta^2}$$

2.3 GRAPHICAL CAPABILITIES

Mathematica has won many advocates because of its graphical capabilities. This section discusses numerical equation solving, numerical extremum searching, interactive graphics drawing and manipulating, and animation as well as two- and three-dimensional plotting.

2.3.1 Two-Dimensional Graphics

2.3.1.1 Basic Plots

Plot[f, {x, x_{min}, x_{max}}] plots f as a function of x from x_{min} to x_{max}.

Example 2.3.1 The normalized eigenfunction for the first excited state of the one-dimensional harmonic oscillator is

$$u_1(x) = \left(\frac{4}{\pi}\right)^{1/4} x e^{-x^2/2}$$

where the displacement x is in units of $(\hbar/m\omega)^{1/2}$, in which m is the mass, ω is the oscillator frequency, and $\hbar$ is Plank's constant divided by 2π. Plot $u_1(x)$ for x from -4 to 4.

```
In[1]:= Plot[((4/π)^(1/4)) x Exp[-(x^2)/2], {x, -4, 4}]
```

Out[1]=

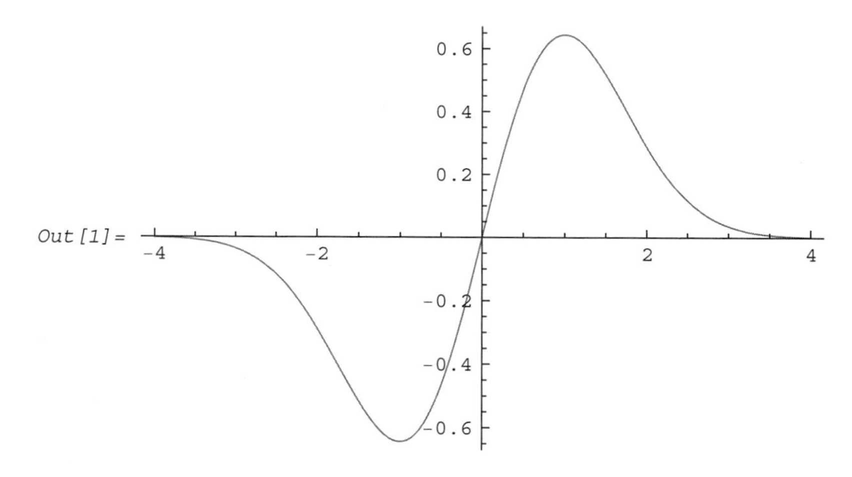

(For a discussion of the one-dimensional harmonic oscillator in quantum mechanics, see [Gos03] or [Lib03].) ■

With **Plot**[f, {x, x_{min}, x_{max}}], *Mathematica* evaluates f from scratch for each needed value of x. In some cases, it is better to evaluate f first into a more explicit symbolic expression before it is subsequently evaluated numerically for those values of x. For the latter approach, we can use **Plot[Evaluate[**f**]**, {x, x_{min}, x_{max}}].

Example 2.3.2 Plot the radial probability density $P(r) = r^2 R_{nl}^2$ for the 1s state of the hydrogen atom. The radial functions R_{nl} are given in Example 2.2.13 in Section 2.2.16.

From Example 2.2.13, we have

```
In[2]:= F[n_, l_, x_] :=
          ((x^l)Exp[-x/2]((n+l)!)LaguerreL[n-l-1, 2l+1, x])
```

```
In[3]:= R[n_, l_, r_]:= ((a^(-3/2))(2/(n^2))
            Sqrt[((n-l-1)!)/(((n+l)!)^3)]F[n, l, 2 r/(n a)])
```

In what follows, let $a = 1$; that is, r is in units of the Bohr radius a.

In this case, it is better to evaluate $r^2R_{10}^2$ first into a more explicit symbolic expression. Otherwise, for each needed value of r, *Mathematica* must invoke the function **R**, which calls the function **F**, which in turn calls the function **LaguerreL**. We now plot the probability distribution $r^2R_{10}^2$:

```
In[4]:= Plot[Evaluate[(r R[1, 0, r])^2 /. a→1], {r, 0, 6}]
```

Out[4]=

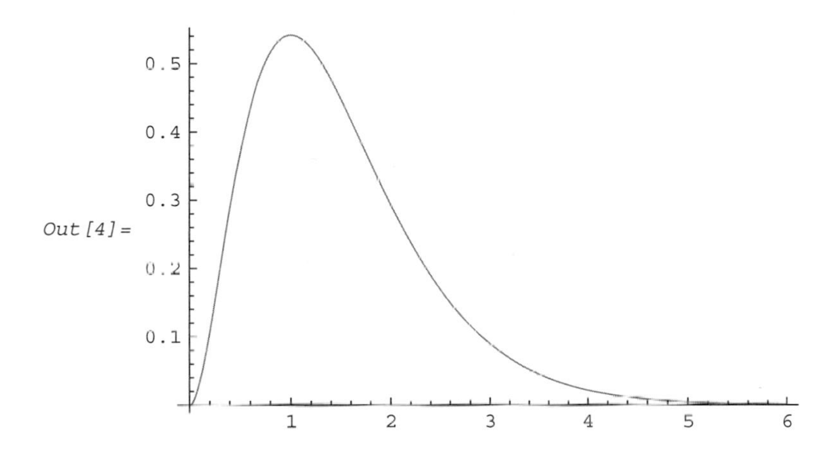

■

In numerical solution of differential equations, results are given as transformation rules in terms of interpolating functions. For plotting the solutions, it is better to use **Plot[Evaluate [***y*[*x*]**/.** *solution*], {*x*, x_{min}, x_{max}}] so that the **InterpolatingFunction** objects can be set up first. For an illustrative instance, refer to Example 2.1.8 in Section 2.1.18. (To see the **InterpolatingFunction** object in the example, enter **InputForm[x /. sol[[1]]]** after the differential equation is solved. **InputForm[***expr***]** prints as a version of *expr* suitable for input to *Mathematica*.)

2.3.1.2 Options

When *Mathematica* generates a graph, it makes many choices, such as the ratio of height to width for the plot, the range of coordinates to include in the plot, and the number of points at which to sample the function. In other words, *Mathematica* specifies the values of the options in the plotting function. The command **Options[***f***]** gives the list of options for the function f and their default values. For the function **Plot**, **Options[Plot]** shows its default options:

```
In[5]:= Options[Plot]

Out[5]= {AlignmentPoint→Center, AspectRatio→ 1/GoldenRatio,
  Axes→True, AxesLabel→None, AxesOrigin→Automatic,
  AxesStyle→{}, Background→None, BaselinePosition→Automatic,
  BaseStyle→{}, ClippingStyle→None, ColorFunction→Automatic,
  ColorFunctionScaling→True, ColorOutput→Automatic,
  ContentSelectable→Automatic, DisplayFunction:→$DisplayFunction,
  Epilog→{}, Evaluated→Automatic, EvaluationMonitor→None,
  Exclusions→Automatic, ExclusionsStyle→None, Filling→None,
  FillingStyle→Automatic, FormatType:→TraditionalForm,
  Frame→False, FrameLabel→None, FrameStyle→{},
  FrameTicks→Automatic, FrameTicksStyle→{},
  GridLines→None, GridLinesStyle→{}, ImageMargins→0.,
  ImagePadding→All, ImageSize→Automatic, LabelStyle→{},
  MaxRecursion→Automatic, Mesh→None, MeshFunctions→{#1&},
  MeshShading→None, MeshStyle→Automatic, Method→Automatic,
  PerformanceGoal:→$PerformanceGoal, PlotLabel→None,
  PlotPoints→Automatic, PlotRange→{Full, Automatic},
  PlotRangeClipping→True, PlotRangePadding→Automatic,
  PlotRegion→Automatic, PlotStyle→Automatic,
  PreserveImageOptions→Automatic, Prolog→{},
  RegionFunction→(True&), RotateLabel→True, Ticks→Automatic,
  TicksStyle→{}, WorkingPrecision→MachinePrecision}
```

Options are specified as transformation rules in the form *OptionName* **->** *OptionValue* or *OptionName* **:>** *OptionValue*. Default values are quite satisfactory for most plots. However, occasionally it is necessary to specify different values for some options. That can be done by giving additional arguments to the **Plot** function:

Plot$[f, \{x, x_{min}, x_{max}\},$

$OptionName_1 \rightarrow OptionValue_1, OptionName_2 \rightarrow OptionValue_2, \ldots]$

Options can be specified in any order after the second argument. If an option is not included, *Mathematica* uses the default value.

Example 2.3.3 Plot the Lennard–Jones potential

$$u(r) = 4\varepsilon\left[\left(\frac{\sigma}{r}\right)^{12} - \left(\frac{\sigma}{r}\right)^{6}\right]$$

where r is the distance between atoms and σ and ε are the empirical parameters.

If $u(r)$ and r are measured respectively in units of 4ε and σ, the potential becomes

$$u(r) = \left[\left(\frac{1}{r}\right)^{12} - \left(\frac{1}{r}\right)^{6}\right]$$

Let us plot the potential:

```
In[6]:= Plot[(1/r^12) - (1/r^6), {r, 0.001, 3.0}]
```

Out[6]=

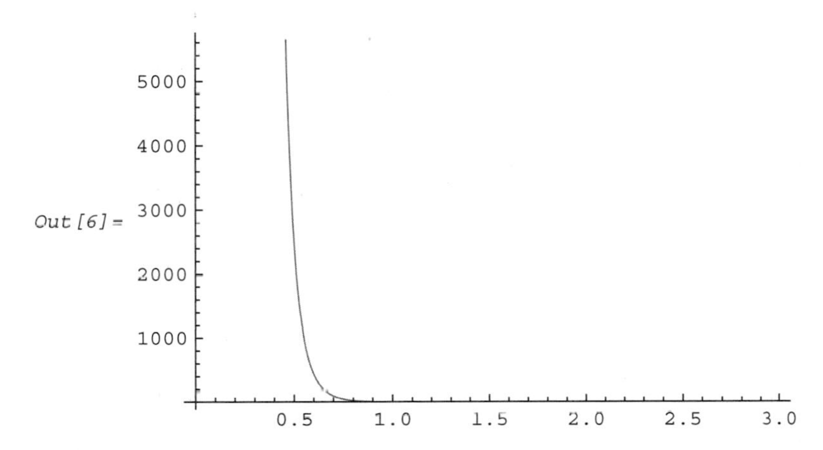

The graph does not show the essential features of the potential. The option **PlotRange** allows us to specify the range of values for $u(r)$ to include in the plot:

```
In[7]:= Plot[(1/r^12) - (1/r^6), {r, 0.001, 3.0}, PlotRange→{-0.3, 0.6}]
```

Out[7]=

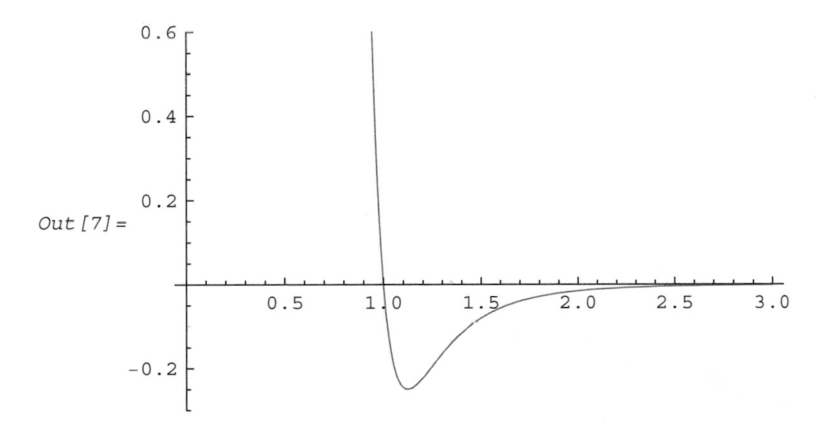

This graph is more interesting. It shows that a minimum occurs at $r/\sigma \approx 1.12$, that the curve is very steep inside the minimum, and that it is flat outside the minimum. (For a discussion of the Lennard–Jones potential, see [AM76].) ■

Example 2.3.4 Two waves of slightly different frequencies travel along a line in the same direction. At a point in space, the displacements, y_1 and y_2, of the waves as functions of time are

$$y_1 = A \cos 90t$$
$$y_2 = A \cos 91t$$

where the time t is in seconds. Plot the resultant displacement at the point as a function of time.

Let the displacements be measured in units of A. The resultant displacement, $y_1 + y_2$, equals $\cos 90t + \cos 91t$. Using **Plot**, we have

```
In[8]:=  Plot[Cos[90 t] + Cos[91 t], {t, -π, π}]
```

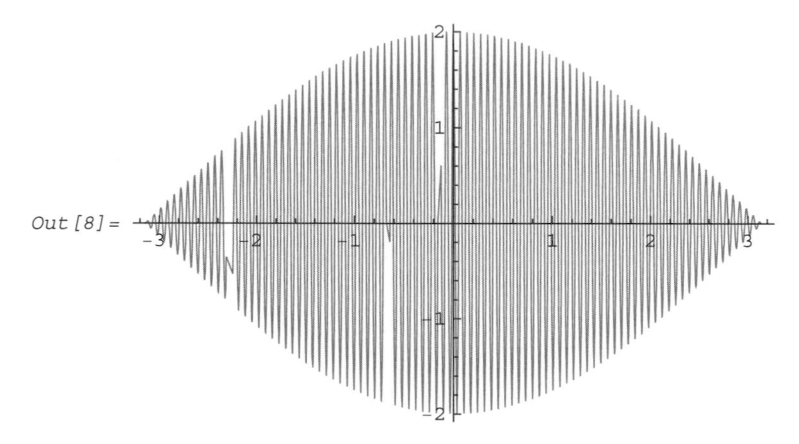

The graph is incorrect. The problem is that *Mathematica* does not sample the function at a sufficient number of points. The default value of the option **PlotPoints**, which specifies how many initial sample points to use, is 50. Let us increase the value to 120:

```
In[9]:= Plot[Cos[90 t] + Cos[91 t], {t, -π, π}, PlotPoints → 120]
```

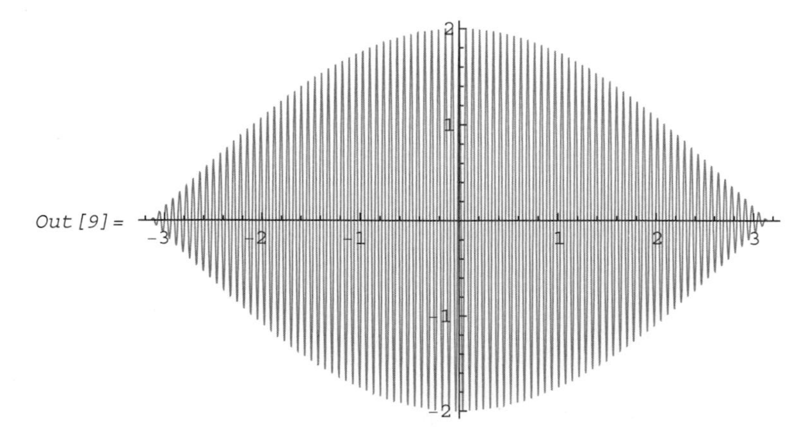

This is the familiar beat pattern. ■

`Show[`*plot*, *options*`]` redraws a plot with the specified options. Since **`Show`** redraws a plot with the original set of points, some options, such as **`PlotPoints`**, cannot be used.

Let us label the axes of the previous plot with **`Show`** by giving the option **`AxesLabel`**, which has the form **`AxesLabel->{"`***xlabel***`", "`***ylabel***`"}`**:

```
In[10]:= Show[%, AxesLabel→{"t (s)", "y (A)"}]
```

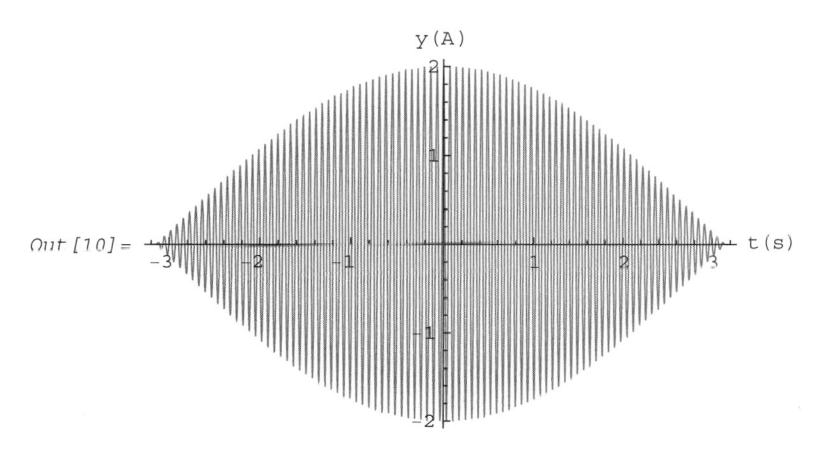

To remove the axes and give the plot a label and also a frame with labels on its two edges but without tick marks, we specify the options **`Axes,PlotLabel,Frame,FrameLabel`**, and **`FrameTicks`**:

```
In[11]:= Show[%, Axes→False, PlotLabel->"The Beat Phenomenon",
           Frame → True, FrameLabel → {"t", "y"}, FrameTicks→None]
```

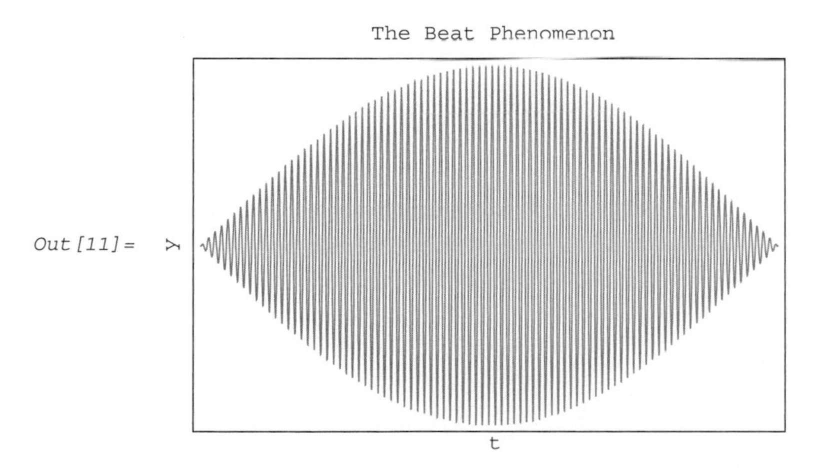

In Example 2.3.2, we plotted the radial probability density without using the numerical value for the Bohr radius a; in Example 2.3.3, we plotted the Lennard–Jones potential without knowing the numerical values for σ and ε. An explanation is in order for the method of plotting graphs without specifying numerical values for the relevant parameters. The method involves making a change of variables into dimensionless variables or, equivalently, choosing convenient units for the variables in terms of the parameters. Let us illustrate this method with an example from thermal physics.

Example 2.3.5 Consider a system that has only three energy eigenstates with nonnegative energies $\epsilon, 0$, and ϵ (again). For this system, the partition function Z is

$$Z = 1 + 2e^{-\frac{\epsilon}{kT}}$$

where k is Boltzmann's constant and T is the absolute temperature. The average energy E can be expressed as

$$E = kT^2 \frac{\partial \ln Z}{\partial T}$$

The heat capacity C is given by

$$C = \frac{\partial E}{\partial T}$$

Plot the average energy E as a function of T, and do the same for the heat capacity C.

We begin with the assignment for the partition function:

```
In[12]:= Z = 1 + 2 Exp[-ϵ/(k T)]
```

$$\text{Out[12]= } 1 + 2\,e^{-\frac{\epsilon}{k\,T}}$$

The average energy E is

```
In[13]:= k(T^2)D[Log[Z], T] // Simplify
```

$$\text{Out[13]= } \frac{2\,\epsilon}{2 + e^{\frac{\epsilon}{k\,T}}}$$

Let us make a change of variables, from E and T to ε and t:

$$\varepsilon = E/\epsilon$$

$$t = kT/\epsilon$$

which can also be written as

$$E = \varepsilon\epsilon$$

$$T = t\epsilon/k$$

In terms of the dimensionless variables ε and t, the preceding *Mathematica* output implies

$$\varepsilon = \frac{2}{2 + e^{\frac{1}{t}}}$$

We can now plot ε as a function of t—that is, E/ϵ as a function of kT/ϵ or, equivalently, E as a function of T, when E and T are measured in units of ϵ and ϵ/k, respectively:

```
In[14]:= Plot[2/(2 + Exp[1/t]), {t, 0, 4}, AxesLabel → {"T (ϵ/k)", "E (ϵ)"}]
```

Out[14]=

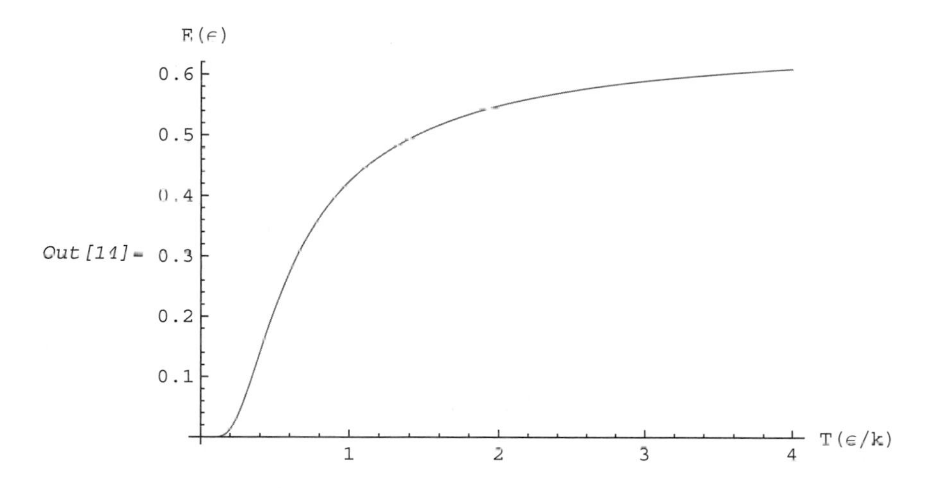

The heat capacity C is

```
In[15]:= D[%%, T]
```

$$\text{Out[15]}= \frac{2\, e^{\frac{\epsilon}{k T}}\, \epsilon^2}{\left(2 + e^{\frac{\epsilon}{k T}}\right)^2 k\, T^2}$$

In terms of the dimensionless variables

$$c = C/k$$

$$t = kT/\epsilon$$

the preceding *Mathematica* output implies

$$c = \frac{2e^{1/t}}{\left(2 + e^{1/t}\right)^2 t^2}$$

We can plot c as a function of t or, equivalently, C as a function of T, provided the units of C and T are k and ϵ/k, respectively:

```
In[16]:= Plot[(2 Exp[1/t])/(((2 + Exp[1/t])^2) (t^2)),
           {t, 0, 4}, AxesLabel → {"T (ϵ/k)", "C (k)"}]
```

Out[16]=

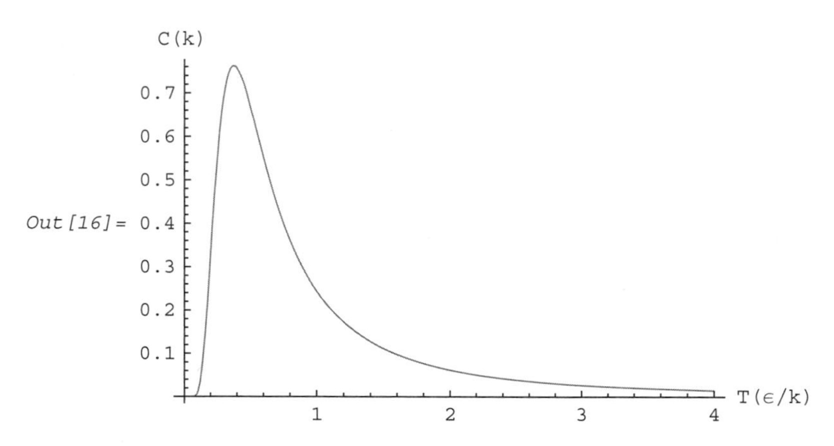


```
In[17]:= Clear[z]
```

■

2.3.1.3 Multiple Plots

Plot[{f_1, f_2, ...}, {x, x_{min}, x_{max}}] and **Plot[Evaluate[{f_1, f_2, ...}], {x, x_{min}, x_{max}}]** plot functions $f_1, f_2, \ldots$ together.

Example 2.3.6 Plot the radial probability density $P(r) = r^2 R_{nl}^2$ for the 1s, 2s, and 3s states of the hydrogen atom. The radial functions R_{nl} are given in Example 2.3.2 in Section 2.3.1.1.

```
In[18]:= Plot[Evaluate[Table[(r R[i, 0, r])^2, {i, 3}] /. a → 1], {r, 0, 25},
          PlotRange → {0, 0.55}, AxesLabel → {"r(a)", "P(r)(1/a)"}]
```

Out[18]=

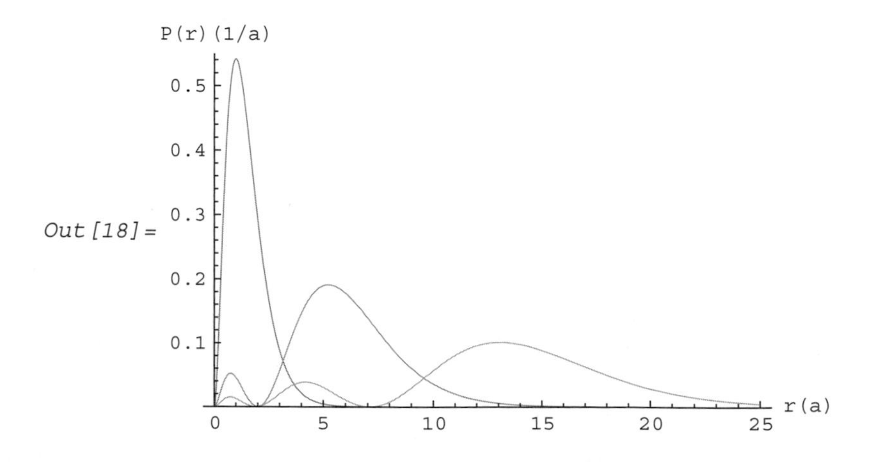

■

To make the plot easier to comprehend, we can assign a different style to each curve with the option **PlotStyle**. The option has the form **PlotStyle->{{***styles for curve*$_1$**}, {***styles for curve*$_2$**},...}**. Some styles are

Thickness[*r***]**	give curve a thickness r as a fraction of the total width of the plot
Dashing[{*r*$_1$**,***r*$_2$**,...}]**	show a dashed curve where successive drawn and undrawn segments are of lengths $r_1, r_2, \ldots$, given as fractions of the total width of the plot; lengths are cyclical—that is, **{***r*$_1$**,***r*$_2$**}** has the same effect as **{***r*$_1$**,***r*$_2$**,***r*$_1$**,***r*$_2$**,...}**
GrayLevel[*i***]**	display curve with a gray level i between 0 (black) and 1 (white)
RGBColor[*red***,***green***,***blue***]**	display curve in a color with specified *red*, *green*, and *blue* components, each between 0 and 1: **RGBColor[1,0,0]** or equivalently **Red** for red, **RGBColor[0,1,0]** or **Green** for green, **RGBColor[0,0,1]** or **Blue** for blue, and **RGBColor[1,1,0]** or **Yellow** for yellow. For more **RGBColor** specifications, see Section 3.5.2.2.

Let us specify different styles for the curves in the preceding graph:

```
In[19]:= Plot[Evaluate[Table[(r R[i, 0, r])^2, {i, 3}] /. a→1], {r, 0, 25},
           PlotRange→{0, 0.55}, AxesLabel→{"r (a)", "P(r) (1/a)"},
           PlotStyle→{{Thickness[0.006]}, {Thickness[0.0075],
              Dashing[{0.01, 0.02}], RGBColor[1, 0, 0]}, {GrayLevel[0.5],
              Thickness[0.009], Dashing[{0.02}], RGBColor[0, 0, 1]}}]
```

Out[19]=

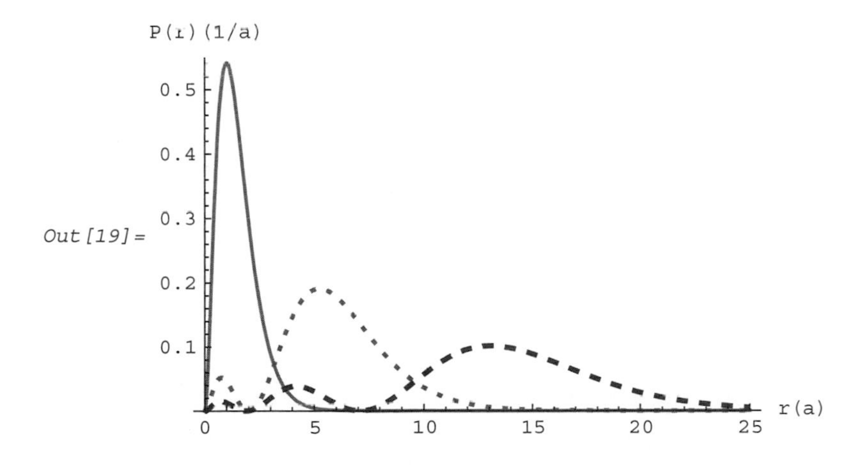


```
In[20]:= Clear[F,R]
```

2.3.1.4 FindRoot

Section 2.1.16 discussed numerical equation solving. The function **NSolve** considered there only finds the roots of polynomial equations. It cannot find solutions to more general equations such as transcendental equations.

Example 2.3.7 Find the roots of the equation

$$\sin x = x^2$$

```
In[1]:= NSolve[Sin[x] == x^2, x]

        Solve::tdep : The equations appear to involve the
            variables to be solved for in an essentially non–algebraic way. ≫

Out[1]= NSolve[Sin[x] == x^2, x]
```

NSolve fails to find the roots and returns the input unevaluated. ■

The function **FindRoot** can find numerical solutions to arbitrary equations. **FindRoot**[*lhs* == *rhs*, {*x*, x_0}] searches for a numerical solution to the equation *lhs* == *rhs*, starting with $x = x_0$. The success in finding a particular root depends heavily on the choice of the starting value. If the starting value is sufficiently close to a solution, **FindRoot** always returns that solution. Mac OS X and Windows front ends provide the means for determining the appropriate starting value graphically.

For determining a starting value and finding a solution, the procedure is

1. Evaluate

   ClickPane[Plot[{*lhs*, *rhs*}, {*x*, x_{min}, x_{max}}], (*xycoord* = #) &]

   ClickPane[*image*, *func*] generates a clickable pane that displays as *image* and applies *func* to the *x*, *y* coordinates of each click within the pane. The function (*xycoord* = #) & is a pure function that assigns its argument (i.e., the *x*, *y* coordinates here) to the variable *xycoord*. (Section 3.3.1 will discuss pure functions.) Rather than using a pure function, we can define

   assigncoord[*lis*_] := (*xycoord* = *lis*)

   and evaluate

   ClickPane[Plot[{*lhs*, *rhs*}, {*x*, x_{min}, x_{max}}], *assigncoord*]

   Pure functions are, perhaps, more elegant than named functions.

2. If the curves do not intersect, there is no solution. If there are intersections, click an intersection with the mouse pointer. The list of x, y coordinates of the intersection becomes the value of *xycoord*.
3. Evaluate

```
FindRoot[lhs == rhs, {x, xycoord[[1]]}]
```

to find the solution.

For determining *several* starting values and finding the corresponding solutions, the procedure is

1. Evaluate

```
ClickPane[Plot[{lhs, rhs}, {x, xmin, xmax}], (xycoord = #) &]
```

2. If the curves do not intersect, there is no solution. If there are intersections, evaluate

```
Dynamic[xycoord]
```

`Dynamic[expr]` displays the dynamically updated current value of *expr*.

3. With the mouse pointer, click an intersection. The output of `Dynamic[xycoord]` in step 2 will immediately display the list of x, y coordinates of the intersection. Copy and paste the x coordinate of the list as an element of a list named *startvalue*.
4. Repeat step 3 for as many intersections as needed.
5. Evaluate

```
startvalue = {x1, x2, ...}
```

6. Evaluate

```
Table[FindRoot[lhs == rhs, {x, startvalue[[i]]}], {i, Length[startvalue]}]
```

to find the solutions.

For *Mathematica* 6.0.2, Mac OS X and Windows front ends also provide a coordinate-picking tool for determining the appropriate starting values graphically:

1. Plot *lhs* and *rhs* of the equation together. If the curves do not intersect, there is no solution.
2. Choose **Drawing Tools** in the Graphics menu, and click the Get Coordinates button (i.e., second button from the top in the right column) in the 2D Drawing Palette.

3. Over the graphic, the pointer turns into a coordinates pointer that resembles a cross; the coordinates of the center of the pointer appear. Move the center of the pointer over an intersection. Without moving the pointer, click the mouse. A point is drawn over the intersection. Repeat for the other intersections, if desired.
4. Select **Copy** in the Edit menu. Place the text insertion point below the graphics cell, and select **Paste** in the Edit menu. A nested list of the coordinates of the intersections is pasted there.

Example 2.3.8 Find the nonzero root of the equation

$$\sin x = x^2$$

Let us follow the procedure outlined previously for determining a starting value and finding a solution.

1. Evaluate

```
In[2]:= Clickpane[Plot[{Sin[x], x^2}, {x, -0.5, 1.5}], (xycoord = #)&]
```

Out[2]=

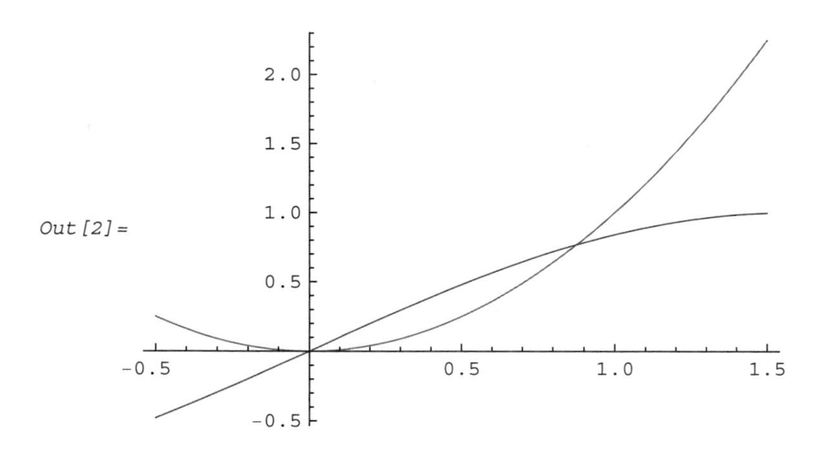

2. For the nonzero root, click with the mouse pointer the intersection that is not the origin.
3. Evaluate

```
In[3]:= FindRoot[Sin[x] == x^2, {x, xycoord[[1]]}]
Out[3]= {x→0.876726}
```

The solution is $x = 0.876726$.

```
In[4]:= Clear[xycoord]
```

With *Mathematica* 6.0.2, we can also determine the starting value with the coordinate-picking tool:

1. Plot sin x and x^2 together:

```
In[5]:= Plot[{Sin[x], x^2}, {x, -0.5, 1.5}]
```

Out[5]=

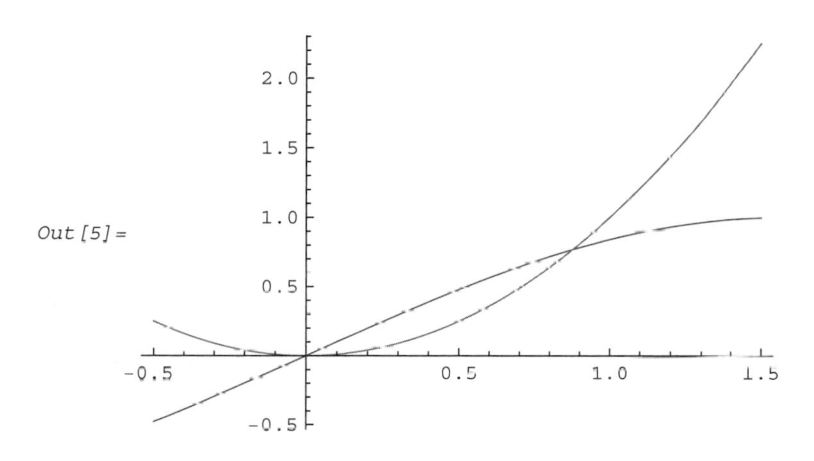

The curves intersect at two points.

2. Choose **Drawing Tools** in the Graphics menu, and click the Get Coordinates button in the 2D Drawing Palette.
3. For the nonzero root, move the center of the pointer over the intersection that is not the origin. Without moving the pointer, click the mouse. A point is drawn over the intersection:

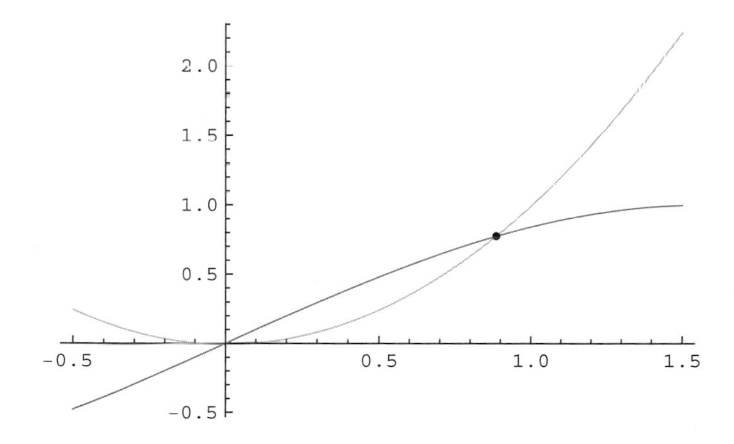

4. Select **Copy** in the Edit menu. Place the text insertion point below the graphics cell, and select **Paste** in the Edit menu. A nested list of the coordinates of the intersection is pasted below the graphics cell:

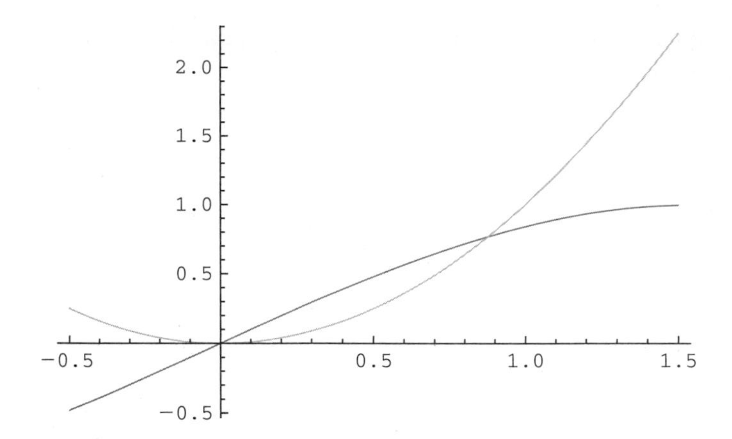


```
{{0.8887, 0.783}}
```

To find the solution, use 0.8887 as the starting value for **FindRoot**:

```
In[6]:= FindRoot[Sin[x] == x^2, {x, 0.8887}]
Out[6]= {x→0.876726}
```

The solution is $x = 0.876726$. ■

Example 2.3.9 Consider a particle of mass m in a square well potential

$$V(x) = \begin{cases} V_0 & |x| > a/2 \\ 0 & |x| < a/2 \end{cases}$$

The energy of the particle is given as

$$E = \frac{h^2}{2\pi^2 ma^2} y^2$$

where h is Planck's constant. For energy eigenvalues associated with even eigenfunctions, the values for y are the solutions to the equation

$$y \tan y = \sqrt{d - y^2}$$

with

$$d = \frac{2\pi^2 m a^2}{h^2} V_0$$

For the special case $d = 16$, determine the energy spectrum for the even eigenstates.

Let us follow the procedure outlined previously for determining several starting values and finding the corresponding solutions.

1. Evaluate

```
In[7]:= ClickPane[Plot[{y Tan[y], Sqrt[16 - y^2]},
            {y, 0, 8}, PlotRange→{{0, 5}, {0, 5}}], (yzcoord = #) &]
```

Out[7]=

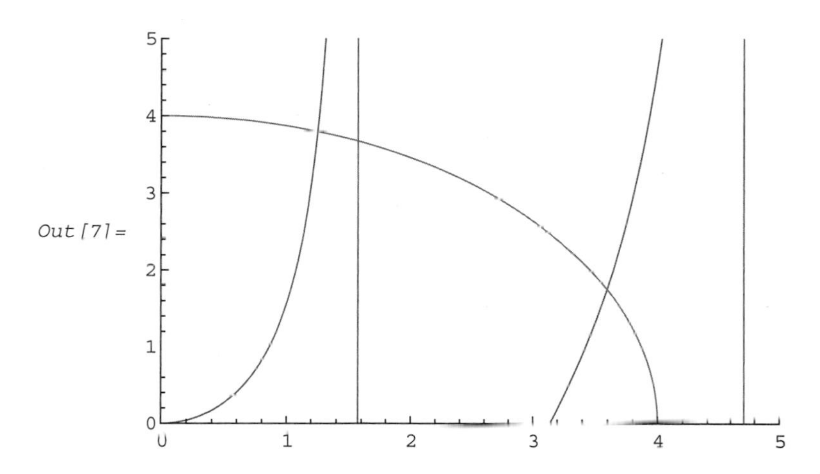

(The vertical lines are vertical asymptotes of the graph of the tangent function. We can remove these lines by including in the **Plot** function the option specification **Exclusions ->{Cos[y] == 0}**.)

2. Evaluate

```
In[8]:= Dynamic[yzcoord]
Out[8]= yzcoord
```

3. With the mouse pointer, click an intersection. Then copy and paste the first element of the list now displayed in step 2 (i.e., Out[8]) to a list named **startvalue**.

```
startvalue = {1.2721070114013822`}
```

4. With the mouse pointer, click the other intersection. Then copy and append the first element of the list displayed now in step 2 to **startvalue**.

```
startvalue = {1.2721070114013822`, 3.606248316725019`}
```

With *Mathematica* 6.0.2, we can also obtain the starting values with the coordinate-picking tool. Exercise 55 in Section 2.3.5 considers this approach.

5. Evaluate

```
In[9]:= startvalue = {1.2721070114013822`, 3.606248316725019`};
```

6. Evaluate

```
In[10]:= Table[FindRoot[y Tan[y] == Sqrt[16 - y^2], {y, startvalue[[i]]}],
           {i, Length[startvalue]}]
Out[10]= {{y→1.25235}, {y→3.5953}}
```

The solutions are $y = 1.25235$ and 3.5953.

In terms of d, the energy E can be expressed as

$$E = \frac{y^2 V_0}{d}$$

With the values 1.25235 and 3.5953 for y, the energies are

```
In[11]:= (1.25235^2)Vo/16
Out[11]= 0.0980238 Vo
```

```
In[12]:= (3.5953^2)Vo/16
Out[12]= 0.0807886 Vo
```

These are the energy eigenvalues of the ground state and the second excited state. (For further discussion of the square well potential, see Section 6.4.)

```
In[13]:= Clear[yzcoord, startvalue]
```

■

2.3.1.5 FindMinimum and FindMaximum

For finding local minima and maxima numerically, *Mathematica* provides the functions **FindMinimum** and **FindMaximum**:

`FindMinimum[`f`, {`x`,`x_0`}]`	search for a local minimum of f, starting at $x = x_0$
`FindMaximum[`f`, {`x`,`x_0`}]`	search for a local maximum of f, starting at $x = x_0$

A value x_0 can serve as the starting value if $f(x_0)$ is sufficiently close to the local minimum or maximum. To determine starting values and find the corresponding local minima or maxima, we only need to modify the procedures described in Section 2.3.1.4 for determining starting values and finding roots.

Example 2.3.10 Find the second smallest local minimum of the Airy function $Ai(z)$.

Let us modify the procedure outlined in Example 2.3.8 of Section 2.3.1.4:

1. Evaluate

```
In[1]:= ClickPane[Plot[AiryAi[z], {z, -9, 4},
          AxesLabel→{"z", "Ai(z)"}], (zycoord=#)&]
```

Out[1]=

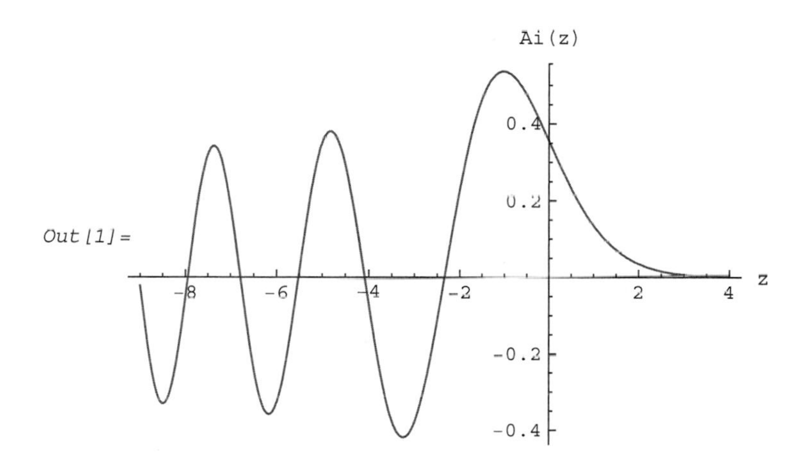

2. With the mouse pointer, click the point of the second smallest local minimum.
3. Evaluate

```
In[2]:= FindMinimum[AiryAi[z], {z, zycoord[[1]]}]
Out[2]= {-0.357908, {z→-6.16331}}
```

The Airy function $Ai(z)$ has a local minimum at $z = -6.16331$, and the local minimum value is -0.357908. (For the use of Airy functions in quantum mechanics, see [Gri05] and [Gos03].)

```
In[3]:= Clear[zycoord]
```

■

Example 2.3.11 For a harmonic oscillator subject to a damping force proportional to the square of its velocity, the equation of motion is

$$\frac{d^2x}{dt^2} + 2\gamma\frac{dx}{dt}\left|\frac{dx}{dt}\right| + \omega_0^2 x = 0$$

Let $\gamma = 0.20\,\text{m}^{-1}$, $\omega_0 = 2.00\,\text{rad/s}$, $x(0) = 1.00\,\text{m}$, and $x'(0) = 0$. For $t > 0$, find the third largest local maximum of x.

In SI units, **NDSolve** finds a numerical solution to the equation of motion for t ranging from 0 to 15:

```
In[4]:= NDSolve[{x''[t] + 0.4 x'[t] Abs[x'[t]] + 4.0 x[t] == 0,
          x[0] == 1.0,  x'[0] == 0}, x, {t, 0, 15}]

Out[4]= {{x→InterpolatingFunction[{{0., 15.}}, <>]}}
```

The position x of the harmonic oscillator as a function of time t is given by

```
In[5]:= x[t_] = x[t] /. %[[1]];
```

To find the third largest local maximum of x, let us modify the procedure outlined in Example 2.3.8 of Section 2.3.1.4:

1. Evaluate

```
In[6]:= ClickPane[Plot[x[t], {t, 0, 15}, PlotRange -> {-1, 1},
          AxesLabel → {"t", "x"}], (txcoord = #) &]
```

Out[6]=

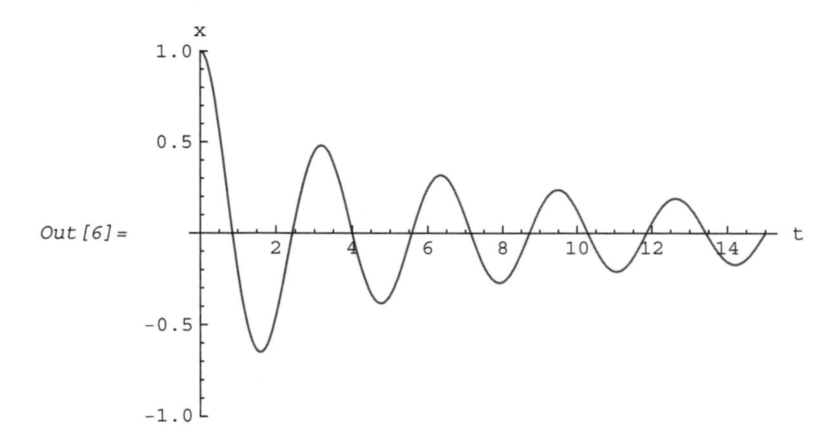

2. With the mouse pointer, click the point of the third largest local maximum of x for $t > 0$.
3. Evaluate

```
In[7]:= FindMaximum[x[t], {t, txcoord[[1]]}]

Out[7]= {0.236875, {t→9.48505}}
```

The output states that x has a local maximum at $t_m = 9.48505$ and that the local maximum value is 0.236875. With units, $t_m = 9.48505$ s and $x(t_m) = 0.236875$ m. The Second Derivative Test confirms that there is indeed a local maximum at t_m:

```
In[8]:= Chop[{x'[t], x''[t]} /. Last[%], 10^-9]

Out[8]= {0, -0.947502}
```

As expected, $x'(t_m) = 0$ and $x''(t_m) < 0$. `Chop[`*expr*, *delta*`]` replaces in *expr* approximate real numbers that are smaller in absolute magnitude than *delta* by the exact integer 0. (For a discussion of the Second Derivative Test, see [Ste03].)

```
In[9]:= Clear[x,txcoord]
```

■

2.3.1.6 Data Plots

`ListPlot[{{`x_1`,`y_1`}, {`x_2`,`y_2`},...}]` plots a list of points with specified x and y coordinates.

Example 2.3.12 Consider the hot Volkswagen example in Section 2.1.13. (a) Plot the data. (b) Plot the least-squares fit curve. (c) Superpose the two plots to examine how well the curve fits the data.

Here are the data:

```
In[1]:= vwdata = {{0, 0}, {1, 1.5}, {2, 6.0}, {3, 13.5}, {4, 24.0},
          {5, 37.5}, {6, 54.0}, {7, 73.5}, {8, 96.0}, {9, 121.5}};
```

Let us plot the data:

```
In[2]:= ListPlot[vwdata, PlotStyle→PointSize[0.025],
          AxesLabel→{"t (s)", "d (m)"}]
```

Out[2]=

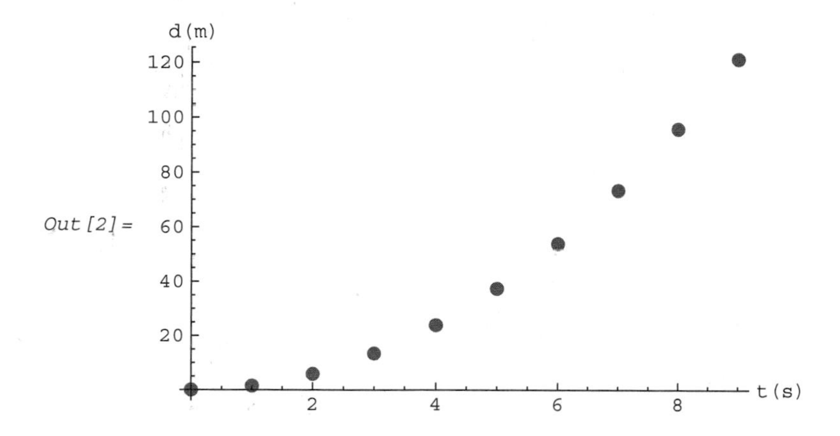

We then plot the least-squares fit curve, namely $1.5\,t^2$:

```
In[3]:= Plot[1.5t^2, {t, 0, 10},
          PlotRange→{-5, 125}, AxesLabel→{"t (s)", "d (m)"}]
```

Out[3]=

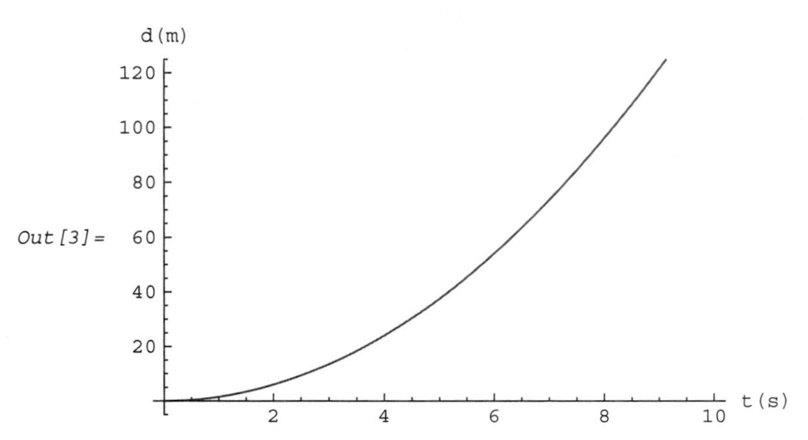

The function **Show** in the form **Show**[$plot_1$, $plot_2$, ..., *options*] allows us to superpose the two plots to see how well the curve fits the data. (Since **Show** redraws the plots with the original points, some options, such as **PlotPoints** and **PlotStyle**, cannot be used.)

```
In[4]:= Show[%, %%]
```

Out[4]=

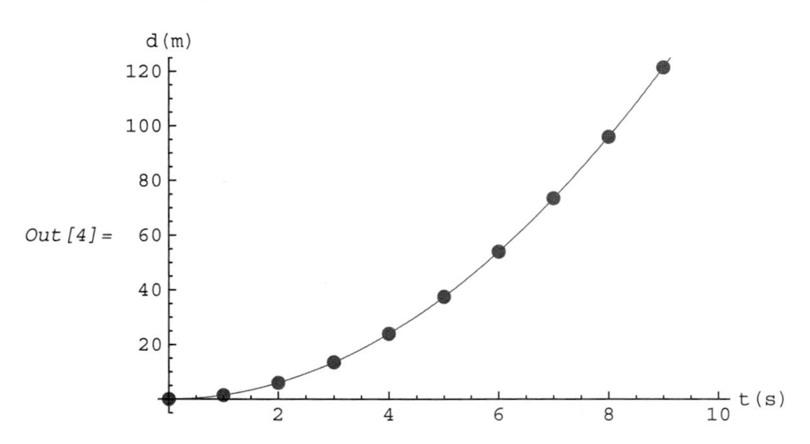

The least-squares fit is excellent. ■

Example 2.3.13 Consider again the hot Volkswagen example. This time, display the last plot without first showing the other two.

Suppose we are only interested in seeing how well the least-squares curve fits the data. There is no reason for taking time to render the plots for the data and the least-squares curve separately. To suppress the display of a plot, add a semicolon at the end of the plotting command.

Let us produce the graphics objects without displaying them:

```
In[5]:= ListPlot[vwdata, PlotStyle→PointSize[0.025]];
```

```
In[6]:= Plot[1.5t^2, {t, 0, 10}, PlotRange→{-5, 125}];
```

To display the two plots together, again use the function **Show**:

```
In[7]:= Show[%, %%, AxesLabel→{"t (s)", "d (m)"}]
```

Out[7]=

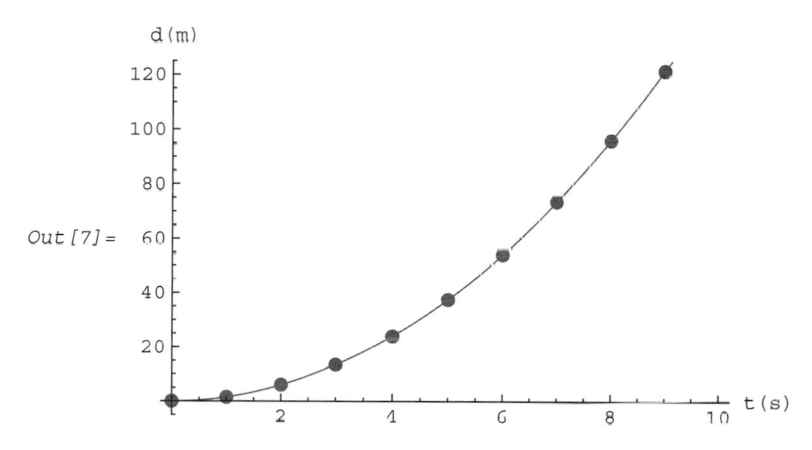

■

ListLinePlot[{{x_1,y_1}, {x_2,y_2}, ...}] plots a line through a list of positions with specified x and y coordinates. Consider once more the hot Volkswagen example. This time, plot a line through the positions specified by the data.

```
In[8]:= ListLinePlot[vwdata, AxesLabel→{"t (s)", "d (m)"}]
```

Out[8]=

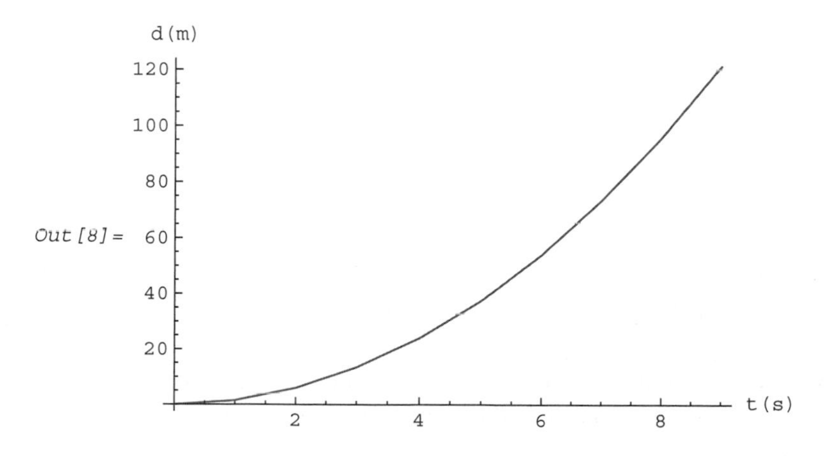

The default value of the option **InterpolationOrder** is **1**, and **ListLinePlot** simply joins the positions with straight lines. To obtain a smooth line, include the option specification **InterpolationOrder -> 3**:

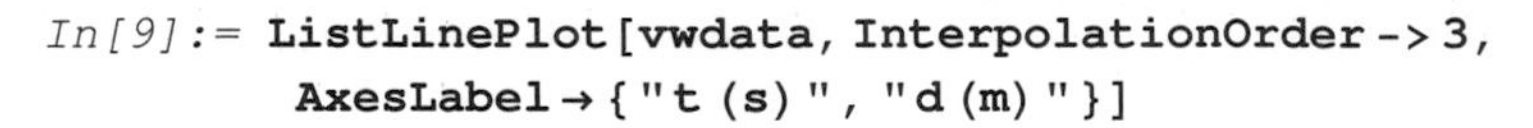

```
In[9]:= ListLinePlot[vwdata, InterpolationOrder -> 3,
         AxesLabel → {"t (s)", "d (m)"}]
```

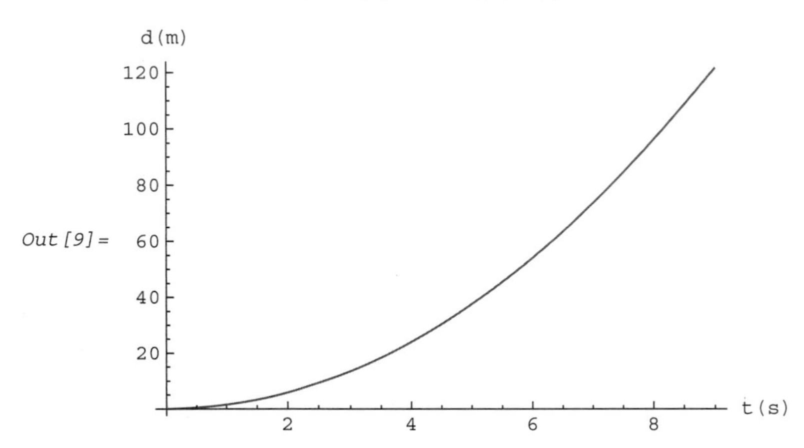

To show the smooth line together with the data points, include the option **Mesh**:

```
In[10]:= ListLinePlot[vwdata, InterpolationOrder -> 3,
          Mesh -> Full, AxesLabel → {"t (s)", "d (m)"}]
```

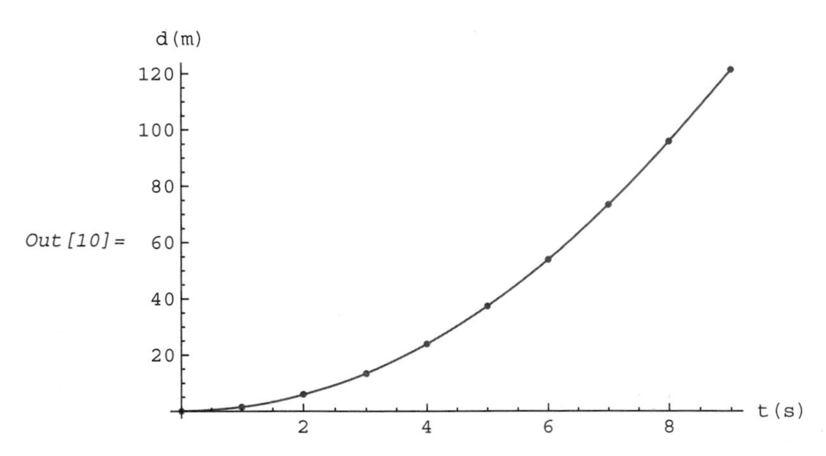

To give the points a different size, include the option **PlotMarkers**:

```
In[11]:= ListLinePlot[vwdata, InterpolationOrder -> 3, Mesh -> Full,
          PlotMarers -> {Automatic, 12}, AxesLabel → {"t (s)", "d (m)"}]
```

Out[11]=

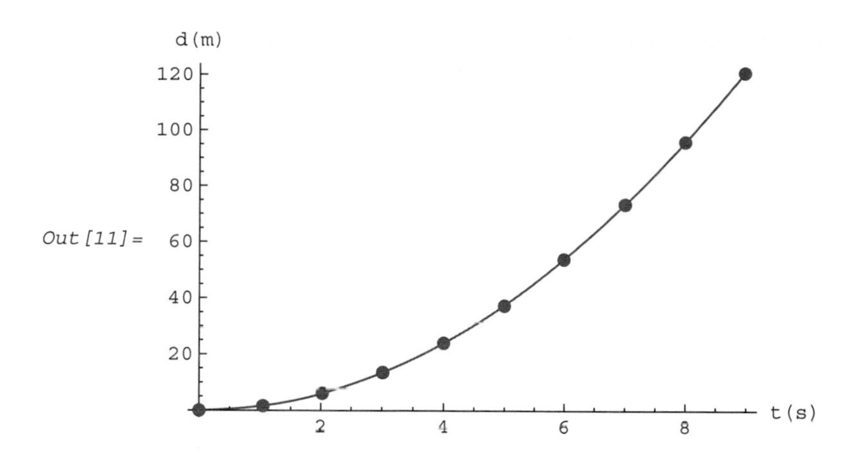

`PlotMarkers -> {Automatic, s}` draws a standard sequence of markers with size s at the points plotted. `Tiny`, `Small`, `Medium`, and `Large` can be used for s.

```
In[12]:= Clear[vwdata]
```

2.3.1.7 Parametric Plots

`ParametricPlot[{fx, fy}, {t, tmin, tmax}]` makes a parametric plot with x and y coordinates f_x and f_y that are functions of the parameter t.

Example 2.3.14 Electrons in an oscilloscope are deflected by two mutually perpendicular electric fields in such a way that at any time t the displacement is given by

$$x = A_x \cos(\omega_x t + \phi_x)$$

$$y = A_y \cos(\omega_y t + \phi_y)$$

where A_x, ω_x, and ϕ_x are, respectively, the amplitude, angular frequency, and phase constant for the horizontal deflection, and A_y, ω_y, and ϕ_y are those for the vertical deflection. Plot the path of the electrons when (a) $A_x = A_y = A$, $\omega_x = \omega_y = \omega$, $\phi_x = 0$, and $\phi_y = \pi/2$; (b) $A_x = A_y = A$, $\omega_x = \omega_y/3 = \omega$, $\phi_x = 0$, and $\phi_y = \pi/2$; (c) $A_x = A_y = A$, $\omega_x = \omega_y/3 = \omega$, $\phi_x = 0$, and $\phi_y = -\pi/4$; and (d) $A_x = A_y = A$, $\omega_x/5 = \omega_y/4 = \omega$, $\phi_x = 0$, and $\phi_y = \pi/2$.

Let the displacements and the time be measured in units of A and $1/\omega$, respectively.

(a) $A_x = A_y = 1$, $\omega_x = \omega_y = 1$, $\phi_x = 0$, and $\phi_y = \pi/2$

```
In[1]:= ParametricPlot[{Cos[t], Cos[t + π/2]},
          {t, 0, 2π}, AxesLabel → {"x (A)", "y (A)"}]
```

Out[1]=

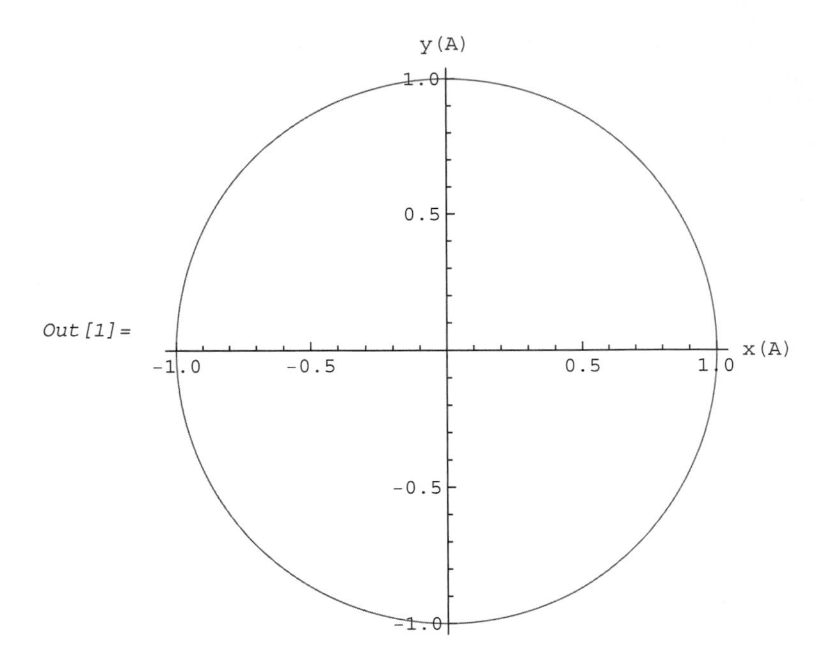

(b) $A_x = A_y = 1$, $\omega_x = \omega_y/3 = 1$, $\phi_x = 0$, and $\phi_y = \pi/2$

```
In[2]:= ParametricPlot[{Cos[t], Cos[3t+π/2]},
          {t, 0, 2π}, AxesLabel→{"x (A) ", "Y (A) "}]
```

Out[2]=

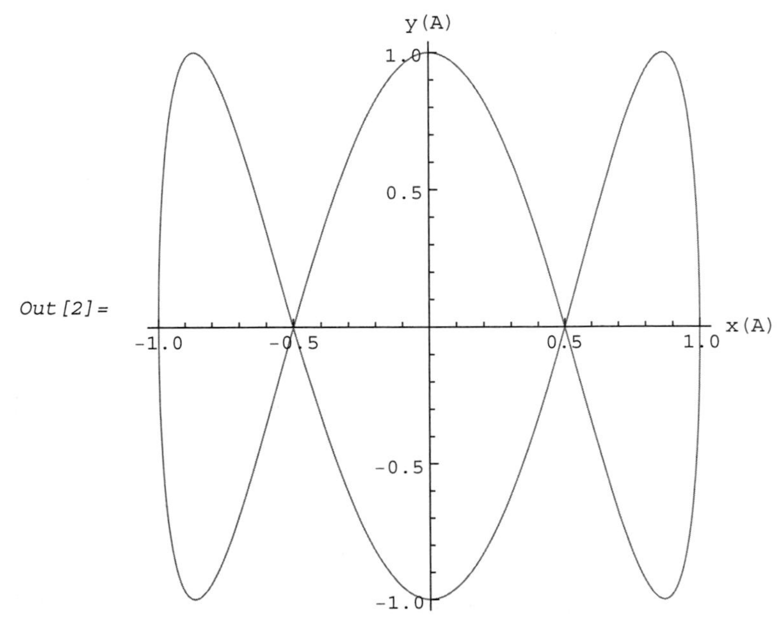

This is a Lissajous curve.

(c) $A_x = A_y = 1, \omega_x = \omega_y/3 = 1, \phi_x = 0$, and $\phi_y = -\pi/4$

```
In[3]:= ParametricPlot[{Cos[t], Cos[3t - π/4]},
        {t, 0, 2π}, AxesLabel→{"x (A)", "y (A)"}]
```

Out[3]=

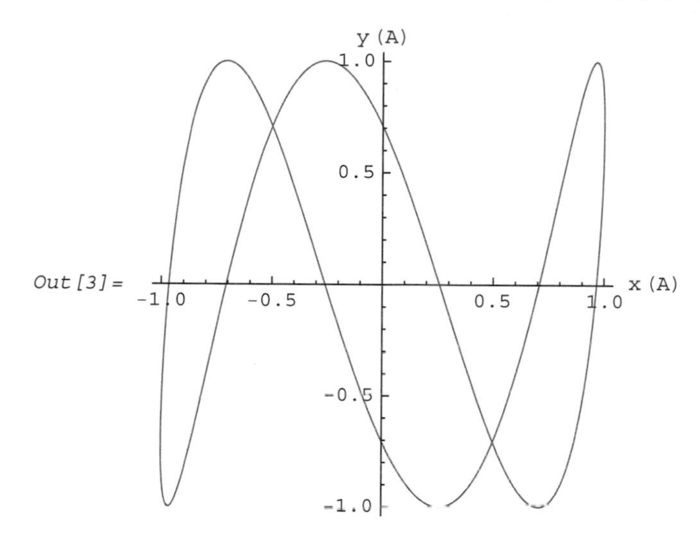

The shape of the Lissajous curve depends on the phase difference $\phi_x - \phi_y$, which is $\pi/4$ in this case, whereas it was $-\pi/2$ in part (b).

(d) $A_x = A_y = 1, \omega_x/5 = \omega_y/4 = 1, \phi_x = 0$, and $\phi_y = \pi/2$

```
In[4]:= ParametricPlot[{Cos[5t], Cos[4t + π/2]},
        {t, 0, 2π}, AxesLabel→{"x (A)", "y (A)"}]
```

Out[4]=

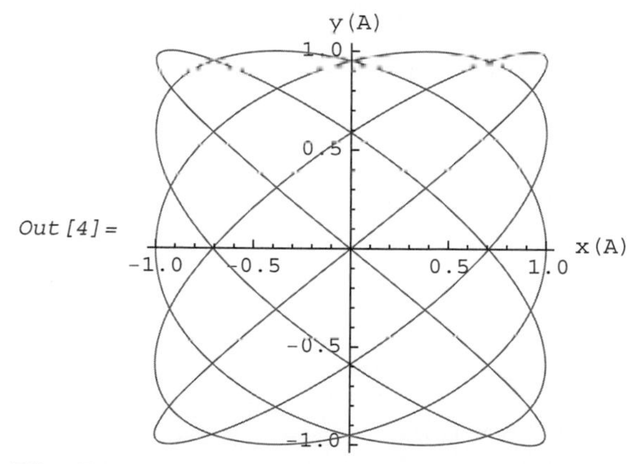

This Lissajous curve is a little more complex than the previous ones. ■

Example 2.3.15 Plot the phase diagram—that is, $x'(t)$ versus $x(t)$—for the damped harmonic oscillator in Example 2.3.11.

As shown in Example 2.3.11, **NDSolve** finds a numerical solution to the equation of motion:

```
In[5]:= NDSolve[{x''[t] + 0.4 x'[t] Abs[x'[t]] + 4.0 x[t] == 0,
          x[0] == 1.0, x'[0] == 0},  x, {t, 0, 25}]
Out[5]:= {{x→InterpolatingFunction[{{0., 25.}}, <>]}}
```

ParametricPlot plots the phase diagram—that is, $x'(t)$ versus $x(t)$:

```
In[6]:= ParametricPlot[Evaluate[{x[t], x'[t]}/.%], {t, 0, 25},
          AxesLabel→{"x (m)", "x' (m/s)"}, PlotRange -> All]
```

Out[6]=

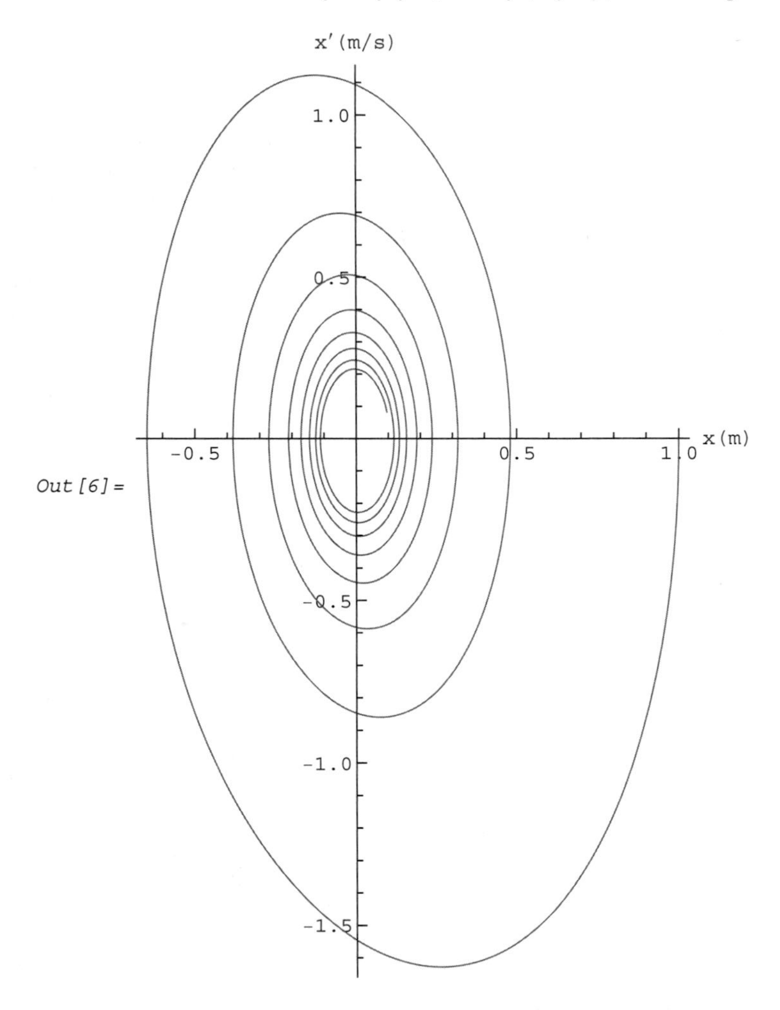

PlotRange -> All includes all points in the plot, whereas **PlotRange -> Automatic**, the default, drops the outlying points. ■

2.3.1.8 Interactive Graphics Drawing

Mathematica allows the interactive drawing and enhancing of objects such as points, lines, circles, polygons, and texts in existing plots or new graphics. The techniques for drawing and enhancing these objects in *Mathematica* notebooks are similar to those for drawing and enhancing the drawing objects in Microsoft Word documents. They consist mainly of pointing, clicking, and dragging with a mouse. What follows is an example illustrating the innovative mixing of programmatic and interactive creation of graphics.

Example 2.3.16 The curve traced out by a fixed point on a circle rolling along a straight line is called a cycloid, whose parametric equations are

$$x = r(\theta - \sin\theta)$$

$$y = r(1 - \cos\theta)$$

where r is the radius of the circle and the straight line is along the x-axis. Also, the parameter θ is the angle of rotation of the circle such that the fixed point and the origin coincide when $\theta = 0$. Create the graphic showing how the point traces out a cycloid:

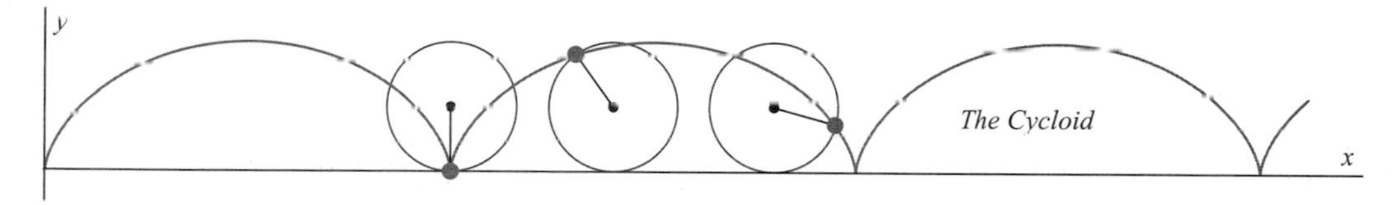

If x and y are measured in units of r, the parametric equations become

$$x = \theta - \sin\theta$$

$$y = 1 - \cos\theta$$

The steps for creating the graphic are

1. Plot the cycloid:

```
In[1]:= ParametricPlot[{θ - Sin[θ], 1 - Cos[θ]},
          {θ, 0, 6.55π}, PlotRange -> {{0, 6.5π}, {-0.5, 2.5}},
          PlotStyle -> {Thickness[0.0025], Red},
          Ticks→ {Table[iπ, {i, 0, 6}], Range[0, 3]}, ImageSize -> 72 × 7.85]
```

Out[1]=

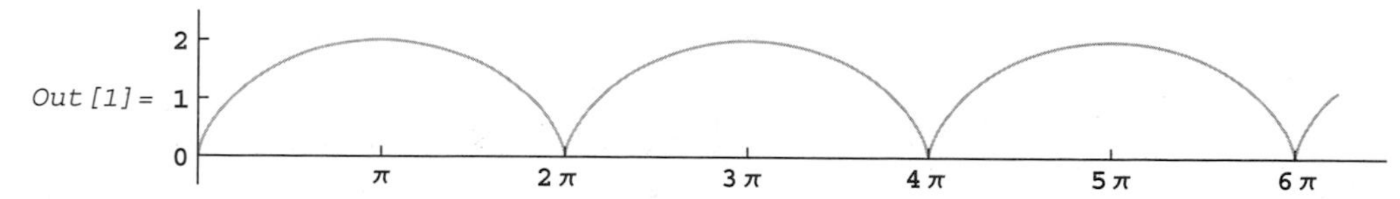

2. Insert auxiliary lines for aligning the objects:

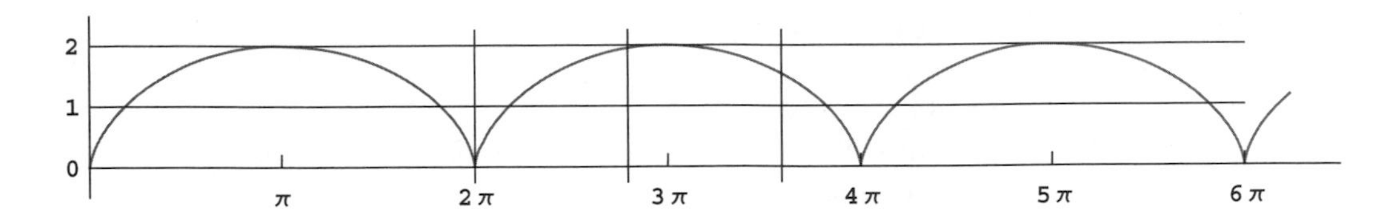

To draw a line, choose **Drawing Tools** in the Graphics menu, click the Line button (i.e., fourth button from the bottom in the left column) in the 2D Drawing Palette, position at one end of the line the pointer that has now turned into a cross, and drag to the other end of the line.

3. Draw the circles:

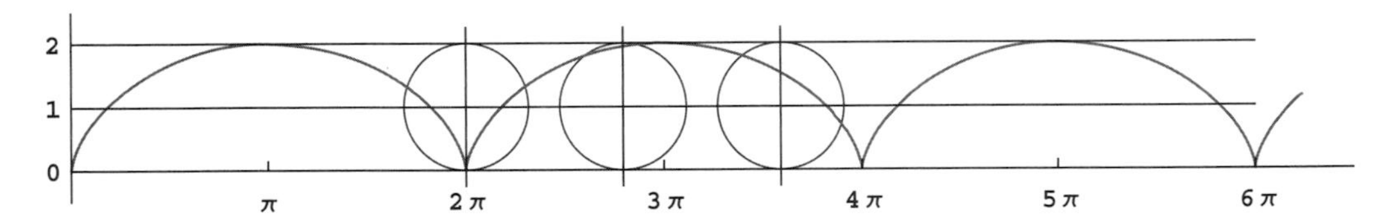

To draw a circle, click (i.e., select) the plot, click the Disk/Circle button (i.e., second button from the bottom in the left column) in the 2D Drawing Palette, position at the intended center of the circle the pointer that has now turned into a cross, hold down the Shift key, drag to draw a circle of radius 1, and release the Shift key. To enhance the circle, double-click it (if it is not already selected), click the Graphics Inspector button (i.e., button at the bottom for *Mathematica* 6.0.0 as well as 6.0.1 or first button from the top in the right column for *Mathematica* 6.0.2) in the 2D Drawing Palette, double-click the color bar in the 2D Graphics Inspector, for Mac OS X only—click the Color Palettes button (i.e., the middle button at the top in the Colors window), click the white button, click the OK button, click the button with a square within another square immediately to the left of the color bar, double-click the color bar, click the blue button, click the OK button, close the 2D Graphics Inspector, and choose **Graphics ▸ Operations ▸ Move to Back**.

4. Add the points and lines (i.e., fixed points on the circles, centers, and radii):

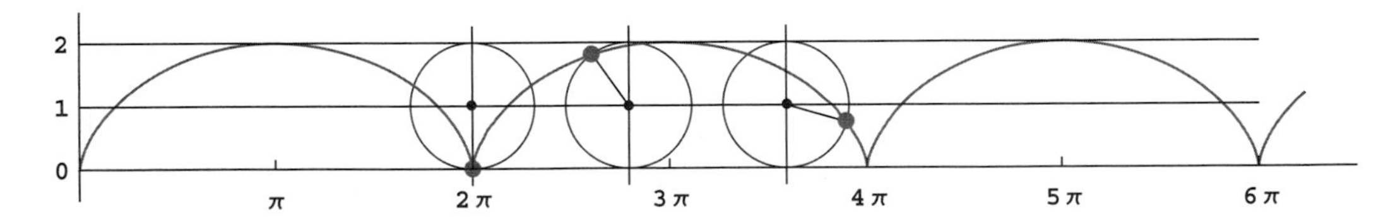

To draw a line for a radius, see step 2. To draw a point for a center or the fixed point on a circle, click the Point button (i.e., fifth button from the bottom in the left column) in the 2D Drawing Palette, and click the intended location with the pointer that has now turned into a cross. To enhance the point, click anywhere outside the graphic, double-click the point, click the Graphics Inspector button in the 2D Drawing Palette, double-click the color bar in the 2D Graphics Inspector, for Mac OS X only—click the Color Palettes button (i.e., the middle button at the top in the Colors window), click a color button, click the OK button, click the Point Size right pointer, select a point size with the slider, and close the 2D Graphics Inspector.

5. Enter the axes and plot labels:

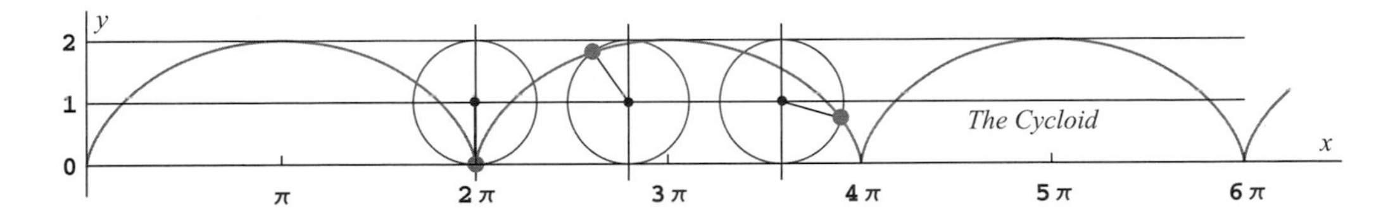

To enter a label, click the TraditionalForm Text button (i.e., bottom button in the left column) in the 2D Drawing Palette, click the intended label location with the pointer that has now turned into an I-beam, and type the label. To enhance the label, click a blank space in the plot; click the label; choose Font (for Mac OS X, Show Fonts) in the Format menu, select a family, typeface, size, and color in the Font window; and click the OK button for Windows or close the window for Mac OS X. (For Mac OS X only, use the Text Color submenu in the Format menu to change the color of the label if the Text Color button in the Font window is broken.)

6. Delete the auxiliary lines:

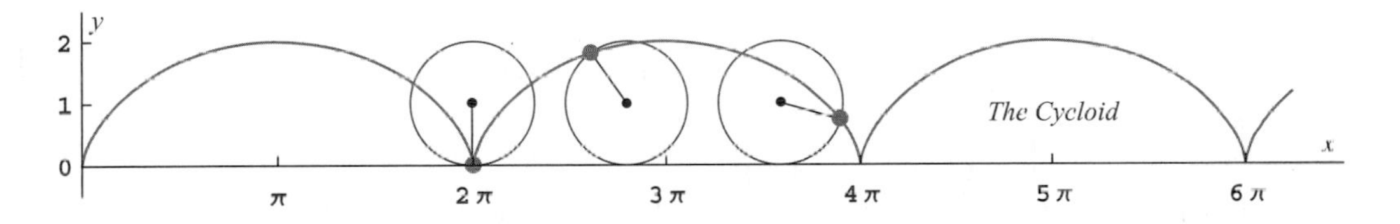

To delete a line, double-click it and press the Backspace key for Windows or the *delete* key for Mac OS X.

7. Delete the auxiliary tick marks:

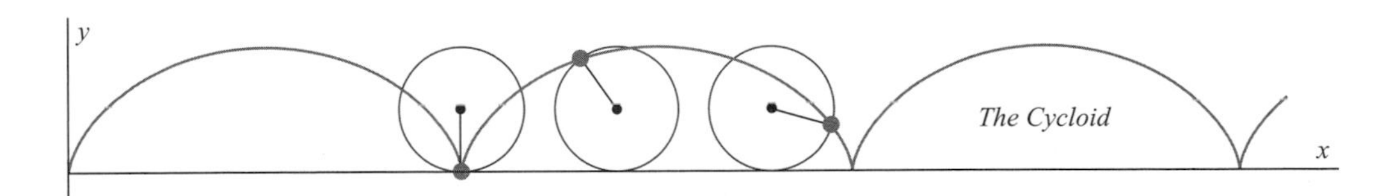

To delete the tick marks, change the current specification of **`Ticks`** to **`None`**: Select the graphics cell bracket, choose **Show Expression** in the Cell menu, scroll to near the bottom of the graphics cell, position the cursor right before *CurrentSpecification* (i.e., right before the first left curly bracket) in `Ticks ->`*CurrentSpecification*, choose **Check Balance** in the Edit menu, type **`None`**, select the graphics cell bracket, and deselect **Show Expression** in the Cell menu.

8. If necessary, refine the positions, shapes, and sizes of the objects by selecting the graphics cell bracket as well as choosing **Show Expression** in the Cell menu; by editing with *extreme* care the numbers in expressions such as **`PointBox`**`[{{`$coord_x$`, `$coord_y$`}}]`, **`LineBox`**`[{{`$coord1_x$`, `$coord1_y$`}, {`$coord2_x$`, `$coord2_y$`}}]`, **`DiskBox`**`[{`$center_x$`, `$center_y$`}, {`r_x`, `r_y`}]`, **`AbsolutePointSize`**`[`d`]`, and **`AbsoluteThickness`**`[`d`]`; and by selecting the graphics cell bracket as well as deselecting **Show Expression** in the Cell menu:

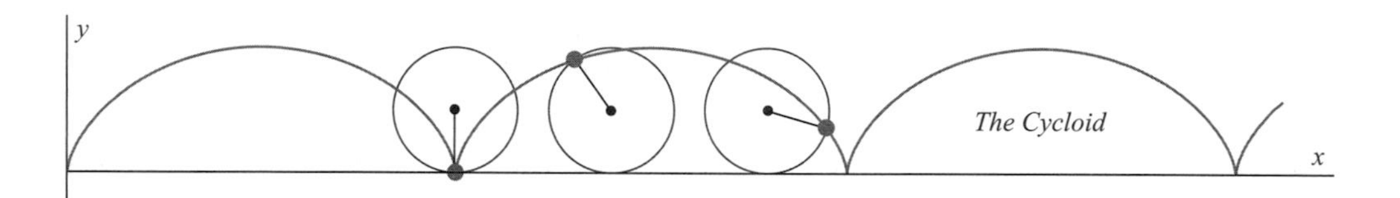

Because this step can be perilous, avoid it if you can or execute it with *extreme* caution if you must. (For more information on two-dimensional graphics, see Section 3.5.2.) ■

2.3.2 Three-Dimensional Graphics

2.3.2.1 Surface Plots

`Plot3D``[`f`, {`x`, `x_{min}`, `x_{max}`}, {`y`, `y_{min}`, `y_{max}`}]` generates a three-dimensional plot of f as a function of the variables x from x_{min} to x_{max} and y from y_{min} to y_{max}.

Example 2.3.17 Two point charges, $+Q$ and $-Q$, are placed at $(-b, 0)$ and $(b, 0)$, respectively. For points on the xy plane, the electric potential V produced by these charges is

$$V = kQ\left(\frac{1}{\sqrt{(x+b)^2+y^2}} - \frac{1}{\sqrt{(x-b)^2+y^2}}\right)$$

where k is the Coulomb constant. Plot V as a function of x and y.

In units of kQ/b, V can be written as

$$V = \left(\frac{1}{\sqrt{(x+1)^2+y^2}} - \frac{1}{\sqrt{(x-1)^2+y^2}}\right)$$

where x and y are in units of b.

Let us plot V with the function **Plot3D**:

```
In[1]:= Plot3D[1/Sqrt[(x+1)^2+y^2]-1/Sqrt[(x-1)^2+y^2], {x, -15, 15},
         {y, -15, 15}, Ticks→{Automatic, Automatic, {-0.04, 0, 0.04}},
         AxesLabel→{"x (b)", "y (b)", "V (kQ/b)"}]
```

Out[1]=

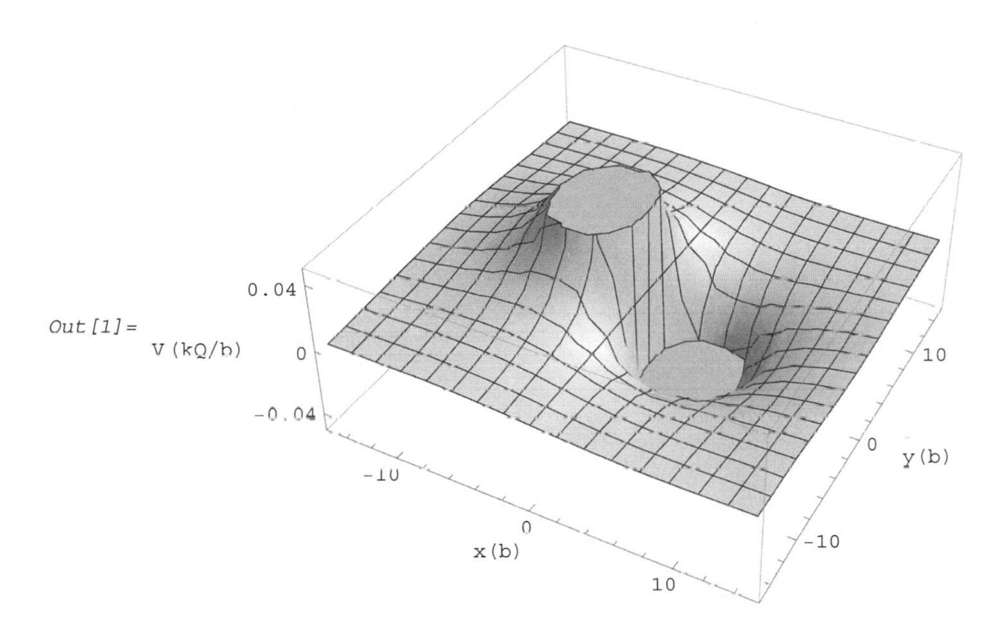

Mathematica has no problem handling the singularities of the potential at $(-b,0)$ and $(b,0)$. The option **Ticks** specifies tick marks for axes:

Ticks→None	draw no tick marks
Ticks→Automatic	place tick marks automatically
Ticks→{*xtick*, *ytick*, *ztick*}	specify tick mark options separately for each axis; tick mark option $\{coord_1, coord_2, \ldots\}$ draws tick marks at the specified positions

To show the essential features of the dipole potential, we specify the following options for **Plot3D**:

PlotRange→{$V_{\min}$, $V_{\max}$}	specify the range of values for V
PlotPoints→*n*	change the number of initial points in each direction, from the default value of 15 to n, for sampling the function
BoxRatios→{r_x, r_y, r_z}	change the ratios of side lengths for the bounding box from the default values of {1,1,0.4} to $\{r_x, r_y, r_z\}$
Boxed→False	remove the bounding box
Axes→False	remove the axes

```
In[2]:= Plot3D[1/Sqrt[(x+1)^2+y^2]-1/Sqrt[(x-1)^2+y^2],
         {x, -15, 15}, {y, -15, 15}, PlotRange→{-0.125, 0.125},
         PlotPoints→40, BoxRatios→{1, 1, 1}, Boxed→False, Axes→False]
```

Out[2]=

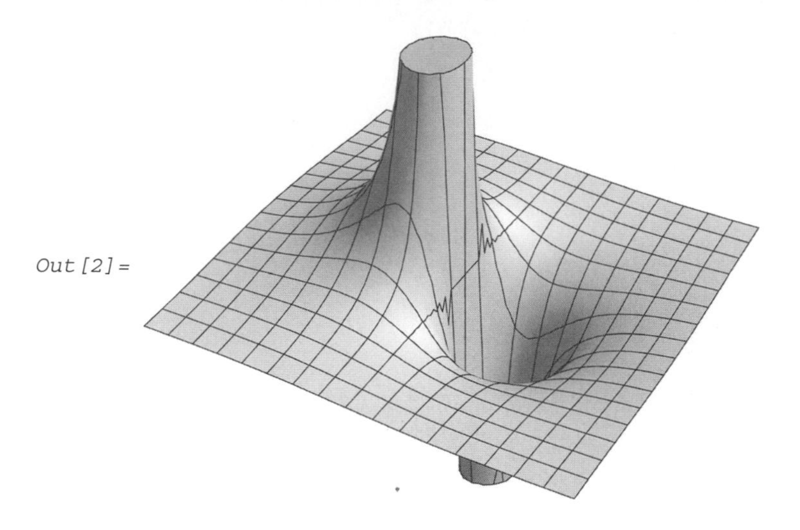

■

2.3.2.2 Viewpoint

ViewPoint is an option for **Plot3D**. **ViewPoint -> {x,y,z}** specifies the coordinates of the point from which three-dimensional graphic objects are viewed. The origin of the right-handed coordinate system is at the center of the bounding box. The axes are parallel to the edges of the box and scaled so that the longest side of the box equals one. Some special values for **ViewPoint** are

{1.3, -2.4, 2}	default
{0, -2, 0} or **Front**	in front, along the negative y-axis
{0, -2, 2}	in front and up
{0, -2, -2}	in front and down
{-2, -2, 0}	in front and left
{2, -2, 0}	in front and right
{0, 2, 0} or **Back**	at back, along the positive y-axis
{2, 0, 0} or **Right**	right, along the positive x-axis
{-2, 0, 0} or **Left**	left, along the negative x-axis
{0, 0, 2} or **Above**	above, along the positive z-axis
{0, 0, -2} or **Below**	below, along the negative z-axis

Example 2.3.18 View the dipole potential in Example 2.3.17 with the values **{-1.770, 2.838, 0.514}** and **{0.806, -3.072, 1.168}** for **ViewPoint**.

To look at the dipole potential with these option specifications, use **Show**:

In[3]:= **Show[%, ViewPoint→{-1.770, 2.838, 0.514}]**

Out[3]=

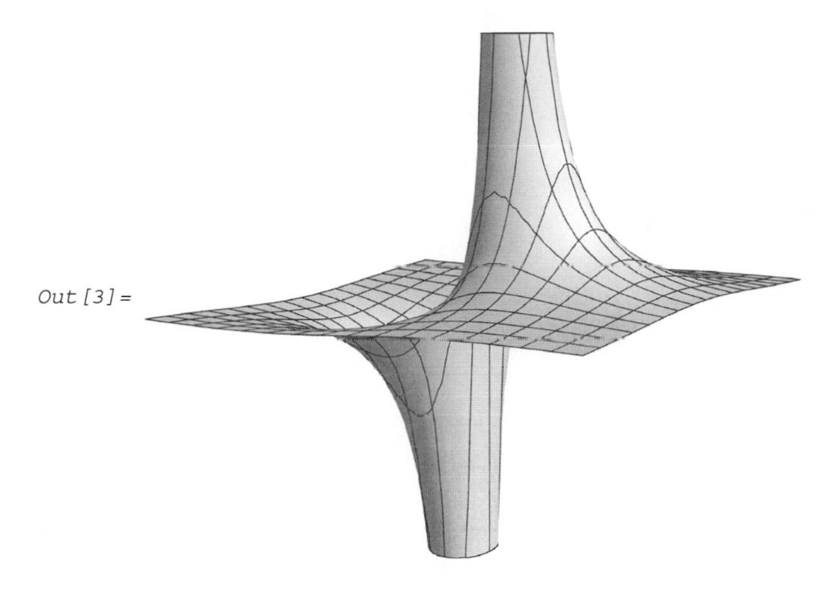

In[4]:= **Show[%, ViewPoint→{0.806, -3.072, 1.168}]**

Out[4]=

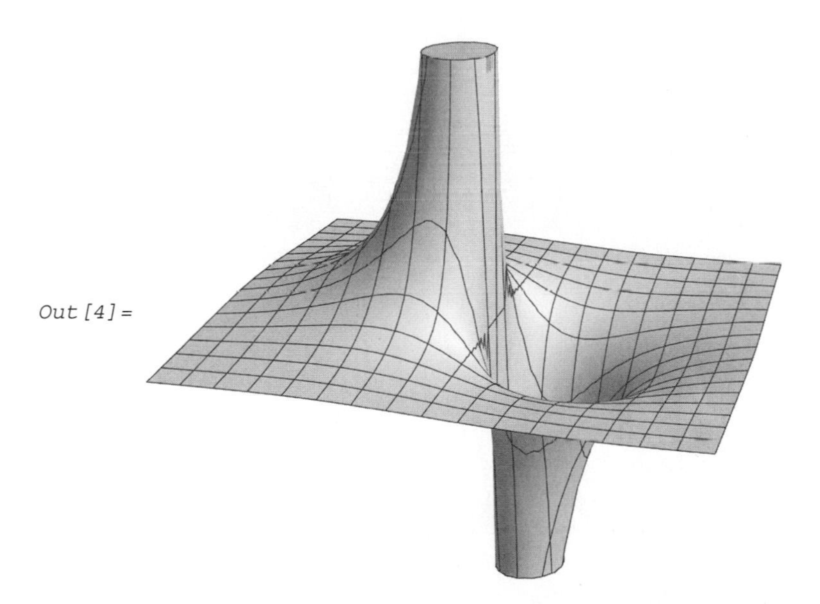

■

To interactively rotate a three-dimensional plot, drag the surface to the desired orientation with the mouse:

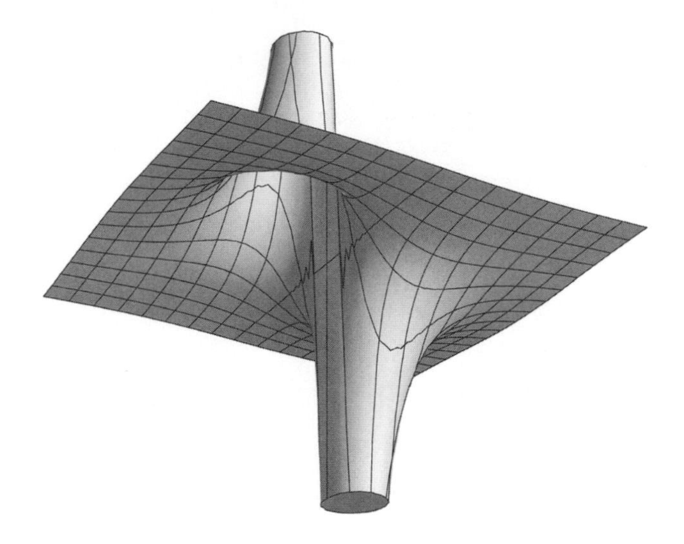

where we have used a plot from Example 2.3.17. (Dragging with the mouse while pressing Ctrl or Alt for Windows or *option* for Mac OS X zooms in for close-up or zooms out for wide views of the plot.) To see the **ViewPoint** specification for this orientation, use **Options**:

In[5]:= **Options** [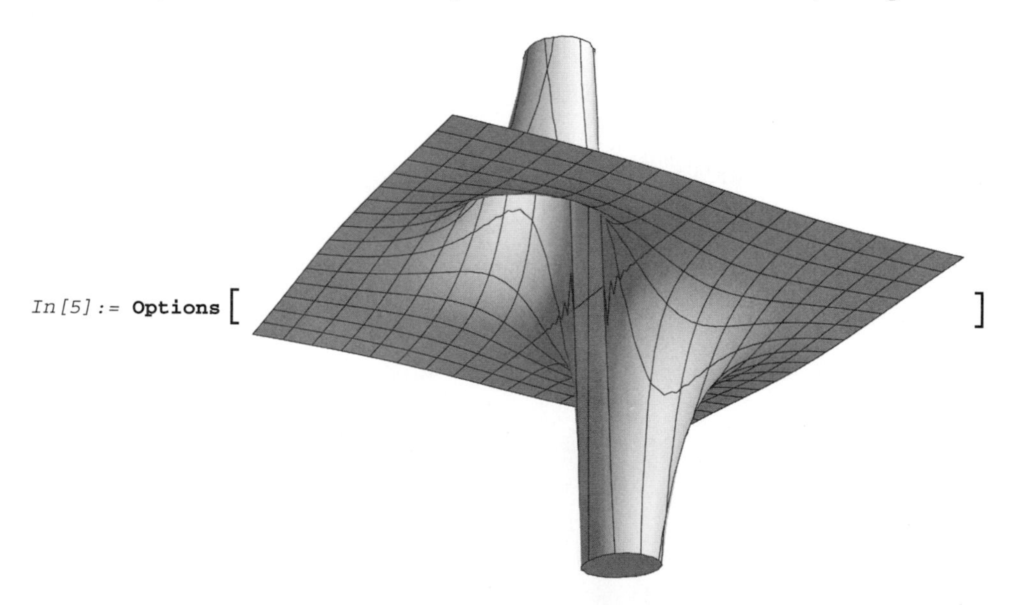]

```
Out[5]= {BoxRatios→{1, 1, 1}, Boxed→False,
        ImageSize→{360., 381.728}, Method→{RotationControl→Globe,
          SpherePoints→Automatic, CylinderPoints→Automatic},
```

```
PlotRange→{{-15, 15}, {-15, 15}, {-0.125, 0.125}},
PlotRangePadding→{Scaled[0.02], Scaled[0.02], Automatic},
ViewPoint→{-1.41493, -2.73862, -1.39605},
ViewVertical→{0., 0., 1.}}
```

where we have copied and pasted the plot as the argument for **Options**. For this orientation, the value of the **ViewPoint** option is **{-1.41493, -2.73862, -1.39605}**. To generate a another plot with this **ViewPoint** specification use **Show**:

```
In[6]:= Show[%%, ViewPoint -> {-1.41493, -2.73862, -1.39605}]
```

Out[6] =

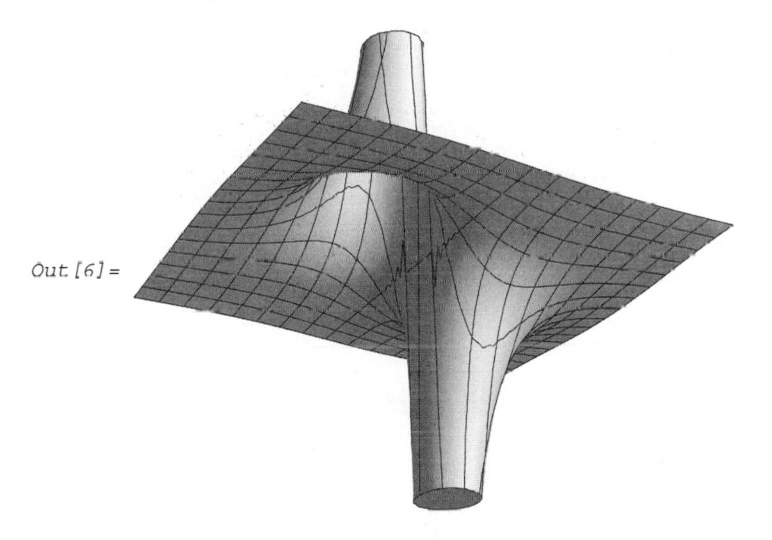

2.3.3 Interactive Manipulation of Graphics

Manipulate[*expr*, {u, u_{min}, u_{max}}]	generate a version of *expr* with controls added to allow interactive manipulation of the value of u, where *expr* can be any two- or three-dimensional graphic or other expression
Manipulate[*expr*, {u, u_{min}, u_{max}, du}]	allow the value of u to vary between u_{min} and u_{max} in steps du
Manipulate[*expr*, {{u, u_{init}}, u_{min}, u_{max}, ...}]	take the initial value of u to be u_{init}
Manipulate[*expr*, {{u, u_{init}, u_{lbl}}, ...}]	label the controls for u with u_{lbl}
Manipulate[*expr*, {u, {u_1, u_2,...}}]	allow u to take on discrete values $u_1, u_2,...$
Manipulate[*expr*, {u,...}, {v,...}, ...]	provide controls to manipulate each of the u, v,...

Example 2.3.19 For a harmonic oscillator subject to a damping force proportional to its velocity, the equation of motion is

$$m\frac{d^2x}{dt^2}+b\frac{dx}{dt}+kx=0$$

where m is the mass, k is the force constant, and b is a positive constant. This equation can be written as

$$\frac{d^2x}{dt^2} + 2\beta\frac{dx}{dt} + \omega_0^2 x = 0$$

where $\beta = b/2m$ is the damping coefficient and $\omega_0 = \sqrt{k/m}$ is the natural angular frequency of the undamped oscillator. Let $\omega_0 = 2.00$ rad/s, $x(0) = 1.00$ m, and $x'(0) = 0$. Using **Manipulate**, determine graphically the values of β for which the motion of the oscillator is (a) underdamped, (b) critically damped, and (c) overdamped.

Manipulate generates the graphic together with controls for interactively varying the value of the damping coefficient β:

```
In[1]:= Manipulate[
  Plot[Evaluate[x[t]/.NDSolve[{x''[t] + 2 β x'[t] + 4 x[t] == 0,
      x[0] == 1, x'[0] == 0}, x, {t, 0, 10}]], {t, 0, 10},
   AxesLabel -> {"t (s)", "x (m)"}, PlotRange -> {-1, 1}],
  {{β, 0, "damping coefficient β (s^-1)"}, 0, 4, 0.1}]
```

Out[1]=

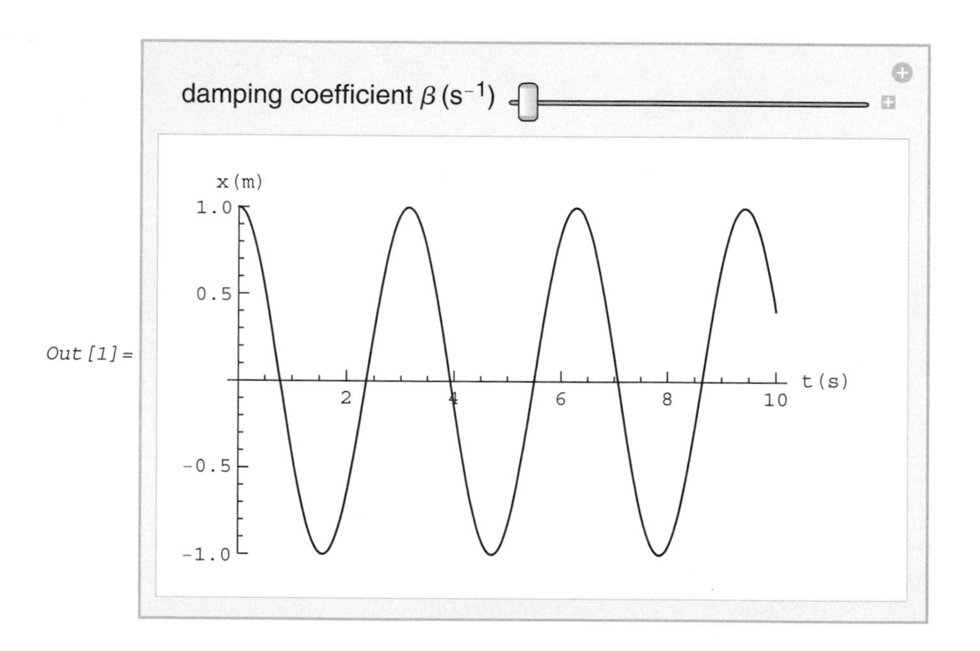

Clicking the button ⊞ near the right end of the slider shows the controls:

Out[1]=

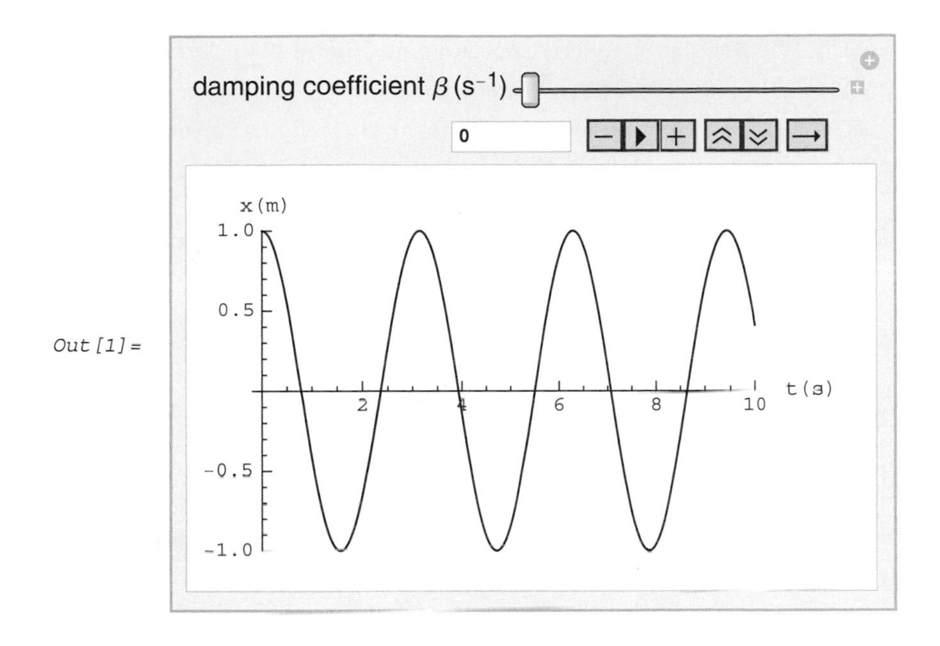

Clicking the button ⊞ or ⊟ increases or decreases the value of β in steps of 0.1. (The other buttons are the animation controls, and Section 2.3.4 will discuss animation.)

(a) For $\beta < 2.0\ \mathrm{s}^{-1}$ (i.e., $\beta < \omega_0$), the motion of the oscillator is underdamped:

Out[1]=

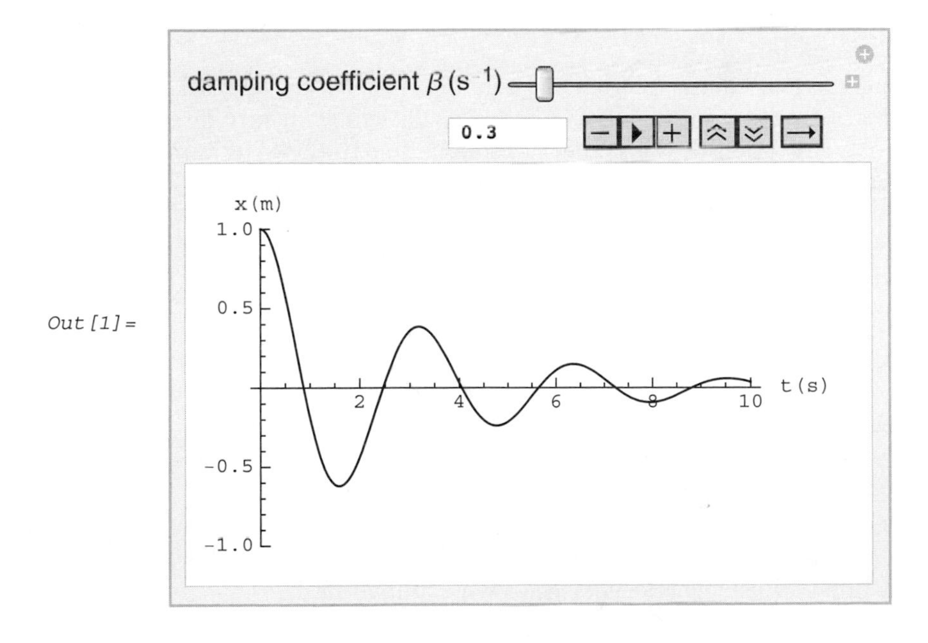

Here, the damping coefficient $\beta = 0.3\,\mathrm{s}^{-1}$. With underdamping, the motion is oscillatory but not periodic because the amplitude decreases with time.

(b) For $\beta = 2.0\,\mathrm{s}^{-1}$ (i.e., $\beta = \omega_0$), the motion of the oscillator is critically damped:

Out[1]=

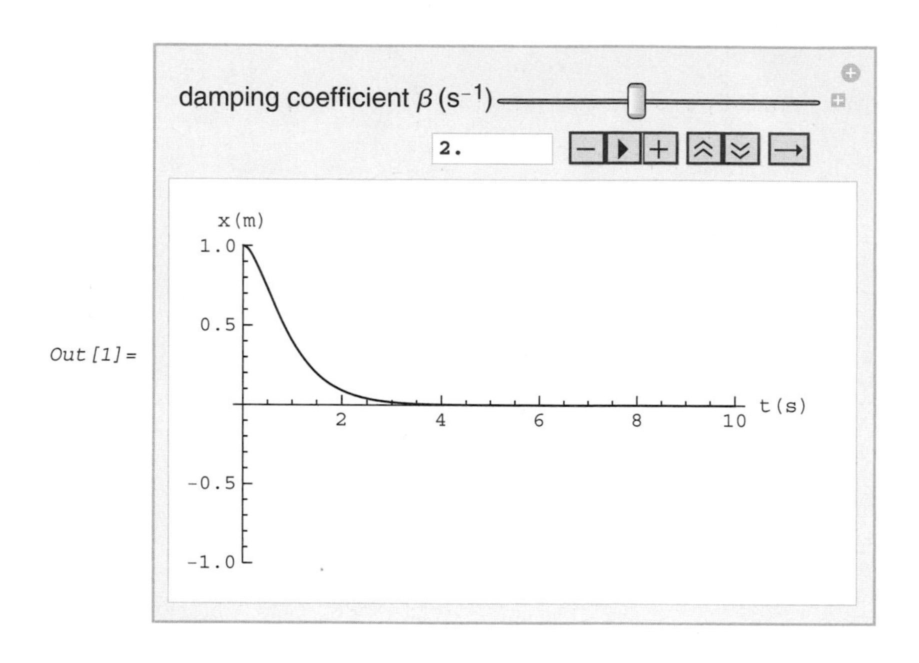

With critical damping, the oscillator reaches the equilibrium position as quickly as possible without overshooting. (To ascertain that the damping coefficient is indeed $2.0\,\mathrm{s}^{-1}$ for critical damping, increase the resolution of the graph by reevaluating the preceding input with the specification **{-0.0001, 0.0001}** instead of **{-1, 1}** for **PlotRange** and then step β through $1.9\,\mathrm{s}^{-1}$, $2.0\,\mathrm{s}^{-1}$, and $2.1\,\mathrm{s}^{-1}$.)

(c) For $\beta > 2.0\ \mathrm{s}^{-1}$ (i.e., $\beta > \omega_0$), the motion of the oscillator is overdamped:

(see the image on the next page)

Here, the damping coefficient $\beta = 3.5\ \mathrm{s}^{-1}$. With overdamping, the oscillator also returns to the equilibrium position without overshooting but not as quickly as for critical damping. (For more information on the damped harmonic oscillator, see [TM04] or [Sym71].) ■

2.3.4 Animation

Mathematica simulates motion by displaying a sequence of graphics, like frames of a movie, in rapid succession. It provides two functions for generating animations:

Out [1]=

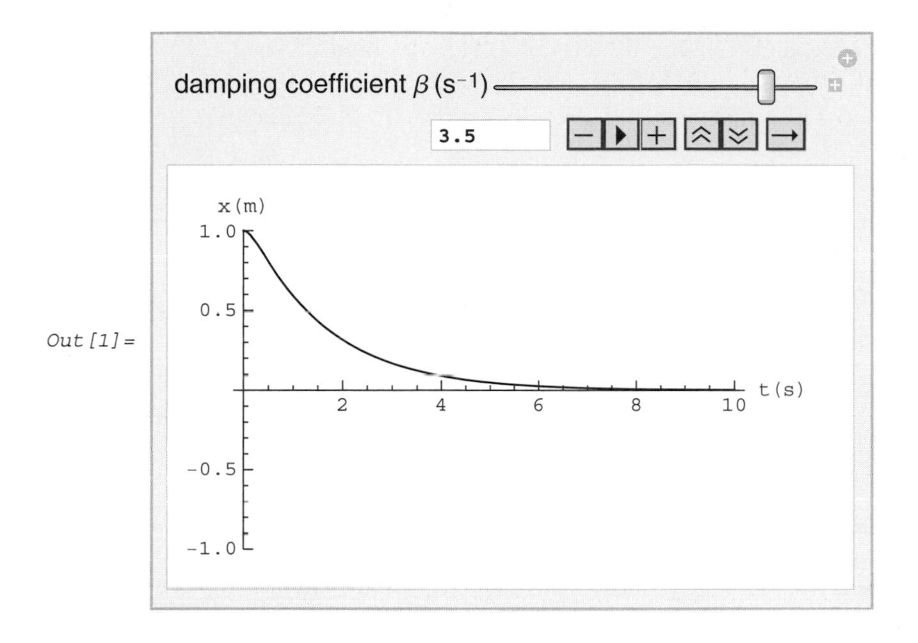

ListAnimate[{$expr_1$, $expr_2$, ...}]	generate an animation whose frames are the successive $expr_i$, which can be any two- or three-dimensional graphics or other expressions
Animate[*expr*, {u, $u_{\min}$, $u_{\max}$}]	generate an animation of *expr* in which u varies continuously from $u_{\min}$ to $u_{\max}$, where *expr* can be any two- or three-dimensional graphic or other expression
Animate[*expr*, {u, $u_{\min}$, $u_{\max}$, du}]	take u to vary in steps du
Animate[*expr*, {u,...}, {v,...}, ...]	vary all the variables u, v, ...

Example 2.3.20 Demonstrate the principle of superposition by simulating the interference of two Gaussian pulses traveling along a string in opposite directions.

Let the displacements y_1 and y_2 of the Gaussian pulses be

$$y_1 = Ae^{-16\alpha(x-vt)^2}$$

$$y_2 = 1.5Ae^{-\alpha(x+vt)^2}$$

These pulses have the same speed v but different amplitudes and widths. The numbers 1.5 and 16 are chosen to make the pulses distinguishable and their interference visible.

If displacement, distance, and time are in units of A, $(1/\alpha)^{1/2}$, and $\left(1/\alpha v^2\right)^{1/2}$, respectively, then y_1 and y_2 can be written as

$$y_1 = e^{-16(x-t)^2}$$

$$y_2 = 1.5e^{-(x+t)^2}$$

The resultant displacement y equals $y_1 + y_2$.

Animate generates the animation, where y_1, y_2, and y are at the top, middle, and bottom, respectively:

```
In[1]:= Animate[
          Plot[
           {Exp[-16(x-t)^2] + 1.5Exp[-(x+t)^2],
            1.5Exp[-(x+t)^2] + 3, Exp[-16(x-t)^2] + 5}, {x, -3.0, 3.0},
           PlotStyle→{{Thickness[0.01], Black},
             {Thickness[0.005], Red}, {Thickness[0.005], Blue}},
           PlotRange→{-0.1, 6.25}, Axes→False], {t, -2.0, 3.5}]
```

Out[1]=

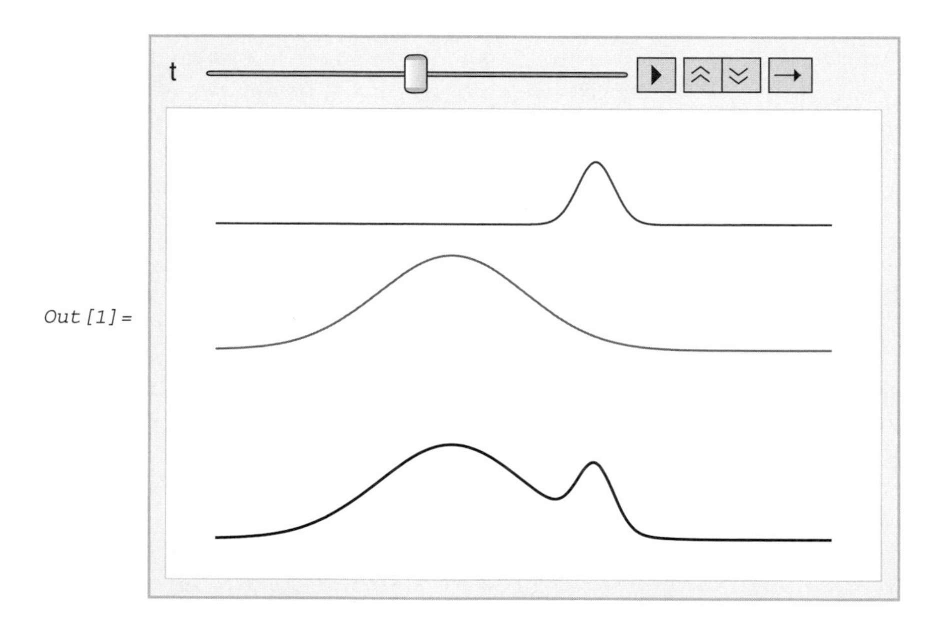

The four buttons at the top of the animation window are the controls. From left to right, the first toggles between Play and Pause, the next two affect the animation speed, and the last specifies the animation direction. To start the animation if it is not already in progress, click the Play button; to stop the animation, click the Pause button.

There is another way to generate the animation:

```
In[2]:= Do[
         Print[
          Plot[
           {Exp[-16(x-t)^2] + 1.5Exp[-(x+t)^2],
            1.5Exp[-(x+t)^2] + 3, Exp[-16(x-t)^2] + 5}, {x, -3.0, 3.0},
           PlotStyle→{{Thickness[0.01], Black},
             {Thickness[0.005], Red}, {Thickness[0.005], Blue}},
           PlotRange→{-0.1, 6.25}, Axes→False, Frame -> True,
           FrameTicks->None]], {t, -2.0, 3.5, 0.5}]
```

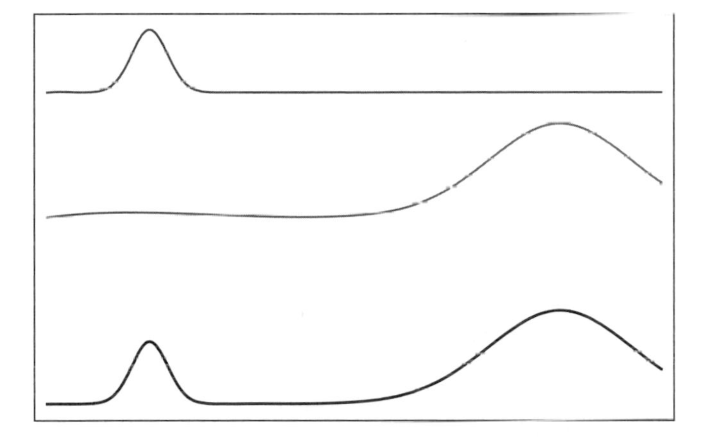

(Only the first of 12 Print cells is shown here. Also, `Print[`*expr*`]` prints *expr* in a Print cell, where *expr* can be any graphic or other expression.) To simulate the interference of the two pulses, select all the Print cells (i.e., SHIFT-click all the Print cell brackets) and choose **Graphics ▸ Rendering ▸ Animate Selected Graphics**. The panel of buttons at the bottom of the notebook window can be used to adjust the speed and direction of the animation. Slow down or speed up the animation by clicking the two buttons on the right. To stop the animation, click anywhere in the notebook window. When *Mathematica* comes to the end of a sequence of frames during an animation, it begins immediately again with the first frame of the sequence. To separate the two sequences, if desired, create a blank Input or Text cell after the last Print cell, select the blank cell and the last Print cell, choose **Graphics ▸ Rendering ▸ Animate Selected Graphics**, click six or more times the two down arrowheads for Windows or the two up arrows for Mac OS X in the panel of buttons at the bottom of the notebook window, and delete the blank cell. To animate, select all the Print cells and choose **Graphics ▸ Rendering ▸ Animate Selected Graphics**. ■

Example 2.3.21 Animate the $n = 2$ normal mode of vibration of the acoustic membrane in Example 1.2.3.

For the animation, we let time t take on the values $i(2\pi/m)$ with $i = 0, 1, \ldots, m-1$. A reasonable value for m is 8; a better one is 16. **Animate** generates the animation, where i varies from 0 to 15 in steps of 1:

```
In[3]:= Animate[
          ParametricPlot3D[{r Cos[θ], r Sin[θ],
            BesselJ[2, r]Sin[2θ]Cos[(2π/16)i]},
           {r, 0, 5.13562}, {θ, 0, 2π},
           PlotRange→{{5, -5}, {5, -5}, {-0.5, 0.5}}, PlotPoints→40,
           BoxRatios→{1, 1, 0.4}, ViewPoint→{2.340, -1.795, 1.659},
           Boxed→False, Axes→False], {i, 0, 15, 1}]
```

Out[3]=

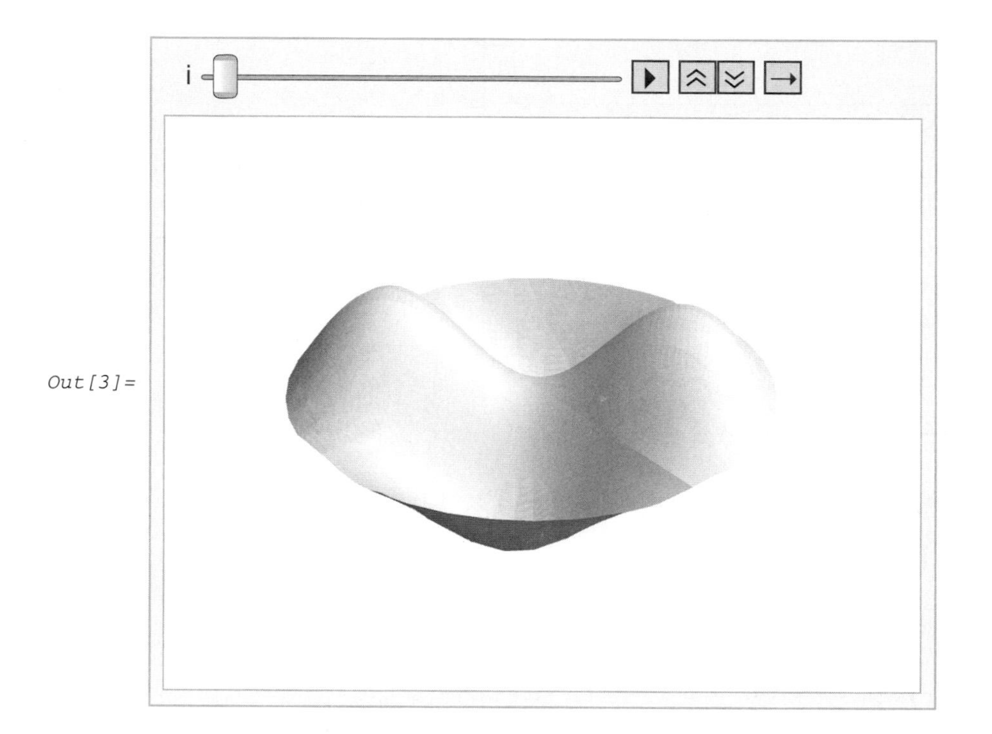

To start the animation if it is not already in progress, click the Play button; to stop the animation, click the Pause button.

Note that we have specified a value for the option **PlotRange** to ensure that the scale remains the same for all the frames. With the default option setting **PlotRange -> Automatic**, the animation is often distorted and jittery. (For an example, see Exercise 46 in Section 2.3.5.) In generating graphics for animating a three-dimensional rotation, we usually include the option setting **SphericalRegion -> True** as well as a specification for **PlotRange**. (For more information on the option **SphericalRegion**, see Section 3.5.3.3.) ■

Example 2.3.22 In an elastic medium, two point sources S_1 and S_2 are at a distance d apart along the x-axis. Each source emits a wave that travels out uniformly in all directions. The two waves have the same frequency and amplitude and start in phase at the sources. For a point P at a distance r_1 from S_1 and r_2 from S_2, the displacement z_1 of the medium due to the wave from S_1 is

$$z_1 = \frac{Y}{r_1} \sin(k\,r_1 - \omega t)$$

and the displacement z_2 due to the wave from S_2 is

$$z_2 = \frac{Y}{r_2} \sin(k\,r_2 - \omega t)$$

where t is the time, Y is a constant, k is the wave number, and ω is the angular frequency. Let the distance d between the sources be $20/k$. For points on the xy-plane, simulate the interference of these two waves.

The resultant displacement z at P is

$$z = z_1 + z_2$$

If displacement, distance, and time are measured, respectively, in units of $Yk, 1/k$, and $1/\omega$, z_1 and z_2 become

$$z_1 = \frac{1}{r_1} \sin(r_1 - t)$$

$$z_2 = \frac{1}{r_2} \sin(r_2 - t)$$

and the distance d between the sources becomes 20. If the coordinates on the xy-plane of S_1, S_2, and P are, respectively, (10, 0), (−10, 0), and (x, y), then

$$r_1 = \sqrt{(x - 10)^2 + y^2}$$

$$r_2 = \sqrt{(x + 10)^2 + y^2}$$

Let us make several assignments:

```
In[4]:= Clear["Global`*"]

In[5]:= r1[x_, y_] := Sqrt[(x-10)^2 + y^2]

In[6]:= r2[x_, y_] := Sqrt[(x+10)^2 + y^2]

In[7]:= z1[r1_, t_] := Sin[r1 - t]/r1

In[8]:= z2[r2_, t_] := Sin[r2 - t]/r2

In[9]:= z[x_, y_, t_] := z1[r1[x, y], t] + z2[r2[x, y], t]
```

ListAnimate together with **Table** and **Plot3D** generates the animation:

```
In[10]:= ListAnimate[
          Table[
           Plot3D[Evaluate[z[x, y, t]], {x, -25, 25},
            {y, 1, 75}, PlotPoints→50, PlotRange→{0, 1.5},
            ViewPoint -> {1.181, 3.090, 0.713}, Boxed→False,
            Axes→False, Mesh -> None], {t, 0, 2Pi - 2Pi/16, 2Pi/16}]]
```

Out[10]=

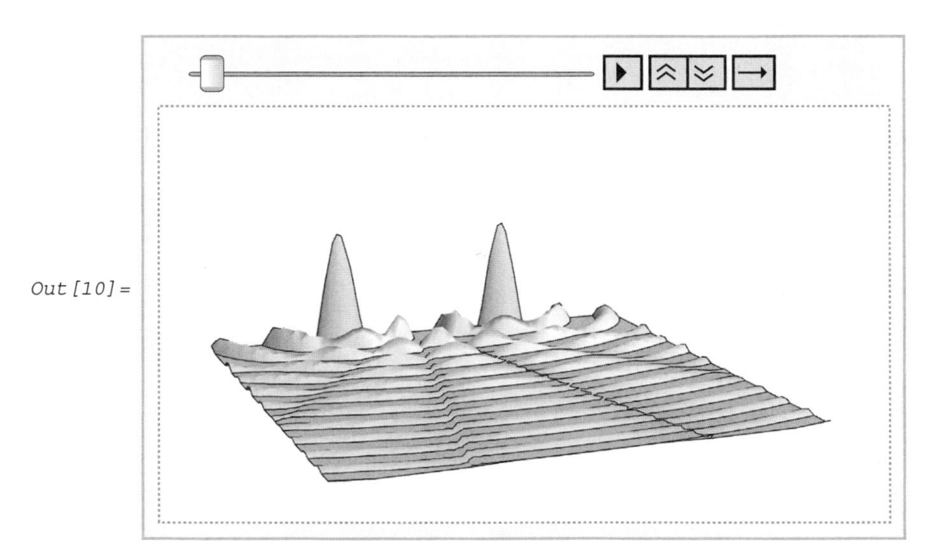

Here is the animation from another viewpoint:

```
In[11]:= ListAnimate[
          Table[
           Plot3D[Evaluate[z[x, y, t]], {x, -25, 25},
            {y, 1, 75}, PlotPoints→50, PlotRange→{0, 1.5},
            ViewPoint -> {1.298, 2.720, 1.540}, Boxed→False,
            Axes→False, Mesh -> None], {t, 0, 2Pi - 2Pi/16, 2Pi/16}]]
```

Out[11]=

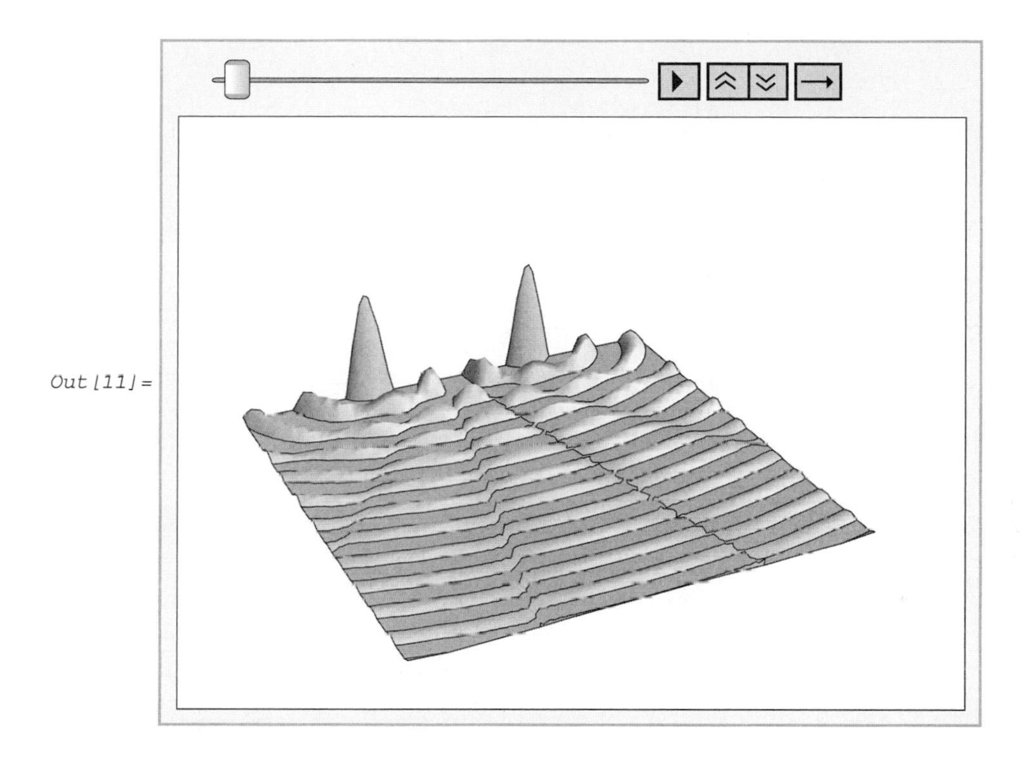

We let the y coordinate vary from 1 to 75 to avoid the singularities of the waves at the sources and to scale the graphics in order to render the interference of the waves visible. To start the animation if it is not already in progress, click the Play button. Note the alternating regions of constructive and destructive interference.

```
In[12]:= Clear[r1, r2, z1, z2, z]
```

■

2.3.5 Exercise

In this book, straightforward, intermediate-level, and challenging exercises are unmarked, marked with one asterisk, and marked with two asterisks, respectively. For solutions to most odd-numbered exercises, see Appendix C.

1. Plot (i.e., graph) the function

$$f(x) = \frac{x}{x^2 + 1}$$

Answer:

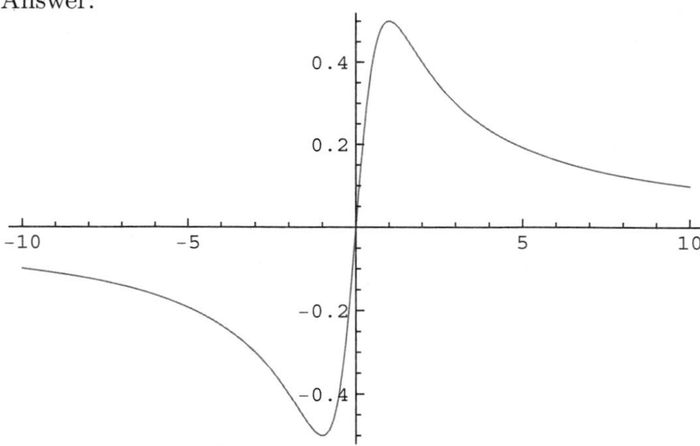

2. Plot (i.e., graph) the function

$$f(x) = x\sqrt{x+3}$$

Answer:

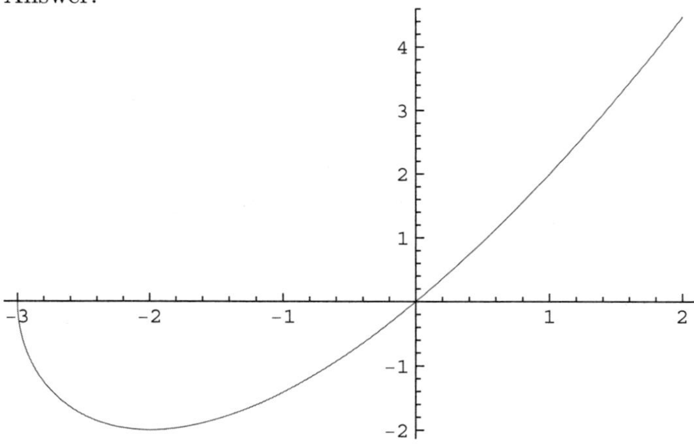

3. Solve the differential equation

$$\frac{d^2y}{dt^2} = -t^3y + 1$$

with $y(0) = 1$ and $y'(0) = 0$. Plot $y(t)$ with t in the range 0 to 8, and label the axes.

Answers:

```
{{y→InterpolatingFunction[{{0., 8.}}, <>]}}
```

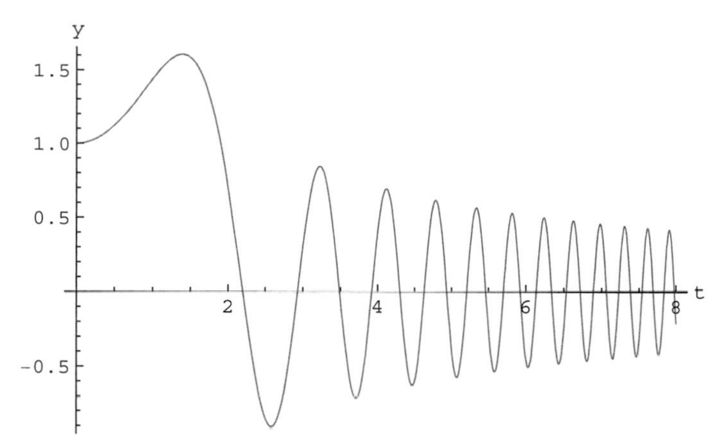

4. Solve the differential equation

$$y\frac{dy}{dt} = t \sin t$$

with the initial condition $y(0) = 1$ for t from 0 to 4.6. Plot the solution with t varying from 0 to 4.6, and label the axes.

Answers:

```
{{y→InterpolatingFunction[{{0., 4.6}}, <>]}}
```

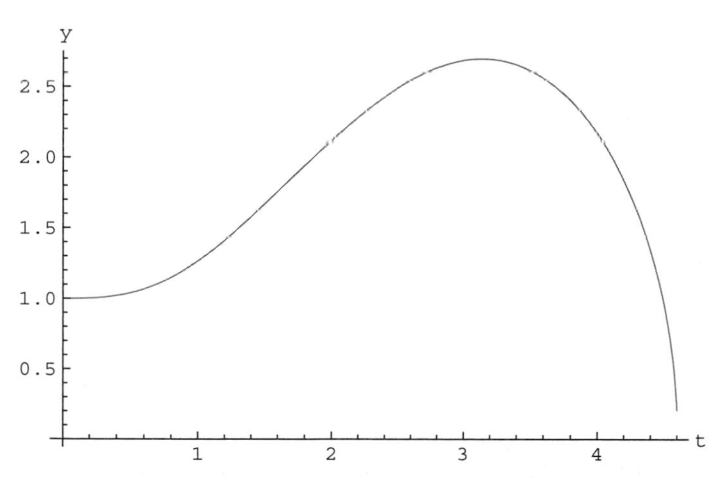

5. For the one-dimensional harmonic oscillator, the time-independent Schrödinger equation is

$$-\frac{\hbar^2}{2m}\frac{d^2\psi}{dx^2} + \frac{1}{2}m\omega^2x^2\psi = E\psi$$

In terms of the dimensionless variable

$$\xi = \sqrt{\frac{m\omega}{\hbar}}\, x$$

the Schrödinger equation can be written as

$$\frac{d^2\psi}{d\xi^2} = \left(\xi^2 - 2n - 1\right)\psi$$

where

$$n = \frac{E}{\hbar\omega} - \frac{1}{2}$$

is a nonnegative integer. Let $n = 5$, $\psi(0) = 0$, and $\psi'(0) = 1$. (a) Solve this equation for ξ ranging from -5 to 5, and evaluate $\psi(-5)$ as well as $\psi(5)$. (b) Plot $\psi(\xi)$.

Answers:

```
{{ψ→InterpolatingFunction[{{-5., 5.}}, <>]}}
{-0.00250397, 0.00250397}
```

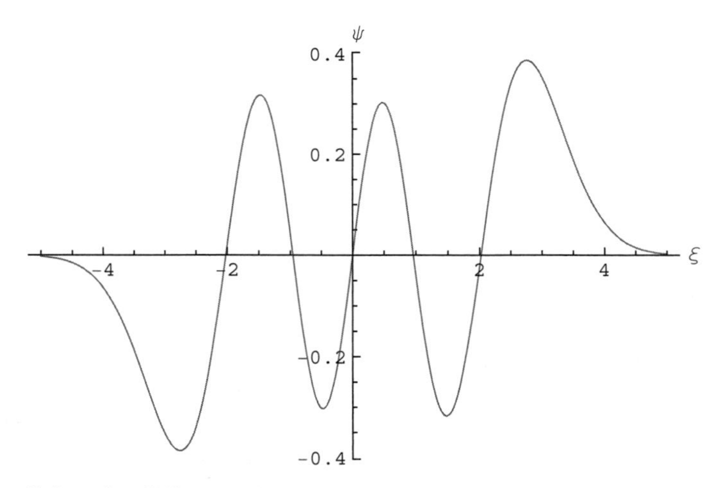

6. Solve the differential equation

$$\frac{d^2x}{dt^2} = \frac{12}{x^{13}} - \frac{6}{x^7}$$

for $x(t)$ with t ranging from 0 to 10. Take as initial conditions $x(0) = 1.02$ and $x'(0) = 0$. Plot $x(t)$ versus t from 0 to 10, and label the axes.

Answer:

```
{{x→InterpolatingFunction[{{0., 10.}}, <>]}}
```

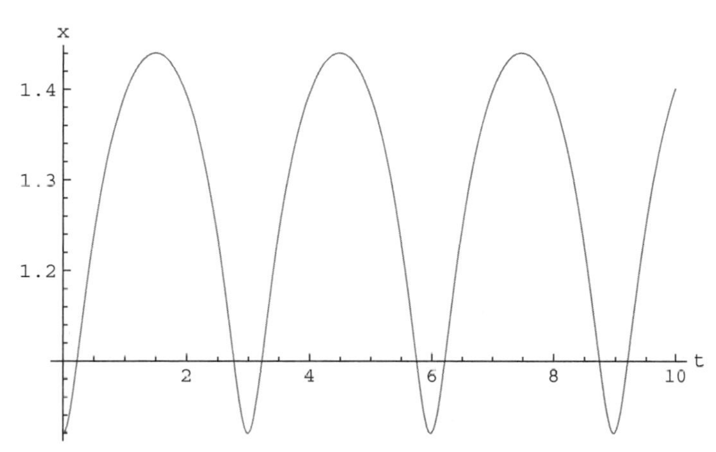

7. In Exercise 6, plot together $x(t)$ and $x'(t)$ versus t from 0 to 10. Show $x'(t)$ as a dashed curve, and label the plot with "$x(t)$ and $x'(t)$".

Answer:

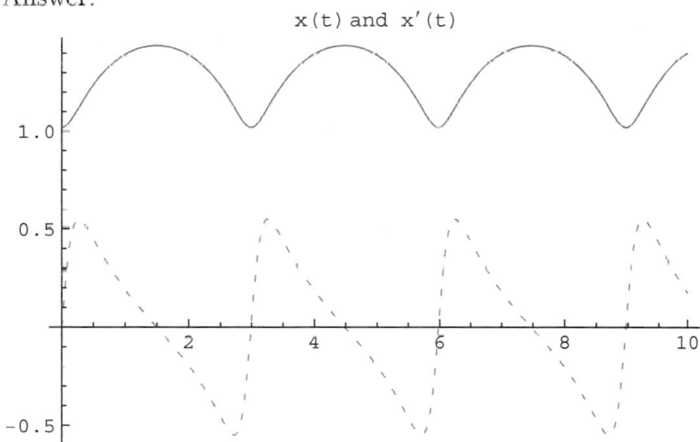

8. In Exercise 6, plot together $x(t)$ and $x'(t)$ versus t from 0 to 10. Show $x(t)$ in **`Orange`** and **`Thick`**, $x'(t)$ in **`Blue`** and **`Thick`**, and the background in **`Lighter[Black, 0.7]`**. Also, label the plot with "$x(t)$ and $x'(t)$".
9. Plot the function

$$f(x) = x^2 + \cos 32x$$

with x varying from −5 to 5. Label the axes. *Hint:* Specify a value for the option **`PlotPoints`**.

Answer:

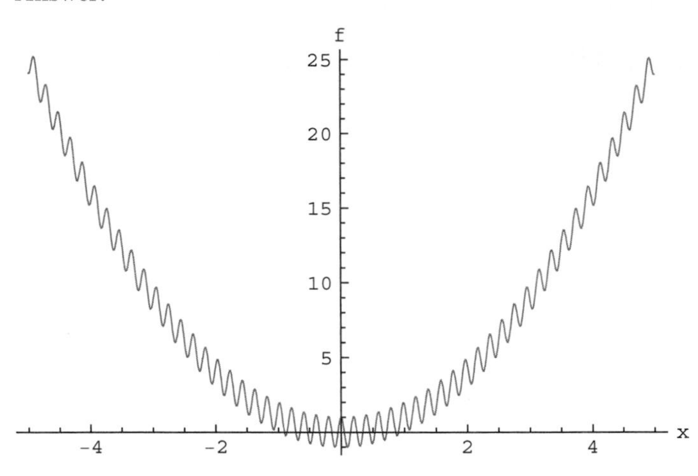

10. According to the Debye theory, the heat capacity of a solid is

$$C_V = 9kN\left(\frac{T}{\theta}\right)^3 \int_0^{\theta/T} \frac{x^4 e^x}{(e^x - 1)^2} dx$$

where θ is the Debye temperature, N is the number of atoms, k is Boltzmann's constant, and T is the absolute temperature. Plot $C_V/3Nk$ as a function of T/θ from 0.001 to 1.4; label the axes. *Hint:* Use **`NIntegrate`**.

Answer:

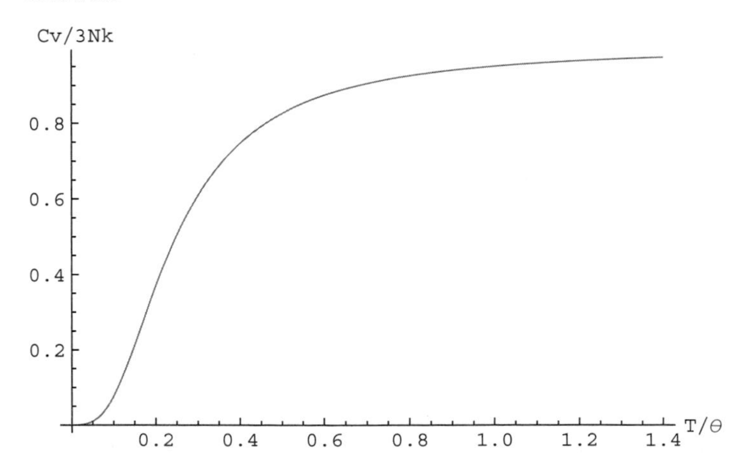

11. Consider a system that has only two energy eigenstates having nonnegative energies 0 and ϵ. For this system, the partition function Z is

$$Z = 1 + e^{-\frac{\epsilon}{kT}}$$

where k is Boltzmann's constant and T is the absolute temperature. The average energy E can be expressed as

$$E = kT^2 \frac{\partial \ln Z}{\partial T}$$

The heat capacity C is given by

$$C = \frac{\partial E}{\partial T}$$

The entropy S can be obtained from

$$S = k \ln Z + \frac{E}{T}$$

Calculate and then plot (as functions of temperature) the following quantities: (a) average energy, (b) heat capacity, and (c) entropy. *Hint:* In the plots, let E, C, S, and T be measured in units of ϵ, k, k, and ϵ/k, respectively.

Answers:

$$\frac{\epsilon}{1 + e^{\frac{\epsilon}{kT}}}$$

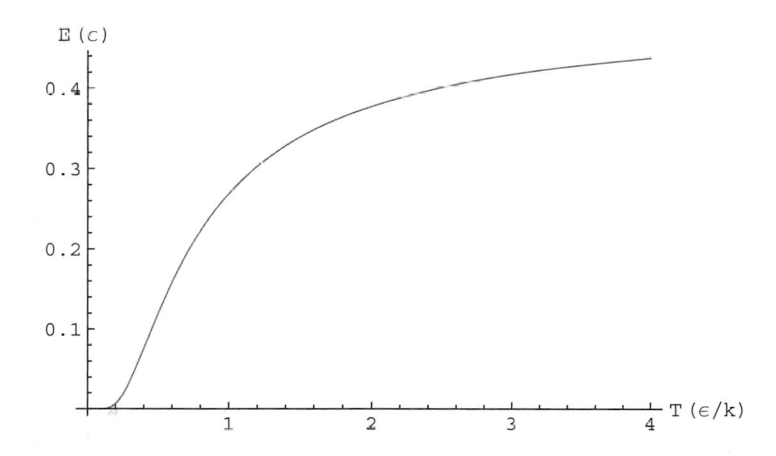

$$\frac{e^{\frac{\epsilon}{kT}} \epsilon^2}{\left(1 + e^{\frac{\epsilon}{kT}}\right)^2 k T^2}$$

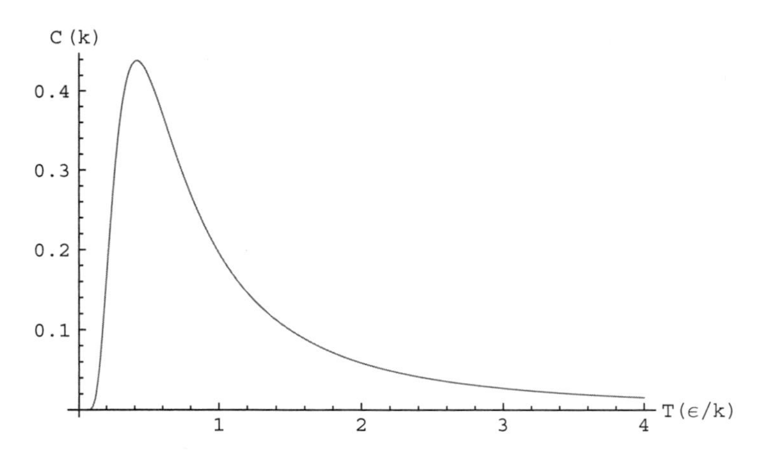

$$\frac{\epsilon}{\left(1+e^{\frac{\epsilon}{kT}}\right)T}+k\,\mathrm{Log}\left[1+e^{-\frac{\epsilon}{kT}}\right]$$

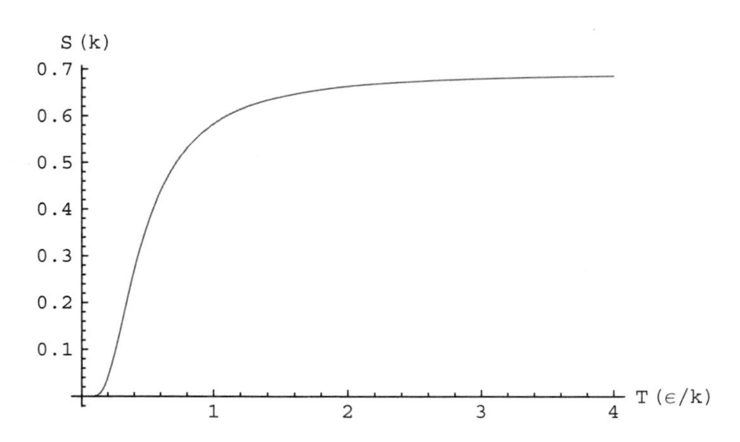

12. (a) Find the power series expansion (representation) at 0 to x^7 (i.e., seventh-degree Taylor polynomial at 0 or seventh-degree Maclaurin polynomial) of

$$f(x) = \tan^{-1} x$$

(b) Plot f together with the series, and examine the graph to determine the region in which the series is a good approximation of f. *Hint*: Use the function **`Normal`**.

Answers:

$$x-\frac{x^3}{3}+\frac{x^5}{5}-\frac{x^7}{7}$$

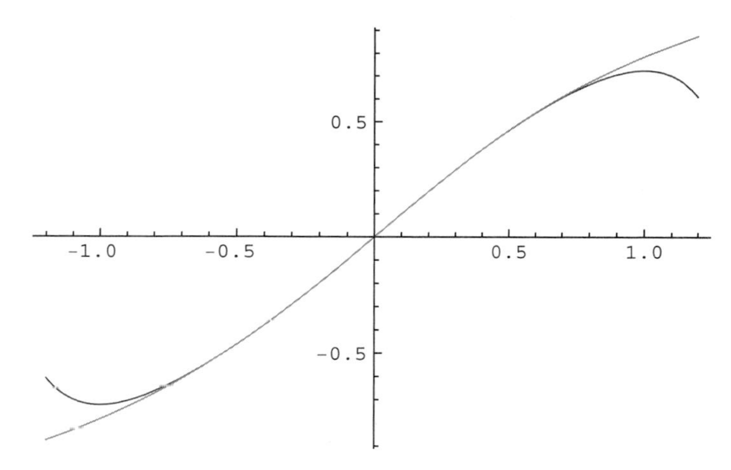

The series is a good approximation of $\tan^{-1} x$ for $|x| \lesssim 0.8$.

13. (a) Find the power series expansion (representation) at 0 up to x^7 (i.e., seventh-degree Taylor polynomial at 0 or seventh-degree Maclaurin polynomial) of

$$f(x) = \sin x \cos x$$

(b) Plot f together with the series, and examine the graph to determine the region in which the series is a good approximation of f. *Hint*: Use the function **Normal**.

Answers:

$$x - \frac{2\,x^3}{3} + \frac{2\,x^5}{15} - \frac{4\,x^7}{315}$$

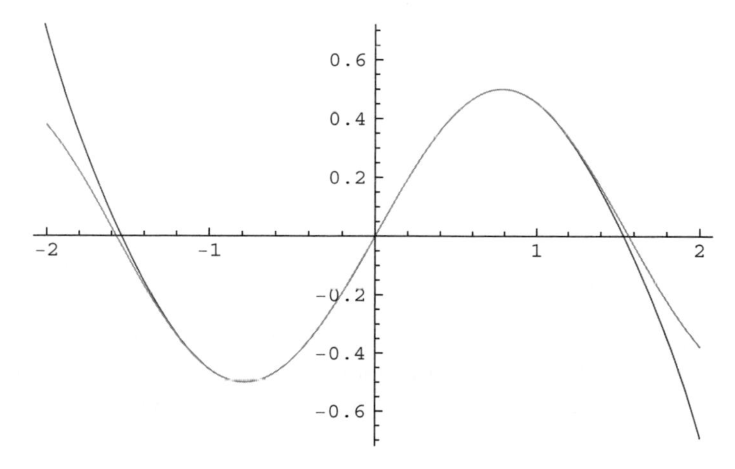

The series is a good approximation of $f(x)$ for $|x| \lesssim 1.3$.

14. Consider the function

$$f(x) = \sin x$$

(a) Define functions T_1, T_3, and T_5, where T_n is the power series expansion of f at 0 up to x^n (i.e., the nth-degree Taylor polynomial of f at 0 or the nth-degree Maclaurin polynomial of f).

(b) Plot f, T_1, T_3, and T_5 together. Show f in {`Black, Thick, Dashing[Small]`} and the background in `Lighter[Yellow, 0.75]`; give each of the other curves the thickness `Thick` and a distinct color; label the plot with "sin(x) vs. x". Using **Find Selected Function** in the Help menu, access information on `Thick, Small,` and `Lighter`. *Hint*: Include `Evaluate` in `Plot,` and specify values for the options `PlotRange, PlotStyle, Background,` and `PlotLabel`.

15. A differential equation that arises in many physical problems is the Bessel equation:

$$x^2\frac{d^2y}{dx^2} + x\frac{dy}{dx} + \left(x^2 - p^2\right)y = 0$$

A solution of the Bessel equation is $J_p(x)$, called the Bessel function of the first kind and order p:

$$J_p(x) = \sum_{k=0}^{\infty} \frac{(-1)^k}{k!\,\Gamma(k+p+1)}\left(\frac{x}{2}\right)^{2k+p}$$

where Γ is the gamma function. The Bessel function of the first kind and order 0 can be expressed as

$$J_0(x) = \sum_{k=0}^{\infty} \frac{(-1)^k}{(k!)^2}\left(\frac{x}{2}\right)^{2k}$$

The nth partial sum of J_0 is

$$S_n(x) = \sum_{k=0}^{n} \frac{(-1)^k}{(k!)^2}\left(\frac{x}{2}\right)^{2k}$$

(a) For $S_n(x)$, define a *Mathematica* function that has two arguments: the first for n and the second for x.

(b) Plot J_0, S_1, S_2, S_3, and S_4 together. Show J_0 in {`Black, Thickness[0.0125], Dashing[Medium]`} and the background in `Lighter[Blue, 0.9]`; give each of the other curves the thickness `Thick` and a distinct color; label the plot with "Partial sums of the Bessel function". Using **Find Selected Function** in the Help menu, access information on `Medium`, `Lighter`, and `Thick`. *Hint*: `BesselJ[`*p*`,`*x*`]` gives $J_p(x)$.

16. The normalized eigenfunctions of the one-dimensional harmonic oscillator are

$$\psi_n(x) = 2^{-n/2}(n!)^{-1/2}\left(\frac{m\omega}{\hbar\pi}\right)^{1/4} H_n\left(\sqrt{m\omega/\hbar}\, x\right) \exp\left(-\frac{m\omega}{2\hbar}x^2\right)$$

where m, ω, $\hbar$, n, and H_n are oscillator mass, oscillator frequency, Planck's constant divided by 2π, nonnegative integers, and Hermite polynomials, respectively. Plot ψ_0, ψ_2, and ψ_4 together. Give a different style to each curve, and label the plot. *Hint*: To locate the $\hbar$ button, choose **Palettes ▸ SpecialCharacters ▸ Technical Symbols** for *Mathematica* 6.0.0 and 6.0.1 or **Palettes ▸ SpecialCharacters ▸ Symbols ▸ Technical Symbols** (i.e., ∞ button or tab) for *Mathematica* 6.0.2.

Answer:

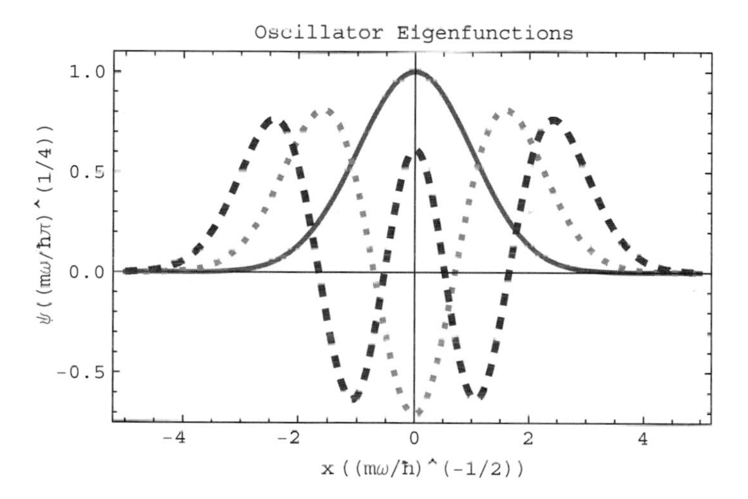

*17. For an ideal Fermi gas in equilibrium at temperature T, the average number of fermions in a single-particle state of energy ε is given by the Fermi–Dirac distribution

$$n(\varepsilon) = \frac{1}{e^{(\varepsilon-\mu)/kT} + 1}$$

where k is Boltzmann's constant and μ is the chemical potential. Assuming that the discrete single-particle energy spectrum can be approximated by a continuous spectrum, generate the following graphs of the distribution:

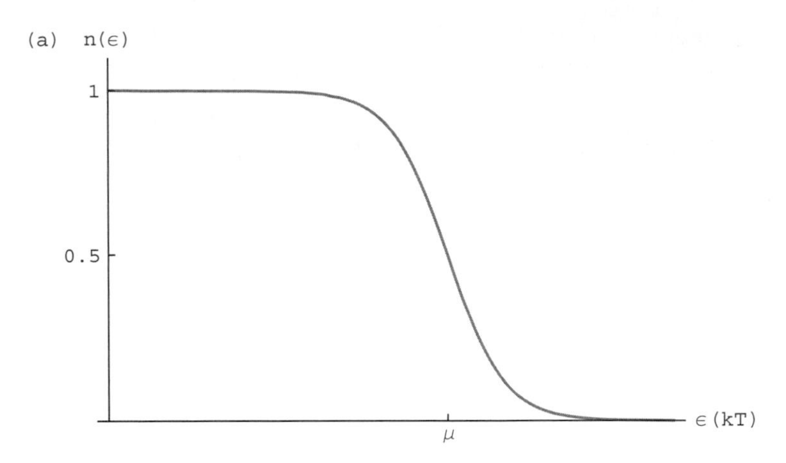

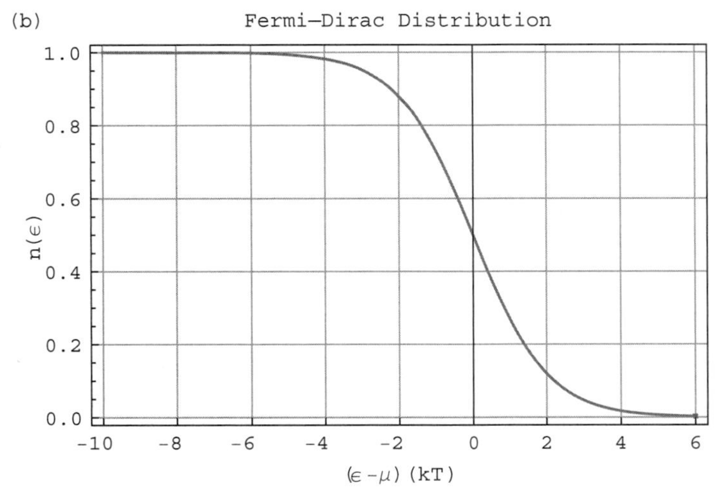

*18. For an ideal Fermi gas in equilibrium at temperature T, the average number of fermions in a single-particle state of energy ε is given by the Fermi–Dirac distribution

$$n(\varepsilon) = \frac{1}{e^{(\varepsilon-\mu)/kT} + 1}$$

where k is Boltzmann's constant and μ is the chemical potential. For an ideal Bose gas, the average number of bosons in a single-particle state of energy ε is given by the Bose–Einstein distribution

$$n(\varepsilon) = \frac{1}{e^{(\varepsilon-\mu)/kT} - 1}$$

Assuming that discrete single-particle energy spectrums can be approximated by continuous spectrums, generate the following graph of the two distributions together:

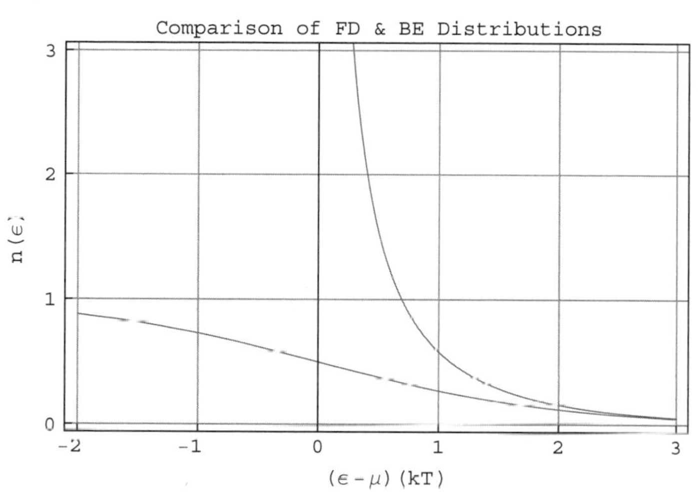

19. Consider the equation

$$6\,\mathrm{sech}^2(x) - \sqrt{5 - x^2} = 0$$

Using **FindRoot**, determine all of its roots to six decimal digits of accuracy.

Answers:

```
{{x→-2.21882}, {x→-1.17705}, {x→1.17705}, {x→2.21882}}
```

20. The *Mathematica* expression for $j_\nu(x)$, the spherical Bessel function of order ν, is

$$\sqrt{\frac{\pi}{2\,x}}\ \mathbf{BesselJ}\left[\nu + \frac{1}{2},\ x\right]$$

(a) Plot $j_3(x)$ for x ranging from 0 to 15.

(b) Using the capability of the notebook front end to obtain a list of coordinates from a graph, determine the first three (from the origin) positive zeroes of $j_3(x)$ to two decimal digits of accuracy.

(c) Using **FindRoot** with the zeroes obtained in part (b) as starting values, determine the first three positive zeroes of $j_3(x)$ to six decimal digits of accuracy.

Answers:

```
{{x→6.98793}, {x→10.4171}, {x→13.698}}
```

21. The Bessel function of the first kind and order 0 is often denoted by $J_0(x)$. For $x > 0$, find (a) its first five (from the origin) zeroes and (b) its first two local minima. Give the answers to six decimal digits of accuracy.

Answers:

```
{{x→2.40483}, {x→5.52008},
 {x→8.65373}, {x→11.7915}, {x→14.9309}}
 {{-0.402759, {x→3.83171}}, {-0.249705, {x→10.1735}}}
```

22. The *Mathematica* expression for $P_n(x)$, the Legendre polynomial of order n, is

```
LegendreP[n, x]
```

(a) Plot $P_5(x)$.
(b) Using **FindRoot**, determine all the zeroes of $P_5(x)$ to six decimal digits of accuracy.
(c) Using **FindMaximum**, determine all the local maxima of $P_5(x)$ to six decimal digits of accuracy.

Answers:

```
{{x→ -0.90618}, {x→ -0.538469},
 {x→ -1.3976 × 10^-20}, {x→0.538469}, {x→0.90618}}
 {{0.419697, {x→ -0.765055}}, {0.346628, {x→0.285232}}}
```

*23. The Einstein model assumes that a solid is equivalent to an assembly of independent one-dimensional harmonic oscillators all vibrating with the same frequency ν. The heat capacity at constant volume C_V is given as

$$C_V = 3Nk\left(\frac{\Theta}{T}\right)^2 \frac{e^{\Theta/T}}{(e^{\Theta/T} - 1)^2}$$

where N is the number of atoms in the solid, k is Boltzmann's constant, and Θ is the Einstein temperature defined by the equation $k\Theta \equiv h\nu$, in which h is Planck's constant. For diamond, the experimental value for $C_V/3Nk$ is 0.3552 at $T = 358.5\,\text{K}$. Use **FindRoot** to determine Θ of diamond. (For a more sophisticated method for determining Θ, see Example 2.4.18.) *Hint:* Make a change of variables from C_V and T to c and t:

$$c = C_V/3Nk$$
$$t = T/\Theta$$

Then, plot c as a function of t to determine graphically for **FindRoot** a starting value for t at which $c = 0.3552$.

Answer:

```
1327.72K
```

24. Using **FindRoot**, justify the approximation

$$\ln(1+x) \approx x$$

for $x \ll 1$. That is, using **FindRoot**, determine the value of x at which the percentage error

$$\left|\frac{\ln(1+x) - x}{\ln(1+x)}\right| \times 100\%$$

first exceeds (a) 5% and (b) 1%.

Answers:

```
{x→0.10164}
{x→0.0200664}
```

25. Using **FindRoot**, justify the approximation

$$\sin x \approx x$$

for $x \ll 1$, where x is in radians. That is, using **FindRoot**, determine the value of x at which the percentage error

$$\left|\frac{\sin x - x}{\sin x}\right| \times 100\%$$

first exceeds (a) 1% and (b) 0.5%. What are these values of x in degrees?

Answers:

(a) 0.244097 radians or 13.9857 degrees
(b) 0.172903 radians or 9.9066 degrees

26. (a) Plot the curve representing the best fit obtained in Exercise 19 of Section 2.1.20. (b) Plot the given points. (c) Superpose the two graphs to see if the fit is acceptable.

Answers:

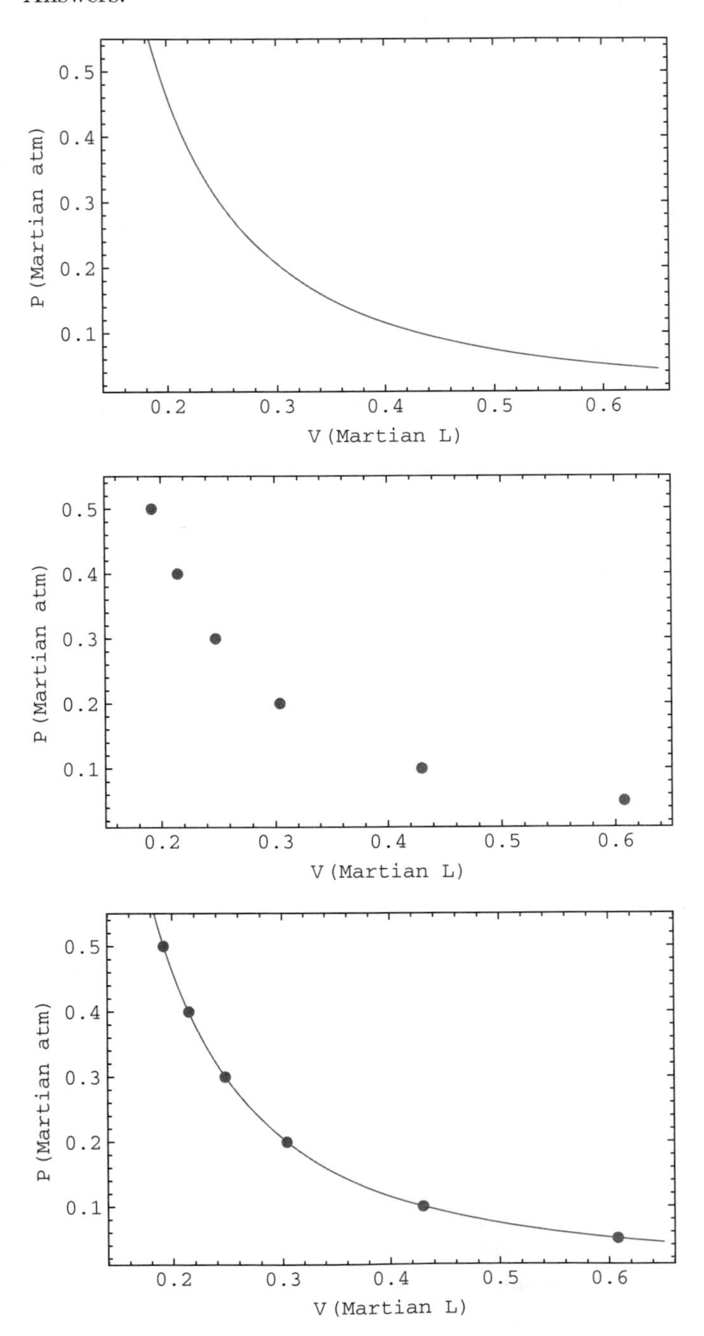

27. (a) Plot the curve representing the best fit obtained in Exercise 20 of Section 2.1.20. (b) Plot the given points. (c) Superpose the two graphs to see if the fit is acceptable.

Answers:

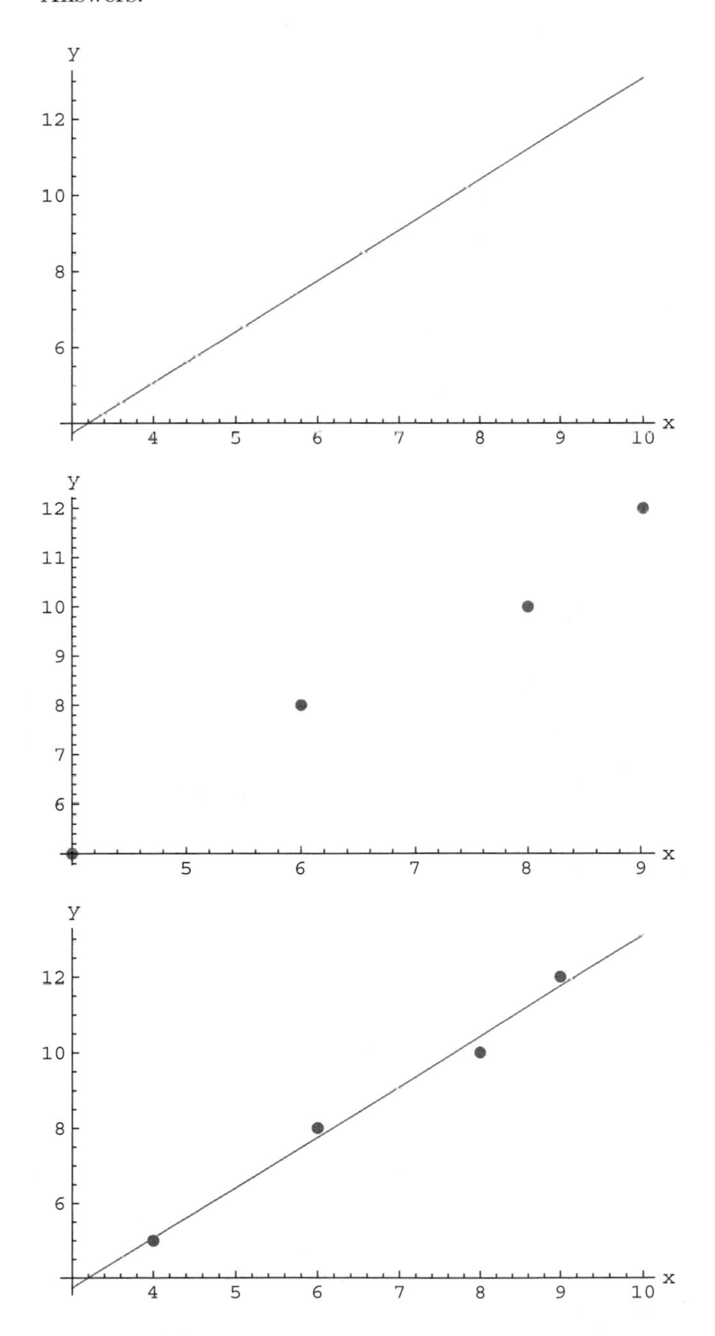

28. A stone falls from rest near the earth's surface. The following table provides the data for distance d traveled versus time t:

t(s)	d(ft)
0	0
0.5	4.2
1	16.1
1.5	35.9
2.0	64.2

(a) Using the least-squares method, determine the magnitude of the free-fall acceleration. (b) Without *first* plotting the data and the curve representing the best fit separately, plot the data and the best-fit curve together.
Answers:

The magnitude of the free-fall acceleration is $32.1\,\mathrm{ft/s^2}$.

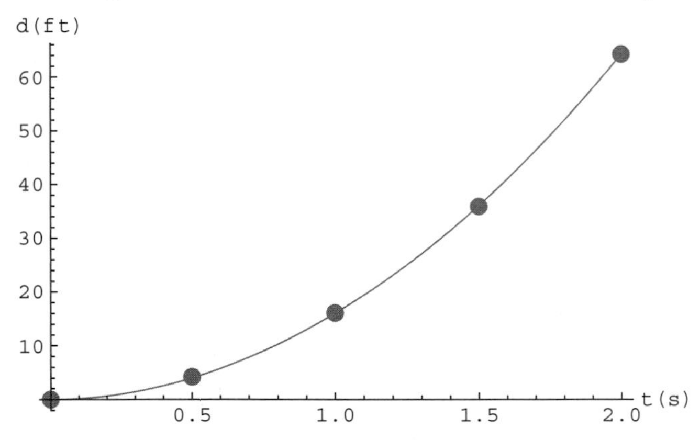

29. (a) Using **Table**, create a list of 10 equally spaced numbers from 0 to 2π, and name it **xlist**. *Hint:* Use steps $2\pi/9$. (b) Create a list of the sines of the numbers in **xlist**, and name it **ylist**. (c) Generate the list of $\{xlist_i, ylist_i\}$. *Hint:* Use **Transpose**. (d) Plot the points $\{xlist_i, ylist_i\}$, and generate the graph shown. *Hint:* Use **ListPlot**, and specify values for the options **PlotStyle** and **AxesLabel**.

30. The surface tension σ of water against air is a linear function of temperature T:

$$\sigma = a + bT$$

where a and b are constants. An experiment provides the following data:

T(°C)	σ(dyne/cm)
−8	77.0
−5	76.4
0	75.6
5	74.9
10	74.22
15	73.49
18	73.05
20	72.75
30	71.18
40	69.56
50	67.91
60	66.18
70	64.4
80	62.6
100	58.9

(a) Using the least-squares method, determine the constants a and b. (b) Without *first* plotting the data and the line representing the best fit *separately*, plot the data and the best-fit line together.

Answers:

$a = 75.9$ dyne/cm and $b = -0.165$ dyne/cm°C.

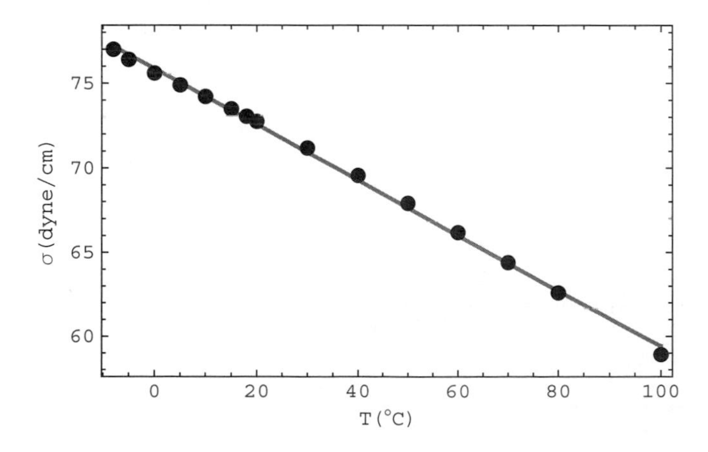

31. (a) Using the method of least squares, fit the function

$$f(x) = a_0 + a_1 x + a_2 x^2 + a_3 x^3$$

to the data

```
{{0, 1}, {0.1, 1.05409}, {0.2, 1.11803}, {0.3, 1.19523},
 {0.4, 1.29099}, {0.5, 1.41421}, {0.6, 1.58114},
 {0.7, 1.82574}, {0.8, 2.23607}, {0.9,  3.16228}}
```

(b) Plot the curve representing the best fit. (c) Plot the data. (d) Superpose the two graphs to see if the fit is acceptable.

Answers:

```
0.95346 + 1.97667 x - 5.87188 x^2 + 7.00395 x^3
```

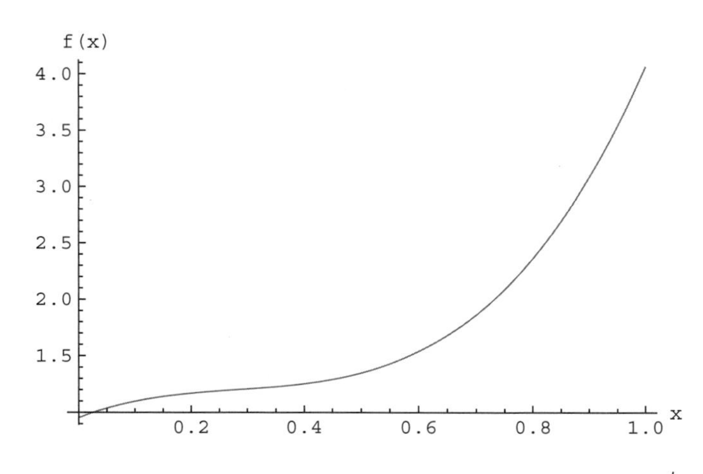

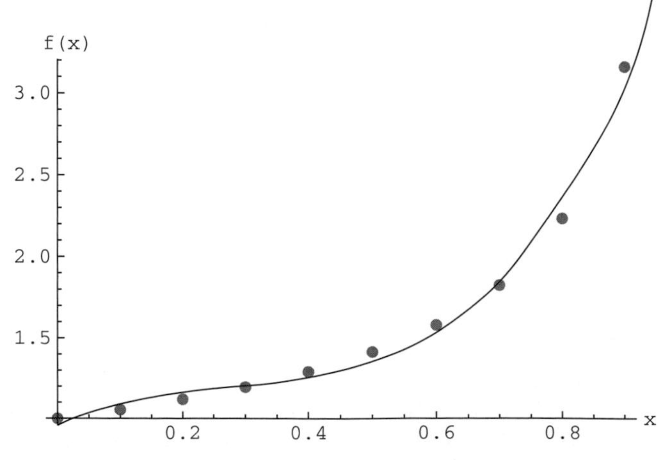

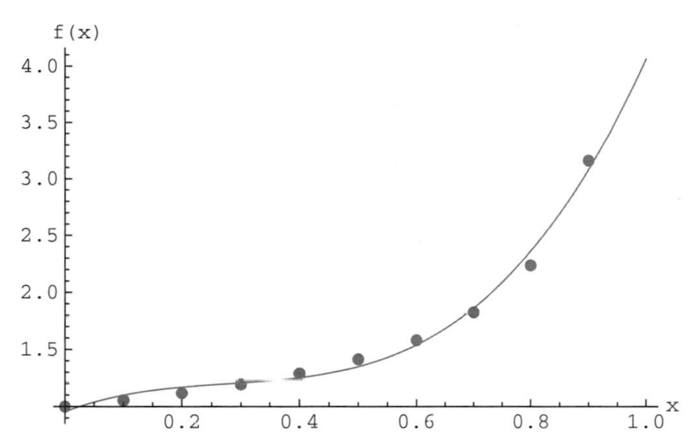

32. An astroid or a hypocycloid of four cusps is a curve traced out by a fixed point on a circle rolling inside another circle when the ratio of the radii is 1 to 4. Plot the astroid whose parametric equations are

$$x = 4\cos^3\theta$$
$$y = 4\sin^3\theta$$

Answer:

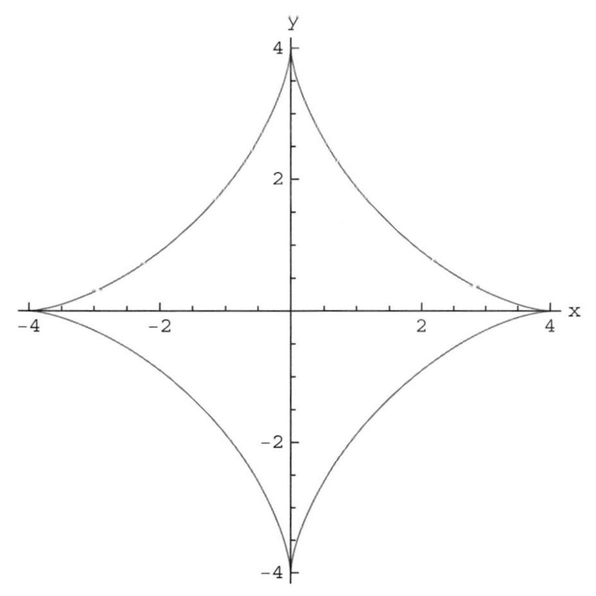

33. The curve traced out by a point on the plane of a circle rolling along a straight line is called a trochoid. If the straight line is along the x-axis and the point is at a

distance d from the center of the circle of radius r, the parametric equations of the trochoid are

$$x = r\theta - d\sin\theta$$

$$y = r - d\cos\theta$$

Plot the trochoid for (a) $d = 0.5r$ and (b) $d = 1.5r$. (The curves in parts (a) and (b) are also known as a curtate cycloid and a prolate cycloid, respectively.) *Hint:* Specify appropriate values for **PlotRange**, **Ticks**, **AxesLabel**, **ImageSize**, and **PlotStyle**.

Answers:

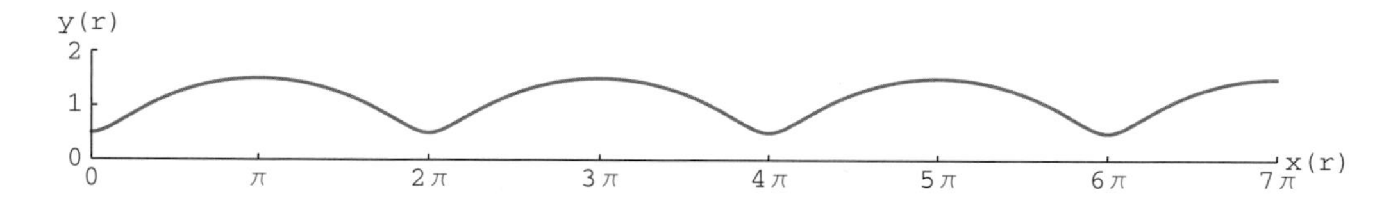

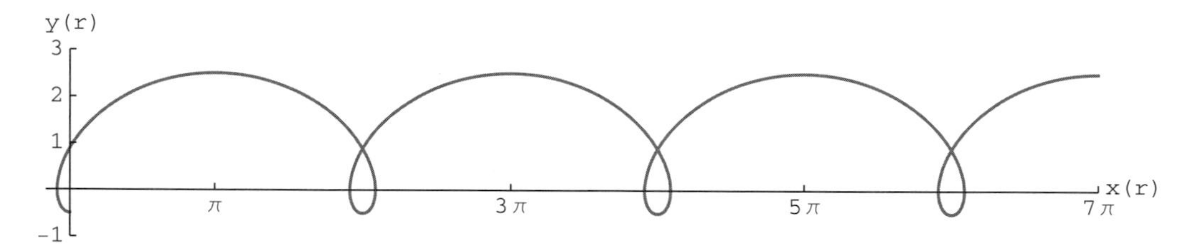

34. The equation of an ellipse is

$$\frac{x^2}{a^2} + \frac{y^2}{b^2} = 1$$

where $a = 2b$. Plot the ellipse exhibiting its true shape.

Answer:

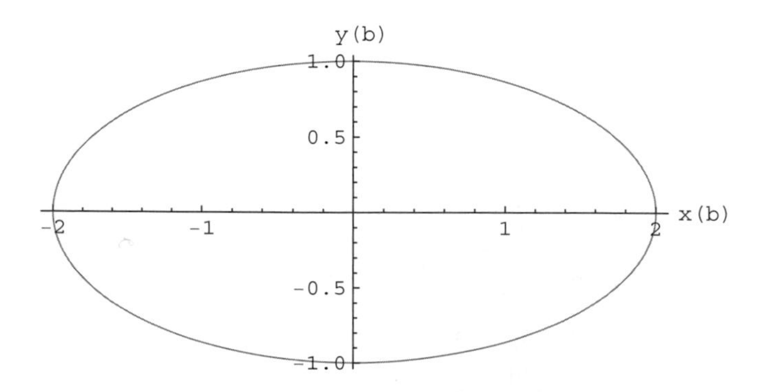

35. Plot

$$\frac{x^2}{3} + \frac{y^2}{5} = 1$$

Hint: Use **ParametricPlot** and be mindful to preserve the true shape of the curve.

Answer:

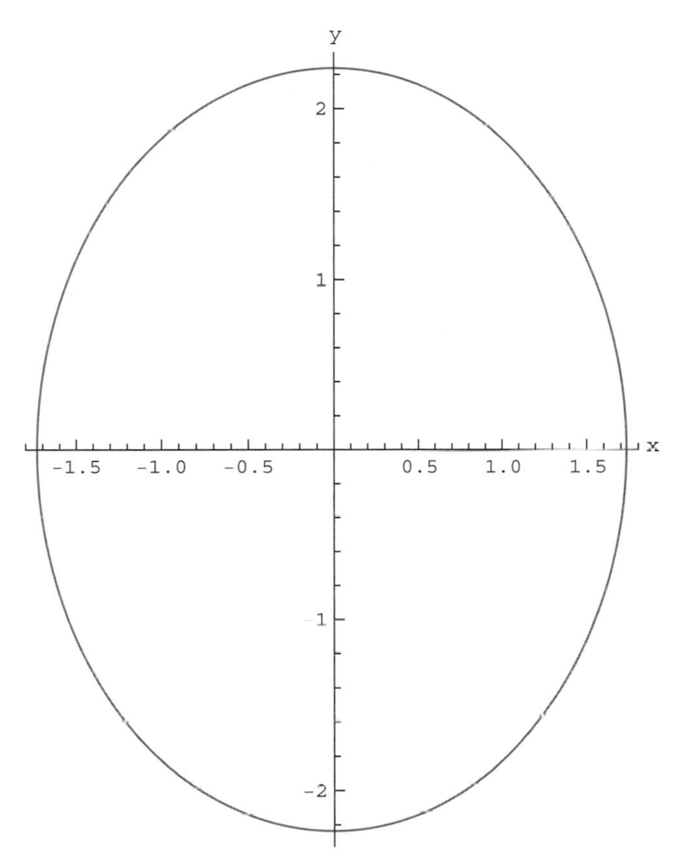

36. The van der Pol equation can be written as

$$\frac{d^2x}{dt^2} + \varepsilon(x^2 - a^2)\frac{dx}{dt} + \omega_0^2 x = 0$$

(For more information on the van der Pol oscillator, see [TM04]) Let $a = 1$, $\omega_0 = 1$, $\varepsilon = 0.05$, and $x'(0) = 0$. Solve the equation and plot the phase diagram (i.e., $x'(t)$ versus $x(t)$) for t from 0 to 100 when (a) $x(0) = 1.0$ and (b) $x(0) = 3.0$.

Answers:

```
{{x→InterpolatingFunction[{{0., 100.}}, <>[]}}
```

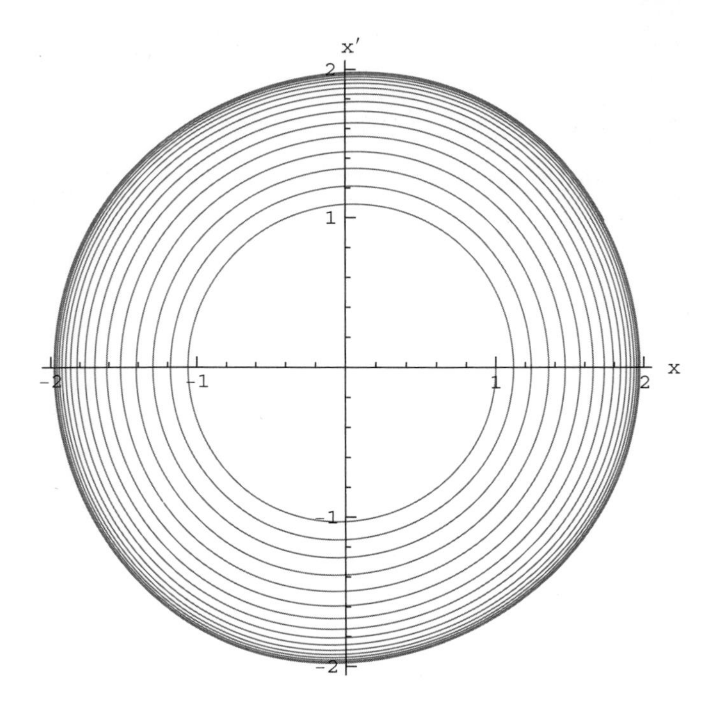


```
{{x→InterpolatingFunction[{{0., 100.}}, <>]}}
```

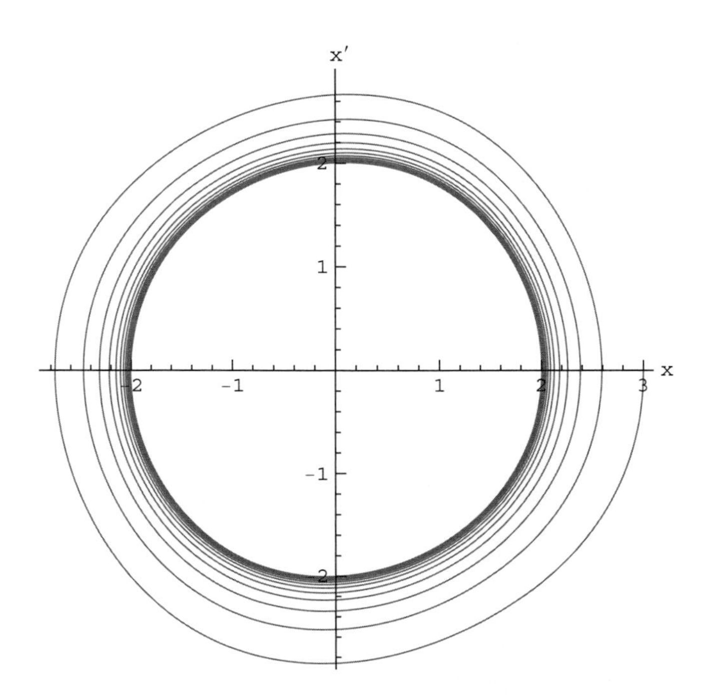

*37. The equation of motion of the van der Pol oscillator can be written as

$$\frac{d^2x}{dt^2} - \varepsilon(1 - x^2)\frac{dx}{dt} + x = 0$$

where ε is a positive parameter. This second-order differential equation can be reduced to two first-order equations:

$$\frac{dx}{dt} = v$$

$$\frac{dv}{dt} = \varepsilon(1 - x^2)v - x$$

Consider the case in which $\varepsilon = 0.5$. Solve these coupled differential equations numerically and plot the phase diagrams (i.e., $v(t)$ versus $x(t)$) with t ranging from 0 to 50 and for five sets of initial conditions: $v(0) = 0$ and $x(0)$ varies from 1 to 3 in steps of 0.5. Repeat with several other sets of randomly selected initial conditions. From these phase diagrams, infer the meaning of "limit cycle."

Answer:

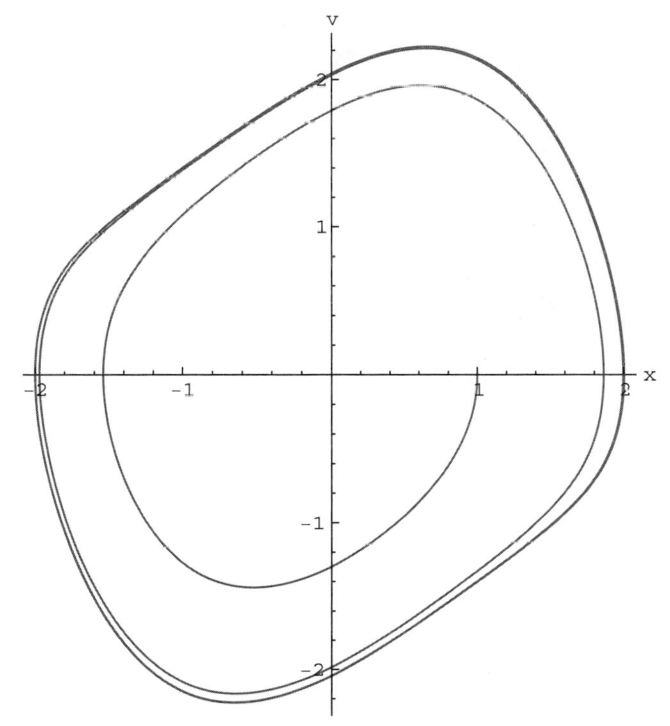

Only one of the five graphs is shown here.

38. Plot the hyperbolic paraboloid given by the equation

$$\frac{x^2}{a^2} - \frac{y^2}{b^2} = 2cz$$

with $a = b = 1$ and $c = 1/2$.

Answer:

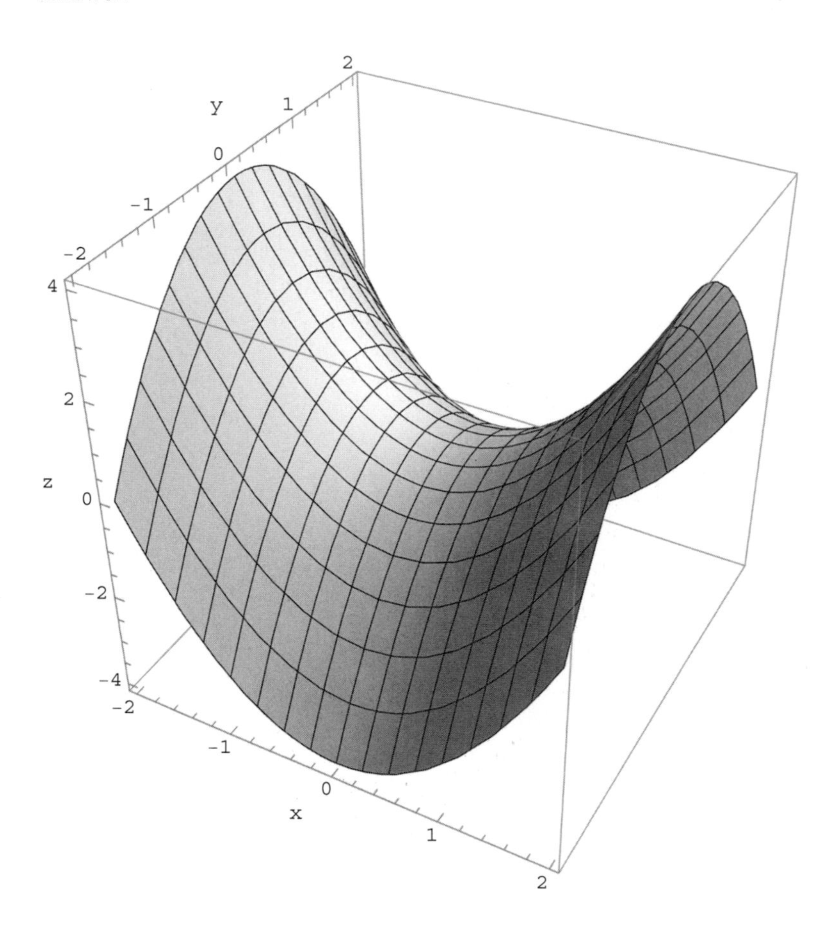

39. Plot the elliptic paraboloid given by the equation

$$\frac{x^2}{a^2} + \frac{y^2}{b^2} = 2cz$$

with $a = b = 1$ and $c = 1/2$.

Answer:

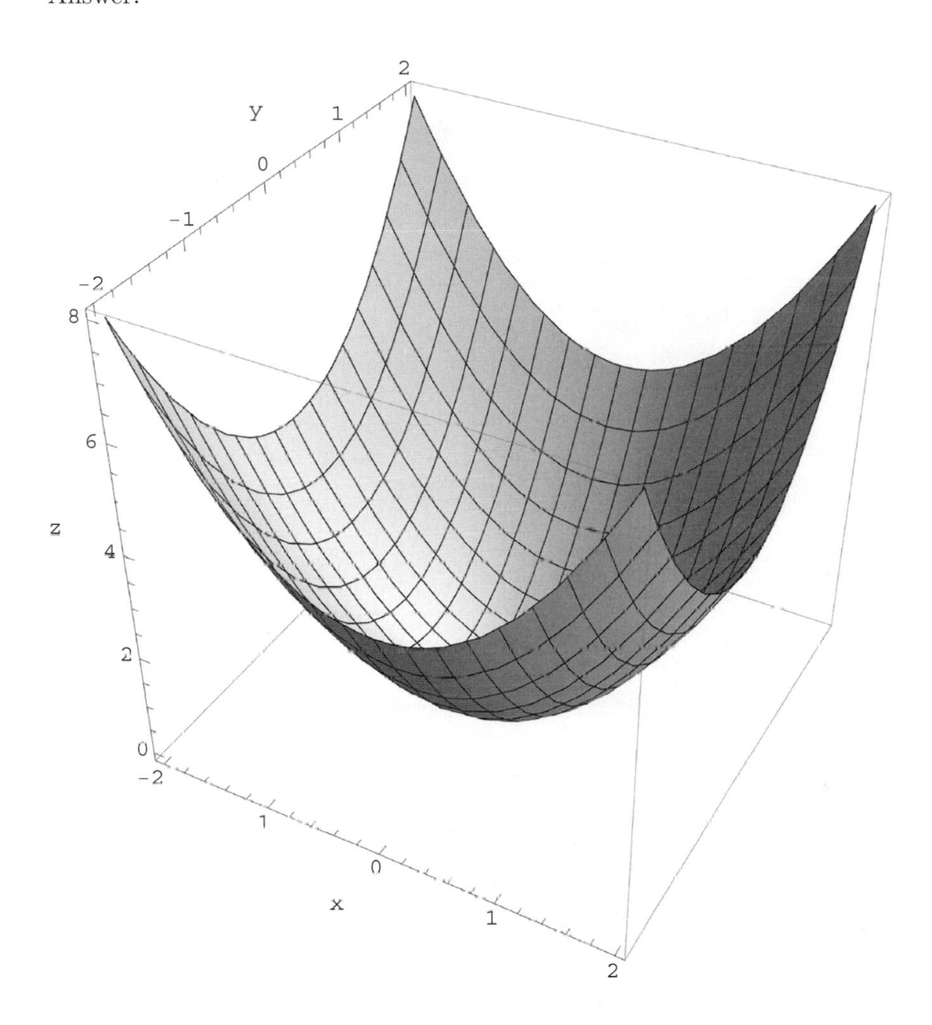

*40. (a) Plot (i.e., graph) the surface defined by the equation

$$z = \ln(x^2 + y^2)$$

with both x and y varying from -15 to 15. Label the axes. *Hint:* Specify values for the options **`PlotRange`**, **`PlotPoints`**, **`BoxRatios`**, **`AxesLabel`**, and **`ViewPoint`**. (b) The surface generated in part (a) lacks the rotational symmetry about the z-axis that the defining equation implies. Using **`ParamertricPlot3D`**, plot the surface showing the rotational symmetry.

Answers:

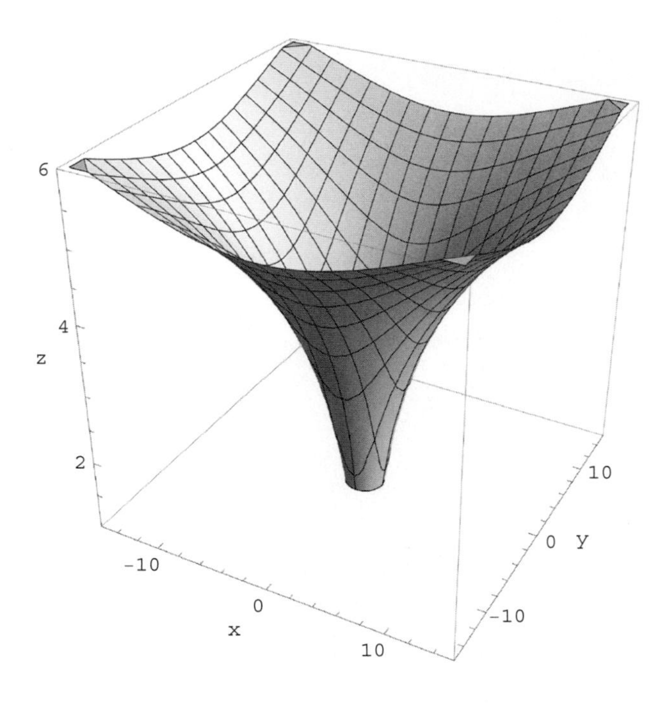

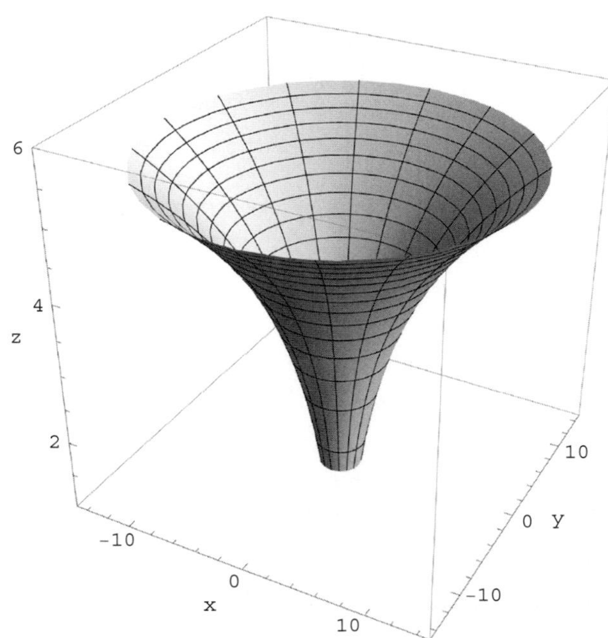

41. Plot (i.e., graph) the function

$$f(x, y) = \sin(x + y)$$

Label the axes.

Answer:

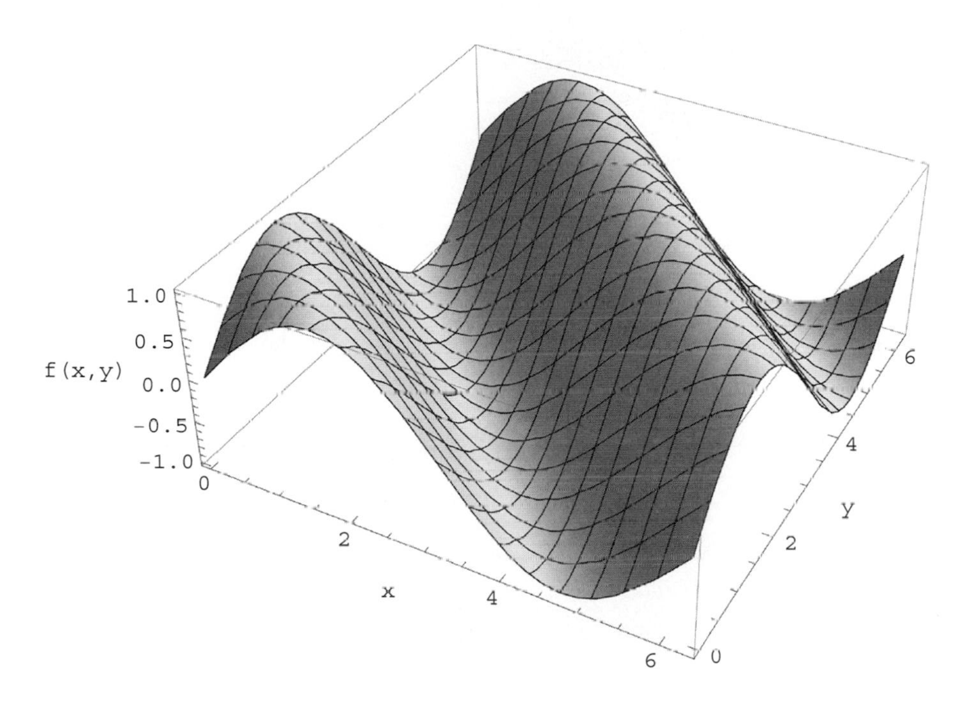

42. In terms of the reduced dimensionless variables, the van der Waals equation can be written as

$$p = \frac{8T}{3V - 1} - \frac{3}{V^2}$$

Plot p versus V from 0.3 to 2 and T from 0.8 to 1.2. Specify the following options:

```
PlotRange→{0, 1.75}, AxesLabel→{"V", "T", "p"},
PlotPoints→25, Mesh→20, ViewPoint→{1.437, -2.653, 1.531}
```

Answer:

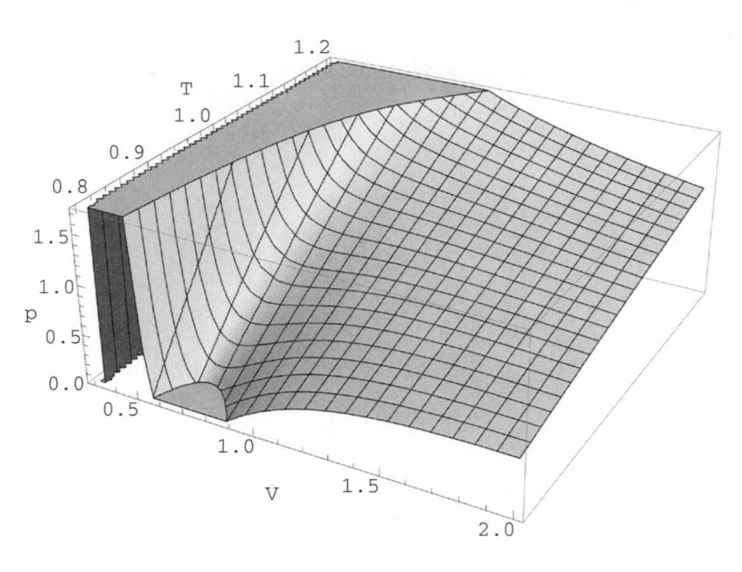

43. Experimenting with various domains and viewpoints, plot the "monkey saddle" defined by the function

$$f(x, y) = xy^2 - x^3$$

to obtain the view

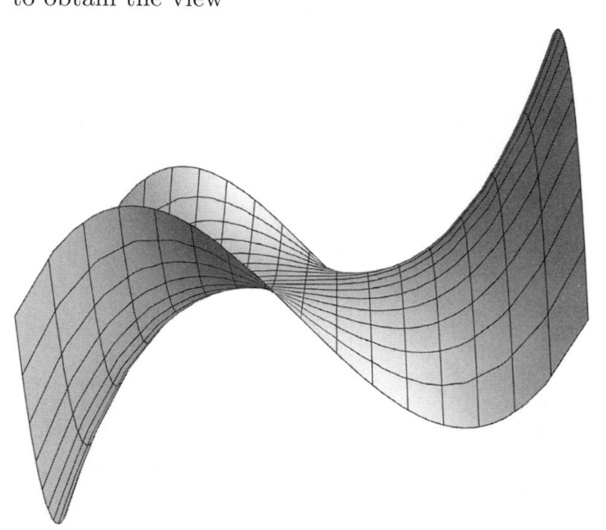

Hint: Specify values for the options **`BoxRatios`**, **`Boxed`**, **`Axes`**, and **`ViewPoint`**.

44. Experimenting with various domains and viewpoints, plot the "dog saddle" defined by the function

$$f(x, y) = xy^3 - yx^3$$

to obtain the view

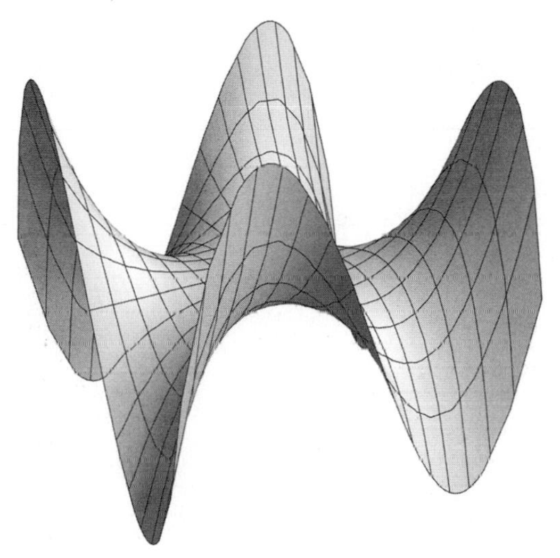

Hint: Specify values for the options **BoxRatios**, **Boxed**, **Axes**, and **ViewPoint**.

45. Plot the function

$$f(x, y) = e^{-(x+y)^2}$$

with both x and y varying from −2 to 2. Label the axes: x, y, and f. Show no tick marks on the f-axis, and put tick marks only at −2, 0, and 2 along the x- and y-axes.

Answer:

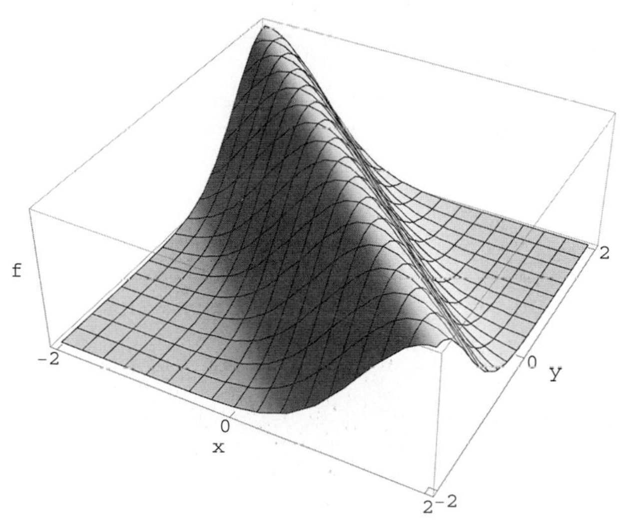

46. In Example 2.3.21 of Section 2.3.4, omit specification for the option **PlotRange**, generate the graphics, and observe the animation. What conclusion can be drawn on the omission of the **PlotRange** specification in generating graphics for animation?
47. The "monkey saddle" is defined by the equation

$$f(x, y) = x(x^2 - 3y^2)$$

Using **Plot3D**, we can plot the surface:

```
Plot3D[x (x^2 - 3 y^2), {x, -2, 2}, {y, -2, 2},
 BoxRatios → {1, 1, 1}, Boxed → False, Axes → False]
```

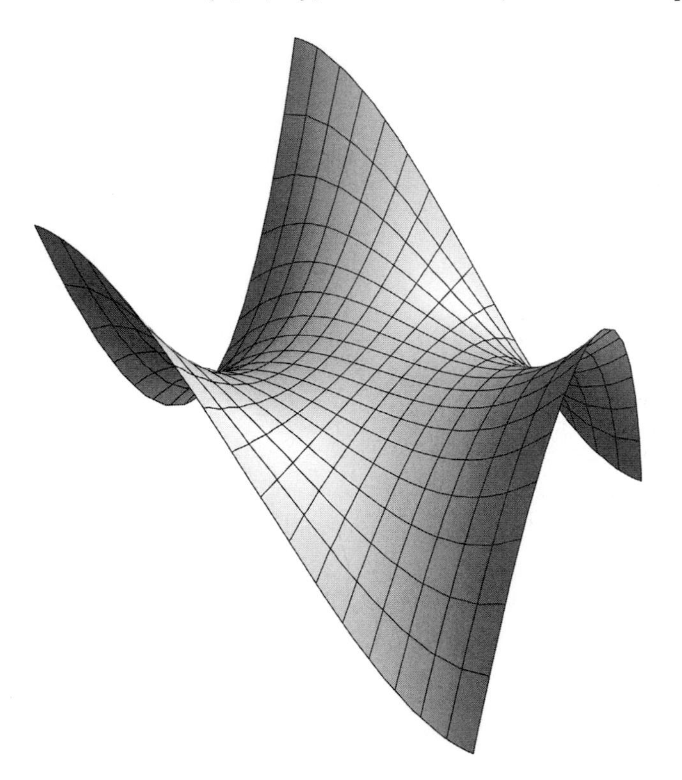

Animate this surface. (*Hint:* Using **Table**, generate a sequence of graphs showing the surfaces defined by the equation

$$f(x, y) = (x + t)(x^2 - 3y^2)$$

where the value of t for each graph varies from -12 to 12 in steps of 1; to animate, use **ListAnimate**.)
48. Two waves of equal frequency and amplitude are moving along a string in opposite directions. If A, k, and ω are, respectively, the amplitude, wave number, and angular

frequency of the waves, the resultant wave is a standing wave described by the equation

$$y(x, t) = 2A \cos kx \cos \omega t$$

Animate the standing wave. (*Hint:* Using **`Table`**, generate a number of graphs for different values of time t; to animate, use **`ListAnimate`**.)

Answer:

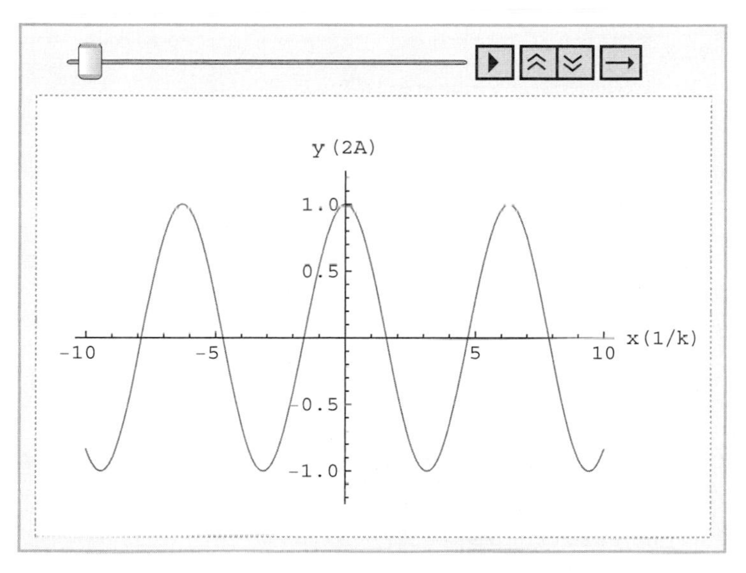

49. Two waves of equal frequency but different amplitudes are moving along a string in opposite directions; their displacements as functions of position x and time t are

$$y_1(x, t) = A \sin(kx - \omega t)$$
$$y_2(x, t) = 0.75A \sin(kx + \omega t)$$

where A is the amplitude of the wave traveling in the positive x direction, k is the wave number, and ω is the angular frequency. The displacement of the resultant wave is

$$y(x, t) = y_1(x, t) + y_2(x, t)$$

This can be written as

$$y(x, t) = \sin(x - t) + 0.75 \sin(x + t)$$

if y, x, and t are in units of A, $1/k$, and $1/\omega$, respectively. Animate the motion of the string. That is, generate a sequence of 16 frames (i.e., plots) corresponding to appropriate values of t, and then animate them to show the wave motion. Let x vary from 0 to 8π. Does the resultant wave have nodes and antinodes?

Answer:

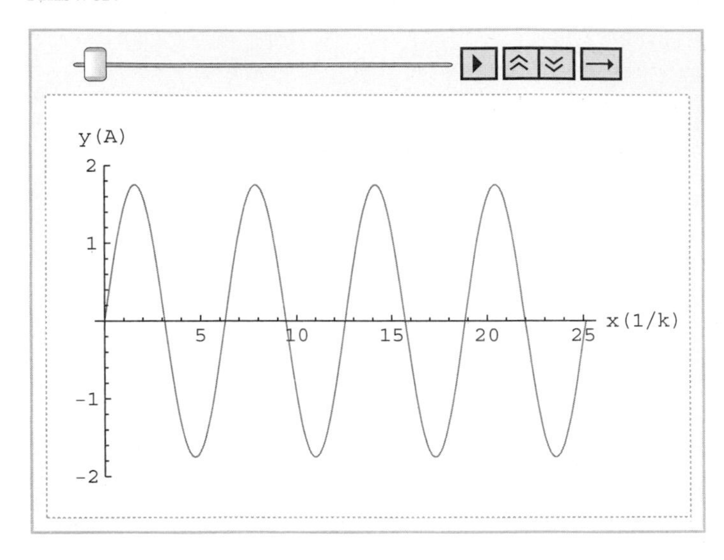

There are antinodes but no nodes.

*50. Plot the ellipsoid given by the equation

$$\frac{x^2}{a^2} + \frac{y^2}{b^2} + \frac{z^2}{c^2} = 1$$

where $a = 2b = 2c$. *Hint:* Use the "front end help" discussed in Section 1.6.3 to obtain information on the function **ParametricPlot3D**.

Answer:

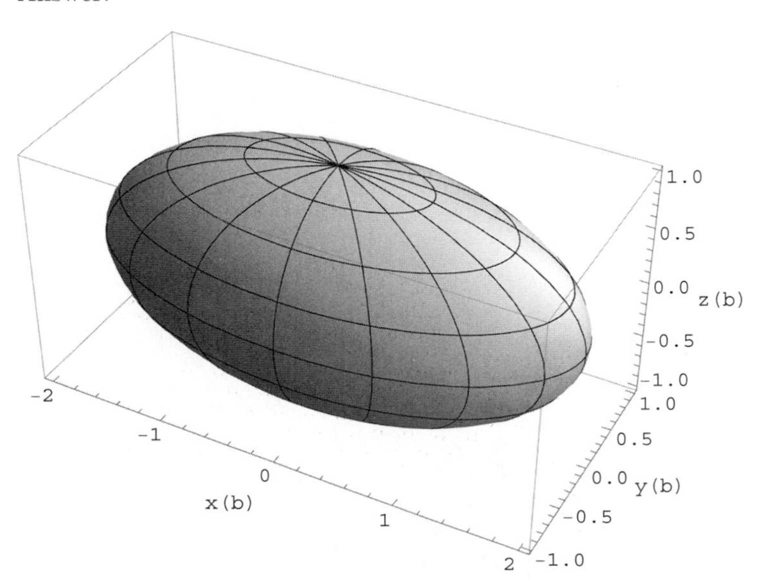

*51. Using **ParametricPlot3D**, plot the elliptic paraboloid

$$x^2 + y^2 = z$$

and the plane

$$x + 3y - 2z + 3 = 0$$

together to create the image

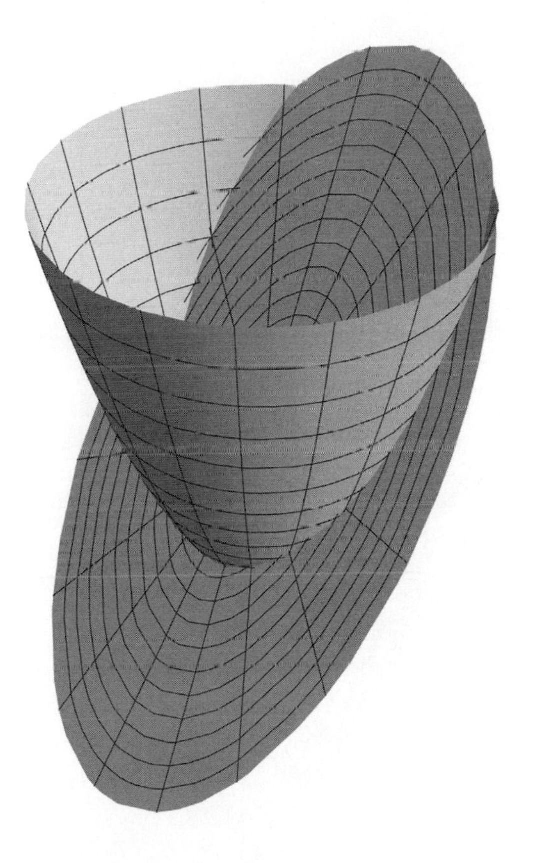

Do *not* use **Show**.

52. Using **ContourPlot3D**, plot a hyperboloid of two sheets whose equation is

$$4x^2 - y^2 - 2z^2 + 4 = 0$$

Specify the options

```
AxesLabel -> {"x", "y", "z"},
ViewPoint → {2.738, 1.865, 0.685}, PlotPoints -> 25
```

Hint: Using the "front end help" discussed in Section 1.6.3, access information on the function **ContourPlot3D**.

Answer:

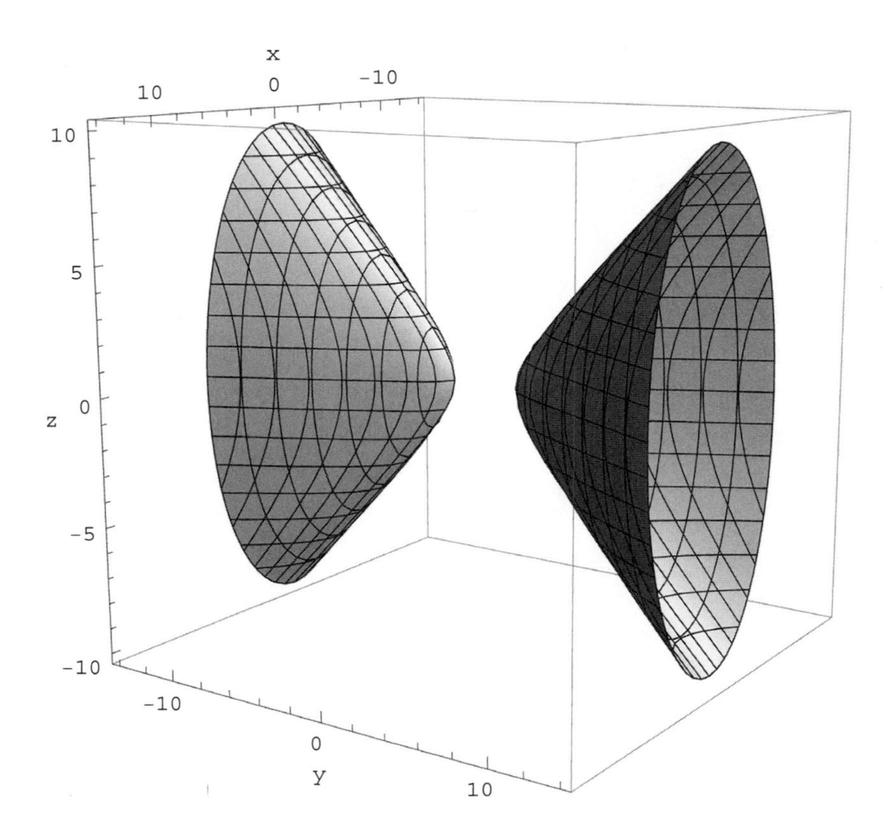

*53. Using **Grid**, display the frames of the animation in Example 2.3.20 in a rectangular array. *Hint:* Use the "front end help" to obtain information on **Grid**, and define

```
g[i_]:=
 Plot[
  {Exp[-16(x-i)^2]+1.5Exp[-(x+i)^2],
   1.5Exp[-(x+i)^2]+3, Exp[-16(x-i)^2+5}, {x, -3.0, 3.0},
  Plotstyle→{{Thickness[0.01], Black},
    {Thickness[0.005], Red}, {Thickness[0.005], Blue}},
  PlotRange→{-0.1, 6.25}, Axes→False,
  Frame -> True, FrameTicks -> None]
```

Answer:

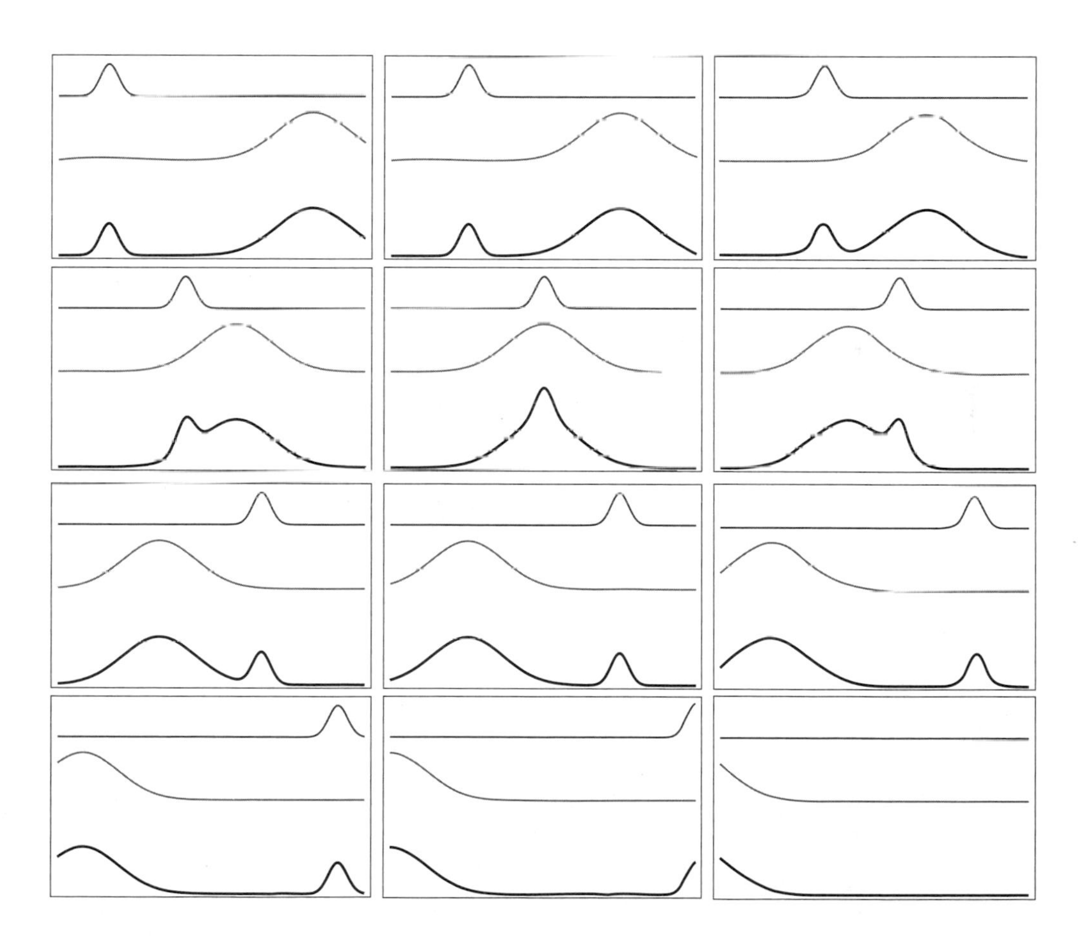

54. Choose **Drawing Tools** and also **Graphics Inspector** in the Graphics menu. Interactively create the following graphic:

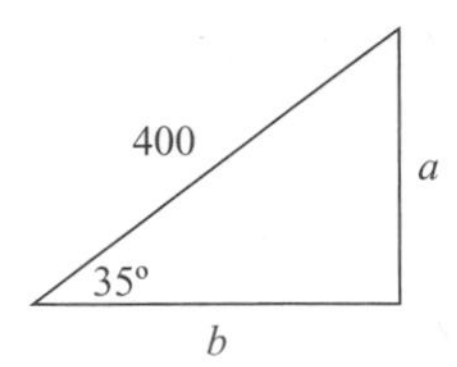

55. This exercise is for users with *Mathematica* 6.0.2. Choose **Drawing Tools** in the Graphics menu. Using the coordinate-picking tool in the 2D Drawing Palette, redo Example 2.3.9. That is, do not use **ClickPane**. *Hint:* Give a name to the nested list of coordinates of the intersections; then use **Table** together with **FindRoot** to determine the solutions.

2.4 LISTS

Section 2.1.11 discussed representations of vectors and matrices by lists and lists of lists, respectively. *Mathematica* employs lists in many other ways. This section provides a general definition of lists and shows how they are created, used, and manipulated.

2.4.1 Defining Lists

Lists are sequences of *Mathematica* objects separated by commas and enclosed by curly brackets. In Section 2.1.11, the objects were numbers or lists of numbers. In general, the objects can be any expressions, such as symbols, character strings, functions, equations, or transformation rules. In a list, the objects are usually, but not necessarily, of the same type.

Built-in *Mathematica* functions provide many examples of lists. These functions often take lists as arguments and return values in terms of lists. For instance, consider using the function **Solve** to solve the system of equations

$$x^2 - 5x - y + 4 = 0$$

$$x - 4y = 1$$

```
In[1]:= Solve[{x^2 - 5 x - y + 4 == 0, x - 4 y == 1}, {x, y}]

Out[1]= {{y → 0, x → 1}, {y → 13/16, x → 17/4}}
```

The first argument of **Solve** is a list of equations; the second argument is a list of variables. The function produces a list of lists of transformation rules for the variables x and y. As another example, **Options[Plot3D]** returns a list of transformation rules for the default options of the function **Plot3D**:

```
In[2]:= Options[Plot3D]
Out[2]= {AlignmentPoint → Center, AspectRatio → Automatic,
         Axes → True, AxesEdge → Automatic, AxesLabel → None,
```

```
AxesStyle→{}, Background→None, BaselinePosition → Automatic,
BaseStyle→{}, BoundaryStyle→GrayLevel[0],Boxed→True,
BoxRatios→{1, 1, 0.4}, BoxStyle→{}, ClippingStyle→Automatic,
ColorFunction→Automatic, ColorFunctionScaling→True,
ColorOutput→Automatic, ContentSelectable→Automatic,
ControllerLinking→Automatic, ControllerMethod→Automatic,
ControllerPath→Automatic, DisplayFunction :→ $DisplayFunction,
Epilog→{}, Evaluated→Automatic, EvaluationMonitor→None,
Exclusions→Automatic, ExclusionsStyle→None,
FaceGrids→None, FaceGridsStyle→{}, Filling→None,
FillingStyle→Opacity[0.5], FormatType :→ TraditionalForm,
ImageMargins→0., ImagePadding→All, ImageSize→Automatic,
LabelStyle→{}, Lighting→Automatic, MaxRecursion→Automatic,
Mesh→Automatic, MeshFunctions→{#1&, #2&}, MeshShading→None,
MeshStyle→Automatic, Method→Automatic, NormalsFunction→Automatic,
PerformanceGoal :→ $PerformanceGoal, PlotLabel→None,
PlotPoints→Automatic, PlotRange→{Full, Full, Automatic},
PlotRangePadding→Automatic, PlotRegion→Automatic,
PlotStyle→Automatic, PreserveImageOptions→Automatic,
Prolog→{}, RegionFunction→(True &), RotationAction→Fit,
SphericalRegion→False, Ticks→Automatic, TicksStyle→{},
ViewAngle→Automatic, ViewCenter→{1/2, 1/2, 1/2}, ViewMatrix→Automatic,
ViewPoint→{1.3, -2.4, 2.}, ViewRange→All, ViewVector→Automatic,
ViewVertical→{0, 0, 1}, WorkingPrecision→MachinePrecision}
```

2.4.2 Generating and Displaying Lists

In addition to the function **Table** introduced in Section 2.1.19, several other functions for generating lists are

Range[*imax*]	generate the list {1, 2, ..., *imax*}
Range[*imin*, *imax*]	generate the list {*imin*, *imin* + 1, ..., *imax*}
Range[*imin*, *imax*, *di*]	generate the list {*imin*, *imin* + *di*, ..., *imax*}
Array[*a*, *n*]	create the list {*a*[1], *a*[2], ..., *a*[*n*]}
Array[*a*, {n_1, n_2, ...}]	create a $n_1 \times n_2 \times \ldots$ nested list, with elements *a*[i_1, i_2, ...]

The functions for displaying *lists* are

ColumnForm[*list*]	display the elements of *list* in a column
MatrixForm[*list*]	display the elements of *list* in a regular array of square cells of the same size
TableForm[*list*]	display the elements of *list* in an array of rectangular cells that are not necessarily of the same size

We can, for example, generate a list of even numbers from 10 to 21 with the function **Range** and display the elements in a column with the function **ColumnForm**:

```
In[1]:= numlist = Range[10, 21, 2]
Out[1]= {10, 12, 14, 16, 18, 20}
```

```
In[2]:= ColumnForm[numlist]
Out[2]= 10
        12
        14
        16
        18
        20
```

For another example, let us use the function **Table** to create the list $\{1 + \sin x, 1 + \sin^3 x, 1 + \sin^5 x, 1 + \sin^7 x\}$:

```
In[3]:= Table[1 + sin [x]^n, {n, 1, 7, 2}]
Out[3]= {1 + Sin[x], 1 + Sin[x]^3, 1 + Sin[x]^5, 1 + Sin[x]^7}
```

We can also create a list of the cubes of the first 10 integers. First define a function for returning the cube of its argument:

```
In[4]:= c[x_] := x^3
```

The function **Array** then gives the list:

```
In[5]:= Array[c, 10]
Out[5]= {1, 8, 27, 64, 125, 216, 343, 512, 729, 1000}
```

For the final example, we generate a 3 × 4 nested list with elements $x^i + y^j$ and display the elements in a rectangular array. The function **Table** generates the nested list:

```
In[6]:= nestlist = Table[x^i + y^j, {i, 3}, {j, 4}]
Out[6]= {{x + y, x + y^2, x + y^3, x + y^4},
         {x^2 + y, x^2 + y^2, x^2 + y^3, x^2 + y^4}, {x^3 + y, x^3 + y^2, x^3 + y^3, x^3 + y^4}}
```

The function **TableForm** displays the nested list in tabular form:

```
In[7]:= TableForm[nestlist]
Out[7]//Tableform=
        x + y      x + y^2      x + y^3      x + y^4
        x^2 + y    x^2 + y^2    x^2 + y^3    x^2 + y^4
        x^3 + y    x^3 + y^2    x^3 + y^3    x^3 + y^4
```

2.4.3 Counting List Elements

The functions for counting list elements are

`Length[`*list*`]`	give the number of elements in *list*
`Count[`*list*`, `*elem*`]`	give the number of times *elem* appears as an element of *list*
`Dimensions[`*list*`]`	give a list of the dimensions of *list*

With the function **`Length`**, we can determine the number of elements in the list **`numlist`** introduced in the preceding section:

```
In[8]:= Length[numlist]
Out[8]= 6
```

The function **`Dimensions`** gives the dimensions of the nested list **`nestlist`** defined in the preceding section:

```
In[9]:= Dimensions[nestlist]
Out[9]= {3, 4}
```

Example 2.4.1 Use the function **`RandomInteger`** to simulate the throwing of a die n times. (a) Plot a bar chart of the number of times each number turns up when $n = 10^2$. (b) Then try $n = 10^3$, 10^4, 10^5, and 10^6. (c) What do these simulations tell us about the meaning of probability?

`RandomInteger[{1, 6}]` returns with equal probability one of the integers from 1 to 6. In other words, it returns the result of a throw with one die. When n is specified, **`RandomInteger[{1, 6},`***n***`]`** returns a list of the results of n throws.

(a) For $n = 10^2$, we have

```
In[10]:= results = RandomInteger[{1, 6}, 10^2]
Out[10]= {2,4,4,5,1,5,2,3,1,1,4,6,3,2,6,2,1,4,3,6,4,2,4,6,1,1,
          1,6,2,5,2,6,5,4,6,2,4,1,1,1,2,1,2,1,1,5,6,5,3,4,2,
          6,6,1,5,2,1,5,1,2,3,1,4,3,4,5,6,5,3,1,5,1,1,6,2,2,
          1,4,6,4,4,4,6,3,3,3,2,1,3,4,2,4,5,5,5,1,1,5,3,5}
```

The function **`Count`** gives, for example, the number of times that ace turned up:

```
In[11]:= Count[results,1]
Out[11]= 24
```

We can make a list of the number of times each number from 1 to 6 turned up:

```
In[12]:= distribution = Table[Count[results, i],{i, 1, 6}]
Out[12]= {24,17,12,17,16,14}
```

To produce a bar chart of this distribution, let us use the function **BarChart** in the package **BarCharts`**. **BarChart[***list***]** generates a bar chart of the data in *list*:

```
In[13]:= Needs["BarCharts`"]

In[14]:= BarChart[distribution]
```

Out[14]=

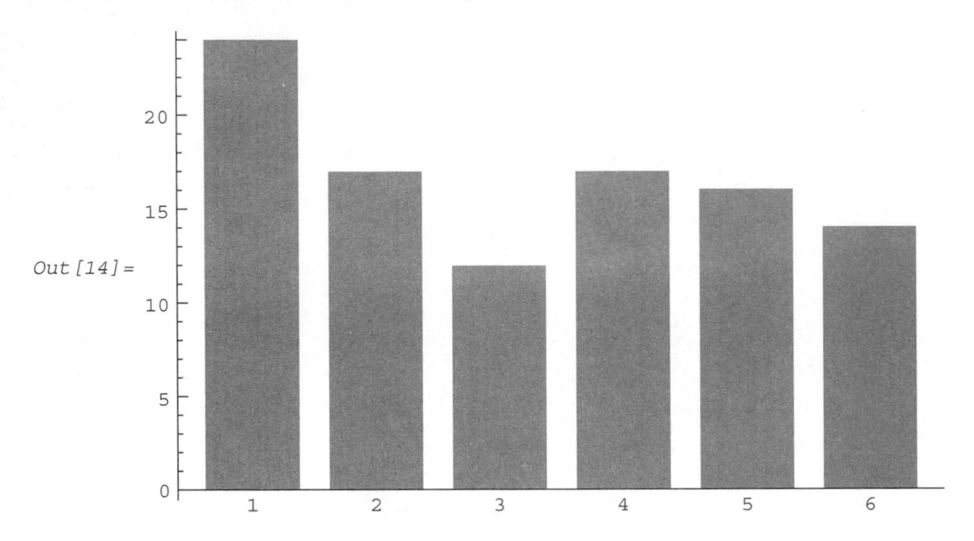

(b) For $n = 10^3$, 10^4, 10^5, and 10^6, we have

```
In[15]:= Do[Print[
           BarChart[Table[Count[RandomInteger[{1, 6}, 10^j], i], {i, 1, 6}],
            PlotLabel -> "n = " <> ToString[10^j]]], {j, 3, 6}]
```

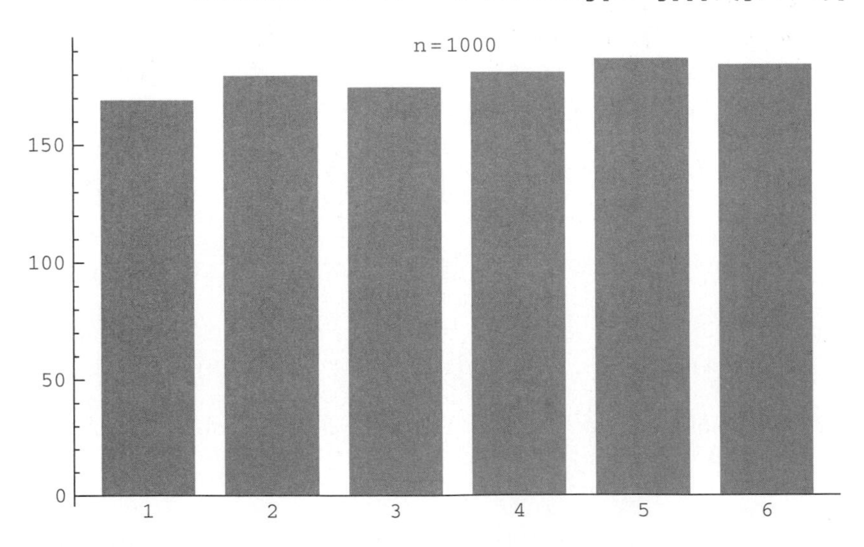

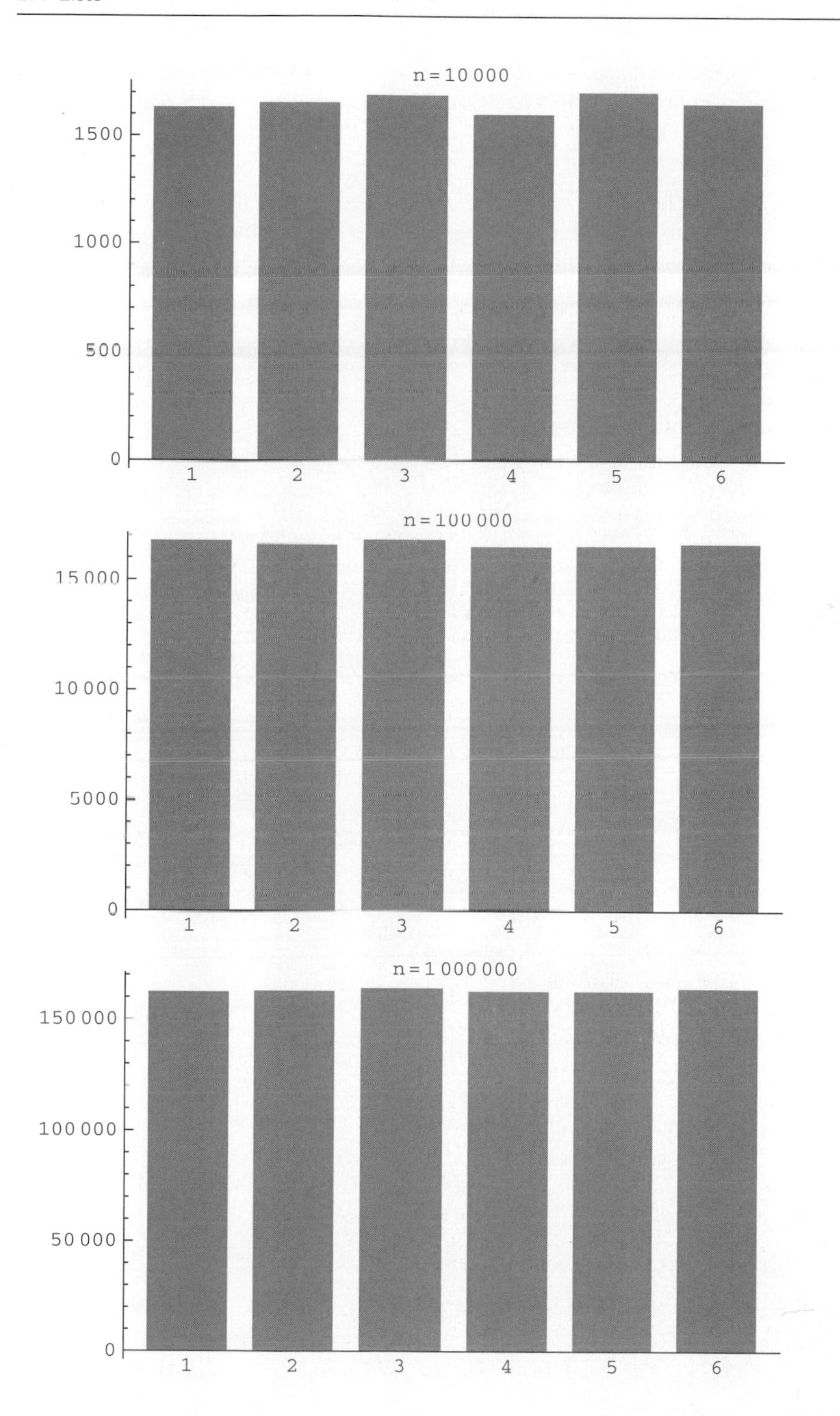
n = 10 000
1500
1000
500
0
1 2 3 4 5 6
n = 100 000
15 000
10 000
5000
0
1 2 3 4 5 6
n = 1 000 000
150 000
100 000
50 000
0
1 2 3 4 5 6

In the `Do` loop, the iteration variable j varies from 3 to 6 in steps of 1. For each value of the variable j, `Print` prints a bar chart of the distribution. `ToString[`*expr*`]` gives a string corresponding to the printed form of *expr*; `"`s_1`" <> "`s_2`" <> ...` or `StringJoin["`s_1`", "`s_2`", ...]` yields a string consisting of a concatenation of the s_i.

(c) One meaning of probability is that it indicates the relative frequency with which a particular event would occur in the long run. Here, for a very great number of throws (approximately 1 million), the distribution became fairly uniform and the fractional number of times that each number turned up approached the value 1/6, the probability of getting a particular number in a single throw.

```
In[16]:= Clear[numlist, c, nestlist, results, distribution]
```

■

2.4.4 Obtaining List and Sublist Elements

The functions for obtaining elements of lists are

`First[`*list*`]`	the first element
`Last[`*list*`]`	the last element
`Part[`*list*`, `*n*`]` or *list*`[[`*n*`]]`	the *n*th element
`Part[`*list*`, -`*n*`]` or *list*`[[-`*n*`]]`	the *n*th element from the end
`Part[`*list*`, {`n_1`, `n_2`, ...}]` or *list* `[[{`n_1`, `n_2`, ...}]]`	the list of the n_1th, n_2th, ... elements
`Take[`*list*`, `*n*`]`	the list of the first *n* elements
`Take[`*list*`, -`*n*`]`	the list of the last *n* elements
`Take[`*list*`, {`*m*`, `*n*`}]`	the list of the *m*th through *n*th elements
`Rest[`*list*`]`	*list* without the first element
`Most[`*list*`]`	*list* without the last element
`Drop[`*list*`, `*n*`]`	*list* without the first *n* elements
`Drop[`*list*`, -`*n*`]`	*list* without the last *n* elements
`Drop[`*list*`, {`*m*`, `*n*`}]`	*list* without the *m*th through *n*th elements

The functions for selecting elements of sublists are

`Part[`*list*`, `*i*`, `*j*`, ...]` or *list*`[[`*i*`, `*j*`, ...]]`	the element at position $\{i, j, \ldots\}$ in *list*
`Extract[`*list*`, {`*i*`, `*j*`, ...}]`	the element at position $\{i, j, \ldots\}$ in *list*—that is, `Part[`*list*`, `*i*`, `*j*`, ...]`
`Extract[`*list*`, {{`i_1`, `j_1`, ...}, {`i_2`, `j_2`, ...}, ...}]`	the list of elements at positions $\{i_1, j_1, \ldots\}$, $\{i_2, j_2, \ldots\}$, ...,—that is, `{Part[`*list*`, `i_1`, `j_1`, ...], Part[`*list*`, `i_2`, `j_2`, ...], ...}`

To illustrate the use of these functions, consider several examples.

Example 2.4.2 (a) Create the list $\{x\cos x, x^2\cos 2x, x^3\cos 3x, x^4\cos 4x, x^5\cos 5x, x^6\cos 6x\}$. (b) Pick out the fourth element. (c) Obtain the list of the third and fifth elements. (d) Get the list with the first three elements. (e) Obtain the list without the second through fourth elements.

(a) Let us generate the list with **Table** and name it **mylist**:

```
In[1]:= mylist = Table[(x^n) cos [n x], {n, 6}]
Out[1]= {x Cos[x], x^2 Cos[2 x], x^3 Cos[3 x], x^4 Cos[4 x], x^5 Cos[5 x], x^6 Cos[6 x]}
```

(b) Now select the fourth element:

```
In[2]:= mylist[[4]]
Out[2]= x^4 Cos[4 x]
```

(c) Here is the list of the third and fifth elements:

```
In[3]:= mylist[[{3, 5}]]
Out[3]= {x^3 Cos[3 x], x^5 Cos[5 x]}
```

(d) The function **Take** returns the list with the first three elements:

```
In[4]:= Take[mylist, 3]
Out[4]= {x Cos[x], x^2 Cos[2 x], x^3 Cos[3 x]}
```

(e) The function **Drop** gives the list without the second through fourth elements:

```
In[5]:= Drop[mylist, {2, 4}]
Out[5]= {x Cos[x], x^5 Cos[5 x], x^6 Cos[6 x]}
In[6]:= Clear[mylist]
```

■

Example 2.4.3 (a) Create a 3 × 4 list of lists with elements c_{ij}. (b) Display the nested list in standard matrix form. (c) Select the third element of the list. (d) Pick out the second element of the first sublist.

(a) The function **Array** generates the nested list with symbolic elements:

```
In[7]:= nestedlist = Array[c, {3, 4}]
Out[7]= {{c[1, 1], c[1, 2], c[1, 3], c[1, 4]},
         {c[2, 1], c[2, 2], c[2, 3], c[2, 4]},
         {c[3, 1], c[3, 2], c[3, 3], c[3, 4]}}
```

(b) The function **MatrixForm** displays the nested list in standard matrix form:

```
In[8]:= MatrixForm[nestedlist]
Out[8]//MatrixForm=

   ( c[1, 1]   c[1, 2]   c[1, 3]   c[1, 4] )
   ( c[2, 1]   c[2, 2]   c[2, 3]   c[2, 4] )
   ( c[3, 1]   c[3, 2]   c[3, 3]   c[3, 4] )
```

(c) Here is the third element of the list:

```
In[9]:= nestedlist[[3]]
Out[9]= {c[3, 1], c[3, 2], c[3, 3], c[3, 4]}
```

(d) Finally, select the second element of the first sublist:

```
In[10]:= nestedlist[[1, 2]]
Out[10]= c[1,2]
```

An equivalent command is

```
In[11]:= Extract[nestedlist,{1, 2}]
Out[11]= C[1, 2]

In[12]:= Clear[nestedlist]
```

■

Example 2.4.4 Using the function **Extract**, generate the list

$$\{\{a, b, c, d\}, e, i, \{i,j\}\}$$

from the list

$$\{\{a, b, c, d\}, \{e,f\}, \{g, h, \{i,j\}\}\}$$

Let us name the second list **seclist**:

```
In[13]:= seclist = {{a, b, c, d}, {e, f}, {g, h, {i, j}}}
Out[13]= {{a, b, c, d}, {e, f}, {g, h, {i, j}}}
```

Extract generates the desired list:

```
In[14]:= Extract[seclist, {{1}, {2, 1}, {3, 3, 1}, {3, 3}}]
Out[14]= {{a, b, c, d}, e, i, {i, j}}
```

Even though it is a bit cumbersome, we can also use the function **Part** to generate the list:

```
In[15]:= {seclist[[1]], seclist[[2, 1]], seclist[[3, 3, 1]], seclist[[3, 3]]}
Out[15]= {{a, b, c, d}, e, i, {i, j}}

In[16]:= Clear[seclist]
```

■

(To explore additional features of the function **Part**, see Exercises 7 and 8 of Section 2.4.12.)

The functions in this section pick out elements according to their positions in the list. The function **Position** allows us to determine the positions of the elements. **Position**[*list*, *elem*] gives a list of the positions at which *elem* appears. For example, we can determine the positions of the symbol a in the list $\{a, \{a, b, a\}, b, \{\{a, b\}, c\}\}$:

```
In[17]:= Position[{a, {a, b, a}, b, {{a, b}, c}}, a]
Out[17]= {{1}, {2, 1}, {2, 3}, {4, 1, 1}}
```

Note that **Extract** and **Position** use the same form for position specifications, whereas **Part** uses a different one.

So far, we have considered functions that relate to the positions of elements in lists. We now examine a function that selects list elements based on their properties rather than positions:

Select[*list*, *criterion*]	pick out all elements e_i of list for which *criterion*[e_i] is **True**
Select[*list*, *criterion*, *n*]	pick out the first n elements for which *criterion*[e_i] is **True**

Built-in predicates can be used as criteria for the **Select** function. A predicate is a function that returns **True** or **False** in testing an element. Some built-in predicates are

ArrayQ	**NumericQ**	**IntegerQ**	**PolynomialQ**
EvenQ	**OrderedQ**	**ListQ**	**PrimeQ**
LetterQ	**Positive**	**MachineNumberQ**	**UpperCaseQ**
LowerCaseQ	**StringQ**	**Negative**	**VectorQ**
MatrixQ	**ValueQ**	**NumberQ**	
NonNegative	**DigitQ**	**OddQ**	

To find out the properties that they test, use on-line help:

```
In[18]:= ?EvenQ
```

EvenQ[*expr*] gives True if *expr* is an even integer, and False otherwise. >>

Let us illustrate the use of **Select** with several examples.

```
In[19]:= Select[{1, 2, 3, 4, 5, 6}, EvenQ]
Out[19]= {2, 4, 6}
```

```
In[20]:= Select[{1, 2, 3, 4, 5, 6}, EvenQ, 2]
Out[20]= {2, 4}
```

Positive[x] gives **True** if x is a positive number.

```
In[21]:= Select[{a, 4 + 5 I, x^2, -2, 3, -5, 2/3}, Positive]
```

Out[21]= $\{3, \frac{2}{3}\}$

VectorQ[*expr*] gives **True** if *expr* is a list, none of whose elements are themselves lists, and gives **False** otherwise.

```
In[22]:= Select[{{a, 2 - 3 I, 7}, x, {-2, {1, -5}}, 2/3}, VectorQ]
Out[22]= {{a, 2 - 3 i, 7}}
```

The only element that is a list without sublists is **{a, 2 - 3 I, 7}**.

2.4.5 Changing List and Sublist Elements

Some functions for inserting, deleting, and replacing list and sublist elements are

Prepend[*list*, *elem*]	insert *elem* at the beginning of *list*
Append[*list*, *elem*]	insert *elem* at the end of *list*
Insert[*list*, *elem*, *i*]	insert *elem* at position *i* in *list*
Insert[*list*, *elem*, {*i*, *j*, ...}]	insert *elem* at position {*i*, *j*, ...} in *list*
Insert[*list*, *elem*, {{i_1, j_1, ...}, {i_2, ...}, ...}]	insert *elem* at positions {i_1, j_1, ... },{i_2, ...}, ... in *list*
Delete[*list*, *i*]	delete the element at position *i* in *list*
Delete[*list*, {*i*, *j*, ...}]	delete the element at position {*i*, *j*, ...} in *list*
Delete[*list*, {{i_1, j_1, ...}, {i_2, ...}, ...}]	delete elements at positions {i_1, j_1, ...},{i_2, ... }, ... in *list*
ReplacePart[*list*, *elem*, *i*]	replace the element at position *i* in *list* with *elem*
ReplacePart[*list*, *elem*, {*i*, *j*, ...}]	replace the element at position {*i*, *j*, ...} with *elem*
ReplacePart[*list*, *elem*, {{i_1,j_1, ...}, {i_2, ...}, ...}]	replace elements at positions {i_1, j_1, ...}, {i_2, ...}, ... with *elem*

Let us illustrate the use of these functions with an example.

Example 2.4.5 Consider the list $\{a, \{a, b, a\}, b, \{\{a, b\}, c\}, d\}$. In this list, (a) add x at the beginning; (b) delete d; (c) delete c; (d) insert y at positions {2}, {2, 2}, and {4, 1, 2}; and (e) replace a with z.

Let us name the list **ourlist**:

```
In[1]:= ourlist = {a, {a, b, a}, b, {{a, b}, c}, d}
Out[1]= {a, {a, b, a}, b, {{a, b}, c}, d}
```

(a) Prepend **x** to the list:

```
In[2]:= Prepend[ourlist, x]
Out[2]= {x, a, {a, b, a}, b, {{a, b}, c}, d}
```

(b) Delete **d** from the list:

```
In[3]:= Delete[ourlist, 5]
Out[3]= {a, {a, b, a}, b, {{a, b}, c}}
```

(c) Delete **c** from the list:

```
In[4]:= Delete[ourlist, {4, 2}]
Out[4]= {a, {a, b, a}, b, {{a, b}}, d}
```

(d) Insert **y** at several positions:

```
In[5]:= Insert[ourlist, y, {{2}, {2, 2}, {4, 1, 2}}]
Out[5]= {a, y, {a, y, b, a}, b, {{a, y, b}, c}, d}
```

(e) Replace **a** with **z** in the list:

```
In[6]:= ReplacePart[ourlist, z, Position[ourlist, a]]
Out[6]= {z, {z, b, z}, b, {{z, b}, c}, d}
```

Position[ourlist, a] gives a list of positions of the symbol **a** in **ourlist**. Of course, we can also replace **a** with **z** in the list with a transformation rule:

```
In[7]:= ourlist /. a→z
Out[7]= {z, {z, b, z}, b, {{z, b}, c}, d}
```

```
In[8]:= Clear[ourlist]
```

■

2.4.6 Rearranging Lists

Some functions for rearranging lists are

Sort[*list*]	sort the elements of *list* into canonical order
Union[*list*]	give a sorted version of *list*, in which all duplicated elements have been dropped
Reverse[*list*]	reverse the order of the elements in *list*
RotateLeft[*list*]	cycle the elements in *list* one position to the left
RotateLeft[*list*, *n*]	cycle the elements in *list* *n* positions to the left
RotateRight[*list*]	cycle the elements in *list* one position to the right

`RotateRight[`*list*`, `*n*`]`	cycle the elements in *list* *n* positions to the right
`Permutations[`*list*`]`	generate a list of all possible permutations of the elements in *list*

For example, we can sort the elements of the list $\{abd, abc, ab, a, d, 10, a, 10, 5.5, 1/4, 23/2, ada\}$ into a standard order:

```
In[1]:= Sort[{abd, abc, ab, a, d, 10, a, 10, 5.5, 1/4, 23/2, ada}]
```

Out[1]= $\{\frac{1}{4}, 5.5, 10, 10, \frac{23}{2}, a, a, ab, abc, abd, ada, d\}$

Mathematica returns a list in which numbers and symbols are arranged in increasing and alphabetical order, respectively. Furthermore, numbers precede symbols. We can also sort the elements of the list and remove duplicated elements:

```
In[2]:= Union[{abd, abc, ab, a, d, 10, a, 10, 5.5, 1/4, 23/2, ada}]
```

Out[2]= $\{\frac{1}{4}, 5.5, 10, \frac{23}{2}, a, ab, abc, abd, ada, d\}$

For another example, let us rotate the elements of the list of the first 10 integers by one position to the left,

```
In[3]:= RotateLeft[Range[10]]
Out[3]= {2, 3, 4, 5, 6, 7, 8, 9, 10, 1}
```

or three places to the right,

```
In[4]:= RotateRight[Range[10], 3]
Out[4]= {8, 9, 10, 1, 2, 3, 4, 5, 6, 7}
```

For the final example, we give all possible permutations of the list $\{a, b, c\}$:

```
In[5]:= Permutations[{a, b, c}]
Out[5]= {{a, b, c}, {a, c, b}, {b, a, c}, {b, c, a}, {c, a, b}, {c, b, a}}
```

2.4.7 Restructuring Lists

Some functions for restructuring lists are

`Partition[`*list*`, `*n*`]`	partition *list* into nonoverlapping sublists of length *n*
`Partition[`*list*`, `*n*`, `*d*`]`	generate sublists with offset *d*
`Split[`*list*`]`	split *list* into sublists consisting of runs of identical elements
`Transpose[`*list*`]`	transpose the first two levels in *list*
`Flatten[`*list*`]`	flatten out nested lists

`Flatten[list, n]`	flatten out the top n levels
`FlattenAt[list, i]`	flatten out a sublist that appears as the ith element of *list*
`FlattenAt[list, {i, j, ...}]`	flatten out the element of *list* at position `{i, j, ...}`
`FlattenAt[list, {{i1, j1, ...}, {i2, j2, ...}, ...}]`	flatten out elements of *list* at several positions

Let us illustrate the use of these functions with several examples.

Example 2.4.6 Consider the list $\{a, b, c, d, e\}$. (a) Partition the list into sublists of two elements. (b) Form sublists of two elements with an offset of one.

(a) Form sublists of two elements:

```
In[1]:= Partition[{a, b, c, d, e}, 2]
Out[1]= {{a, b}, {c, d}}
```

The unpaired element at the end is discarded.

(b) Use an offset of one for successive sublists:

```
In[2]:= Partition[{a, b, c, d, e}, 2, 1]
Out[2]= {{a, b}, {b, c}, {c, d}, {d, e}}
```

■

Example 2.4.7 Convert the list $\{a, a, a, b, c, c, c, c, d, e, e\}$ into the list $\{\{a, a, a\}, \{b\}, \{c, c, c, c\}, \{d\}, \{e, e\}\}$.

The function **`Split`** effects the desired conversion:

```
In[3]:= Split[{a, a, a, b, c, c, c, c, d, e, e}]
Out[3]= {{a, a, a}, {b}, {c, c, c, c}, {d}, {e, e}}
```

■

Example 2.4.8 (a) Give the list $\{a, \{b, c\}, d, \{a, \{b, \{c, \{e, f\}\}\}\}\}$ a name. (b) Flatten out the list. (c) Flatten out the sublist at position 2. (d) Flatten out the sublist at position $\{4, 2, 2, 2\}$.

(a) Assign the name **`demolist`** to the list:

```
In[4]:= demolist = {a, {b, c}, d, {a, {b, {c, {e, f}}}}}
Out[4]= {a, {b, c}, d, {a, {b, {c, {e, f}}}}}
```

(b) Eliminate nested lists:

```
In[5]:= Flatten[demolist]
Out[5]= {a, b, c, d, a, b, c, e, f}
```

The inner brackets of **`demolist`** are removed.

(c) Flatten out the sublist at position 2 of the list:

```
In[6]:= FlattenAt[demolist, 2]
Out[6]= {a, b, c, d, {a, {b, {c, {e, f}}}}}
```

(d) Flatten out the sublist at position {4, 2, 2, 2} of the list:

```
In[7]:= FlattenAt[demolist, {4, 2, 2, 2}]
Out[7]= {a, {b, c}, d, {a, {b, {c, e, f}}}}
```

```
In[8]:= Clear[demolist]
```

■

Example 2.4.9 We can label the energy eigenstates of the hydrogen atom with four quantum numbers: the principal quantum number n, azimuthal or orbital quantum number l, magnetic quantum number m_l, and spin magnetic quantum number m_s. The allowed values of these quantum numbers are

$$n = 1, 2, 3, \ldots$$

$$l = 0, 1, 2, \ldots, (n-1)$$

$$m_l = -l, -l+1, \ldots, 0, 1, 2, \ldots, +l$$

$$m_s = -\frac{1}{2}, +\frac{1}{2}$$

(a) Using the functions **Table** and **Flatten**, generate a list of possible lists of quantum numbers $\{n, l, m_l, m_s\}$ for $n \leq 3$. (b) Using the function **Select**, pick out those lists in part (a) with $l = 1$.
(a) **Table** creates a nested list of quantum numbers for $n \leq 3$:

```
In[9]:= Table[{n, l, ml, ms}, {n, 3}, {l, 0, n-1}, {ml, -l, l}, {ms, -1/2, 1/2}]
```

$$\text{Out[9]= } \{\{\{\{1, 0, 0, -\tfrac{1}{2}\}, \{1, 0, 0, \tfrac{1}{2}\}\}\}\},$$
$$\{\{\{\{2, 0, 0, -\tfrac{1}{2}\}, \{2, 0, 0, \tfrac{1}{2}\}\}\}, \{\{\{2, 1, -1, -\tfrac{1}{2}\}, \{2, 1, -1, \tfrac{1}{2}\}\},$$
$$\{\{2, 1, 0, -\tfrac{1}{2}\}, \{2, 1, 0, \tfrac{1}{2}\}\}, \{\{2, 1, 1, -\tfrac{1}{2}\}, \{2, 1, 1, \tfrac{1}{2}\}\}\}\},$$
$$\{\{\{\{3, 0, 0, -\tfrac{1}{2}\}, \{3, 0, 0, \tfrac{1}{2}\}\}\}, \{\{\{3, 1, -1, -\tfrac{1}{2}\}, \{3, 1, -1, \tfrac{1}{2}\}\},$$
$$\{\{3, 1, 0, -\tfrac{1}{2}\}, \{3, 1, 0, \tfrac{1}{2}\}\}, \{\{3, 1, 1, -\tfrac{1}{2}\}, \{3, 1, 1, \tfrac{1}{2}\}\}\},$$
$$\{\{\{3, 2, -2, -\tfrac{1}{2}\}, \{3, 2, -2, \tfrac{1}{2}\}\}, \{\{3, 2, -1, -\tfrac{1}{2}\}, \{3, 2, -1, \tfrac{1}{2}\}\},$$

$$\{\{3,2,0,-\tfrac{1}{2}\}\{3,2,0,\tfrac{1}{2}\}\},\ \{\{3,2,1,-\tfrac{1}{2}\},\ \{3,2,1,\tfrac{1}{2}\}\},$$
$$\{\{3,2,2,-\tfrac{1}{2}\},\ \{3,2,2,\tfrac{1}{2}\}\}\}\}\}$$

To create a list of possible lists of quantum numbers, we use **`Flatten`** to flatten out the top three levels of the preceding list:

```
In[10]:= qnlists = Flatten[%, 3]
```

$$\text{Out[10]=}\ \{\{1,0,0,-\tfrac{1}{2}\},\ \{1,0,0,\tfrac{1}{2}\},\ \{2,0,0,-\tfrac{1}{2}\},\ \{2,0,0,\tfrac{1}{2}\},$$
$$\{2,1,-1,-\tfrac{1}{2}\},\ \{2,1,-1,\tfrac{1}{2}\},\ \{2,1,0,-\tfrac{1}{2}\},\ \{2,1,0,\tfrac{1}{2}\},$$
$$\{2,1,1,-\tfrac{1}{2}\},\ \{2,1,1,\tfrac{1}{2}\},\ \{3,0,0,-\tfrac{1}{2}\},\ \{3,0,0,\tfrac{1}{2}\},$$
$$\{3,1,-1,-\tfrac{1}{2}\},\ \{3,1,-1,\tfrac{1}{2}\},\ \{3,1,0,-\tfrac{1}{2}\},\ \{3,1,0,\tfrac{1}{2}\},$$
$$\{3,1,1,-\tfrac{1}{2}\},\ \{3,1,1,\tfrac{1}{2}\},\ \{3,2,-2,-\tfrac{1}{2}\},\ \{3,2,-2,\tfrac{1}{2}\},$$
$$\{3,2,-1,-\tfrac{1}{2}\},\ \{3,2,-1,\tfrac{1}{2}\},\ \{3,2,0,-\tfrac{1}{2}\},\ \{3,2,0,\tfrac{1}{2}\},$$
$$\{3,2,1,-\tfrac{1}{2}\},\ \{3,2,1,\tfrac{1}{2}\},\ \{3,2,2,-\tfrac{1}{2}\},\ \{3,2,2,\tfrac{1}{2}\}\}$$

(b) Because none of the built-in predicates introduced in Section 2.4.4 can test whether the second element of a list is equal to 1, we need to define our own predicate:

```
In[11]:= mytest[x_] := x[[2]] == 1
```

`Select` gives a list of lists of quantum numbers with $l = 1$:

```
In[12]:= Select[qnlists, mytest]
```

$$\text{Out[12]=}\ \{\{2,1,-1,-\tfrac{1}{2}\},\ \{2,1,-1,\tfrac{1}{2}\},\ \{2,1,0,-\tfrac{1}{2}\},\ \{2,1,0,\tfrac{1}{2}\},$$
$$\{2,1,1,-\tfrac{1}{2}\},\ \{2,1,1,\tfrac{1}{2}\},\ \{3,1,-1,-\tfrac{1}{2}\},\ \{3,1,-1,\tfrac{1}{2}\},$$
$$\{3,1,0,-\tfrac{1}{2}\},\ \{3,1,0,\tfrac{1}{2}\},\ \{3,1,1,-\tfrac{1}{2}\},\ \{3,1,1,\tfrac{1}{2}\}\}$$

```
In[13]:= Clear[qnlists, mytest]
```

■

2.4.8 Combining Lists

Some functions for combining lists are

`Join``[`$list_1$`, `$list_2$`, ...]`	concatenate lists together
`Union``[`$list_1$`, `$list_2$`, ...]`	give a sorted list of all the distinct elements that appear in any of the $list_i$

Intersection[*list*$_1$, *list*$_2$, ...]	give a sorted list of the elements common to all the *list*$_i$
Complement[*listall*, *list*$_1$, *list*$_2$, ...]	give a sorted list of the elements in *listall* that are not in any of the *list*$_i$

Let us consider an example.

Example 2.4.10 Name the lists $\{1,2,3\}$, $\{2,3,4,5\}$, and $\{1,2,3,4,5,6,7,8\}$ **listA**, **listB**, and **listC**, respectively. Evaluate (a) **Join[listA, listB]**, (b) **Union[listA, listB]**, (c) **Intersection[listA, listB]**, and (d) **Complement[listC, listA, listB]**.

Create and name the lists:

```
In[1]:= listA = Range[3]
Out[1]= {1, 2, 3}
```

```
In[2]:= listB = Range[2, 5]
Out[2]= {2, 3, 4, 5}
```

```
In[3]:= listC = Range[8]
Out[3]= {1, 2, 3, 4, 5, 6, 7, 8}
```

(a)

```
In[4]:= Join[listA, listB]
Out[4]= {1, 2, 3, 2, 3, 4, 5}
```

The lists are joined and the order is preserved.

(b)

```
In[5]:= Union[listA, listB]
Out[5]= {1, 2, 3, 4, 5}
```

(c)

```
In[6]:= Intersection[listA, listB]
Out[6]= {2, 3}
```

(d)

```
In[7]:= Complement[listC, listA, listB]
Out[7]= {6, 7, 8}
```

The functions **Union**, **Intersection**, and **Complement** return sorted lists with duplicated elements removed.

```
In[8]:= Clear[listA, listB, listC]
```

■

2.4.9 Operating on Lists

Built-in mathematical functions in *Mathematica* operate separately on each element of a list. For instance, **D[{x, x^2, x^3, x^4}, x]** differentiates each element of the list:

```
In[1]:= Table[x^n, {n, 4}]
Out[1]= {x, x^2, x^3, x^4}

In[2]:= D[%, x]
Out[2]= {1, 2 x, 3 x^2, 4 x^3}
```

With the list $\{a, b, c\}$ as its argument, the function **Sin** takes the sine of each element:

```
In[3]:= Sin[{a, b, c}]
Out[3]= {Sin[a], Sin[b], Sin[c]}
```

Also, the function **Sqrt** when applied to the list $\{3, 4, 6, 8\}$ returns a list of the square roots:

```
In[4]:= Sqrt[{3, 4, 6, 8}]
```
Out[4]= $\{\sqrt{3}, 2, \sqrt{6}, 2\sqrt{2}\}$

Functions that operate separately on each element of a list have the attribute **Listable**. Not all built-in functions have this attribute. For example, the function **Print** does not have the attribute **Listable**:

```
In[5]:= Print[{a, b, c}]
        {a, b, c}

In[6]:= {Print[a], Print[b], Print[c]}
        a
        b
        c
Out[6]= {Null, Null, Null}
```

Null is a *Mathematica* symbol representing "nothing." **Print[{a, b, c}]** does not return the same result as **{Print[a], Print[b], Print[c]}**. Thus, it is not **Listable**.

User-defined functions are, in general, not **Listable** unless they are assigned this attribute, in a manner to be discussed in Section 3.2.5. For instance, define the function

```
In[7]:= g[x_] := 2
```

and evaluate

```
In[8]:= g[{a, b, c}]
Out[8]= 2
```

The function g does not act on each element separately; it treats the list as a single object; it returns the number 2 instead of the list $\{2,2,2\}$.

```
In[9]:= Clear[g]
```

To apply a function that is not **Listable** to each element of a list, we use the function **Map**. **Map**[*f*, *list*] or *f*/@*list* applies f to each element of *list*. For example, we can apply f to each element of $\{a,b,c\}$:

```
In[10]:= Map[f, {a, b, c}]
Out[10]= {f[a], f[b], f[c]}
```

We can also use the special input form "**/@**":

```
In[11]:= f /@ {a, b, c}
Out[11]= {f[a], f[b], f[c]}
```

Sometimes, it is desirable to apply separately a function that is not **Listable** to only some elements of a list. The function **MapAt**[*f*, *list*, *n*] applies f to the element at position n of *list*. **MapAt**[*f*, *list*, {{n_1}, {n_2}, ...}] applies f to elements at positions $n_1, n_2, \ldots$ of *list*. For instance, apply f separately to elements a and d of the list $\{a,b,c,\{d,e\}\}$:

```
In[12]:= MapAt[f, {a, b, c, {d, e}}, {{1}, {4, 1}}]
Out[12]= {f[a], b, c, {f[d], e}}
```

The positions of **a** and **d** are {1} and {4, 1}, respectively.

It is often necessary to supply the elements of a list as arguments of a function. **Apply**[*f*, {*a*, *b*, ...}] or *f*@@{*a*, *b*, ...} gives *f*[*a*, *b*, ...]. For example, let us add the elements of the list $\{a,b,c,d\}$:

```
In[13]:= Plus @@ {a, b, c, d}
Out[13]= a + b + c + d
```

Plus[a_1, a_2, ..., a_n] returns the sum of $a_1, a_2, \ldots,$ and a_n.

2.4.10 Using Lists in Computations

This section illustrates how lists are used in interactive calculations.

Example 2.4.11 The change in entropy between states a and b of a system is given by

$$S_b - S_a = \int_a^b \frac{dQ}{T}$$

where T is the absolute temperature of the system and dQ is the small amount of heat that enters or leaves the system. The integral is evaluated over any reversible path connecting the two states. For a process in which the pressure is held constant and the temperature varies from T_a to T_b, the equation can be written as

$$S_b - S_a = \int_{T_a}^{T_b} \frac{C_p(T)dT}{T}$$

where $C_p(T)$ is the heat capacity at constant pressure of the system. The following data were obtained for rhombic sulfur:

T(K)	C_p(cal/K)	T(K)	C_p(cal/K)
15	0.311	160	4.123
20	0.605	170	4.269
25	0.858	180	4.404
30	1.075	190	4.526
40	1.452	200	4.639
50	1.772	210	4.743
60	2.084	220	4.841
70	2.352	230	4.927
80	2.604	240	5.010
90	2.838	250	5.083
100	3.060	260	5.154
110	3.254	270	5.220
120	3.445	280	5.286
130	3.624	290	5.350
140	3.795	298.1	5.401
150	3.964		

Determine the change in entropy, and plot C_p versus T.

A reasonable approximation to the preceding integral is given by the trapezoidal rule

$$\sum_i \left(\frac{C_p(T_{i+1})}{T_{i+1}} + \frac{C_p(T_i)}{T_i} \right) \left(\frac{T_{i+1} - T_i}{2} \right)$$

Assign the name **T** to the list of temperatures in units of K:

```
In[1]:= T = {15, 20, 25, 30, 40, 50, 60, 70, 80, 90,
        100, 110, 120, 130, 140, 150, 160, 170, 180, 190, 200,
        210, 220, 230, 240, 250, 260, 270, 280, 290, 298.1};
```

Give the name **Cp** to the corresponding list of heat capacities in cal/K:

```
In[2]:= Cp = {0.311, 0.605, 0.858, 1.075, 1.452, 1.772, 2.084,
          2.352, 2.604, 2.838, 3.060, 3.254, 3.445, 3.624, 3.795,
          3.964, 4.123, 4.269, 4.404, 4.526, 4.639, 4.743, 4.841,
          4.927, 5.010, 5.083, 5.154, 5.220, 5.286, 5.350, 5.401};
```

Let us create the list $\{Cp_i/T_i\}$:

```
In[3]:= CpoverT = Cp/T
Out[3]= {0.0207333, 0.03025, 0.03432, 0.0358333, 0.0363, 0.03544, 0.0347333,
         0.0336, 0.03255, 0.0315333, 0.0306, 0.0295818, 0.0287083,
         0.0278769, 0.0271071, 0.0264267, 0.0257688, 0.0251118, 0.0244667,
         0.0238211, 0.023195, 0.0225857, 0.0220045, 0.0214217, 0.020875,
         0.020332, 0.0198231, 0.0193333, 0.0188786, 0.0184483, 0.0181181}
```

The change in entropy evaluated with the trapezoidal rule is

```
In[4]:= Sum[(CpoverT[[i + 1]] + CpoverT[[i]]) ((T[[i + 1]] - T[[i]])/2),
         {i, Length[T] - 1}]
Out[4]= 7.50452
```

with the units cal/K. (Instead of using the trapezoidal rule, we can also compute an approximation to the integral with the functions **Interpolation** and **Integrate**. See Exercise 31 in Section 2.4.12.)

To obtain the list $\{T_i, Cp_i\}$ from the lists **T** and **Cp**, use the function **Transpose**:

```
In[5]:= Transpose[{T, Cp}]
Out[5]= {{15, 0.311}, {20, 0.605}, {25, 0.858}, {30, 1.075}, {40, 1.452},
         {50, 1.772}, {60, 2.084}, {70, 2.352}, {80, 2.604}, {90, 2.838},
         {100, 3.06}, {110, 3.254}, {120, 3.445}, {130, 3.624},
         {140, 3.795}, {150, 3.964}, {160, 4.123}, {170, 4.269},
         {180, 4.404}, {190, 4.526}, {200, 4.639}, {210, 4.743},
         {220, 4.841}, {230, 4.927}, {240, 5.01}, {250, 5.083}, {260, 5.154},
         {270, 5.22}, {280, 5.286}, {290, 5.35}, {298.1, 5.401}}
```

ListPlot gives a plot of C_p versus T:

```
In[6]:= ListPlot[%, PlotStyle → PointSize[0.015],
         AxesLabel → {"T (K)", "Cp (cal/K)"}]
```

Out[6]=

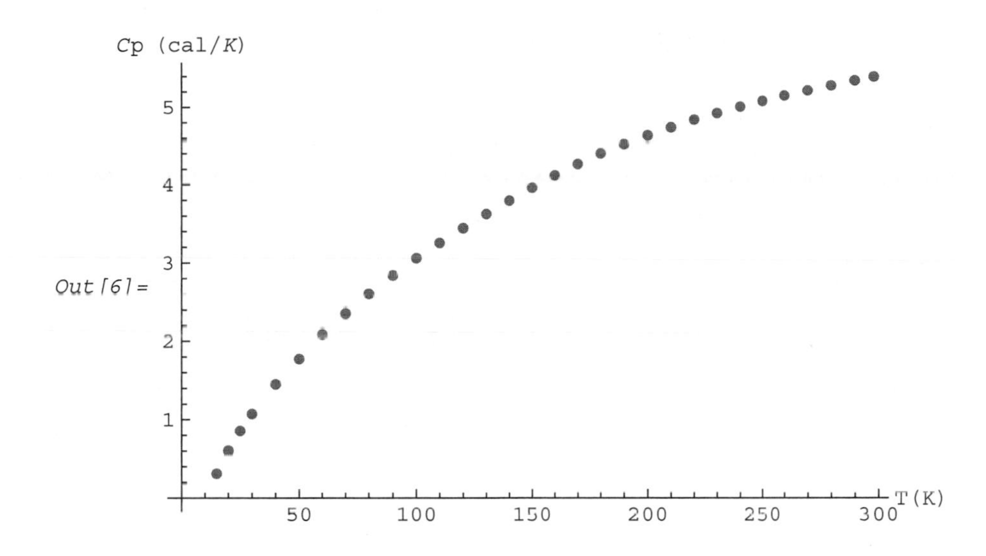


```
In[7]:= Clear[T,Cp]
```

■

Example 2.4.12 Consider the differential equation

$$\frac{dx}{dt} + x = e^{4t}$$

with $x(0) = 1$. (a) Determine $x(t)$ numerically for t in the range 0 to 0.5. (b) Determine $x(t)$ analytically. (c) Compare the numerical and analytical solutions for t from 0 to 0.5 in steps of 0.1.

(a) To solve the differential equation numerically, use the function **NDSolve** discussed in Section 2.1.18:

```
In[8]:= NDSolve[{x'[t] + x[t] == Exp[4t], x[0] == 1}, x, {t, 0, 0.5}]
Out[8]= {{x→InterpolatingFunction[{{0., 0.5}}, <>]}}
```

NDSolve caters to multiple solutions and returns a list of lists of transformation rules. In this case, there is only one solution and one rule. The rule replaces the variable x by the solution that is given in terms of an **InterpolatingFunction**. Let us assign the name **numsoln** to the **InterpolatingFunction**:

```
In[9]:= numsoln = x /. %[[1]]
Out[9]= InterpolatingFunction[{{0., 0.5}}, <>]
```

(b) The function **DSolve** discussed in Section 2.2.14 solves the differential equation analytically:

```
In[10]:= DSolve[{x'[t] + x[t] == Exp[4t], x[0] == 1}, x[t], t]
```

Out[10]= $\left\{\left\{x[t] \to \frac{1}{5} e^{-t}\left(4 + e^{5t}\right)\right\}\right\}$

DSolve also allows multiple solutions and returns a nested list of transformation rules. Here again, there is only one solution and one rule. The solution is given as the value of the replacement rule. Let the solution be the body of the definition for a function named **ansoln**:

```
In[11]:= ansoln[t_] = x[t] /. %[[1]]
```

Out[11]= $\frac{1}{5} e^{-t}\left(4 + e^{5t}\right)$

(c) Now, generate a list of $\{t_i$, numsoln(t_i), ansoln$(t_i)\}$ with the function **Table**:

```
In[12]:= NumberForm[Table[{t, numsoln[t], ansoln[t]}, {t, 0, 0.5, 0.1}], 10]
Out[12]//NumberForm=
        {{0., 1., 1.}, {0.1, 1.022234855, 1.022234874},
         {0.2, 1.100092808, 1.100092788}, {0.3, 1.25667762, 1.256677961},
         {0.4, 1.526862302, 1.526862522}, {0.5, 1.963035787, 1.963035748}}
```

where the numbers are given to 10-digit precision. We can use **TableForm** to display the result in a more familiar format:

```
In[13]:= NumberForm[TableForm[%,
           TableHeadings -> {None, {"t", "numerical", "analytical"}},
           TableSpacing -> {3, 8}], 10]
Out[13]//NumberForm=
```

t	numerical	analytical
0.	1.	1.
0.1	1.022234855	1.022234874
0.2	1.100092808	1.100092788
0.3	1.25667762	1.256677961
0.4	1.526862302	1.526862522
0.5	1.963035787	1.963035748

```
In[14]:= Clear[numsoln, ansoln]
```

■

Example 2.4.13 The plane pendulum consists of a particle of mass m constrained by a rigid and massless rod to move in a vertical circle of radius L, as in Figure 2.4.1.

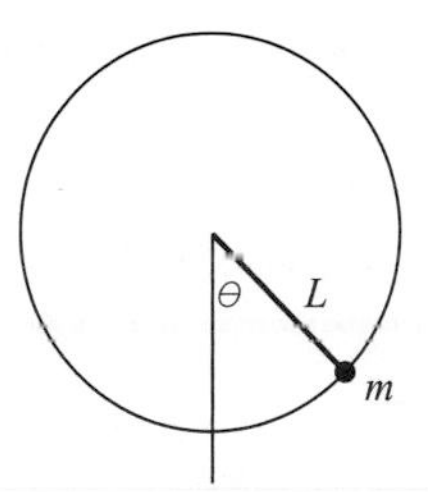

Figure 2.4.1. The plane pendulum.

The equation of motion is

$$\frac{d^2\theta}{dt^2} + \omega_0^2 \sin\theta = 0 \tag{2.4.1}$$

where θ is the angle that the rod makes with the vertical and

$$\omega_0^2 = \frac{g}{L} \tag{2.4.2}$$

In Equation 2.4.2, g is the magnitude of the acceleration due to gravity. If the time t is in units of $1/\omega_0$, Equation 2.4.1 becomes

$$\frac{d^2\theta}{dt^2} + \sin\theta = 0 \tag{2.4.3}$$

Let $\theta(0) = \pi/15$ and $\theta'(0) = 0$. Plot the positions of the particle for t varying from 0 to π in steps of $\pi/8$.

Let us begin by solving Equation 2.4.3 numerically:

```
In[15]:= NDSolve[{θ''[t] + Sin[θ[t]] == 0, θ[0] == π/15, θ'[0] == 0}, θ, {t, 0, π}]
Out[15]= {{θ→InterpolatingFunction[{{0., 3.14159}}, <>]}}
```

As explained in Example 2.4.12, the function **θ** that is the numerical solution of the differential equation can be obtained with the assignment

```
In[16]:= θ = θ /. %[[1]]
Out[16]= InterpolatingFunction[{{0., 3.14159}}, <>]
```

If lengths are in units of L and the fixed end of the rod is the origin, the x and y coordinates of the particle for t from 0 to π in steps of $\pi/8$ are

```
In[17]:= Table[{Sin[θ[t]], -Cos[θ[t]]}, {t, 0, π, π/8}]
Out[17]= {{0.207912, -0.978148}, {0.192402, -0.981316}, {0.147937, -0.988997},
         {0.0807484, -0.996735}, {0.000902549, -1.}, {-0.079086, -0.996868},
         {-0.146674, -0.989185}, {-0.19172, -0.98145}, {-0.207904, -0.978149}}
```

The function **ListPlot** introduced in Section 2.3.1.6 plots the positions of the particle:

```
In[18]:= ListPlot[%, AspectRatio → Automatic, PlotStyle → PointSize[0.03],
          PlotRange → {{-0.5, 0.5}, {0, -1.1}}, AxesLabel → {"x (L)", "y (L)"}]
```

Out[18] =

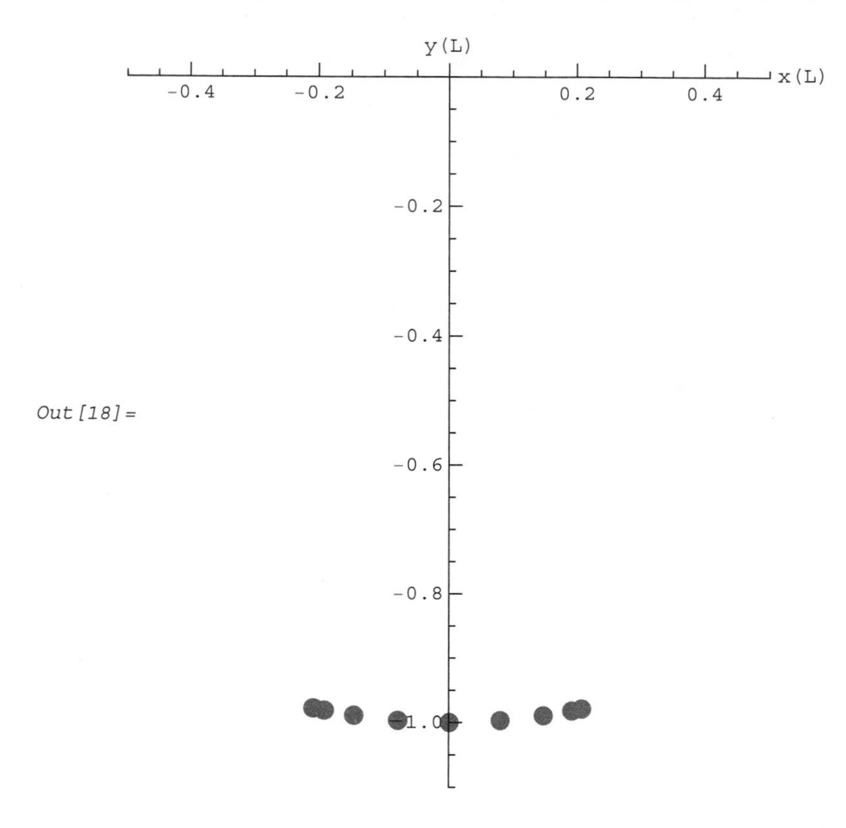


```
In[19]:= Clear[θ]
```

■

Example 2.4.14 Prove that

$$\text{(a)} \quad \mathbf{A} \times (\mathbf{B} \times \mathbf{C}) + \mathbf{B} \times (\mathbf{C} \times \mathbf{A}) + \mathbf{C} \times (\mathbf{A} \times \mathbf{B}) = \mathbf{0}$$

$$\text{(b)} \quad \nabla \cdot (f\mathbf{A}) = f(\nabla \cdot A) + \mathbf{A} \cdot (\nabla f)$$

$$\text{(c)} \quad \nabla\times (\nabla\times \mathbf{A}) = \nabla\, (\nabla\cdot \mathbf{A}) - \nabla^2\mathbf{A}$$

where f is a scalar point function and **A**, **B**, and **C** are vector point functions.

The package **VectorAnalysis`** contains many functions for vector analysis. Some of them are

CoordinateSystem	give the name of the default coordinate system
Coordinates[]	give a list of the default names of the coordinate variables in the default coordinate system

`SetCoordinates[`*coordsys*`]`	set the default coordinate system to be *coordsys* with default variable names
`SetCoordinates[`*coordsys*`[`*varnames*`]]`	set the default coordinate system to be *coordsys* with variable names *varnames*
`DotProduct[`v_1`,` v_2`]`	compute the dot product (sometimes called inner product or scalar product) of the three-dimensional vectors v_1 and v_2 in the default coordinate system
`CrossProduct[`v_1`,` v_2`]`	compute the cross product (sometimes called vector product) of the three-dimensional vectors v_1 and v_2 in the default coordinate system
`Div[`*v*`]`	give the divergence of the vector point function *v* in the default coordinate system
`Curl[`*v*`]`	give the curl of the vector point function *v* in the default coordinate system
`Grad[`*f*`]`	give the gradient of the scalar point function *f* in the default coordinate system
`Laplacian[`*f*`]`	give the Laplacian of the scalar point function *f* in the default coordinate system

```
In[20]:= Needs["VectorAnalysis`"]
```

Unless otherwise set, the Cartesian coordinate system is the default system:

```
In[21]:= CoordinateSystem
Out[21]= Cartesian
```

The default names of the coordinate variables are **Xx**, **Yy**, and **Zz**:

```
In[22]:= Coordinates[]
Out[22]= {Xx, Yy, Zz}
```

To conform with traditional notation, let us change the names of the coordinate variables to **x**, **y**, and **z**:

```
In[23]:= SetCoordinates[Cartesian[x, y, z]]
Out[23]= Cartesian[x, y, z]
```

We can verify the new names of the coordinate variables:

```
In[24]:= Coordinates[]
Out[24]= {x, y, z}
```

Mathematica represents three-dimensional vectors by three-element lists. Because the letter **C** has a built-in meaning, we let **a**, **b**, and **c** be aliases of the vector point functions **A**, **B**, and **C**, respectively, and

```
In[25]:= a = {ax[x, y, z], ay[x, y, z], az[x, y, z]};
         b = {bx[x, y, z], by[x, y, z], bz[x, y, z]};
         c = {cx[x, y, z], cy[x, y, z], cz[x, y, z]};
         f = func[x, y, z];
```

(a) To prove vector identity (a), show that *lhs* of the identity evaluates to the null vector $\{0, 0, 0\}$:

```
In[29]:= CrossProduct[a, CrossProduct[b, c]] +
          CrossProduct[b, CrossProduct[c, a]] +
          CrossProduct[c, CrossProduct[a, b]]
Out[29]= {0, 0, 0}
```

We have proved the vector identity for any point (x, y, z). The proof remains valid for the special case in which the vector functions **A**, **B**, and **C** are constant vectors independent of the coordinates x, y, and z.

(b) To prove vector identity (b), show that *lhs* – *rhs* evaluates to 0:

```
In[30]:= (Div[f a] - f Div[a] - DotProduct[a, Grad[f]]) // Simplify
Out[30]= 0
```

(c) To prove vector identity (c), show that *lhs* – *rhs* evaluates to $\{0, 0, 0\}$:

```
In[31]:= Curl[Curl[a]] - Grad[Div[a]] + Laplacian /@ a
Out[31]= {0, 0, 0}
```

We have proved the vector identity in Cartesian coordinates. With curvilinear coordinates such as spherical or cylindrical coordinates, the components of the Laplacian of a vector are not, in general, the Laplacians of the corresponding components of the vector. For curvilinear coordinates, vector identity (c) often serves as the definition for the Laplacian of a vector.

```
In[32]:= Clear[a, b, c, f]
```

■

Example 2.4.15 An uncharged conducting sphere of radius a is placed in an initially uniform electric field $\mathbf{E}_0$ aligned in the z direction. (a) Show that the potential outside the sphere is

$$V(r, \theta) = -E_0\left(r - \frac{a^3}{r^2}\right)\cos\theta$$

where r is the distance from the origin located at the center of the sphere and θ is the polar angle—that is, the angle down from the z-axis. (b) Determine the components of the electric field outside the sphere. (c) Find the induced charge density on the surface of the sphere and verify that the total charge on the sphere is zero.

The package **VectorAnalysis`** contains many functions for vector analysis. Those relevant to this problem were defined in Example 2.4.14 of this section.

```
In[33]:= Needs["VectorAnalysis`"]
```

Let us set the spherical coordinate system to be the default system:

```
In[34]:= SetCoordinates[Spherical];
```

Coordinates[] gives the default names of coordinate variables in the default coordinate system:

```
In[35]:= Coordinates[]
Out[35]= {Rr, Ttheta, Pphi}
```

Let us change the names of the coordinate variables to the customary **r**, **θ**, and **ϕ**:

```
In[36]:= SetCoordinates[Spherical[r, θ, ϕ]]
Out[36]= Spherical[r, θ, ϕ]
```

(a) To show that $V(r, \theta)$ is the electric potential outside the sphere, we invoke the uniqueness theorem: The solution to Laplace's equation with specified boundary conditions is unique. That is, we must show that V satisfies Laplace's equation and matches the appropriate boundary conditions.

Laplace's equation is

$$\nabla^2 V = 0$$

We can prove that the given potential satisfies this equation. Let **E0** be the alias of E_0.

```
In[37]:= Clear[a, E0]

In[38]:= V = -E0 (1 - (a/r)^3) r Cos[θ];

In[39]:= Laplacian[V] // Simplify
Out[39]= 0
```

The boundary conditions are

$$\begin{array}{ll} V = 0 & r = a \\ V = -E_0\, z = -E_0\, r \cos\theta & r \gg a \end{array}$$

We can verify that V matches these conditions:

```
In[40]:= V/.r→a
Out[40]= 0
```

```
In[41]:= V/.((a/r)^3→0)
Out[41]= -E0 r Cos[θ]
```

where we have translated $r >> a$ into `(a/r)^3->0`. (For questions raised by this translation, see Exercise 34 of Section 2.4.12.) Hence, the given $V(r, \theta)$ is the potential outside the sphere.

(b) The electric field is given by

$$\mathbf{E} = -\nabla V$$

Thus, `-Grad[v]` yields a list of the radial, polar, and azimuthal components of $\mathbf{E}$, namely $\{E_r, E_\theta, E_\phi\}$:

```
In[42]:= -Grad[V]//Simplify
```

$$\text{Out[42]= } \left\{\text{E0}\left(1+\frac{2\,\text{a}^3}{\text{r}^3}\right)\text{Cos}[\theta],\ \frac{\text{E0}\left(\text{a}^3-\text{r}^3\right)\text{Sin}[\theta]}{\text{r}^3},\ 0\right\}$$

(c) From Gauss's law, the surface charge density on the sphere is given by

$$\sigma = \varepsilon_0\, E_r|_{r=a}$$

where ε_0 is the permittivity of free space. Therefore, the induced surface charge density is

```
In[43]:= ε0 %[[1]]/.r→a
Out[43]= 3 E0 ε0 Cos[θ]
```

with `ε0` being the alias of ε_0. The total charge on the sphere is

$$Q = \int_{\text{sphere}} \sigma dS = 2\pi a^2 \int_0^{\pi} \sigma(\theta) \sin\theta d\theta$$

which must be zero because the sphere was initially uncharged:

```
In[44]:= (2π(a^2))Integrate[% Sin[θ], {θ, 0, π}]
Out[44]= 0
```

(For a discussion of Laplace's equation as well as this problem of an uncharged metal sphere placed in an otherwise uniform electric field, see [Gri99].)

```
In[45]:= Clear[V]
```

■

2.4.11 Analyzing Data

For sophisticated analyses of data, *Mathematica* provides a number of statistics packages, such as **LinearRegression`**, **NonlinearRegression`**, **MultivariateStatistics`**, and **StatisticalPlots`**. For locating the packages, see Section 1.5. Here, we cover the rudiments of data analysis.

2.4.11.1 Basic Error Analysis

This section gives a brief introduction to error analysis. Errors, also called uncertainties, are intrinsic to measurements. Even with great skills and extreme care, experimenters can never eliminate errors in measurements. Errors due to random fluctuations in measurements are called random errors. Errors caused by blunders of the experimenters or defects of the equipment are not random errors and are not covered here.

Random errors are described by the Gaussian or normal distribution

$$P_G(x) = \frac{1}{\sigma_G\sqrt{2\pi}} \exp\left[-\frac{1}{2}\left(\frac{x-\mu_G}{\sigma_G}\right)^2\right] \tag{2.4.4}$$

which has two parameters, μ_G and σ_G. Here is a plot of the Gaussian distribution:

```
In[1]:= PG[x_] := (1/(σG Sqrt[2π]))Exp[-(1/2)(((x-μG)/σG)^2)]
```

```
In[2]:= Plot[Evaluate[σG PG[x] /. x → μG + σG z], {z, -3, 3}, PlotRange → {0, 0.5},
          Frame → True, FrameLabel → {" x - μG (σG) ", "PG (1/σG) "}, Axes → False]
```

Out[2] =

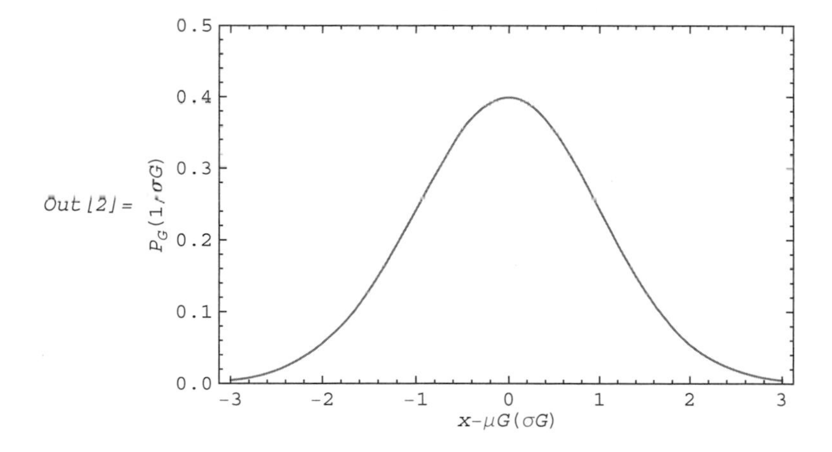

where we have plotted $\sigma_G P_G(x)$ versus $(x-\mu_G)/\sigma_G$—that is, $P_G(x)$ in units of $(1/\sigma_G)$ versus $(x-\mu_G)$ in units of σ_G. The distribution has a maximum at $x=\mu_G$ and a width of approximately $2\sigma_G$.

If we make a measurement of a quantity, the probability of finding the value between x and $x+dx$ is $P_G(x)dx$. In other words, if we make N measurements of a quantity and obtain the values $x_1, x_2, \ldots,$ and x_N, the fractional number of x_i with values between x and $x+dx$ is

$P_G(x)dx$, as N approaches infinity. The mean or average value of the measurements is

$$\mu = \int_{-\infty}^{\infty} x P_G(x)\, dx \tag{2.4.5}$$

The standard deviation, which is the root-mean-square deviation from the mean, is

$$\sigma = \sqrt{\int_{-\infty}^{\infty} (x-\mu)^2 P_G(x)\, dx} \tag{2.4.6}$$

Equations 2.4.4 and 2.4.5 imply that $\mu = \mu_G$:

```
In[3]:= Integrate[x PG[x], {x, -∞, ∞}, Assumptions → σG > 0]
Out[3]= μG
```

Equations 2.4.4 and 2.4.6 together with $\mu = \mu_G$ imply that $\sigma = \sigma_G$:

```
In[4]:= Simplify[Sqrt[Integrate[((x - μ)^2) PG[x] /. μG → μ,
          {x, -∞, ∞}, Assumptions -> σG > 0]], σG > 0]
Out[4]= σG
```

Since $\mu = \mu_G$ and $\sigma = \sigma_G$, we will, as is customary, drop the subscript G from the Gaussian parameters μ_G and σ_G. Since N approaches infinity, it is reasonable to expect that μ, which is the mean, approaches the true value of the measured quantity.

In practice, N is finite and we do not know μ and σ. It can be shown that the *best* estimate of μ from our data is the sample mean $\bar{x}$:

$$\mu \approx \bar{x} = \frac{\sum_{i=1}^{N} x_i}{N} \tag{2.4.7}$$

The *best* estimate of σ is the sample standard deviation s:

$$\sigma \approx s = \sqrt{\frac{1}{N-1}\sum_{i=1}^{N}(x_i - \bar{x})^2} \tag{2.4.8}$$

For each measurement, the uncertainty of the result is σ. For N measurements, the uncertainty of each measurement contributes to the determination of the uncertainty of the mean. It can be shown that the uncertainty of the mean, also known as the standard error, is

$$\sigma_\mu = \frac{\sigma}{\sqrt{N}} \tag{2.4.9}$$

From Equations 2.4.8 and 2.4.9, the *best* estimate of the uncertainty of the mean is

$$\sigma_\mu \approx \frac{s}{\sqrt{N}} \tag{2.4.10}$$

From Equations 2.4.7 and 2.4.10, the experimental value of the measured quantity is

$$\overline{x} + \frac{s}{\sqrt{N}}$$

For counting experiments such as counting the decays from radioactive nuclei, the experimental errors are statistical rather than random. We do not know for sure if a nucleus decays in a given time interval; we only know the probability of its decay. Statistical errors are described by the Poisson distribution

$$P_P(x) = \frac{\mu_P{}^x}{x!} e^{-\mu_P} \tag{2.4.11}$$

which has only one parameter μ_P. Because the Poisson distribution describes results of counting experiments, it is a discrete distribution. Although Equation 2.4.11 gives the Poisson distribution as a continuous function of x, the only physically significant values of x are the nonnegative integers, $0, 1, 2, \ldots$. Here is a plot of the Poisson distribution for $\mu_P = 1.50$:

```
In[5]:= PP[x_]:=((μP^x)/(x!)) Exp[-μP]
```

```
In[6]:= Plot[Evaluate[PP[x] /. μP→1.50], {x, 0, 8}, PlotRange→{0, 0.4},
          Frame→True, FrameLabel→{"x", "Pₚ"}, Axes→False]
```

Out[6]=

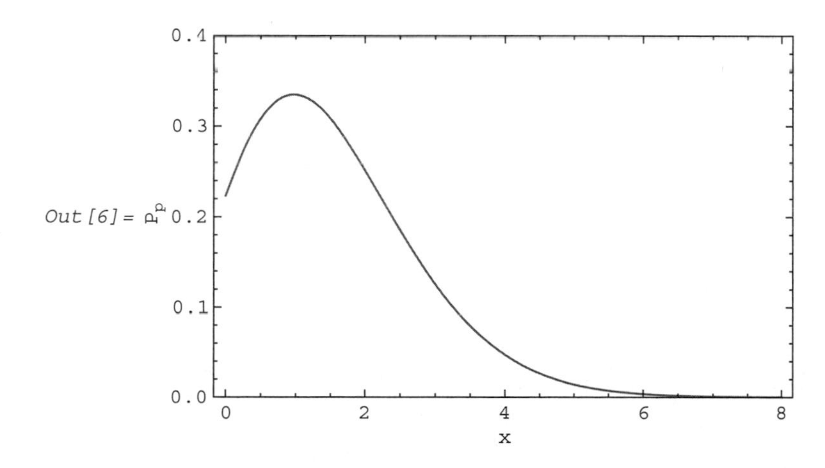

where the maximum occurs at $x < \mu_P$.

The probability of observing x items, events, or "successes" in a given time interval Δt is $P_P(x)$. In other words, if we count the number of items, events, or "successes" in a given time interval N times and obtain the values $x_1, x_2, \ldots,$ and x_N, the fractional number of these with the value x is $P_P(x)$, as N approaches infinity. The mean or average number of counts in the time interval Δt is

$$\mu = \sum_{x=0}^{\infty} x P_P(x) \tag{2.4.12}$$

The standard deviation, which is the root-mean-square deviation from the mean, is

$$\sigma = \sqrt{\sum_{x=0}^{\infty} (x-\mu)^2 P_P(x)} \tag{2.4.13}$$

Equations 2.4.11 and 2.4.12 imply that $\mu = \mu_P$:

```
In[7]:= Sum[x PP[x], {x, 0, ∞}]
Out[7]= μP
```

Equations 2.4.11 and 2.4.13 together with $\mu = \mu_P$ imply that $\sigma = \sqrt{\mu}$:

```
In[8]:= Sqrt[Sum[((x - μ)^2) PP[x] /. μP → μ, {x, 0, ∞}]]
Out[8]= √μ
```

Since $\mu = \mu_P$, we will, as is customary, drop the subscript P from the Poisson parameter μ_P.

In practice, N is finite and we do not know μ. It can be shown that the *best* estimate of μ from our data is the sample mean $\bar{x}$:

$$\mu \approx \bar{x} = \frac{\sum_{i=1}^{N} x_i}{N} \tag{2.4.14}$$

Because $\sigma = \sqrt{\mu}$, the *best* estimate of the standard deviation is

$$\sigma \approx \sqrt{\bar{x}} \tag{2.4.15}$$

From Equations 2.4.9 and 2.4.15, the *best* estimate of the uncertainty of the mean is

$$\sigma_\mu \approx \sqrt{\frac{\bar{x}}{N}} \tag{2.4.16}$$

From Equations 2.4.14 and 2.4.16, the experimental mean number of counts in the time interval Δt is

$$\bar{x} \pm \sqrt{\frac{\bar{x}}{N}}$$

or the experimental mean number of count per unit time is

$$\frac{\bar{x}}{\Delta t} \pm \sqrt{\frac{(\bar{x}/\Delta t)}{N\Delta t}}$$

(This section made a number of assertions and provided a brief introduction to error analysis. For justification of the assertions or for more information on error analysis, see [Tay82] or [BR03].)

```
In[9]:= Clear[PG, PP]
```

Several of the *Mathematica* functions for basic error analysis are

`Total[{x1, x2, ..., xN}]`	total, $\sum_{i=1}^{N} x_i$
`Mean[{x1, x2, ..., xN}]`	sample mean $\bar{x}$, $(1/N)\sum_{i=1}^{N} x_i$
`Variance[{x1, x2, ..., xN}]`	sample variance s^2, $\frac{1}{N-1}\sum_{i=1}^{N}(x_i - \bar{x})^2$
`StandardDeviation[{x1, x2, ..., xN}]`	sample standard deviation s, $\sqrt{s^2}$

Example 2.4.16 A student made 30 measurements of the time of flight, in seconds, of an object: 8.16, 8.14, 8.12, 8.16, 8.18, 8.10, 8.18, 8.18, 8.18, 8.24, 8.16, 8.14, 8.17, 8.18, 8.21, 8.12, 8.12, 8.17, 8.06, 8.10, 8.12, 8.10, 8.14, 8.09, 8.16, 8.16, 8.21, 8.14, 8.16, 8.13. Determine the sample mean and the sample standard deviation. Assuming all errors are random, give the best estimate for the time of flight and its uncertainty.

Let us begin with the assignment:

```
In[10]:= tlist = {8.16, 8.14, 8.12, 8.16, 8.18, 8.10, 8.18, 8.18, 8.18, 8.24,
          8.16, 8.14, 8.17, 8.18, 8.21, 8.12, 8.12, 8.17, 8.06, 8.10,
          8.12, 8.10, 8.14, 8.09, 8.16, 8.16, 8.21, 8.14, 8.16, 8.13};
```

The sample mean (in seconds) is

```
In[11]:= NumberForm[Mean[tlist], 4]
Out[11]//NumberForm=
        8.149
```

The sample standard deviation (in seconds) is

```
In[12]:= NumberForm[StandardDeviation[tlist], 2]
Out[12]//NumberForm=
        0.039
```

The standard deviation of the mean (in seconds) is

```
In[13]:= NumberForm[%/Sqrt[Length[tlist]], 1]
Out[13]//NumberForm=
        0.007
```

Thus, the best estimates for the time of flight and its uncertainty are 8.149 s and 0.007 s, respectively. The experimental result is usually stated as 8.149 ± 0.007 s.

```
In[14]:= Clear[tlist]
```

■

Example 2.4.17 The following counts were recorded in 1-minute intervals from a radioactive source:

Number of decays	0	1	2	3	4	5	6	7	8	9	10
Times observed	1	9	20	24	19	11	11	0	3	1	1

(a) Determine the best estimate of the mean number of decays in 1 minute and its uncertainty.
(b) Plot together the bar histogram of the fractional number of "times observed" versus the "number of decays" and the Poisson distribution with μ being the best estimate of the mean number of decays in 1 minute.

(a) Let us begin with the assignment for the list of numbers of decays observed in 1 minute:

```
In[15]:= dpm = Range[0, 10]
Out[15]= {0, 1, 2, 3, 4, 5, 6, 7, 8, 9, 10}
```

For each number of decays observed in 1 minute, there is a number of times observed. The list of numbers of times observed is

```
In[16]:= freq = {1, 9, 20, 24, 19, 11, 11, 0, 3, 1, 1}
Out[16]= {1, 9, 20, 24, 19, 11, 11, 0, 3, 1, 1}
```

and the list of fractional numbers of times observed is

```
In[17]:= frac = freq/Total[freq]
```

Out[17]= $\left\{\frac{1}{100}, \frac{9}{100}, \frac{1}{5}, \frac{6}{25}, \frac{19}{100}, \frac{11}{100}, \frac{11}{100}, 0, \frac{3}{100}, \frac{1}{100}, \frac{1}{100}\right\}$

The best estimate of the mean number of decays in 1 minute is the sample mean:

```
In[18]:= μ = Total[dpm frac] // N
Out[18]= 3.61
```

The best estimate of the standard deviation of the mean σ_μ is given by

```
In[19]:= Sqrt[μ/Total[freq]]
Out[19]= 0.19
```

Thus, the best estimates of the mean number of decays in 1 minute and its uncertainty are 3.61 and 0.19, respectively. The experimental result is usually stated as 3.61 ± 0.19 decays per minute.

(b) To plot the bar histogram, we use the function **GeneralizedBarChart** in the package **BarCharts`**. **GeneralizedBarChart**[*datalist*] generates a bar chart of the data, where *datalist* specifies the positions x_i, heights h_i, and widths w_i of the bars in the form

$$\{\{x_1, h_1, w_1\}, \{x_2, h_2, w_2\}, \ldots\}$$

Here is the bar histogram of the fractional number of "times observed" versus the "number of decays" together with the plot of the Poisson distribution with $\mu = 3.61$:

```
In[20]:= Needs["BarCharts`"]

In[21]:= Show[GeneralizedBarChart[Transpose[{dpm, frac, Table[0.2, {11}]}]],
           Plot[(μ^x Exp[-μ])/((x)!), {x, 0, 10}, PlotStyle→Thickness[0.0125]],
           Frame→True, FrameLabel→{"Decay Number", "frac times observed"},
           FrameTicks→{Automatic, Automatic, None, None}, Axes→False]
```

Out[21]=

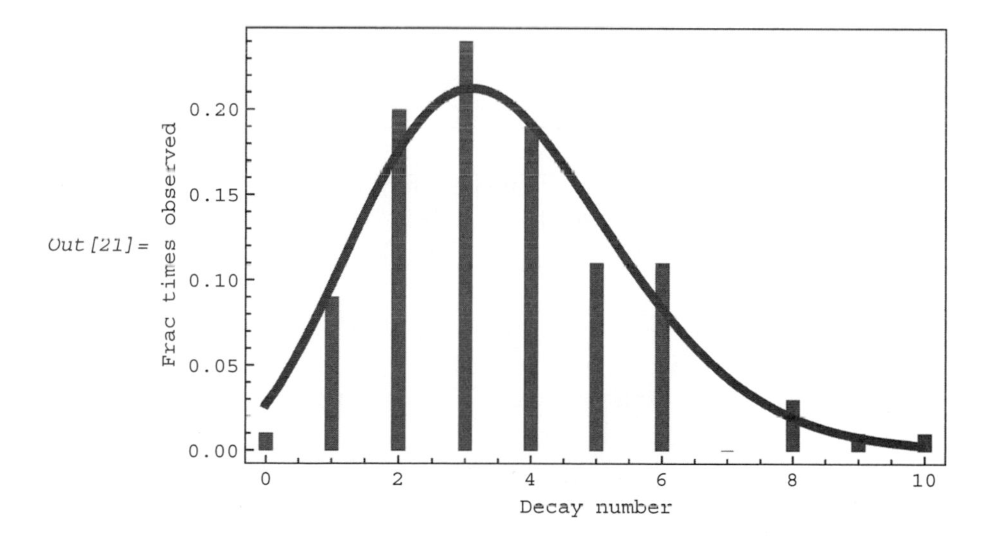


```
In[22]:= Clear[dpm, freq, frac, μ]
```

■

2.4.11.2 Nonlinear Least-Squares Fit

Section 2.1.13 discussed linear least-squares fit when the fitting function has the form

$$y(x_1, x_2, \ldots, x_n) = \sum_{k=1}^{m} a_k f_k(x_1, x_2, \ldots, x_n)$$

where x_i are the independent variables, a_k are the parameters, and f_k are the specified functions. With the specified functions, **Fit** determines the parameter values for the least-squares fit. Often, the fitting function, also known as the model, is not linear in the parameters. In these cases, we can use the function **FindFit**.

FindFit[*data*, *model*, *pars*, *vars*] searches for a least-squares fit to a list of data according to the model containing the variables in the list *vars* and the parameters in the list *pars*. **FindFit**[*data*, *model*, {{par_1, p_1}, {par_2, p_2}, ...}, *vars*] starts the search for a fit with {par_1 -> p_1, par_2 -> p_2, ...}. Like that for the function **Fit**, the data argument has the form

$$\{\{x_{11}, x_{12}, \ldots, y_1\}, \{x_{21}, x_{22}, \ldots, y_2\}, \ldots\}$$

When the list *vars* has only one variable, the data argument takes the form

$$\{\{x_1, y_1\}, \{x_2, y_2\}, \ldots\}$$

Furthermore, if $x_1 = 1, x_2 = 2, \ldots$—that is, $x_i = i$—this can be written as

$$\{y_1, y_2, \ldots\}$$

If the list *vars* or *pars* has only one element, its enclosing curly brackets may be omitted. **FindFit** returns a list of replacements for the parameters for the least-squares fit. (For a discussion of the method of least squares, see [Won92]. In addition to the least-squares method, **FindFit** supports the use of other methods for curve fitting. For details, use the "front end help.")

Example 2.4.18 The Einstein model assumes that the N atoms in a crystal is equivalent to an assembly of $3N$ independent one-dimensional harmonic oscillators, all vibrating with the same frequency ν. The heat capacity at constant volume C_V is given as

$$C_V = 3Nk\left(\frac{\theta_E}{T}\right)^2 \frac{e^{\theta_E/T}}{\left(e^{\theta_E/T} - 1\right)^2}$$

where k is Boltzmann's constant, T is the absolute temperature, and θ_E is a characteristic parameter, called the Einstein temperature and defined by the equation $k\theta_E \equiv h\nu$, in which h

is Planck's constant. The following table displays the data for diamond:

T(K)	$C_V/(3Nk)$	T(K)	$C_V/(3Nk)$
222.4	0.1278	413.0	0.4463
262.4	0.1922	479.2	0.5501
283.7	0.2271	520.0	0.6089
306.4	0.2653	879.7	0.8871
331.3	0.3082	1,079.7	0.9034
358.5	0.3552	1,258.0	0.9235

Use **FindFit** to find θ_E of diamond. How well does Einstein's formula for the specific heat agree with the data?

Here are the list of temperatures T_i in kelvins and the corresponding list of $C_{V_i}/(3Nk)$ for diamond:

```
In[1]:= Tlist = {222.4, 262.4, 283.7, 306.4, 331.3,
          358.5, 413.0, 479.2, 520.0, 879.7, 1079.7, 1258}
Out[1]= {222.4, 262.4, 283.7, 306.4, 331.3,
          358.5, 413., 479.2, 520., 879.7, 1079.7, 1258}
```

```
In[2]:= Clist = {0.1278, 0.1922, 0.2271, 0.2653, 0.3082,
          0.3552, 0.4463, 0.5501, 0.6089, 0.8871, 0.9034, 0.9235}
Out[2]= {0.1278, 0.1922, 0.2271, 0.2653, 0.3082,
          0.3552, 0.4463, 0.5501, 0.6089, 0.8871, 0.9034, 0.9235}
```

Therefore, the list of data for diamond is

```
In[3]:= data = Transpose[{Tlist, Clist}]
Out[3]= {{222.4, 0.1278}, {262.4, 0.1922}, {283.7, 0.2271}, {306.4, 0.2653},
          {331.3, 0.3082}, {358.5, 0.3552}, {413., 0.4463}, {479.2, 0.5501},
          {520., 0.6089}, {879.7, 0.8871}, {1079.7, 0.9034}, {1258, 0.9235}}
```

Let us evaluate **FindFit** with the data for diamond and the model for $C_V/(3Nk)$ from the Einstein formula:

```
In[4]:= FindFit[data, ((θ/T)^2)Exp[θ/T]/((Exp[θ/T] - 1)^2), θ, T]
Out[4]= {θ→1296.21}
```

where the independent variable is T and the parameter is θ, the alias of θ_E. Thus, θ_E for the best fit is 1296.21 K.

To compare Einstein's theory with experiment, we now superimpose the experimental points and the theoretical curve for $C_V/(3Nk)$ as a function of T/θ_E. Let us first generate from the data the list of T_i/θ_E and the nest list of $\{T_i/\theta_E, C_{Vi}/(3Nk)\}$:

```
In[5]:= tlist = Tlist/1296.21
Out[5]= {0.171577, 0.202436, 0.218869, 0.236381, 0.255591, 0.276576,
         0.318621, 0.369693, 0.40117, 0.678671, 0.832967, 0.970522}
```

```
In[6]:= newdata = Transpose[{tlist, Clist}]
Out[6]= {{0.171577, 0.1278}, {0.202436, 0.1922}, {0.218869, 0.2271},
         {0.236381, 0.2653}, {0.255591, 0.3082}, {0.276576, 0.3552},
         {0.318621, 0.4463}, {0.369693, 0.5501}, {0.40117, 0.6089},
         {0.678671, 0.8871}, {0.832967, 0.9034}, {0.970522, 0.9235}}
```

Note that the equation for heat capacity at constant volume can be written as

$$C_V/3Nk = \left(\frac{1}{t}\right)^2 \frac{e^{1/t}}{\left(e^{1/t} - 1\right)^2}$$

where $t \equiv T/\theta_E$. We can now plot the experimental points together with the theoretical curve for $C_V/(3Nk)$ as a function of t—that is, $C_V/(3Nk)$ as a function of T/θ_E:

```
In[7]:= Show[Plot[(1/(t^2))Exp[1/t]/((Exp[1/t] - 1)^2), {t, 0, 1}],
         ListPlot[newdata, PlotStyle→PointSize[0.02]],
         AxesLabel→{"T/θE", "CV/3Nk"}]
```

Out[7]=

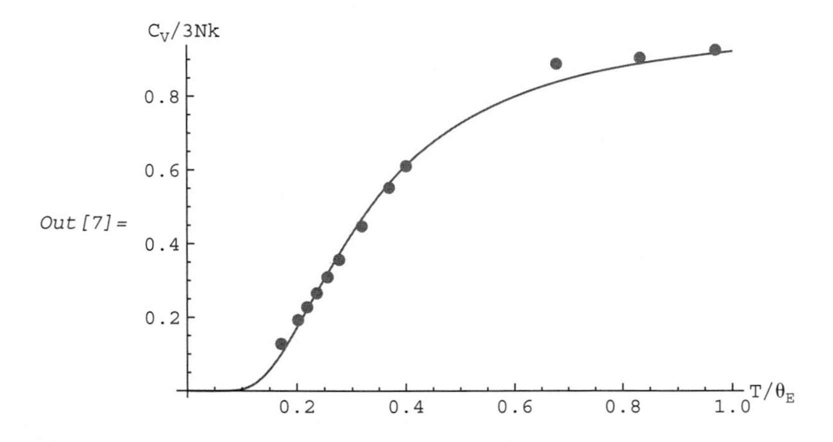

The fit is reasonably good. However, for temperatures below approximately $0.24\theta_E$, the curve drops off to zero too fast as T -> 0. The Debye model gives a better description of the heat capacities of solids. (For a discussion of the Debye model, see [Bai99] or [Sch00].)

```
In[8]:= Clear[Tlist, Clist, data, tlist, newdata]
```

■

In Example 2.4.18, we invoked **FindFit** without specifying a starting value for the parameter. A word of caution is in order. For some invocations of **FindFit**, specifications of starting values for the parameters are imperative. For an example, see Exercise 38 in Section 2.4.12.

2.4.11.3 Interpolation

Given a set of data points, we would like to find the underlying function that corresponds to them. Section 2.4.11.2 discussed curve fitting. With a fitting function containing parameters, **FindFit** determines the values of the parameters for the least-squares fit. What if we do not have a fitting function? The function **Interpolation** returns an approximate function corresponding to the data.

Interpolation[*data*] constructs an **InterpolatingFunction** object that represents an approximate function that interpolates *data*. The data argument has the form

$$\{\{x_1, y_1\}, \{x_2, y_2\}, \ldots\}$$

If $x_1 = 1, x_2 = 2, \ldots$—that is, $x_i = i$—this can be written as

$$\{y_1, y_2, \ldots\}$$

Interpolation finds the approximate function by fitting polynomial curves between successive data points. The degree of the polynomials can be specified by including the option **InterpolationOrder**, which has a default value of 3. (To explore the option, see Exercise 40 of Section 2.4.12.)

We first saw **InterpolatingFunction**[*domain*, <>] as the values of the replacement rules returned by **NDSolve**. **InterpolatingFunction**[*domain*, <>][*x*] returns the value of the approximate function for any argument *x* within the *domain* of the data. We can differentiate, integrate, plot, and perform other numerical operations on it.

Let us illustrate the use of **Interpolation** and **InterpolatingFunction**. Here is a list of carbon dioxide concentrations collected at Mauna Loa Observatory, Hawaii:

```
In[1]:= data = {355.98, 356.72, 357.81, 359.15, 359.66, 359.25, 357.03, 355.,
          353.01, 353.31, 354.16, 355.4, 356.7, 357.16, 358.38, 359.46, 360.28,
          359.6, 357.57, 355.52, 353.7, 353.98, 355.33, 356.8, 358.36, 358.91,
          359.97, 361.26, 361.68, 360.95, 359.55, 357.49, 355.84, 355.99,
          357.58, 359.04, 359.96, 361., 361.64, 363.45, 363.79, 363.26, 361.9,
          359.46, 358.06, 357.75, 359.56, 360.7, 362.05, 363.25, 364.03,
          364.72, 365.41, 364.97, 363.65, 361.49, 359.46, 359.6, 360.76,
          362.33, 363.18, 364., 364.57, 366.35, 366.79, 365.62, 364.47, 362.51,
          360.19, 360.77, 362.43, 364.28, 365.32, 366.15, 367.31, 368.61,
          369.29, 368.87, 367.64, 365.77, 363.9, 364.23, 365.46, 366.97,
          368.15, 368.87, 369.59, 371.14, 371., 370.35, 369.27, 366.94,
          364.63, 365.12, 366.67, 368.01, 369.14, 369.46, 370.52, 371.66,
          371.82, 371.7, 370.12, 368.12, 366.62, 366.73, 368.29, 369.53,
          370.28, 371.5, 372.12, 372.87, 374.02, 373.3, 371.62, 369.55,
```

```
367.96, 368.09, 369.68, 371.24, 372.43, 373.09, 373.52, 374.86,
375.55, 375.41, 374.02, 371.49, 370.71, 370.24, 372.08, 373.78};
```

(For details about the sampling methods, see [KW04]. For information on importing and exporting data, see Exercises 47–50 in Section 2.4.12.) The list gives consecutive monthly carbon dioxide concentration in parts per million (ppm) from January 1992 to December 2002. **Interpolation** finds the approximate function for the data:

```
In[2]:= concentration = Interpolation[data]
Out[2]= InterpolatingFunction[{{1., 132.}}, <>]
```

where the name **concentration** has been given to the approximate function. We can plot the function:

```
In[3]:= Plot[concentration[x], {x, 1, 132},
         Axes → False, Frame → True, FrameTicks →
          {Transpose[{Table[i, {i, 1, 132, 24}], Table[1991 + i, {i, 1, 11, 2}]}],
           Automatic, Automatic, Automatic},
         FrameLabel → {"Year", "Concentration (ppm)"}]
```

Out[3]=

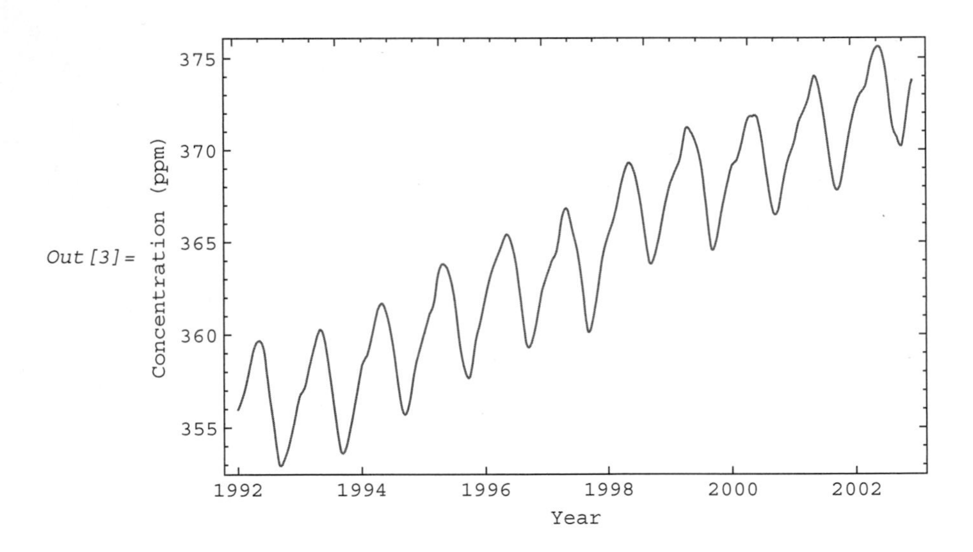

Here is the plot showing the concentration versus the number of months from January 1992:

```
In[4]:= Plot[concentration[x], {x, 1, 132}, Axes → False,
         Frame → True, FrameLabel → {"Month", "Concentration (ppm)"}]
```

Out[4]=

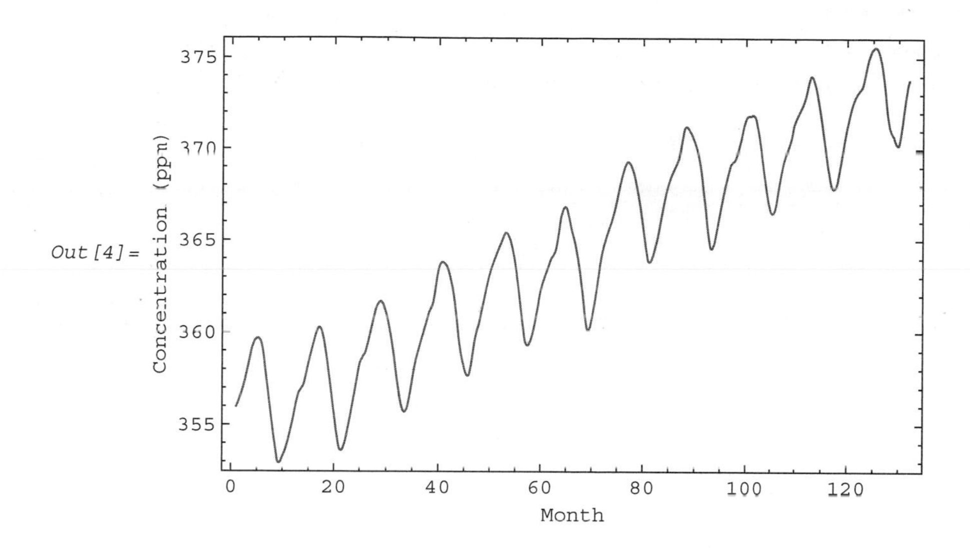

We can find the local maximum, for example, near the 77th month:

```
In[5]:= NumberForm[FindMaximum[concentration[x], {x, 77}], 5]
Out[5]//NumberForm=
        {369.29, {x→77.075}}
```

As discussed in Section 2.3.1.5, **FindMaximum**[f, {x, x_0}] searches for a local maximum in f, starting from the point $x = x_0$. The Second Derivative Test confirms that a local maximum indeed exists at $x = 77.075$:

```
In[6]:= {concentration'[x], concentration''[x] // Negative} /. %[[2]] // Chop
Out[6]= {0, True}
```

The result shows that for the approximate function the first derivative is zero and the second derivative is negative at $x = 77.075$. (For a discussion of the Second Derivative Test, see [Ste03].) The maximum value is

```
In[7]:= NumberForm[concentration[x],5] /. %%[[2]]
Out[7]//NumberForm=
        369.29
```

which is what **FindMaximum** found.

```
In[8]:= Clear[data, concentration]
```

2.4.12 Exercises

In this book, straightforward, intermediate-level, and challenging exercises are unmarked, marked with one asterisk, and marked with two asterisks, respectively. For solutions to most odd-numbered exercises, see Appendix C.

1. (a) Using the function **RandomInteger**, generate a nested list for a 5 × 5 matrix with elements that are random integers from 0 to 20. (b) Display the nested list in standard matrix form. *Hint:* Use the "front end help" to access information on **RandomInteger**.
2. Determine the length and dimensions of the list {{{3,4,5},{4,5,6}},{{4,5,6},{5,6,7}}}.

   Answers:

```
2
{2, 2, 3}
```

3. Use the function **RandomInteger** to simulate tossing a coin n times. (a) Plot a bar chart of the distribution of getting heads and tails when $n = 10^2$. (b) Then try $n = 10^3, 10^4, 10^5$, and 10^6. (c) What do these simulations tell us about the meaning of probability? *Hint:* For the function **BarChart** in the package **BarCharts`**, specify here the option

```
BarLabels → {"heads", "tails"}
```

*4. Consider a volume V partitioned into 10 bins of equal volume. Use the function **Random-Integer** to simulate tossing N molecules randomly into these bins. (a) With the function **BarChart** in the package **BarCharts`**, plot a bar chart of the number of molecules in each bin when $N = 10^2$. (b) Then try $N = 10^3, 10^4, 10^5$, and 10^6. (c) What conclusion about the distribution of molecules in these bins emerges from these simulations? (This exercise illustrates macroscopic regularity in an ideal gas. For a discussion of macroscopic regularity and the source of this exercise, see Chapter 2 of [Bai99].)

5. Create the list $\{x-1, x^3-1, x^5-1, x^7-1, x^9-1, x^{11}-1\}$ with the function **Table**. With the function **Drop**, obtain the list without the second and third elements.

   Answer:

```
{-1 + x, -1 + x^7, -1 + x^9, -1 + x^11}
```

6. Consider the list $\{r, \{s,t\}, \{w, \{x, \{y,z\}\}\}\}$. Using **Part**, pick out the pieces: (a) r, (b) $\{s,t\}$, (c) w, (d) $\{y,z\}$, and (e) z.

   Answers:

```
r
{s, t}
w
```

```
{y, z}
z
```

*7. Using the "front end help" discussed in Section 1.6.3, access information on the function **Part**. Then consider the matrix

$$\begin{pmatrix} a[1,1] & a[1,2] & a[1,3] & a[1,4] \\ a[2,1] & a[2,2] & a[2,3] & a[2,4] \\ a[3,1] & a[3,2] & a[3,3] & a[3,4] \\ a[4,1] & a[4,2] & a[4,3] & a[4,4] \end{pmatrix}$$

Using **Part**, pick out (a) the column

$$\begin{pmatrix} a[1,3] \\ a[2,3] \\ a[3,3] \\ a[4,3] \end{pmatrix}$$

and (b) the submatrix

$$\begin{pmatrix} a[2,3] & a[2,4] \\ a[3,3] & a[3,4] \\ a[4,3] & a[4,4] \end{pmatrix}$$

Hint:

list[[{a, b, ...}, {α, β, ...}]]

gives

{{*list*[[a, α]], *list*[[a, β]], ...}, {*list*[[b, α]], *list*[[b, β]], ...}, ...}.

Answers:

$$\begin{pmatrix} \texttt{a[1, 3]} \\ \texttt{a[2, 3]} \\ \texttt{a[3, 3]} \\ \texttt{a[4, 3]} \end{pmatrix}$$

$$\begin{pmatrix} \texttt{a[2, 3]} & \texttt{a[2, 4]} \\ \texttt{a[3, 3]} & \texttt{a[3, 4]} \\ \texttt{a[4, 3]} & \texttt{a[4, 4]} \end{pmatrix}$$

8. We can use the function **Part** to change pieces of matrices:

matrix[[*i*, *j*]] = *new*	replace the (i,j)th element of *matrix* with *new*
matrix[[*i*]] = *new*	replace the *i*th row of *matrix* with *new*
matrix[[**All**, *i*]] = *new*	replace the *i*th column of *matrix* with *new*

Consider the matrix

$$\begin{pmatrix} a[1,1] & a[1,2] & a[1,3] & a[1,4] \\ a[2,1] & a[2,2] & a[2,3] & a[2,4] \\ a[3,1] & a[3,2] & a[3,3] & a[3,4] \end{pmatrix}$$

Using **Part**, replace (a) $a[3,4]$ with $b[3,4]$, (b) the second row with $\{b[2,1], b[2,2], b[2,3], b[2,4]\}$, and (c) the fourth column with $\{b[1,4], b[2,4], b[3,4]\}$.

Answers:

$$\begin{pmatrix} a[1,1] & a[1,2] & a[1,3] & a[1,4] \\ a[2,1] & a[2,2] & a[2,3] & a[2,4] \\ a[3,1] & a[3,2] & a[3,3] & b[3,4] \end{pmatrix}$$

$$\begin{pmatrix} a[1,1] & a[1,2] & a[1,3] & a[1,4] \\ b[2,1] & b[2,2] & b[2,3] & b[2,4] \\ a[3,1] & a[3,2] & a[3,3] & a[3,4] \end{pmatrix}$$

$$\begin{pmatrix} a[1,1] & a[1,2] & a[1,3] & b[1,4] \\ a[2,1] & a[2,2] & a[2,3] & b[2,4] \\ a[3,1] & a[3,2] & a[3,3] & b[3,4] \end{pmatrix}$$

9. Consider the list of 17 integers:

{14, 29, 30, 35, 53, 86, 42, 76, 16, 98, 87, 54, 100, 69, 20, 101, 3}

With the function **Select**, pick out from the list the integers that are odd.

Answer:

```
{29, 35, 53, 87, 69, 101, 3}
```

10. (a) Use **RandomInteger** to create lists **ourlistA** and **ourlistB** each containing 10 random integers in the range 1 to 300. (b) Using the functions **ReplacePart** and **Part**, replace the third element of **ourlistA** with the seventh element of **ourlistB**.
11. (a) Use **RandomInteger** to create lists **mylistA** and **mylistB** each containing 20 random integers in the range 301 to 600. (b) Using the functions **ReplacePart** and **Part**, replace the second element of **mylistA** with the 12th element of **mylistB**.

12. Consider the list of 20 integers:

```
{10, 56, 13, 88, 64, 1, 58, 39, 65,
 85, 44, 67, 73, 23, 89, 43, 10, 57, 3, 67}
```

(a) Using **Partition**, partition the list into five nonoverlapping sublists of four integers. (b) Using **Part**, pick out the third sublist created in part (a).

Answers:

```
{{10,56,13,88},{64,1,58,39},
 {65,85,44,67},{73,23,89,43},{10,57,3,67}}
{65,85,44,67}
```

13. Consider the list $\{a, \{b, c\}, a, \{d, \{e, \{f, \{a, c, g\}\}\}\}\}$. (a) Flatten out the list. (b) Flatten out sublists $\{b, c\}$ and $\{a, c, g\}$.

Answers:

```
{a, b, c, a, d, e, f, a, c, g}
{a, b, c, a, {d, {e, {f, a, c, g}}}}
```

14. Consider the list $\{a, \{b, c\}, \{a, \{\{a, f\}, g\}\}, \{d, \{e, \{f, \{a, c, g\}\}\}\}\}$. (a) Flatten out the list. (b) Flatten out sublists $\{a, f\}$ and $\{a, c, g\}$.

Answers:

```
{a, b, c, a, a, f, g, d, e, f, a, c, g}
{a, {b, c}, {a, {a, f, g}}, {d, {e, {f, a, c, g}}}}
```

15. Flatten out sublists $\{g, h\}$ and $\{r, s\}$ in the list $\{a, \{d, \{e, \{f, \{g, h\}\}\}\}, \{\{i, \{j, k\}\}, \{p, q, \{r, s\}\}\}\}$.

Answer:

```
{a, {d, {e, {f, g, h}}}, {{i, {j, k}}, {p, q, r, s}}}
```

16. Consider the list $\{a, \{\{b, f\}, c\}, \{a, \{a, \{f, d\}\}\}, \{d, \{e, \{f, a, c, g\}\}\}\}$. (a) Flatten out the list. (b) Flatten out sublists $\{b, f\}$ and $\{f, d\}$.

Answers:

```
{a, b, f, c, a, a, f, d, d, e, f, a, c, g}
{a, {b, f, c}, {a, {a, f, d}}, {d, {e, {f, a, c, g}}}}
```

17. Use **RandomInteger** to create a list of 30 random integers in the range 1 to 200, and name the list **someRandoms**. With **Select**, pick out from **someRandoms** the integers that

are greater than 50. *Hint:* What follows is a function that returns **True** if its argument exceeds 50:

```
bigger[x_] := x > 50
```

18. We can label the energy eigenstates of the hydrogen atom with four quantum numbers: the principal quantum number n, azimuthal or orbital quantum number l, magnetic quantum number m_l, and spin magnetic quantum number m_s. The allowed values of these quantum numbers are specified in Example 2.4.9. Consider $n = 4$. Use **Table** to generate a nested list of possible lists of quantum numbers $\{n, l, m_l, m_s\}$. Use **Flatten[*list*, 2]** to flatten out the top two levels of the nested list. Then, use **Length** to determine the number of elements in the resulting list and, therefore, the number of eigenstates for $n = 4$.

    Answer:

    ```
    32
    ```

19. We can label the energy eigenstates of the hydrogen atom with four quantum numbers: the principal quantum number n, azimuthal or orbital quantum number l, magnetic quantum number m_l, and spin magnetic quantum number m_s. The allowed values of these quantum numbers are specified in Example 2.4.9. (a) Using the functions **Table** and **Flatten**, generate a nested list of possible lists of quantum numbers $\{n, l, m_l, m_s\}$ for $2 \le n \le 4$. (b) Using the function **Select**, pick out those lists in part (a) with $m_l = -2$.

    Answer:

    $$\{\{3, 2, -2, -\tfrac{1}{2}\}, \{3, 2, -2, \tfrac{1}{2}\}, \{4, 2, -2, -\tfrac{1}{2}\}, \{4, 2, -2, \tfrac{1}{2}\}, \{4, 3, -2, -\tfrac{1}{2}\}, \{4, 3, -2, \tfrac{1}{2}\}\}$$

*20. Consider the lists $\{n_1, n_2, n_3\}$ of three quantum numbers, where n_1, n_2, and n_3 are nonnegative integers—that is, $0, 1, 2, 3 \ldots$. (a) Using the functions **Table** and **Flatten**, generate a single list of all possible lists of these three quantum numbers for $n_1 \le 5, n_2 \le 5$, and $n_3 \le 5$. (b) Using the function **Select**, pick out from this list those lists in which $n_1 + n_2 + n_3 = 4$.

    Answer:

    ```
    {{0, 0, 4}, {0, 1, 3}, {0, 2, 2}, {0, 3, 1}, {0, 4, 0},
     {1, 0, 3}, {1, 1, 2}, {1, 2, 1}, {1, 3, 0}, {2, 0, 2},
     {2, 1, 1}, {2, 2, 0}, {3, 0, 1}, {3, 1, 0}, {4, 0, 0}}
    ```

*21. Using **Union**, **Sort**, and **SameQ**, write a function that determines whether a list has only distinct elements. For example,

```
noduplicates[{d, c, a, b}]
True
noduplicates[{d, d, d, a}]
False
noduplicates[{a, a, a, a}]
False
```

Hint: From the "front end help," obtain information on **SameQ**.

*22. Using **Union**, **Sort**, and **UnSameQ**, write a function that determines whether a list has duplicate elements. For example,

```
duplicates[{d, c, a, b}]
False
duplicates[{d, d, d, a}]
True
duplicates[{a, a, a, a}]
True
```

Hint: From the "front end help," obtain information on **UnSameQ**.

*23. Using **Union**, **Length**, and **Equal**, write a function that determines whether all the elements of a list are the same. For example,

```
onlyduplicates[{d, c, a, b}]
False
onlyduplicates[{a, a, a, d}]
False
onlyduplicates[{a, a, a, a}]
True
onlyduplicates[{x^2 -1, x^2 -1, x^2 -1, x^2 -1, (x -1) (x + 1) }]
False
onlyduplicates[{{-1, -1, 2}, {-1, -1, 2}, {-1, -1, 2}, {-1, -1, 2}}]
True
```

24. The function f is defined by

$$f(x) = \sin^2 x$$

Using **MapAt**, apply f to several parts of

```
{a, {b, c}, a, {d, {e, {a, c, g}}}}
```

in order to generate

```
{a, {b, Sin[c]^2}, Sin[a]^2, {d, {e, {Sin[a]^2, c, g}}}}
```

25. (a) Using **Range**, create a list of the squares of the first eight even integers. (b) Using **Apply**, **Plus**, and **Length**, determine the mean value of the elements of the list created in part (a).

    Answers:

```
{4, 16, 36, 64, 100, 144, 196, 256}
102
```

26. (a) Using **Range**, create a list of the cubes of the first four positive integers (i.e., natural numbers). (b) Repeat for the next four positive integers. (c) Merge these two lists and name the new list **mylist**. (d) Using **Apply**, **Plus**, and **Length**, find the average of the elements in **mylist**. (e) Create a new list in which each element is the square of the difference between the corresponding element and the average of the elements in **mylist**.

    Answers:

```
{1, 8, 27, 64}
{125, 216, 343, 512}
{1, 8, 27, 64, 125, 216, 343, 512}
162
{25 921, 23 716, 18 225, 9604, 1369, 2916, 32 761, 122 500}
```

27. Consider the list of 16 integers:

```
ourlist = {84, 79, 30, 45, 51, 86, 42, 57, 6, 98, 3, 87, 14, 100, 69, 20}
```

  (a) Using **Apply**, **Plus**, and **Length**, determine the mean (average) of the 16 integers.
  (b) Using **Total** and **Length**, determine the mean of the 16 integers.
  (c) Using **Mean**, determine the mean of the 16 integers.
  (d) Using **Plus**, **Take**, and **Sort**, find the median of the 16 integers. The median is the middle element of a sorted list or the average of the middle two elements, if the list has an even number of elements.
  (e) Using **Median**, find the median of the 16 integers.
  (f) Using the function **Select**, pick out from the list the integers that are larger than 50. *Hint:* Define a function as a predicate (criterion).
  (g) Using **Partition**, partition **ourlist** into four lists, each of four integers.
  (h) List the third row of the 4×4 matrix created in part (g).
  (i) List the third column of the 4×4 matrix created in part (g).
  (j) Using **Flatten**, convert the nested list created in part (g) back into a **ourlist**.

Answers:

```
54.4375
54.4375
54.4375
54
54
{84, 79, 51, 86, 57, 98, 87, 100, 69}
{{84, 79, 30, 45}, {51, 86, 42, 57}, {6, 98, 3, 87}, {14, 100, 69, 20}}
{6, 98, 3, 87}
{30, 42, 3, 69}
{84, 79, 30, 45, 51, 86, 42, 57, 6, 98, 3, 87, 14, 100, 69, 20}
```

28. (a) Create a list of the integers from 0 to 10, and name it **xlist**. (b) Create a list of the squares of the integers from 0 to 10, and name it **ylist**. (c) Generate the list of $\{xlist_i, ylist_i\}$. (d) Plot the points $\{xlist_i, ylist_i\}$.

    Answers:

```
{0, 1, 2, 3, 4, 5, 6, 7, 8, 9, 10}
{0, 1, 4, 9, 16, 25, 36, 49, 64, 81, 100}
{{0, 0}, {1, 1}, {2, 4}, {3, 9}, {4, 16},
 {5, 25}, {6, 36}, {7, 49}, {8, 64}, {9, 81}, {10, 100}}
```

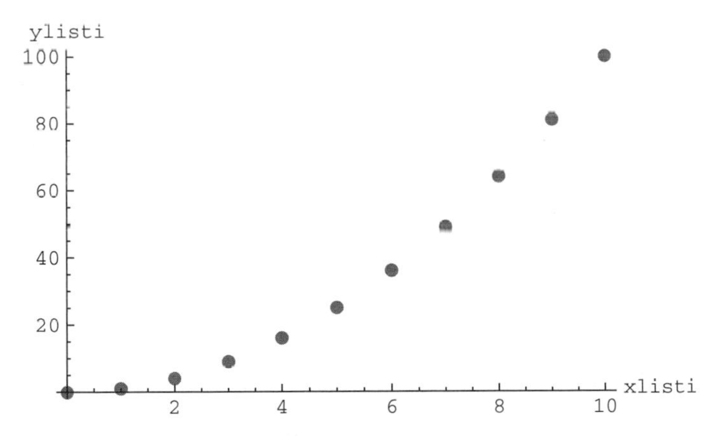

29. Consider the list of the first 100 integers. (a) Partition the list into two sublists. (b) Reverse the order of the second sublist. (c) Add the corresponding elements of the two sublists. (d) What can be inferred about the sum and mean of the first 100 integers?

    Answers:

```
{{1, 2, 3, 4, 5, 6, 7, 8, 9, 10, 11, 12, 13, 14, 15, 16, 17, 18,
  19, 20, 21, 22, 23, 24, 25, 26, 27, 28, 29, 30, 31, 32, 33, 34,
```

```
 35, 36, 37, 38, 39, 40, 41, 42, 43, 44, 45, 46, 47, 48, 49, 50},
{51, 52, 53, 54, 55, 56, 57, 58, 59, 60, 61, 62, 63, 64, 65, 66, 67,
 68, 69, 70, 71, 72, 73, 74, 75, 76, 77, 78, 79, 80, 81, 82, 83, 84,
 85, 86, 87, 88, 89, 90, 91, 92, 93, 94, 95, 96, 97, 98, 99, 100}}
{{1, 2, 3, 4, 5, 6, 7, 8, 9, 10, 11, 12, 13, 14, 15, 16, 17, 18,
  19, 20, 21, 22, 23, 24, 25, 26, 27, 28, 29, 30, 31, 32, 33, 34,
  35, 36, 37, 38, 39, 40, 41, 42, 43, 44, 45, 46, 47, 48, 49, 50},
 {100, 99, 98, 97, 96, 95, 94, 93, 92, 91, 90, 89, 88, 87, 86, 85,
  84, 83, 82, 81, 80, 79, 78, 77, 76, 75, 74, 73, 72, 71, 70, 69, 68,
  67, 66, 65, 64, 63, 62, 61, 60, 59, 58, 57, 56, 55, 54, 53, 52, 51}}
{101, 101, 101, 101, 101, 101, 101, 101, 101, 101, 101, 101,
 101, 101, 101, 101, 101, 101, 101, 101, 101, 101, 101, 101,
 101, 101, 101, 101, 101, 101, 101, 101, 101, 101, 101, 101, 101,
 101, 101, 101, 101, 101, 101, 101, 101, 101, 101, 101, 101, 101}
5050
```

$$\frac{101}{2}$$

30. The function g is defined by

$$g(x) = a + b \ln x$$

Apply g to the integers 3 and 27 in the list $\{1, 2, 3, \{3, 4, 5, \{27, 9, 3\}\}\}$.

Answer:

```
{1, 2, a + b Log[3], {a + b Log[3], 4, 5, {a + b Log[27], 9, a + b Log[3]}}}
```

*31. Using **`Interpolation`** and **`Integrate`**, compute an approximation to the integral in Example 2.4.11 and thus determine the change in entropy of the system. Compare the result with that calculated with the trapezoidal rule.

Answer:

```
7.51218
```

with the units cal/K.

32. Prove that

$$\nabla \cdot (\mathbf{A} \times \mathbf{B}) = \mathbf{B} \cdot (\nabla \times \mathbf{A}) - \mathbf{A} \cdot (\nabla \times \mathbf{B})$$

where **A** and **B** are any two vector point functions.

33. Show that

$$\nabla \times (f\,\mathbf{A}) = f\,(\nabla \times \mathbf{A}) - \mathbf{A} \times (\nabla f)$$

where f is a scalar point function and **A** is a vector point function.

34. In Example 2.4.15, we evaluated **V /. ((a/r)^3 -> 0)** in order to verify that the potential V matches the boundary condition at $r >> a$. (a) Why don't we apply the transformation rule **(a/r) -> 0** instead of the rule **(a/r)^3 -> 0**? (b) Why don't we simply apply the rule **r -> ∞**? (c) Is it more motivating to evaluate **V /. a -> (f r) /. f -> 0** or **V //. {a -> (f r), f -> 0}** rather than **V /. ((a/r)^3 -> 0)**? (d) How does the operator "**//.**", which we have not introduced, perform replacements?

35. The following data

```
data = {{0.46, 0.19}, {0.69, 0.27}, {0.71, 0.28}, {1.04, 0.62},
  {1.11, 0.68}, {1.14, 0.70}, {1.14, 0.74}, {1.20, 0.81},
  {1.31, 0.93}, {2.03, 2.49}, {2.14, 2.73}, {2.52, 3.57},
  {3.24, 3.90}, {3.46, 3.55}, {3.81, 2.87}, {4.06, 2.24},
  {4.93, 0.65}, {5.11, 0.39}, {5.26, 0.33}, {5.38, 0.26}}
```

follow roughly a normal distribution

$$y(x) = a_1 \exp\left\{-\frac{1}{2}\left(\frac{x-a_2}{a_3}\right)^2\right\}$$

(a) Using **FindFit**, determine the values of a_1, a_2, and a_3. (b) To see how well the least-squares curve fits the data, plot the curve and data together.

Answers:

```
{a1 → 3.97924, a2 → 2.99725, a3 → 1.00152}
```

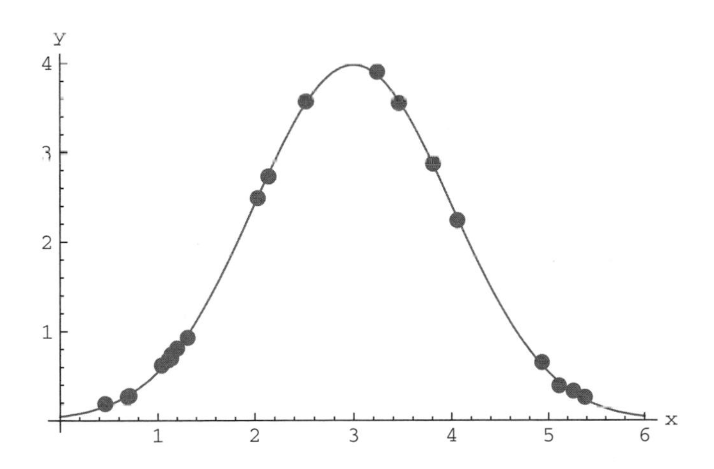

36. Here is a list of coordinates $\{x_i, y_i\}$ of a freely hanging string:

```
data = {{0, 16.3935}, {0.1, 14.1192}, {0.2, 12.1632},
  {0.3, 10.4814}, {0.4, 9.03585}, {0.5, 7.79399}, {0.6, 6.72783},
  {0.7, 5.81333}, {0.8, 5.02988}, {0.9, 4.3598}, {1., 3.78801},
```

```
{1.1, 3.30161}, {1.2, 2.88963}, {1.3, 2.5428}, {1.4, 2.25328},
{1.5, 2.01456}, {1.6, 1.82125}, {1.7, 1.66899}, {1.8, 1.55436},
{1.9, 1.47476}, {2., 1.42841}, {2.1, 1.41426}, {2.2, 1.43199},
{2.3, 1.482}, {2.4, 1.56542}, {2.5, 1.68413}, {2.6, 1.8408},
{2.7, 2.03897}, {2.8, 2.2831}, {2.9, 2.57869}, {3., 2.93242},
{3.1, 3.35224}, {3.2, 3.84764}, {3.3, 4.42977}, {3.4, 5.11175},
{3.5, 5.90897}, {3.6, 6.83939}, {3.7, 7.92398}, {3.8, 9.1872},
{3.9, 10.6575}, {4., 12.3681}, {4.1, 14.3574}, {4.2, 16.6705}}
```

(a) Using **`FindFit`**, determine the values for a, b, and c of the fitting function

$$y(x) = a \cosh(b\,x + c)$$

(b) To see how well the least-squares curve fits the data, plot the curve and data together.

Answers:

```
{a → 1.41421, b → 1.5, c → -3.14159}
```

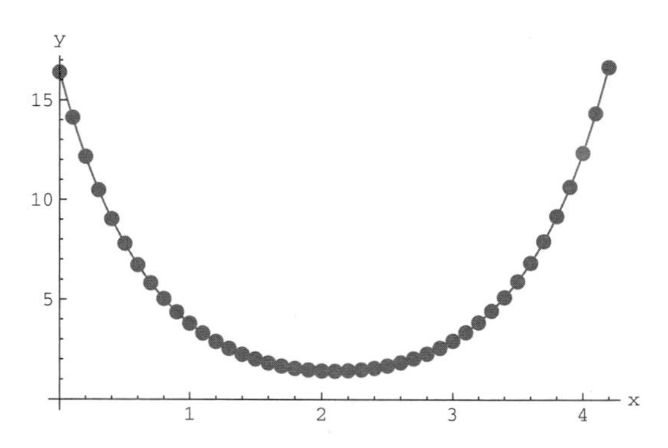

37. Here are the data for beta-particle counts by a Geiger counter in 15-second intervals for an irradiated silver coin. (The data are taken from [BRO3].)

Time (s)	Counts	Time (s)	Counts
15	775	450	35
30	479	465	24
45	380	480	30
60	302	495	26
75	185	510	28
90	157	525	21
105	137	540	18
120	119	555	20
135	110	570	27
150	89	585	17

(*Continued*)

Time (s)	Counts	Time (s)	Counts
165	74	600	17
180	61	615	14
195	66	630	17
210	68	645	24
225	48	660	11
240	54	675	22
255	51	690	17
270	46	705	12
285	55	720	10
300	29	735	13
315	28	750	16
330	37	765	9
345	49	780	9
360	26	795	14
375	35	810	21
390	29	825	17
405	31	840	13
420	24	855	12
435	25	870	18

(a) Using **FindFit**, determine the values of the parameters a_1, a_2, a_3, a_4, and a_5 in the fitting function

$$y = a_1 + a_2\, e^{-t/a_3} + a_4\, e^{-t/a_5}$$

Specify the option **PrecisionGoal→6** for **FindFit**. (**PrecisionGoal** stipulates how many effective digits of precision should be sought in the final result.) (b) To see how well the function fits the data, plot the data and function together.

Answers:

```
{a1 → 14.4281, a2 → 178.456, a3 → 157.075, a4 → 965.105, a5 → 30.0404}
```

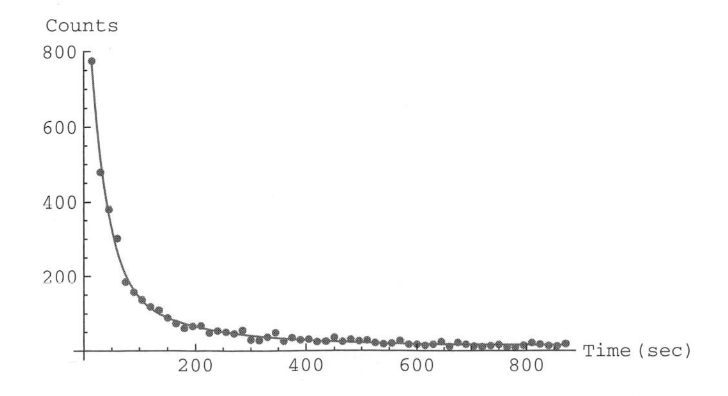

38. Newton's law of cooling is an empirical law stating that the rate of cooling of an object is proportional to the difference between its temperature and the ambient temperature, provided that the temperature difference is not too large. Mathematically, the law can be expressed as

$$\frac{dT}{dt} = -r\,(T - Ta)$$

where T, T_a, and t are the temperature of the object, ambient temperature, and time, respectively. The positive constant r depends on the heat transfer mechanism and the geometric and thermal properties of the object. (a) Let T_0 be the initial temperature of the object. Solve the differential equation for $T(t)$. (b) Here are the water-cooling data taken from [WP05]:

```
{{0, 156}, {29, 154}, {64, 152}, {81, 150}, {109, 150}, {119, 148},
 {139, 148}, {149, 147}, {159, 146}, {174, 145}, {189, 144},
 {199, 144}, {209, 144}, {219, 143}, {229, 143}, {236, 142},
 {249, 142}, {259, 141}, {269, 141}, {279, 140}, {289, 140},
 {299, 139.5}, {309, 139}, {319, 139}, {329, 138}, {339, 138},
 {349, 137}, {359, 137}, {367, 136}, {379, 136}, {389, 136},
 {399, 135}, {409, 135}, {419, 134}, {429, 134}, {439, 133},
 {449, 133}, {459, 133}, {469, 132.5}, {479, 132}, {489, 132},
 {499, 131}, {509, 131}, {519, 131}, {529, 130}, {539, 130},
 {549, 129.5}, {559, 129}, {569, 129}, {579, 128.5}, {589, 128},
 {599, 128}, {609, 128}, {619, 127.5}, {629, 127}, {639, 127},
 {649, 126}, {659, 126}, {669, 126}, {679, 125.5}, {689, 125},
 {699, 125}, {709, 124.5}, {719, 124}, {729, 124}, {739, 123.5},
 {749, 123}, {759, 123}, {769, 122.5}, {779, 122}, {789, 122},
 {799, 122}, {809, 121.5}, {819, 121}, {829, 121}, {839, 121}}
```

where the elements of the list are $\{t_i, T_i\}$ for $i = 1, 2, \ldots, 76$, and the time t_i is in seconds and the temperature T_i is in degrees Fahrenheit. The ambient temperature is 79°F. Using the solution $T(t)$ obtained in part (a) as the fitting function and **FindFit**, determine the value of the constant r. Then, plot the data and function together. How well does the function fit the data? (c) Plot the sum of the squares of the deviations $(T_i - T(t_i))$ for $i = 1, 2, \ldots, 76$ versus r from 0 to 0.003. Graphically determine the value of r at which the sum of the squares of the deviations is a minimum. (d) Using **FindFit**, the solution $T(t)$ in part (a) as the fitting function, and the value of r determined in part (c) as the starting value for r, determine the value of the constant r. Then, plot the data and function together. How well does the function fit the data? What conclusion can be drawn from this exercise?

Answers:

$e^{-rt}(T0 - Ta) + Ta$

```
{r → 0.}
```

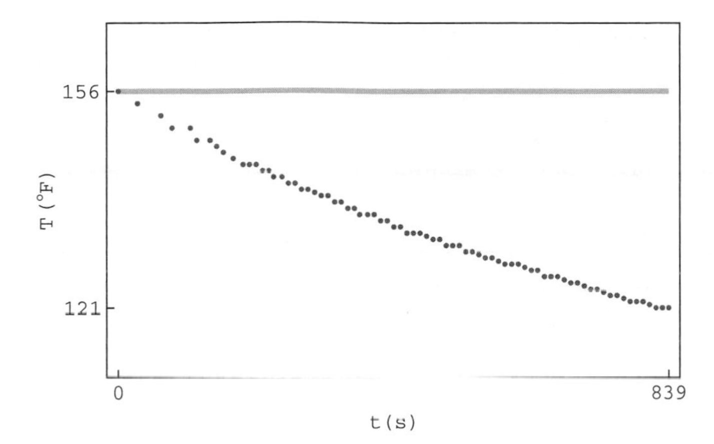

The fit is terrible.

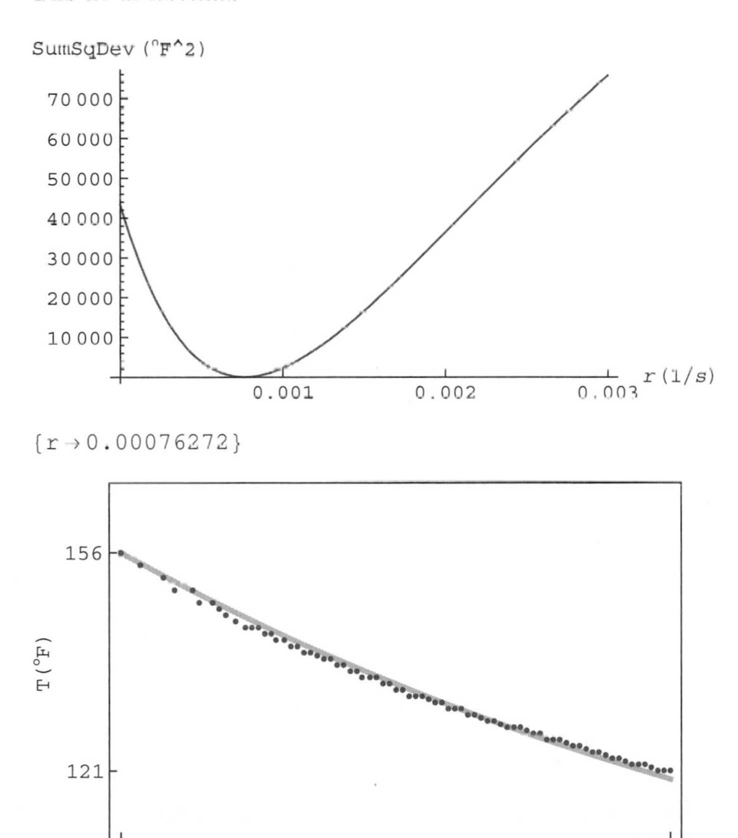

The fit is good. Here the specification of a starting value for the parameter r is crucial.

39. Consider the following data for nitrogen (N_2):

T(K)	100	200	300	400	500	600
B(cm^3/mol)	-160	-35	-4.2	9.0	16.9	21.3

where T is the temperature and B is the second virial coefficient. (a) Using **Interpolation**, find the approximate function corresponding to the data. (b) Plot the function and data together. (c) Determine the second virial coefficient at 450 K.

Answers:

```
InterpolatingFunction[{{100., 600.}}, <>]
```

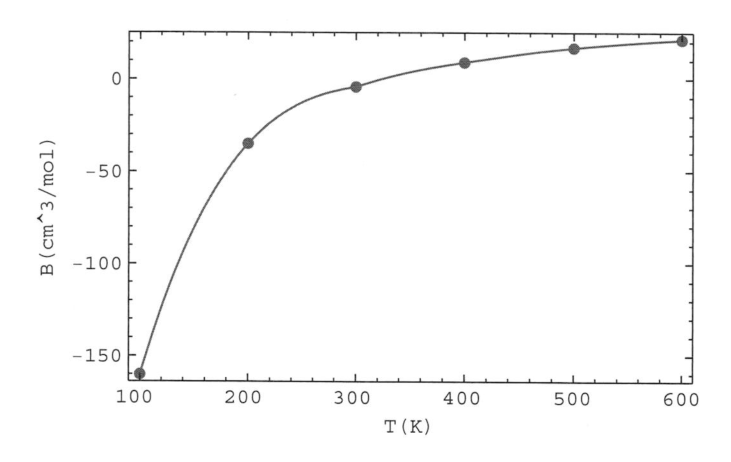

$$\frac{13.5\ \mathrm{cm}^3}{\mathrm{mol}}$$

40. Here are some data for water:

T(°C)	50	60	65	75	80
ρ(kg/m^3)	988.0	985.7	980.5	974.8	971.6

where T is the temperature and ρ is the density. (a) Using **Interpolation** with the option **InterpolationOrder ->** n, find the approximate functions for $n = 1, 2, 3$, and 4. (b) Plot together the approximate functions in part (a). Give each curve a different color. (c) What conclusions about the option **InterpolationOrder** can be drawn from the graph?

Answer:

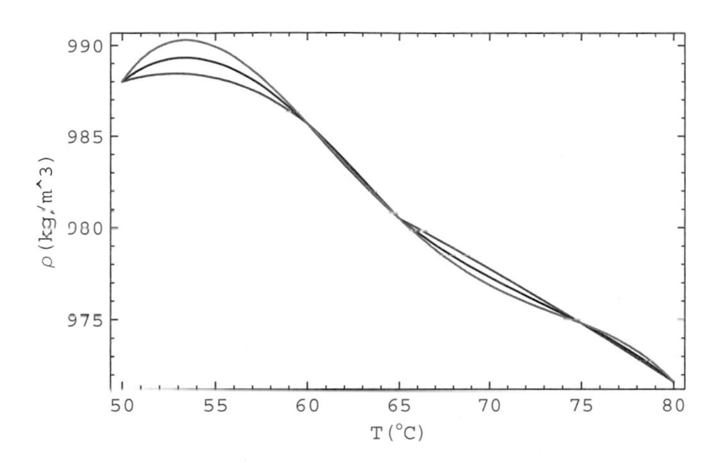

41. A surveyor made 10 measurements of the distance (in feet) between two points: 1023.56, 1023.47, 1023.51, 1023.49, 1023.51, 1023.48, 1023.50, 1023.53, 1023.48, 1023.52. (a) Determine the sample mean and the sample standard deviation. (b) Assuming all errors are random, give the best estimate for the distance between the two points and its uncertainty.

Answers:

```
1023.505
0.027
0.009
```

42. What is the probability that a measurement of a quantity falls within one standard deviation of the true value of the quantity? Assume that all errors are random.

Answer:

```
0.682689
```

43. A student recorded the following counts in 1-minute intervals from a radioactive source:

Number of decays	9	11	15	16	19	22	24
Times observed	2	2	2	1	1	1	1

(a) Determine the best estimates of the mean number of decays in 1 minute, the standard deviation, and the uncertainty of the mean. (b) Plot together the bar histogram of the fractional number of "times observed" versus the "number of decays" and the Poisson distribution with μ being the best estimate of the mean number of decays in 1 minute.

Answers:

```
15.1
3.9
1.2
```

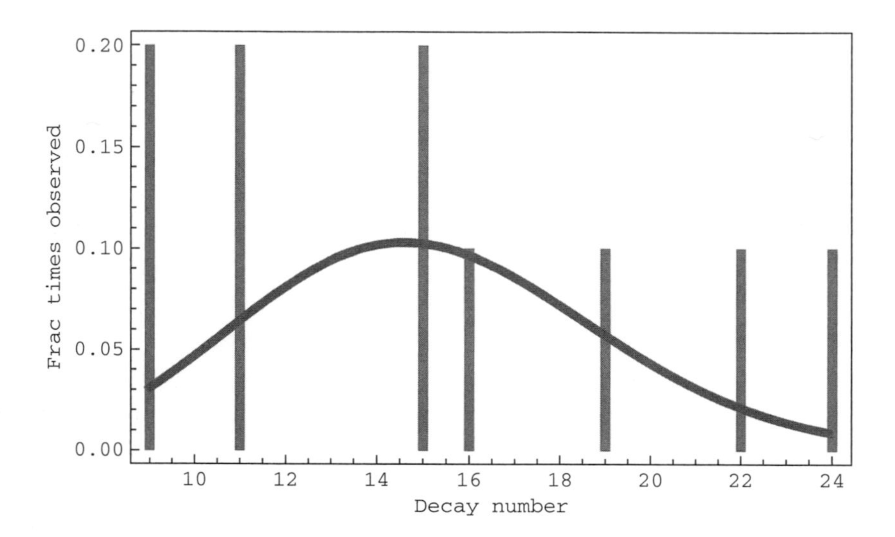

*44. If the average count of neutrinos in a detector is 2 per day, what is the probability of having a count of 8 or more in one (a) day and (b) 10-minute interval? *Hint:* Using the "front end help," access information on **NSum**.

Answers:

```
0.0011
3.4×10^-20
```

*45. **Accumulate**[*list*] gives a list of the successive accumulated totals of elements in *list*. For example,

```
Accumulate[{a, b, c, d}]
{a, a + b, a + b + c, a + b + c + d}
Accumulate[{{a, b}, {c, d}, {e, f}}]
{{a, b}, {a + c, b + d}, {a + c + e, b + d + f}}
```

Using **Table**, **Sum**, **Part**, and **Length**, write a function named **myaccumulate** that is a replicate of the built-in function **Accumulate**. For example,

```
myaccumulate[{a, b, c, d}]
{a, a+b, a+b+c, a+b+c+d}
myaccumulate[{a, b, c, d, e}]
{a, a+b, a+b+c, a+b+c+d, a+b+c+d+e}
myaccumulate[{{a, b}, {c, d}, {e, f}}]
{{a, b}, {a+c, b+d}, {a+c+e, b+d+f}}
myaccumulate[{{a, b}, {c, d}, {e, f}, {g, h}}]
{{a, b}, {a+c, b+d}, {a+c+e, b+d+f}, {a+c+e+g, b+d+f+h}}
```

46. **Accumulate**[*list*] gives a list of the successive accumulated totals of elements in *list*. For example,

```
Accumulate[{a, b, c, d}]
{a, a+b, a+b+c, a+b+c+d}
```

Differences[*list*] gives the successive differences of elements in *list*. For example,

```
Differences[{a, b, c, d}]
{-a+b, -b+c, -c+d}
```

(a) Generate a list named **mylist** whose elements are **a**[*i*] with *i* ranging from 1 to *n*, where *n* is a random integer in the range 10 to 20. (b) Apply **Accumulate** to **mylist** (c) Apply **Differences** to the list generated in part (b). What can be said about the relation between **Accumulate** and **Differences**?

47. This exercise is for Windows users only. (a) Create a new folder at a convenient location, name the folder MyFolder, and determine its address (i.e., full path). *Hint:* To determine the address of a folder, open the folder; the address bar displays its address. (b) Evaluate.

```
MyData = {{"t (s)", "d (ft)"},
   {0.5, 4.2}, {1.0, 16.1}, {1.5, 35.9}, {2.0, 64.2}};
```

(c) Using **Export**, export the data in part (b) in the form of a table to a text file named Grav in MyFolder. *Hint:* **Export**["*fullfilename*", *data*, "**Table**"] exports *data* in the table format to a file; the *fullfilename* of the file Grav is *FolderAddress***Grav.txt**, where **txt** is the text file extension and *FolderAddress* is the address of the folder determined in part (a). (d) Using **Export**, export the data to an Excel file named Gravl in MyFolder. *Hint:* **Export**["*fullfilename*", *data*] exports *data* to a file; the *fullfilename* of the Excel file Gravl is *FolderAddress***Gravl.xls**.

48. This exercise is for Mac OS X users only. (a) Create a new folder at a convenient location, name the folder MyFolder, and determine its location. *Hint:* To determine the location of a folder, open the folder and choose **File ▸ Get Info**; General/Where displays its location. (b) Evaluate

```
MyData = {{"t (s)", "d (ft)"},
   {0.5, 4.2}, {1.0, 16.1}, {1.5, 35.9}, {2.0, 64.2}};
```

(c) Using **Export**, export the data in part (b) in the form of a table to a text file named Grav in MyFolder. *Hint:* **Export["***fullfilename***",** *data*, **"Table"]** exports *data* in the table format to a file; the *fullfilename* of the file Grav is *FolderLocation*/**MyFolder/Grav.txt**, where **txt** is the text file extension and *FolderLocation* is the location of the folder determined in part (a). (d) Using **Export**, export the data to an Excel file named Gravl in MyFolder. *Hint:* **Export["***fullfilename***",** *data***]** exports *data* to a file; the *fullfilename* of the Excel file Gravl is *Folderlocation*/**MyFolder/Gravl.xls**.

*49. This exercise is for Mac OS X users only. (a) Create a new folder at a convenient location, name the folder OurFolder, and determine its location. *Hint:* To determine the location of a folder, open the folder and choose **File ▸ Get Info**; General/Where displays its location. (b) Visit the website http://cdiac.esd.ornl.gov/trends/co2/sio-mlo.htm. Click the Digital Data link near the top of the page. Copy the data including the table headings and paste the data into a blank Microsoft Word document. Choose **File ▸ Save**, select Text Only in the drop-down Format menu, name the document ACDC.txt, and save it in OurFolder. (c) Import the data in part (b) to a *Mathematica* notebook and name it **data**. *Hint:* **Import["***fullfilename***", "Table"]** imports a table of data from a file; the *fullfilename* of the file ACDC.txt is *FolderLocation*/**OurFolder/ACDC.txt**, where **txt** is the text file extension and *FolderLocation* is the location of the folder determined in part (a). (d) Using **Drop**, delete the first nine elements from **data** (i.e., delete the table headings as well as data for years 1958–1964 and retain the data for years 1965–2004) and name the result **mydata**. (e) Using **Take** and **Map**, delete the first and last two columns of **mydata** (i.e., delete the columns for Year, Annual, and Annual-Fit) and name the result **ourdata**. *Hint:* The following function definition may be useful:

```
func[x_] := Take[x, {2, 13}]
```

(f) Using **Flatten**, flatten out **ourdata** and name it **newdata**. (g) Using **Interpolation** and **Plot**, generate for **newdata** the plot

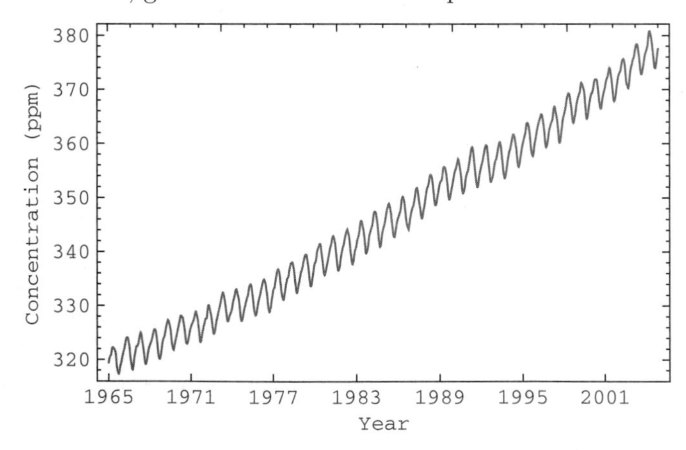

Hint: See Section 2.4.11.3.

*50. This exercise is for Windows users only. (a) Create a new folder at a convenient location, name the folder OurFolder, and determine its address (i.e., full path). *Hint:* To determine the address of a folder, open the folder; the address bar displays its address. (b) Visit the web site http://cdiac.esd.ornl.gov/trends/co2/sio-mlo.htm. Click the Digital Data link near the top of the page. Copy the data including the table headings and paste the data into a blank Notepad document. Choose **File ▸ Save**, select Text Documents in the drop-down "Save as type" menu, name the document ACDC.txt, and save it in OurFolder. (c) Import the data in part (b) to a *Mathematica* notebook and name it **data**. *Hint:* **Import["***full file name***", "Table"]** imports a table of data from a file; the *full file name* of the file ACDC.txt is *FolderAddress***\ACDC.txt**, where **txt** is the text file extension and *FolderAddress* is the address of the folder determined in part (a). (d) Using **Drop**, delete the first nine elements from **data** (i.e., delete the table headings as well as data for years 1958–1964 and retain the data for years 1965–2004) and name the result **mydata**. (e) Using **Take** and **Map**, delete the first and last two columns of **mydata** (i.e., delete the columns for Year, Annual, and Annual-Fit) and name the result **ourdata**. *Hint:* The following function definition may be useful:

```
func[x_] := Take[x, {2, 13}]
```

(f) Using **Flatten**, flatten out **ourdata** and name it **newdata**. (g) Using **Interpolation** and **Plot**, generate for **newdata** the plot

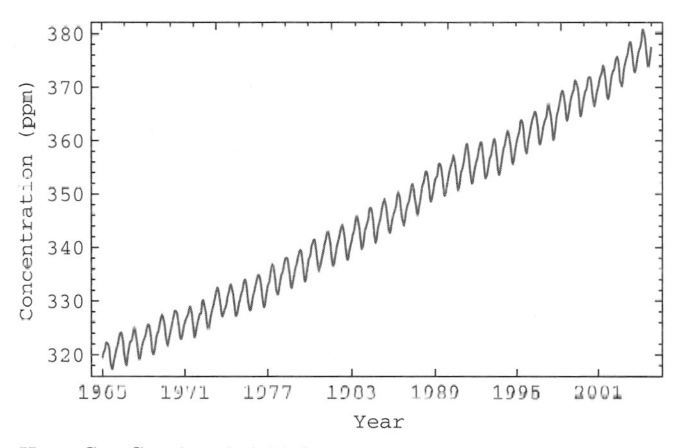

Hint: See Section 2.4.11.3.

2.5 SPECIAL CHARACTERS, TWO-DIMENSIONAL FORMS, AND FORMAT TYPES

We have postponed to this section the introduction of *Mathematica*'s marvelous typesetting capabilities that feature special characters, two-dimensional forms, and various format types. Section 2.5.3 elucidates the reasons for the delay. To reproduce input and output that resemble those in this section, click **Mathematica ▸ Preferences ▸ Evaluation** for Mac OS X or **Edit ▸ Preferences ▸ Evaluation** for Windows (i.e., click the Evaluation tab in the Preferences window of the Application (Mathematica) menu for Mac OS X or the Edit menu for Windows) and then select **StandardForm** and **TraditionalForm** in the drop-down menus of "Format type of new input cells:" and "Format type of new output cells:", respectively.

2.5.1 Special Characters

Section 1.6.2 introduced the notion of special characters in *Mathematica*, focused only on the Greek letters, and showed how to enter them in a notebook. There are many other special characters. In addition to ordinary keyboard characters, *Mathematica* recognizes nearly 1000 special characters. We can use the special characters just like ordinary keyboard characters. For a complete list of these named characters, choose **Help ▸ Document Center ▸ Notebooks and Documents/Special Characters ▸ More About/Listing of All Special Characters**; clicking a character displays its reference page showing its description and relevant links.

There are four kinds of special characters: letters and letterlike forms, operators, structural elements, and spacing characters. This section shows several ways to enter these characters, gives examples of them, and cautions that there are many similar-looking but different characters.

2.5.1.1 Ways to Enter Special Characters

There are three ways to enter special characters in a notebook:

Palettes — *Mathematica* 6.0.0 and 6.0.1 : Choose **Palettes ▸ SpecialCharacters**, and select an item such as **General Operators** in the drop-down menu; click a character such as ⊕, and click the Insert button to paste it to the notebook.
Mathematica 6.0.2 : Choose **Palettes ▸ SpecialCharacters**, and click a button (or tab) such as Symbols; click one of the buttons/tabs such as General Operators (i.e., second one from the left in the row of seven buttons/tabs); click a character such as ⊕ to paste it to the notebook.

Full names — Enter from the keyboard the full name of a character such as\[CirclePlus] for ⊕; *Mathematica* automatically converts the full name to the character; to display full names, if desired, click the cell bracket, select **Option Inspector** in the Format menu, choose **selection** and **alphabetically** at the top of the Option Inspector window, and change the setting of **ShowSpecialCharacters** to False.

Aliases — Enter from the keyboard an alias of a character such as ESCc+ESC for ⊕, where ESC represents the Esc key for Windows and the *esc* key for Mac OS X.

2.5.1.2 Letters and Letterlike Forms

Here are some letters and letterlike forms together with their full names as well as aliases:

Character	Full Name	Aliases
α	`\[Alpha]`	⋮a⋮, ⋮alpha⋮
β	`\[Beta]`	⋮b⋮, ⋮beta⋮
γ	`\[Gamma]`	⋮g⋮, ⋮gamma⋮
δ	`\[Delta]`	⋮d⋮, ⋮delta⋮

(*Continued*)

Character	Full Name	Aliases
ϵ	\[Epsilon]	⋮e⋮, ⋮epsilon⋮
ε	\[CurlyEpsilon]	⋮ce⋮, ⋮cepsilon⋮
θ	\[Theta]	⋮q⋮, ⋮th⋮, ⋮theta⋮
λ	\[Lambda]	⋮l⋮, ⋮lambda⋮
π	\[Pi]	⋮p⋮, ⋮pi⋮
ϕ	\[Phi]	⋮f⋮, ⋮ph⋮, ⋮phi⋮
ψ	\[Psi]	⋮y⋮, ⋮ps⋮, ⋮psi⋮
ω	\[Omega]	⋮o⋮, ⋮w⋮, ⋮omega⋮
Γ	\[CapitalGamma]	⋮G⋮, ⋮Gamma⋮
Δ	\[CapitalDelta]	⋮D⋮, ⋮Delta⋮
Φ	\[CapitalPhi]	⋮F⋮, ⋮Ph⋮, ⋮Phi⋮
Ψ	\[CapitalPsi]	⋮Y⋮, ⋮Ps⋮, ⋮Psi⋮
Ω	\[CapitalOmega]	⋮O⋮, ⋮W⋮, ⋮Omega⋮
ℓ	\[ScriptL]	⋮scl⋮
ℰ	\[ScriptCapitalE]	⋮scE⋮
ℋ	\[ScriptCapitalH]	⋮scH⋮
ℒ	\[ScriptCapitalL]	⋮scL⋮
ℝ	\[DoubleStruckCapitalR]	⋮dsR⋮
°	\[Degree]	⋮deg⋮
Å	\[Angstrom]	⋮Ang⋮
ℏ	\[HBar]	⋮hb⋮
∞	\[Infinity]	⋮inf⋮
ⅇ	\[ExponentialE]	⋮ee⋮
ⅈ	\[ImaginaryI]	⋮ii⋮
ⅉ	\[ImaginaryJ]	⋮jj⋮
∠	\[Angle]	
•	\[Bullet]	⋮bu⋮
†	\[Dagger]	⋮dg⋮

In aliases, ⋮ stands for the key ESC.

We can use letters and letterlike forms in symbol names just like ordinary keyboard letters. For example, the energy eigenvalues of the one-dimensional harmonic oscillator are

```
In[1]:= ℰ[n_] := (n+1/2)ℏω
```

where n takes on any nonnegative integer, and the energy for $n = 5$ is

```
In[2]:= ℰ[5]
```

Out[2]= $\frac{11\,\omega\,\hbar}{2}$

```
In[3]:= Clear[ℰ]
```

Although most of the letters and letterlike forms do not have built-in meanings, several do:

Character	Equivalent to
π	**Pi**
ⅇ	**E**
∞	**Infinity**
°	**Degree**
ⅈ	**I**
ⅉ	**I**

For example,

```
In[4]:= {α, ψ, Γ, βπ, π, ⅇ^-2, Sin[60°], ⅈ^2} //N
Out[4]= {α, ψ, Γ, βπ, 3.14159, 0.135335, 0.866025, -1.}
```

Note that π has built-in meaning, but π in βπ assumes no special meaning.

2.5.1.3 Operators

What follow are some operators grouped according to functionality.

Common Mathematical Operators

Character	Full Name	Alias	Example	Built-in Meaning
×	\[Times]	⋮*⋮	$x \times y$	**Times** (*)
÷	\[Divide]	⋮div⋮	$x \div y$	**Divide** (/)
√	\[Sqrt]	⋮sqrt⋮	$\sqrt{x}$	**Sqrt**
×	\[Cross]	⋮cross⋮	$x \times y$	**Cross**
∂	\[PartialD]	⋮pd⋮	$\partial_x y$	**D**
ⅆ	\[DifferentialD]	⋮dd⋮	$\int f\, dx$	for use in integrals
∫	\[Integral]	⋮int⋮	$\int f\, dx$	integral sign

The operators ×, ÷, and × are infix operators that go between their operands; the operators √ and ∂ are prefix operators that precede their operands; and the characters ⅆ and ∫ are elements of compound operators that evaluate integrals. These operators follow built-in evaluation rules:

```
In[5]:= (3×√4)÷6
Out[5]= 1
```

```
In[6]:= {Ax, Ay, Az}×{Bx, By, Bz}
```

Out[6]= $\{A_y B_z - A_z B_y, A_z B_x - A_x B_z, A_x B_y - A_y B_x\}$

```
In[7]:= ∫Sin[x] ⅆx
Out[7]= -cos(x)
```

Note that × is different from × and that the special character ⅆ is different from the ordinary keyboard character d. (Section 2.5.2.1 will discuss ways to enter subscripts.)

Logical Operators

<table>
<tr><th>Character</th><th>Full Name</th><th>Alias</th><th>Example</th><th>Built-in Meaning</th></tr>
<tr><td>∧</td><td>\[And]</td><td>⋮&&⋮, ⋮and⋮</td><td>$x \wedge y$</td><td>And (&&)</td></tr>
<tr><td>∨</td><td>\[Or]</td><td>⋮||⋮</td><td>$x \vee y$</td><td>Or (||)</td></tr>
<tr><td>¬</td><td>\[Not]</td><td>⋮!⋮</td><td>$\neg x$</td><td>Not (!)</td></tr>
<tr><td>⇒</td><td>\[Implies]</td><td>⋮=>⋮</td><td>$x \Rightarrow y$</td><td>Implies</td></tr>
</table>

The operator ¬ is a prefix operator and the others are infix operators. These operators have built-in meanings:

```
In[8]:= ¬((5 > 3)∧(7 < 5))
Out[8]= True
```

Bracketing Operators

<table>
<tr><th>Character</th><th>Full Name</th><th>Alias</th><th>Example</th><th>Built in Meaning</th></tr>
<tr><td>〚</td><td>\[LeftDoubleBracket]</td><td>⋮[[⋮</td><td>$m〚i〛$</td><td>for use in Part ([[)</td></tr>
<tr><td>〛</td><td>\[RightDoubleBracket]</td><td>⋮]]⋮</td><td>$m〚i,j〛$</td><td>for use in Part (]])</td></tr>
<tr><td>〈</td><td>\[LeftAngleBracket]</td><td>⋮<⋮</td><td>$\langle x, y, z\rangle$</td><td></td></tr>
<tr><td>〉</td><td>\[RightAngleBracket]</td><td>⋮>⋮</td><td>$\langle x\rangle$</td><td></td></tr>
<tr><td>|</td><td>\[LeftBracketingBar]</td><td>⋮l|⋮</td><td>$|x, y|$</td><td></td></tr>
<tr><td>|</td><td>\[RightBracketingBar]</td><td>⋮r|⋮</td><td>$|x|$</td><td></td></tr>
</table>

The first two characters are elements of a compound operator with built-in evaluation rules:

```
In[9]:= {2, 3, 4, 5}〚2〛
Out[9]= 3
```

The other characters do not have built-in meanings and are matchfix operators, which come in matching pairs that enclose or delimit their operands:

```
In[10]:= {〈x, y〉, |2 + 3 I|}
Out[10]= {〈x,y〉, |2 + 3 ⅈ|}
```

The operators 〈 〉 and | |, formed with pairs of matchfix operators, are named **AngleBracket** and **BracketingBar**, respectively. We can assign meanings to these operators. For example,

```
In[11]:= |x_| := Abs[x]

In[12]:= |2 + 3 I|
Out[12]= √13
```

For another example, let us define a scalar product of two functions of the variable ζ:

```
In[13]:= ⟨φ_, ψ_⟩ :=
           Integrate[(φ /. Complex[u_, v_] -> Complex[u, -v]) ψ, {ζ, -∞, ∞}]
```

(Section 3.1.2 will explain the necessity of using the rule with the function **Complex** rather than the simpler rule **I -> -I** to transform φ to φ^*.) The energy eigenfunctions of the one-dimensional harmonic oscillator are

```
In[14]:= u[n_, ζ_] := (Sqrt[1/(Sqrt[π] n! 2^n)] * HermiteH[n, ζ] Exp[-(ζ^2)/2])
```

where the dimensionless coordinate $\zeta = \sqrt{m\omega/\hbar}\, x$. Thus,

```
In[15]:= ⟨u[1, ζ], u[1, ζ]⟩
Out[15]= 1

In[16]:= ⟨u[1, ζ], ζ^2 u[3, ζ]⟩
```

Out[16]= $\sqrt{\frac{3}{2}}$

```
In[17]:= Clear[u, BracketingBar, AngleBracket]
```

Other Operators

Character	Full Name	Alias	Example	Built-in Meaning
⊕	\[CirclePlus]	⋮c+⋮	$x \oplus y$	
⊗	\[CircleTimes]	⋮c*⋮	$\otimes x$, $x \otimes y$	
·	\[CenterDot]	⋮.⋮	$x \cdot y$	
⋃	\[Union]	⋮un⋮	$x \bigcup y$	**Union**
⋂	\[Intersection]	⋮inter⋮	$x \bigcap y$	**Intersection**
∇	\[Del]	⋮del⋮	∇f	
□	\[Square]	⋮sq⋮	$\Box x$	
≡	\[Congruent]	⋮===⋮	$x \equiv y$	
~	\[Tilde]	⋮~⋮	$x \sim y$	
≈	\[TildeTilde]	⋮~~⋮	$x \approx y$	
≃	\[TildeEqual]	⋮~=⋮	$x \simeq y$	
∝	\[Proportional]	⋮prop⋮	$x \propto y$	
≠	\[NotEqual]	⋮!=⋮	$x \neq y$	**Unequal** (!=)
≥	\[GreaterEqual]	⋮>=⋮	$x \geq y$	**GreaterEqual** (>=)

(*Continued*)

Character	Full Name	Alias	Example	Built-in Meaning
≤	\[LessEqual]	⋮<=⋮	$x \leq y$	**LessEqual** (<=)
≯	\[NotGreater]	⋮!>⋮	$x \ngtr y$	
⊂	\[Subset]	⋮sub⋮	$x \subset y$	
∈	\[Element]	⋮el⋮	$x \in y$	**Element**
→	\[Rule]	⋮->⋮	$x \to y$	**Rule** (->)
:→	\[RuleDelayed]	⋮:>⋮	$x :\to y$	**RuleDelayed** (:>)
⇀	\[RightVector]	⋮vec⋮	$x \rightharpoonup y, \overrightarrow{x}$	

These operators are infix operators with the following exceptions: □ and ∇ are prefix operators, ⊗ can be a prefix operator, and ⇀ can be an overfix operator that goes over its operand. Several operators have built-in evaluation rules:

```
In[18]:= {3 ≥ 3, 5 ≠ 5, {a, b, c}⋃{b, c, d}} /. {a → 1}
Out[18]= {True, False, {1, b, c, d}}
```

Many of these operators do not have special meanings:

```
In[19]:= {x⊕y, x⊗y, x∝y, 3≃2, ∇x^2}
Out[19]= {x⊕y, x⊗y, x∝y, 3≃2, ∇x²}
```

We can assign meanings to them. For instance, let us define the Kronecker product of two matrices:

```
In[20]:= ρ_⊗σ_ := Outer[Times, ρ, σ]
```

Consider, for example,

```
In[21]:= α = {{α11, α12, α13}, {α21, α22, α23}, {α31, α32, α33}};

In[22]:= β = {{β11, β12}, {β21, β22}};
```

Thus,

```
In[23]:= α⊗β
```

Out[23]=
$$\begin{pmatrix} \begin{pmatrix} \alpha 11\,\beta 11 & \alpha 11\,\beta 12 \\ \alpha 11\,\beta 21 & \alpha 11\,\beta 22 \end{pmatrix} & \begin{pmatrix} \alpha 12\,\beta 11 & \alpha 12\,\beta 12 \\ \alpha 12\,\beta 21 & \alpha 12\,\beta 22 \end{pmatrix} & \begin{pmatrix} \alpha 13\,\beta 11 & \alpha 13\,\beta 12 \\ \alpha 13\,\beta 21 & \alpha 13\,\beta 22 \end{pmatrix} \\ \begin{pmatrix} \alpha 21\,\beta 11 & \alpha 21\,\beta 12 \\ \alpha 21\,\beta 21 & \alpha 21\,\beta 22 \end{pmatrix} & \begin{pmatrix} \alpha 22\,\beta 11 & \alpha 22\,\beta 12 \\ \alpha 22\,\beta 21 & \alpha 22\,\beta 22 \end{pmatrix} & \begin{pmatrix} \alpha 23\,\beta 11 & \alpha 23\,\beta 12 \\ \alpha 23\,\beta 21 & \alpha 23\,\beta 22 \end{pmatrix} \\ \begin{pmatrix} \alpha 31\,\beta 11 & \alpha 31\,\beta 12 \\ \alpha 31\,\beta 21 & \alpha 31\,\beta 22 \end{pmatrix} & \begin{pmatrix} \alpha 32\,\beta 11 & \alpha 32\,\beta 12 \\ \alpha 32\,\beta 21 & \alpha 32\,\beta 22 \end{pmatrix} & \begin{pmatrix} \alpha 33\,\beta 11 & \alpha 33\,\beta 12 \\ \alpha 33\,\beta 21 & \alpha 33\,\beta 22 \end{pmatrix} \end{pmatrix}$$

The product is not commutative:

```
In[24]:= β⊗α
```

$$\text{Out[24]=}\begin{pmatrix} \begin{pmatrix} \alpha 11\,\beta 11 & \alpha 12\,\beta 11 & \alpha 13\,\beta 11 \\ \alpha 21\,\beta 11 & \alpha 22\,\beta 11 & \alpha 23\,\beta 11 \\ \alpha 31\,\beta 11 & \alpha 32\,\beta 11 & \alpha 33\,\beta 11 \end{pmatrix} & \begin{pmatrix} \alpha 11\,\beta 12 & \alpha 12\,\beta 12 & \alpha 13\,\beta 12 \\ \alpha 21\,\beta 12 & \alpha 22\,\beta 12 & \alpha 23\,\beta 12 \\ \alpha 31\,\beta 12 & \alpha 32\,\beta 12 & \alpha 33\,\beta 12 \end{pmatrix} \\ \begin{pmatrix} \alpha 11\,\beta 21 & \alpha 12\,\beta 21 & \alpha 13\,\beta 21 \\ \alpha 21\,\beta 21 & \alpha 22\,\beta 21 & \alpha 23\,\beta 21 \\ \alpha 31\,\beta 21 & \alpha 32\,\beta 21 & \alpha 33\,\beta 21 \end{pmatrix} & \begin{pmatrix} \alpha 11\,\beta 22 & \alpha 12\,\beta 22 & \alpha 13\,\beta 22 \\ \alpha 21\,\beta 22 & \alpha 22\,\beta 22 & \alpha 23\,\beta 22 \\ \alpha 31\,\beta 22 & \alpha 32\,\beta 22 & \alpha 33\,\beta 22 \end{pmatrix} \end{pmatrix}$$

(For a discussion of tensor products, see [Hal58].)

```
In[25]:= Clear[α, β, CircleTimes]
```

2.5.1.4 Structural Elements and Spacing Characters

Structural elements specify structures. Two examples are

Character	Full Name	Alias
⋱	`\[Continuation]`	`⋮cont⋮`
	`\[InvisibleComma]`	`⋮,⋮`

In *Mathematica* output, the character ⋱ indicates the continuation of an expression onto the next line:

```
In[26]:= 80!
Out[26]= 71 569 457 046 263 802 294 811 533 723 186 532 165 584 657 342 365 752 577 109 445 058 227 ⋱
         039 255 480 148 842 668 944 867 280 814 080 000 000 000 000 000 000
```

Invisible commas are interpreted on input as ordinary commas:

```
In[27]:= ⟨αβδγ⟩
Out[27]= ⟨α, β, δ, γ⟩
```

Spacing characters indicate spaces. Two examples are

Character	Full Name	Alias
␣	`\[SpaceIndicator]`	`⋮space⋮`
	`\[InvisibleSpace]`	`⋮is⋮`

The character ␣ represents the space key on a keyboard. (For examples, see Section 2.5.1.5.) Also, *Mathematica* treats it on input as an ordinary space:

```
In[28]:= 10␣25
Out[28]= 250
```

Invisible spaces of zero width are interpreted on input as ordinary spaces:

```
In[29]:= xy /. {x → 3, y → 5}
Out[29]= 15
```

where the invisible space placed between **x** and **y** is interpreted as multiplication.

2.5.1.5 Similar-Looking Characters

Beware of characters that look alike but have disparate meanings. Some examples are

Character	Full Name	Alias
Σ	\[CapitalSigma]	⋮S⋮, ⋮Sigma⋮
∑	\[Sum]	⋮sum⋮
Π	\[CapitalPi]	⋮P⋮, ⋮Pi⋮
∏	\[Product]	⋮prod⋮
U	keyboard U	
⋃	\[Union]	⋮un⋮
ϵ	\[Epsilon]	⋮e⋮, ⋮epsilon⋮
∈	\[Element]	⋮el⋮
d	keyboard d	
ⅆ	\[DifferentialD]	⋮dd⋮
e	keyboard e	
ⅇ	\[ExponentialE]	⋮ee⋮
i	keyboard i	
ⅈ	\[ImaginaryI]	⋮ii⋮
μ	\[Mu]	⋮m⋮, ⋮mu⋮
µ	\[Micro]	⋮mi⋮
A	keyboard A	
Α	\[CapitalAlpha]	⋮A⋮, ⋮Alpha⋮
Å	\[CapitalARing]	⋮Ao⋮
Å	\[Angstrom]	⋮ang⋮
×	\[Times]	⋮*⋮
⨯	\[Cross]	⋮cross⋮
∧	\[And]	⋮&&⋮, ⋮and⋮
∧	\[Wedge]	⋮^⋮

(*Continued*)

Character	Full Name	Alias
→	`\[Rule]`	`⋮->⋮`
→	`\[RightArrow]`	`⋮␣->⋮`
=	keyboard =	
=	`\[LongEqual]`	`⋮l=⋮`
*	keyboard*	
⋆	`\[Star]`	`⋮star⋮`
\	keyboard \	
\	`\[Backslash]`	`⋮\⋮`
|	keyboard |	
|	`\[VerticalSeparator]`	`⋮|⋮`
|	`\[VerticalBar]`	`⋮␣|⋮`
⎸	`\[LeftBracketBar]`	`⋮l|⋮`
⎹	`\[RightBracketBar]`	`⋮r|⋮`

As mentioned in Section 2.5.1.4, the character ␣ stands for Space entered by pressing the space bar.

2.5.2 Two-Dimensional Forms

Mathematica supports two-dimensional as well as one-dimensional input. This section shows several ways to enter two-dimensional forms, describes a number of two-dimensional forms that have built-in meanings, and illustrates the use of two-dimensional forms in solving physics problems.

2.5.2.1 Ways to Enter Two-Dimensional Forms

Palettes

To enter a two-dimensional form with a palette, highlight or select an expression, choose **BasicMathInput** in the Palettes menu, and click a two-dimensional form. The two-dimensional form is pasted in the notebook with the current selection already inserted in the primary or selection placeholder "■". Other expressions can be entered into the remaining placeholders "□". To move from one placeholder to another, use the Tab key or the mouse.

Let us create, for example, the two-dimensional input $\mathbf{x}^{\mathbf{a}}$:

1. Enter

   x

2. Highlight or select **x**, choose **BasicMathInput** in the Palettes menu, and click the button at the upper left-hand corner:

   $\mathbf{x}^{\square}$

3. Enter **a** in the placeholder:

   $\mathbf{x}^{\mathbf{a}}$

Except for a few special superscripts, such as $x^+, x^\dagger$, and x^*, most superscripts are interpreted as powers by *Mathematica.* However, subscripts have no built-in meanings.

For another example, let us enter the two-dimensional form

$$\sum_{n=1}^{k} x_n + \int_1^4 x^2 \, dx$$

1. Enter

   $$x + x^2$$

2. Select the first **x** from the left, choose **BasicMathInput** in the Palettes menu, and click the subscript button:

   $$x_\square + x^2$$

3. Enter **n** in the placeholder:

   $$x_n + x^2$$

4. Select x_n, and click the fifth button from the top of the left column in the palette:

   $$\sum_{\square=\square}^{\square} x_n + x^2$$

5. Enter **n**, **1**, and **k** in the appropriate placeholders, using the Tab key to move from one placeholder to another:

   $$\sum_{n=1}^{k} x_n + x^2$$

6. Select x^2, and click the fourth button from the top in the left column of buttons in the palette:

   $$\sum_{n=1}^{k} x_n + \int_\square^\square x^2 \, d\square$$

7. Type **1**, **4**, and **x** in the respective placeholders, using the Tab key to go from one placeholder to another:

   $$\sum_{n=1}^{k} x_n + \int_1^4 x^2 \, dx$$

Control Keys

We can use control characters to enter two-dimensional forms:

Control Character	Built-in Meaning
CTRL [^] or CTRL [6]	go to the superscript position
CTRL [_] or CTRL [-]	go to the subscript position
CTRL [@] or CTRL [2]	go into a square root
CTRL [%] or CTRL [5]	go from subscript to superscript or vice versa, from underscript to overscript or vice versa, or to the exponent position in a root
CTRL [/]	go to the denominator for a fraction
CTRL [+] or CTRL [=]	go to the underscript position
CTRL [&] or CTRL [7]	go to the overscript position
CTRL [␣]	return from a special position

CTRL [*Key*] stands for Control-*Key*—that is, hold down the Control key and press *Key*. (The Control key is labeled *control* for Mac OS X and Ctrl for Windows.) The character ␣ represents Space—that is, the space bar.

Here are several useful characters for entering two-dimensional forms:

Character	Full Name	Alias	Built-in Meaning
∑	`\[Sum]`	`⋮sum⋮`	summation sign
∏	`\[Product]`	`⋮prod⋮`	product sign
∫	`\[Integral]`	`⋮int⋮`	integral sign
ⅆ	`\[DifferentialD]`	`⋮dd⋮`	for use in integrals
∂	`\[PartialD]`	`⋮pd⋮`	partial derivative operator

Let us create again the two-dimensional input $\mathbf{x}^{\mathbf{a}}$. The key sequence is

x CTRL **[^] a** CTRL **[␣]**

The key sequences for entering

$$\sum_{n=1}^{k} x_n + \int_1^4 x^2 \, dx$$

are

⋮sum⋮ CTRL **[+] n = 1** CTRL **[%] k** CTRL **[␣] x** CTRL **[_] n** CTRL **[␣] +**
⋮int⋮ CTRL **[_] 1** CTRL **[%] 4** CTRL **[␣] x** CTRL **[^] 2** CTRL **[␣] ⋮dd⋮x**

Ordinary Characters

We can use only ordinary printable characters to enter two-dimensional forms:

Characters	Form
\!\(x\^y\)	superscript x^y
\!\(x_y\)	subscript x_y
\!\(x\^y\%z\)	subscript and superscript x_z^y
\!\(\@x\)	square root $\sqrt{x}$
\!\(x\/y\)	built–up fraction $\frac{x}{y}$
\!\(x\+y\)	underscript $\underset{y}{x}$
\!\(x\&y\)	overscript $\overset{y}{x}$
\!\(x\+y\%z\)	underscript and overscript $\overset{y}{\underset{z}{x}}$

Note that each expression is enclosed on the left by the characters "\!\(" and on the right by the characters "\)". We can use the characters "\(" and "\)" within an expression to generate invisible parentheses for grouping elements.

Let us create once more the two-dimensional input $\mathbf{x^a}$. The sequence of characters is

```
\!\(x\^a\)
```

To display the two-dimensional form, select the expression and choose **Cell ▸ Convert To ▸ StandardForm**:

$$x^a$$

The character sequences for entering

$$\sum_{n=1}^{k} x_n + \int_1^4 x^2 \, dx$$

are

```
\!\(\(\[Sum]\+\(n=1\)\%k\(x\_n\)\) +
    \(\[Integral]\_1\%4\(x\^2\)\[DifferentialD]x\)\)
```

To display the two-dimensional form, select the expression and choose **Cell ▸ Convert To ▸ StandardForm**:

$$\sum_{n=1}^{k} x_n + \int_1^4 x^2 \, dx$$

Create Table/Matrix

To create matrices, we can use the **BasicMathInput** palette, or choose **Insert ▸ Table/Matrix ▸ New** and enter the necessary specifications in the dialog box. The Tab key and several control characters are useful in editing matrices:

Key or Control Character	Built-in Meaning
CTRL [,]	add a column
CTRL [↵] (Control – Return)	add a row
TAB	go to the next placeholder
CTRL [␣]	move out of the matrix

Consider, for example, the matrix

$$\begin{pmatrix} -2 & 1 & 3 \\ 0 & -1 & 1 \\ 1 & 2 & 0 \end{pmatrix}$$

To enter the matrix:

1. Choose **Insert ▸ Table/Matrix ▸ New**, click the Matrix button at the top of the dialog box, enter the number 3 in the fields for "Number of rows:" and "Number of columns:", and click the OK button:

$$\begin{pmatrix} \square & \square & \square \\ \square & \square & \square \\ \square & \square & \square \end{pmatrix}$$

2. Enter the matrix elements, using the Tab key to move from one placeholder to the next:

$$\begin{pmatrix} -2 & 1 & 3 \\ 0 & -1 & 1 \\ 1 & 2 & 0 \end{pmatrix}$$

We can find the inverse of this matrix:

In[1]:= **Inverse@** $\begin{pmatrix} -2 & 1 & 3 \\ 0 & -1 & 1 \\ 1 & 2 & 0 \end{pmatrix}$

Out[1]= $\begin{pmatrix} -\frac{1}{4} & \frac{3}{4} & \frac{1}{2} \\ \frac{1}{8} & -\frac{3}{8} & \frac{1}{4} \\ \frac{1}{8} & \frac{5}{8} & \frac{1}{4} \end{pmatrix}$

2.5.2.2 Some Two-Dimensional Forms with Built-in Meanings

The following are some two-dimensional forms together with their corresponding one-dimensional forms and built-in meanings:

Two-Dimensional Form	One-Dimensional Form	Built-in Meaning
x^y	`x^y`	power
$\frac{x}{y}$	`x/y`	division
$\sqrt{x}$	`Sqrt[x]`	square root
$\sqrt[n]{x}$	`x^(1/n)`	nth root
$\sum_{i=imin}^{imax} f$	`Sum[f, {i, imin, imax}]`	sum
$\prod_{i=imin}^{imax} f$	`Product[f, {i, imin, imax}]`	product
$\int f\, dx$	`Integrate[f,x]`	indefinite integral
$\int_{xmin}^{xmax} f dx$	`Integrate[f, {x, xmin, xmax}]`	definite integral
$\partial_x f$	`D[f,x]`	partial derivative
$\partial_{x,y} f$	`D[f, x, y]`	multivariate partial derivative

2.5.2.3 Two-Dimensional Notation in Physics

To illustrate the use of two-dimensional forms for input, this section gives several physics examples.

Example 2.5.1 Define the gradient of a scalar point function in traditional mathematical notation, and apply it to the electric potential produced by a single point charge placed at the origin of a Cartesian coordinate system in order to obtain the corresponding electric field.

In terms of the Cartesian coordinates, the gradient of f can be defined as

```
∇f_ := ∂x f î + ∂y f ĵ + ∂z f k̂
```

To display this definition in traditional mathematical notation, click the cell bracket and choose **TraditionalForm** in the Convert To submenu of the Cell menu:

In[2]:= $\nabla f\_ := \frac{\partial f}{\partial x}\hat{i} + \frac{\partial f}{\partial y}\hat{j} + \frac{\partial f}{\partial z}\hat{k}$

(Section 2.5.3 will relate the necessary precaution for using TraditionalForm input.)

The electric potential due to a point charge q located at the origin is

In[3]:= $\phi[x\_, y\_, z\_] := k\frac{q}{\sqrt{x^2 + y^2 + z^2}}$

In electrostatics, the electric field can be written as

$$\mathbf{E} = -\nabla\phi$$

Thus, the electric field produced by the point charge is

```
In[4]:= -∇ϕ[x, y, z] // Factor
```

$$\text{Out[4]:=} \frac{k\,q\left(x\hat{i}+y\hat{j}+z\hat{k}\right)}{(x^2+y^2+z^2)^{3/2}}$$

```
In[5]:= Clear[ϕ, Del]
```

■

Example 2.5.2 For the one-dimensional harmonic oscillator, determine the matrix element of the kinetic energy operator between the first and third excited energy eigenstates.

In terms of the dimensionless coordinate $\zeta = \sqrt{m\omega/\hbar}\,x$, the kinetic energy operator of the one-dimensional harmonic oscillator is

$$T = \frac{p^2}{2m} = -\frac{\hbar^2}{2m}\frac{d^2}{dx^2} = -\frac{\hbar\omega}{2}\frac{d^2}{d\zeta^2}$$

The energy eigenfunctions are

$$\text{In[6]:=}\ \mathbf{u}_{n\_}[\xi\_] := \sqrt{\frac{1}{\sqrt{\pi}\,n!\,2^n}}\ \mathbf{H}_n[\xi]\ e^{-\frac{\xi^2}{2}}$$

where n can assume any nonnegative integer and

```
In[7]:= Hn_[ξ_] := HermiteH[n, ξ]
```

As shown in Section 2.5.1.3, the scalar product of two functions of the variable ζ can be defined as

$$\text{In[8]:=}\ \langle\varphi\_, \psi\_\rangle := \int_{-\infty}^{\infty} \varphi^*\,\psi\,d\xi$$

in which φ^* is the complex conjugate of φ and is given by

```
In[9]:= φ_* := φ /. Complex[u_, v_] → Complex[u, -v]
```

where the keyboard character "*" is entered as the superscripts of φ and $\varphi\_$. Thus, the matrix element of the kinetic energy operator between the first and third excited states is

$$\text{In[10]:=}\left\langle \mathbf{u}_1[\xi],\ -\frac{\hbar\,\omega}{2}\partial_{\xi,\xi}\mathbf{u}_3[\xi]\right\rangle$$

$$\text{Out[10]=} -\frac{1}{2}\sqrt{\frac{3}{2}}\,\omega\,\hbar$$

```
In[11]:= Clear[Subscript, AngleBracket, SuperStar]
```

■

Example 2.5.3 The normalized energy eigenfunctions of the hydrogen atom can be written as

$$\psi_{n\,l\,m}(r,\theta,\phi) = R_{nl}(r)Y_{lm}(\theta,\phi)$$

where R_{nl} and Y_{lm} are the radial functions and the normalized spherical harmonics, respectively. In terms of the associated Laguerre polynomials L_p^q and the Bohr radius a, the radial functions are given by

$$R_{nl}(r) = a^{-3/2}\frac{2}{n^2}\sqrt{\frac{(n-l-1)!}{[(n+l)!]^3}}F_{nl}\left(\frac{2r}{na}\right)$$

with

$$F_{nl}(x) = x^l e^{-x/2}L_{n-l-1}^{2l+1}(x)$$

The associated Laguerre polynomials $L_p^q(x)$ are related to the generalized Laguerre polynomials **LaguerreL[p, q, x]** of *Mathematica* by

$$L_p^q(x) = (p+q)!\ \texttt{LaguerreL[p, q, x]}$$

The allowed quantum numbers n, l, and m are given by the rules

$$n = 1, 2, 3, \ldots$$
$$l = 0, 1, 2, \ldots, (n-1)$$
$$m = -l, -l+1, \ldots, 0, 1, 2, \ldots, +l$$

Plot the probability densities for the states specified by the quantum numbers $(n, l, m) = (4, 3, 1), (4, 2, 1)$, and $(4, 2, 0)$. Explain the apparent lack of spherical symmetry in these plots for the hydrogen atom, which has a spherically symmetric Hamiltonian.

We cannot really generate a three-dimensional plot for the probability density $\psi^*\psi$ because it depends on three coordinates (r, θ, ϕ) and, therefore, its plot must be four-dimensional. We can, however, make a density plot effectively using the gray level at each point (more precisely, at each cell) as the fourth coordinate or dimension. That is, the gray-level intensity, varying from 0 (black) to 1 (white), at a point is an increasing function of and, hence, represents the probability density at that point.

As it turns out, the probability densities of the hydrogen atom are independent of the angle ϕ. Only two-dimensional density plots on a plane containing the z- or polar axis are necessary because the three-dimensional plots can readily be obtained by rotating the two-dimensional ones through 360° about the z-axis.

The wave functions can be defined by

```
ψn_l_m_[r_, θ_, ϕ_] := Rnl[r] Ylm[θ, ϕ]
```

```
In[12]:= Rn_l_[r_] := a^(-3/2) (2/n^2) Sqrt[(n - l - 1)!/((n + l)!)^3] Fnl[2r/(n a)]

In[13]:= Fn_l_[x_] := x^l e^(-x/2) (n + l)! LaguerreL[n - l - 1, 2l + 1, x]

In[14]:= Yl_m_[θ_, ϕ_] := SphericalHarmonicY[l, m, θ, ϕ]
```

where invisible commas :,: have been inserted between all quantum numbers in the foregoing definitions in order to mimic traditional physics notation. Although this definition for the wave function is perfectly sound, Section 3.2.3 will show us how to write a better definition that complies with the rules on the quantum numbers n, l, and m:

```
In[15]:= ψn_Integer?Positivel_Integer?NonNegativem_Integer[r_, θ_, ϕ_] :=
           Rnl[r] Ylm[θ, ϕ] /; l ≤ n - 1 ∧ -l ≤ m ≤ l
```

where, as before, invisible commas :,: have been inserted between all quantum numbers in this definition, including between **n_Integer?Positive** and **l_Integer?NonNegative** as well as between **l_Integer?NonNegative** and **m_Integer**. The definitions for R, F, and Y remain the same.

The complex conjugate of the wave function φ can be obtained from

```
In[16]:= φ_* := φ /. Complex[u_, v_] → Complex[u, -v]
```

Let us define a function for generating the density plots:

```
In[17]:= hydrogenPlot[n_, l_, m_] :=
           (DensityPlot[Evaluate[ψnlm[r, θ, ϕ]* ψnlm[r, θ, ϕ] //.
               {a → 1, r -> Sqrt[x^2 + z^2], θ → ArcCos[z/r]}], {x, -10 (1.5 n + 1)/2,
             10 (1.5 n + 1)/2}, {z, -10 (1.5 n + 1)/2, 10 (1.5 n + 1)/2},
            Mesh → False, PlotPoints → 200, FrameLabel → {"x", "z"},
            PlotLabel → {n, l, m}, ImageSize → 144, FrameTicks → None])
```

DensityPlot[f, {x, $xmin$, $xmax$}, {z, $zmin$, $zmax$}] makes a density plot of f as a function of x and z. The option **ImageSize** → w specifies that the width of image to display for an object is w printer's points. In the definition for the function **hydrogenPlot**, the rules for r and θ effect a coordinate transformation from (r, θ) to (x, z) and the rule for a makes the Bohr radius the unit of length. Note that invisible or ordinary commas must be inserted between the quantum numbers n, l, and m in the preceding definition.

Here are the density plots:

```
In[18]:= Row[{hydrogenPlot[4, 3, 1],
            hydrogenPlot[4, 2, 1], hydrogenPlot[4, 2, 0]}, Spacer[36]]
```

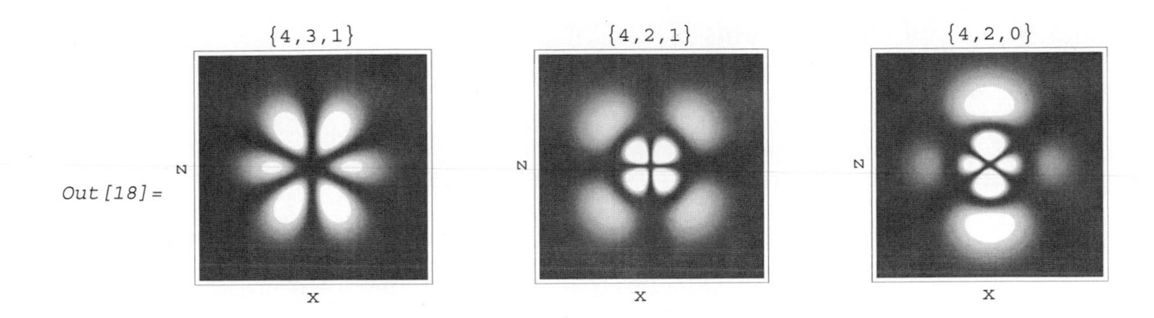

`Row[{`$expr_1$`, `$expr_2$`, ...}, `s`]` is an object that formats with the $expr_i$ arranged in a row and with s as a separator between successive $expr_i$. `Spacer[`w`]` specifies a spacer w printer's points wide.

The probability densities exhibit rotational symmetry about the z- or polar axis but lack the full spherical symmetry of the central potential. However, we cannot observe these probability densities experimentally. Otherwise, we can establish a preferred or z direction in space in the absence of external fields. All we can actually measure is the average probability density of the n^2 degenerate states in a shell identified by the principal quantum number n. It can be verified that the average probability density of a shell is spherically symmetric:

```
In[19]:= shell[n_] := 1/n^2 Sum[Sum[ψnlm[r, θ, ϕ]* ψnlm[r, θ, ϕ], {m, -l, l}], {l, 0, n-1}] // Simplify
```

$$\texttt{In[19]:= shell[n\_]} := \frac{1}{n^2}\sum_{l=0}^{n-1}\sum_{m=-l}^{l}\psi_{nlm}[r,\theta,\phi]^*\,\psi_{nlm}[r,\theta,\phi]\ \texttt{// Simplify}$$

```
In[20]:= shell[4]
```

$$\texttt{Out[20]=}\ \frac{e^{-\frac{r}{2a}}\left(9216a^6-13\,824ra^5+10\,368r^2a^4-3264r^3a^3+504r^4a^2-36r^5a+r^6\right)}{9\,437\,184a^9\pi}$$

The average probability density for the $n = 4$ shell is indeed spherically symmetric because it depends only on the coordinate r. We can easily verify that the average probabilities for other values of n are also spherically symmetric.

The rotational invariance of the Hamiltonian actually imposes only the m-degeneracy, called essential degeneracy, in a subshell labeled by the quantum numbers n and l. The additional l-degeneracy, called accidental degeneracy, is specific to the Coulomb potential. We can show that the average probability densities of the subshells are spherically symmetric—that is, independent of the angle θ:

$$\texttt{In[21]:= subshell[n\_, l\_]} := \frac{1}{2l+1}\sum_{m=-l}^{l}\psi_{nlm}[r,\theta,\phi]^*\,\psi_{nlm}[r,\theta,\phi]\ \texttt{// Simplify}$$

```
In[22]:= FreeQ[Table[subshell[n, l], {n, 1, 4}, {l, 0, n-1}], θ]
Out[22]:= True

In[23]:= Clear[Subscript, SuperStar, hydrogenPlot, shell, subshell]
```

■

2.5.3 Input and Output Forms

Mathematica supports several forms or format types for input and output:

InputForm	a one-dimensional form using keyboard characters
OutputForm	a form using keyboard characters; for output only
StandardForm	a form utilizing special characters and two-dimensional forms; the default form for both input and output
TraditionalForm	a form that attempts to imitate traditional mathematical notation; primarily for output

To illustrate these format types, consider the mathematical expression

$$\frac{x^3}{2+\sin^2 x} + \int f(x)\,dx$$

The functions **InputForm**, **OutputForm**, **StandardForm**, and **TraditionalForm** display the expression in their respective forms:

- **InputForm**

```
In[1]:= x^3/(2 + Sin[x]^2) + ∫f[x] dx // InputForm

Out[1]//InputForm=
        Integrate[f[x], x] + x^3/(2 + Sin[x]^2)
```

- **OutputForm**

```
In[2]:= x^3/(2 + Sin[x]^2) + ∫f[x] dx // OutputForm

Out[2]//OutputForm=
                               3
                              x
        Integrate[f[x], x] + -----------
                                        2
                             2 + Sin[x]
```

- **StandardForm**

```
In[3]:= x^3/(2 + Sin[x]^2) + ∫f[x] dx // StandardForm

Out[3]//StandardForm=
                     3
                    x
        ∫f[x]dx + -----------
                            2
                  2 + Sin[x]
```

- **TraditionalForm**

In[4]:= $\frac{x^3}{2+\text{Sin}[x]^2}+\int f[x]\,dx$ // TraditionalForm

Out[4]//TraditionalForm=

$$\frac{x^3}{\sin^2(x)+2}+\int f(x)\,dx$$

The current choice of **Format type of new input cells** in the Evaluation tab of the Preferences submenu in the Mathematica (Application) menu for Mac OS X or the Edit menu for Windows specifies globally the default form of input. However, we can enter expressions in InputForm with **Format type of new input cells** set at either **InputForm** or **StandardForm**. For example, consider again the mathematical expression

$$\frac{x^3}{2+\sin^2 x}+\int f(x)\,dx$$

Leave **Format type of new input cells** set at **StandardForm**, which is the default, and type the expression in InputForm:

```
x^3/(2+Sin[x]^2)+Integrate[f[x],x]
```

The three choices of **Format type of new output cells** in the Evaluation tab of the Preferences submenu in the Mathematica (Application) menu for Mac OS X or the Edit menu for Windows correspond to three output forms. The current choice specifies globally the form in which output will be displayed. Now evaluate the previous input with **Format type of new output cells** at each of the three settings. Here are the outputs:

- **OutputForm**

$$\text{Integrate}[f[x],x]+\frac{x^3}{2+\sin[x]^2}$$

- **StandardForm**

$$\int f[x]\,dx+\frac{x^3}{2+\text{Sin}[x]^2}$$

- **TraditionalForm**

$$\frac{x^3}{\sin^2(x)+2}+\int f(x)\,dx$$

The appeal of StandardForm with the special characters and two-dimensional forms notwithstanding, we must still learn InputForm en route to mastery of *Mathematica.*

As will be explained in Section 3.1, the *Mathematica* kernel sees only FullForm, and the one-dimensional InputForm is its close cousin, whereas StandardForm bears little resemblance to it. Consequently, it is much easier to debug the code in InputForm than in StandardForm. For interactive *Mathematica* sessions, this advantage may be insignificant. For substantial programs, consider writing the code in InputForm, fix the bugs, and then convert it to StandardForm. Conversion of the code from InputForm to StandardForm can be done with the **Convert To** command in the Cell menu; minor edits of the result may be necessary. Through Section 2.4, most of the code in this book has been written in InputForm that does not utilize special characters and two-dimensional forms.

Traditional mathematical notation is discernible, intuitive, natural, and elegant to those who are mathematically proficient, and many mathematicians, scientists, and engineers have long awaited a computer algebra system that would utilize such a notation. The main difficulty lies in its lack of the precision required for computer input and output. For example, $a(1+x)$ can mean a times $1+x$ or the function a with the argument $1+x$. Also, $\tan^{-1}\theta$ may be interpreted as $(\tan\theta)^{-1}$ or $1/\tan\,\theta$ rather than the customary inverse tangent or arctangent of θ, as $\tan^2\theta$ is normally taken to be $(\tan\theta)^2$. StandardForm is *Mathematica*'s answer to the call because it has the required precision for input as well as output and resembles somewhat the familiar mathematical notation. For those who want "the real thing," *Mathematica* offers TraditionalForm, which gives a fair rendition of the traditional mathematical notation, but with the caveat that it is primarily for output only.

TraditionalForm lacks the necessary precision for input because it tries to imitate the traditional mathematical notation. For example,

```
In[5]:= {S[1 + α], FresnelS[1 + α], S(1 + α)} // StandardForm
Out[5]//StandardForm=
        {S[1 + α], FresnelS[1 + α], S(1 + α)}

In[6]:= {S[1 + α], FresnelS[1 + α], S(1 + α)} // TraditionalForm
Out[6]//TraditionalForm=
```

$$\{S(\alpha+1),\ S(\alpha+1),\ S\,(\alpha+1)\}$$

where the first element is a function S with the argument $1+\alpha$, the second is a Fresnel integral, and the third is S times $(1+\alpha)$. The three elements appear distinct in StandardForm but look identical in TraditionalForm except for the tiny extra space in the third element. If we input them in TraditionalForm, how can we tell what they actually represent, and how can *Mathematica* distinguish between the first two?

If the lure of TraditionalForm for input becomes irresistible, consider entering the expression in StandardForm, clicking the cell bracket, and then choosing **TraditionalForm** in the Convert To submenu of the Cell menu. *Mathematica* will insert hidden tags in the converted expression, now in TraditionalForm, so as to make it sufficiently precise for input. For example, enter

$$\mathcal{E} := \sum_{k=1}^{f} \dot{q}_k\, \partial_{\dot{q}_k} L - L$$

in StandardForm and convert this to

$$\mathcal{E} := \sum_{k=1}^{f} \dot{q}_k \frac{\partial L}{\partial \dot{q}_k} - L$$

which is a legitimate expression in TraditionalForm for *Mathematica* input. For another example, enter

```
DSolve[D[x[t], {t, 2}] == -γ D[x[t], t] - ω₀² x[t], x[t], t]
```

and convert it to

$$\text{In[7]:=}\ \mathbf{DSolve}\left[\frac{\partial^2 x(t)}{\partial t^2} = -x(t)\,\omega_0^2 - \gamma\,\frac{\partial x(t)}{\partial t},\ x(t),\ t\right]$$

$$\text{Out[7]=}\ \left\{\left\{x(t) \to e^{\frac{1}{2}t\left(-\gamma-\sqrt{\gamma^2-4\,\omega_0^2}\right)} c_1 + e^{\frac{1}{2}t\left(\sqrt{\gamma^2-4\,\omega_0^2}-\gamma\right)} c_2\right\}\right\}$$

where we have evaluated the converted expression to verify its legitimacy as an expression for input.

2.5.4 Exercises

For these exercises, choose **StandardForm** for both **Format type of new input cells** and **Format type of new output cells** in the Evaluation tab of the Preferences submenu under the Mathematica (Application) menu for Mac OS X or the Edit menu for Windows. In this book, straightforward, intermediate-level, and challenging exercises are unmarked, marked with one asterisk, and marked with two asterisks, respectively. For solutions to most odd-numbered exercises, see Appendix C.

1. Utilizing special characters and two-dimensional forms, redo Examples (i) 2.1.2, (ii) 2.1.6, (iii) 2.2.9, (iv) 2.2.14, (v) 2.3.17, (vi) 2.4.11, and (vii) 2.4.15.
2. Utilizing special characters and two-dimensional forms, redo Exercises (i) 47 of Section 2.2.20, (ii) 48 of Section 2.2.20, and (iii) 11 of Section 2.3.5.
3. Using special characters and two-dimensional forms, enter the following mathematical expressions in StandardForm, then convert them to TraditionalForm with the Convert To submenu in the Cell menu, and finally evaluate the converted expressions:

   (a)
   $$\sum_{n=0}^{\infty} n \frac{1}{n!} \left(\frac{t}{\tau_0}\right)^n e^{-t/\tau_0}$$

   (b)
   $$\frac{\sum_{n=0}^{\infty} n \exp\left(-\frac{n\varepsilon}{kT}\right)}{\sum_{n=0}^{\infty} \exp\left(-\frac{n\varepsilon}{kT}\right)}$$

Answers:

$$\frac{t}{\tau_0}$$

$$\frac{1}{-1 + e^{\frac{\epsilon}{k T}}}$$

4. Using TraditionalForm, solve the differential equation

$$\gamma \frac{d^2 n}{dz^2} + \frac{z}{2} \frac{dn}{dz} = 0$$

That is, *create* and evaluate the input

$$\mathbf{DSolve}\left[\frac{1}{2} z \frac{\partial n(z)}{\partial z} + \gamma \frac{\partial^2 n(z)}{\partial z^2} == 0,\ n(z),\ z\right] \mathbf{//TraditionalForm}$$

Answer:

$$\left\{\left\{n(z) \to c_2 + \sqrt{\pi}\sqrt{\gamma}\, c_1 \operatorname{erf}\left(\frac{z}{2\sqrt{\gamma}}\right)\right\}\right\}$$

5. Reduce the integral

$$-\frac{1}{\omega_1} \int_{\omega_1 t}^{0} \exp(-\gamma t) \exp\left(\left[(\gamma - \beta)/\omega_1\right] z\right) \sin z \, dz$$

to

$$-\frac{e^{-t\gamma}\left(e^{t(\gamma-\beta)}(\beta - \gamma)\sin(t\,\omega_1) + \left(e^{t(\gamma-\beta)}\cos(t\,\omega_1) - 1\right)\omega_1\right)}{(\beta-\gamma)^2 + \omega_1^2}$$

6. In the $n = 1$, $l = 0$ state, the radial probability density for the hydrogen atom is

$$r^2 \frac{4}{a^3} e^{-2r/a}$$

where a is the Bohr radius. Using special characters and two-dimensional forms, (a) enter the probability density in StandardForm and (b) calculate the probability of finding the electron within a Bohr radius from the nucleus. *Hint:* Integrate the probability density from 0 to a.

Answer:

```
0.323324
```

7. In the $n = 2$, $l = 1$ state, the radial probability density for the hydrogen atom is

$$r^2 \frac{1}{24a^3} \frac{r^2}{a^2} e^{-r/a}$$

where a is the Bohr radius. Using special characters and two-dimensional forms, (a) enter the probability density in StandardForm and (b) find the most likely distance of the electron from the nucleus. *Hint:* Set the first derivative of the probability density to zero and solve for r.

Answer:

```
{r → 4 a}
```

*8. The Maxwell distribution for the speeds of molecules in an ideal gas at a temperature T is

$$f(v) = 4\pi \left(\frac{m}{2\pi kT}\right)^{3/2} v^2 e^{-mv^2/2kT}$$

where m is the mass of a molecule, k is Boltzmann's constant, and $f(v)\,dv$ is the probability that a molecule has a speed in the range v to $v + dv$. Using special characters and two-dimensional forms, (a) enter the definition for **`f[v_]`** in StandardForm and (b) determine the root-mean-square (rms) speed, the average speed, and the most probable speed.

Answer:

$$\sqrt{3}\,\sqrt{\frac{k\,T}{m}}$$

$$2\,\sqrt{\frac{2}{\pi}}\,\sqrt{\frac{k\,T}{m}}$$

$$\left\{v \to \frac{\sqrt{2}\,\sqrt{k}\,\sqrt{T}}{\sqrt{m}}\right\}$$

9. The Fermi distribution function is

$$\frac{1}{e^{(\varepsilon-\mu)/kT} + 1}$$

Using special characters and two-dimensional forms, enter the function in StandardForm and evaluate it for $\varepsilon = 7.20\,\text{eV}$. Use the following values for the constants:

$$\mu = 7.11\,\text{eV}$$

$$k = 8.617 \times 10^{-5}\,\text{eV/K}$$

$$T = 300\,\text{K}$$

Answer:

```
0.0298435
```

*10. In the *RLC* series circuit, a resistor with resistance R, a capacitor with capacitance C, and an inductor with inductance L are connected in series with an AC source whose frequency and rms voltage are f and V_{rms}, respectively. The rms current I_{rms} is given by

$$I_{\text{rms}} = \frac{V_{\text{rms}}}{Z}$$

where the impedance Z of the circuit is defined as

$$Z = \sqrt{R^2 + (X_L - X_C)^2}$$

with the inductive reactance

$$X_L = 2\pi f L$$

and the capacitive reactance

$$X_C = \frac{1}{2\pi f C}$$

(a) Using special characters and two-dimensional forms, define a function that takes five arguments—R, C, L, V_{rms}, and f—and that returns the rms current I_{rms}. (b) Evaluate the function for a circuit with

$$R = 148\,\Omega$$

$$C = 1.5 \times 10^{-6}\,\text{F}$$

$$L = 35.7 \times 10^{-3}\,\text{H}$$

$$V_{\text{rms}} = 35.0\,\text{V}$$

$$f = 512\,\text{Hz}$$

Hint: `I` and `C` have predefined meanings. Also, pattern names must be symbols (i.e., pattern names cannot contain subscripts). Furthermore, avoid intermediate definitions for X_L, X_C, and Z. Finally, load the package **`Units``**.

Answer:

```
0.200609 Amp
```

*11. Consider a single-loop circuit that consists of a resistor R, a capacitor C, and an inductor L connected in series across an AC source whose instantaneous voltage is

$$V(t) = V_0 \sin 2\pi f t$$

where V_0 is the maximum voltage of the source and f is its frequency. The rms current can be expressed as

$$I_{\mathrm{rms}} = \frac{V_{\mathrm{rms}}}{\sqrt{R^2 + (X_L - X_C)^2}}$$

where

$$V_{\mathrm{rms}} = \frac{V_0}{\sqrt{2}}$$

$$X_C = \frac{1}{2\pi f C}$$

$$X_L = 2\pi f L$$

The average power delivered by the AC source in this circuit is

$$\overline{P} = I_{\mathrm{rms}} V_{\mathrm{rms}} \cos\phi$$

The quantity $\cos\phi$ is called the power factor and is given by

$$\cos\phi = \frac{R}{\sqrt{R^2 + (X_L - X_C)^2}}$$

The circuit is said to be in resonance when the rms current has its maximum value. This occurs when $X_C = X_L$, or equivalently, when the frequency of the source equals the resonance frequency

$$f_0 = \frac{1}{2\pi\sqrt{LC}}$$

Let

$$C = 5.10\,\mu\mathrm{F}$$

$$V_{\mathrm{rms}} = 11.0\,\mathrm{V}$$

$$f_0 = 1.30\,\mathrm{kHz}$$

$$\overline{P}_0 = 25.0\,\mathrm{W}$$

where $\overline{P}_0$ is the average power delivered by the source at resonance. Using special characters, two-dimensional forms, and the package **`Units``**, find the values of the inductance and the resistance, and calculate the power factor when the frequency of the source is 2.31 kHz. *Hint:* For a discussion on obtaining parts of expressions, see Section 3.1.3.1.

Answers:

(2.93889×10^{-3}) Henry

4.84 Ohm

0.163817

2.6 PROBLEMS

In this book, straightforward, intermediate-level, and challenging problems are unmarked, marked with one asterisk, and marked with two asterisks, respectively. For solutions to selected problems, see Appendix D.

1. Determine the absolute value of

$$\frac{1.735\times10^{-7}(\sin 23°)e^{35.7}}{\sqrt{3+7i}\,(2+9i)^2}$$

and express the result in scientific notation with three significant figures.

Answer:

9.23×10^{5}

2. Determine the sum of the infinite series

$$1+\frac{1}{3^4}+\frac{1}{5^4}+\frac{1}{7^4}+\cdots$$

Answer:

$$\frac{\pi^4}{96}$$

3. Determine

$$\sum_{k=1}^{n} k^4$$

Answer:

$$\frac{1}{30} n (1 + n) (1 + 2 n) \left(-1 + 3 n + 3 n^2 \right)$$

4. Invert the 5×5 Hilbert matrix, whose matrix elements m_{ij} are equal to $1/(i + j - 1)$, with i and j each ranging from 1 to 5.

Answer:

$$\begin{pmatrix} 25 & -300 & 1050 & -1400 & 630 \\ -300 & 4800 & -18\,900 & 26\,880 & -12\,600 \\ 1050 & -18\,900 & 79\,380 & -117\,600 & 56\,700 \\ -1400 & 26\,880 & -117\,600 & 179\,200 & -88\,200 \\ 630 & -12\,600 & 56\,700 & -88\,200 & 44\,100 \end{pmatrix}$$

5. Solve the system of equations and verify the solution.

$$2x + y = 3z$$
$$6x + 24 + 4y = 0$$
$$20 - 5x + 2z = 0$$

Answers:

$$\left\{\left\{x \to \frac{24}{7}, y \to -\frac{78}{7}, z \to -\frac{10}{7}\right\}\right\}$$

```
{{True, True, True}}
```

6. Solve the system of equations and verify the solutions.

$$x^2 + y^2 = 1$$
$$x + 3y = 0$$

Answers:

$$\left\{\left\{x \to -\frac{3}{\sqrt{10}}, y \to \frac{1}{\sqrt{10}}\right\}, \left\{x \to \frac{3}{\sqrt{10}}, y \to -\frac{1}{\sqrt{10}}\right\}\right\}$$

```
{{True, True}, {True, True}}
```

7. Applying Kirchhoff's rules to a certain circuit (not shown here) results in the system of equations

$$5I_1 + 7I_2 = 4$$
$$5I_1 - 2I_3 = 32$$
$$I_1 + I_3 = I_2$$

Find I_1, I_2, and I_3, and express the result in decimal notation to three significant figures. *Hint:* The symbol **I** has a built-in meaning in *Mathematica.*

Answers:

$$\left\{\left\{i_1 \to \frac{232}{59}, i_2 \to -\frac{132}{59}, i_3 \to -\frac{364}{59}\right\}\right\}$$

```
{{i1.→3.93, i2.→-2.24, i3.→-6.17}}
```

8. Solve the system of equations

$$3x = y + z$$

$$2by + 3ax = h$$

$$g + cz - 2by = f$$

for x, y, and z in terms of a, b, c, f, g, and h.

Answer:

$$\left\{\left\{x \to -\frac{-2bf + 2bg - 2bh - ch}{3(2ab + ac + 2bc)},\right.\right.$$

$$\left.\left. y \to -\frac{af - ag - ch}{2ab + ac + 2bc}, z \to -\frac{-af - 2bf + ag - 2bg - 2bh}{2ab + ac + 2bc}\right\}\right\}$$

9. Solve the system of equations

$$x^2 + 2xy + y^2 = 1$$

$$x^3 + x^2y + y^2 + y^3 = 4$$

Answers:

```
{{x→3.30278, y→-2.30278}, {x→-3.09455, y→2.09455},
{x→2., y→-1.}, {x→0.0472757-1.13594 i, y→-1.04728+1.13594i},
{x→0.0472757+1.13594 i, y→-1.04728-1.13594 i},
{x→-0.302776, y→1.30278}}
```

10. By specifying the option **Assumptions → -1 < Re[p] < 1** in the function **Integrate**, show that the integral

$$\int_0^{\pi/2} \tan^p(x)\, dx$$

evaluates to

$$\frac{1}{2}\pi \sec\left(\frac{p\pi}{2}\right)$$

11. Consider the integral

$$\int \frac{x^3}{\sqrt{a^2 - x^2}}\, dx$$

(a) Evaluate the integral. (b) Differentiate the result and recover the integrand.

Answers:

$$-\frac{1}{3}\sqrt{a^2 - x^2}\left(2\, a^2 + x^2\right)$$

$$\frac{x^3}{\sqrt{a^2 - x^2}}$$

12. Consider the integral

$$\int \frac{1}{x^2 \sqrt{(a^2 - x^2)^3}}\, dx$$

(a) Evaluate the integral. (b) Differentiate the result and recover the integrand.

Answers:

$$\frac{-a^4 + 3\, a^2 x^2 + 2\, x^4}{a^4\, x \sqrt{\left(a^2 - x^2\right)^3}}$$

$$\frac{1}{x^2 \sqrt{\left(a^2 - x^2\right)^3}}$$

13. The current i in an electric circuit is given by

$$\frac{di}{dt} + 2i = \sin t$$

with $i(0) = 0$. (a) Determine $i(t)$ numerically for t in the range 0 to 1.1. (b) Determine $i(t)$ analytically. (c) Compare the numerical and analytical solutions for t from 0 to 1 in steps of 0.1.

Answers:

```
{{i→InterpolatingFunction[{{0., 1.1}}, <>]}}
```

$$\left\{\left\{\mathtt{i[t]} \to -\frac{1}{5}\, e^{-2t}\left(-1 + e^{2t}\,\mathtt{Cos[t]} - 2\, e^{2t}\,\mathtt{Sin[t]}\right)\right\}\right\}$$

t	numerical	analytical
0	0	0
0.1	0.0046787	0.00467868
0.2	0.0175184	0.0175184
0.3	0.0369031	0.0369031
0.4	0.0614209	0.0614209
0.5	0.0898296	0.0898296
0.6	0.121029	0.121029
0.7	0.154038	0.154038
0.8	0.18798	0.18798
0.9	0.222069	0.222069
1.	0.255595	0.255595

14. Consider the differential equation

$$\frac{dx}{dt} + x + e^t x^2 = 0$$

with $x(0) = 1$. (a) Determine $x(t)$ numerically for t in the range 0 to 0.4. (b) Determine $x(t)$ analytically. (c) Compare the numerical and analytical solutions for t from 0 to 0.4 in steps of 0.1.

Answers:

```
{{x→InterpolatingFunction[{{0., 0.4}}, <>]}}
```

$$\left\{\left\{\mathtt{x[t]} \to \frac{e^{-t}}{1 + t}\right\}\right\}$$

t	numerical	analytical
0.	1.	1.
0.1	0.822579	0.822579
0.2	0.682276	0.682276
0.3	0.56986	0.56986
0.4	0.4788	0.4788

15. Solve the differential equation

$$\frac{d^2y}{dt^2} - 2y = 3e^{-t^2}$$

Answer:

$$\left\{\left\{y[t] \to e^{\sqrt{2}\,t}\,C[1] + e^{-\sqrt{2}\,t}\,C[2] + \frac{3}{4}\,e^{\frac{1}{2}-\sqrt{2}\,t}\sqrt{\frac{\pi}{2}}\left(\text{Erf}\left[\frac{1}{\sqrt{2}}-t\right] + e^{2\sqrt{2}\,t}\,\text{Erf}\left[\frac{1}{\sqrt{2}}+t\right]\right)\right\}\right\}$$

16. Find the functions $x(t)$ and $y(t)$ that satisfy the differential equations

$$\frac{dx}{dt} + \frac{dy}{dt} - 4y = 1$$

$$\frac{dy}{dt} - 3y + x = t^2$$

and the initial conditions $x(0) = y(0) = 0$.

Answer:

$$\left\{\left\{x[t] \to \frac{1}{4}\left(1 - e^{2t} + 6t + 4t^2\right),\ y[t] \to \frac{1}{4}\left(1 - e^{2t} + 2t\right)\right\}\right\}$$

17. Consider the differential equation

$$-\frac{d^2\psi}{dx^2} + V(x)\psi = E\psi$$

with

$$V(x) = -6\ \text{sech}^2(x)$$

Also, $\psi(0) = 1$, $\psi'(0) = 0$, and $E = -4$. Determine numerically and plot $\psi(x)$ for x in the range -5 to 5.

Answers:

```
{{ψ→InterpolatingFunction[{{-5.,5.}},<>]}}
```

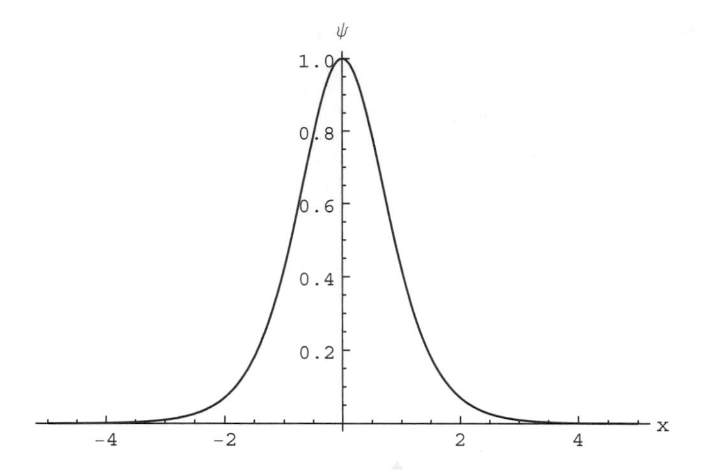

18. In an experiment, the motion of a block down a smooth incline is observed and the following data are obtained:

$\sin\theta$	$a(\mathrm{m/s^2})$
0.271	2.78
0.282	2.66
0.302	2.65
0.319	3.11
0.336	3.27
0.350	3.31
0.360	3.55
0.376	3.60
0.389	3.68
0.400	3.92
0.418	3.90

where θ is the angle the incline makes with the horizontal and a is the acceleration of the block down the incline. (a) Find the equation of the straight line that best fits the data. (b) Determine the gravitational acceleration g. (c) Calculate the percentage error of the experimental value for g from the accepted value of 9.80 $\mathrm{m/s^2}$. (d) Superpose the best fit curve on the data to examine the fit.

Answers:

$a = 9.57482\,\mathrm{m/s^2}\,\sin(\theta)$

$9.57482\,\mathrm{m/s^2}$

2.3%

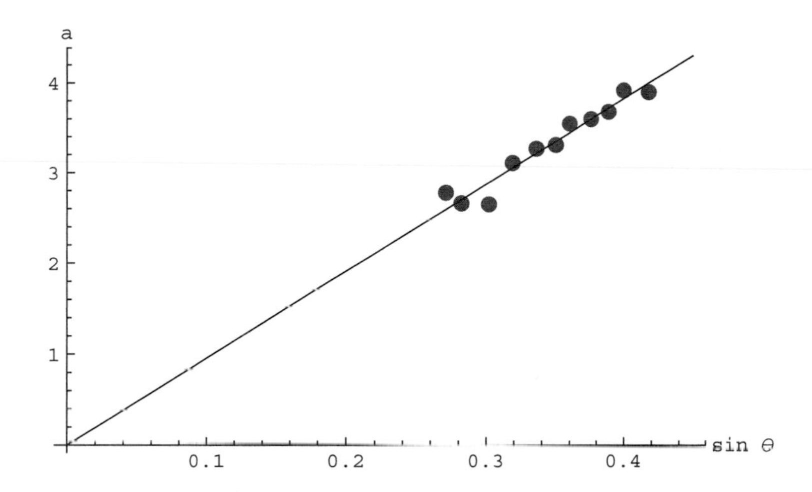

19. The decay rate R (often referred to as activity) of a radioactive sample can be expressed as

$$R = R_0 e^{-\lambda t}$$

where R_0, λ, and t are the initial decay rate, decay constant, and time, respectively. The half-life $T_{1/2}$ of the nuclei in the sample can be obtained from

$$T_{1/2} = \frac{\ln 2}{\lambda}$$

Determine R_0, λ, and $T_{1/2}$ from the data:

t(h)	R(decays/min)
1.00	3100
2.00	2450
4.00	1480
6.00	910
8.00	545
10.0	330
12.0	200

Answers:

```
{3985.53 decays/min, 0.00411946 min^-1, 168.262 min}
```

20. Find the solution for $x > 0$ of the following equation to five significant figures:

$$e^{-x} \sin x = x^2 - \frac{1}{2}$$

Answer:

```
{x→0.90453}
```

21. Find the three roots of the equation

$$\sin^2(x) = \tanh\left(\frac{x}{2}\right) + \frac{1}{5}$$

Give your answers to six significant figures. *Hint:* Use **FindRoot**.

Answers:

```
{{x→-0.264672}, {x→0.914801}, {x→1.84447}}
```

22. $J_n(x)$ is called the Bessel function of the first kind and order n. Plot J_0, J_1, and J_2 together. Choose a different style for each curve, specify labels for the left and bottom frame edges, and give the label "The first three Bessel functions" to the plot.

Answer:

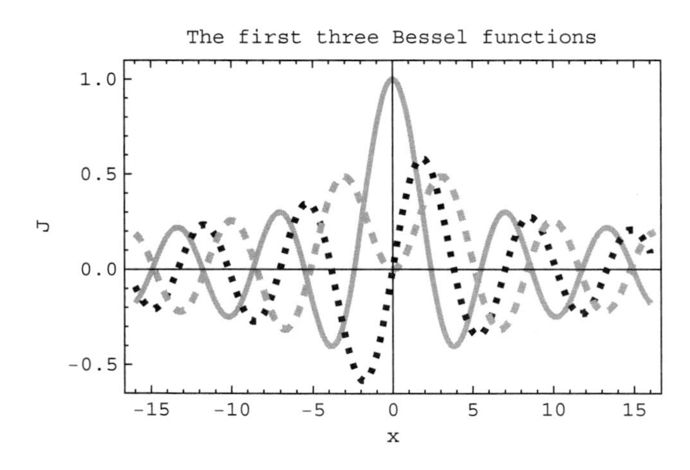

23. Plot the elliptic paraboloid given by the equation

$$\frac{x^2}{a^2} + \frac{y^2}{b^2} = 2cz$$

where $a = b = 1$ and $c = 1/2$. Using the **ViewPoint** option, show the picture from several viewpoints.

Answer: (Only one of the four elliptic paraboloids is included here.)

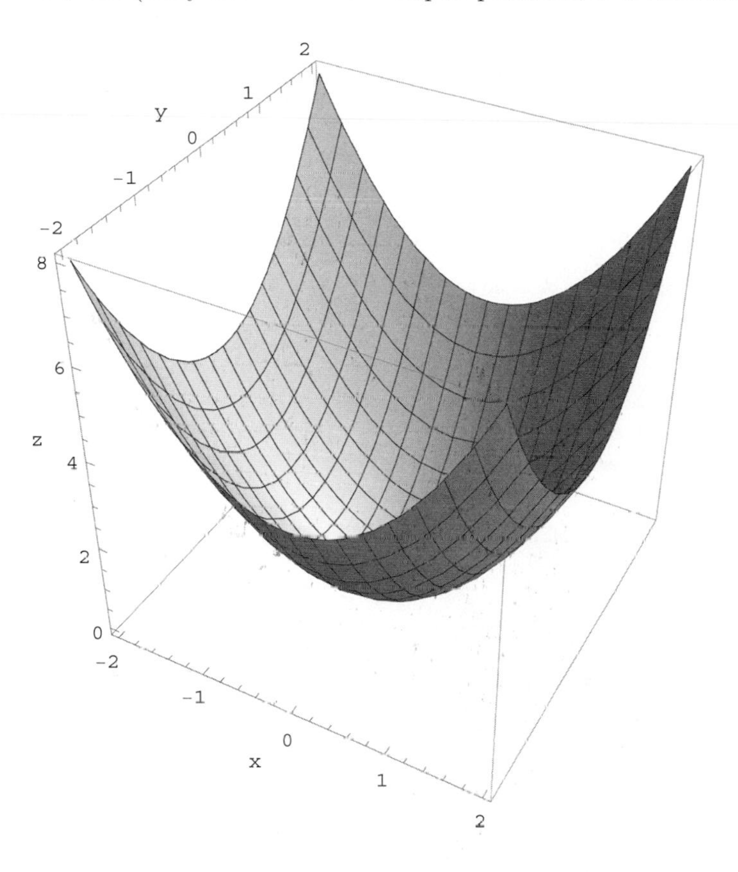

*24. The graphs in Problem 23 do not look like elliptic paraboloids, which have rotational symmetry about the z-axis. Use the **ParametricPlot3D** function to replot the graphs.

Answer: (Only one of the four elliptic paraboloids is included here.)

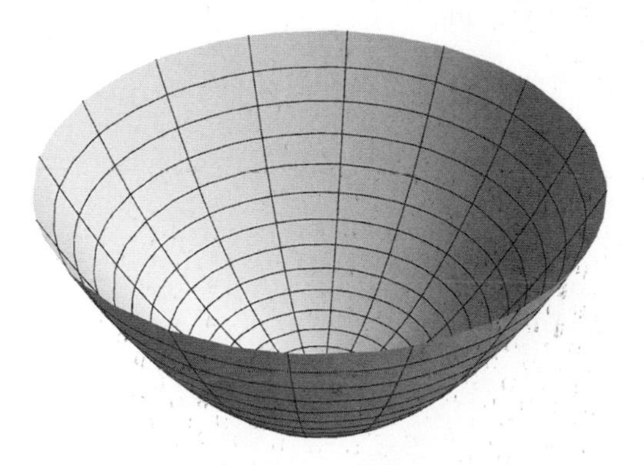

25. Consider again Problem 23 on the elliptic paraboloid. Let a plane be defined by the equation

$$ax + by + cz + d = 0$$

with $a = 0.25$, $b = 0.25$, $c = -1$, and $d = 2.5$. Superpose this plane on the elliptic paraboloid and produce the graph

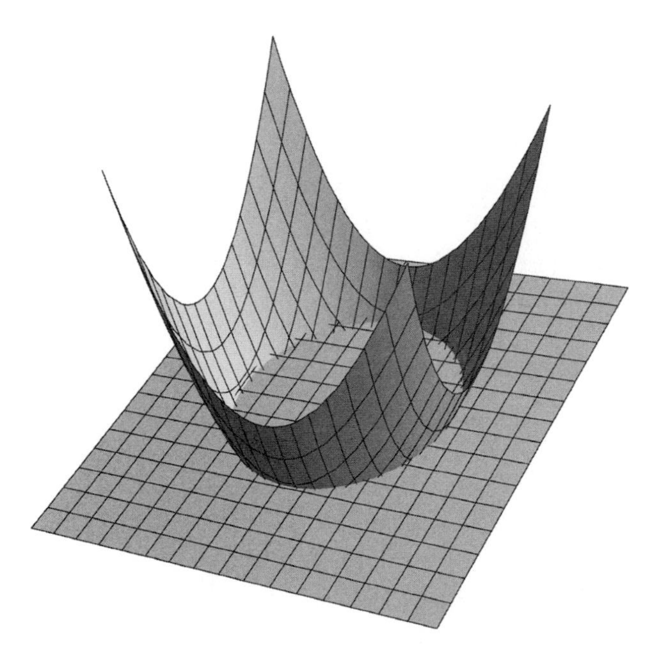

26. Using **`Manipulate`**, demonstrate the effects of varying λ from 3π to 9π on the graph of three functions:

$$y_1(x) = 1$$

$$y_2(x) = -1$$

$$y_3(x) = \cos x + \frac{\lambda}{2}\frac{\sin x}{x}$$

Answer:

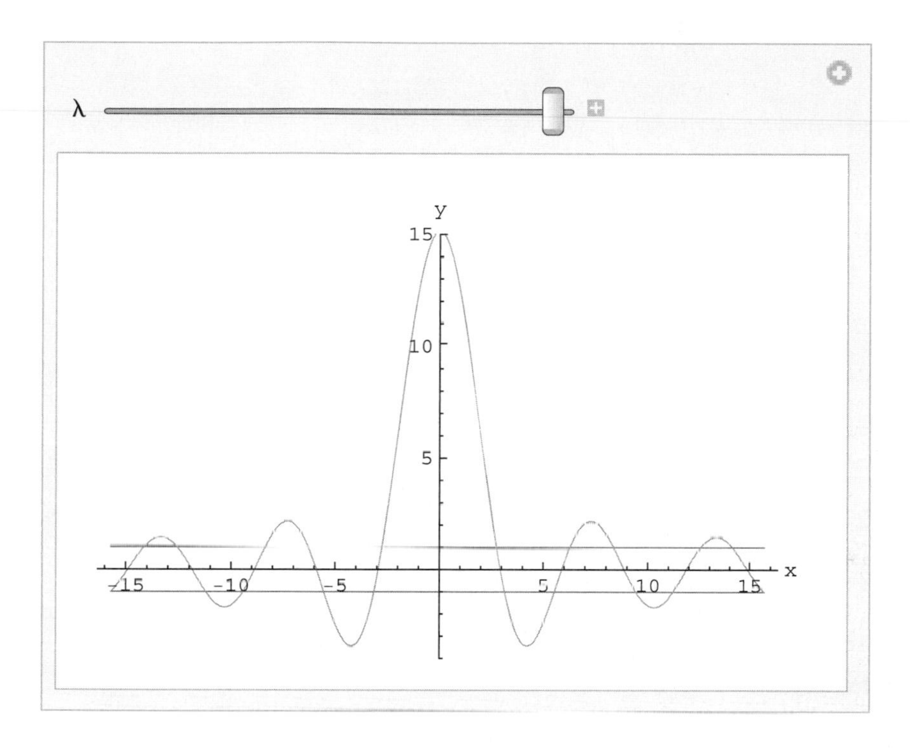

27. Let $a = 1$ in Equation 2.2.1 and

$$f(t) = \begin{cases} \sin \omega_0 t & |t| < T \\ 0 & |t| > T \end{cases}$$

with $T > 0$. Show that the Fourier transform of $f(t)$ is

$$g(\omega) = -\frac{i}{\sqrt{2\pi}} \left\{ \frac{\sin(\omega - \omega_0)T}{(\omega - \omega_0)} - \frac{\sin(\omega + \omega_0)T}{(\omega + \omega_0)} \right\}$$

Verify the uncertainty relation

$$\Delta\omega\Delta t \approx 4\pi$$

*28. If the wave function in configuration space is the Gaussian wave packet

$$\psi(x) = A \exp\left[\frac{i}{\hbar} p_0\, x - \frac{(x - x_0)^2}{2b^2}\right]$$

where A and b are positive real constants, show that the wave function in momentum space is

$$\varphi(p) = A \frac{b}{\sqrt{\hbar}} \exp\left[\frac{i}{\hbar} x_0 \,(p_0 - p) - \frac{b^2\,(p - p_0)^2}{2\hbar^2}\right]$$

Verify that the product of the uncertainty in position Δx and the uncertainty in momentum Δp is of order h, which is Planck's constant. *Hint: Mathematica* can use a little bit of help in evaluating the Fourier transform.

29. A one-dimensional relativistic particle of mass m under the action of a constant force F starts from rest at the origin at time $t = 0$. The equation of motion is

$$F = \frac{dp}{dt}$$

where p is the relativistic momentum. (a) Show that the velocity of the particle is

$$v = \frac{(F/m)t}{\sqrt{1 + (Ft/mc)^2}}$$

and its displacement is

$$x = \frac{mc^2}{F}\left[\sqrt{1 + \left(\frac{Ft}{mc}\right)^2} - 1\right]$$

(b) Plot the relativistic velocity together with the classical one as a function of time.
(c) Plot the relativistic displacement together with the classical one as a function of time. *Hint:* Adopt a convenient system of units.

Answers:

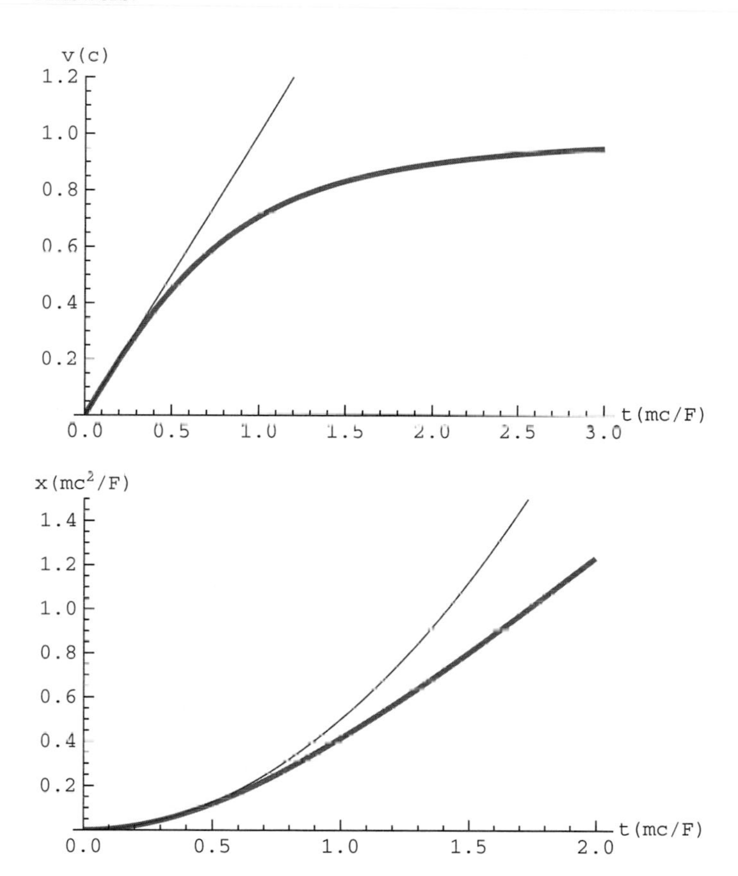

30. A sphere centered at the origin has radius a and charge density

$$\rho(r, \theta) = \rho_0 \frac{a}{r^2}(a - 2\,r)\sin\theta$$

where ρ_0 is a constant and r and θ are the usual spherical coordinates. The potential for points on the z-axis can be expressed as

$$V(z) = \int_0^1 \int_0^{\pi} \frac{(1 - 2\,r)\sin^2\theta}{\sqrt{r^2 + z^2 - 2\,z\,r\cos\theta}} dr d\theta$$

where z and V are in units of a and $\rho_0 a^2/(2\epsilon_0)$, respectively. Plot $V(z)$ for z from 1 to 3. *Hint:* Use **`NIntegrate`**.

Answer:

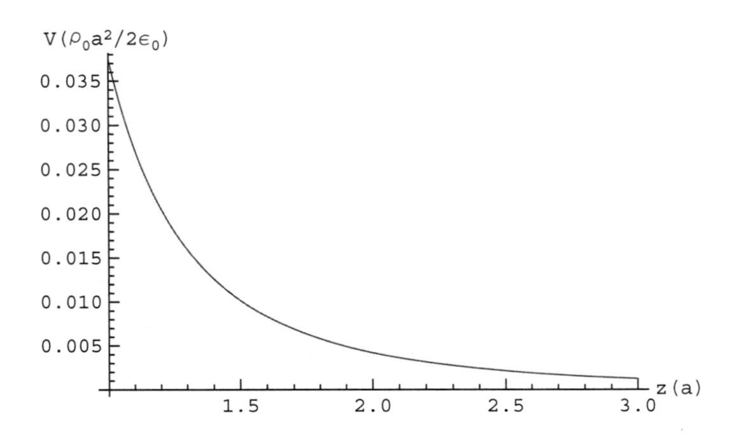

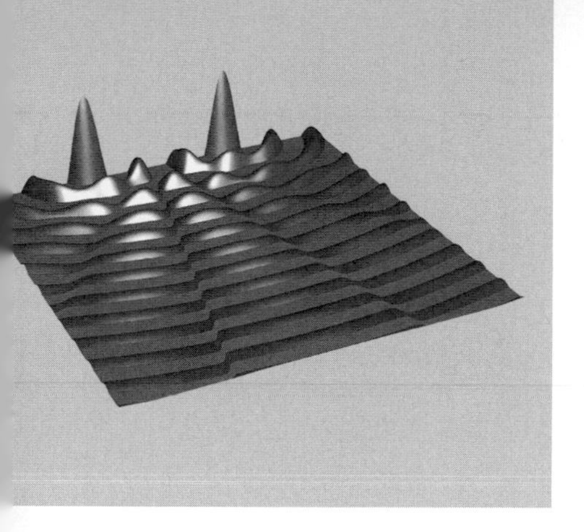

Chapter 3
Programming in *Mathematica*

We have examined the dazzling numerical, symbolic, and graphical capabilities of *Mathematica*. Yet *Mathematica*'s real power rests on its programming capabilities. Many programming concepts, such as assignments, transformation rules, and lists, were introduced in Chapter 2. This chapter covers five programming topics: expressions, patterns, functions, procedures, and graphics. It also discusses three programming styles and the writing of packages.

3.1 EXPRESSIONS

Chapter 2 showed that *Mathematica* works with many kinds of objects, such as algebraic expressions, mathematical functions, equations, functions for specifying operations, assignments, transformation rules, graphics, and lists. They are all expressions. This section discusses the building blocks and internal representation of expressions. It also considers their manipulation and evaluation.

3.1.1 Atoms

Most *Mathematica* objects are composed of parts. For example, the list **{a, b, c}** has four parts. We can extract these parts with the function **Part** introduced in Section 2.4.4:

```
In[1]:= Table[Part[{a, b, c}, i],{i,0,3}]
Out[1]= {List, a, b, c}
```

There are several types of special *Mathematica* objects, called atoms, that do not have parts or are indivisible. The following table lists the types, descriptions, and examples of these objects:

Type	Description	Example
Symbol	symbol	**x**, π, $\hbar$, **Sin**
String	character string "*text* "	**"The Charm of Strange Quarks"**
Integer	integer	**25**
Real	approximate real number *n*.*m*, where *n* and *m* are integers	**1.2345**
Rational	rational number *n*/*m*, where *n* and *m* are integers and $m \neq 0$	**2/3**
Complex	complex number *a* + **I** *b*, where *a* and *b* are integers, approximate real numbers, or rational numbers	**4.12 + 3/5 I**

The function **AtomQ** tests whether an object is one of the atomic types. **AtomQ**[*object*] yields **True** if *object* is an atom, and it yields **False** otherwise:

```
In[2]:= AtomQ[e]
Out[2]= True

In[3]:= AtomQ[4.12 + 3/5 I]
Out[3]= True

In[4]:= AtomQ[1 + x^2 == 0]
Out[4]= False

In[5]:= AtomQ[{a, b, c}]
Out[5]= False
```

For an atom, the function **Head** gives its type:

```
In[6]:= Head[x]
Out[6]= Symbol

In[7]:= Head["The Charm of Strange Quarks"]
Out[7]= String

In[8]:= Head[25]
Out[8]= Integer

In[9]:= Head[1.2345]
Out[9]= Real

In[10]:= Head[2/3]
Out[10]= Rational
```

3.1.2 Internal Representation

Mathematica handles many kinds of objects; the following table lists several of them together with examples:

Algebraic expressions	`x + y`
Mathematical functions	`Sin[x]`
Equations	`y''[x] + y[x] + 2 x == 0`
Functions for specifying operations	$D\left[\frac{1}{3}x^3, x\right]$
Assignments	`a = b`
Transformation rules	`a → b`
Lists	`{a, b, c}`

These objects are all expressions.

Mathematica represents expressions internally in a uniform way. The building blocks of expressions are atoms, which are the simplest expressions. There is a rule for obtaining more complicated expressions from simpler ones: If $h, e_1, e_2, \ldots$ are expressions, then h`[`e_1`, `e_2`, `e_3`, ...]` is also an expression, where h is called the head of the expression and the e_i are called the elements of the expression. Expressions may contain no elements and take the form h`[]`. *Mathematica* expressions are the atoms and the objects represented internally in the form h`[`e_1`, `e_2`, `e_3`, ...]`, which are called normal expressions.

The input forms of expressions may differ from their internal forms. `FullForm[`*expr*`]` shows the internal representation of *expr*:

```
In[1]:= FullForm[x + y]
Out[1]//FullForm=
        Plus[x, y]

In[2]:= FullForm[Sin[x]]
Out[2]//FullForm=
        Sin[x]

In[3]:= FullForm[y''[x] + y[x] + 2 x == 0]
Out[3]//FullForm=
        Equal[Plus[Times[2, x], y[x], Derivative[2][y][x]], 0]

In[4]:= FullForm[a → b]
Out[4]//FullForm=
        Rule[a, b]
```

The output forms of expressions may also be different from their internal forms. We can use `FullForm` to show their internal representations. For example, let us plot x for x from 0 to 1:

```
In[5]:= Plot[x, {x, 0, 1}]
```

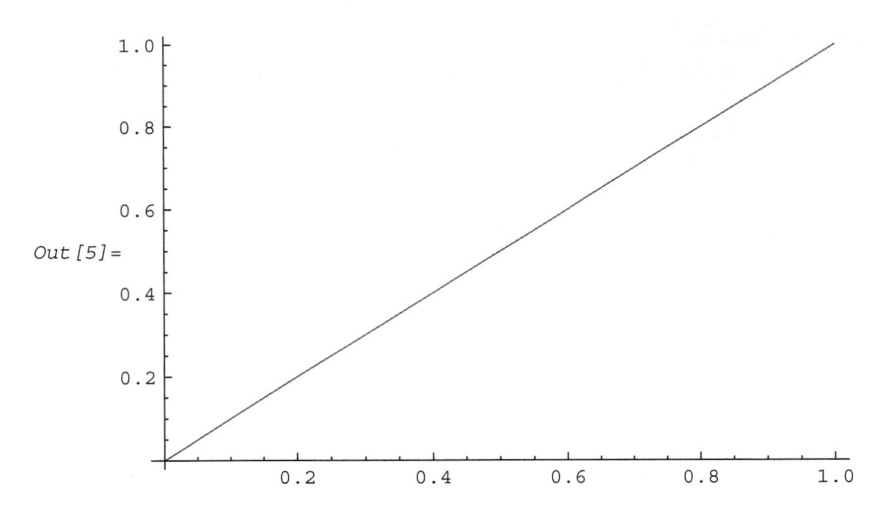

Here is the internal representation of the graphics object:

```
In[6]:= Short[FullForm[%],7]
Out[6]//Short=
        Graphics[List[List[List[],List[],List[Hue[0.67`,0.6`,0.6`],
            Line[List[List[2.040816326530612`*^-8, 2.040816326530612`*^-8],
              List[0.00030671792055962676`, 0.00030671792055962676`],
              \[LeftSkeleton]74\[RightSkeleton],
              List[0.99999002011587`,0.99999002011587`],
              List[0.9999999795918367`,0.9999999795918367`]]]]]],
         List[\[LeftSkeleton]1\[RightSkeleton]]]
```

FullForm returned many lines. The printing of the output has been shortened with **Short**[*expr*, *n*], which reduces the printing of *expr* to approximately *n* lines. **\[LeftSkeleton]***n***\[RightSkeleton]** indicates that *n* elements have been omitted.

To see the full form of an expression, sometimes it is necessary to wrap the function **Hold** around the expression to prevent its evaluation before the internal representation is revealed. For example, consider

```
In[7]:= FullForm[D[1/3x^3, x]]
Out[7]//FullForm=
        Power[x, 2]
```

In this case, *Mathematica* returns the full form of the result of differentiation. To see the internal representation of **D[1/3x^3, x]**, enter

```
In[8]:= FullForm[Hold[D[1/3x^3, x]]]
Out[8]//FullForm=
        Hold[D[Times[Times[1, Power[3, -1]], Power[x, 3]], x]]
```

`Hold[`*expr*`]` maintains *expr* in an unevaluated form.

`FullForm[`*expr*`]` shows the full form of *expr* even if it contains special symbols, such as `->`, `/.`, and `/@`:

```
In[9]:= FullForm[a → b]
Out[9]//FullForm=
        Rule[a, b]
```

```
In[10]:= FullForm[Hold[(f /@expr /. f → g)]]
Out[10]//FullForm=
        Hold[ReplaceAll[Map[f, expr], Rule[f, g]]]
```

Knowing the internal representation of expressions is essential to programming. For example, let us determine the complex conjugate of $(a+ib)(c+id)$, where a, b, c, and d are real. To find the complex conjugate, we simply replace i by $-i$, since a, b, c, and d are real:

```
In[11]:= (a + I b) (c - I d) /. I → -I
Out[11]= (a - i b) (c - i d)
```

Yet *Mathematica* returns an incorrect result. Let us use `FullForm` to reveal the source of error:

```
In[12]:= FullForm[(a + I b) (c - I d)]
Out[12]//FullForm=
        Times[Plus[a, Times[Complex[0, 1], b]],
         Plus[c, Times[Complex[0, -1], d]]]
```

We see that *Mathematica* represents `I` as `Complex[0, 1]` and `-I` as `Complex[0, -1]` internally. To determine the complex conjugate, enter

```
In[13]:= (a + I b) (c - I d) /. Complex[x_, y_] → Complex[x, -y]
Out[13]= (a - i b) (c + i d)
```

This time *Mathematica* yields the correct answer. The notion of "blanks" was introduced in Section 2.2.16. Here, the symbol *arg*_ means any expression to be named *arg* on the right-hand side of the transformation rule.

The function `Head` introduced in Section 3.1.1 can also be used to extract the head of an expression:

```
In[14]:= Head[1 + x + x^2]
Out[14]= Plus
```

`FullForm` shows that `Plus` is indeed the head of the expression:

```
In[15]:= FullForm[1 + x + x^2]
Out[15]//FullForm=
        Plus[1, x, Power[x, 2]]
```

Length[*expr*] gives the index n in the internal form $h[e_1, e_2, e_3, \ldots, e_n]$ of *expr*:

```
In[16]:= Length[1 + x + x^2]
Out[16]= 3
```

For expressions containing no elements, **Length** returns the number 0:

```
In[17]:= Length[h[]]
Out[17]= 0
```

3.1.3 Manipulation

3.1.3.1 Obtaining Parts of Expressions

The positions of the parts of an expression are determined according to the internal form $h[e_1, e_2, \ldots, e_n]$ of the expression: h is part 0 and e_i is the ith part. If the ith part is itself an expression, the position of its jth part is $\{i,j\}$. This indexing scheme continues if a subexpression of a subexpression is also an expression.

The parts of an expression that require exactly n indices for specifying their positions are said to be at level n. Level 0 refers to the entire expression. **Level**[*expr*, *levelspec*] gives a list of all subexpressions of *expr* on levels specified by *levelspec*. The second argument, *levelspec*, of the function is entered according to the level-specification scheme

n	levels 1 through n
Infinity	all levels
$\{n\}$	level n only
$\{n_1, n_2\}$	levels n_1 through n_2

If the option **Heads -> True** is included as its third argument, **Level** returns a list that includes also the heads of subexpressions on levels specified by *levelspec*. Consider the expression **1 + x + 2 (x + 3)^m**. **FullForm** reveals its internal representation:

```
In[1]:= FullForm[1 + x + 2 (x + 3)^m]
Out[1]//Fullform=
        Plus[1, x, Times[2, Power[Plus[3, x], m]]]
```

Here is a list of subexpressions at the level $n = 4$:

```
In[2]:= Level[1 + x + 2 (x + 3)^m, {4}]
Out[2]= {3, x}
```

The function **TreeForm** is helpful in seeing the levels of expressions. **TreeForm**[*expr*] prints out *expr* in a form that shows its level structure:

```
In[3]:= TreeForm[1 + x + 2 (x + 3)^m]
Out[3]//TreeForm=
```

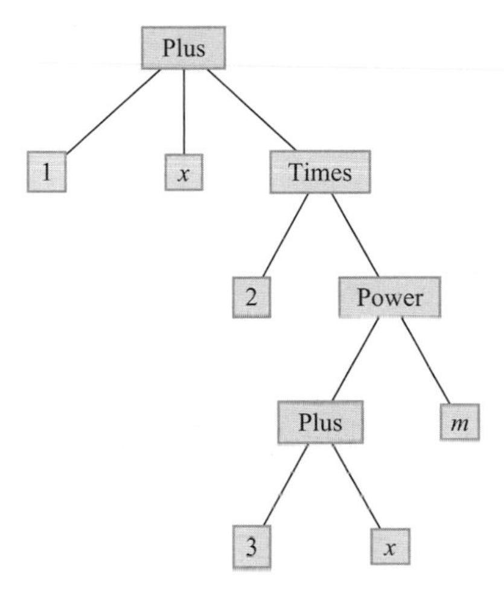

We see that the fourth level indeed consists of the parts **3** and **x**. Here is a list of subexpressions including heads on levels 3 through 4:

```
In[4]:= Level[1 + x + 2 (x + 3)^m, {3, 4}, Heads → True]
Out[4]= {Power, Plus, 3, x, 3 + x, m}
```

Lists are examples of expressions. The functions introduced in Section 2.4.4 for obtaining list and sublist elements can be generalized to functions for obtaining pieces of expressions. The functions for obtaining parts of expressions are

First[*expr*]	the first element
Last[*expr*]	the last element
Part[*expr*, *n*] or *expr*[[*n*]]	the *n*th part
Part[*expr*, -*n*] or *expr*[[-*n*]]	the *n*th part from the end
Part[*expr*, {n_1, n_2, ...}] or *expr*[[{n_1, n_2, ...}]]	*expr* with the n_1th, n_2th, ... elements
Take[*expr*, *n*]	*expr* with the first *n* elements
Take[*expr*, -*n*]	*expr* with the last *n* elements
Take[*expr*, {*m*, *n*}]	*expr* with the *m*th through *n*th elements
Rest[*expr*]	*expr* without the first element
Most[*expr*]	*expr* without the last element
Drop[*expr*, *n*]	*expr* without the first *n* elements
Drop[*expr*, -*n*]	*expr* without the last *n* elements
Drop[*expr*, {*m*, *n*}]	*expr* without the *m*th through *n*th elements

The functions for selecting parts of subexpressions are

Part[*expr*, *i*,*j*,...] or *expr*[[*i*,*j*,...]]	the part at position {*i*,*j*,...} of *expr*
Extract[*expr*, {*i*,*j*,...}]	the part at position {*i*,*j*,...} of *expr*—that is, **Part**[*expr*, *i*,*j*,...]
Extract[*expr*, {{i_1,j_1,...},{i_2,j_2,...},...}]	the list of parts at positions {i_1,j_1,...}, {i_2,j_2,...},...—that is, {**Part**[*expr*, i_1, j_1,...], **Part**[*expr*, i_2,j_2,...], ...}

Let us illustrate the use of these functions with the expression **1 + x + 2 x^2 + 3 x^3 + Sin[x]**. Seeing the full form of this expression can be helpful in understanding the results of function calls:

```
In[5]:= FullForm[1 + x + 2 x^2 + 3 x^3 + Sin[x]]
Out[5]//FullForm=
        Plus[1, x, Times[2, Power[x, 2]], Times[3, Power[x, 3]], Sin[x]]
```

We can pick out one part at a time. Here is the last element:

```
In[6]:= Last[1 + x + 2 x^2 + 3 x^3 + Sin[x]]
Out[6]= Sin[x]
```

Here is the fourth element:

```
In[7]:= Part[1 + x + 2 x^2 + 3 x^3 + Sin[x], 4]
Out[7]= 3 x^3
```

We can obtain expressions with some of the elements. Here is the expression with the first, third, and fourth elements:

```
In[8]:= Part[1 + x + 2 x^2 + 3 x^3 + Sin[x], {1, 3, 4}]
Out[8]= 1 + 2 x^2 + 3 x^3
```

Here is the expression with the first two elements:

```
In[9]:= Take[1 + x + 2 x^2 + 3 x^3 + Sin[x], 2]
Out[9]= 1 + x
```

Here is the expression without the second through fourth elements:

```
In[10]:= Drop[1 + x + 2 x^2 + 3 x^3 + Sin[x], {2, 4}]
Out[10]= 1 + Sin[x]
```

We can select parts of subexpressions as well as expressions. Here is the part at the position $\{4, 2, 1\}$:

```
In[11]:= Part[1 + x + 2 x^2 + 3 x^3 + Sin[x], 4, 2, 1]
Out[11]= x
```

Here is a list of parts at several positions:

```
In[12]:= Extract[1 + x + 2 x^2 + 3 x^3 + Sin[x], {{4, 2, 1}, {3, 0}, {3, 1}, {0}, {5}}]
Out[12]= {x, Times, 2, Plus, Sin[x]}
```

The functions in this section pick out parts according to their positions in expressions. The function **Position** allows us to determine the positions of parts in expressions:

Position[*expr*, *part*]	give a list of the positions at which *part* appears in *expr*
Position[*expr*, *part*, *levelspec*]	give only positions on levels specified by *levelspec*

Let us determine the positions of the symbol **x**:

```
In[13]:= Position[1 + x + 2 x^2 + 3 x^3 + Sin[x], x]
Out[13]= {{2}, {3, 2, 1}, {4, 2, 1}, {5, 1}}
```

We can determine the positions at the third level only:

```
In[14]:= Position[1 + x + 2 x^2 + 3 x^3 + Sin[x], x, {3}]
Out[14]= {{3, 2, 1}, {4, 2, 1}}
```

The function **Select** focuses on the properties rather than the positions of elements in expressions:

Select[*expr*, *criterion*]	*expr* with the elements for which *criterion* gives **True**
Select[*expr*, *criterion*, *n*]	*expr* with the first *n* elements for which *criterion* gives **True**

AtomQ and the built-in predicates described in Section 2.4.4 can be used as criteria for the **Select** function. Let us employ again the expression **1 + x + 2 x^2 + 3 x^3 + Sin[x]** to demonstrate the use of **Select**. We can obtain an expression with the elements that are of the atomic types:

```
In[15]:= Select[1 + x + 2 x^2 + 3 x^3 + Sin[x], AtomQ]
Out[15]= 1 + x
```

Here is the expression with only the first atom:

```
In[16]:= Select[1 + x + 2 x^2 + 3 x^3 + Sin[x], AtomQ, 1]
Out[16]= 1
```

We can also generate an expression with the elements that are odd numbers:

```
In[17]:= Select[1 + x + 2 x^2 + 3 x^3 + Sin[x], OddQ]
Out[17]= 1
```

Note that only the name—that is, without an argument—of a function or predicate is entered as *criterion* for the **Select** function.

Several other built-in functions that test the properties of expressions can also be used to define criteria for the **Select** function:

Function	Yields True If
PolynomialQ[*expr*, *var*]	*expr* is a polynomial in *var*
PolynomialQ[*expr*, {var_1, ...}]	*expr* is a polynomial in the var_i
VectorQ[*expr*, *test*]	*expr* is a vector and *test* gives **True** when applied to each of the elements in *expr*
MatrixQ[*expr*, *test*]	*expr* is a matrix and *test* gives **True** when applied to each of the matrix elements in *expr*
SameQ[*lhs*, *rhs*] or *lhs* === *rhs*	the expression *lhs* is identical to *rhs*
UnsameQ[*lhs*, *lhs*] or *lhs* =!= *lhs*	the expressions *lhs* and *lhs* are not identical
OrderedQ[*h*[e_1, e_2, ...]]	the e_i are in canonical order
MemberQ[*expr*, *form*]	an element of *expr* matches *form*
FreeQ[*expr*, *form*]	no subexpression in *expr* matches *form*
MatchQ[*expr*, *form*]	*form* matches *expr*

Let us demonstrate how to construct criteria with these functions. We define a function that, when applied to an expression, returns **True** if the expression is of the form $a_0 + a_1x + a_2x^2 + \cdots + a_nx^n$, for any integer $n \geq 0$ and any a_i:

```
In[18]:= test1[expr_]:= PolynomialQ[expr,x]
```

With this function, we can obtain an expression with the elements that are polynomials in the variable x:

```
In[19]:= Select[1 + x + 2 x^2 + 3 x^3 + Sin[x] + 4/x^3, test1]
Out[19]= 1 + x + 2 x^2 + 3 x^3
```

Here is a function for testing whether an expression is identical to **Sin[x]**:

```
In[20]:= newtest[expr_]:= (expr === Sin[x])
```

With the function **newtest**, we can pick out the term **Sin[x]**:

```
In[21]:= Select[1 + x + 2 x^2 + 3 x^3 + Sin[x] + 4/x^3, newtest]
Out[21]= Sin[x]
```

The following function tests whether an expression is free of integers:

```
In[22]:= mytest[expr_]:= FreeQ[expr,_Integer]
```

(Section 3.2.3.1 will introduce the pattern _*h*, which stands for any single expression with the head *h*.) This function allows us to generate an expression with the elements that contain no integers:

```
In[23]:= Select[1 + x + 2 x^2 + 3 x^3 + Sin[x] + 4/x^3, mytest]

Out[23]= x + Sin[x]
```

We can, of course, define criteria without using these built-in functions. Here is a function for testing whether an expression is of length greater than or equal to 2:

```
In[24]:= ourtest[expr_]:= (Length[expr] ≥ 2)
```

Let us generate an expression with the elements of appropriate lengths:

```
In[25]:= Select[1 + x + 2 x^2 + 3 x^3 + Sin[x] + 4/x^3, ourtest]

Out[25]= 4/x^3 + 2 x^2 + 3 x^3
```

It is always a good practice to clear symbols that are no longer needed:

```
In[26]:= Clear[test1, newtest, mytest, ourtest]
```

3.1.3.2 Changing Parts of Expressions

The functions introduced in Section 2.4.5 for adding, deleting, and replacing list and sublist elements can be generalized to ones for changing parts of expressions:

`Prepend[`*expr*, *part*`]`	add *part* at the first position of *expr*
`Append[`*expr*, *part*`]`	add *part* at the last position of *expr*
`Insert[`*expr*, *part*, *i*`]`	insert *part* at position *i* in *expr*
`Insert[`*expr*, *part*, `{`*i*, *j*, ...`}]`	insert *part* at position {*i*,*j*,...} in *expr*
`Insert[`*expr*, *part*, `{{`i_1, j_1, ...`}`, `{`i_2, ...`}`, ...`}]`	insert *part* at positions {i_1,j_1,...}, {i_2,...},... in *expr*
`Delete[`*expr*, *i*`]`	delete the part at position *i* in *expr*
`Delete[`*expr*, `{`*i*, *j*, ...`}]`	delete the part at position {*i*, *j*, ...} in *expr*
`Delete[`*expr*, `{{`i_1, j_1, ...`}`, `{`i_2, ...`}`, ...`}]`	delete parts at positions {i_1, j_1,...}, {i_2,...},... in *expr*
`ReplacePart[`*expr*, *part*, *i*`]`	replace the part at position *i* in *expr* with *part*

`ReplacePart[`*expr*, *part*, `{`*i*, *j*, `...}]`	replace the part at position $\{i, j, \ldots\}$ with *part*
`ReplacePart[`*expr*, *part*, `{{`i_1, j_1, `...}, {`i_2, `...}, ...}]`	replace parts at positions $\{i_1, j_1, \ldots\}$, $\{i_2, \ldots\}, \ldots$, with *part*

Let us illustrate the use of these functions with the generic expression `f[a, g[b], c, d]`:

```
In[1]:= Prepend[f[a, g[b], c, d], e]
Out[1]= f[e, a, g[b], c, d]
```

```
In[2]:= Insert[f[a, g[b], c, d], e, 3]
Out[2]= f[a, g[b], e, c, d]
```

Now consider the expression

In[3]:= $\frac{x}{(1+x)^2}$

Out[3]= $\frac{x}{(1+x)^2}$

The function **Insert** allows us to add x^2 after **x** in the denominator:

In[4]:= **Insert**$[\%, x^2, \{2, 1, 3\}]$

Out[4]= $\frac{x}{(1+x+x^2)^2}$

We can remove x^2 with the function **Delete**:

In[5]:= **Delete[%, {2, 1, 3}]**

Out[5]= $\frac{x}{(1+x)^2}$

Let us replace **y** in the expression $\frac{x}{y}$ by $1+x^2$:

In[6]:= **ReplacePart**$\left[\frac{x}{y}, 1+x^2, \{2, 1\}\right]$

Out[6]= $\frac{x}{1+x^2}$

To see that **y** is at the position $\{2, 1\}$, use the function **FullForm** or **Position**:

In[7]:= **FullForm**$\left[\frac{x}{y}\right]$

```
Out[7]//FullForm=
        Times[x, Power[y, -1]]
```

In using these functions, we must remember that *Mathematica* tries to evaluate the expressions whenever it can before their parts are changed. Consider replacing the second element of `D[x + y^2, x]` by `y`:

```
In[8]:= ReplacePart[D[x + y^2, x], y, 2]
        ReplacePart::partw: Part {2} of 1 does not exist. >>
Out[8]= ReplacePart[1, y, 2]
```

Mathematica returns an error message because the derivative was evaluated to **1** before its second argument could be replaced by **y**. To obtain the desired result, enter

```
In[9]:= ReleaseHold[ReplacePart[Hold[D[x + y^2, x]], y, {1, 2}]]
Out[9]= 2 y
```

ReleaseHold[*expr*] removes **Hold** in *expr*.

If the heads of the expressions possess special properties, such as associativity and commutativity, the results of function calls would reflect these properties. For example, consider the function call

```
In[10]:= Prepend[Plus[1, x, x^2], x^3]
Out[10]= 1 + x + x^2 + x^3
```

Contrary to expectation, the element x^3 appears at the end rather than the beginning. This is because *Mathematica* sorts the elements of output into a standard order since **Plus** is both associative and commutative. Section 3.2.5 will discuss assigning attributes to the heads of expressions.

3.1.3.3 Rearranging Expressions

The functions introduced in Section 2.4.6 for rearranging lists can be generalized to ones for rearranging expressions. Several of them are

Sort[*expr*]	sort the elements of *expr* into canonical order
Union[*expr*]	give a sorted version of *expr*, in which all duplicated elements have been dropped
Reverse[*expr*]	reverse the order of the elements in *expr*
Permutations[*expr*]	generate a list of *expr* with all possible permutations of the elements

Let us use the expression **f[d, c, b, a, a]** to illustrate the use of these functions. **Sort** arranges the elements into a standard order:

```
In[1]:= Sort[f[d, c, b, a, a]]
Out[1]= f[a, a, b, c, d]
```

Union sorts the elements and discards the duplicated ones:

```
In[2]:= Union[f[d, c, b, a, a]]
Out[2]= f[a, b, c, d]
```

Permutations gives a list of the expression with all possible permutations of its elements:

```
In[3]:= Short[Permutations[f[d, c, b, a, a]], 4]
Out[3]//Short=
        {f[d, c, b, a, a], f[d, c, a, b, a], f[d, c, a, a, b],
         f[d, b, c, a, a], f[d, b, a, c, a], f[d, b, a, a, c],
         ≪48≫, f[a, a, d, c, b], f[a, a, d, b, c], f[a, a, c, d, b],
         f[a, a, c, b, d], f[a, a, b, d, c], f[a, a, b, c, d]}
```

The function **Short**, introduced in Section 3.1.2, shortened the printing of the output, where **<<**_n_**>>** indicates that _n_ elements have been omitted.

Again, if the heads of expressions possess special properties, the results of function calls would reflect these properties:

```
In[4]:= Permutations[Plus[1, x, x^2]]
Out[4]= {1 + x + x^2, 1 + x + x^2, 1 + x + x^2, 1 + x + x^2, 1 + x + x^2, 1 + x + x^2}
```

The elements of each permutation have been arranged into the same standard order since **Plus** is both associative and commutative.

3.1.3.4 Restructuring Expressions

Most functions introduced in Section 2.4.7 for restructuring lists can be generalized to ones for restructuring expressions:

Partition[*expr*, *n*]	partition without overlap the elements of *expr* into subexpressions with the same head as *expr* and *n* elements
Partition[*expr*, *n*, *d*]	partition with offset *d* the elements of *expr* into subexpressions with the same head as *expr* and *n* elements
Flatten[*expr*]	flatten out all subexpressions with the same head as *expr*
Flatten[*expr*, *n*]	flatten to level *n*
Flatten[*expr*, *n*, *h*]	flatten subexpressions with head *h* to level *n*
FlattenAt[*expr*, *i*]	flatten only the *i*th element of *expr*
FlattenAt[*expr*, {*i*, *j*, ...}]	flatten only the part of *expr* at position {*i*, *j*, ...}
FlattenAt[*expr*, {{i_1, j_1, ...}, {i_2, j_2, ...}, ...}]	flatten parts of *expr* at several positions

We can group without overlap and in pairs the elements of the expression **f[a, b, c, d, e]**:

```
In[1]:= Partition[f[a, b, c, d, e], 2]
Out[1]= f[f[a, b], f[c, d]]
```

Note that the leftover element at the end is dropped! We can also group the elements in triples with an offset of one element between successive triples:

```
In[2]:= Partition[f[a, b, c, d, e], 3, 1]
Out[2]= f[f[a, b, c], f[b, c, d], f[c, d, e]]
```

To demonstrate several properties of the functions **Flatten** and **FlattenAt**, we assign an expression to the variable **myexpr**:

```
In[3]:= myexpr = f[a, h[b, h[c, d]], f[a, h[d]], h[e, f[g]], f[a, f[b, f[c, d]]]]
Out[3]= f[a, h[b, h[c, d]], f[a, h[d]], h[e, f[g]], f[a, f[b, f[c, d]]]]
```

Flatten removes the heads of subexpressions with the same head as the expression:

```
In[4]:= Flatten[myexpr]
Out[4]= f[a, h[b, h[c, d]], a, h[d], h[e, f[g]], a, b, c, d]
```

Observe that none of the elements with the head **h** are affected, including the element **h[e, f[g]]**. We can flatten subexpressions with the head **h** to the second level:

```
In[5]:= Flatten[myexpr, 2, h]
Out[5]= f[a, b, c, d, f[a, h[d]], e, f[g], f[a, f[b, f[c, d]]]]
```

This time none of the elements with the head **f** are affected. Now apply the function **FlattenAt** to the fifth element of **myexpr**:

```
In[6]:= FlattenAt[myexpr, 5]
Out[6]= f[a, h[b, h[c, d]], f[a, h[d]], h[e, f[g]], a, f[b, f[c, d]]]
```

Note that the subexpressions of the fifth element remain unchanged. We can flatten only the part at the position $\{2, 2\}$:

```
In[7]:= FlattenAt[myexpr, {2, 2}]
Out[7]= f[a, h[b, c, d], f[a, h[d]], h[e, f[g]], f[a, f[b, f[c, d]]]]
```

For a more concrete example of restructuring expressions, we define the function

In[8]:= $\mathbf{q[x\_] := \frac{1}{1 + x}}$

With this function, we can generate a nested expression:

```
In[9]:= Nest[q, x, 4]
```

Out[9]= $\frac{1}{1 + \frac{1}{1 + \frac{1}{1 + \frac{1}{1 + x}}}}$

Nest[*f*, *expr*, *n*] gives an expression with *f* applied *n* times to *expr*. The function **FullForm** shows the internal representation of the nested expression:

```
In[10]:= FullForm[%]
Out[10]//FullForm=
        Power[Plus[1,
          Power[Plus[1, Power[Plus[1, Power[Plus[1, x], -1]], -1]], -1]], -1]
```

We can reduce this structure by flattening its subexpressions that have **Power** as their heads:

```
In[11]:= FlattenAt[%, {{1, 2}, {1, 2, 1, 2}, {1, 2, 1, 2, 1, 2}}]
```

Out[11]= $\frac{1}{1+x}$

For another example, consider the polynomial

In[12]:= $\frac{x^2}{3} + x^3 + \frac{5x^5}{3} + x^7 + x^{10} + \frac{4x^{17}}{3}$

Out[12]= $\frac{x^2}{3} + x^3 + \frac{5x^5}{3} + x^7 + x^{10} + \frac{4x^{17}}{3}$

We would like to generate a list of the exponents. First, we flatten those subexpressions with the head **Times** to the first level:

```
In[13]:= Flatten[%, 1, Times]
```

Out[13]= $\frac{10}{3} + x^2 + x^3 + x^5 + x^7 + x^{10} + x^{17}$

The function **Table** gives the desired list:

```
In[14]:= Table[%[[i, 2]], {i, 2, Length[%]}]
Out[14]= {2, 3, 5, 7, 10, 17}
```

Here is another way to generate the list:

```
In[15]:= Exponent[%%, x, List] // Rest
Out[15]= {2, 3, 5, 7, 10, 17}
```

where **Exponent**[*expr*, *form*, *h*] applies *h* to the set of exponents with which *form* appears in *expr*.

There are other functions for restructuring expressions. Several of them are

Thread[*f*[*args*]]	thread *f* over any lists that appear in *args*
Thread[*f*[*args*], *h*]	thread *f* over any objects with head *h* that appear in *args*
Outer[*f*, $list_1$, $list_2$, ...]	give the generalized outer product of the $list_i$
Inner[*f*, $list_1$, $list_2$, *g*]	give the generalized inner product of the $list_i$

Thread applies a function to the corresponding elements of several lists:

```
In[16]:= Thread[f[{a1, a2, a3}, {b1, b2, b3}]]
Out[16]= {f[a1, b1], f[a2, b2], f[a3, b3]}
```

Here is a way to make a list of rules:

```
In[17]:= Thread[Rule[{a1, a2, a3}, {b1, b2, b3}]]
Out[17]= {a1 → b1, a2 → b2, a3 → b3}
```

With a head specified as an additional argument, **Thread** threads a function over the function's arguments that have the specified head:

```
In[18]:= Thread[f[g[a1, a2, a3], g[b1, b2, b3], h[c1, c2, c3]], g]
Out[18]= g[f[a1, b1, h[c1, c2, c3]],
           f[a2, b2, h[c1, c2, c3]], f[a3, b3, h[c1, c2, c3]]]
```

Note that the argument that does not have the head **g** appears in all subexpressions.

Outer applies a function to all possible combinations of the elements from several lists, with each combination consisting of one element from each list:

```
In[19]:= Outer[f, {a1, a2, a3}, {b1, b2, b3}]
Out[19]= {{f[a1, b1], f[a1, b2], f[a1, b3]},
           {f[a2, b1], f[a2, b2], f[a2, b3]}, {f[a3, b1], f[a3, b2], f[a3, b3]}}
```

Inner generates an expression with the specified head and with elements from threading a function over two lists:

```
In[20]:= Inner[f, {a1, a2, a3}, {b1, b2, b3}, g]
Out[20]= g[f[a1, b1], f[a2, b2], f[a3, b3]]
```

3.1.3.5 Operating on Expressions

The functions **Map**, **MapAt**, and **Apply** discussed in Section 2.4.9 for operating on lists can be generalized to functions for operating on expressions. This section also introduces the function **MapAll**.

Map[*f*, *expr*] or *f* **/@** *expr*	apply *f* to each element on the first level in *expr*
Map[*f*, *expr*, *levelspec*]	apply *f* to parts of *expr* specified by *levelspec*
MapAll[*f*, *expr*] or *f* **//@** *expr*	apply *f* to every subexpression in *expr*

Let us illustrate the use of these functions with the expression **h[p[a, b, c],q[r, s, t]]**. **Map**[*f*, *expr*] or *f* **/@** *expr* wraps *f* around each element of *expr*:

```
In[1]:= f /@h[p[a, b, c], q[r, s, t]]
Out[1]= h[f[p[a, b, c]], f[q[r, s, t]]]
```

We can apply f to the parts of *expr* at specified levels:

```
In[2]:= Map[f, h[p[a, b, c], q[r, s, t]], {2}]
Out[2]= h[p[f[a], f[b], f[c]], q[f[r], f[s], f[t]]]
```

Here f is applied at the second level. If the option **Heads -> True** is specified, the function f is also applied to the heads at the specified levels:

```
In[3]:= Map[f, h[p[a, b, c], q[r, s, t]], {2}, Heads → True]
Out[3]= h[f[p][f[a], f[b], f[c]], f[q][f[r], f[s], f[t]]]
```

MapAll[*f*, *expr*] applies f to every subexpression of *expr*, including *expr*:

```
In[4]:= MapAll[f, h[p[a, b, c], q[r, s, t]]]
Out[4]= f[h[f[p[f[a], f[b], f[c]]], f[q[f[r], f[s], f[t]]]]]
```

Sometimes, it is desirable to apply separately a function to only some parts of an expression.

MapAt[*f*, *expr*, *n*]	apply f to the element at position n in *expr*. If n is negative, the position is counted from the end.
MapAt[*f*, *expr*, {*i*, *j*, ...}]	apply f to the part of *expr* at position $\{i, j, \ldots\}$
MapAt[*f*, *expr*, {{i_1, j_1, ...}, {i_2, j_2, ...}, ...}]	apply f to parts of *expr* at several positions

Consider the expression

In[5]:= **myexpr = a +** $\frac{b}{c} + \frac{d + \frac{e}{g}}{r + \frac{s}{t}}$

Out[5]= $a + \frac{b}{c} + \frac{d + \frac{e}{g}}{r + \frac{s}{t}}$

How do we apply the function f to the denominators? We first use **FullForm** to reveal the expression's internal representation upon which the positions of its parts are based:

```
In[6]:= FullForm[myexpr]
Out[6]//FullForm=
        Plus[a, Times[b, Power[c, -1]], Times[Plus[d, Times[e, Power[g, -1]]],
          Power[Plus[r, Times[s, Power[t, -1]]], -1]]]
```

Note that denominators are represented in the form **Power**[*denominator*, **-1**]. We can obtain the positions of **Power** at specified levels of the expression with the function **Position**:

```
In[7]:= Position[myexpr, Power, {3}]
Out[7]= {{2, 2, 0}, {3, 2, 0}}
```

Replacing the `0` in each sublist here by `1` gives the positions of the denominators since in each subexpression of the form `Power[`*denominator*`, -1]`, part 0 is `Power` and part 1 is *denominator*:

```
In[8]:= %/.(0→1)
Out[8]= {{2, 2, 1}, {3, 2, 1}}
```

Let `MapAt` apply f to the denominators at the third level:

```
In[9]:= MapAt[f, myexpr, %]
```

$$\text{Out[9]= } a+\frac{b}{f[c]}+\frac{d+\frac{e}{g}}{f\left[r+\frac{s}{t}\right]}$$

`MapAt` can also apply f to denominators at other levels:

```
In[10]:= MapAt[f, myexpr, Position[myexpr, Power, {5, 6}] /. (0→1)]
```

$$\text{Out[10]= } a+\frac{b}{c}+\frac{d+\frac{e}{f[g]}}{r+\frac{s}{f[t]}}$$

```
In[11]:= Clear[myexpr]
```

The method presented here for mapping at the denominators works well for the given expression. For other algebraic expressions, there is a more general and complex algorithm for applying a function to their numerators or denominators (see [Cop91]).

The function `Apply` changes heads of expressions:

`Apply`[*f*, *expr*] or *f*@@*expr*	replace the head of *expr* by *f*
`Apply`[*f*, *expr*, *levelspec*]	replace heads in parts of *expr* specified by *levelspec*

Let us illustrate the use of `Apply` with the expression `h[p[a, b, c], q[r, s, t]]`, introduced previously in this section:

```
In[12]:= Apply[f, h[p[a, b, c], q[r, s, t]]]
Out[12]= f[p[a, b, c], q[r, s, t]]
```

`Apply` replaces `h` with `f` as the head of the expression. We can also replace the heads of subexpressions by giving a level specification:

```
In[13]:= Apply[f, h[p[a, b, c], q[r, s, t]], {1}]
Out[13]= h[f[a, b, c], f[r, s, t]]
```

3.1.3.6 Manipulating Equations

This section presents two examples on manipulating equations: one taken from introductory mechanics and the other extracted from Section 5.3.

Example 3.1.1 A projectile moves in the xy plane where the y-axis is directed vertically upward. It is initially at the origin, and its initial velocity $\mathbf{v}_0$ makes an angle θ_0 with the positive x direction. Derive an equation relating the x and y coordinates of the projectile, and determine the horizontal range. Neglect air resistance.

The x and y components of the projectile acceleration can be expressed as

$$\frac{d^2x}{dt^2} = 0 \tag{3.1.1}$$

$$\frac{d^2y}{dt^2} = -g \tag{3.1.2}$$

where g is the magnitude of the free-fall acceleration. The components of the initial velocity are

$$v_{x0} = v_0 \cos\theta_0 \tag{3.1.3}$$

$$v_{y0} = v_0 \sin\theta_0 \tag{3.1.4}$$

and the initial coordinates are

$$x_0 = 0 \tag{3.1.5}$$

$$y_0 = 0 \tag{3.1.6}$$

DSolve solves Equations 3.1.1 and 3.1.2 subject to the initial conditions (Equations 3.1.3–3.1.6). Let us do this example in TraditionalForm to demonstrate the typesetting capabilities of *Mathematica*. Heeding the admonition given in Section 2.5.3 about the lack of necessary precision of TraditionalForm for input, we enter in StandardForm

```
DSolve[{D[x[t], {t, 2}] == 0,
  D[y[t], {t, 2}] == -g, (D[x[t], t] /. t→0) == v₀ Cos[θ₀],
  (D[y[t], t] /. t→0) == v₀ Sin[θ₀], x[0] == 0, y[0] == 0}, {x[t], y[t]}, t]
```

To convert this input to TraditionalForm, paste a copy of it below, click the bracket of the new cell, and choose **TraditionalForm** in the Convert To submenu of the Cell menu:

In[1]:= $\mathbf{DSolve}\left[\left\{\dfrac{\partial^2 x(t)}{\partial t^2}=0,\ \dfrac{\partial^2 y(t)}{\partial t^2}=-g,\ \dfrac{\partial x(t)}{\partial t}\ /.\,t\to 0=v_0\cos(\theta_0),\ \dfrac{\partial y(t)}{\partial t}\ /.\,t\to 0=v_0\sin(\theta_0),\ x(0)=0,\ y(0)=0\right\},\ \{x(t),\ y(t)\},\ t\right]$

Out[1]= $\left\{\left\{x(t)\to t\cos(\theta_0)\,v_0,\ y(t)\to \frac{1}{2}\left(2\,t\sin(\theta_0)\,v_0 - g\,t^2\right)\right\}\right\}$

where we have chosen **Mathematica ▸ Preferences ▸ Evaluation ▸ Format type of new output cells ▸ TraditionalForm** for Mac OS X or **Edit ▸ Preferences ▸ Evaluation ▸ Format type of new output cells ▸ TraditionalForm** for Windows before the evaluation of the input. (For the rest of this example, we enter each input in StandardForm first and then convert it to TraditionalForm.) These rules provide the equations relating the x and y coordinates to time t:

In[2]:= $\{x,y\} == \{x(t),y(t)\}$ /.% [[1]]

Out[2]= $\{x,y\} = \left\{t\cos(\theta_0)\,v_0, \frac{1}{2}\left(2\,t\sin(\theta_0)\,v_0 - g\,t^2\right)\right\}$

Solve these equations for y in terms of x by eliminating t:

In[3]:= **Simplify [Solve [%, y, t]]**

Out[3]= $\left\{\left\{y \to x\tan(\theta_0) - \frac{g\,x^2\sec^2(\theta_0)}{2\,v_0^2}\right\}\right\}$

where `Solve[`*eqns*`,`*vars*`,`*elims*`]` attempts to solve the equations for *vars*, eliminating the variables *elims*. Let us turn the preceding rule for y into an equation:

In[4]:= **Equal @@ % [[1, 1]]**

Out[4]= $y = x\tan(\theta_0) - \frac{g\,x^2\sec^2(\theta_0)}{2\,v_0^2}$

where we have used the special input form for `Apply`. This equation gives the relation between the x and y coordinates; this is the equation of the trajectory of the projectile. To determine the horizontal range, solve this equation for x with $y = 0$:

In[5]:= **Simplify [Solve [% /. y → 0, x]]**

Out[5]= $\left\{\{x \to 0\}, \left\{x \to \frac{\sin(2\,\theta_0)\,v_0^2}{g}\right\}\right\}$

Whereas the first rule specifies the initial x coordinate, the second gives the range:

In[6]:= **R == x /. % [[2]]**

Out[6]= $R = \frac{\sin(2\,\theta_0)\,v_0^2}{g}$

Note that `R` is a mathematical variable without an assigned value:

In[7]:= **ValueQ[R]**

Out[7]= False

where `ValueQ[expr]` gives `True` if a value has been defined for *expr*, and it gives `False` otherwise. To turn the last equation into an assignment, replace `Equal (==)` with `Set (=)` or `SetDelayed (:=)`,

```
In[8]:= Set @@ %%
```

$$\text{Out[8]=}\ \frac{\sin(2\,\theta_0)\,v_0^2}{g}$$

Now `R` has an assigned value:

```
In[9]:= R
```

$$\text{Out[9]=}\ \frac{\sin(2\,\theta_0)\,v_0^2}{g}$$

```
In[10]:= Clear[R]
```

Before leaving this example, let us restore the default output format type to StandardForm by choosing **Mathematica ▸ Preferences ▸ Evaluation ▸ Format type of new output cells ▸ StandardForm** for Mac OS X or **Edit ▸ Preferences ▸ Evaluation ▸ Format type of new output cells ▸ StandardForm** for Windows. ■

Example 3.1.2 Without using `DSolve`, obtain the solution to the differential equations

$$\frac{dx}{dt} = by \tag{3.1.7}$$

$$\frac{dy}{dt} = -bx + e \tag{3.1.8}$$

where b and e are constants.

Instead of using `DSolve`, we solve these equations with the "paper and pencil" approach. Although the procedure is rather tedious, it can be argued that this approach may be more appropriate for undergraduate education since it elucidates the fundamentals of mathematics, which normal use of *Mathematica* often obscures.

Let Equations 3.1.7 and 3.1.8 be assigned to a variable:

```
In[11]:= eqn = {x'[t] == b y[t], y'[t] == -b x[t] + e}
Out[11]= {x'[t] == b y[t], y'[t] == e - b x[t]}
```

Now differentiate both sides of the first equation,

```
In[12]:= D[eqn[[1]], t]
Out[12]= x''[t] == b y'[t]
```

solve the resulting equation and the second equation for `x''[t]` by eliminating `y'[t]`,

```
In[13]:= Solve[{%, eqn[[2]]}, x''[t], y'[t]] //ExpandAll
```

Out[13]= $\{\{x''[t] \to b\,e - b^2\,x[t]\}\}$

and turn this rule into an equation,

```
In[14]:= Equal @@ %[[1, 1]]
```

Out[14]= $x''[t] == b\,e - b^2\,x[t]$

This equation has the same form as the equation of motion of a harmonic oscillator acted on by a constant force. The solution to this equation is well-known:

In[15]:= **xsol = C1** $e^{-i b t}$ **+ C2** $e^{i b t}$ **+** $\frac{e}{b}$

Out[15]= $\frac{e}{b} + C1\,e^{-i b t} + C2\,e^{i b t}$

Similarly, differentiate both sides of Equation 3.1.8, solve the resulting equation and Equation 3.1.7 for **y''[t]** with the elimination of **x'[t]**,

```
In[16]:= Solve[{D[eqn[[2]], t], eqn[[1]]}, y''[t], x'[t]]
```

Out[16]= $\{\{y''[t] \to -b^2\,y[t]\}\}$

and turn the rule into an equation,

```
In[17]:= Equal @@ %[[1, 1]]
```

Out[17]= $y''[t] == -b^2\,y[t]$

Again, this equation has the same form as the equation of motion of a harmonic oscillator, and its solution is known:

In[18]:= **ysol = C3** $e^{-i b t}$ **+ C4** $e^{i b t}$

Out[18]= $C3\,e^{-i b t} + C4\,e^{i b t}$

The difficulty is that there are four arbitrary constants **C1**, **C2**, **C3**, and **C4** for two first-order differential equations (Equations 3.1.7 and 3.1.8) that permit only two initial conditions. The problem arises because new solutions that may not satisfy the original equations were introduced when these equations were differentiated. To eliminate these extraneous solutions, substitute **xsol** and **ysol** and their derivatives into Equations 3.1.7 and 3.1.8,

```
In[19]:= eqn /. {x[t] -> xsol, x'[t] -> D[xsol, t], y[t] -> ysol, y'[t] -> D[ysol, t]}
```

Out[19]= $\{-i\,b\,C1\,e^{-i b t} + i\,b\,C2\,e^{i b t} == b\left(C3\,e^{-i b t} + C4\,e^{i b t}\right),$

$-i\,b\,C3\,e^{-i b t} + i\,b\,C4\,e^{i b t} == e - b\left(\frac{e}{b} + C1\,e^{-i b t} + C2\,e^{i b t}\right)\}$

and solve for **C3** and **C4**:

```
In[20]:= Solve[%, {C3, C4}] //Flatten
Out[20]= {C3 → -ⅈ C1, C4 → ⅈ C2}
```

Thus, replacing **C3** and **C4** in **ysol** according to these rules gives the solution to Equations 3.1.7 and 3.1.8:

```
In[21]:= {xsol, ysol /. %}
```

$$Out[21]= \left\{\frac{e}{b} + C1\, e^{-i b t} + C2\, e^{i b t},\ -i\, C1\, e^{-i b t} + i\, C2\, e^{i b t}\right\}$$

which can be written as

```
In[22]:= ExpToTrig[%] /.
            {C1 -> 1/2 (C[1] + ⅈ C[2]), C2 -> 1/2 (C[1] - ⅈ C[2])}//Simplify
```

$$Out[22]= \left\{\frac{e}{b} + C[1]\, Cos[b\, t] + C[2]\, Sin[b\, t],\ C[2]\, Cos[b\, t] - C[1]\, Sin[b\, t]\right\}$$

Of course, **DSolve** can solve Equations 3.1.7 and 3.1.8 with almost no effort from us:

```
In[23]:= DSolve[{x'[t] == b y[t], y'[t] == -b x[t] + e}, {x[t], y[t]}, t]//Simplify
```

$$Out[23]= \left\{\left\{x[t] \to \frac{e}{b} + C[1]\, Cos[b\, t] + C[2]\, Sin[b\, t],\ y[t] \to C[2]\, Cos[b\, t] - C[1]\, Sin[b\, t]\right\}\right\}$$

```
In[24]:= Clear[eqn, xsol, ysol]
```

■

3.1.4 Exercises

In this book, straightforward, intermediate-level, and challenging exercises are unmarked, marked with one asterisk, and marked with two asterisks, respectively. For solutions to most odd-numbered exercises, see Appendix C.

1. Determine the internal forms of the following expressions:
   (a) **x - y**
   (b) **x^2 / y^2**
   (c) **{a, {b, {c, d}}, {e, f}}**
   (d) **(a + I b) / (c - I d)**
   (e) **Integrate[x^2, x]**
   (f) **DSolve[y''[x] - 3 y'[x] - 18 y[x] == x Exp[4 x], y[x], x]**
   (g) **f[a] + g[c, d] /. f[x_] → x^3**
   (h) **Sin /@ (a + b + c)**

(i)
```
CrossProduct[a, CrossProduct[b, c]] +
 CrossProduct[b, CrossProduct[c, a]] +
 CrossProduct[c, CrossProduct[a, b]]
```

Answers:

```
Plus[x, Times[-1, y]]

Times[Power[x, 2], Power[y, -2]]

List[a, List[b, List[c, d]], List[e, f]]

Times[Plus[a, Times[Complex[0, 1], b]],
 Power[Plus[c, Times[Complex[0, -1], d]], -1]]

Hold[Integrate[Power[x, 2], x]]
Hold[DSolve[Equal[
   Plus[Derivative[2][y][x], Times[-1, Times[3, Derivative[1][y][x]]],
    Times[-1, Times[18, y[x]]]], Times[x, Exp[Times[4, x]]]], y[x], x]]

Hold[ReplaceAll[Plus[f[a], g[c, d]],
  Rule[f[Pattern[x, Blank[]]], Power[x, 3]]]]

Hold[Map[Sin, Plus[a, b, c]]]

Plus[CrossProduct[a, CrossProduct[b, c]],
 CrossProduct[b, CrossProduct[c, a]],
 CrossProduct[c, CrossProduct[a, b]]]
```

2. Determine the internal forms of the following expressions.

(a) $\hbar$

(b) $\frac{x}{y}$

(c) a^2

(d) a_β

(e) $\begin{pmatrix} \mathcal{A} & \mathcal{B} \\ \mathcal{C} & \mathcal{D} \end{pmatrix}$

(f) $\int_1^2 x^4 \, dx$

(g) `f[a] × g[c, d] /. c→1`

(h) $\sum_{n=0}^{\infty} \frac{(-1)^n x^{2n}}{2^{2n} (n!)^2}$

(i) $\text{DSolve}\left[\left\{\frac{\partial^2 x(t)}{\partial t^2} == 0, \frac{\partial^2 y(t)}{\partial t^2} == -g, \left(\frac{\partial x(t)}{\partial t} /. t \to 0\right) == v_0 \cos(\theta_0),\right.\right.$
$\left.\left.\left(\frac{\partial y(t)}{\partial t} /. t \to 0\right) == v_0 \sin(\theta_0), x(0) == 0, y(0) == 0\right\}, \{x(t), y(t)\}, t\right]$

Answers:

```
\[HBar]

Times[x, Power[y, -1]]

Power[a, 2]

Subscript[a, \[Beta]]

List[List[\[ScriptCapitalA], \[ScriptCapitalB]],
 List[\[ScriptCapitalC], \[ScriptCapitalD]]]

Hold[Integrate[Power[x, 4],List[x, 1, 2]]]

Hold[ReplaceAll[Cross[f[a], g[c, d]], Rule[c, 1]]]

Hold[Sum[Times[Power[-1, n], Power[x, Times[2, n]],
   Power[Times[Power[2, Times[2, n]], Power[Factorial[n], 2]], -1]],
  List[n, 0, Infinity]]]

Hold[
 DSolve[List[Equal[D[x[t], List[t, 2]], 0], Equal[D[y[t], List[t, 2]],
    Times[-1, g]], Equal[ReplaceAll[D[x[t], t], Rule[t, 0]],
    Times[Subscript[v, 0], Cos[Subscript[\[Theta], 0]]]],
   Equal[ReplaceAll[D[y[t], t], Rule[t, 0]],
    Times[Subscript[v, 0], Sin[Subscript[\[Theta], 0]]]],
   Equal[x[0], 0], Equal[y[0], 0]], List[x[t], y[t]], t]]
```

3. Write a function that returns the complex conjugate of an expression the parts of which, with the exception of the imaginary unit `I`, are all real.
4. Obtain a list of all subexpressions on levels 2 through 4 of the expression

$$1+2\left(-\frac{\pi}{4}+x\right)+2\left(-\frac{\pi}{4}+x\right)^2+\frac{8}{3}\left(-\frac{\pi}{4}+x\right)^3$$

Answer:

$$\left\{2,-\frac{1}{4},\pi,-\frac{\pi}{4},x,-\frac{\pi}{4}+x,2,-\frac{\pi}{4},x,\right.$$
$$\left.-\frac{\pi}{4}+x,2,\left(-\frac{\pi}{4}+x\right)^2,\frac{8}{3},-\frac{\pi}{4},x,-\frac{\pi}{4}+x,3,\left(-\frac{\pi}{4}+x\right)^3\right\}$$

5. Obtain a list of all subexpressions including heads on levels 3 and 4 of the expression

$$\frac{\sqrt{3}}{2}+\frac{1}{2}\left(-\frac{\pi}{3}+x\right)-\frac{1}{4}\sqrt{3}\left(-\frac{\pi}{3}+x\right)^2-\frac{1}{12}\left(-\frac{\pi}{3}+x\right)^3+\frac{\left(-\frac{\pi}{3}+x\right)^4}{16\sqrt{3}}$$

Answer:

$$\left\{\text{Power}, 3, \frac{1}{2}, \text{Plus}, \text{Times}, -\frac{1}{3}, \pi, -\frac{\pi}{3}, \text{x}, \text{Power}, 3, \frac{1}{2}, \text{Power}, \text{Plus}, -\frac{\pi}{3}, \text{x}, -\frac{\pi}{3}+\text{x}, 2, \text{Power}, \text{Plus}, -\frac{\pi}{3}, \text{x}, -\frac{\pi}{3}+\text{x}, 3, \text{Power}, 3, -\frac{1}{2}, \text{Power}, \text{Plus}, -\frac{\pi}{3}, \text{x}, -\frac{\pi}{3}+\text{x}, 4\right\}$$

6. Consider the expression

```
x^2/(x-1) == 1/(x-1)
```

Obtain a list of all subexpressions on levels 1 through 3.

Answer:

$$\left\{-1+\text{x}, -1, \frac{1}{-1+\text{x}}, \text{x}, 2, \text{x}^2, \frac{\text{x}^2}{-1+\text{x}}, -1, \text{x}, -1+\text{x}, -1, \frac{1}{-1+\text{x}}\right\}$$

7. Consider the expression

```
x^3 + (1 + z)^2
```

Obtain a list of all subexpressions on only level 2 of this expression.

Answer:

$\{\text{x}, 3, 1+\text{z}, 2\}$

8. Consider the expression

```
x^3 + (1 + z)^2
```

Obtain a list of all subexpressions on levels 1 and 2 of this expression.

Answer:

$\{\text{x}, 3, \text{x}^3, 1+\text{z}, 2, (1+\text{z})^2\}$

9. Consider the expression

```
a + b/c + (d + e)/(1 + f/(1 - g/h))
```

Using **Part** and **Rest**, pick out the subexpression **g/h**.

Answer:

$$\frac{\text{g}}{\text{h}}$$

10. Determine the position of $\mathbf{a}^3$ at the sixth level of the expression

$$\sqrt{\frac{24\,a^2\,b}{1+3\,a^3}\sqrt{a^3\,b^2}}$$

Answer:

```
{{3, 1, 2, 1, 2, 2}}
```

11. With the function **Part**, obtain the piece **1 + t** from the expression

```
a + b/c + (d + e/g) / (r + s/ (1 + t))
```

Answer:

```
1 + t
```

12. With the function **Extract**, obtain from the expression

$$\frac{\lambda_1}{1+\frac{\lambda_2}{1+\frac{\lambda_3}{1+x}}}$$

the list

$\{\lambda_1, \lambda_2, \lambda_3\}$

Answer:

$\{\lambda_1, \lambda_2, \lambda_3\}$

13. Using **Select**, create a list of all atomic types in the list

$\{\pi, \phi, 2\,\pi, e^x, f_3, 4+\sqrt{7}\,i, 4+2\,i, 7.12, \bar{P}, \text{"Chem/Phyx 340"}, a+b\}$

Answer:

$\{\pi, \phi, 4+2\,i, 7.12, \text{Chem/Phyx 340}\}$

14. Using **Select**, create a list of all atomic types in the list

$\{\pi, \phi, 2\,\alpha, \text{Sin}[\theta], \gamma_n, 4+\sqrt{7}\,i, 2/3, 7.12, \hat{z}, \text{"Happy New Year!"}, a^2+b^2\}$

Answer:

$\{\pi, \phi, \frac{2}{3}, 7.12, \text{Happy New Year!}\}$

15. Using **Select**, create a list of all numeric quantities in the list

```
{x, Pi, ⅇ, ∞, °, "It is a wonderful world.", 44/100, {a, {b, c}}, 4 + 5 ⅈ}
```

Answer:

```
{π, ⅇ, °, 11/25, 4 + 5 ⅈ}
```

16. Use **RandomInteger** to create a list of 50 random integers in the range 1 to 200, and name the list **someRandoms**. Using **Select,** form the list of numbers belonging to **someRandoms**, which are less than 50. *Hint:* Define a function as a predicate (i.e., criterion).

17. Using **Select**, remove all elements containing **y** in the expression

```
1 + 3 x + 3 x^2 + x^3 + 3 y + 6 x y + 3 x^2 y + 3 y^2 + 3 x y^2 + y^3
```

Answer:

$1 + 3x + 3x^2 + x^3$

*18. Consider the list

```
{x, π, "Fortran is archaic", y + z, 99 + 44/100, {y, z}, a + ⅈb, 2.718, 340}
```

Using **Select**, create a list of all elements that are not atomic types.

Answer:

```
{a + ⅈ b, y + z, {y, z}}
```

19. Insert **b** after every element in the expression

$h[e_1, e_2, e_3, e_4, e_5, e_6, e_7]$

Answer:

$h[e_1, b, e_2, b, e_3, b, e_4, b, e_5, b, e_6, b, e_7, b]$

20. Using **ReplacePart**, replace x^2 by y^3 in the expression

$$\frac{x + y}{1 + 3x + 3x^2 + x^3 + 3y + 6xy + 3x^2y + 3y^2 + 3xy^2 + y^3}$$

Answer:

$$\frac{x + y}{1 + 3x + x^3 + 3y + 6xy + 3y^2 + 3xy^2 + 4y^3 + 3y^4}$$

21. Consider the list

```
{{a, {b, c}}, {d, {e, {f, {g, h}}}}, {{i, j}, k}}
```

(a) Using **FlattenAt**, generate the list

```
{{a, {b, c}}, {d, {e, {f, g, h}}}, {{i, j}, k}}
```

(b) Using **Flatten**, generate the list

```
{a, b, c, d, e, {f, {g, h}}, i, j, k}
```

22. From the expression

```
a + b / c + (d + e) / (f + g)
```

obtain the expression

$$a+b+\frac{1}{c}+d+e+\frac{1}{f+g}$$

23. Given two lists:

```
t = {15, 20, 25, 30, 40, 50, 60, 70, 80, 90,
   100, 110, 120, 130, 140, 150, 160, 170, 180, 190, 200,
   210, 220, 230, 240, 250, 260, 270, 280, 290, 298.1};

cp = {0.311, 0.605, 0.858, 1.075, 1.452, 1.772, 2.084,
    2.352, 2.604, 2.838, 3.060, 3.254, 3.445, 3.624, 3.795,
    3.964, 4.123, 4.269, 4.404, 4.526, 4.639, 4.743, 4.841,
    4.927, 5.010, 5.083, 5.154, 5.220, 5.286, 5.350, 5.401};
```

(a) Create a list consisting of elements of **cp** divided by the corresponding elements of **t**. *Hint:* Make it simple! (b) Create the same list with the function **Thread** and the function **f** defined by

```
f[x_, y_] := x / y
```

Answer:

```
{0.0207333, 0.03025, 0.03432, 0.0358333, 0.0363, 0.03544, 0.0347333,
 0.0336, 0.03255, 0.0315333, 0.0306, 0.0295818, 0.0287083,
 0.0278769, 0.0271071, 0.0264267, 0.0257688, 0.0251118, 0.0244667,
 0.0238211, 0.023195, 0.0225857, 0.0220045, 0.0214217, 0.020875,
 0.020332, 0.0198231, 0.0193333, 0.0188786, 0.0184483, 0.0181181}
```

24. Using **Map**, square both sides of the equation

$\sqrt{1 + x} == x^2 + 2$

Answer:

$1 + x == (2 + x^2)^2$

25. Consider the equation

$\frac{x^2}{x - 1} == \frac{1}{x - 1}$

Using **Map**, multiply both sides of the equation by **(x - 1)**.

Answer:

$x^2 == 1$

26. Using **MapAt**, apply **func** to **t** in the expression

a + b/c + (d + e/f) / (r + s/t)

Answer:

$a + \frac{b}{c} + \frac{d + \frac{e}{f}}{r + \frac{s}{\text{func}[t]}}$

*27. Apply **func** to all the numerators of the expression

a + b/c + (d + e/f) / (r + s/t)

Answer:

$a + \frac{\text{func}[b]}{c} + \frac{\text{func}\left[d + \frac{\text{func}[e]}{f}\right]}{r + \frac{\text{func}[s]}{t}}$

28. Using **Apply** and **Equal**, turn the rule

x^2 + x → 1

into an equation.

Answer:

$x + x^2 == 1$

29. Using **Apply**, turn the rule

$$x \to v_{0x}\,t + \frac{1}{2} a_x\,t^2$$

into an equation.

Answer:

$$x == \frac{t^2\,a_x}{2} + t\,v_0$$

30. Turn the rule

$$x -> \frac{-b + \sqrt{b^2 - 4\,a\,c}}{2\,a}$$

into an equation.

Answer:

$$x == \frac{-b + \sqrt{b^2 - 4\,a\,c}}{2\,a}$$

31. Using **Apply** and **Expand**, turn the sum

$$5\,a + 3\,b + \frac{2\,c + 5\,a}{7}$$

into the list

$$\left\{\frac{40\,a}{7},\ 3\,b,\ \frac{2\,c}{7}\right\}$$

*32. Using **Select**, **Flatten**, **Apply**, and **FreeQ**, obtain a list of the variables in the order they appear in the expression

$$x^3 + 3\,x^2\,y^2 + 3\,x\,y^4 + y^6 + 3\,x^2\,z^3 + 6\,x\,y^2\,z^3 + 3\,y^4\,z^3 + 3\,x\,z^6 + 3\,y^2\,z^6 + z^9$$

Answer:

```
{x, x, y, x, y, y, x, z, x, y, z, y, z, x, z, y, z, z}
```

3.2 PATTERNS

The power of *Mathematica* comes from its pattern-matching capabilities. In Section 2.2.16, we introduced the pattern *arg_* in order to define functions. This section shows that _, called a blank, is the basic pattern that stands for any expression. We can construct other patterns with blanks and also name or restrict patterns. Pattern-matching capabilities of *Mathematica* allow us to define functions that accept only arguments matching specific patterns. In pattern matching, *Mathematica* examines structural rather than mathematical equivalence and considers attributes of functions. Patterns can have default values and can contain alternative or repeated patterns. This section concludes with a discussion of multiple blanks. Whereas single blanks stand for single expressions, multiple blanks represent sequences of expressions.

3.2.1 Blanks

The basic pattern is the "blank," which stands for any single expression. Its special input form and internal form are `_` and `Blank[]`, respectively. For example, `h[_]` represents the class of expressions of the form `h[`*element*`]`, where *element* can be any single expression.

```
In[1]:={h[], h[x^2], h[Sin[x]], h[3+4I],
          g[x], h[b, c], h[a, b, c, d]} /. h[_] → "MATCH"
Out[1]= {h[], MATCH, MATCH, MATCH, g[x], h[b, c], h[a, b, c, d]}
```

The expressions `h[x^2]`, `h[Sin[x]]`, `h[3+4I]` match the pattern `h[_]` because each of them has the head `h` and a single element.

The pattern `{_, _}` stands for a list with any two elements. The function `Position` introduced in Section 3.1.3.1 gives a list of the positions at which parts matching a pattern appear in an expression.

```
In[2]:= Position[{{}, {a}, {b, c}, {c, {b, d}}}, {_, _}]
Out[2]= {{3}, {4, 2}, {4}}
```

`Position` returns a list of the positions of `{b, c}`, `{b, d}`, and `{c, {b, d}}`, which match the pattern.

From a list of expressions, we can pick out with the function `Cases` those matching the pattern "`_^_`", which represents any expression to any power. `Cases[`*expr*`, `*pattern*`, `*levelspec*`]` gives a list of parts that match *pattern* and that are on levels of *expr* specified by *levelspec*. If the third argument is omitted, *levelspec* takes the default value `{1}`.

```
In[3]:= Cases[{3, x, a^4, Sqrt[1+x]}, _^_]
Out[3]= {a^4, √(1+x)}
```

Note that `3` and `x`, which are mathematically equivalent to `Sqrt[3]^2` and `x^1`, respectively, do not match the pattern "`_^_`". Section 3.2.4 will show that pattern matching is based on internal forms of expressions and structural equivalence.

From the list of expressions, we can also delete with the function `DeleteCases` those matching the pattern "`_^_`". `DeleteCases[`*expr*`, `*pattern*`, `*levelspec*`]` removes all parts that match *pattern* and that are on levels of *expr* specified by *levelspec*. If the third argument is omitted, *levelspec* takes the default value `{1}`.

```
In[4]:= DeleteCases[{3, x, a^4, Sqrt[1 + x]}, _^_]
Out[4]= {3, x}
```

3.2.2 Naming Patterns

It is often necessary to assign names to patterns. Names of patterns must be symbols; that is, **Head[*name*]** must return **Symbol**. Two-dimensional forms, for example, cannot be names of patterns because they are not symbols:

```
In[1]:= Head /@ {x^y, x_y, x_z^y, Underscript[x, y], x^†, x^+, x^*, x̂, x̄}
Out[1]= {Power, Subscript, Power, Underscript,
         SuperDagger, SuperPlus, SuperStar, OverHat, OverBar}
```

The pattern *x_* stands for any single expression to be named *x* on the right-hand side of a definition or transformation rule. Its full form is **Pattern[x, Blank[]]**.

```
In[2]:= (h[a] + h[b, c] + h[a, a]) h[d, e, f] /. h[x_, y_] → x^y
out[2]= (a^a + b^c + h[a]) h[d, e, f]
```

Both **h[b, c]** and **h[a, a]** match **h[x_, y_]**. Yet only **h[a, a]** matches **h[x_, x_]** because a pattern containing pieces with the same name (**x**, here) can only match expressions the corresponding pieces of which are identical:

```
In[3]:= (h[a] + h[b, c] + h[a, a]) h[d, e, f] /. h[x_, x_] → x^x
Out[3]= (a^a + h[a] + h[b, c]) h[d, e, f]
```

Let **myfunc** be a function that requires an argument matching the pattern "**x_^n_**".

```
In[4]:= myfunc[x_^n_] := {x, n}
```

Apply **myfunc** to the elements of the list introduced at the end of Section 3.2.1:

```
In[5]:= Map[myfunc, {3, x, a^4, Sqrt[1 + x]}]
Out[5]= {myfunc[3], myfunc[x], {a, 4}, {1 + x, 1/2}}
```

Again, only **a^4** and **Sqrt[1 + x]** match the pattern "**x_^n_**".

```
In[6]:= Clear[myfunc]
```

For a pattern more complex than a single blank, *x*:*pattern* represents any expression matching *pattern*, and this expression is named *x* on the right-hand side of a definition or transformation rule:

```
In[7]:= Sin[1 + a^2] /. h : Sin[x_ + y_] → {h, x, y}
Out[7]= {Sin[1 + a^2], 1, a^2}
```

Here, the expression matching the pattern `Sin[x_ + y_]` is named `h` on the right-hand side of the transformation rule.

As mentioned previously, a pattern with pieces of the same name can only match expressions in which the corresponding pieces are identical. Whereas `{Sin[a], Sin[a]}` matches `{h_, h:Sin[x_]}`,

```
In[8]:= {Sin[a], Sin[a]} /. {h_, h : Sin[x_]} → h
Out[8]= Sin[a]
```

`{Cos[a], Sin[a]}` does not match the pattern because the elements are not the same:

```
In[9]:= {Cos[a], Sin[a]} /. {h_, h : Sin[x_]} → h
Out[9]= {Cos[a], Sin[a]}
```

3.2.3 Restricting Patterns

Patterns stand for classes of expressions. How can we construct patterns that represent subclasses of a class of expressions? We can delineate subclasses of expressions with head specifications, conditions, and tests.

3.2.3.1 Types

The pattern *_h* stands for any single expression with the head *h*. The pattern *x_h* represents any single expression with the head *h*, and this expression is named *x* on the right-hand side of a definition or transformation rule. Section 3.1.1 gave the heads of atomic expressions: `Symbol`, `String`, `Integer`, `Real`, `Rational`, and `Complex`. For any list, the head is `List`.

With the function `Union` introduced in Section 2.4.8 and the function `Cases` introduced in Section 3.2.1, we can pick out the variables of an algebraic expression:

```
In[1]:= Expand[(x + y + z)^3]
```

Out[1]= $x^3 + 3x^2 y + 3 x y^2 + y^3 + 3 x^2 z + 6 x y z + 3 y^2 z + 3 x z^2 + 3 y z^2 + z^3$

```
In[2]:= Cases[%, _Symbol, Infinity] // Union
Out[2]= {x, y, z}
```

With `Infinity` as the level specification, `Cases[`*expr*`,`*pattern*`,`*levelspec*`]` returns a list of parts that match *pattern* and appear at any level of *expr*.

We can square the rational numbers in a list:

```
In[3]:= {x, "good", 3, 2/3, 5/7} /. x_Rational → x^2
```

Out[3]= $\left\{x, \text{good}, 3, \frac{4}{9}, \frac{25}{49}\right\}$

The geometric mean of n quantities is defined to be the product of these quantities to the $1/n$ power. Here is a function for computing the geometric mean of the elements of a list:

```
In[4]:= geomean[x_List] := Apply[Times, x]^(1/Length[x])
```

We can determine the geometric mean of a list of four quantities:

```
In[5]:= geomean[{a1, a2, a3, a4}]
Out[5]= (a1 a2 a3 a4)^(1/4)
```

The argument of the function must be a list:

```
In[6]:= geomean[2, 3, 4]
Out[6]= geomean[2, 3, 4]
```

Mathematica returns the function unevaluated since the argument is not a list.

```
In[7]:= Clear[geomean]
```

Consider defining a function that returns the Legendre polynomial of order n with the formula of Rodrigues:

$$P_n(x) = \frac{1}{2^n\, n!} \frac{d^n (x^2-1)^n}{dx^n}$$

where $n = 0, 1, 2, 3, \ldots$, and $-1 \leq x \leq 1$.

```
In[8]:= P[n_Integer, x_] := ((1/(2^n n!)) D[(x^2 - 1)^n, {x, n}]) // Simplify
```

We can tabulate the first few Legendre polynomials:

```
In[9]:= Do[Print["n = ", i, "   ", P[i, x]], {i, 0, 3}]
```

n = 0 $\quad 1$

n = 1 $\quad x$

n = 2 $\quad \frac{1}{2}(-1+3x^2)$

n = 3 $\quad \frac{1}{2}x(-3+5x^2)$

Note that the first argument must match the pattern **n_Integer**—that is, it must be an integer:

```
In[10]:= P[2.5,x]
Out[10]= P[2.5, x]
```

Mathematica returns the function unevaluated since the first argument is not an integer.

A few words about the function **Print** are in order. **Print**[*expr*$_1$, *expr*$_2$, ...] prints the expressions *expr*$_1$, *expr*$_2$, ... concatenated together. The expressions *expr*$_1$, *expr*$_2$, ... can be any expressions including graphics and are evaluated before printing. If an expression has the form "*text*", the string of *text* is printed. The special character `\n` in a text string indicates a newline. For example, let $y = 2$:

```
In[11]:= y = 2;
```

To print $y + 5 = 7$, enter

```
In[12]:= Print["y + 5 = ", y + 5]
         y + 5 = 7
```

Note that `Print[y + 5 = 7]` differs from `Print["y + 5 = ", y + 5]`. `Print[y + 5 = 7]` triggers an error message and prints the number to the right of the assignment operator "`=`":

```
In[13]:= Print[y + 5 = 7]
         Set::write: Tag Plus in 2 + 5 is Protected. >>
         7

In[14]:= Clear[y]
```

3.2.3.2 Tests

We can restrict a pattern with a test. The pattern *pattern*`?`*test* stands for any expression that matches *pattern* and on which the application of *test* gives `True`. The built-in functions introduced in Sections 2.4.4 and 3.1.3.1 for testing properties of expressions can be used with this pattern.

Let us count the number of even integers in a list of 10 random integers from 0 to 100.

```
In[15]:= RandomInteger[100, 10]
Out[15]= {23, 67, 64, 73, 68, 97, 51, 97, 18, 30}
```

`Count[`*expr*`,`*pattern*`,`*levelspec*`]` gives the total number of subexpressions that match *pattern* and that are on levels of *expr* specified by *levelspec*. If the third argument is omitted, *levelspec* takes the default value `{1}`.

```
In[16]:= Count[%, _?EvenQ]
Out[16]= 4
```

In Section 3.2.3.1, there is a flaw in the definition of a function that returns the Legendre polynomial of order n. The first argument must be a nonnegative integer! The flaw can be corrected by introducing the pattern `n_Integer?NonNegative` for the argument.

```
In[17]:= Clear[P]

In[18]:= P[n_Integer?NonNegative, x_] :=
           ((1/(2^n n!)) D[(x^2 - 1)^n, {x, n}]) // Simplify
```

```
In[19]:= P[2,x]
```

$$\text{Out[19]= } \frac{1}{2}\left(-1+3x^2\right)$$

If the first argument is a negative integer, *Mathematica* returns the function unevaluated:

```
In[20]:= P[-2, x]
Out[20]= P[-2, x]
```

3.2.3.3 Conditions

The pattern object *pattern* /; *condition* stands for any expression matching *pattern* provided that the evaluation of *condition* yields **True**. The operator /; can be construed as "such that," "provided that," "only if," or "whenever."

From a nested list, we can pick out sublists that are vectors of numbers:

```
In[21]:= Cases[{a, {a, b}, {a, {b, c}}, 3, {3, 4}, {5, {6, 7}}},
           x_ /; VectorQ[x, NumberQ], Infinity]
Out[21]= {{3, 4}, {6, 7}}
```

The logical operator **&&**, which has the meaning "and," allows imposition of multiple conditions on a pattern. Upon evaluation, $condition_1$ **&&** $condition_2$ returns **True** when both $condition_1$ and $condition_2$ yield **True**. Consider selecting the integers between 10 to 80 from a list of 10 random integers in the range 1 to 100:

```
In[22]:= RandomInteger[{1, 100}, 10]
Out[22]= {2, 81, 9, 80, 17, 47, 32, 51, 87, 13}

In[23]:= Cases[%, x_Integer /; (x > 10 && x < 80)]
Out[23]= {17, 47, 32, 51, 13}
```

where the condition **x > 10 && x < 80** can also be entered as **10 < x < 80**.

In Section 3.2.3.1, there is another flaw in the definition of a function that returns the Legendre polynomial of order *n*. The second argument must be restricted to the interval $[-1, 1]$. We can correct this defect by introducing the pattern **x_ /; Abs[x] <= 1** for the second argument and by modifying the body of the function to prevent the substitution of a numerical value for **x** before the derivative is taken.

```
In[24]:= Clear[P]
In[25]:= P[n_Integer?NonNegative, x_ /; Abs[x] ≤ 1] :=
           ((1/(2^n n!))D[(y^2 - 1)^n, {y, n}] /. y → x) // Simplify
```

(Note that **<=** is equivalent to **≤**.) Let us plot **P[2,x]** as a function of **x** from **-1** to **1**:

```
In[26]:= Plot[P[2, x], {x, -1, 1}, AxesLabel → {"x", "P[2, x]"}]
```

Out[26]=

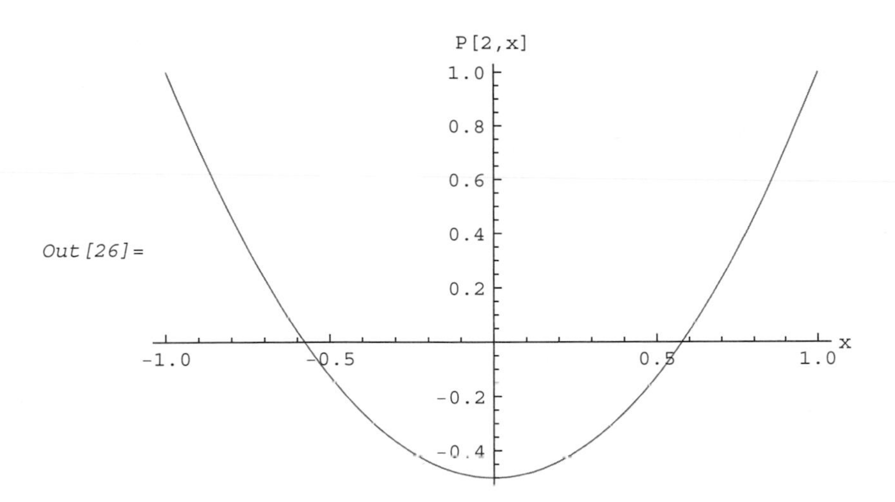

If we try to plot **P[2,x]** beyond the interval $[-1, 1]$, *Mathematica* plots the function only in the interval (i.e., domain) for which it is defined:

```
In[27]:= Plot[P[2, x], {x, -2, 2}, AxesLabel → {"x", "P[2, x]"}]
```

Out[27]=

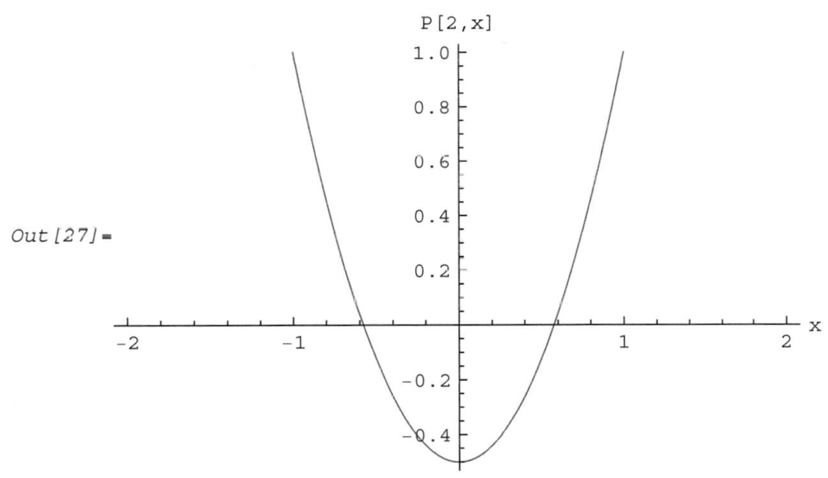

Note that **Abs[x] <= 1** is not evaluated if **x** does not have a numerical value:

```
In[28]:= Abs[x] ≤ 1
Out[28]= Abs[x] ≤ 1
```

Consequently, the function no longer works if **x** is a mathematical variable:

```
In[29]:= P[1, x]
Out[29]= P[1, x]
```

We fixed one flaw but introduced another. The function should work regardless of whether the second argument is a mathematical symbol or a number. We can solve this problem by defining separate functions for arguments matching different patterns: one for the pattern **x_Symbol** and another for the pattern **x_Real /; Abs[x] <= 1**:

```
In[30]:= Clear[P]

In[31]:= P[n_Integer?NonNegative, x_Symbol] :=
           ((1/(2^n n!)) D[(x^2 - 1)^n, {x, n}]) // Simplify

In[32]:= P[n_Integer?NonNegative, x_Real/; Abs[x] ≤ 1] :=
           ((1/(2^n n!)) D[(y^2 - 1)^n, {y, n}] /. y→x) // Simplify
```

Now the function works for both types of second arguments:

```
In[33]:= P[2,x]
```

Out[33]= $\frac{1}{2}\left(-1+3\,x^2\right)$

```
In[34]:= P[2,0.5]
Out[34]= -0.125
```

Of course, we can define a function that returns the Legendre polynomial of order n in terms of the built-in function **LegendreP**:

```
In[35]:= newP[n_Integer?NonNegative, x_Symbol] := LegendreP[n, x]
In[36]:= newP[n_Integer?NonNegative, x_Real /; Abs[x] ≤ 1] := LegendreP[n, x]
```

Here are **newP[2,x]** and **newP[2,0.5]**:

```
In[37]:= newP[2,x]
```

Out[37]= $\frac{1}{2}\left(-1+3\,x^2\right)$

```
In[38]:= newP[2,0.5]
Out[38]= -0.125

In[39]:= Clear[P,newP]
```

Consider plotting the potential well

$$V(x) = \begin{cases} 0 & |x| > a \\ -V_0 & |x| < a \end{cases}$$

Let V be measured in units of V_0 and x in units of a—that is, $V_0 = 1$ and $a = 1$. We give two separate definitions for two regions of **x** values:

```
In[40]:= V[x_ /; Abs[x] > 1] := 0
         V[x_ /; Abs[x] < 1] := -1
```

Here is a plot for the potential well:

```
In[42]:= Plot[V[x], {x, -2.5, 2.5},
          AxesLabel→{"x (a)", "V (V₀)"}, PlotStyle→Thickness[0.01]]
```

Out[42]=

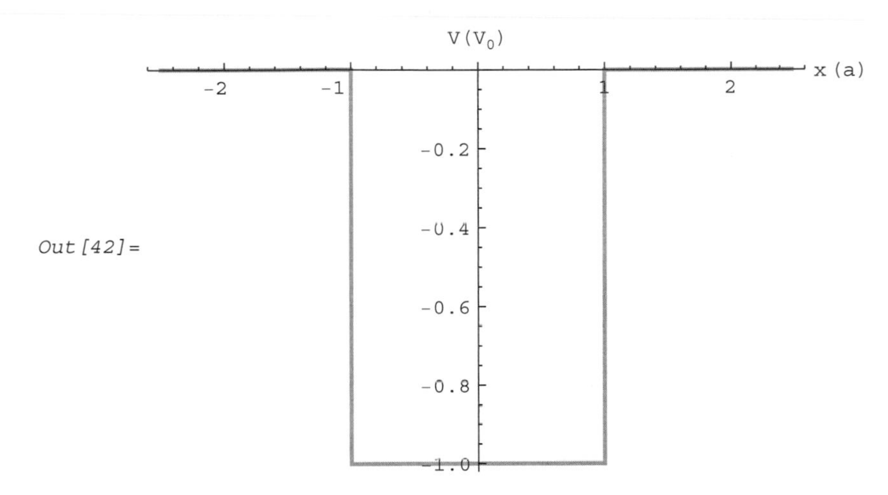

Names in a condition constraining a pattern must appear in the pattern. Consider the following definitions for the potential well:

```
In[43]:= Clear[V]

In[44]:= V[x_ /; Abs[x] > a] := 0
         V[x_ /; Abs[x] < a] := -1
```

The symbol **a** does not appear on the left-hand side of the `/;` operator.

```
In[46]:= V[0.5]
Out[46]= V[0.5]
```

Mathematica returns the function unevaluated unless there is a global value for **a**.

```
In[47]:= Clear[V]
```

In addition to putting conditions on patterns, we can also impose conditions on entire transformation rules and definitions. The transformation rule *lhs* `:>` *rhs* `/;` *condition* and the definition *lhs* := *rhs* `/;` *condition* apply only if the evaluation of *condition* yields **True**.

Let us specify a condition for the application of a transformation rule:

```
In[48]:= {f[2], f[a], g[6], f[5]} /. f[x_] :→ x /; NumberQ[x]
Out[48]= {2, f[a], g[6], 5}
```

Note that the `:>` (equivalent to `:→`) operator is used. Using the operator `->` (equivalent to `→`) causes the right-hand side of the rule—that is, *rhs* `/;` *condition*—to be evaluated immediately:

```
In[49]:= {f[2], f[a], g[6], f[5]}/.f[x_]→x/; NumberQ[x]
Out[49]= {2/; NumberQ[2], a/; NumberQ[a], g[6], 5/; NumberQ[5]}
```

We can put mathematical constraints on the arguments of a function.

```
In[50]:= func[x_, y_] := Sqrt[x - y] /; x > y

In[51]:= func[5, 4]
Out[51]= 1
```

The definition is not used if the first argument is not larger than the second:

```
In[52]:= func[5, 5]
Out[52]= func[5, 5]
```

Note that the operator := rather than the = operator is used in the definition. Using the = operator causes immediate evaluation of the right-hand side of the definition—that is, *rhs* `/;` *condition*:

```
In[53]:= Clear[func]

In[54]:= func[x_, y_] = Sqrt[x - y] /; x > y
Out[54]= √(x - y)/; x > y

In[55]:= func[5, 4]
Out[55]= 1/; 5 > 4

In[56]:= Clear[func]
```

3.2.4 Structural Equivalence

In matching patterns, *Mathematica* looks for structural equivalences between expressions and patterns. Mathematically equivalent expressions match different patterns if their structures are different. For example, `a^2 - b^2` and `(a + b) (a - b)` are mathematically equivalent. The expression `a^2 - b^2` matches the pattern `x_^2 - y_^2` because they have the same structure:

```
In[1]:= MatchQ[a^2 - b^2, x_^2 - y_^2]
Out[1]= True
```

The expression `Sin[a]^2 - Cos[b]^2` also matches the pattern `x_^2 - y_^2` because they are structurally equivalent:

```
In[2]:= MatchQ[Sin[a]^2 - Cos[b]^2, x_^2 - y_^2]
Out[2]= True
```

Yet **(a + b) (a - b)** does not match **x_^2 -y_^2** because their structures are different:

```
In[3]:= MatchQ[(a + b) (a - b), x_^2 - y_^2]
Out[3]= False
```

Although **a^2 - b^2** and **(a + b) (a - b)** are mathematically equivalent, they do not match the same pattern because they have different structures. A pattern that **(a + b) (a - b)** matches is **(x_ + y_) (x_ - y_)**:

```
In[4]:= MatchQ[(a + b) (a - b), (x_ + y_) (x_ - y_)]
Out[4]= True
```

The structures of expressions that *Mathematica* considers are those of the internal forms of the expressions. For example, **(a - b)** and **(a - 2b)** appear to have the same structure as the pattern **(x_ - y_)**. Whereas **(a - b)** matches the pattern, **(a - 2b)** does not:

```
In[5]:= MatchQ[a - b, x_ - y_]
Out[5]= True

In[6]:= MatchQ[a - 2b, x_ - y_]
Out[6]= False
```

An examination of the internal forms reveals that **a - 2b** and **x_ - y_** have different structures:

```
In[7]:= FullForm[a - 2b]
Out[7]//FullForm=
        Plus[a,Times[-2,b]]

In[8]:= FullForm[x_ - y_]
Out[8]//FullForm=
        Plus[Pattern[x, Blank[]], Times[-1, Pattern[y, Blank[]]]]
```

Although **a** can fill in for **Pattern[x,Blank[]]** and **b** can fill in for **Pattern[y,Blank[]]**, the first argument of **Times** is **-2** in the internal form of **a - 2b**, whereas it is **-1** in the internal form of "**x_ - y_**".

3.2.5 Attributes

We can assign general properties, called attributes, to functions. In evaluation and pattern matching, *Mathematica* considers these attributes. Section 2.4.9 discussed the attribute **Listable**. A function with the attribute **Listable** automatically acts on each element of a list that appears as its argument. This section examines two attributes that are important to pattern matching: **Orderless** and **Flat**. Section 3.3.5 will describe the attribute **Protected**. With on-line help, we can access the definitions of the remaining *Mathematica* attributes that are not used as often: **OneIdentity, Constant, Locked, ReadProtected, HoldFirst,**

HoldRest, HoldAll, Temporary, Stub, NumericFunction, HoldAllComplete, NHoldFirst, NHoldRest, NHoldAll, and **SequenceHold**.

The function **Attributes** lists the attributes of functions. **Attributes**[*f*] gives the attributes assigned to *f*. The operator **??** also shows the attributes of functions.

There are two ways to assign attributes to functions:

Attributes[*f*] = {$attr_1$, $attr_2$, ...}	set the attributes of *f*
SetAttributes[*f*, *attr*]	add *attr* to the attributes of *f*

If we are to specify for a function any of the attributes that affect pattern matching, namely **Orderless**, **Flat**, and **OneIdentity**, we must make the assignment before giving definitions or transformation rules to the function. (See Exercise 11 of Section 3.2.9.)

There are two ways to remove the attributes of functions:

Attributes[*f*] = {}	set *f* to have no attributes
ClearAttributes[*f*, *attr*]	remove *attr* from attributes of *f*

We can also use **ClearAll** to remove the attributes of functions. However, in addition to removing the attributes, **ClearAll**[$symbol_1$, $symbol_2$, ...] clears all values, definitions, messages, and defaults associated with the symbols. Be aware that **Clear**[$symbol_1$, $symbol_2$, ...], which we have been using, clears only values and definitions for the specified symbols.

The arguments of a function with the attribute **Orderless** are automatically sorted into standard order. Consequently, if the function **f** has the attribute **Orderless, f[a,c,b], f[b,a,c],** and so forth are equivalent to **f[a,b,c]**. In other words, **f** is a commutative function. *Mathematica* recognizes this property in matching patterns.

Addition and multiplication are commutative; the functions **Plus** and **Times** have the attribute **Orderless**. Let us illustrate how this attribute affects pattern matching:

```
In[1]:= {{a+b, a}, {a+b, b}}/.{x_+y_, x_} :> {x, y}
Out[1]= {{a, b}, {b, a}}
```

The list **{a + b, a}** matches the pattern with **x → a** and **y → b**. The list **{a + b, b}** also matches the pattern because *Mathematica* takes account of the attribute **Orderless** for the function **Plus** and rewrites the list as **{b + a, b}**. In this case, the list matches the pattern with **x → b** and **y → a**.

The arguments of a function with the attribute **Flat** are automatically flattened out. Therefore, if the function **f** has the attribute **Flat, f[f[a], f[b, c]], f[f[a], b, f[f[c]]],** and so forth are equivalent to **f[a,b,c]**. In other words, **f** is an associative function. *Mathematica* takes this property into consideration in matching patterns.

In addition to being commutative, addition and multiplication are also associative; the functions **Plus** and **Times** have the attribute **Flat**. Let us demonstrate with an example the combined effects of the attributes **Orderless** and **Flat**.

```
In[2]:= (a+b+c)/.(x_+y_) :> {x, y}
Out[2]= {a, b+c}
```

The expression matches the pattern with **x → a** and **y → b + c**. Because the function **Plus** has the attributes **Orderless** and **Flat**, there are many other possible matches. *Mathematica* returns the first match it finds. To see the other matches, we attach a condition to the rule. The function **Print**, which is never evaluated to **True**, can be used as a condition to show the possible matches in the order attempted by *Mathematica*.

```
In[3]:= (a + b + c) /. (x_ + y_) :> {x, y} /; Print[{x, y}];
        {a, b + c}
        {b, a + c}
        {c, a + b}
        {a + b, c}
        {a + c, b}
        {b + c, a}
        {a, b}
        {b, a}
        {a, c}
        {c, a}
        {b, c}
        {c, b}
```

(**ReplaceList** also reveals all ways that the entire expression can match the pattern. **ReplaceList**[*expr*, *rules*] attempts to transform the entire expression *expr* by applying a rule or list of rules in all possible ways, and it returns a list of the results obtained.

```
In[4]:= ReplaceList[(a + b + c), (x_ + y_) → {x, y}]
Out[4]= {{a, b + c}, {b, a + c}, {c, a + b}, {a + b, c}, {a + c, b}, {b + c, a}}
```

Why did **ReplaceList** not return the matches {a, b}, {b, a}, {a, c}, {c, a}, {b, c}, and {c, b}?)

3.2.6 Defaults

The pattern *x*_:*v* stands for any single expression that, if omitted, has default value *v*. The expression is to be named *x* on the right-hand side of a definition or transformation rule. For example, the list **{a, b}** matches the pattern **{x_, y_ : d}**,

```
In[1]:= {a, b} /. {x_, y_ : d} :> {x^2, y^2}
Out[1]= {a^2, b^2}
```

and the rule applies. The rule also applies even when the second element of the list is omitted:

```
In[2]:= {a} /. {x_, y_ : d} :> {x^2, y^2}
Out[2]= {a^2, d^2}
```

In this case, the second element is given the default value **d**.

The pattern $x\_h:v$ stands for an expression that, if omitted, has default value v. The expression has the head h and is to be named x on the right-hand side of a definition or transformation rule. For example, the list **{a, 2}** matches the pattern **{x_, y_Integer:10}**,

```
In[3]:= {a, 2}/.{x_, y_Integer: 10} :→ {x^2, y^2}
Out[3]= {a², 4}
```

and the rule applies. What happens if the second element is omitted?

```
In[4]:= {a}/.{x_, y_Integer: 10} :→ {x^2, y^2}
Out[4]= {a², 100}
```

In this case, the second element assumes the default value **10**. For a list to match the pattern **{x_, y_Integer:10}**, it must have one or two elements and, if present, the second element must be an integer.

```
In[5]:= {a, 2, 3}/.{x_, y_Integer: 10} :→ {x^2, y^2}
Out[5]= {a, 2, 3}
```

The list fails to match the pattern because it has three elements.

```
In[6]:= {a, b}/.{x_, y_Integer: 10}:→{x^2, y^2}
Out[6]= {a, b}
```

The list does not match the pattern because its second element is not an integer.

The pattern $x\_.$ represents an expression with a built-in default value. The expression is to be named x on the right-hand side of a definition or transformation rule. We can use this pattern to give a built-in default value to a piece in each of three patterns. In the pattern "$x\_ + y\_.$", the default value for y is 0; in "$x\_y\_.$", the default for y is 1; and in "$x\_\hat{}y\_.$", the default for y is 1. For example, **a + b** matches the pattern "**x_ + y_.**",

```
In[7]:= a + b/. x_ + y_. :→ x^2 + y^2
Out[7]= a² + b²
```

and if the second term of the sum is omitted, the expression still matches the pattern,

```
In[8]:= a/. x_ + y_. :→ x^2 + y^2
Out[8]= a²
```

where the default value 0 is used for the second term that is missing. The expression **a b** matches the pattern "**x_ y_.**",

```
In[9]:= ab/. x_ y_. :→ x^2 + y^2
Out[9]= a² + b²
```

and if the second term of the product is omitted, it still matches the pattern,

```
In[10]:= a/. x_y_. :→ x^2 + y^2
Out[10]= 1 + a²
```

where the default value of 1 is used for the missing term in the product.

Example 3.2.1 Construct a pattern that represents any difference of two terms.

In Section 3.2.4, we found that **a - 2b** does not match the pattern "**x_ - y_**". Neither does **a - 5** match the pattern "**x_ - y_**":

```
In[11]:= MatchQ[a - 5, x_ - y_]
Out[11]= False
```

FullForm reveals the reason for the mismatch.

```
In[12]:= FullForm[x_ - y_]
Out[12]//FullForm=
         Plus[Pattern[x, Blank[]], Times[-1, Pattern[y, Blank[]]]]

In[13]:= FullForm[a - 5]
Out[13]//FullForm=
         Plus[-5, a]
```

The problem is that **-5** is an atom and therefore does not match **Times[-1, Pattern[y, Blank[]]**. Is there a pattern that matches the difference of any two expressions? A pattern is "**x_ + n_?Negative y_.**". The built-in default value for **y** is **1**.

```
In[14]:= Cases[{a - Sqrt[b], a - 2 b, a - b/3, a^2 - 5.2 b^3, a - 5,
           Tan[x] - Sin[y], a + b, Tan[x] + Sin[x]}, x_ + n_?Negative y_.]
Out[14]= {a - √b, a - 2 b, a - b/3, a² - 5.2 b³, -5 + a, -Sin[y] + Tan[x]}
```

Note that **a + b** and **Tan[x] + Sin[x]** do not match the pattern. ■

Example 3.2.2 Construct a pattern that matches any expression of the form $a_n x^n$, where a_n, x, and n are a number, a variable, and a positive integer, respectively.

The pattern **a_ x_^n_** matches the expression **3 x^2**:

```
In[15]:= MatchQ[3 x^2, a_ x_^n_]
Out[15]= True
```

Yet it does not match the expression **x^2**:

```
In[16]:= MatchQ[x^2, a_ x_^n_]
Out[16]= False
```

Neither does it match the expression **2 x**:

```
In[17]:= MatchQ[2 x, a_ x_^n_]
Out[17]= False
```

The pattern **a_. x_Symbol^n_./; NumberQ[a]&&Positive[n]&&IntegerQ[n]** matches any expression of the form $a_n x^n$ where a_n, x, and n are a number, a variable, and a positive integer, respectively:

```
In[18]:= Cases[{3 x^2, x^2, 2 x, x^1.5, 1/y, 3, a + b, 2 y^(-2),
          y^2 + 1, x/y^2, Sqrt[x], Tan[x]},
          a_. x_Symbol^n_./; NumberQ[a]&& Positive[n]&& IntegerQ[n]]
Out[18]= {3 x^2, x^2, 2 x}
```

■

3.2.7 Alternative or Repeated Patterns

The pattern *patt*$_1$ | *patt*$_2$ | ... is a pattern that can have one of several forms. For example, the pattern **_Integer|_Symbol|_Rational** matches any integer, variable, or rational number:

```
In[1]:= Cases[{3, x, 2/3, 2 + 3 I, Tan[x], Sqrt[x], 1.5},
         _Integer|_Symbol|_Rational]
Out[1]= {3, x, 2/3}
```

Cases picks out the integer, symbol, and rational number from the list.

Example 3.2.3 Construct a pattern that stands for any monomial, which is a number or a number times a variable to some positive integer power.

A monomial is a number or an expression of the form $a_n x^n$, where a_n, x, and n are a number, a variable, and a positive integer, respectively. The pattern constructed in Example 3.2.2 matches the latter. We only need to include in the pattern an alternative that represents a number:

```
In[2]:= Cases[{3 x^2, x^2, 2 x, x^1.5, 1/y, 3, a + b,
         1.5, 2 y^(-2), y^2 + 1, x/y^2, Sqrt[x], Tan[x]},
         (a_. x_Symbol^n_./; Positive[n]&& IntegerQ[n]&& NumberQ[a])|a_/;
          NumberQ[a]]
Out[2]= {3 x^2, x^2, 2 x, 3, 1.5}
```

Cases picks out all the monomials in the list. ■

The pattern *expr*.. is a pattern or other expression repeated one or more times. The pattern *expr*... is a pattern or other expression repeated zero or more times. For example, **{a, a, a, a, {a, b}, {a, b}}** matches the pattern **{x_Symbol.., y_List..}**:

```
In[3]:= {a, a, a, a, {a, b}, {a, b}} /.{x_Symbol.., y_List..} :> {x, y}
Out[3]= {a, {a, b}}
```

3.2.8 Multiple Blanks

The "double blank" stands for any sequence of one or more expressions. Its special input form and internal form are __ and `BlankSequence[]`, respectively. The "triple blank" represents any sequence of zero or more expressions. Its special input form and internal form are ___ and `BlankNullSequence[]`, respectively. We can set up other patterns with single, double, and triple blanks and also name or restrict these patterns.

Let us define a function that lists its argument:

```
In[1]:= myfunc[x_] := {x}

In[2]:= myfunc[a]
Out[2]= {a}
```

This function does not accept more than one argument:

```
In[3]:= myfunc[a, b, c]
Out[3]= myfunc[a, b, c]
```

With a double blank, we can define a function that lists one or more arguments:

```
In[4]:= ourfunc[x__] := {x}

In[5]:= ourfunc[a]
Out[5]= {a}

In[6]:= ourfunc[a,b,c,d]
Out[6]= {a, b, c, d}
```

The pattern __*h* stands for any sequence of one or more expressions, each with a head *h*. Thus, `__List` represents any sequence of one or more lists, and `{__List}` matches any list of lists.

```
In[7]:= Cases[{{}, {{}}, {a}, {{a}}, {a, b}, {{a}, {b, c}}}, {__List}]
Out[7]= {{{}}, {{a}}, {{a}, {b, c}}}
```

There are three nested lists.

There are often several possible matches of an expression to a pattern containing multiple blanks:

```
In[8]:= {a, b, c, d}/.{x__, y__} :> "SMILE"/; Print[{{x}, {y}}];
        {{a}, {b, c, d}}
        {{a, b}, {c, d}}
        {{a, b, c}, {d}}
```

To reveal all the possible matches in the order attempted by *Mathematica*, we attached, as a condition to the transformation rule, `Print[{{x}, {y}}]`; this prints each match to the pattern but is never evaluated to `True` for the rule to actually apply.

Example 3.2.4 Evaluate the constant **E** to 60 digits, determine the number of digits between the decimal point and a repeated sequence of numbers, and identify the sequence.

To compute **E** to 60 digits, use the function **N**:

```
In[9]:= N[E, 60]
Out[9]= 2.71828182845904523536028747135266249775724709369995957496697
```

ToString[*expr*] gives a string corresponding to the printed form of *expr*:

```
In[10]:= ToString[%]
Out[10]= 2.71828182845904523536028747135266249775724709369995957496697
```

The function **Characters** gives a list of the characters in a string:

```
In[11]:= charactersOfE = Characters[%]
Out[11]= {2, ., 7, 1, 8, 2, 8, 1, 8, 2, 8, 4, 5, 9, 0, 4, 5, 2, 3,
          5, 3, 6, 0, 2, 8, 7, 4, 7, 1, 3, 5, 2, 6, 6, 2, 4, 9, 7, 7, 5,
          7, 2, 4, 7, 0, 9, 3, 6, 9, 9, 9, 5, 9, 5, 7, 4, 9, 6, 6, 9, 7}
```

Finally, determine how many digits fall between the decimal point and a repeated sequence of numbers:

```
In[12]:= Print["digits      Sequence\n"];
         charactersOfE/.{"2", ".", a___, x__/; Length[{x}] > 1, x__, ___} :>
           "Ha!"/; Print[Length[{a}], "            ", x];
          digits     Sequence

          1          1828
          48          95
```

To search for all the possible matches to the pattern **{"2", ".", a___, x__/; Length[{x}] > 1, x__, ___}**, we attached, as a condition to the rule, the function **Print**, which is never evaluated to **True**. In addition to preventing *Mathematica* from identifying only the first match to the pattern, **Print[Length[{a}], "   ", x]** prints for each match the number of digits between the decimal point and the repeated sequence of numbers and also prints the value of **x**, which is the repeated sequence. There are two repeated sequences: 1828 appearing after 1 digit and 95 appearing after 48 digits. ■

```
In[14]:= Clear[myfunc, ourfunc, charactersOfE]
```

3.2.9 Exercises

In this book, straightforward, intermediate-level, and challenging exercises are unmarked, marked with one asterisk, and marked with two asterisks, respectively. For solutions to most odd-numbered exercises, see Appendix C.

1. The wavelengths of the Balmer series of lines in the hydrogen spectrum can be calculated from the formula

$$\lambda = (364.5\ \text{nm})\frac{n^2}{n^2 - 4}$$

where n can be any integer larger than 2. Define a function for evaluating λ—the function accepts only one argument that must be an integer larger than 2. For example,

```
{λ[-1], λ[2], λ[3, 4], λ[3.5], λ[3]}
{λ[-1], λ[2], λ[3, 4], λ[3.5], 656.1 nm}
```

2. The wavelengths of the Paschen series of lines in the hydrogen spectrum can be calculated from the formula

$$\lambda = (820.1\ \text{nm})\frac{n^2}{n^2 - 3^2}$$

where n can be any integer larger than 3. Define a function for evaluating λ—the function accepts only one argument that must be an integer larger than 3. For example,

```
{λ[-5], λ[3], λ[4, 7], λ[5.5], λ[6]}
{λ[-5], λ[3], λ[4, 7], λ[5.5], 1093.47 nm}
```

3. Using a multiclause definition with two clauses, write the signum function, sgn(x), which equals 1 if x is positive and equals −1 if x is negative. Plot the function for x from −2 to 2.
4. Define a function that accepts only three arguments: The first must be a complex number, the second must be a number smaller than 10, and the third must be an even integer. The function returns the absolute value of the first argument divided by the sum of the second and third arguments. Here are some examples:

```
{ourfunc[2 + ⅈ, -4, 14], ourfunc[4 + 3 ⅈ, 2/7, 10], ourfunc[2 + 7 ⅈ, 1.0, 4]}
```

$\left\{\frac{1}{2\sqrt{5}}, \frac{35}{72}, 1.45602\right\}$

```
{ourfunc[2 + 7 ⅈ, π, 4], ourfunc[2 + ⅈ, 4/5, 13], ourfunc[β, -4, 13],
  ourfunc[2.1, -4, 13], ourfunc[β, η, 13], ourfunc[3 + 5 î, -4, 12, 1],
  ourfunc[β, -4, 11], ourfunc[{1, 3}, 1, 13], ourfunc["a", -4, 13]}
```

$\left\{\text{ourfunc}[2 + 7\,\mathrm{i}, \pi, 4], \text{ourfunc}\left[2 + \mathrm{i}, \frac{4}{5}, 13\right], \text{ourfunc}[\beta, -4, 13],\right.$

$\text{ourfunc}[2.1, -4, 13], \text{ourfunc}[\beta, \eta, 13], \text{ourfunc}\left[3 + 5\,\hat{\mathrm{i}}, -4, 12, 1\right],$

$\left.\text{ourfunc}[\beta, -4, 11], \text{ourfunc}[\{1, 3\}, 1, 13], \text{ourfunc}[a, -4, 13]\right\}$

5. *Mathematica* represents vectors by lists. Thus, two-dimensional vectors have the form {a_1, a_2}. Using a multiclause definition with eight clauses, write a function that classifies nonzero two-dimensional vectors in a rectangular coordinate system according to the following scheme:

Vector Classification	Value of Function
in quadrant I	1
in quadrant II	2
in quadrant III	3
in quadrant IV	4
along the $+x$-axis	5
along the $-x$-axis	6
along the $+y$-axis	7
along the $-y$-axis	8

Here are some examples:

```
{vectorClassify[{-10, 0}], vectorClassify[{2, 0}],
 vectorClassify[{0, -5.2}], vectorClassify[{-2.6, 4}],
 vectorClassify[{3, -3}], vectorClassify[{3π/2, -2.1}]}
```

```
{6, 5, 8, 2, 4, 4}
```

```
{vectorClassify[{1, 1, 2}],
 vectorClassify[{0, 0}], vectorClassify[1, 0]}
```

```
{vectorClassify[{1, 1, 2}],
 vectorClassify[{0, 0}], vectorClassify[1, 0]}
```

6. In terms of spherical coordinates, the rectangular coordinates are given by

$$x = r \sin\theta \cos\phi$$

$$y = r \sin\theta \sin\phi$$

$$z = r \cos\theta$$

Define a function that accepts three arguments r, θ, and ϕ: r must be nonnegative, θ (in degrees) must be greater than or equal to 0 and less than or equal to 180, and ϕ (in degrees) must be greater than or equal to 0 and less than or equal to 360. The function returns a list of the corresponding rectangular coordinates. Some examples are

```
{convert[2, 0, 0], convert[10.0, 30.0, 55.0], convert[10, 30, 55]}
{{0., 0., 2.}, {2.86788, 4.09576, 8.66025}, {2.86788, 4.09576,
  8.66025}}
```

```
{convert[-10, 30.0, 55.0], convert[d, 30, 55],
 convert[2 + 3 i, 30.0, 55.0], convert[10.0, 30.0, a],
 convert[10, 200.0, 55.0], convert[10, 30, 450]}
```

```
{convert[-10, 30., 55.], convert[d, 30, 55], convert[2+3 i, 30., 55.],
 convert[10., 30., a], convert[10, 200., 55.], convert[10, 30, 450]}
```

Hint: Arguments of the trigonometric functions of *Mathematica* are assumed to be in radians; use the constant **Degree**.

7. Consider the list

```
{a, {a}, {a, a}, {b, c}, {d, {e, f}},
 {Sin[a x], Sin[a x]}, {a, a, a}, {Sin[b x], Tan[c x]}}
```

Using **Cases**, pick out from this list *all* sublists of two *different* elements. That is, generate the list

```
{{b, c}, {e, f}, {d, {e, f}}, {Sin[b x], Tan[c x]}}
```

Hint: Note the difference between **Unequal**(**!=** or ≠) and **UnsameQ**(**=!=**).

8. Write a function that expands only the operand of the **Log** function in an expression. Using the function, transform

```
((1 - x)^3 + Log[(a - b)^2] + Log[(c + d)^2])^2
```

to

```
((1 - x)^3 + Log[a^2 - 2 a b + b^2] + Log[c^2 + 2 c d + d^2])^2
```

Hint: Use **Expand**; also, note the difference between **->** (equivalent to →) and **:>** (equivalent to :→).

9. Using a transformation rule that expands the operand of the **Sin** function, transform

```
{(1 - x + Sin[(a - b)^2])^2, Sqrt[(a - b)^3 Cos[(4 - x)^3]],
 (a - 3) ArcTan[Sin[π((x - 1) (x + 1) - x^2)]]}
```

to

```
{(1 - x + Sin[a^2 - 2 a b + b^2])^2, Sqrt[(a - b)^3 Cos[(4 - x)^3]], 0}
```

Hint: Use **Expand**; also, note the difference between **->** (equivalent to →) and **:>** (equivalent to :→).

***10.** Consider the associated Legendre polynomials $P_n^m(x)$, where n must be a nonnegative integer; m can take on only the values $-n, -n+1, \ldots, n-1, n$; and x is confined to the interval $[-1, 1]$. P_n^0 equals P_n, which is the Legendre polynomial of order n. For positive

m, P_n^m can be determined from the equation

$$P_n^m(x) = (-1)^m(1-x^2)^{m/2}\frac{d^m P_n(x)}{dx^m}$$

For negative m, P_n^m can be obtained from the formula

$$P_n^{-m}(x) = (-1)^m \frac{(n-m)!}{(n+m)!} P_n^m(x)$$

Define a function for generating the associated Legendre polynomials. Plot $P_2^1(x)$. Determine $P_n^m(x)$ and $P_n^m(0.5)$ for $n = 0, 1, 2$, and all possible m. Verify the results with the built-in function **LegendreP**.

11. Define a function that takes three arguments: a symbol, an integer, and a complex number. However, the order in which arguments are specified is not important. In other words, the function should accept a symbol, an integer, or a complex number as an argument at any position, but no two arguments can be of the same type. The function returns an ordered list of the arguments: the symbol followed by the integer and then the complex number. For example, **f[2 + I, g, 5]** gives **{g, 5, 2 + I}**. *Hint:* For the function, assignment of attributes must occur before the definition.

***12.** Define a function that takes four arguments: an integer, an approximate real number, a rational number, and a complex number. However, the order in which arguments are specified is not important. In other words, the function should accept an integer, an approximate real number, a rational number, or a complex number as an argument at any position, but no two arguments can be of the same type. The function returns a list of the arguments in the same order as they appear in the function. Here are some examples:

```
h[3 + 5 i, 7/3, 5.23, 11]
```

$\left\{3+5i, \frac{7}{3}, 5.23, 11\right\}$

```
h[11, 7/3, 3 + 5 i, 5.23]
```

$\left\{11, \frac{7}{3}, 3+5i, 5.23\right\}$

```
h[2, 3, 4/3, 5.5]
```

$h\left[2, 3, \frac{4}{3}, 5.5\right]$

13. Define a function that accepts only two arguments: The first must be a symbol and the second must be an integer. The function returns the product of the symbol and the integer. If the second argument is omitted, it takes on the default value **7**. For example,

```
{func[x, 10], func[x], func[2, 5], func[x, y, 4], func[{x, 6}]}
{10 x, 7 x, func[2, 5], func[x, y, 4], func[{x, 6}]}
```

14. Define a function that accepts only two arguments: The first must be a symbol and the second must be an integer. The function returns the symbol to the power of the integer. If the second argument is omitted, it takes on the default value **3**. For example,

```
{func[x, 10], func[x], func[2, 5], func[x, y, 4], func[{x, 6}]}
```

$\{x^{10}, x^3, \text{func}[2, 5], \text{func}[x, y, 4], \text{func}[\{x, 6\}]\}$

15. Without using the function **Range**, write a function **myRange** that accepts only one *or* two integer arguments. When given a positive integer n as the only argument, it returns the list of consecutive integers from 1 to n; when given n as the first argument and m as the second such that $n < m$, it returns the list of consecutive integers from n to m. Some examples are

```
{myRange[5], myRange[1], myRange[0], myRange[-7]}
{{1, 2, 3, 4, 5}, {1}, myRange[0], myRange[-7]}
{myRange[-3, 2], myRange[2, 4], myRange[5, 2], myRange[3.5, 5.5]}
{{-3, -2, -1, 0, 1, 2}, {2, 3, 4}, myRange[5, 2], myRange[3.5, 5.5]}
```

Hint: Use **Table**.

16. Construct a pattern **myPattern** that matches the quotient of any two expressions. For example,

$$\text{Cases}\left[\left\{\frac{3x^3}{1+x^2}, x^3, \frac{2}{3}x, \frac{1}{b}, 2+3i, 2/7, \frac{\sqrt{1+a}}{\sqrt{1+b^2}}, c\,d\right\}, \text{myPattern}\right]$$

$$\left\{\frac{3x^3}{1+x^2}, \frac{1}{b}, \frac{\sqrt{1+a}}{\sqrt{1+b^2}}\right\}$$

Hint: In *Mathematica*, rational numbers are atoms rather than quotients.

17. Construct a pattern **ourPattern** that matches any expression of the form $a_n x^n$, where n is an integer, x and a are different symbols, and the subscript of a and the exponent of x are identical. For example,

$$\text{Cases}\left[\left\{\frac{b_{-3}}{x^3}, \frac{b_{-2}}{x^2}, \frac{b_{-1}}{x}, \frac{x_{-3}}{x^3}, \frac{x_{-2}}{x^2}, \frac{x_{-1}}{x}, b_0, x\,b_1, x^2\,b_2, x^3\,b_3,\right.\right.$$

$$\left.\left. x_2\,x^2, 3y^4, a+b, \sqrt{1+x}, \text{Sin}[x]\,x, x_{11}\,y^{11}, \frac{1}{z}\right\}, \text{ourPattern}\right]$$

$$\left\{\frac{b_{-3}}{x^3}, \frac{b_{-2}}{x^2}, \frac{b_{-1}}{x}, b_0, x\,b_1, x^2\,b_2, x^3\,b_3, y^{11}\,x_{11}\right\}$$

18. Construct a pattern **yourPattern** that matches any quadratic expressions in the variable x: $a + bx + cx^2$, where a, b, and c are *any* coefficients independent of x and $c \neq 0$. For example,

```
Cases[{1, x, 1 + x, Sin[x] x^2, 1 + x^2, x + x^2, x^2, a + Cos[x] x + c x^2,
   f[x] + b x + c x^2, b x + c x^2, a + c x^2, x^3, m x + b, {a, b},
   a + b y + c y^2, y + 2 y x + y^2 x^2, c x^2, a + b x + c x^2}, yourPattern]
{1 + x^2, x + x^2, x^2, b x + c x^2, a + c x^2, y + 2 x y + x^2 y^2, c x^2, a + b x + c x^2}
```

Hint: Use built-in defaults, **FreeQ**, and alternative patterns.

19. Define a function that only accepts either a rational number or an integer as the argument and that returns the absolute value of the number. For example,

```
{ourfunc[-3/5], ourfunc[-5], ourfunc[7],
 ourfunc[x], ourfunc[-3/5, -5], ourfunc[1.2]}
{3/5, 5, 7, ourfunc[x], ourfunc[-3/5, -5], ourfunc[1.2]}
```

20. Define a function that converts only a product of integers into a list of the integers. Here are some examples:

```
{prod2list[2 × 3 × 4 × 5 × 6 × 7], prod2list[2], prod2list[2.0 × 6 × 20]}
{{2, 3, 4, 5, 6, 7}, prod2list[2], prod2list[2. × 6 × 20]}
prod2list[22
          3
          4
          55
          6
          9]
{22, 3, 4, 55, 6, 9}
```

Hint: With "front end help" discussed in Section 1.6.3, obtain information about the *Mathematica* attribute **HoldAll**. Also, examine the internal forms of products.

***21.** Produce the decimal expansion of π out to 770 places, and determine the number of digits past the decimal point before six 9's appear in a row.

Answer:

```
761
```

***22.** Evaluate the constant **EulerGamma** to 1000 digits, and determine the number of digits past the decimal point before three 9's appear in a row.

Answer:

```
889
```

***23.** Evaluate $\sqrt{5}$ to 6200 digits. For each sequence of three *or* more 7's appearing in a row, determine the number of 7's in the sequence as well as the number of digits between the decimal point and the sequence.

Answers:

Number of 7's	Number of Digits
3	3714
3	4317
3	4861
3	6150
4	6150
3	6151

24. Write a function that takes only an expression with two or more elements as an argument and that returns a list of the first and last elements of the expression, using (a) **First**, **Last**, **Part** (including **[[]]**), **Extract**, **Take**, **Rest**, or **Drop** and (b) none of the built-in functions mentioned in part (a). Some examples are

```
{firstandlast[{a, b, {c, d}}],
 firstandlast[a + b + c + d^2], firstandlast[h[e1, e2, e3, e4, e5]]}
```

$\{\{a, \{c, d\}\}, \{a, d^2\}, \{e_1, e_5\}\}$

```
{firstandlast[{a}], firstandlast[3], firstandlast[Sin[x]]}
{firstandlast[{a}], firstandlast[3], firstandlast[Sin[x]]}
```

Hint: Note the form of normal expressions, and use multiple blanks.

25. Write a function that computes the mean of a list of one or more numbers. Determine the mean of a list of n random integers in the range 0 to 1000, where n is a random integer in the range 1 to 100.

26. Construct a pattern **newPattern** that matches a list of two or more numbers if the sum of the first element squared and the rest of the elements is less than 7. That is, the pattern matches the list $\{e_1, e_2, \ldots, e_n\}$ if $e_1{}^2 + e_2 + \cdots + e_n < 7$. For example,

$$\text{Cases}\left[\left\{\{1, 2, 3\}, \{2, 3, 4, 5\}, \{a, b\}, \{1, x, x^2, x^3\}, 2 + 3i, 2/7, \frac{\sqrt{1 + a}}{\sqrt{1 + b^2}}, \{2/3, 2.75, 1, 0.2\}\right\}, \text{newPattern}\right]$$

$$\left\{\{1, 2, 3\}, \left\{\frac{2}{3}, 2.75, 1, 0.2\right\}\right\}$$

27. Define a function **myfunc** that accepts only a sequence of one or more lists as its arguments and that returns a list of the lengths of the lists. For example,

```
{myfunc[{1}, {2, 3, 4, 5}, {6, 7}, {8, 9}],
 myfunc[{a, b}, {c, 5}], myfunc[x, 3, {1}]}
{{1, 4, 2, 2}, {2, 2}, myfunc[x, 3, {1}]}
```

3.3 FUNCTIONS

3.3.1 Pure Functions

Sections 2.2.7 and 2.2.16 stressed the hazards of named functions and programming variables with global values and promoted the practice of clearing them as soon as they are no longer needed. Elegant *Mathematica* programs minimize the number of temporary programming variables and named functions. To reduce the number of temporary variables, these programs prefer replacements with transformation rules over assignments. To reduce the number of temporary named functions, they make frequent use of pure or anonymous functions, which are functions without names. Unless a function is to be invoked repeatedly or its arguments are to be constrained, consider using a pure function. **Function**[*var*, *body*] is a pure function with a single formal parameter *var*. **Function**[*var*, *body*][*arg*] returns the result from evaluating *body* in which the formal variable *var* is replaced everywhere by the argument *arg*.

To specify an operation such as multiplying an expression by 3, we can define a named function

```
In[1]:= f[var_]:= 3 var
```

Let us apply this function to an argument **y**:

```
In[2]:= f[y]
Out[2]= 3 y
```

We can also use the pure function **Function[var, 3 var]**:

```
In[3]:= Function[var, 3 var][y]
Out[3]= 3 y
```

The pure function returns the same result.

Since the name of the formal parameter is irrelevant, *Mathematica* allows the use of two short forms: **Function**[*body*] or *body***&**. In these forms, the formal parameter is **#**. Let us go back to the example of multiplying an expression by 3. The pure function can now take the forms **Function[3 #]** or **3 # &**. With **y** as the argument again, we have

```
In[4]:= Function[3 #][y]
Out[4]= 3 y

In[5]:= 3 #&[y]
Out[5]= 3 y
```

The function **Select** together with a pure function allows us to select numbers greater than 4 from a list:

```
In[6]:= Select[{1, a, x^2, 3, 5, 1 + x, 7}, # > 4&]
Out[6]= {5, 7}
```

In Section 3.1.3.1, we defined the function **test1** for testing whether an expression is a polynomial in the variable x:

```
In[7]:= test1[expr_]:= PolynomialQ[expr,x]
```

With this function, we can obtain an expression with the elements that are polynomials in the variable x:

```
In[8]:= Select[(1+x+2x^2+3x^3+Sin[x]), test1]
```

$$Out[8]= 1 + x + 2x^2 + 3x^3$$

Using a pure function makes **test1** unnecessary:

```
In[9]:= Select[(1+x+2x^2+3x^3+Sin[x]),
          Function[var, PolynomialQ[var, x]]]
```

$$Out[9]= 1 + x + 2x^2 + 3x^3$$

We can also use the short form for the pure function:

```
In[10]:= Select[(1+x+2x^2+3x^3+Sin[x]), PolynomialQ[#, x]&]
```

$$Out[10]= 1 + x + 2x^2 + 3x^3$$

In Section 3.1.3.4, we defined a function to illustrate the restructuring of expressions:

$$In[11]:= q[x\_]:= \frac{1}{1 + x}$$

With this function, we can generate a nested expression:

```
In[12]:= Nest[q, x, 4]
```

$$Out[12]= \frac{1}{1 + \frac{1}{1 + \frac{1}{1 + \frac{1}{1 + x}}}}$$

Using a pure function, we no longer need the function **q**:

$$In[13]:= Nest\left[Function\left[\frac{1}{1 + \#}\right], x, 4\right]$$

$$Out[13]= \frac{1}{1 + \frac{1}{1 + \frac{1}{1 + \frac{1}{1 + x}}}}$$

We can also use the ampersand notation for the pure function:

```
In[14]:= Nest[1/(1 + #) &, x, 4]
```

$$\text{Out[14]=}\quad \frac{1}{1+\frac{1}{1+\frac{1}{1+\frac{1}{1+x}}}}$$

A pure function can have more than one parameter. `Function[{`x_1`, `x_2`,...}, `*body*`]` is a pure function with a list of formal parameters $x_1, x_2, \ldots$. In the short forms, the formal parameters are `#` (or `#1`), `#2`, ..., and `#`n for the first, second, ..., and nth variables in the pure function, respectively. The expression `##` stands for the sequence of all variables in a pure function, and `##`n represents the sequence of variables starting with the nth one.

Here is a function for computing the sum of the squares of its two arguments:

```
In[15]:= myfunc[x_, y_] := x^2 + y^2

In[16]:= myfunc[a, b]

Out[16]= a^2 + b^2
```

The same operation can be done with a pure function:

```
In[17]:= #1^2 + #2^2 & [a, b]

Out[17]= a^2 + b^2
```

In Section 3.2.8, we defined a function that lists one or more arguments:

```
In[18]:= ourfunc[x__] := {x}

In[19]:= ourfunc[a]

Out[19]= {a}

In[20]:= ourfunc[a, b, c, d]

Out[20]= {a, b, c, d}
```

The same operations can be performed with the pure function `{##}&`:

```
In[21]:= {##}&[a]

Out[21]= {a}

In[22]:= {##}&[a, b, c, d]

Out[22]= {a, b, c, d}

In[23]:= ClearAll[f, test1, q, myfunc, ourfunc]
```

By including an additional argument, we can assign attributes to a pure function. **Function[{**x_1**,**x_2**,...}**, *body*, **{***attributes***}]** is a pure function that has the specified attributes. Consider a pure function that is equivalent to the function **p**:

```
In[24]:= Function[x, p[x]][{a, b, c}]
Out[24]= p[{a, b, c}]
```

With the attribute **Listable**, the pure function is automatically applied to each element of the list that appears as its argument:

```
In[25]:= Function[x, p[x], Listable][{a, b, c}]
Out[25]= {p[a], p[b], p[c]}
```

(Note that we cannot assign attributes to a pure function in the short forms.)

We conclude this section on pure functions with two examples: one from vector analysis and another from quantum mechanics.

Example 3.3.1 Show that

$$\nabla(\mathbf{A}\cdot\mathbf{B}) = \mathbf{A}\times(\nabla\times\mathbf{B}) + \mathbf{B}\times(\nabla\times\mathbf{A}) + (\mathbf{A}\cdot\nabla)\mathbf{B} + (\mathbf{B}\cdot\nabla)\mathbf{A}$$

where **A** and **B** are any two vector point functions. In terms of the basis vectors **i**, **j**, and **k** of the Cartesian coordinate system and the components of the vector functions, the third term on the *rhs* of the vector identity can be written as

$$\begin{aligned}(\mathbf{A}\cdot\nabla)\mathbf{B} &= \left(A_x\frac{\partial B_x}{\partial x} + A_y\frac{\partial B_x}{\partial y} + A_z\frac{\partial B_x}{\partial z}\right)\mathbf{i} \\ &+ \left(A_x\frac{\partial B_y}{\partial x} + A_y\frac{\partial B_y}{\partial y} + A_z\frac{\partial B_y}{\partial z}\right)\mathbf{j} \\ &+ \left(A_x\frac{\partial B_z}{\partial x} + A_y\frac{\partial B_z}{\partial y} + A_z\frac{\partial B_z}{\partial z}\right)\mathbf{k}\end{aligned}$$

and a similar expression can be given for the fourth term.

The package **VectorAnalysis`** contains many functions for vector analysis, and those relevant to this example were defined in Example 2.4.14 in Section 2.4.10.

```
In[26]:= Needs["VectorAnalysis`"]
```

CoordinateSystem gives the name of the default coordinate system:

```
In[27]:= CoordinateSystem
Out[27]= Cartesian
```

Coordinates[] returns the default names of coordinate variables in the default coordinate system:

```
In[28]:= Coordinates[]
Out[28]= {Xx, Yy, Zz}
```

To conform with traditional notation, let us change the names of the coordinate variables to **x**, **y**, and **z**:

```
In[29]:= SetCoordinates[Cartesian[x,y,z]]
Out[29]= Cartesian[x,y,z]
```

We can verify the new names of the coordinate variables:

```
In[30]:= Coordinates[]
Out[30]= {x, y, z}
```

Mathematica represents three-dimensional vectors by three-element lists. For the vector functions **A** and **B**, we let

```
In[31]:= A = {Ax[x, y, z], Ay[x, y, z], Az[x, y, z]};
         B = {Bx[x, y, z], By[x, y, z], Bz[x, y, z]};
```

To prove the vector identity, show that *lhs* − *rhs* evaluates to the null vector $\{0,0,0\}$:

```
In[33]:= Grad[DotProduct[A, B]] - CrossProduct[A, Curl[B]] -
           CrossProduct[B, Curl[A]] - (DotProduct[A, {∂x#, ∂y#, ∂z#}]& /@B) -
            (DotProduct[B, {∂x#, ∂y#, ∂z#}]& /@A)
Out[33]= {0, 0, 0}
```

We have used pure functions in the last two terms.

```
In[34]:= Clear[A, B]
```

■

Example 3.3.2 The Hamiltonian of the three-dimensional isotropic harmonic oscillator is the sum of three terms:

$$H = H_x + H_y + H_z$$

$$H_i = \frac{p_i^2}{2m} + \frac{1}{2} k\, q_i{}^2, \quad i = x, y \text{ or } z$$

where m is the mass, and k is the spring constant. Also, p_i and q_i are, respectively, the momentum and displacement from the equilibrium position of the mass along the i direction. For each direction, we can write the time-independent Schrödinger equation as

$$H_i u_{n_i}(q_i) = e_i u_{n_i}(q_i)$$

and express the energy eigenfunctions and eigenvalues as

$$u_{n_i}(q_i) = 2^{-n_i/2}(n_i!)^{-1/2}\left(\frac{m\omega}{\hbar\pi}\right)^{1/4} H_{n_i}\left(\sqrt{m\omega/\hbar}\, q_i\right)\exp\left(-\frac{m\omega}{2\hbar}{q_i}^2\right)$$

$$e_i = \hbar\omega\left(n_i + \frac{1}{2}\right), \qquad n_i = 0, 1, 2, \ldots, \infty$$

where H_{n_i} are the Hermite polynomials, and $\omega = \sqrt{k/m}$. The three-dimensional eigenfunctions are products of the one-dimensional ones:

$$\psi_{n_x n_y n_z}(q_x, q_y, q_z) = u_{n_x}(q_x)u_{n_y}(q_y)u_{n_z}(q_z)$$

Note that the three-dimensional eigenfunctions are specified by three quantum numbers $\{n_x, n_y, n_z\}$ that are nonnegative integers. Direct substitution verifies that these eigenfunctions satisfy the three-dimensional time-independent Schrödinger equation

$$H\psi_{n_x n_y n_z}(q_x, q_y, q_z) = E_n\psi_{n_x n_y n_z}(q_x, q_y, q_z)$$

where the energy eigenvalues are

$$E_n = (n + 3/2)\hbar\omega$$

with

$$n = n_x + n_y + n_z = 0, 1, 2, \ldots, \infty$$

(a) For a given n, how many distinct triples $\{n_x, n_y, n_z\}$ are possible? In other words, derive a formula for the degree of degeneracy (also known simply as degeneracy) of the energy eigenvalue E_n. (b) Generate the nested list of quantum numbers $\{n_x, n_y, n_z\}$ up to $n_x = n_y = n_z = 6$. (c) For each n ranging from 0 to 6, determine in the nested list generated in part (b) the number of elements $\{n_x, n_y, n_z\}$ that satisfy the relation $n = n_x + n_y + n_z$. That is, determine the degree of degeneracy of the energy eigenvalue E_n for $n = 0, 1, \ldots, 6$. Do these results substantiate the degeneracy formula derived in part (a)?

(a) For each n, the first quantum number n_x can be $0, 1, \ldots, n$. If the value of n_x is k, then the second quantum number n_y can be $0, 1, \ldots, n - k$; that is, there are $n - k + 1$ possible quantum numbers. If n_x and n_y are specified, the third quantum number n_z is uniquely determined by the relation $n = n_x + n_y + n_z$. Thus, for each n, the number of distinct $\{n_x, n_y, n_z\}$ is

```
In[35]:= degeneracy[n_] = Sum[(n - k + 1), {k, 0, n}] // Factor
```

$$\text{In[35]:= degeneracy[n\_]} = \sum_{k=0}^{n}(n - k + 1) \;//\; \text{Factor}$$

$$\text{Out[35]=}\ \frac{1}{2}(1 + n)(2 + n)$$

(b) Here is the nested list of quantum numbers up to $n_x = n_y = n_z = 6$:

```
In[36]:= (qnumbers = Flatten[
            Table[{nx, ny, nz}, {nx, 0, 6}, {ny, 0, 6}, {nz, 0, 6}], 2]) // Short
Out[36]//Short=
        {{0, 0, 0}, {0, 0, 1}, {0, 0, 2}, << 337 >>, {6, 6, 4}, {6, 6, 5}, {6, 6, 6}}
```

where we have limited the length of the output with **Short**. (**Short[***expr***]** prints as a short form of *expr*, less than approximately one line long.)

(c) With n ranging from 0 to 6, the degeneracy for each E_n is

```
In[37]:= Do[Print["n = " <> ToString[n] <> "  degeneracy = " <>
           ToString[Length[Select[qnumbers, Plus@@ # == n&]]]], {n, 0, 6}]

        n = 0     degeneracy = 1
        n = 1     degeneracy = 3
        n = 2     degeneracy = 6
        n = 3     degeneracy = 10
        n = 4     degeneracy = 15
        n = 5     degeneracy = 21
        n = 6     degeneracy = 28
```

Plus@@ # == *n***& [{**n_x**,** n_y**,** n_z**}]** returns **True** if $n_x + n_y + n_z = n$. **Select[qnumbers, Plus@@ # ==** *n***&]** picks out from the list **qnumbers** all elements $\{n_x, n_y, n_z\}$ that satisfy the criterion $n_x + n_y + n_z = n$. **ToString[***expr***]** gives a string corresponding to the printed form of *expr* in OutputForm, and the operator **<>** with the full name **StringJoin** concatenates two strings. The formula for the degree of degeneracy derived in part (a) gives the same results:

```
In[38]:= Do[Print["n = " <> ToString[n] <>
           "  degeneracy = " <> ToString[degeneracy[n]]], {n, 0, 6}]

        n = 0     degeneracy = 1
        n = 1     degeneracy = 3
        n = 2     degeneracy = 6
        n = 3     degeneracy = 10
        n = 4     degeneracy = 15
        n = 5     degeneracy = 21
        n = 6     degeneracy = 28

In[39]:= ClearAll[degeneracy, qnumbers]
```

■

3.3.2 Selecting a Definition

In *Mathematica*, we can give a function more than one definition. When the function is called—that is, invoked—how does *Mathematica* select which definition to use? It starts with the most specific definition, tries one at a time—the more specific definitions before the more general ones—and without going any further uses the first definition that matches the call. For example, consider the definitions

```
In[1]:= ClearAll[f]

In[2]:= f[y_]:= 3y

In[3]:= f[x_Integer]:= x^2

In[4]:= f[5]:= 1/25
```

and the function call `f[7]`. *Mathematica* first tries the definition for `f[5]`, which is the most specific because `5` is a subclass of the class of expressions represented by `x_Integer`, which in turn stands for a subclass of the class of expressions represented by `y_`. Since `f[7]` does not match `f[5]`, *Mathematica* proceeds to the definition for `f[x_Integer]`, where it finds a match. The second definition is then used, and the result is **49**:

```
In[5]:= f[7]
Out[5]= 49
```

The **?** operator shows how *Mathematica* orders these definitions:

```
In[6]:= ?f

Global`f

f[5]:= 1/25

f[x_Integer]:= x^2

f[y_]:= 3y
```

Definitions often lack definite order in generality. In such cases, *Mathematica* gives higher priority to definitions that are entered earlier. For instance, consider the definitions

```
In[7]:= ClearAll[f]

In[8]:= f[{x_, y_}]:= x^3 - y^3

In[9]:= f[x_Real?Positive]:= Sqrt[x]

In[10]:= f[x_Integer]:= x^2
```

Since none of these definitions can be considered to be more specific, *Mathematica* simply stores them in the order they are given:

```
In[11]:= ?f
```

```
Global`f

f[{x_, y_}] := x^3 - y^3
f[x_Real?Positive] := √x
f[x_Integer] := x^2
```

Beware! *Mathematica* does not always observe the rule of generality in prioritizing or storing definitions. For example, consider the definitions

```
In[12]:= ClearAll[f]

In[13]:= f[x_Integer] := x^2

In[14]:= f[x_Integer?Positive] := 1/x

In[15]:= f[x_Integer /; x > 50] := -x
```

These definitions have a definite order in generality: **f[x_Integer /; x > 50]** is a special case of **f[x_Integer?Positive]**, which in turn is a special case of **f[x_Integer]**. Therefore, we expect *Mathematica* to store them in the same order. Well, it does not:

```
In[16]:= ?f
```

```
Global`f

f[x_Integer?Positive] := 1/x
f[x_Integer/;x > 50] := -x
f[x_Integer] := x^2
```

Although *Mathematica* stores the definition for **f[x_Integer]** last as expected, it stores the other two definitions according to the relative order in which they are entered. Therefore, it is often a good idea to check how *Mathematica* orders the definitions with the **?** operator to ensure that the appropriate definition is used.

```
In[17]:= ClearAll[f]
```

3.3.3 Recursive Functions and Dynamic Programming

Consider, for example, a recursion relation

$$H_n(z) = 2z H_{n-1}(z) - 2(n-1)H_{n-2}(z)$$

for the Hermite polynomials, where n can be any integer greater than 1 and the initial conditions are

$$H_0(z) = 1$$

$$H_1(z) = 2z$$

(The recursion relation [Gos03, Mes00, Gri05] for the Hermite polynomials is also known as a recurrence formula [MF53] or a recurrence relation [Has91]. Some authors use the terms recursion and recurrence interchangeably [Lib03, Vve93]. See [Tho92] for an interesting discussion of recursion versus recurrence. For a contrasting view, see [MF53].)

In programming, a function is recursive if it is defined in terms of itself. We can determine the Hermite polynomials with the recursive function

```
In[1]:= hermite[n_Integer /; n > 1, z_] :=
         Expand@(2 z hermite[n - 1, z] - 2 (n - 1) hermite[n - 2, z])
```

and the initial conditions

```
In[2]:= hermite[0, z_] := 1
```

```
In[3]:= hermite[1, z_] := 2 z
```

where **hermite** is an alias for H. Unless n is small, generating Hermite polynomials with the recursive function is computation-intensive because a call to **hermite[*n*, z]** invokes two calls, one to **hermite[*n* - 1, z]** and another to **hermite[*n* - 2, z]**. Each of these calls in turn triggers two additional ones, and so on until each branch eventually terminates with a call to **hermite[1, z]** or **hermite[0, z]**. Evaluating **hermite[20, z]**, for instance, requires a total of 21,891 calls to the function **hermite** (see Exercise 13a in Section 3.3.7). The following table shows the number of calls to each **hermite[*k*, z]**, for $0 \leq k \leq 20$, in the evaluation of **hermite[20, z]** (see Exercise 14 in Section 3.3.7):

k	Number of Function Calls	k	Number of Function Calls
0	4181	11	55
1	6765	12	34
2	4181	13	21
3	2584	14	13
4	1597	15	8
5	987	16	5
6	610	17	3
7	377	18	2
8	233	19	1
9	144	20	1
10	89		

Most of these function calls, however, are unnecessary if the function is so written that *Mathematica* remembers, in the form of rules or definitions, the values that it has computed for the function with specific arguments. The method of creating and storing new rules for a function during an evaluation is called dynamic programming. The syntax is rather simple:

func[*arg*$_1$_, *arg*$_2$_, ...] := *func*[*arg*$_1$, *arg*$_2$, ...] = *rhs*

For the Hermite polynomials, the definition becomes

```
In[4]:= ClearAll[hermite]

In[5]:= hermite[n_Integer/; n>1, z_]:=
         hermite[n, z] = Expand@(2 z hermite[n-1, z] - 2(n-1)hermite[n-2, z])
```

while the initial conditions remain the same:

```
In[6]:= hermite[0, z_]:= 1

In[7]:= hermite[1, z_]:= 2 z
```

Mathematica stores the value of the function as soon as it is computed, and consequently there is now only one computation for each **hermite[*k*, z]**, for $k \leq n$, in the evaluation of **hermite[*n*, z]** (see Exercise 12b in Section 3.3.7). When *Mathematica* needs a value for a specific **hermite[*k*, z]**, it simply looks up the appropriate definition stored in the global rule base in a manner described in Section 3.3.2. Evaluating **hermite[20, z]** with dynamic programming, for example, only requires a total of 39 instead of 21,891 calls (see Exercise 13c in Section 3.3.7) to the function **hermite**. The following table shows the number of calls to each **hermite[*k*, z]**, for $0 \leq k \leq 20$, in the evaluation of **hermite[20, z]** (see Exercise 13c in Section 3.3.7):

k	Number of Function Calls	k	Number of Function Calls
0	1	11	2
1	2	12	2
2	2	13	2
3	2	14	2
4	2	15	2
5	2	16	2
6	2	17	2
7	2	18	2
8	2	19	1
9	2	20	1
10	2		

The **?** operator reveals why there is such a reduction of function calls.

```
In[8]:= ClearAll[hermite]

In[9]:= hermite[n_Integer/; n > 1, z_]:=
         hermite[n, z] = Expand@(2 z hermite[n - 1, z] - 2(n - 1) hermite[n - 2, z])

In[10]:= hermite[0, z_]:= 1

In[11]:= hermite[1, z_]:= 2 z

In[12]:= ?hermite
```

```
Global`hermite

hermite[n_Integer/;n>1, z_]:=
 hermite[n, z] = Expand[2 z hermite[n - 1, z] - 2 (n - 1) hermite[n - 2, z]]
hermite[0, z_]:= 1
hermite[1, z_]:= 2 z
```

Before an evaluation of **hermite[*n*, z]**, there are just three definitions for **hermite** in the global rule base that *Mathematica* searches during computations.

```
In[13]:= hermite[8, z]
Out[13]= 1680 - 13440 z^2 + 13440 z^4 - 3584 z^6 + 256 z^8

In[14]:= ?hermite
```

```
Global`hermite

hermite[2, z] = -2 + 4 z^2
hermite[3, z] = -12 z + 8 z^3
hermite[4, z] = 12 - 48 z^2 + 16 z^4
hermite[5, z] = 120 z - 160 z^3 + 32 z^5
hermite[6, z] = -120 + 720 z^2 - 480 z^4 + 64 z^6
hermite[7, z] = -1680 z + 3360 z^3 - 1344 z^5 + 128 z^7
hermite[8, z] = 1680 - 13440 z^2 + 13440 z^4 - 3584 z^6 + 256 z^8
hermite[n_Integer/;n>1, z_]:=
 hermite[n, z] = Expand[2 z hermite[n - 1, z] - 2 (n - 1) hermite[n - 2, z]]
hermite[0, z_]:= 1
hermite[1, z_]:= 2 z
```

After an evaluation of, for instance, **hermite[8, z]**, there are seven additional rules in the global rule base, one for each **hermite[*k*, z]**, with k ranging from 2 to 8. Evaluating

hermite[9, z] now requires only 3 (instead of 109) function calls: **hermite[9, z]**, **hermite[8, z]**, and **hermite[7, z]**, since the rules for the latter two are already stored in the global rule base and no recomputations are necessary. (For further discussion of dynamic programming, see [WGK05, Gra98].)

```
In[15]:= ClearAll[hermite]
```

3.3.4 Functional Iterations

We sometimes wish to apply a function repeatedly to an expression a specified number of times or until a criterion is satisfied. Several built-in functions are available for this purpose:

Nest[*f*, *expr*, *n*]	give the result of applying *f* nested *n* times to *expr*
NestList[*f*, *expr*, *n*]	return a list of the results of applying *f* nested 0 through *n* times to *expr*; that is, generate the list {*expr*, *f*[*expr*], *f*[*f*[*expr*]], ...} with $n+1$ elements
FixedPoint[*f*, *expr*]	start with *expr*, then apply *f* repeatedly until the result no longer changes
FixedPoint[*f*, *expr*, *n*]	stop after at most *n* steps; the default value for *n* is 65,536
FixedPoint[*f*, *expr*, **SameTest** -> *comp*]	stop when the function *comp* applied to two successive results yields **True**; the default setting is **SameTest -> SameQ**
FixedPointList[*f*, *expr*]	generate a list giving the results of applying *f* repeatedly, starting with *expr*, until the results no longer change; that is, give the list {*expr*, *f*[*expr*], *f*[*f*[*expr*]], ...}, stopping when the elements no longer change
FixedPointList[*f*, *expr*, *n*]	stop after at most *n* steps—that is, the list having at most $n+1$ elements
FixedPointList[*f*, *expr*, **SameTest** -> *comp*]	terminate the list when the function *comp* applied to two successive elements yields **True**; the default setting is **SameTest -> SameQ**

For an example, we generate a geometric progression to nine terms:

```
In[1]:= Plus @@ NestList[r#&, a, 8]
```

Out[1]= $a + a r + a r^2 + a r^3 + a r^4 + a r^5 + a r^6 + a r^7 + a r^8$

With **FixedPoint**, we can implement the Newton–Raphson algorithm for evaluating square roots. To determine the square root of a number r, start with an initial guess x_0 and obtain successively better approximations with the formula

$$x_{n+1} = \frac{1}{2}\left(x_n + \frac{r}{x_n}\right)$$

Starting with $x_0 = 3.0$, we compute, for example, the square root of 11:

```
In[2]:= FixedPoint[1/2 (# + 11/#) &, 3.0]
Out[2]= 3.31662
```

Let us verify that the computed value is indeed correct:

```
In[3]:= % == Sqrt[11.0]
Out[3]= True
```

To further illustrate the use of these built-in functions, we introduce the logistic map, a simple population growth model exhibiting chaotic behavior.

Example 3.3.3 Consider the logistic map given by the one-dimensional difference equation

$$x_{n+1} = f(x_n)$$

where $n = 0, 1, \ldots$, and

$$f(x) = \mu x(1 - x)$$

with $0 \le x \le 1$ and $0 \le \mu \le 4$. (a) Show that the points $x = 0.6$ and $x = 0$ are stable fixed points when the map parameter μ assumes the values 2.5 and 0.24, respectively. (b) Plot on the same graph x_n versus n with $\mu = 3.64$ and for two slightly different initial points $x_0 = 0.500$ and 0.501.
(a) A point x satisfying the condition

$$x = f(x)$$

is called a fixed point. A fixed point is stable, or attracting, if upon repeated application of the map f all points converge to it.

Consider the case $\mu = 2.5$:

```
In[4]:= μ = 2.5;
```

The point $x = 0.6$ is a fixed point of the map

```
In[5]:= f[x_] := μ x (1 - x)
```

because

```
In[6]:= 0.6 == f[0.6]
Out[6]= True
```

To show that $x = 0.6$ is a stable or attracting fixed point, we apply f repeatedly to a number of randomly selected initial points:

```
In[7]:= Table[FixedPoint[f, RandomReal[]], {10}]
Out[7]= {0.6, 0.6, 0.6, 0.6, 0.6, 0.6, 0.6, 0.6, 0.6, 0.6}
```

The arbitrary starting points all converge to $x = 0.6$. Note that the points $x = 0$ and 1 are exceptions; they do not converge to $x = 0.6$ because $f(0) = f(1) = 0$.

Now consider the case $\mu = 0.24$:

```
In[8]:= μ = 0.24;
```

The point $x = 0$ is a fixed point since

```
In[9]:= 0 == f[0]
Out[9]= True
```

Let us apply f repeatedly to an arbitrary initial point—for example, $x_0 = 0.612$:

```
In[10]:= FixedPoint[f, 0.612]
Out[10]= $Aborted
```

The computation was manually aborted because *Mathematica* kept running for a long time without returning any result. **FixedPointList** reveals the problem.

```
In[11]:= FixedPointList[f, 0.612, 39]
Out[11]= {0.612, 0.0569894, 0.012898, 0.00305559,
          0.000731101, 0.000175336, 0.0000420733, 0.0000100972,
          2.42329 × 10^-6, 5.81589 × 10^-7, 1.39581 × 10^-7, 3.34995 × 10^-8,
          8.03988 × 10^-9, 1.92957 × 10^-9, 4.63097 × 10^-10, 1.11143 × 10^-10,
          2.66744 × 10^-11, 6.40186 × 10^-12, 1.53645 × 10^-12, 3.68747 × 10^-13,
          8.84993 × 10^-14, 2.12398 × 10^-14, 5.09756 × 10^-15, 1.22341 × 10^-15,
          2.93619 × 10^-16, 7.04686 × 10^-17, 1.69125 × 10^-17, 4.05899 × 10^-18,
          9.74158 × 10^-19, 2.33798 × 10^-19, 5.61115 × 10^-20, 1.34668 × 10^-20,
          3.23202 × 10^-21, 7.75686 × 10^-22, 1.86165 × 10^-22, 4.46795 × 10^-23,
          1.07231 × 10^-23, 2.57354 × 10^-24, 6.17649 × 10^-25, 1.48236 × 10^-25}
```

FixedPoint and **FixedPointList** stop when the comparison function *comp* in the option specification **SameTest -> *comp*** yields **True** upon applying to two successive results; the default option setting is **SameTest -> SameQ**.

```
In[12]:= SameQ[%[[-1]], %[[-2]]]
Out[12]= False
```

where the negative indices **-1** and **-2** refer, respectively, to the first and second elements from the end of the list. The problem is that **SameQ** returns **False** even though the two numbers being compared are practically zero. We can specify a comparison function less stringent than **SameQ**.

```
In[13]:= Table[FixedPoint[f, RandomReal[],
              SameTest → (Abs[#1 - #2] < 10^-11 &)] // Chop, {10}]
Out[13]= {0, 0, 0, 0, 0, 0, 0, 0, 0, 0}
```

The pure function **Abs[#1 - #2] < 10^-11 &** returns **True** when the absolute value of the difference between two successive iterates is less than 10^{-11}. The point $x = 0$ is indeed a stable fixed point because randomly selected initial points all converge to it upon repeated application of the map f. With the function **Chop**, we have replaced small approximate real numbers by zeros. (**Chop[***expr*, *tol***]** replaces in *expr* the numbers whose absolute magnitudes differ from zero by less than *tol* by the exact integer **0**; **Chop** uses a default tolerance (i.e., *tol*) of 10^{-10}.)

(b) **ListLinePlot[{{{**x_{11}**,** y_{11}**}, {**x_{12}**,** y_{12}**}, ...}, {{**x_{21}**,** y_{21}**}, {**x_{22}**,** y_{22}**}, ...}, ...}]** plots on the same graph several lines, each through a list of points **{{**x_{i1}**,** y_{i1}**}, {**x_{i2}**,** y_{i2}**}, ...}**.

Consider the case with $\mu = 3.64$ and two slightly different initial points, $x_0 = 0.500$ and 0.501:

```
In[14]:= μ = 3.64;

In[15]:= ListLinePlot[{Transpose[{Range[0, 44], NestList[f, 0.500, 44]}],
           Transpose[{Range[0, 44], NestList[f, 0.501, 44]}]},
          PlotRange → {0, 1}, AxesLabel → {"n", "xn"}]
```

Out[15]=

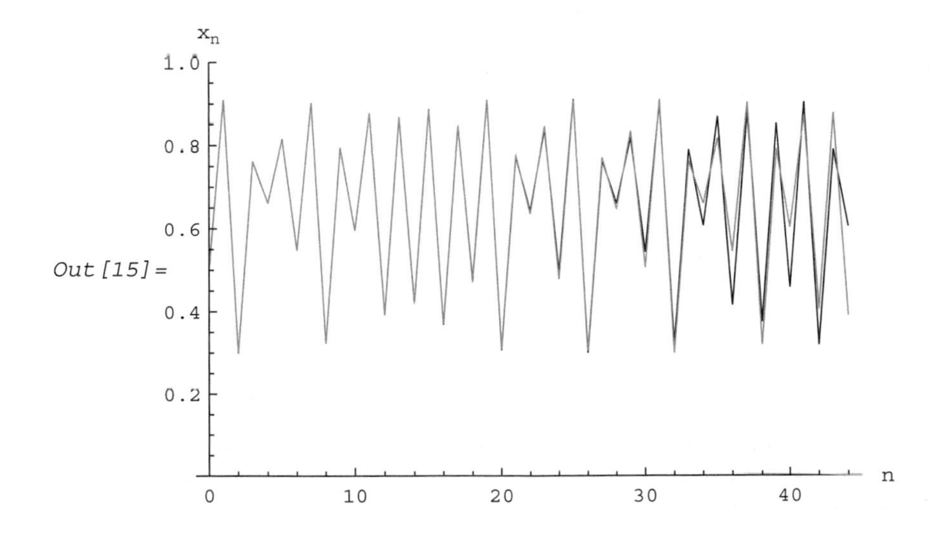

Whereas `NestList[f,x0,n]` returns the list `{x0,f[x0],f[f[x0]],...}` with $n+1$ elements, `Transpose[{Range[0,n], NestList[f,x0,n]}]` gives the list `{{0, x0}, {1,f[x0]}, {2,f[f[x0]]},...}` with the same number of elements. Let us zoom in on the portion of the graph for n ranging from 30 to 44:

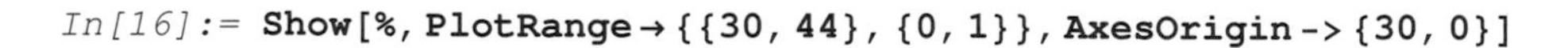

```
In[16]:= Show[%, PlotRange→{{30, 44}, {0, 1}}, AxesOrigin -> {30, 0}]
```

Out[16]=

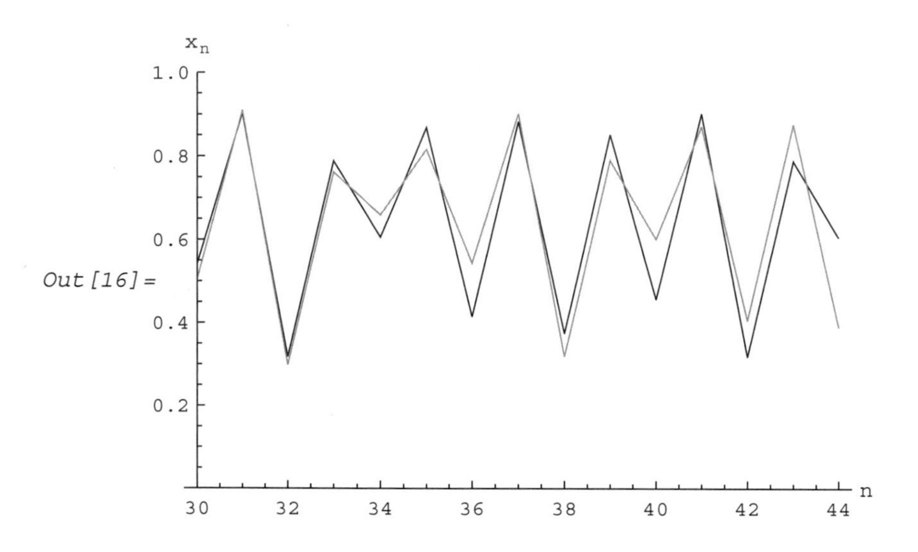

There is significant divergence between the two curves. The values of x_{44} for $x_0 = 0.500$ and 0.501 are

```
In[17]:= Nest[f, 0.500, 44]
Out[17]= 0.605808
```

```
In[18]:= Nest[f, 0.501, 44]
Out[18]= 0.390978
```

A 0.2% separation between the initial points leads to a 35.5% spread of the iterates after only 44 iterations! Sensitive dependence of the iterates x_n on the initial condition is a characteristic of chaotic behavior.

```
In[19]:= ClearAll[f, μ]
```

■

3.3.5 Protection

As discussed in Section 3.2.5, we can assign general properties, called attributes, to functions. `SetAttributes[f, attr]` adds *attr* to the attributes of function f and `ClearAttributes[f, attr]` removes *attr* from the attributes of f.

Many built-in *Mathematica* functions have the attribute **Protected**, which prevents us from inadvertently giving any definitions for these functions. For example, consider the function **D** for computing partial derivatives. **Attributes[D]** returns a list of its attributes:

```
In[1]:= Attributes[D]
Out[1]= {Protected, ReadProtected}
```

Because **D** has the attribute **Protected**, giving it any definition, be it consistent or inconsistent with built-in rules, triggers an error message:

```
In[2]:= D[u_ + v_, x_] := D[u, x] + D[v, x]
        SetDelayed::write: Tag D in ∂x_(u_ + v_) is Protected. >>
Out[2]= $Failed
```

Mathematica, however, allows us to specify rules for built-in functions provided that we first remove the protection. To clear **Protected**, use either **ClearAttributes** or **Unprotect**. **Unprotect[**s_1**,**s_2**,...]** removes the attribute **Protected** for the symbols s_i.

Section 3.3.6 will show that there are occasions when altering the default rule base for built-in functions is appropriate. We, however, modify built-in functions at our peril because we may add new rules that are mathematically invalid. Furthermore, we may override built-in rules with erroneous ones because *Mathematica* applies user-defined rules before built-in ones. Let us, for example, evaluate

```
In[3]:= D[x Sin[x], x]
Out[3]= x Cos[x] + Sin[x]
```

Now unprotect **D**,

```
In[4]:= Unprotect[D];
```

and enter the definition

```
In[5]:= D[u_ v_, x_] := D[v, x] - v D[u, x]
```

After giving the rule, we should restore the protection from accidental definitions with either **SetAttributes** or **Protect**. **Protect[**s_1**,**s_2**,...]** sets the attribute **Protected** for the symbols s_i.

```
In[6]:= Protect[D];
```

The erroneous rule that we just gave for **D** overrides the built-in product rule and can lead to disastrous results. Let us reevaluate the preceding derivative:

```
In[7]:= D[x Sin[x], x]
Out[7]= Cos[x] - Sin[x]
```

The answer is no longer $x\cos x + \sin x$; it is now mathematically incorrect!

We can assign the attribute **Protected** to user-defined functions. For example,

```
In[8]:= myfunc[x_] := x^2 + 1

In[9]:= Protect[myfunc];

In[10]:= Attributes[myfunc]
Out[10]= {Protected}
```

Giving another definition for **myfunc** triggers an error message:

```
In[11]:= myfunc[x_, y_] := x^2 + y^2
         SetDelayed::write: Tag myfunc in myfunc [x_ , y_] is Protected. >>
Out[11]= $Failed
```

After unprotecting **myfunc**, we may give the definition that was rejected:

```
In[12]:= Unprotect[myfunc];
In[13]:= myfunc[x_, y_] := x^2 + y^2
```

To verify that *Mathematica* recognizes this definition, let us call **myfunc** with two arguments:

```
In[14]:= myfunc[a, b]
Out[14]= a^2 + b^2

In[15]:= ClearAll[myfunc]
```

3.3.6 Upvalues and Downvalues

Familiar to us are definitions of the forms

```
f[args] = rhs
f[args] := rhs
```

These definitions are associated with the symbol f and are called downvalues for f. As examples,

```
f[0] = 1;
f[x_, y_] := x^2 + y^3
f[g[x_], h[y_]] := p[x, y]
```

define downvalues for **f**.

Definitions of the forms

```
f /: g[..., f, ...] = rhs
f /: g[..., f, ...] := rhs
```

```
f /: g[..., f [args], ...] = rhs
f /: g[..., f [args], ...] := rhs
```

are also associated with the symbol f, but they are called upvalues for f. As examples,

```
f /: Re[f] = 0;
f /: Log[f[x_]] := q[x]
f /: g[f[x_], h[y_]] := w[x + y]
```

define upvalues for **f**.

Upvalues for f can be given in compact forms

```
g[f] ^= rhs
g[f] ^:= rhs
g[f [args]] ^= rhs
g[f [args]] ^:= rhs
```

The first two definitions in the previous group of examples can be expressed as

```
Re[f] ^= 0;
Log[f[x_]] ^:= q[x]
```

For functions with more than one argument,

```
g[arg1, arg2, ...] ^= rhs
g[arg1, arg2, ...] ^:= rhs
```

define upvalues for the heads of all arg_i. The third definition in the earlier examples can be given as

```
g[f[x_], h[y_]] ^:= w[x + y]
```

Whereas the earlier definition is associated only with **f**, this definition is associated with both **f** and **h**.

To illustrate the use of downvalues and upvalues, consider adding new rules for the built-in function **Abs**. **Abs**[z] gives the absolute value of the real or complex number z. If *expr* is not a number, *Mathematica* does not always simplify **Abs**[*expr*]. For instance,

```
In[1]:= ClearAll[a, b]

In[2]:= Abs[a + ⅈ b]
Out[2]= Abs[a + ⅈ b]
```

where **a** and **b** are mathematical variables—that is, variables without assigned values. *Mathematica* does not transform $|a + ib|$ to $\sqrt{a^2 + b^2}$. We can add new rules for simplifying

Abs[*expr*] when *expr* consists of numbers and variables some of which are assumed real and either positive or negative. To declare variable **a** to be real and positive, use the definitions

```
In[3]:= Im[a] ^= 0;
        Positive[a] ^= True;
```

which define upvalues for **a**. Similarly, to declare variable **b** to be real and negative, use the definitions

```
In[5]:= Im[b] ^= 0;
        Negative[b] ^= True;
```

which define upvalues for **b**. Before specifying new rules for **Abs**, we must first unprotect the symbol since it has the attribute **Protected**:

```
In[7]:= Unprotect[Abs];
```

For simplifying **Abs**[*expr*] where *expr* is a product or quotient of numbers, variables, and expressions of the forms u^x and $x + yi$ in which u may be imaginary and x and y are real, add the definitions

```
In[8]:= Abs[x_/;(Im[x] == 0 && Positive[x])] := x

In[9]:= Abs[x_/;(Im[x] == 0 && Negative[x])] := -x

In[10]:= Abs[u_^x_/;Im[x] == 0]:= Abs[u]^x

In[11]:= Abs[u_ v_] := Abs[u] Abs[v]

In[12]:= Abs[(x_ + y_.Complex[0, n_])/;
            (MatchQ[Im/@Level[{x, y}, {-1}], {(0)..}] && FreeQ[{x, y}, w_^z_/;
              (!IntegerQ[z] && !(Positive[w] === True))])] := Sqrt[x^2 + (n y)^2]

In[13]:= Abs[u_] := Sqrt[Re[u]^2 + Im[u]^2]
```

which define downvalues for **Abs**. The fifth rule is valid only if x and y are real because **Complex[0, n_]** stands for any pure imaginary number ni. The two conditions restricting the pattern **(x_ + y_. Complex[0, n_])** ensure that x and y are indeed real. **Level[{x, y}, {-1}]** gives a list of the atoms in **{x, y}**—that is, in x and y. The condition **MatchQ[Im /@ Level[{x, y}, {-1}], {(0)..}]** requires that these atoms are all real because the pattern **{(0)..}** represents a list of one or more zeros. The second condition with **FreeQ** prohibits combinations of these real atoms from becoming imaginary. **FreeQ**[*expr*,*form*] yields **True** if no subexpression in *expr* matches *form*, and it yields **False** otherwise. The condition **FreeQ[{x, y}, w_^z_/;(!IntegerQ[z] && !(Positive[w] === True))]** insists that, for an expression to match the pattern **(x_ + y_. Complex(0, n_])**, x and y must not contain any subexpression of the form w^z unless z is an integer or w is a positive. To avoid unintended definitions, we should restore the protection for **Abs**:

```
In[14]:= Protect[Abs];
```

Let us apply these definitions to several examples:

```
In[15]:= Abs[b]
Out[15]= -b
```

In[16]:= $\text{Abs}\left[\frac{3^{3/2}}{a^{2/3}} + \frac{a^2}{b^3}\, i\right]$

Out[16]= $\sqrt{\frac{27}{a^{4/3}} + \frac{a^4}{b^6}}$

In[17]:= $\text{Abs}\left[\left(\frac{2a}{3} - 3b^2 a^{-7}\, i\right)\frac{b u^2}{v^{2/3}}\right]$

Out[17]= $-\frac{b\sqrt{\frac{4a^2}{9} + \frac{9b^4}{a^{14}}}\left(\text{Im}[u]^2 + \text{Re}[u]^2\right)}{\left(\text{Im}[v]^2 + \text{Re}[v]^2\right)^{1/3}}$

In[18]:= $\text{Abs}\left[\left(a^3 - \frac{b^2}{a^4}\, i\right)\frac{(u+a)^2}{(2+3a\, i)^5 v^3}\right]$

Out[18]= $\frac{\sqrt{a^6 + \frac{b^4}{a^8}}\left(\text{Im}[u]^2 + \text{Re}[a+u]^2\right)}{\left(4+9a^2\right)^{5/2}\left(\text{Im}[v]^2 + \text{Re}[v]^2\right)^{3/2}}$

Despite the negative signs in two results, they are positive since **b** was declared negative. (See Exercise 22 of Section 3.3.7 for simplifying **Abs**[*expr*] with the built-in function **ComplexExpand**.)

For a class of objects of a particular type, we often have to decide whether to define special arithmetic operators for them or augment the definitions of built-in operators in order to extend their domains to include these objects. Furthermore, we need to choose between giving the definitions as upvalues or downvalues. Consider, for example, three-dimensional vectors. *Mathematica* usually represents them as three-element lists. We can represent them as the objects **vector**[A_x, A_y, A_z], where the arguments are the Cartesian components of a vector. Rather than defining a special function, such as **vectorPlus**, for adding vectors, we may find it more convenient to give a new definition to **Plus** for vector addition. The rule can be specified as a downvalue for **Plus** because it is the head of the expression **vector[Ax_, Ay_, Az_] + vector[Bx_, By_, Bz_]**:

```
In[19]:= Head[vector[Ax_, Ay_, Az_] + vector[Bx_, By_, Bz_]]
Out[19]= Plus
```

Since *Mathematica* tries user-defined rules before built-in ones, the new rule will be tried in every invocation of **Plus**, even when the computations do not involve vectors, and consequently will slow *Mathematica* down. For efficiency, we should give a rule for addition as an upvalue for **vector**:

```
In[20]:= vector[Ax_,Ay_,Az_] + vector[Bx_,By_,Bz_]^:=
          vector[Ax + Bx, Ay + By, Az + Bz]
```

This rule is associated with **vector** rather than **Plus**. Let us add, for example, three vectors:

```
In[21]:= vector[a1, a2, a3] + vector[b1, b2, b3] + vector[c1, c2, c3]
Out[21]= vector[a1 + b1 + c1, a2 + b2 + c2, a3 + b3 + c3]

In[22]:= ClearAll[a, b, f, h, vector]
```

3.3.7 Exercises

In this book, straightforward, intermediate-level, and challenging exercises are unmarked, marked with one asterisk, and marked with two asterisks, respectively. For solutions to most odd-numbered exercises, see Appendix C.

1. Write named functions that perform the operations specified by the pure functions:
   (a) `(1 + #^3)&`
   (b) `{ #, #^2 }&`
   (c) `( # /.x→y)&`
   (d) `(1/ # 1^#2)&`
   (e) `{ #1, #2^#3}&`
   (f) `( - ⅈD[ #, x])&`
   (g) `(( #^3 )&/@ # )&`
   (h) `Function[x, Apply[And, Map[OddQ, x]]]`
   (i) `Function[x, Delete[x, RandomInteger[{1, Length[x]}]]]`

2. Define pure functions corresponding to the named functions:
   (a) `f[x_]:= 1/(1+x)`
   (b) `g[x_, y_] := (x+y)^2`
   (c) `mytest[expr_]:= FreeQ[expr, _Integer]`
   (d) `ourtest[expr_]:= Length[expr] ≥ 2`
   (e) `test1[x_, y_]:= x>y`
   (f) `ourfunc[x_]:= f[x] + g[x]`
   (g) `func[x_]:= μ x(1 - x)`
   (h) `myfunc[x_]:= {Re[(x - 1)/(x + 1)], Im[(x - 1)/(x + 1)]}`

3. Use **Cases** and a pure function to obtain from a list those elements that are integers greater than 3. For example, from the list

```
{1, a, 2.0, 5.0, 4, x^2, Sin[x]}
```

obtain the list

```
{4}
```

4. Use **Select** and a pure function to pick out from a list of pairs those pairs in which the sum of the elements is smaller than 5. For example, from the list

```
{{1, 2}, {20, 1}, {a, 2}}
```

obtain the list

```
{{1, 2}}
```

Hint: Use also **Plus** and **Apply**.

5. Use **Select** and a pure function to pick out from the list of triples those triples whose products of the elements are larger than 20:

```
{{4.294, 3.757, 7.222}, {9.240, 3.008, 1.001}, {0.696, 3.826, 0.375},
 {1.931, 4.814, 5.422}, {7.161, 3.665, 0.212}, {1.809, 4.298, 7.333},
 {5.745, 3.287, 1.215}, {1.901, 3.335, 0.022}, {8.769, 3.246, 0.137}}
```

6. Use **RandomInteger** to create a list of 50 random integers in the range 1 to 200. Then, use **Select** and a pure function to pick out from the list the numbers that are less than 50.

7. Generate a list of triples $\{x, y, z\}$, where x, y, z, and the length of the list are random integers from 0 to 10. From this list, obtain a list of those triples satisfying the condition that $(x + y + z)$ is an even number. Also, obtain the list of those triples with $(x + y + z)$ being an odd number. For example, from the generated list

```
{{9, 1, 6}, {1, 9, 10}, {3, 3, 2},
 {5, 0, 10}, {6, 4, 1}, {9, 6, 8}, {7, 7, 8}}
```

obtain the list

```
{{9, 1, 6}, {1, 9, 10}, {3, 3, 2}, {7, 7, 8}}
```

and also the list

```
{{5, 0, 10}, {6, 4, 1}, {9, 6, 8}}
```

8. Using only built-in functions and pure functions, write a function that takes as its arguments a list

```
{a1, a2, ..., an}
```

as well as a nested list

```
{{{b11, c11}, {b12, c12},...,{b1m, c1m}},
 {{b21, c21}, {b22, c22},...,{b2m, c2m}},
 ...,
 {{bn1, cn1}, {bn2, cn2},...,{bnm, cnm}}}
```

and returns the list

```
{{a1, {b11, c11}, {b12, c12},...,{b1m, c1m}},
 {a2, {b21, c21}, {b22, c22},...,{b2m, c2m}},
 ...,
 {an, {bn1, cn1}, {bn2, cn2},...,{bnm, cnm}}}
```

for any positive integers n and m. For example,

```
list1 = Table[ai, {i, 1, 3}]
{a1, a2, a3}
list2 = Table[
  HoldForm[{bnm, cnm}]/.{n→ToString[i], m -> j}, {i, 1, 3}, {j, 1, 2}]
{{{b11, c11}, {b12, c12}},
 {{b21, c21}, {b22, c22}}, {{b31, c31}, {b32, c32}}}
func[list1, list2]
{{a1, {b11, c11}, {b12, c12}},
 {a2, {b21, c21}, {b22, c22}}, {a3, {b31, c31}, {b32, c32}}}
```

9. Show that

$$\nabla \times (\mathbf{A} \times \mathbf{B}) = (\mathbf{B} \cdot \nabla)\mathbf{A} - (\mathbf{A} \cdot \nabla)\mathbf{B} + \mathbf{A}(\nabla \cdot \mathbf{B}) - \mathbf{B}(\nabla \cdot \mathbf{A})$$

where **A** and **B** are any two vector point functions.

*10. Write a function for tabulating the number of times each distinct element appears in a list. For example, `tabulate[{b, 1, b, 1, a, a, c, a, Sin[x], Sin[x]}]` should return

```
1       2
a       3
b       2
c       1
Sin[x]  2
```

Hint: Use `Union`, `Count`, `Map`, `Transpose`, and `TableForm`.

11. (a) What is the internal form of `{v₀, θ₀}`? (b) Make the assignment

```
{v₀, θ₀} = {55 m/s, 30°};
```

Applying the function **Clear** to v_0 and θ_0 generates a couple of error messages:

```
Clear[v₀, θ₀]
Clear::ssym: v₀ is not a symbol or a string. ≫
Clear::ssym: θ₀ is not a symbol or a string. ≫
```

How are the values for v_0 and θ_0 stored in the global rule base? How can we clear these values without using the **Unset** operator "`=.`"? *Hint:* Use the operator "`??`".

*12. `Trace[`*expr*`]` generates a list of all expressions used in the evaluation of *expr*. Using the function **Trace**, describe how *Mathematica* evaluates the Hermite polynomials `hermite[`*n*`, z]` discussed in Section 3.3.3 when (a) dynamic programming *is not* used and (b) dynamic programming *is* used. Consider only the case with $n = 4$.

*13. Whereas `Trace[`*expr*`]`, introduced in Problem 12, generates a list of all expressions used in the evaluation of *expr*, `Trace[`*expr*`, `*form*`]` includes only expressions that match the pattern *form*. (a) Write a function `totalCalls[n_, z_]` that returns the total number of function calls to **hermite** in the evaluation of `hermite[`*n*`, z]` when dynamic programming *is not* used. Evaluate `totalCalls[20, z]`. (b) Write a function `partialCalls[n_, k_, z_]` that returns the number of function calls to `hermite[`*k*`, z]` in the evaluation of `hermite[`*n*`, z]`, with $k < n$, when dynamic programming *is not* used. Evaluate `partialCalls[20,3,z]`. (c) Repeat parts (a) and (b) concerning the evaluation of `hermite[`*n*`,z]` but for the cases in which dynamic programming *is* used.

*14. Using the function **partialCalls** defined in Problem 13b in the forms `Table[partialCalls[20,k,z],{k,0,20}]` or `Do[Print[partialCalls[20,k,z]], {k, 0, 20}]` for the task of listing the number of function calls to `hermite[`*k*`, z]` in the evaluation of `hermite[20, z]` with k ranging from 0 to 20 is rather expensive because there is a fresh evaluation of `Trace[hermite[20, z], hermite[`*k*`, z]]` for each k. Find a frugal method for accomplishing this task. *Hint:* With **FullForm**, examine the internal representation of the result of evaluating `Trace[hermite[20, z], hermite[`*k*`, z]]` when k is 10, for example. Instead of **FullForm**, **InputForm** may also be used.

*15. A 3.00-g leaf falls from rest at a height of 2.00 m near the surface of the earth. The effect of air resistance may be approximated by a frictional force proportional to the velocity. That is, if the x-axis is directed upward, the net force on the leave is $F = -mg - bv$, where m is the mass, g is the magnitude of the acceleration due to gravity, and v is the velocity. Let the constant $b = 0.0300\,\text{kg/s}$. (a) Determine the terminal speed of the leaf. (b) Use Euler's method of numerical analysis and dynamic programming to find the velocity and position of the leaf as functions of time, from the instant it is released until 99% of the terminal speed is reached. Plot the velocity as a function of time and also the position as a function of time. *Hint:* Try the step size $\Delta t = 0.005\,\text{s}$. (c) Use **NDSolve** to do part (b) above—that is, to find the velocity and position of the leaf as functions of time, from the instant it is released until 99% of the terminal speed is reached. (For information on Euler's method, see [Ste03, SJ04].)

*16. A hailstone falls from rest near the surface of the earth. The effect of air resistance may be approximated by a frictional force proportional to the square of the speed. If the x-axis is directed upward, the net force on the hailstone is given by

$$F = -mg + Cv^2$$

where g is the magnitude of the acceleration due to gravity, and v is the speed. Let the mass $m = 4.80 \times 10^{-4}$ kg and the constant $C = 2.50 \times 10^{-5}$ kg/m. (a) Determine the terminal speed of the hailstone. (b) Use Euler's method of numerical analysis and dynamic programming to find the velocity and position of the hailstone at 0.1-s intervals until the hailstone reaches 99% of the terminal speed. Plot the velocity as a function of time and also the position as a function of time. (c) Use **NDSolve** to do part (b) above—that is, to find the velocity and position of the hailstone at 0.1-s intervals until the hailstone reaches 99% of the terminal speed. (For information on Euler's method, see [Ste03, SJ04].)

17. **Nest**, **NestList**, **FixedPoint**, and **FixedPointList** apply repeatedly a function of one argument to an expression. **Fold** and **FoldList** work with functions of two arguments. **FoldList[***f*, *x*, {*a*, *b*, ...}**]** gives {*x*, *f*[*x*, *a*], *f*[*f*[*x*, *a*], *b*], ...} and **Fold[***f*, *x*, *list***]** yields the last element of **FoldList[***f*, *x*, *list***]**. Without using *Mathematica*, determine (a) **FoldList[Plus, 0,** *list***]**, (b) **FoldList[Times, 1,** *list***]**, and (c) **Fold[10 #1 + #2&, 0,** *digits***]**, where *list* and *digits* stand for an arbitrary list and any list of integers from 0 to 9, respectively. Verify your conclusions with *Mathematica*.

18. Implement the Newton–Raphson method for finding roots (more precisely, zeros) of functions. To determine a root of a function $f(x)$, start with an approximate value x_0 and obtain successively better approximations with the formula

$$x_{i+1} = x_i - f(x_i)/f'(x_i)$$

Determine the first five positive roots of the Bessel function of the first kind $J_0(x)$.

19. Consider the tent map given by the one-dimensional difference equation

$$x_{n+1} = f(x_n)$$

where $n = 0, 1, \ldots,$ and

$$f(x) = \mu\left(1 - 2\left|x - \frac{1}{2}\right|\right) = 2\mu \begin{cases} x & 0 \le x \le \frac{1}{2} \\ 1 - x & \frac{1}{2} \le x \le 1 \end{cases}$$

with $0 \le \mu \le 1$. (a) Verify that the point $x = 0$ is a stable fixed point when the map parameter μ is less than 1/2. (b) Show that there are two fixed points when the map parameter μ is greater than 1/2 and that neither point is stable or attracting.

20. Add rules for the built-in function **Sqrt** to simplify

```
Sqrt[x^2]
```

when **x** is declared real and either positive or negative.

21. (a) After entering the additional rules defined in Exercise 20 for the built-in function **Sqrt**, we have, for example,

```
Im[a] ^= 0;
Positive[a] ^= True;

Im[b] ^= 0;
Negative[b] ^= True;

{Sqrt[a^2], Sqrt[b^2]}
{a, -b}
```

Yet *Mathematica* does not simplify

$\{\sqrt{a^2}, \sqrt{b^2}\}$

as expected. Why? Explain. *Hint.* Use **FullForm**.

(b) Add the necessary rules for **Power** (*without* adding *any* rules for **Sqrt**) to simplify both

```
{Sqrt[a^2], Sqrt[b^2]}
```

and

$\{\sqrt{a^2}, \sqrt{b^2}\}$

22. **ComplexExpand[***expr***]** expands *expr* assuming that all variables are real. **ComplexExpand[***expr***, {***x*₁**,** *x*₂**, ...}]** expands *expr* assuming that variables matching any of the x_i are complex. Let **a** be declared real and positive and **b** be declared real and negative. Simplify the expressions

```
Abs[b]
```

$$\text{Abs}\left[\frac{3^{3/2}}{a^{2/3}} + \frac{a^2}{b^3}\, i\right]$$

$$\text{Abs}\left[\left(\frac{2\,a}{3} - 3\,b^2 a^{-7}\, i\right)\frac{b\,u^2}{v^{2/3}}\right]$$

$$\text{Abs}\left[\left(a^3 - \frac{b^2}{a^4}\, i\right)\frac{(u+a)^2}{(2+3\,a\,i)^5\, v^3}\right]$$

where **u** an **v** are complex variables. Compare the results with those in Section 3.3.6 and explain any discrepancies. *Hint:* Apply **ComplexExpand** to **Abs**[*expr*].

3.4 PROCEDURES

A procedure, or compound expression, is a sequence of expressions separated by semicolons. *Mathematica* evaluates the expressions consecutively and returns the result from evaluating the last expression. For instance,

```
In[1]:= r = 3; s = √4 + r; r^2 + s^2
Out[1]= 34
```

If a procedure ends with a semicolon, *Mathematica* returns **Null**, which is a symbol representing "nothing" and is not printed as an output. For example, define the function **func** to be a procedure ending with a semicolon:

```
In[2]:= func[y_] := func[y] = (Print["What a wonderful world!"];
          Print[Plot[Sin[x], {x, 0, 2Pi}]];
          x = y;)
```

Note that the parentheses are necessary for delimiting the body of the function and that the syntax for the definition allows the function to remember values that it finds, as explained in Section 3.3.3. Let us call the function

```
In[3]:= func[6]
        What a wonderful world!
```

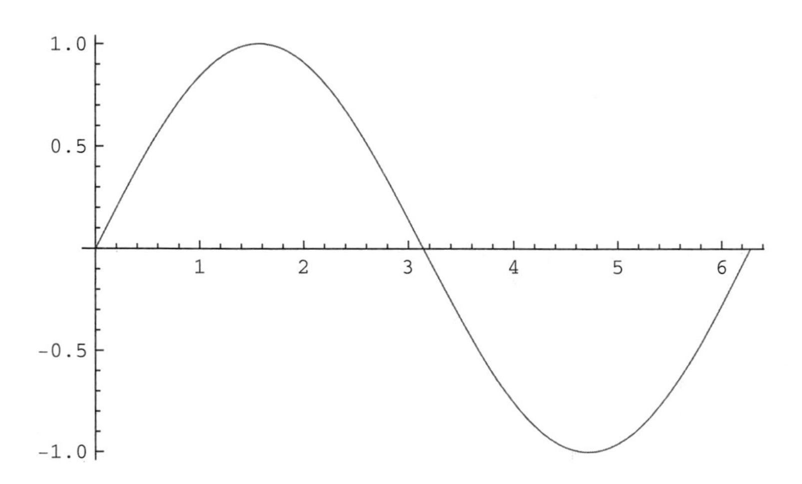

Mathematica displayed no output—that is, **Out[*n*]**. The value of **func[6]** is **Null**:

```
In[4]:= ?func
```

```
Global`func

func[6] = Null
func[y_]:= func[y] = (Print[What a wonderful world!];
   Print[Plot[Sin[x],{x, 0, 2π}]]; x=y;)
```

If the last semicolon is removed from the definition of **func**, the value of **func[6]** would be **6**. Printing an expression as well as a plot and assigning value to a variable are just side effects. To retrieve the value of the last element in the compound expression, use the operator **%*n***:

```
In[5]:= %%^2 + 3
Out[5]= 39
```

Often, an unfortunate side effect of procedures is the introduction of global variables such as **r**, **s**, and **x** with assigned values:

```
In[6]:= {r,s,x}
Out[6]= {3,5,6}
```

Section 3.4.1 will present a construct for avoiding global variables by declaring variables to be local to a particular procedure.

Mathematica normally evaluates the expressions in a procedure one after another. However, we can use conditional or iterative constructs to alter this sequential flow of control. Section 3.4.2 will examine conditionals for branching, and Section 3.4.3 will describe loops for iteration.

```
In[7]:= ClearAll[func,r,s,x]
```

3.4.1 Local Symbols

Sections 2.2.7 and 2.2.16 warned against conflicts of global symbols, which have assigned values or function definitions, with those of the same names elsewhere in a *Mathematica* session and exalted the merit of clearing them as soon as they are no longer needed. To minimize the collisions of symbols, we can often use transformation rules and pure functions instead of temporary variables and auxiliary functions. *Mathematica* supports another construct for avoiding symbol collisions. **Module**[{x_1, x_2, y_1 = y_{10}, x_3, y_2 = y_{20}, ...}, *body*] specifies that occurrences of the symbols x_i, y_i, ... in *body* should be treated as local and that y_i should be assigned the initial value y_{i0}. **Module** declares symbols as local to the module by giving them new and unique names within its body.

As an example, let us create a global variable **p** with the value **10**:

```
In[1]:= p = 10
Out[1]= 10
```

This variable has no effects on evaluations within **Module** if **p** is specified as a local variable within the module:

```
In[2]:= Module[{q, p}, Print[q]; q = p^2; Print[q]; p = 50; Print[p, "  ", q]]
        q$344
        p$344^2
        50   2500
```

Also, evaluations within **Module** have not affected the global variable **p**:

```
In[3]:= p
Out[3]= 10
```

Furthermore, the value of the local variable **q** is not visible outside the module:

```
In[4]:= q
Out[4]= q
```

The local variables **p** and **q** within **Module** are given, respectively, the names **p$***n* and **q$***n* to distinguish them from the corresponding global variables and variables in other modules. Thus, local variables are completely independent of the global variables and variables in other modules with identical names. The positive integer *n* in *symbol***$***n* increases each time **Module** is called in a *Mathematica* session. Thus, names of local variables are always unique to a particular module.

```
In[5]:= Module[{q, p}, Print[q]; q = p^2; Print[q]; p = 50; Print[p, "  ", q]]
        q$348
        p$348^2
        50   2500
```

Note that **p$344** and **q$344** have become **p$348** and **q$348**.

A few words of caution are in order. Even if variables are specified as local to a module, their global values will be captured if they appear in the expressions of initial values:

```
In[6]:= u = 5;

In[7]:= Module[{u = 120, v = u^2 + 1}, Print[v]]
        26

In[8]:= ClearAll[p,u]
```

3.4.2 Conditionals

A conditional function returns the value of one of several alternative expressions only if a condition is met. The conditional functions of *Mathematica* are **If**, **Which**, **Switch**, and **Piecewise**:

If[*condition*, *t*]	give *t* if *condition* evaluates to **True**, and **Null** if it evaluates to **False**
If[*condition*, *t*, *f*]	give *t* if *condition* evaluates to **True**, and *f* if it evaluates to **False**
If[*condition*, *t*, *f*, *u*]	give *u* if *condition* evaluates to neither **True** nor **False**
Which[$condition_1$, $value_1$, $condition_2$, $value_2$, ...]	evaluate each of the $condition_i$ in turn, returning the value of the $value_i$ corresponding to the first one that yields **True**
Switch[*expr*, $form_1$, $value_1$, $form_2$, $value_2$, ...]	evaluate *expr*, then compare it with each of the $form_i$ in turn, evaluating and returning the $value_i$ corresponding to the first match found
Piecewise[{{val_1, $cond_1$}, {val_2, $cond_2$}, ...}]	give the first val_i for which $cond_i$ is **True**
Piecewise[{{val_1, $cond_1$}, ...}, *val*]	give *val* if all $cond_i$ are **False**; the default for *val* is **0**

The relational and logical expressions described in Section 2.2.17 and the built-in predicates introduced in Sections 2.4.4 and 3.1.3.1 can serve as conditions for these conditional expressions.

Section 3.2.3.3 showed that we can put conditions on patterns, transformation rules, and definitions. The pattern object *pattern* /; *condition* stands for any expression matching *pattern* provided that the evaluation of *condition* yields **True**. The transformation rule *lhs* :> *rhs* /; *condition* and the definition *lhs* := *rhs* /; *condition* apply only if the evaluation of *condition* yields **True**.

To illustrate the use of conditional constructs, let us consider three examples.

Example 3.4.1 Write several functions for determining the maximum of a list of numbers.

We begin by defining the function **ourMax** for finding the maximum of a list of numbers to be a procedure containing the conditional function **If**:

```
In[1]:= ourMax[x_List] := Module[{maxNum = x[[1]]},
          Do[If[x[[i]] > maxNum, maxNum = x[[i]]], {i, 2, Length[x]}];
          maxNum]
```

The function **Do** was introduced in Section 2.1.19 and will be explained further in Section 3.4.3.2.

To extract the maximum, we can also define the recursive function **myMax** with a multiclause definition using patterns:

```
In[2]:= myMax[{x_,y_, z___}/;x>y] := myMax[{x, z}]
```

```
In[3]:= myMax[{x_,y_,z___}/;x≤y] := myMax[{y,z}]
```

```
In[4]:= myMax[{x_}] := x
```

Conditional constructs may not be necessary. Here is a simple function for determining the maximum of a list of numbers:

```
In[5]:= newMax[x_List] := Last[Sort[x]]
```

Of course, we can just use the built-in function **Max**.

Let us apply these functions to a list of random integers from 0 to 100 and of random length from 2 to 15:

```
In[6]:= RandomInteger[100,RandomInteger[{2,15}]]
Out[6]= {88,27,77,94,61,8}
```

```
In[7]:= {ourMax[%],myMax[%],newMax[%],Max[%]}
Out[7]= {94,94,94,94}
```

All these functions return the same result.

```
In[8]:= ClearAll[ourMax,myMax,newMax]
```

■

Example 3.4.2 Write several functions for the Kronecker delta δ_{nm} defined by

$$\delta_{nm} = \begin{cases} 1 & n = m \\ 0 & n \neq m \end{cases}$$

where n and m are integers.

We can write a single-clause definition with the built-in function **KroneckerDelta**:

```
In[9]:= ourkDel[n_Integer,m_Integer] := KroneckerDelta[n, m]
```

where **KroneckerDelta**[n_1, n_2, ...] gives the Kronecker delta $\delta_{n_1 n_2 \ldots}$ that is equal to 1 if all the n_i are equal and is equal to 0 otherwise.

We can write another single-clause definition with the function **Boole**:

```
In[10]:= mykDel[n_Integer,m_Integer] := Boole[n == m]
```

where **Boole**[*expr*] yields **1** if *expr* is **True** and **0** if it is **False**.

We can write yet another single-clause definition with the function **If**:

```
In[11]:= kDel[n_Integer,m_Integer] := If[n == m,1,0]
```

We can also give a multiclause definition:

```
In[12]:= newkDel[n_Integer,m_Integer] := 1/;n == m
```

```
In[13]:= newkDel[n_Integer,m_Integer] := 0/;n≠m
```

Often, multiclause definitions are easier to understand and modify than single-clause definitions.

As an example, let us apply these functions to a list of eight pairs of random integers from 0 to 3:

```
In[14]:= Table[{RandomInteger[3],RandomInteger[3]}, {8}]
Out[14]= {{1, 1}, {0, 0}, {1, 1}, {3, 2}, {3, 1}, {0, 2}, {3, 1}, {3, 2}}
```

```
In[15]:= {ourkDel @@#, mykDel @@#, kDel @@#, newkDel @@#}&/@%
Out[15]= {{1, 1, 1, 1}, {1, 1, 1, 1}, {1, 1, 1, 1}, {0, 0, 0, 0},
          {0, 0, 0, 0}, {0, 0, 0, 0}, {0, 0, 0, 0}, {0, 0, 0, 0}}
```

The functions give the same results.

```
In[16]:= ClearAll[ourkDel,mykDel,kDel,newkDel]
```

■

Example 3.4.3 Write several functions for the step potential

$$V(x) = \begin{cases} 0 & x < -3a \\ -2b & -3a \leq x < -a \\ -b & -a \leq x < a \\ -2b & a \leq x < 3a \\ 0 & 3a \leq x \end{cases}$$

Let $a = 1$ and $b = 1$.

The conditional function **Which** allows us to give a single-clause definition:

```
In[17]:= V[x_] := Which[x< -3||x≥3,0, -3≤x< -1||1≤x<3, -2, -1≤x < 1, -1]
```

In the definition, we have used the logical operator **||**. Let us plot this function:

```
In[18]:= Plot[V[x],{x, -5,5},
           PlotStyle→Thickness[0.01],AxesLabel→{" x","v"}]
```

Out[18]=

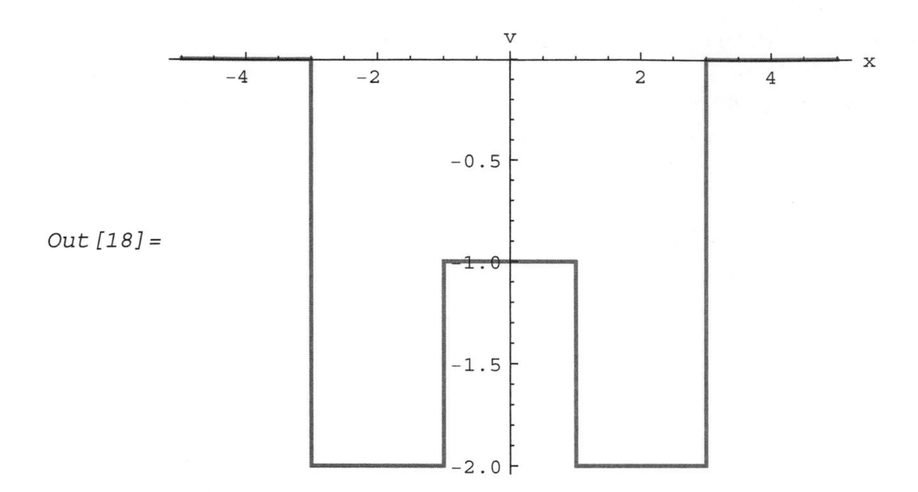

The function **Piecewise** allows us to give another single-clause definition:

```
In[19]:= myV[x_] := Piecewise[{{0,x < -3},
          {-2, -3 ≤ x <-1},{-1, -1≤x<1},{-2,1≤x<3},{0,3≤x}}]
```

Here is a graph of this function:

```
In[20]:= Plot[myV[x],{x, -5,5},
          PlotStyle→Thickness[0.01],AxesLabel→{" x","v"}]
```

Out[20]=

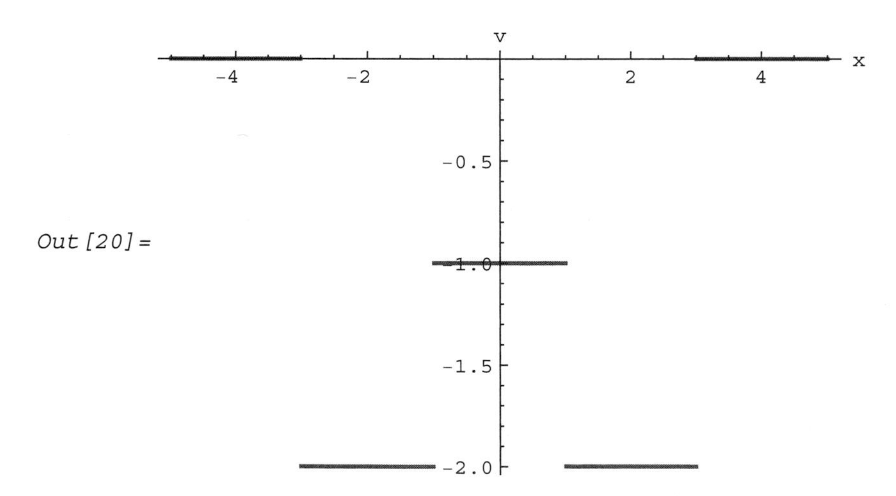

Without the vertical lines at the discontinuities, this is a more authentic graph of the function than the preceding one because the function has a definite value at each discontinuity. The preceding graph implies that the function has no definite values at the discontinuities. (We can add the vertical lines to this graph by including in the **Plot** function the option specification **Exclusions -> None**; we can remove the vertical lines in the preceding graph by including in the **Plot** function the option specification **Exclusions -> {-3, -1, 1, 3}**.)

We can also give a multiclause definition:

```
In[21]:= newV[x_/;x< -3||x≥3]:=0

In[22]:= newV[x_/;(x≥ -3&&x< -1)||(x≥1&&x<3)]:= -2

In[23]:= newV[x_/;x≥ -1&&x<1]:= -1
```

Plotting this function produces the same graph as the first one:

```
In[24]:= Plot[newV[x],{x, -5,5},
           PlotStyle→Thickness[0.01],AxesLabel→{" x","v"}]
```

Out[24]=

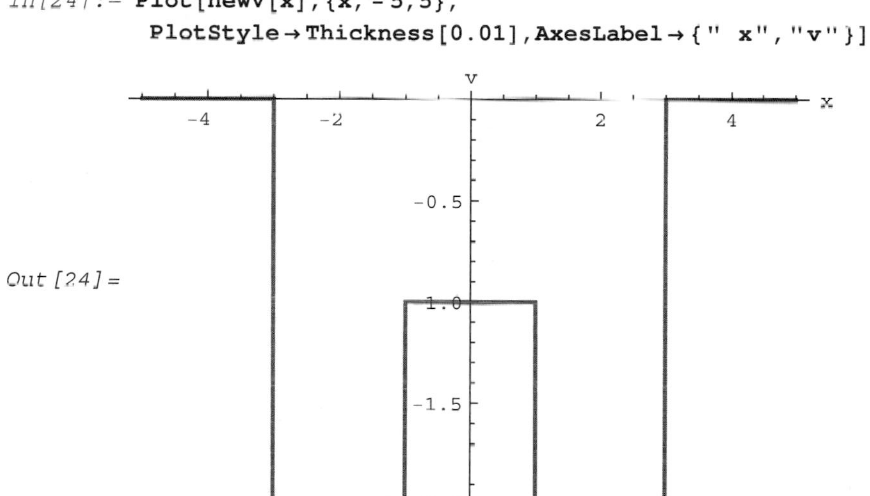

Whereas multiclause definitions are not differentiable, we can differentiate single-clause definitions involving conditional functions. For example,

```
In[25]:= ∂x newV[x]
Out[25]= newV′[x]

In[26]:= ∂x V[x]
Out[26]= Which[x< -3||x≥3, 0, -3 ≤x< -1||1≤x<3, 0, -1≤x<1, 0]

In[27]:= ∂x myV[x]

Out[27]= { 0                x< -3|| -3<x< -1|| -1<x<1||1<x<3||x>3
           Indeterminate    True
```

We must exercise caution in using this feature of conditional functions because *Mathematica* may return flawed derivatives. For instance,

```
In[28]:= {∂x myV[x],∂x V[x]}/.x→1
Out[28]= {Indeterminate,0}
```

Whereas the result for **myV** is correct, the result for **V** is wrong because the potential V is in fact not differentiable at 1.

```
In[29]:= ClearAll[V,myV,newV]
```

■

If an evaluated condition yields neither **True** nor **False** in a call to the function **Which** or **Piecewise**, *Mathematica* returns all or part of the conditional expression:

```
In[30]:= ClearAll[x,y]

In[31]:= x=2;

In[32]:= Which[x < 1, 1, y < 1, 2, x > 1, 3, y > 1, 4, True, "Wonderful!"]
Out[32]= Which[y < 1,2,x > 1,3,y > 1,4,True,Wonderful!]

In[33]:= Piecewise[
           {{1,x < 1},{2,y < 1},{3,x > 1},{4,y > 1},{"Wonderful!",True}}]

Out[33]= { 2   y<1
         { 3   True
```

The evaluation of **y < 1** yielded neither **True** nor **False**. Also, in StandardForm, **Piecewise[{{2, y < 1}, {3, True}}]** is output as

```
{ 2   y<1
{ 3   True
```

```
In[34]:= ClearAll[x]
```

When a function with more than one definition (i.e., with a multiclause definition) is called, *Mathematica*, in principle, starts with the most specific definition, proceeds from the more specific definitions to the more general ones, and uses the first definition that matches the call. As noted in Section 3.3.2, *Mathematica*'s notion of generality is not always scrutable. Consider the example cited in that section:

```
In[35]:= ClearAll[f]

In[36]:= f[x_Integer] := x^2

In[37]:= f[x_Integer?Positive] := 1/x

In[38]:= f[x_Integer/; x > 50] := -x
```

What would *Mathematica* return for `f[60]`? Since the last definition is the most specific, we expect `-60` to be the value of `f[60]`.

```
In[39]:= f[60]
Out[39]= 1/60
```

Contrary to expectation, *Mathematica* used the second definition. The function `Switch` allows us to give a single-clause definition in which the patterns are always tested in the specified order:

```
In[40]:= ClearAll[f]

In[41]:= f[y_] := Switch[y, x_Integer /; x > 50,
              -y, x_Integer?Positive, 1/y, x_Integer, y^2]

In[42]:= f[60]
Out[42]= -60
```

There is no ambiguity this time. Since `60` matches the first pattern, *Mathematica* returns `-60` for `f[60]`, as expected.

```
In[43]:= ClearAll[f]
```

3.4.3 Loops

3.4.3.1 Changing Values of Variables

In using looping constructs, we often wish to change the values of variables. *Mathematica* provides special notations for such modifications:

x++	increase the value of *x* by 1, returning the old value of *x*
++*x*	increase the value of *x* by 1, returning the new value of *x*
x--	decrease the value of *x* by 1, returning the old value of *x*
--*x*	decrease the value of *x* by 1, returning the new value of *x*
x += *dx*	add *dx* to *x* and return the new value of *x*
x -= *dx*	subtract *dx* from *x* and return the new value of *x*
x *= *c*	multiply *x* by *c* and return the new value of *x*
x /= *c*	divide *x* by *c* and return the new value of *x*
{*x*, *y*} = {*y*, *x*}	interchange the values of *x* and *y*
PrependTo[*s*, *elem*]	prepend *elem* to the value of *s*, and reset *s* to the result
AppendTo[*s*, *elem*]	append *elem* to the value of *s*, and reset *s* to the result

For example, consider

```
In[1]:= x = 3b
Out[1]= 3b
```

```
In[2]:= x *= 4 b^2
Out[2]= 12 b^3
```

```
In[3]:= x
Out[3]= 12 b^3
```

Note that prior to the modification of a variable, it must already have an assigned value.

After increasing the value of x by 1, `++`x returns the new value of x, whereas x`++` yields the old one. For instance,

```
In[4]:= x = 3
Out[4]= 3
```

```
In[5]:= ++x
Out[5]= 4
```

```
In[6]:= x
Out[6]= 4
```

```
In[7]:= x++
Out[7]= 4
```

```
In[8]:= x
Out[8]= 5
```

The functions **PrependTo** and **AppendTo** modify lists. For example, consider

```
In[9]:= s = {1,2,3,4}
Out[9]= {1,2,3,4}
```

```
In[10]:= AppendTo[s,r]
Out[10]= {1,2,3,4,r}
```

```
In[11]:= s
Out[11]= {1,2,3,4,r}
```

```
In[12]:= ClearAll[x,s]
```

3.4.3.2 Do, While, and For

The functions **Do**, **While**, and **For** evaluate expressions or procedures repeatedly. The value returned by these functions is **Null**, which, as mentioned previously, is simply a symbol indicating the absence of an expression or a result and is not printed as an output. These functions are used only for their side effects, such as assigning values to variables and printing expressions as well as graphics.

The function **Do** evaluates an expression or a procedure repetitively a specified number of times. The iterative constructs are

`Do[`$expr$`, {`i_{max}`}]`	evaluate $expr$ i_{max} times
`Do[`$expr$`, {`i, i_{max}`}]`	evaluate $expr$ with the variable i successively taking on the values 1 through i_{max} (in steps of 1)
`Do[`$expr$`, {`i, i_{min}, i_{max}`}]`	start with $i = i_{min}$
`Do[`$expr$`, {`i, i_{min}, i_{max}, di`}]`	use steps di
`Do[`$expr$`, {`i, i_{min}, i_{max}`}, {`j, j_{min}, j_{max}`}, ...]`	evaluate $expr$ looping over different values of j, etc. for each i

Do has no output because it always returns **Null**. For instance, consider

```
In[13]:= ClearAll[f]
```

```
In[14]:= f[x_] := x^2
```

```
In[15]:= Do[f[i],{i,3}]
```

To display **f[i]** with **i = 1**, **2**, and **3**, we include the function **Print**:

```
In[16]:= Do[Print[f[i]],{i,3}]
          1
          4
          9
```

The number of times **Do**[$expr$, {i, $imin$, $imax$, di}] evaluates $expr$ is

$$\frac{i_{max} - i_{min}}{di} + 1$$

rounded off to the nearest smaller integer. Consider determining

$$\sum_{n=1}^{\infty} \frac{1}{n^k}$$

for $k = 2, 4, 6$, and 8.

```
In[17]:= Do[Print["k = ", k, "  ",
            Sum[1/(n^k), {n, 1, Infinity}]//N],{k, 2, 8, 2}]
          k = 2    1.64493
          k = 4    1.08232
          k = 6    1.01734
          k = 8    1.00408
```

The number of iterations is $(8 - 2)/2 + 1 = 4$. We can obtain the same result with, for example, the iterator **{k, 2, 9.3, 2}**.

Instead of constructing a **Do** loop, we can often use the function **Nest** discussed in Section 3.3.4. Example 3.3.3 introduced the logistic map defined by the equation

$$x_{n+1} = f(x_n)$$

where $n = 0, 1, \ldots,$ and

$$f(x) = \mu x(1 - x)$$

with $0 \leq x \leq 1$ and $0 \leq \mu \leq 4$. With a **Do** loop, we can determine x_{20} for the map parameter $\mu = 3.64$ and the initial point $x_0 = 0.500$:

```
In[18]:= x = 0.500; Do[x = 3.64 x (1 - x), {20}]; x
Out[18]= 0.305517
```

We can also compute x_{20} with the function **Nest**:

```
In[19]:= Nest[(3.64 # (1 - #)) &, 0.500, 20]
Out[19]= 0.305517
```

Nest is perhaps more polished than **Do**.

```
In[20]:= ClearAll[f, x]
```

The functions **While** and **For** evaluate expressions or procedures repeatedly as long as certain conditions are satisfied. The looping constructs are

While[*condition*, *body*]	evaluate *condition*, then *body*, repetitively, until *condition* first fails to give **True**
For[*start*, *condition*, *incr*, *body*]	execute *start*, then repeatedly evaluate *body* and *incr* until *condition* fails to give **True**

To illustrate the use of **While**, determine x such that $x = 1 + 1/x$:

```
In[21]:= x = 1.0;
         While[1 + 1/x =!= x,  x = 1 + 1/x];
         x
Out[23]= 1.61803
```

where **UnsameQ**[*lhs*, *rhs*], with the special input form *lhs* **=!=** *rhs*, yields **True** if the expression *lhs* is not identical to *rhs*, and it yields **False** otherwise. The number $(1 + \sqrt{5})/2 \approx 1.61803$ is called the golden ratio. Using the function **FixedPoint** introduced in Section 3.3.4 to determine the golden ratio is perhaps more elegant than using the looping construct:

```
In[24]:= FixedPoint[(1 + 1/#) &, 1.0]
Out[24]= 1.61803
```

Of course, this result is in agreement with the approximate numerical value of the built-in mathematical constant **GoldenRatio**:

```
In[25]:= N[GoldenRatio] == % == %%
Out[25]= True
```

```
In[26]:= ClearAll[x]
```

As an example on using **For**, generate a list of the prime numbers less than 40:

```
In[27]:= For[{n, v} = {1, {}}, Prime[n] < 40, n++, AppendTo[v, Prime[n]]];
         v
Out[28]= {2,3,5,7,11,13,17,19,23,29,31,37}
```

Prime[*n*] gives the *n*th prime number, the special form *n*++ increments the value of *n* by 1, and **AppendTo**[*v*, *elem*] is equivalent to *v*=**Append**[*v*, *elem*].

```
In[29]:= ClearAll[n, v]
```

After executing *start*, the order of evaluation in a **For** loop is *condition*, *body*, and *incr*. For instance,

```
In[30]:= For[n = 1, n < 3, n++, Print[n]]
         1
         2
```

Here, **n** is first set to **1**. The test **n < 3** gives **True**, **1** is printed, and **n** is increased to **2**; **n < 3** again yields **True**, **2** is printed, and **n** is increased to **3**; **n < 3** now gives **False** and the loop stops. A side effect is that **n** retains the value **3**:

```
In[31]:= n
Out[31]= 3
```

For both **While** and **For**, *body* may not be evaluated at all if *condition* yields **False** at the outset.

```
In[32]:= For[n = 1, n < 1, n++, Print[n]]
```

Here, **n** is first set to **1**, **n < 1** gives **False**, the loop terminates immediately, and **n** still has the value **1**:

```
In[33]:= n
Out[33]= 1
```

```
In[34]:= ClearAll[n]
```

3.4.4 Named Optional Arguments

Mathematica functions allow two kinds of arguments: positional arguments and named optional arguments or, simply, options. Whereas the meanings of the positional arguments are determined by their positions in the argument sequence, those of the named optional arguments are specified by their names.

Built-in *Mathematica* plotting functions, for example, accept many options that, as we have seen, are expressed as transformation rules and can be entered in any order after the positional arguments in the argument sequences of the functions. To illustrate how to set up options in user-defined functions, consider a problem from introductory electrostatics: For points on a plane, write a function that returns the electric potential and electric field produced by a system of point charges confined to the plane.

At position $\mathbf{r}$, the electric potential due to a system of N point charges $q_1, q_2, \ldots, q_N$ at positions $\mathbf{r}_1, \mathbf{r}_2, \ldots, \mathbf{r}_N$, respectively, can be written as

$$\varphi(\mathbf{r}) = k \sum_{i=1}^{N} \frac{q_i}{|\mathbf{r} - \mathbf{r}_i|} \tag{3.4.1}$$

and the electric field is given by

$$\mathbf{E}(\mathbf{r}) = -\nabla\varphi(\mathbf{r}) \tag{3.4.2}$$

where k is the Coulomb constant. In the SI system of units, $k = 8.98755 \times 10^9$ volt meter/coulomb and the units of charge, electric potential, and electric field are coulomb, volt, and volt/meter, respectively. In the Gaussian system, $k = 1$ and the units of charge, electric potential, and electric field are statcoulomb (or esu), statvolt, and statvolt/centimeter, respectively. In a two-dimensional Cartesian coordinate system, Equations 3.4.1 and 3.4.2 imply

$$\varphi(x, y) = k \sum_{i=1}^{N} \frac{q_i}{\sqrt{(x - x_i)^2 + (y - y_i)^2}} \tag{3.4.3}$$

$$E_x = -\frac{\partial\varphi}{\partial x} \tag{3.4.4}$$

$$E_y = -\frac{\partial\varphi}{\partial y} \tag{3.4.5}$$

where E_x and E_y are the x and y components of the electric field, respectively.

For determining the electric potential and electric field at a point (x, y), the function **`electrostaticField`** takes two positional arguments. The first argument is for a list of lists, each of which has three elements: the charge, the x coordinate, and the y coordinate of a charge in the charge distribution producing the field. The pattern *patt*`..` is a pattern repeated one or more times. The pattern *q*`:{{_, _, _}..}` stands for a list of trios, and this list is to be named *q* in the body of the function. The second argument is for a list of x and y coordinates of the point at which the electric potential and electric field are evaluated. The pattern *patt*`:`*v* or `Optional[`*patt*`,`*v*`]`, introduced in Section 3.2.6, represents an expression that, if omitted, has default value *v*. The pattern `Optional[`*p*`:{_,_},{0,0}]` matches any two-element list to be named *p* on the *rhs* of the function definition; if we do not explicitly give a second positional argument to the function **electrostaticField** when it is called, *p* assumes the default value **{0,0}**. That is, unless otherwise specified, the origin is taken as the point at which the electric potential and electric field are determined. With the inclusion of **opts__Rule**, options may be given as trailing arguments for specifying the system of units and the unit for the angle that the electric field makes counterclockwise from the +*x*-axis. The value for the option **systemOfUnits** can be either **SI** or **Gaussian**, and that for the option **angle** can be either **radian** or **degree**. Comments on the function are included in the body of the function.

With the function **Options**, we first name the options and specify their default values. **Options[*f*]** gives the list of default options assigned to *f*.

```
In[1]:= Options[electrostaticField] = {systemOfUnits→SI,angle→radian};
```

Let us present the definition for the function and then analyze how options work.

```
electrostaticField[q:{{_,_,_}..},
  Optional[p:{_,_},{0,0}],opts___Rule] :=
 Module[
  {(* declare local variables *)
   phi,x,y,ex,ey,e,electricPotential,electricField,theta,

   (* determine option values and
    assign them to local variables as initial values *)
    units = systemOfUnits /.{opts}/.Options[electrostaticField],
    dir = angle /.{opts} /. Options[electrostaticField]},

   (* from Equation 3.4.3; electric potential at a point
    (x,y) in units of (k * unit of charge/unit of length) *)
   phi = Sum[q[[i,1]]/Sqrt[(x-q[[i,2]])^2 + (y-q[[i,3]])^2],
    {i,1,Length[q]}];

   (* from Equations 3.4.4 and 3.4.5;
   x and y components of the electric field at the point
    p in units of (k*unit of charge/unit of length^2) *)
   ex = N[-D[phi,x]/.{x→p[[1]],y→p[[2]]}];
   ey = N[-D[phi,y]/.{x→p[[1]],y→p[[2]]}];

   (* magnitude of the electric field at the point
    p in units of (k*unit of charge/unit of length^2) *)
   e = Sqrt[ex^2 + ey^2];

   (* the electric potential at the point
    p in units of (k*unit of charge/unit of length) *)
   phi = N[phi/.{x→p[[1]],y→p[[2]]}];

   (* determine the specified system of units,
   electric potential,and magnitude of the electric field *)
   Which[
    units === SI,
```

```
electricPotential = ((phi * 8.98755 * 10^9 volts) // ScientificForm);
electricField = ((e * 8.98755 * 10^9 volts/meter) // ScientificForm),
units === Gaussian,
electricPotential = ((phi statvolts) //ScientificForm);
electricField = ((e statvolts/centimeter) //ScientificForm),True,
Return["The value of the option systemOfUnits
        must be either SI or Gaussian."]
    ];

(* print the values of the electric
 potential and magnitude of the electric field *)
Print["electric potential = ",electricPotential, "\n\n"];
Print["magnitude of electric field = ",electricField,"\n\n\n"];

(* determine the direction that the electric
field makes counterclockwise from the +x axis *)
If[
 (ex == 0) && (ey == 0),
 Return[],
 theta = ArcTan[Abs[ey/ex]]
  ];
Which[
 (ex > 0) && (ey ≥ 0) || (ex == 0) && (ey > 0),Null,
 (ex < 0) && (ey ≥ 0),theta = (Pi - theta),
 (ex ≤ 0) && (ey < 0),theta = (Pi + theta),
 (ex > 0) && (ey < 0),theta = (2 Pi - theta),
 True,Return["Determination of angle failed!"]
    ];

(* assign appropriate unit to the angle *)
Which[
 dir === radian,theta = (N[theta] rad),
 dir === degree,theta = (N[theta/Degree] deg),
 True,Return["The value of the option angle
          must be either radian or degree."]
   ];
(* print the value for the angle *)
Print["angle of electric field from +x axis = ",theta];
   ]
```

The function **Return** appears at several places in the function definition. **Return[***expr***]** returns the value *expr* from a function and exits all procedures and loops in the function; **Return[]** returns the value **Null**, which is not printed, and exits the function.

We can embellish the previous definition with special characters and two-dimensional forms. To proceed safely and efficiently, paste a copy of the definition below, click the bracket of the new cell, choose **StandardForm** in the Convert To submenu of the Cell menu, and edit the result. An adorned definition for the function **electrostaticField** is

```
In[2]:= electrostaticField[q:{{_,_,_}..},p:{_,_}:{0,0},opts___Rule]:=
   Module[
   {(* declare local variables *)
    φ,x,y,E,ℰ,electricPotential,electricField,θ,

   (* determine option values and
      assign them to local variables as initial values *)
     units = systemOfUnits/.{opts}/.Options[electrostaticField],
     dir = angle/.{opts}/.Options[electrostaticField]},

   (* from Equation 3.4.3; electric potential at a point
    (x,y) in units of (k × unit of charge/unit of length) *)
```

$$\varphi = \sum_{i=1}^{\text{Length}[q]} \frac{q[\![i,1]\!]}{\sqrt{(x-q[\![i,2]\!])^2+(y-q[\![i,3]\!])^2}};$$

```
   (* from Equations 3.4.4 and 3.4.5;
   x and y components of the electric field at the point p
    in units of (k × unit of charge/(unit of length)^2) *)
```

$E_x = N[-\partial_x\varphi /. \{x \to p[\![1]\!], y \to p[\![2]\!]\}];$
$E_y = N[-\partial_y\varphi /. \{x \to p[\![1]\!], y \to p[\![2]\!]\}];$

```
   (* magnitude of the electric field at the point p
    in units of (k × unit of charge/(unit of length)^2) *)
```

$\mathcal{E} = \sqrt{E_x{}^2 + E_y{}^2};$

```
   (* the electric potential at the point p
    in units of (k × unit of charge/unit of length) *)
   φ = N[φ/.{x→p[[1]],y→p[[2]]}];

   (* determine the specified system of units,
   electric potential,and magnitude of the electric field *)
   Which[
    units === SI,
    electricPotential = ScientificForm[φ 8.98755 10^9 volts];
```

$\text{electricField} = \text{ScientificForm}\left[\frac{\mathcal{E}\ 8.98755\ 10^9\ \text{volts}}{\text{meter}}\right],$

```
    units === Gaussian,
    electricPotential = ScientificForm[φ statvolts];
```

$\text{electricField} = \text{ScientificForm}\left[\frac{\mathcal{E}\ \text{statvolts}}{\text{centimeter}}\right],$

```
    True,
    Return["The value of the option
       systemOfUnits must be either SI or Gaussian."]
```

```
        ];
    (* print the values of the electric
     potential and magnitude of the electric field *)
    Print["electric potential = ", electricPotential, "\n\n"];
    Print["magnitude of electric field = ", electricField, "\n\n\n"];

    (* determine the direction that the electric
      field makes counterclockwise from the +x axis *)
    If[
     Ex == 0 ∧ Ey == 0,
     Return[],
     θ = ArcTan[Abs[Ey/Ex]]
      ];
    Which[
      (Ex > 0 ∧ Ey ≥ 0) ∨ (Ex == 0 ∧ Ey > 0), Null,
      (Ex < 0) ∧ (Ey ≥ 0), θ = π - θ,
      (Ex ≤ 0) ∧ (Ey < 0), θ = π + θ,
      (Ex > 0) ∧ (Ey < 0), θ = 2π - θ,
      True, Return["Determination of angle failed!"]
         ];

    (* assign appropriate unit to the angle *)
    Which[
     dir === radian, θ = N[θ] rad,
     dir === degree, θ = N[θ/°] deg,
     True, Return["The value of the option
        angle must be either radian or degree."]
        ];

    (* print the value for the angle *)
    Print["angle of electric field from +x axis = ", θ];
       ]
```

where φ, $\mathbf{E_x}$, $\mathbf{E_y}$, $\mathcal{E}$, and θ have replaced **phi**, **ex**, **ey**, **e**, and **theta**, respectively, and we have reentered all the comments that were lost in the conversion. Beware that the Greek letter **E** used here appears similar to, but is in fact different from, the ordinary keyboard letter **E**, which represents the exponential constant in *Mathematica*. Furthermore, **E**, rather than $\mathbf{E_x}$ and $\mathbf{E_y}$, is included in the list of local variables, which must be symbols. Whereas **E** is a symbol, E_x and E_y are not because their full forms are **Subscript[E,x]** and **Subscript[E,y]**, respectively. (Could we have replaced **e** with the Greek letter **E** instead of the script letter $\mathcal{E}$?)

Let us examine how *Mathematica* extracts the option values. With **Options**, we name the options and specify their default values:

```
Options[electrostaticField] = {systemOfUnits → SI, angle → radian};
```

The option **systemOfUnits** allows the choice between the SI and Gaussian systems of units, and the option **angle** permits the selection between radian and degree for the unit of angle. The default settings are **SI** and **radian** for **systemOfUnits** and **angle**, respectively. When the function **electrostaticField** is called, *Mathematica* evaluates

```
systemOfUnits/.{opts}/.Options[electrostaticField]
```

and

```
angle/.{opts}/.Options[electrostaticField]
```

to determine the option values. Consider, for example, the evaluation of

```
systemOfUnits/.{opts}/.Options[electrostaticField]
```

The replacement operator **/.** associates to the left. That is, the rules in **{opts}** are tried before those in **Options[electrostaticField]**. If the rule **systemOfUnits -> *optionValue*** appears explicitly in the actual argument sequence of the function **electrostaticField** when it is called, this rule is passed to **{opts}**, and **systemOfUnits/.{opts}** becomes *optionValue* after the replacement. The rules in **Options[electrostaticField]** are not used because *optionValue* does not match the *lhs* of any of the rules. Thus, **systemOfUnits/.{opts}/. Options[electrostaticField]** is evaluated to *optionValue*. If a rule for **systemOfUnits** has not been specified explicitly in the actual argument sequence of the function **electrostaticField**—that is, in **{opts}**—**systemOfUnits/.{opts}** simply becomes **systemOfUnits**. The rule **systemOfUnits → SI** in **Options[electrostaticField]** is then applied, and the entire expression **systemOfUnits/.{opts}/.Options[electrostaticField]** is evaluated to **SI**. In both cases, the evaluation results in an appropriate option value that is then assigned as an initial value to the local variable **units**. Similarly, proper option value is determined and assigned as an initial value to the local variable **dir**. Instead of recycling the symbols **systemOfUnits** and **angle**, we assign the option values to the new symbols **units** and **dir** because assigning values to symbols with the option names can be perilous. (See Exercises 9 and 10 of Section 3.4.6.) It is not necessary to assign the option values as initial values to the local variables **units** and **dir**. We can make the assignments to these local variables later in the execution of the function after the local variables declaration. If we choose to assign the option values as initial values in the manner indicated in the local variable list of **Module**, we should remember that initial values are always evaluated before the module is executed. That is, the global values of local variables are captured if they appear in the expressions of initial values, as explained in Section 3.4.1.

To illustrate the use of the function **electrostaticField**, consider the charge distribution: A charge of +500 esu is located at $(-10, 30)$ and another charge of -200 esu is located at $(30, 20)$, where the coordinates are in centimeters. Let us determine the electric potential and the electric field at the origin:

```
In[3]:= electrostaticField[{{500, -10, 30}, {-200, 30, 20}},
          systemOfUnits → Gaussian, angle → degree]
        electric potential = (1.02644 × 10^1) statvolts
                                      (4.82897 × 10^-1) statvolts
        magnitude of electric field = ---------------------------
                                              centimeter
        angle of electric field from +x axis = 306.335 deg
```

For another example, consider a column of 101 equally spaced point charges, each of 2 μC, distributed along the y-axis from $y = -50$ to $+50$, where lengths are in meters. To determine the electric potential and electric field at the point (100, 0), call **electrostaticField**:

```
In[4]:= electrostaticField[Table[{2 × 10^-6, 0, i}, {i, -50, 50}], {100, 0}]
       electric potential = (1.74603×10^4) volts
                                     (1.62058 × 10^2) volts
       magnitude of electric field = ----------------------
                                            meter
       angle of electric field from +x axis = 0. rad
```

Now consider a uniform line charge of length b placed along the y-axis, as in Figure 3.4.1, where the center of the line charge is at the origin.

The electric potential at a point P on the x-axis at a distance d from the origin is given by

$$\varphi = \frac{2kQ}{b} \ln \left(\frac{(b/2) + \sqrt{(b/2)^2 + d^2}}{d} \right) \tag{3.4.6}$$

where Q is the total charge. The magnitude of the electric field at point P can be expressed as

$$E = \frac{2kQ}{d} \frac{1}{\sqrt{b^2 + 4d^2}} \tag{3.4.7}$$

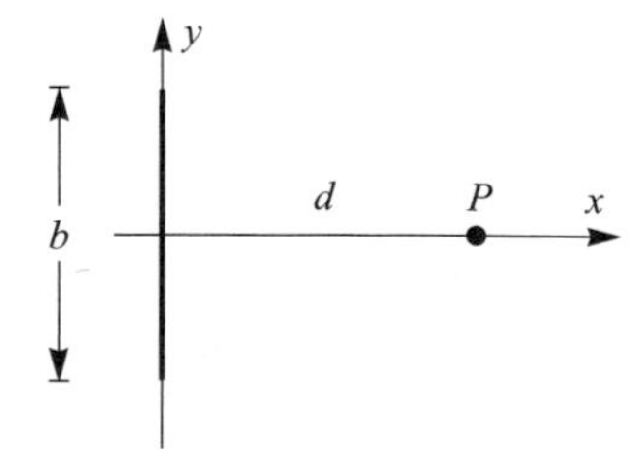

Figure 3.4.1. A uniform line charge of length b positioned along the y-axis.

and the electric field is in the positive x direction. (For derivations of Equations 3.4.6 and 3.4.7, see [HRK02].) For $Q = 101 \times 2\,\mu C$, $b = 100\,\text{m}$, $d = 100\,\text{m}$, Equations 3.4.6 and 3.4.7 give

$$\varphi = 1.74727\,10^4 \text{ volts}$$

$$E = 1.62382\,10^2 \frac{\text{volts}}{\text{meter}}$$

Comparing these results with those obtained previously for the column of point charges suggests that a column of closely spaced (i.e., distance between two charges $\ll d$) equal point charges approximates rather well a continuous uniform line charge. The difference between the values for the electric potential is less than 0.1%, and that for the electric field is less than 0.2%.

The function **SetOptions** allows us to change the default values for options. **SetOptions**[*f*, *opt*$_1$ -> *value*$_1$, *opt*$_2$ -> *value*$_2$, ...] resets the specified default options for f. Let **Gaussian** be the default value for **systemOfUnits**:

```
In[5]:= SetOptions[electrostaticField, systemOfUnits→Gaussian];
```

Consider again the previous example with a distribution of two point charges: A charge of +500 esu at $(-10, 30)$ and another charge of -200 esu at $(30, 20)$, where the coordinates are in centimeters. In calling **electrostaticField** to determine the electric potential and electric field at the origin, we do not need to specify the **systemOfUnits** option explicitly anymore:

```
In[6]:= electrostaticField[{{500, -10, 30}, {-200, 30, 20}}, angle→degree]
```

electric potential = (1.02644×10^{1}) statvolts

magnitude of electric field = $\dfrac{(4.82897 \times 10^{-1})\ \text{statvolts}}{\text{centimeter}}$

angle of electric field from +x axis = 306.335 deg

In setting up the function **electrostaticField**, we made the assignment

```
Options[electrostaticField] = {systemOfUnits→ SI, angle→ radian};
```

before defining the function. If this expression is located within the body of the function, we cannot assign the option values as initial values to local variables. Furthermore, we cannot use **SetOptions** to change the default options because every call to **electrostaticField** returns immediately the default options to their original settings and overrides the effects of **SetOptions**.

```
In[7]:= ClearAll[electrostaticField]
```

3.4.5 An Example: Motion of a Particle in One Dimension

The equation of motion of a particle with mass m moving along the x-axis can be written as

$$x''(t) = \frac{1}{m} F(x(t), x'(t), t) \tag{3.4.8}$$

where $F(x(t), x'(t), t)$ is the resultant force acting on the particle and $x(t)$ is its position at time t. This equation simply states that the acceleration of the particle equals the net force per unit mass.

To illustrate the use of several constructs presented previously in this section, we write a function for plotting the position, velocity, and acceleration versus time of the particle and for displaying these plots together. The function takes four arguments: the net force per unit mass, the initial position, the initial velocity, and the time interval from $t = 0$. Options can be entered as trailing arguments for specifying the plots to be omitted and the labels to be placed at the ends of the axes or on the edges of the frame of each plot. Two-dimensional graphics options that affect all the plots can also be included for passing to the functions **Plot** and **Show** with the exceptions **PlotLabel**, **AxesLabel**, **Ticks**, **FrameLabel**, and **FrameTicks**. With the specifications **velocityPlot -> False** and **combinationPlot -> False**, for example, the function does not produce the velocity versus time plot nor display together the plots that are generated. With the inclusion of **accelerationAFLabel -> {***xlabel***,** *ylabel***}**, for instance, the function places the specified labels at the ends of the x- and y-axes of the acceleration versus time plot if the option value of **Axes** is **True** and on the bottom and left-hand edges of the frame around the plot if the option value of **Frame** is **True**; the default settings of **Axes** and **Frame** are **False** and **True**, respectively. The description of the function is embedded in its body.

```
In[1]:= (* name the options and specify their default values *)
        Options[motion1DPlot] =
          {positionPlot → True,
           velocityPlot → True,
           accelerationPlot → True,
           combinationPlot → True,
           positionAFLabel → { "t (s)", "x (m)"},
           velocityAFLabel → { "t (s)", "v (m/s)"},
           accelerationAFLabel → { "t (s)", "a (m/s²)"},
           combinationAFLabel → { "t (s)", None}};

In[2]:= motion1DPlot[a_, x0_, v0_, tmax_, opts___Rule] :=
          Module[
           {(* declare local variables *)
            sol, curves = {}, plotx, plotv, plota,

            (* determine option values and
             assign them as initial values to local variables *)
            position = positionPlot/.{opts}/.Options[motion1DPlot],
            velocity = velocityPlot/.{opts}/.Options[motion1DPlot],
            acceleration = accelerationPlot/.{opts}/.Options[motion1DPlot],
            combination = combinationPlot/.{opts}/.Options[motion1DPlot],
            positionLabel = positionAFLabel/.{opts}/.Options[motion1DPlot],
            velocityLabel = velocityAFLabel/.{opts}/.Options[motion1DPlot],
            accelerationLabel = accelerationAFLabel/.{opts}/.
              Options[motion1DPlot], combinationLabel =
             combinationAFLabel/.{opts}/.Options[motion1DPlot],
```

```
(* select valid options for Plot and Show and
 assign them as initial values to local variables *)
optPlot = Sequence @@FilterRules[{opts},Options[Plot]],
optShow = Sequence @@FilterRules[{opts},Options[Graphics]]},

(* set text of a warning message *)
motion1DPlot::argopt =
 "Each of the values for the options positionPlot,
   velocityPlot, accelerationPlot, and
   combinationPlot must be either True or False.";

(* verify option specifications *)
If[Count[{position,velocity,
    acceleration,combination},True|False] =!= 4,
 Message[motion1DPlot::argopt];Return[$Failed]];

(* solve the equation of motion numerically *)
sol = NDSolve[{x''[t] == a,x[0] == x0,x'[0] == v0},x,{t,0,tmax}];

(* plot position vs. time *)
If[position,
 plotx = Plot[Evaluate[x[t]/.sol],{t,0,tmax},
   PlotLabel→ "position vs.time",AxesLabel→ positionLabel,
   Ticks→ Automatic,FrameLabel→ positionLabel,
   FrameTicks→ Automatic, Evaluate[optPlot],
   PlotRange→ All, Axes→ False, Frame→ True];
 Print[plotx];
 AppendTo[curves,plotx]];

(* plot velocity vs. time *)
If[velocity,
 plotv = Plot[Evaluate[x'[t]/.sol],{t,0,tmax},
   PlotLabel→ "velocity vs. time", AxesLabel→ velocityLabel,
   Ticks→ Automatic,FrameLabel→ velocityLabel, FrameTicks→
    Automatic, Evaluate[optPlot],PlotStyle→ Dashing[{0.03,0.03}],
   PlotRange→ All, Axes→ False,Frame→ True];
 Print[plotv];
 AppendTo[curves,plotv]];

(* plot acceleration vs. time *)
If[acceleration,
 plota = Plot[Evaluate[a/.sol],{t,0,tmax},
   PlotLabel→ "acceleration vs. time",
   AxesLabel→ accelerationLabel, Ticks→ Automatic,
   FrameLabel→ accelerationLabel, FrameTicks→ Automatic,
   Evaluate[optPlot], PlotStyle→ RGBColor[1,0,0],
   PlotRange→ All, Axes→ False, Frame→ True];
```

```
    Print[plota];
    AppendTo[curves,plota]];

   (* combine the plots *)
   If[(combination)&&(Length[curves]>1),
    Show[curves,
     PlotLabel→ "combination", AxesLabel→ combinationLabel,
     Ticks→{Automatic,None},FrameLabel→ combinationLabel,
     FrameTicks→{Automatic, None}, optShow]]]
```

In verifying option specifications, **motion1DPlot** uses the function **Message** to print a warning message if any of the options **position**, **velocity**, **acceleration**, and **combination** have values other than **True** or **False**. **Message**[*symbol*::*tag*] prints the message previously assigned to *symbol*::*tag*.

For passing options to **Plot** and **Show**, **motion1DPlot** employs the function **FilterRules** because **Plot** and **Show** do not accept user-defined options such as **positionPlot** and **velocityAFLabel**. **Sequence@@FilterRules[{**opt_1**,**opt_2**,...}, Options[**f**]]** returns a sequence of options from $opt_1, opt_2, \ldots$ that are valid options for f. We assign the sequence of specified options that are valid for **Plot** as an initial value to the local variable **optPlot**, and the sequence of specified options that are valid for **Show** as an initial value to **optShow**.

To illustrate the use of **motion1DPlot**, let us consider several examples. For the first example, a pebble is thrown vertically upward with an initial speed of 25 m/s. If air resistance is negligible, the force acting on the pebble is

$$F = -mg$$

where g is the magnitude of the acceleration due to gravity and the positive x direction is upward. The function **motion1DPlot** generates the graphs:

```
In[3]:= motion1DPlot[-9.80,0,25,5.1]
```

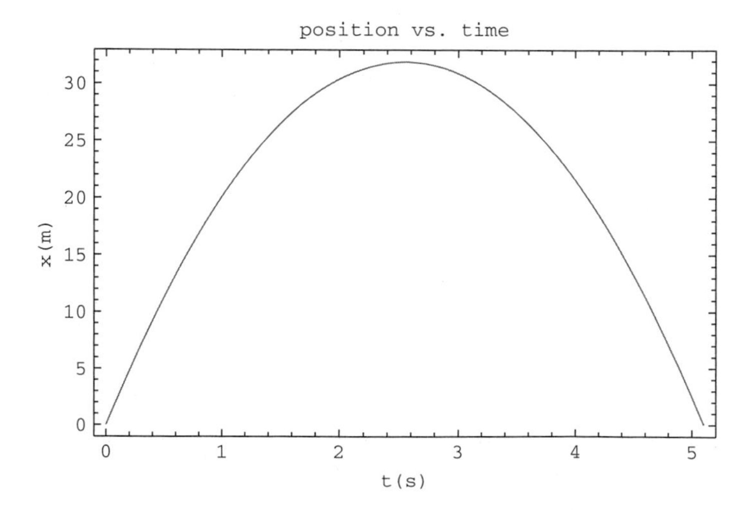

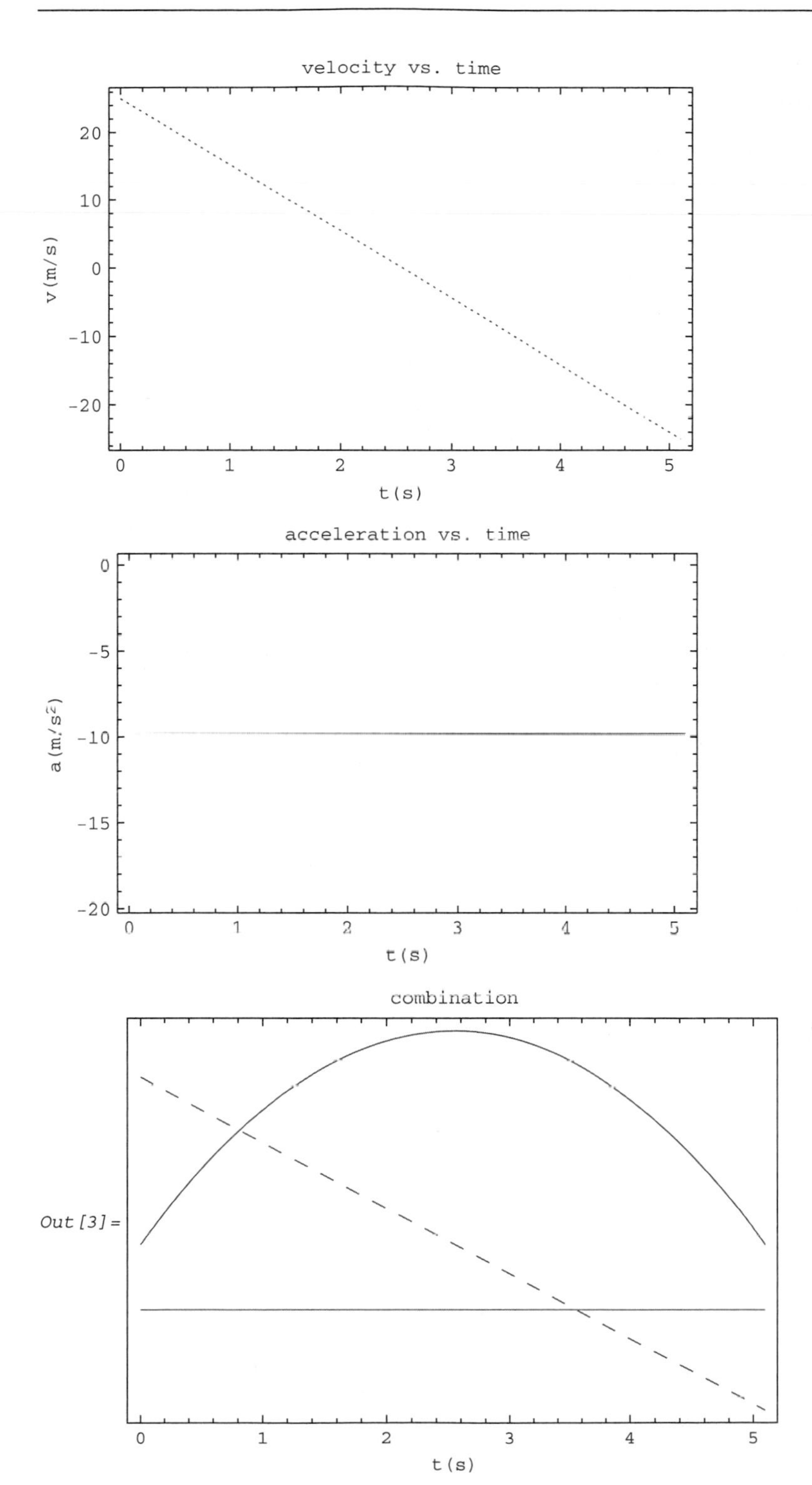
velocity vs. time
v(m/s)
20
10
0
-10
-20
0
1
2
3
4
5
t(s)
acceleration vs. time
a(m/s^2)
0
-5
-10
-15
-20
0
1
2
3
4
5
t(s)
combination
Out[3]=
0
1
2
3
4
5
t(s)

To approximate the effect of air resistance, we include a retarding drag force proportional to the square of the velocity and express the resultant force as

$$F = -mg - k\,x'(t)|x'(t)|$$

where the parameter k is approximately 10^{-4} kg/m for a pebble of radius 0.01 m (see [GTC07] for more information). Let the pebble's mass $m = 10^{-2}$ kg. The function **motion1DPlot** produces a different set of graphs:

```
In[4]:= motion1DPlot[-9.80 - 10^-2 x'[t] Abs[x'[t]], 0, 25, 4.49]
```

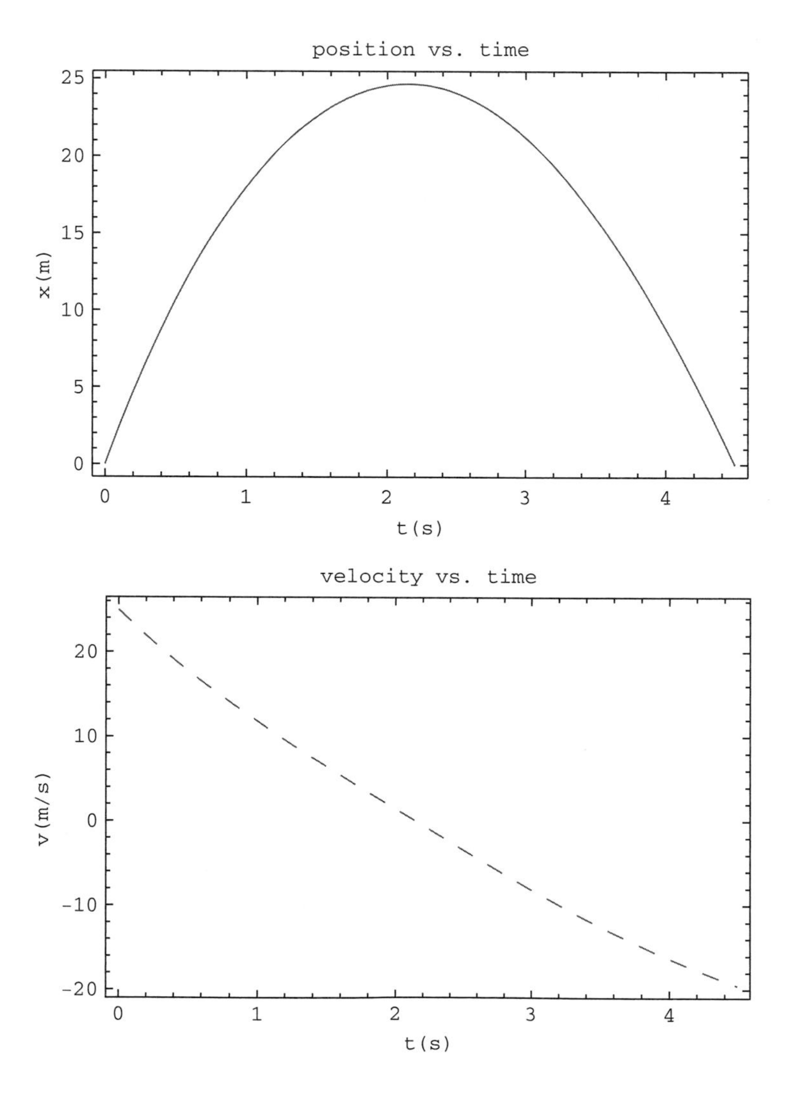

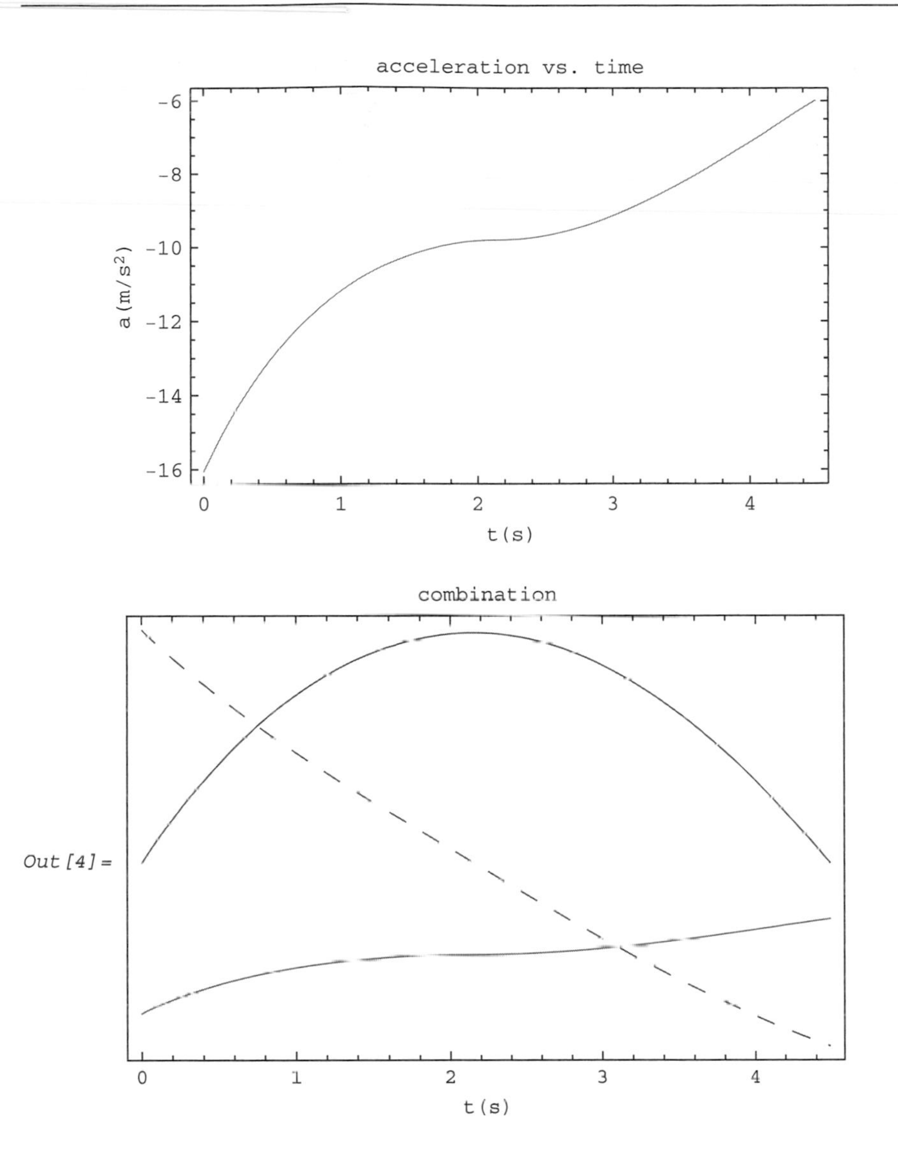

The maximum height is approximately 32 m and the acceleration is constant if air resistance is absent, whereas the actual height reached is less than 25 m and acceleration in fact increases during the flight. Furthermore, when the pebble returns to its initial position, the speed is the same as the initial speed of 25 m/s if air resistance is neglected, whereas the realistic speed is less than 20 m/s.

For another example, a particle initially at rest has mass $m = 1$ kg and is subjected to a force

$$F(t) = k\,t\,e^{-\alpha t}$$

where $k = 1\,\mathrm{N/s}$ and $\alpha = 0.5\,\mathrm{s}^{-1}$. (This example is extracted from a problem in [TM04].) Let us call **motion1DPlot**:

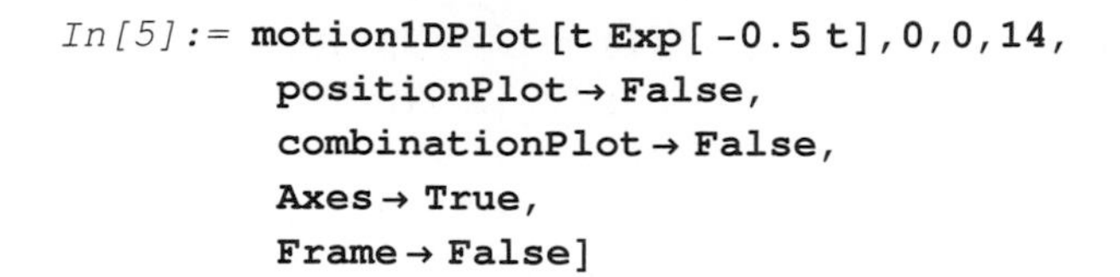

```
In[5]:= motion1DPlot[t Exp[-0.5 t],0,0,14,
          positionPlot → False,
          combinationPlot → False,
          Axes → True,
          Frame → False]
```

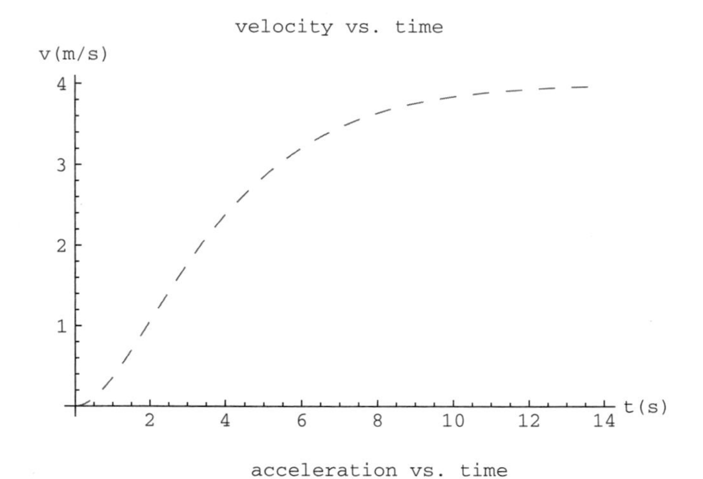

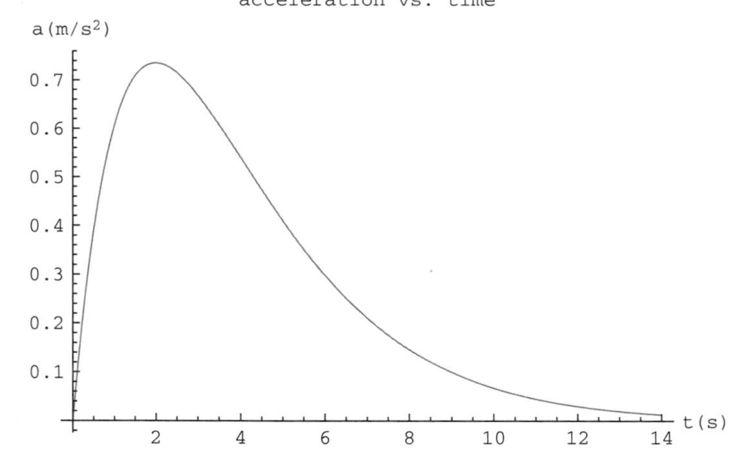

We conclude this section with an example from molecular physics. Solution of the time-independent Schrödinger equation for a diatomic molecule is difficult. For an approximation, we can separate the equation into three equations for different kinds of motion: motion of the electrons, rotation of the nuclei, and vibration of the nuclei. The equation for the electronic motion is still complex because it must include the interactions of the electrons with the nuclei as well as their mutual interactions [Gas96]. However, the solution of the equation for rotational motion is straightforward because the nuclei rotate as a rigid body about the center of mass [ER85]. The two-body problem of the vibrational motion, which is our focus here, can be transformed into two one-body problems: the motion of the center of mass and the one-dimensional motion of either nucleus as viewed from the other. The motion of the center of mass is just that of a free particle. The analysis of the relative motion can be further simplified with the observation that semiclassical approximation is often justified because of the large mass of the nuclei. In Equation 3.4.8, which is the classical equation of motion, the mass m is now the reduced mass and the resultant force is the force on the moving nucleus due to the observing one [Sym71]. The vibrational motion is reduced to the motion of a particle in a one-dimensional potential that is often taken to be the Lennard–Jones or 6–12 potential

$$V(x) = 4b\left[\left(\frac{d}{x}\right)^{12} - \left(\frac{d}{x}\right)^{6}\right]$$

where b and d are the empirical parameters. If $V(x)$ and x are measured, respectively, in units of $4b$ and d, this potential becomes

$$V(x) = \frac{1}{x^{12}} - \frac{1}{x^6}$$

Let us plot this potential:

```
In[6]:= Plot[(1/x^12 - 1/x^6), {x, 0.4, 3},
          PlotRange → {-0.27, 0.3}, AxesLabel → {"x (d)", "V (4b)"}]
```

Out[6]=

It is steep and repulsive for $x < 1$, has a minimum of $-1/4$ at $x = 2^{1/6}$, and approaches zero as x goes to infinity. The force on the moving nucleus is

```
In[7]:= -∂x (1/x^12 - 1/x^6)
```

$$\text{Out[7]=} \quad \frac{12}{x^{13}} - \frac{6}{x^7}$$

For $x_0 = 1.32$ and $v_0 = 0$, the energy is between $-1/4$ and 0 and the motion is bounded and periodic but asymmetric about $x = 2^{1/6}$. Let the unit of mass be m. The function **motion1DPlot** gives the graphs describing the motion of either nucleus as viewed from the other:

```
In[8]:= motion1DPlot[12/x[t]^13 - 6/x[t]^7, 1.32, 0, 6.8,
        positionAFLabel → {"t ((d/2) √(m/b))", "x (d)"},
        velocityAFLabel → {"t ((d/2) √(m/b))", "v (√(4b/m))"},
        accelerationAFLabel → {"t ((d/2) √(m/b))", "a (4b/(md))"},
        combinationAFLabel → {"t ((d/2) √(m/b))", None}]
```

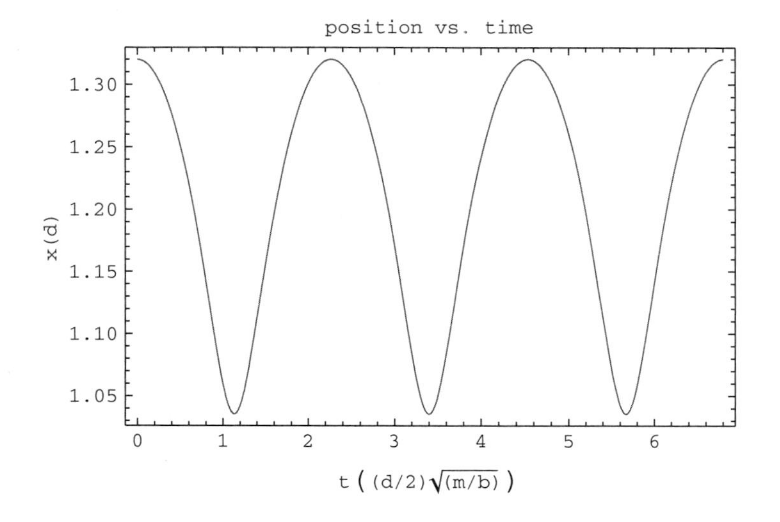

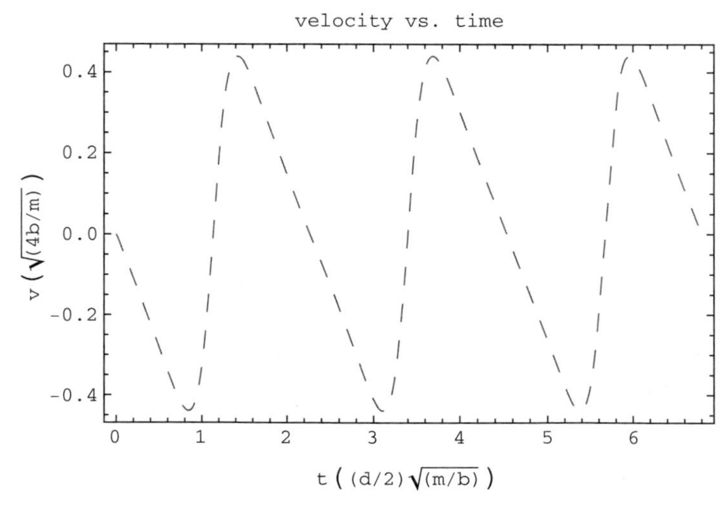

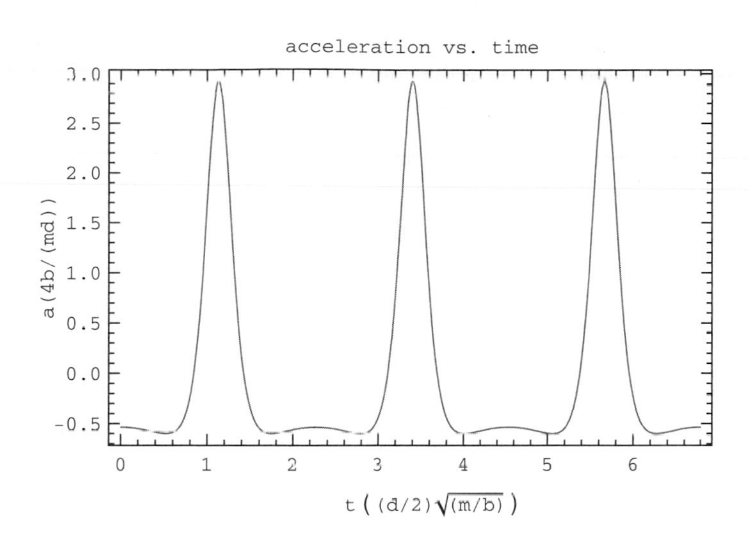

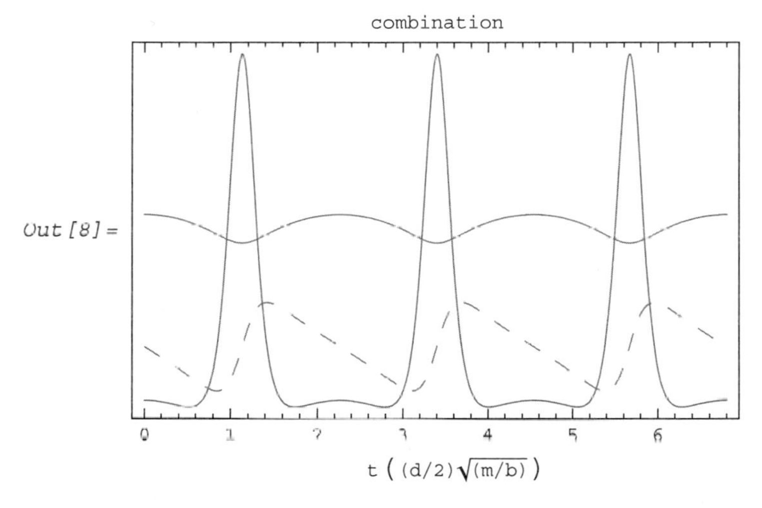

Although these graphs, based on Newton's laws, offer insight into the vibration of the nuclei, we must be leery of the Newtonian notion that all energies between $-1/4$ and 0 are possible for the bound states. The vibrational energy is in fact quantized. In a semiclassical approximation, we impose on the classical results the Sommerfeld–Wilson quantization rule in order to isolate the discrete energies. (Refer to [KM90] for a discussion of semiclassical quantization of molecular vibrations.)

For the H_2 molecule, the Lennard–Jones potential unfortunately yields vibrational energies that are inconsistent with experimental observations. A better choice leading to results that are in reasonable agreement with experimental data is the Morse potential

$$V(x) = V_0 \left[\left(1 - e^{-\alpha(x/x_{\min}-1)} \right)^2 - 1 \right]$$

where the dissociation energy $V_0 = 4.73\,\mathrm{eV}$, the equilibrium internuclear separation $x_{\mathrm{min}} = 0.74\,\text{Å}$, and the parameter $\alpha = 1.44$ for the H_2 molecule. Let us plot the potential in units of V_0 as a function of x in units of x_{min}:

```
In[9]:= Plot[(1 - e^(-1.44 (x - 1)))^2 - 1, {x, 0, 6}, AxesLabel → {"x (xmin)", "V (V0)"}]
```

Out[9]=

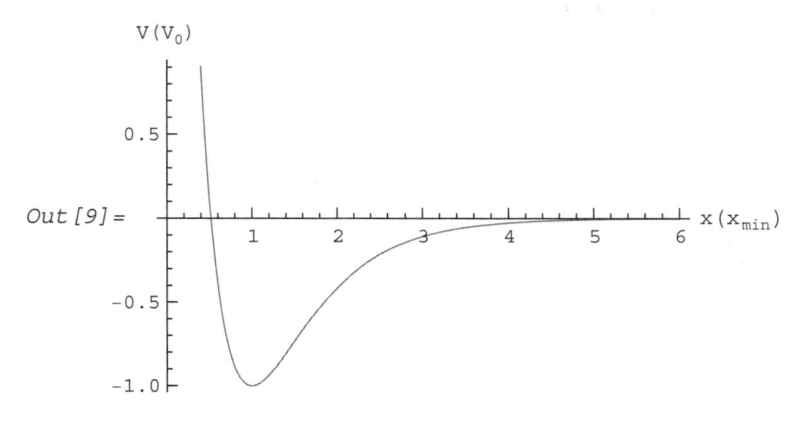

Although the Lennard–Jones and Morse potentials have the same essential features, the latter allows an additional parameter α for adjusting the curvature of the minimum. We will not invoke the function **motion1DPlot** to generate the graphs because they resemble those produced with the Lennard–Jones potential.

```
In[10]:= ClearAll[motion1DPlot]
```

3.4.6 Exercises

In this book, straightforward, intermediate-level, and challenging exercises are unmarked, marked with one asterisk, and marked with two asterisks, respectively. For solutions to most odd-numbered exercises, see Appendix C.

1. Write a single-clause definition and also a multiclause definition for the function

$$f(x) = |x|$$

   What is the derivative of f at 0?

2. (a) Write a single-clause definition and also a multiclause definition for the potential

$$V(x) = \begin{cases} V_0 & x > a \\ V_1 & 0 \le x \le a \\ 0 & x < 0 \end{cases}$$

   where $V_1 = 1.6V_0$. (b) Using the single-clause definition, plot $V(x)$ for x from $-a$ to $3a$. Repeat, using the multiclause definition. (c) Determine the derivatives of V at $x = a$ and $x = 2a$. *Hint:* Let V and x be measured in units of V_0 and a, respectively.

3. (a) Write a single-clause definition and also a multiclause definition for the potential

$$V(x) = \begin{cases} 0 & x < -2a \\ -2b & -2a \le x < -a \\ 0 & -a \le x < a \\ -b & a \le x < 3a \\ b & 3a \le x \end{cases}$$

(b) Using the single-clause definition, plot $V(x)$ for x from $-4a$ to $4a$. Repeat, using the multiclause definition. *Hint:* Let V and x be measured in units of b and a, respectively.

4. The Elliot Wave Theory for stock market forecast is based on Fibonacci numbers. Each Fibonacci number is defined as the sum of the previous two. The first and second are defined to be one. The sequence begins with $1, 1, 2, 3, 5, 8, 13, \ldots$. Here is a function defined to be a procedure for determining the nth Fibonacci number:

```
ourfibonacci[n_Integer?Positive] :=
 (fn1 = 1; fn2 = 1; Do[{fn1, fn2} = {fn2, fn1 + fn2}, {n - 2}]; fn2)
```

For example,

```
ourfibonacci[200]
280 571 172 992 510 140 037 611 932 413 038 677 189 525
```

Of course, the built-in *Mathematica* function also returns the same number:

```
Fibonacci[200]
280 571 172 992 510 140 037 611 932 413 038 677 189 525
```

The problem with the function definition is that **fn1** and **fn2** are global variables that can cause symbol collisions.

```
{fn1, fn2}
{173 402 521 172 797 813 159 685 037 284 371 942 044 301,
 280 571 172 992 510 140 037 611 932 413 038 677 189 525}
```

(a) Using **Module**, modify the function definition so that **fn1** and **fn2** are declared local variables with initial values equal to 1. With the new function definition, we have

```
myfibonacci[200]
280 571 172 992 510 140 037 611 932 413 038 677 189 525
```

```
{fn1, fn2}
{fn1, fn2}
```

That is, **fn1** and **fn2** are no longer global variables with assigned values.

(b) Using dynamic programming introduced in Section 3.3.3, write a function for determining the nth Fibonacci number.

5. Write a single-clause definition for a function that takes a list of numbers as its argument and returns a list of their inverses, except the function should only report the number of zeros present when there are zeros in the list. As examples,

```
func[{1,2,3,4,a}]
func[{1,2,3,4,a}]

func[{2,3,4,10}]
```

$\left\{\frac{1}{2}, \frac{1}{3}, \frac{1}{4}, \frac{1}{10}\right\}$

```
func[{0,3,0,0}]
3 zeroes in list
```

Hint: Use **Map**, **If**, **Count**, **VectorQ**, **NumberQ**, **Print**, and a pure function.

*6. The median of a list of real numbers is the middle element in the sorted version of the list if the length of the list is odd and the average of the two middle elements otherwise.
(a) Using **Module**, **If**, **Sort**, and **OddQ**, write a function **yourmedian** that accepts only a list of real numbers as the argument and returns the median of the list:

```
yourmedian[{76,56,23,78,34}]
56

yourmedian[{3.5,5.6,3/2,78,34,310/3}]
19.8

yourmedian[{2.5,2 + 3i, 6}]
yourmedian::arg : List of real numbers is
     expected at position 1 in yourmedian[{2.5, 2+3i, 6}].
yourmedian[{2.5,2+3 I,6}]

yourmedian[{a,b,f,c,g,e}]
yourmedian::arg : List of real numbers is
     expected at position 1 in yourmedian[{a,b,f,c,g,e}].
yourmedian[{a,b,f,c,g,e}]
```

The function must not introduce any unnecessary global variables. *Hint:* Using Find Selected Function in the Help menu, obtain information about **VectorQ**, **Message**, **ToString**, and **StringJoin**.
(b) Using the built-in function **Median**, determine again the medians of

```
{76,56,23,78,34}
{3.5,5.6,3/2,78,34,310/3}
{2.5,2 + 3i,6}
{a,b,f,c,g,e}
```

7. (a) Using **Module** and **While**, write a function that accepts a positive integer as the argument and returns the sum of the first n prime numbers. With **Timing**, determine the time required to evaluate the function for $n = 10^6$. (b) Using **Table** and **Apply**, repeat part (a). (c) Using **Sum**, repeat part (a) again.

8. Using a **For** loop, print

```
The square root of 3 is 1.73205
The square root of 6 is 2.44949
The square root of 9 is 3.
The square root of 12 is 3.4641
The square root of 15 is 3.87298
```

*9. Consider the function **binomialExpansion**:

```
Options[binomialExpansion] = {caption→False, exponent→2};
binomialExpansion[x_Symbol, y_Symbol, opts___Rule] :=
 Module[{caption, exponent},
  caption = caption/.{opts}/.Options[binomialExpansion];
  exponent = exponent/.{opts}/.Options[binomialExpansion];
  If[caption === True, Print["Expansion of ", (x + y)^exponent]];
  Expand[(x + y)^exponent]]
```

(a) Why doesn't this function work as expected? As examples,

```
binomialExpansion[a,b]
(a + b)^exponent$70

binomialExpansion[a,b,caption→True,exponent→10]
(a + b)^exponent$74
```

(b) Correct the errors in the function definition so that

```
binomialExpansion[a,b]
a^2 + 2 a b + b^2

binomialExpansion[a,b,exponent→10,caption→True]
Expansion of (a + b)^10
a^10 + 10 a^9 b + 45 a^8 b^2 + 120 a^7 b^3 + 210 a^6 b^4 +
 252 a^5 b^5 + 210 a^4 b^6 + 120 a^3 b^7 + 45 a^2 b^8 + 10 a b^9 + b^10
```

*10. Consider the function **myBinomialExpansion**:

```
Options[myBinomialExpansion] = {caption→False};
myBinomialExpansion[x_Symbol, y_Symbol, exp_, opts___Rule] :=
 Module[{},
```

```
caption = caption/.{opts}/.Options[myBinomialExpansion];
If[caption === True, Print["Expansion of ", (x + y)^exp]];
Expand[(x + y)^exp]]
```

(a) Why do two identical calls to the function give different results? Whereas the first call prints the caption, the second does not:

```
myBinomialExpansion[a,b,2,caption→True]
Expansion of (a + b)^2
a^2 + 2ab + b^2

myBinomialExpansion[a,b,2,caption→True]
a^2 + 2ab + b^2
```

(b) Correct the errors in the function definition so that two identical calls to the function give the same result:

```
myBinomialExpansion[a,b,2,caption→True]
Expansion of (a + b)^2
a^2 + 2ab + b^2

myBinomialExpansion[a,b,2,caption→True]
Expansion of (a + b)^2
a^2 + 2ab + b^2
```

*11. A particle of mass m moves in the potential

$$V(x) = -\frac{V_0 a^2 \left(a^2 + x^2\right)}{8a^4 + x^4}$$

(a) Plot the potential and the force on the particle as a function of the position of the particle. (b) Choose three sets of initial conditions: one resulting in the energy of the particle greater than zero, one the energy of the particle between $-V_0/8$ and zero, and one the energy of the particle between $-V_0/4$ and $-V_0/8$. For each set of initial conditions, plot the particle's position, velocity, and acceleration as a function of time. *Hint:* Adopt an appropriate system of units.

*12. A particle of mass m moves in the potential

$$V(x) = V_0 \left(\left(\frac{x}{a}\right)^2 - \beta \left(\frac{x}{a}\right)^3 \right)$$

where V_0 and a are positive constants. Adopt a system of units in which $m = 1$, $V_0 = 1$, and $a = 1$. Let $\beta = 1/12$. (a) Plot the potential and also the force on the particle. (b) Choose two sets of initial conditions: $x_0 = -3$ and v_0 resulting in the energy of the particle being greater than $64/3$, and $x_0 = 5$ and v_0 resulting in the energy of the particle lying between $64/3$ and zero. For each set of initial conditions, plot the particle's position, velocity, and acceleration as a function of time.

**13. Modify the function `motion1DPlot` in Section 3.4.5 so that it takes as its first argument the entire Equation 3.4.8 rather than just the *rhs* of the equation. Also, the function should print an error message if any expression other than an equation is entered as the first argument, the initial position and velocity are not real numbers, or the time interval is not a positive real number. Real numbers include all approximate real numbers, integers, and rational numbers. Implement this error message feature of the function two ways: (a) with the function `If` and (b) with the pattern object *pattern* `/;` *condition*. (That is, define two different functions.)

14. (a) Write a function **code that accepts two arguments: a string and a positive integer. The function, defined to be a procedure, returns another string resulting from each argument string uppercase letter in the alphabet sequence being replaced by the corresponding letter in the alphabet sequence shifted n places to the right. Consider, for example, $n = 7$. The alphabet sequence is

$$A, B, C, D, E, F, G, H, I, J, K, L, M, N, O, P, Q, R, S, T, U, V, W, X, Y, Z$$

The sequence shifted seven places to the right is

$$T, U, V, W, X, Y, Z, A, B, C, D, E, F, G, H, I, J, K, L, M, N, O, P, Q, R, S$$

Thus, for instance, A is replaced by T, L by E, and Y by R. If

```
PROVERB =
  "TO EVERY MAN IS GIVEN THE KEY TO THE
     GATES OF HEAVEN; THE SAME KEY OPENS THE GATES OF
     HELL.  THE VALUE OF SCIENCE. RICHARD P. FEYNMAN.";
```

then

```
code[PROVERB,7]
```

```
MH XOXKR FTG BL ZBOXG MAX DXR MH MAX
  ZTMXL HY AXTOXG; MAX LTFX DXR HIXGL MAX ZTMXL HY
  AXEE.  MAX OTENX HY LVBXGVX. KBVATKW I. YXRGFTG.
```

(b) Write a function **decode** that accepts two arguments: a string and a positive integer. The function, defined to be a procedure, decodes the string returned by the function **code** and recovers the original string. For example,

```
decode[%,7]
```

```
TO EVERY MAN IS GIVEN THE KEY TO THE
  GATES OF HEAVEN; THE SAME KEY OPENS THE GATES OF
  HELL.  THE VALUE OF SCIENCE. RICHARD P. FEYNMAN.
```

3.5 GRAPHICS

Section 2.3 discussed the interactive use of *Mathematica*'s graphical capabilities. This section considers graphics programming that extends these capabilities. Section 3.5.1 introduces the concept of graphics objects, which are represented by *Mathematica* expressions containing graphics elements and options. There are two kinds of graphics elements: graphics primitives and graphics directives. Graphics primitives represent the building blocks of graphics objects, and graphics directives determine how they are rendered. Whereas graphics directives affect the rendering of individual building blocks, graphics options influence the overall appearance of the graphics. Sections 3.5.2 and 3.5.3 describe how to produce two- and three-dimensional graphics, respectively.

The availability of color is one reason why *Mathematica* graphics are so enticing. Unfortunately, this book is not printed in color; in the printing of this book, all color specifications are converted to gray levels. Consequently, some of the pictures are less than appealing. To be awed by the wonderful colors of the graphics, see the notebook for Section 3.5 on the CD accompanying this book.

3.5.1 Graphics Objects

Built-in plotting functions generate plots of lists of data or mathematical functions. To make a plot, the function creates an expression representing the graphics object and then calls for the display of its graphical image. The function **InputForm** allows us to see the expression:

```
In[1]:= InputForm[ListPlot[Table[Random[], {5}], PlotStyle→PointSize[0.02]]]
Out[1]//InputForm=
        Graphics[{Hue[0.67, 0.6, 0.6], PointSize[0.02], Point[{{1.,
        0.8302589107544965}, {2., 0.22154204924586535}, {3.,
        0.7112612179161621}, {4., 0.9194336617304331}, {5.,
        0.05481438263742046}}]},
         {AspectRatio -> GoldenRatio^(-1), Axes -> True, AxesOrigin -> {0,
        Automatic}, PlotRange -> Automatic, PlotRangeClipping -> True}]
```

The expression has the head **Graphics** and elements consisting of graphics elements and graphics options. The graphics elements are the graphics primitive **Point[{{**x_1**,**y_1**},...]** and the graphics directives **Hue[**h**,**s**,**b**]** and **PointSize[**n**]**. The graphics options have the form *OptionName* **->** *OptionValue*; Section 2.3.1.2 introduced several of them.

Although the plotting functions are excellent for plotting lists of data and mathematical functions, graphics programming must be invoked to produce more complex graphics. In these cases, we need to construct the expressions representing the graphics objects. The expressions have the forms

Graphics[*elements*, *options***]**	represent a two-dimensional graphics object that is displayed in StandardForm as a graphical image
Graphics3D[*elements*, *options***]**	represent a three-dimensional graphics object that is displayed in StandardForm as a graphical image

Unless there is only one graphics primitive, the first argument of **Graphics** and **Graphics3D** must be a list or nested list of graphics elements that are either graphics primitives or graphics directives. The first argument takes the form

$\{prim_1, dir_1, dir_2, prim_2, sublist_1, prim_3, dir_3, prim_4, prim_5, sublist_2, \ldots\}$

where the $prim_j$ are the graphics primitives, the dir_j are the graphics directives, and the $sublist_j$ have a similar form as that of the list. Examples of graphics primitives are **Point**, **Line**, **Polygon**, and **Text**; examples of graphics directives are **GrayLevel**, **RGBColor**, **PointSize**, and **Thickness**.

In a list, each graphics directive modifies the subsequent graphics primitives in the list or sublists of the list:

```
In[2]:= Graphics[{Point[{1, 1}], PointSize[0.15],
          {GrayLevel[0.5], Point[{2, 2}]}, Point[{3, 3}]},
         AxesOrigin→{0, 0}, Axes→True, PlotRange→{{0, 4}, {0, 4}}]
```

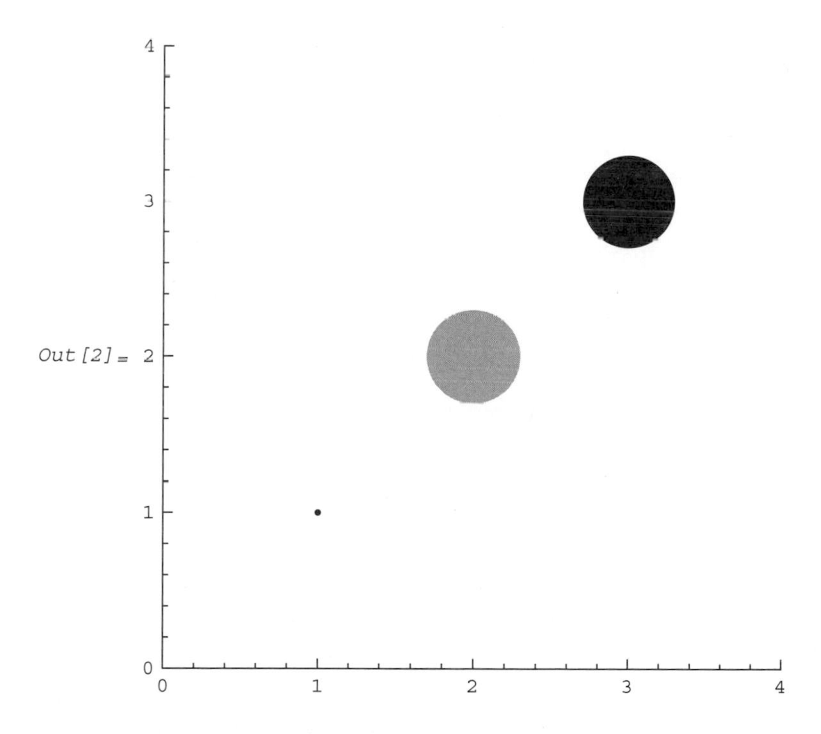

In the preceding graphic, **GrayLevel** affects only the gray-level intensity of the point at {2, 2} and **PointSize** specifies the diameter of the points at {2, 2} and {3, 3}. The graphics directives have no influence on the almost invisible point at {1, 1}.

If there is another graphics directive with the same name but different specification further down the list, the latter prevails:

```
In[3]:= Graphics[
         {PointSize[0.05], Point[{1, 1}],
          {PointSize[0.10], GrayLevel[0.5], Point[{2, 2}]},
          Point[{3, 3}], PointSize[0.175], Point[{4, 4}]},
         AxesOrigin → {0, 0}, Axes → True, PlotRange → {{0, 5}, {0, 5}}]
```

Out[3]=

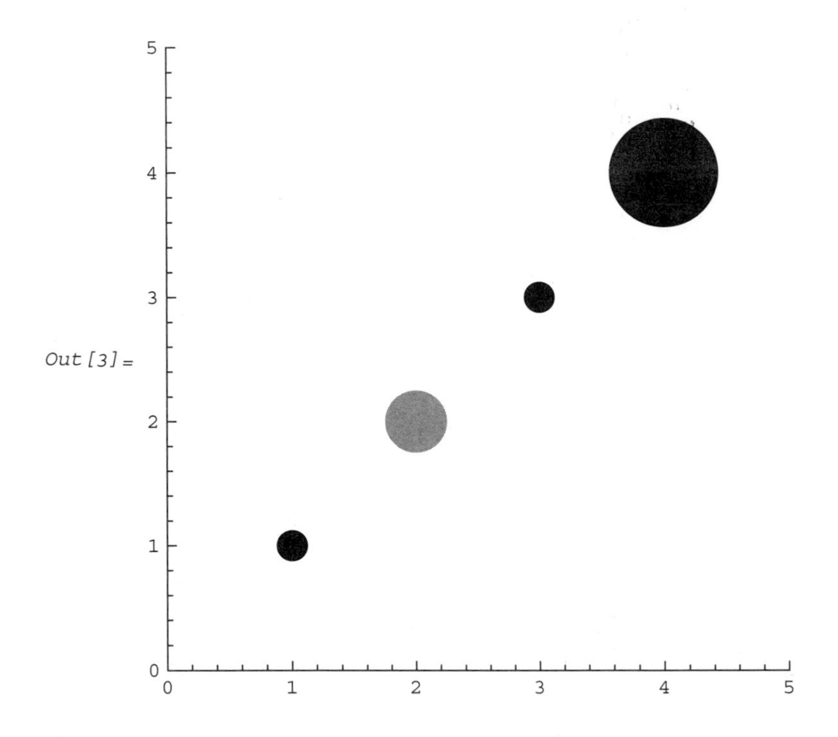

Here, **PointSize[0.05]** specifies the diameter of the points at $\{1,1\}$ and $\{3,3\}$. **PointSize[0.10]** determines the diameter of the point at $\{2, 2\}$, and **PointSize[0.175]** stipulates the size of the point at $\{4, 4\}$. Note that **PointSize[0.10]** and **GrayLevel[0.5]** do not affect points outside their list.

Whereas graphics directives modify the rendering of building blocks represented by graphics primitives in the list, graphics options have global effects on the plot. Specifications of graphics options cause *Mathematica* to construct a collection of graphics elements. The function **FullGraphics** allows us to see these elements. **InputForm[FullGraphics[*g*]]** takes an expression representing a graphics object and generates a new one in which graphics options, if any, are given as explicit lists of graphics elements:

```
In[4]:= Short[InputForm[FullGraphics[Graphics[Point[{1, 1}], Axes → True]]], 7]
```

```
Out[4]//Short=
      Graphics[{Point[{1, 1}], {{GrayLevel[0.], AbsoluteThickness[0.25],
        Line[{{0.25, 0.}, {0.25, 0.0125}}]}, Text[0.25, {0.25, -0.025},
        {0., 1.}], {GrayLevel[0.], AbsoluteThickness[0.25], Line[{{0.5,
        0.}, {0.5, 0.0125}}]}, Text[0.5, {0.5, -0.025}, {0., 1.}],
        {GrayLevel[0.], AbsoluteThickness[0.25], Line[{{0.75, 0.}, {0.75,
        0.0125}}]}, Text[0.75, {0.75, -0.025}, {0., 1.}], {GrayLevel[0.],
        AbsoluteThickness[0.25], Line[{{1., 0.}, {1., 0.0125}}]}, Text[1,
        {1., -0.025}, {0., 1.}], {GrayLevel[0.], AbsoluteThickness[0.25],
        Line[{{1.25, 0.}, {1.25, 0.0125}}]}, Text[1.25, {1.25, -0.025},
        {0., 1.}], {GrayLevel[0.], AbsoluteThickness[0.25], Line[{{1.5,
        0.}, {1.5, 0.0125}}]}, Text[1.5, {1.5, -0.025}, {0., 1.}],
        {GrayLevel[0.], AbsoluteThickness[0.25], Line[{{1.75, 0.}, {1.75,
        0.0125}}]}, <<78>>, {GrayLevel[0.], <<1>>, Line[{<<2>>}]},
        {GrayLevel[0.], AbsoluteThickness[0.125], Line[{{0., 1.7},
        {0.0075, 1.7}}]}, {GrayLevel[0.], AbsoluteThickness[0.125],
        Line[{{0., 1.8}, {0.0075, 1.8}}]}, {GrayLevel[0.],
        AbsoluteThickness[0.125], Line[{{0., 1.85}, {0.0075, 1.85}}]},
        {GrayLevel[0.], AbsoluteThickness[0.125], Line[{{0., 1.9},
        {0.0075, 1.9}}]}, {GrayLevel[0.], AbsoluteThickness[0.125],
        Line[{{0., 1.95}, {0.0075, 1.95}}]}, {GrayLevel[0.],
        AbsoluteThickness[0.25], Line[{{0., 0.}, {0., 2.}}]}}}]
```

With the function **Short**, we have reduced the length of the output.

The function **Show**, introduced in Section 2.3, can display graphics objects:

Show[*g*, *options*]	display the graphics object represented by the expression g with the specified options added
Show[g_1, g_2, ..., *options*]	display several graphics objects combined; add the options, if specified

Section 2.3.1.2 considered graphics options. The lists of nondefault options in the g_i are concatenated. Options explicitly specified in **Show** override those included in the g_i, and they remain with the graphics expression that **Show** returns. (To see the graphics expression, again use **InputForm**.)

3.5.2 Two-Dimensional Graphics

3.5.2.1 Two-Dimensional Graphics Primitives

Mathematica recognizes many two-dimensional graphics primitives:

Arrow[{{x_1, y_1}, {x_2, y_2}}]	arrow from $\{x_1, y_1\}$ to $\{x_2, y_2\}$
Circle[{x, y}, r]	circle of radius r centered at the point $\{x, y\}$
Circle[{x, y}, {r_x, r_y}]	ellipse with semiaxes r_x and r_y

Circle[{x, y}, r, {θ_1, θ_2}]	circular arc where θ_1 and θ_2 are, respectively, the starting and finishing angles measured counterclockwise in radians from the positive x direction
Circle[{x, y}, {r_x, r_y}, {θ_1, θ_2}]	elliptical arc
Disk[{x, y}, r]	filled disk of radius r centered at the point $\{x, y\}$
Disk[{x, y}, {r_x, r_y}]	elliptical disk with semiaxes r_x and r_y
Disk[{x, y}, r, {θ_1, θ_2}]	segment of a disk
Inset[*obj*, {x, y}]	object *obj* inset at position $\{x, y\}$ in a graphic, where *obj* can be a graphic, string, or any other expression
GraphicsComplex[{pt_1, pt_2, ...}, *data*]	graphics complex in which coordinates given as integers i in graphics primitives in *data* are taken to be pt_i; *data* can be any nested list of graphics primitives and directives
Line[{{x_1, y_1}, {x_2, y_2}, ...}]	line joining points $\{x_1, y_1\}, \{x_2, y_2\}, \ldots$
Line[{{{x_{11}, y_{11}}, {x_{12}, y_{12}}, ...}, ...}]	collection of lines
Point[{x, y}]	point at coordinates $\{x, y\}$
Point[{{x_1, y_1}, {x_2, y_2}, ...}]	collection of points
Polygon[{{x_1, y_1}, {x_2, y_2}, ...}]	filled polygon with the specified list of corners
Rectangle[{x_{min}, y_{min}}, {x_{max}, y_{max}}]	filled rectangle, oriented parallel to the axes
Text[*expr*, {x, y}]	text corresponding to the textual form of *expr*, centered at $\{x, y\}$; the options **Background** and **FormatType** can be given (see Section 3.5.2.3 for information on options)
Text[*expr*, {x, y}, {-1, 0}]	text with its left-hand end at $\{x, y\}$
Text[*expr*, {x, y}, {1, 0}]	text with its right-hand end at $\{x, y\}$
Text[*expr*, {x, y}, {0, -1}]	text centered above $\{x, y\}$
Text[*expr*, {x, y}, {0, 1}]	text centered below $\{x, y\}$

Coordinates of graphics primitives may be given in terms of the ordinary coordinates $\{x, y\}$ that we have been using in this book or the display coordinates relative to the display area for the graphic. The display, or scaled, coordinates are given in the form **Scaled**[{*sx*, *sy*}]. These coordinates, $\{sx, sy\}$, vary from 0 to 1 along the x and y directions, and the origin is located at the lower left corner of the display area. As an example, let us display two rectangles—one specified with the ordinary coordinates and another specified with the scaled or display coordinates:

```
In[1]:= Graphics[
          {Rectangle[{0.5, 0.5}, {1, 1}],
           Rectangle[Scaled[{0.5, 0.5}], Scaled[{1, 1}]]},
          PlotRange→{{0, 2}, {0, 2}}, Frame→True,
          FrameTicks→{{0, 0.5, 1, 2}, {0, 0.5, 1, 2}}]
```

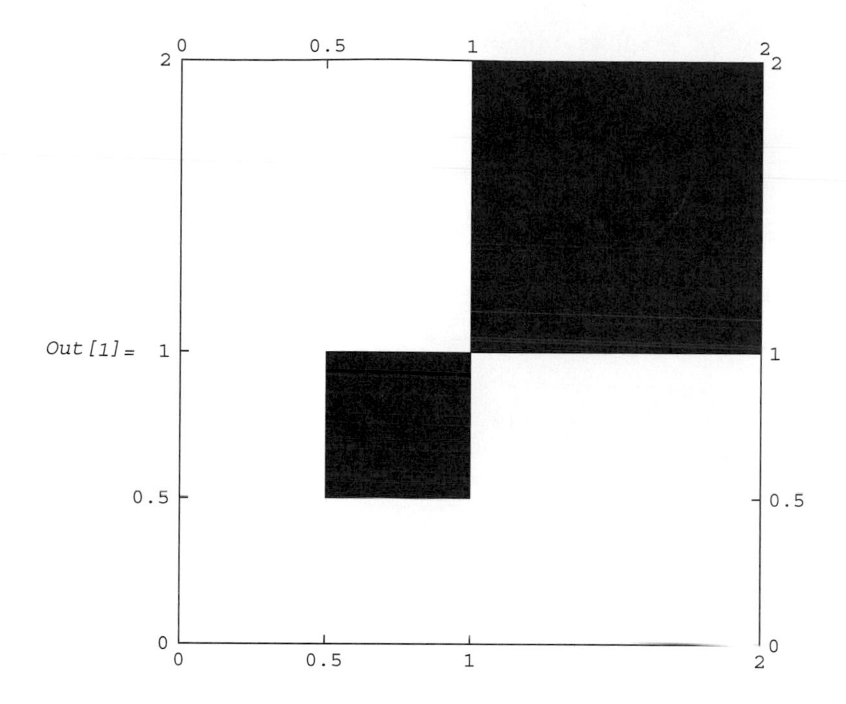

In the expression representing the graphics object, the lower-left and upper-right corners of the smaller rectangle are specified in the ordinary coordinates, whereas those of the larger rectangle are given in the scaled coordinates.

To illustrate the use of some of the graphics primitives in setting up graphics expressions representing graphics objects, we select an example from geometric optics.

Example 3.5.1 Construct a ray diagram for a thin converging lens.

To begin, we generate a list of graphics primitives representing the lines. The function **Map** with the special input form **/@** applies the function **Line** to each list of points joined by a line:

```
In[2]:= g1 = Line /@
          {{{-13, 0}, {17, 0}}, {{0, -4}, {0, 4}},
           {{-10, 2}, {0, 2}, {15, -3}}, {{-10, 2}, {15, -3}}}
Out[2]= {Line[{{-13, 0}, {17, 0}}], Line[{{0, -4}, {0, 4}}],
         Line[{{-10, 2}, {0, 2}, {15, -3}}], Line[{{-10, 2}, {15, -3}}]}
```

Map and **Arrow** give a list of graphics primitives for the arrows representing the object and the image:

```
In[3]:= g2 = Arrow /@ {{{-10, 0}, {-10, 2}}, {{15, 0}, {15, -3}}}
Out[3]= {Arrow[{{-10, 0}, {-10, 2}}], Arrow[{{15, 0}, {15, -3}}]}
```

Map and **Polygon** give a list of graphics primitives for the triangles depicting the arrowheads:

```
In[4]:= g3 = Polygon /@ {{{-5, 2}, {-5.4, 2.4}, {-5.4, 1.6}},
            {{-5, 1}, {-5.314, 1.471}, {-5.471, 0.6862}}}
Out[4]= {Polygon[{{-5, 2}, {-5.4, 2.4}, {-5.4, 1.6}}],
          Polygon[{{-5, 1}, {-5.314, 1.471}, {-5.471, 0.6862}}]}
```

The functions **Thread** and **Circle** provide a list of two graphics primitives representing the two circular arcs of the lens:

```
In[5]:= g4 = Thread[
          Circle[{{13, 0}, {-13, 0}}, 13.60, {{2.843, 3.440}, {5.985, 6.582}}]]
Out[5]= {Circle[{13, 0}, 13.6, {2.843, 3.44}],
          Circle[{-13, 0}, 13.6, {5.985, 6.582}]}
```

Using **Thread** is elegant but optional. We may enter the list of graphics primitives directly.

Thread and **Disk** yield a list of two graphics primitives for the two focal points:

```
In[6]:= g5 = Thread[Disk[{{-6, 0}, {6, 0}}, 0.25]]
Out[6]= {Disk[{-6, 0}, 0.25], Disk[{6, 0}, 0.25]}
```

Finally, we produce a list of graphics primitives representing the labels:

```
In[7]:= g6 = {
            Text["F", {-6, -0.275}, {0, 1}],
            Text["F", {6.05, 0.2}, {0, -1}],
            Text[Style["Ray Diagram for a\nThin Converging Lens",
              FontFamily → "Helvetica", Bold, 9], {3.5, 2.8}, {-1, 0}]};
```

In a text string, **\n** indicates a new line. Also, the function **Style** determines the style of the title for the diagram. **Style[***expr***,** *options***]** displays with *expr* formatted using the specified option settings. Some common options are **FontSize**, **FontWeight**, **FontSlant**, **FontFamily**, **FontColor**, and **Background**. Also, **Bold** is equivalent to **FontWeight -> "Bold"**, **Italic**, to **FontSlant -> "Italic"**, **RGBColor[***r***,** *g***,** *b***]**, to **FontColor -> RGBColor[***r***,** *g***,** *b***]**, and any positive number *n*, to **FontSize ->** *n*. **Style[***expr***, "***style***"]** displays with *expr* formatted using option settings for the specified named style in the current notebook. Some typical named styles are **Title**, **Section**, **Subsection**, **Text**, **Input**, and **Output**. (For more information on **Style**, see the notes on **BaseStyle** in Section 3.5.2.3.)

The graphics expression is displayed in StandardForm as the ray diagram:

```
In[8]:= Graphics[Join[g1, g2, g3, g4, g5, g6]]
```

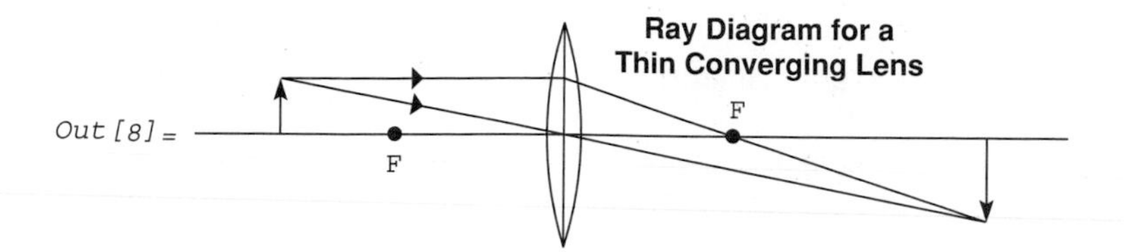


```
In[9]:= ClearAll[g1, g2, g3, g4, g5, g6]
```

■

3.5.2.2 Two-Dimensional Graphics Directives

Mathematica supports many two-dimensional graphics directives:

GrayLevel[*level*]	gray level between 0 (black) and 1 (white)
RGBColor[*r*,*g*,*b*]	color with specified red, green, and blue components, each between 0 and 1
Hue[*h*]	color with hue *h* between 0 and 1
Hue[*h*,*s*,*b*]	color with specified hue, saturation, and brightness, each between 0 and 1
Opacity[*a*]	opacity *a* between 0 (perfectly transparent) and 1 (opaque)
PointSize[*d*]	give all points a diameter *d* as a fraction of the total width of the graphic
AbsolutePointSize[*d*]	give all points a diameter *d* measured in absolute units—that is, in units of printer's points—each approximately equal to 1/72 of an inch
Thickness[*d*]	give all lines a thickness *d* as a fraction of the total width of the graphic; line thickness can also be specified using the graphics directives **Thick** and **Thin**
AbsoluteThickness[*d*]	give all lines a thickness *d* measured in absolute units
Dashing[{d_1, d_2, ...}]	show all lines as a sequence of dashed segments of lengths $d_1, d_2, \ldots$ (repeated cyclically) with d_i given as a fraction of the total width of the graphic; line dashing can also be specified using the graphics directives **Dashed**, **Dotted**, and, **DotDashed**
AbsoluteDashing[{d_1, d_2, ...}]	use absolute units to measure dashed segments
Arrowheads[*spec*]	arrowheads of arrows are drawn using sizes, positions, and forms specified by *spec*; *spec* can be *s* (default arrowhead with scaled size *s*), **{Automatic**, *pos*} (default arrowhead at position *pos*), {*s*,*pos*} (scaled default arrowhead at position *pos*), or {*s*,*pos*,*g*} (arrowhead drawn as graphic *g*); *s* is given as a fraction of the total width of the graphic; *s* can also

	assume the symbolic value **Tiny**, **Small**, **Medium**, or **Large**; *pos* runs from 0 to 1 from the tail to the head of the arrow
Arrowheads[{$spec_1$, $spec_2$,...}]	arrows are drawn with several arrowhead elements; the argument {s_0,s_1,...,s_n} specifies arrowheads with scaled sizes s_i at positions i/n; {$-s$, s} specifies double-headed arrows
EdgeForm[*g*]	edges of polygons, disks, and rectangles are drawn according to the specified graphics directive or list of directives *g*; no edges are drawn is the default
FaceForm[*g*]	faces of polygons, disks, and rectangles are drawn according to the specified graphics directive or list of directives *g*
Directive[g_1, g_2, ...]	single graphics directive composed of the directives $g_1, g_2, \ldots$

In the preceding table, d and d_i can also assume the symbolic value **Tiny, Small, Medium,** or **Large,** each of which is specified in absolute units independent of the total width of the graphic. Furthermore, **RGBColor** specifications for some common colors are

Beige	**RGBColor[0.640004, 0.580004, 0.5]**
Black	**Black**, **GrayLevel[0]**, or **RGBColor[0, 0, 0]**
Blue	**Blue** or **RGBColor[0, 0, 1]**
Brown	**Brown** or **RGBColor[0.6, 0.4, 0.2]**
Chocolate	**RGBColor[0.823496, 0.411802, 0.117603]**
Cyan	**Cyan** or **RGBColor[0, 1, 1]**
Gold	**RGBColor[1, 0.843104, 0]**
Green	**Green** or **RGBColor[0, 1, 0]**
Ivory	**RGBColor[1, 1, 0.941206]**
Lavender	**RGBColor[0.902005, 0.902005, 0.980407]**
Magenta	**Magenta** or **RGBColor[1, 0, 1]**
Olive	**RGBColor[0.230003, 0.370006, 0.170003]**
Orange	**Orange** or **RGBColor[1, 0.5, 0]**
Peach	**RGBColor[0.44, 0.26, 0.26]**
Pink	**Pink** or **RGBColor[1, 0.5, 0.5]**
Purple	**Purple** or **RGBColor[0.5, 0, 0.5]**
Red	**Red** or **RGBColor[1, 0, 0]**
Tomato	**RGBColor[1, 0.388195, 0.278405]**
Turquoise	**RGBColor[0.250999, 0.878399, 0.815699]**
Violet	**RGBColor[0.559999, 0.370006, 0.599994]**
White	**White**, **GrayLevel[1]**, or **RGBColor[1, 1, 1]**
Yellow	**Yellow** or **RGBColor[1, 1, 0]**

The following example from a favorite lecture demonstration illustrates the use of some of the graphics directives.

Example 3.5.2 Construct a diagram showing a lecture demonstration in which a projectile is aimed at a target that falls from rest at the moment when the projectile is fired.

Plot generates the expression representing the trajectory of the projectile:

```
In[1]:= gp1 = Plot[10 x/15 - 7 x^2/225, {x, 0, 15}, PlotStyle → Thickness[0.01]];
```

(To see the expression generated by **Plot**, enter **InputForm[gp1]**.)

Here is the nested list of graphics elements for the lines, two of which are dashed:

```
In[2]:= gp2 = {Line[{{0, 0}, {5, 0}}], {Dashing[{0.02, 0.01}],
            Line[{{0, 0}, {15, 10}}], Line[{{15, 10}, {15, 3}}]}};
```

The nested list of graphics elements for the arrows in the diagram is

```
In[3]:= gp3 =
          {{Arrowheads[Large], Thick,
            Arrow[{{0, 0}, {5, 50/15}}], Arrow[{{15, 10}, {15, 7}}],
            Arrow[{{11.8956, 3.5332}, {12.3375, 3.48956}}]},
           Arrow[{{12.5, 1}, {14.5, 2.5}}]};
```

The nested list of graphics elements for the projectile, target, and point of collision is

```
In[4]:= gp4 =
          {{Red, Opacity[0.6], Disk[{15, 3}, 0.4]}, Tooltip[Disk[{0, 0}, 0.4],
            "projectile"], Tooltip[Disk[{15, 10}, 0.4], "target"]};
```

where **Tooltip[***expr*, *label***]** displays *label* as a tooltip while the mouse pointer is in the area where *expr* is displayed.

Finally, the list of **Text** primitives representing the texts is

```
In[5]:= gp5 =
          {Text[Style[v0, "Subsubsection"], {3.3, 3.8}],
           Text[θ0, {2.5, 0.6}],
           Text[Style["Point of\ncollision",
             FontFamily → "Helvetica"], {10.75, 1}]};
```

The function **Show** displays the combined graphics objects:

```
In[6]:= Show[gp1, Graphics[Join[gp2, gp3, gp4, gp5]],
         PlotRange → {{-1, 17}, {-1, 12}}, Axes → False, AspectRatio → Automatic]
```

Out[6]=

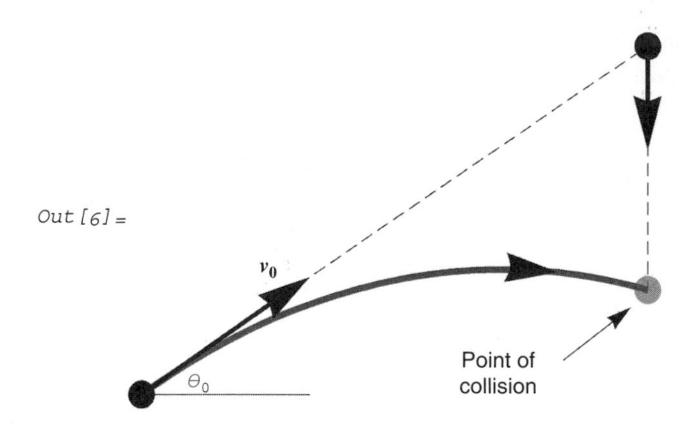

where moving the mouse pointer over a black disk displays its label. Note that **Show** can combine expressions created by built-in plotting functions with those constructed explicitly with graphics elements. Furthermore, **Show** renders graphics objects in the order their expressions appear as its arguments, and later objects cover earlier ones in the final display. Objects represented by graphics primitives in **Graphics** are also rendered in the order their primitives appear in the list, and later objects are similarly drawn over earlier ones.

```
In[7]:= ClearAll[gp1, gp2, gp3, gp4, gp5]
```

■

3.5.2.3 Two-Dimensional Graphics Options

Section 2.3.1 introduced many options for two-dimensional plotting functions. **Options[Plot]** lists the options and their default settings for **Plot**:

```
In[1]:= Options[Plot]

Out[1]= {AlignmentPoint → Center, AspectRatio → 1/GoldenRatio,
    Axes → True, AxesLabel → None, AxesOrigin → Automatic,
    AxesStyle → {}, Background → None, BaselinePosition → Automatic,
    BaseStyle → {}, ClippingStyle → None, ColorFunction → Automatic,
    ColorFunctionScaling → True, ColorOutput → Automatic,
    ContentSelectable → Automatic, DisplayFunction :→ $DisplayFunction,
    Epilog → {}, Evaluated → Automatic, EvaluationMonitor → None,
    Exclusions → Automatic, ExclusionsStyle → None, Filling → None,
    FillingStyle → Automatic, FormatType :→ TraditionalForm, Frame → False,
    FrameLabel → None, FrameStyle → {}, FrameTicks → Automatic,
    FrameTicksStyle → {}, GridLines → None, GridLinesStyle → {},
    ImageMargins → 0., ImagePadding → All, ImageSize → Automatic,
    LabelStyle → {}, MaxRecursion → Automatic, Mesh → None,
    MeshFunctions → {#1&}, MeshShading → None, MeshStyle → Automatic,
    Method → Automatic, PerformanceGoal :→ $PerformanceGoal,
```

```
PlotLabel→None, PlotPoints→Automatic,
PlotRange→{Full, Automatic}, PlotRangeClipping→True,
PlotRangePadding→Automatic, PlotRegion→Automatic,
PlotStyle→Automatic, PreserveImageOptions→Automatic, Prolog→{},
RegionFunction→(True &), RotateLabel→True, Ticks→Automatic,
TicksStyle→{}, WorkingPrecision→MachinePrecision}
```

The options for **Graphics** form a subset of that for **Plot**, and some of them are

Option Name	Default Value	Description
AlignmentPoint	**Center**	how objects should by default be aligned when they appear in **Inset**; *opos* specifies that **Inset**[*obj*, *pos*] is equivalent to **Inset**[*obj*, *pos*, *opos*]– that is, align the inset so that position *opos* in the object lies at position *pos* in the enclosing graphic; {*x*, *y*} represents ordinary coordinates in a graphic; **Center** corresponds to the center of the whole image; see the notes on **Inset** in Section 3.5.2.1
AspectRatio	**Automatic**	the ratio of height to width for a plot; **Automatic** sets it from the actual coordinate values in the plot, namely, one coordinate unit in the *x* direction has the same size in the display area as one coordinate unit in the *y* direction
Axes	**False**	whether to include axes; **True** draws all axes and **False** draws no axes
AxesLabel	**None**	labels to be put on axes; **None** gives no axis labels; *ylabel* specifies a label for the *y*-axis; {*xlabel*, *ylabel*} puts labels on both axes
AxesOrigin	**Automatic**	the point at which axes cross; {*x*, *y*} specifies the point where they cross; **Automatic** uses an internal algorithm to determine where the axes should cross
AxesStyle	**{}**	how axes should be rendered; *style*, all axes are to be generated with the specified style; *style* can be a rule for an option such as **FontSize** or **FontFamily**; *style* can be a graphics directive such as **Thick**, **Red**, **Dashed**, **Thickness**, **Dashing**, or **Directive** (i.e., combination of directives and options); {*xstyle*, *ystyle*}, *xstyle* for *x*-axis and *ystyle* for *y*-axis

(Continued)

Option Name	Default Value	Description
Background	**None**	background color for the graphic; the setting can be a **GrayLevel**, **Hue**, **RGBColor**, or **Opacity** (in the form **Opacity[***a*, *color***]**) directive; **None**, no background should be used
BaselinePosition	**Automatic**	the baseline of a graphics object for purposes of alignment with surrounding text or other expressions; *pos*, position *pos* in an object should align with the baseline of surrounding text or other expressions; *pos* can be **Automatic**, **Bottom**, **Top**, **Center**, **Axis** (axis of the object), and **Scaled[***y***]** (fraction *y* of the height of the object)
BaseStyle	**{}**	base style specifications for the graphics object *obj*; *spec*, specifies that *obj* should always be displayed as **Style[***obj*, *spec***]**; we introduced **Style** in Example 3.5.1; what follows is some additional information; *spec* can be any list of a named style in the current notebook, option settings, and graphics directives as well as **Underlined** (fonts underlined), **Larger** (fonts larger), **Smaller** (fonts smaller), **Large** (fonts large), **Medium** (fonts medium-sized), **Small**(fonts small), and **Tiny** (fonts tiny); explicit option settings always override those defined by the named style
ContentSelectable	**Automatic**	whether and how content objects should be selectable; **True**, single click immediately selects a content object; **False**, content objects cannot be selected—that is, single click selects the complete graphics object; **Automatic**, double click selects a content object
DisplayFunction	**$DisplayFunction**	an option for **Show** (and the plotting functions); specifies a function to apply to the graphics object before returning it; for Windows and Mac OS X, **$DisplayFunction** is **Identity**
Epilog	**{}**	a list of graphics elements to be rendered after the main part of the graphics is rendered
FormatType	**TraditionalForm**	the default format type for text; typical settings include **Automatic**, **InputForm**, **OutputForm**, **StandardForm**, and **TraditionalForm**
Frame	**False**	whether to draw a frame around the plot
FrameLabel	**None**	labels to be placed on the edges of the frame around a plot; *label* specifies a label for the

Option Name	Default Value	Description
		bottom edge of the frame; `{`*bottom*`,` *left*`}` specifies labels for the bottom and left-hand edges of the frame; `{{`*left*`,`*right*`},` `{`*bottom*`,`*top*`}}` specifies labels for each of the edges of the frame
FrameStyle	`{}`	style for the frame; *style*, all edges of the frame are drawn with the specified style; *style* can be a graphics directive such as **Thick**, **Red**, **Dashed**, **Thickness**, **Dashing**, or **Directive** (i.e., combination of directives); `{{`*left*`,`*right*`},` `{`*bottom*`,`*top*`}}`, different edges are drawn with different styles
FrameTicks	**Automatic**	what tick marks to draw if there is a frame; **None** gives no tick marks; **Automatic** places tick marks automatically; **True** places tick marks automatically on bottom and left edges; **All** places tick marks automatically on all edges; `{{`*left*`,`*right*`},{`*bottom*`,`*top*`}}` specifies tick mark options separately for each edge of the frame; for each edge, tick marks can be specified as described in this table in the notes for **Ticks**
FrameTicksStyle	`{}`	how frame ticks (tick marks and tick labels) should be rendered; *style*, all ticks are to be rendered with the specified style; *style* can be a graphics directive such as **Thick**, **Red**, **Dashed**, **Thickness**, **RGBColor**, **Dashing**, or **Directive** (i.e., combination of directives); *style* can be a rule for an option such as **FontSize** or **FontFamily**; *style* can also be a style name from the current stylesheet; `{{`*left*`,`*right*`},` `{`*bottom*`,`*top*`}}`, ticks on different edges should use different styles; explicit style specifications for **FrameTicks** override those for **FrameTicksStyle**
GridLines	**None**	what grid lines to include; **None** gives no grid lines; **Automatic** places a grid line for every major tick mark; `{`*xgrid*`,`*ygrid*`}` specifies separately grid lines in each direction; *xgrid* can be **None**, **Automatic**, `{`x_1`,`x_2`, ...}` (i.e., at the specified positions), or `{{`x_1`,`*style*$_1$`}, ...}` (at the specified positions with the specified styles); similarly for *ygrid*; see **GridLinesStyle** for style specifications
GridLinesStyle	`{}`	how grid lines should be rendered; *style*, all grid lines should be rendered with the specified style; *style* can be a graphics directive such as **Thick**, **Red**, **Dashed**, **Thickness**, **RGBColor**, **Dashing**, or **Directive** (i.e., combination of directives); *style* can also be a

(Continued)

Option Name	Default Value	Description
		style name from the current stylesheet; {*xstyle*, *ystyle*}, *x* and *y* grid lines should use different styles; explicit style specifications for **GridLines** override those for **GridLinesStyle**
ImageMargins	0	the margins to leave around the graphic; *m*, same margins in printer's points on all sides; {{*left*, *right*}, {*bottom*, *top*}}, different margins in printer's points on different sides; margins specified by **ImageMargins** appear outside the region defined by **ImageSize**
ImagePadding	**All**	what extra padding should be left for extended objects such as thick lines as well as tick and axis labels; *m*, same padding in printer's points on all sides; {{*left*, *right*}, {*bottom*, *top*}}, different paddings in printer's points on different sides; **All**, enough padding for all objects; **None**, no padding; padding specified by **ImagePadding** is left inside the region defined by **ImageSize**
ImageSize	**Automatic**	the size at which to render the graphic; *w* specifies a width of *w* printer's points; {*w*, *h*} specifies a width of *w* printer's points and a height of *h* printer's points; *w* and *h* can also be **Automatic**, **Tiny**, **Small**, **Medium**, or **Large**
LabelStyle	{}	the style for labels; *spec*, replace labels by **Style**[*label*, *spec*]; for information on **Style**, see Example 3.5.1
Method	**Automatic**	what graphics methods to use; "*name*", use the method with the specified name; **Automatic**, pick the method automatically
PlotLabel	**None**	an overall label for a plot; any expression can be used as a label; arbitrary strings of text can be given as "*text*"
PlotRange	**All**	the range of coordinates to include in a plot; **All** includes all points; **Automatic** drops outlying points; **Full** includes full range of original data; {*min*, *max*} specifies limits for *y*; {{x_{min}, x_{max}}, {y_{min}, y_{max}}} gives limits for each coordinate; *s* is equivalent to {{-*s*, *s*}, {-*s*, *s*}}
PlotRangeClipping	**False**	whether graphics objects should be clipped at the edge of the region defined by **PlotRange** or

Option Name	Default Value	Description
		should be allowed to extend to the actual edge of the image
PlotRangePadding	**Automatic**	how much further axes etc. should extend beyond the range of coordinates specified by **PlotRange**; p, the same padding in all directions; `{`p_x`,`p_y`}`, different padding in x and y directions; `{{`p_{xL}`,`p_{xR}`},{`p_{yB}`,`p_{yT}`}}`, different padding on left and right as well as bottom and top; p and p_i can be s (s coordinate units), **Scaled**[s] (a fraction s of the plot), **Automatic**, or **None**
PlotRegion	**Automatic**	the region that the plot should fill in the final display area; `{{`sx_{min}`,`sx_{max}`},{`sy_{min}`,`sy_{max}`}}` specifies the region in scaled coordinates; **Automatic** fills the final display area with the plot; when the plot does not fill the entire display area, the rest of the area is rendered according to the setting for the option **Background**
Prolog	`{}`	a list of graphics elements to be rendered before the main part of the graphics is rendered
RotateLabel	**True**	whether labels on vertical frame axes should be rotated to be vertical
Ticks	**Automatic**	what tick marks (and labels) to draw if there are axes; **None** gives no tick marks; **Automatic** places tick marks automatically; `{`*xticks*`,` *yticks*`}` specifies tick mark options separately for each axis; for the x-axis, **None** draws no tick marks, **Automatic** places tick marks automatically, `{`x_1`,` x_2`, ...}` draws tick marks at the specified positions, `{{`x_1`,` *label*$_1$`}, ...}` draws tick marks with the specified labels, `{{`x_1`,` *label*$_1$`,` *len*$_1$`}, ...}` draws tick marks with the specified scaled lengths, `{{`x_1`,` *label*$_1$`, {`*plen*$_1$`,` *mlen*$_1$`}}, ...}` draws tick marks with the specified lengths in the positive and negative directions, and `{{`x_1`,` *label*$_1$`,` *len*$_1$`,` *style*$_1$`}, ...}` draws tick marks with the specified styles; similarly for the y-axis
TicksStyle	`{}`	how axis ticks (tick marks and tick labels) should be rendered; *style*, all ticks are to be rendered with the specified style; *style* can be a rule for an option such as **FontSize** or **FontFamily**; *style* can be a graphics directive such as **Thick**, **Red**, **Dashed**, **Thickness**, **RGBColor**, **Dashing**, or **Directive** (i.e., a combination of directives and options); *style* can also

(Continued)

Option Name	Default Value	Description
		be a style name from the current stylesheet; {*xstyle*, *ystyle*}, ticks on different axes should use different styles; explicit style specifications for **Ticks** override those for **TicksStyle**

Note that **Graphics** and **Plot** have several different default option settings:

	Default for **Plot**	Default for **Graphics**
AspectRatio	**1/GoldenRatio**	**Automatic**
Axes	**True**	**False**
PlotRange	**{Full, Automatic}**	**All**
PlotRangeClipping	**True**	**False**

Section 2.3.1.2 discussed many of these options. In what follows, we provide examples for several new possibilities.

As mentioned in Example 3.5.2, **Show** can combine expressions created by built-in plotting functions with those constructed explicitly with graphics primitives and directives. **Epilog** and **Prolog** provide other ways to effect such combinations.

```
In[2]:= Plot[Sin[x], {x, 0, 2π},
          Epilog→{{GrayLevel[0.8], Polygon[{{2, 0}, {3, 1}, {4, 0}}]}}]
```

Out[2]=

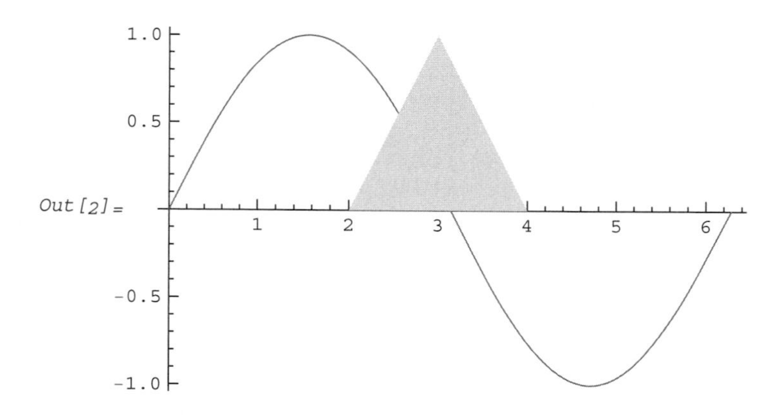

With **Epilog**, the main part of the graphic is rendered first and, therefore, the triangle covers part of the curve.

```
In[3]:= Plot[Sin[x], {x, 0, 2π},
          Prolog→{{GrayLevel[0.8], Polygon[{{2, 0}, {3, 1}, {4, 0}}]}}]
```

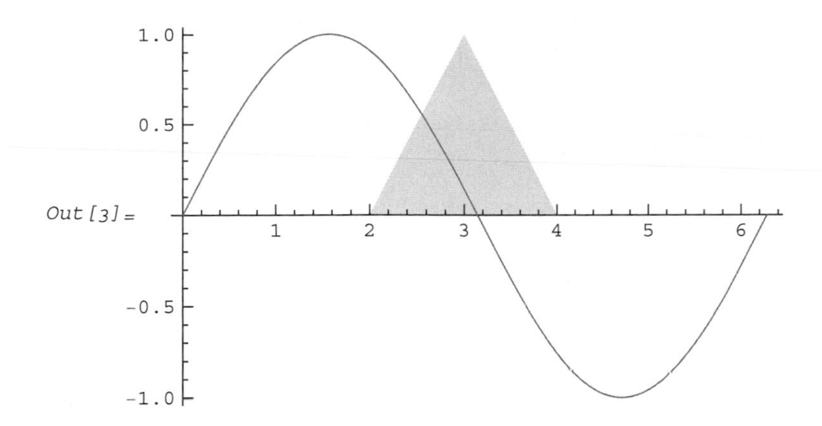

With **Prolog**, the triangle is behind the curve, which is rendered last. We can use **Graphics** rather than **Plot** to generate these plots:

```
In[4]:= Graphics[
    {GrayLevel[0.8], Polygon[{{2, 0}, {3, 1}, {4, 0}}]},
    Prolog -> List @@ Plot[Sin[x], {x, 0, 2π}],
    Axes -> True, AspectRatio -> 1/GoldenRatio,
    PlotRange -> {{0, 2π}, {-1, 1}},
    Ticks -> {Automatic, {-1.0, -0.5, 0.5, 1.0}}]
```

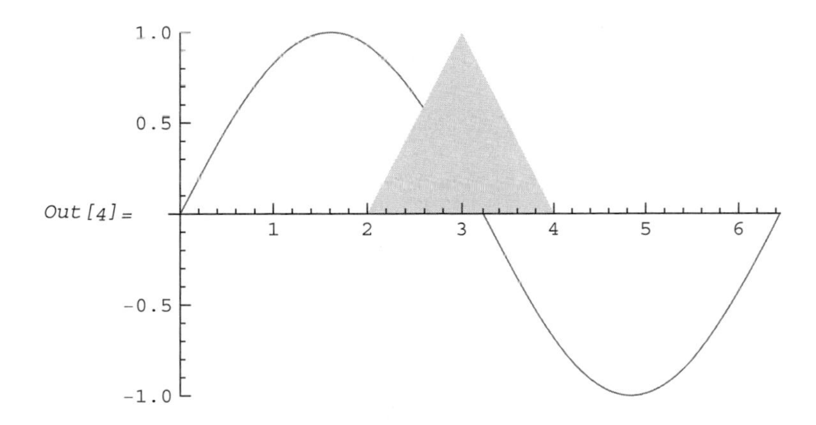


```
In[5]:= Graphics[
    {GrayLevel[0.8], Polygon[{{2, 0}, {3, 1}, {4, 0}}]},
    Epilog -> List @@Plot[Sin[x], {x, 0, 2π}],
    Axes -> True, AspectRatio -> 1/GoldenRatio,
    PlotRange -> {{0, 2π}, {-1, 1}},
    Ticks -> {Automatic, {-1.0, -0.5, 0.5, 1.0}}]
```

Out[5]=

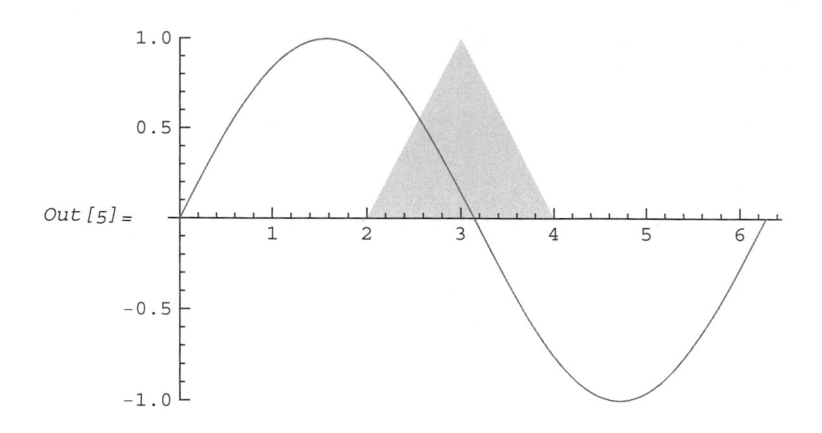

There are several ways to change the formats for texts in graphics: options **BaseStyle**, **FrameTicksStyle**, **LabelStyle**, **TicksStyle**, and **FormatType**, and functions **Style** and **Text**. Setting the options affects the default style and format type for a particular graphic; using the functions imposes the style and format type for a specific piece of text. Also, specifications for the functions take precedence over those for the options. Here is a plot of the charge on the capacitor versus time for the *RC* circuit:

```
In[6]:= Plot[{1, 1 - e^-t}, {t, 0, 4},
          PlotRange → {0, 1.2},
          PlotStyle → {Dashing[{0.015, 0.015}], Thickness[0.009]},
          AxesLabel → {"t", "q"}, PlotLabel → "Charge vs. time",
          Ticks → {{{1, "RC"}}, {{0.632, "Q'"}, {1, "Q"}}},
          Epilog → {Dashing[{0.015, 0.015}],
            Line[{{1, 0}, {1, 0.632}}], Line[{{0, 0.632}, {1, 0.632}}]}]
```

Out[6]=

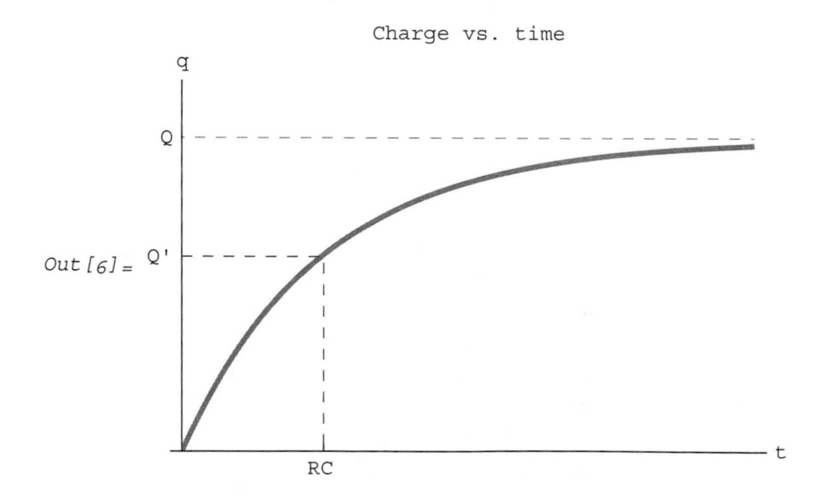

where after one time constant, RC, the charge reaches the value Q', which is 0.632 of the maximum value Q. Let us change the text styles in the plot:

```
In[7]:= Show[%,
          LabelStyle→{FontFamily→"Times", Italic, 12},
          PlotLabel→
           Style["Charge vs. time", FontFamily→"Helvetica", Plain, Red, 14]]
```

Out[7]=

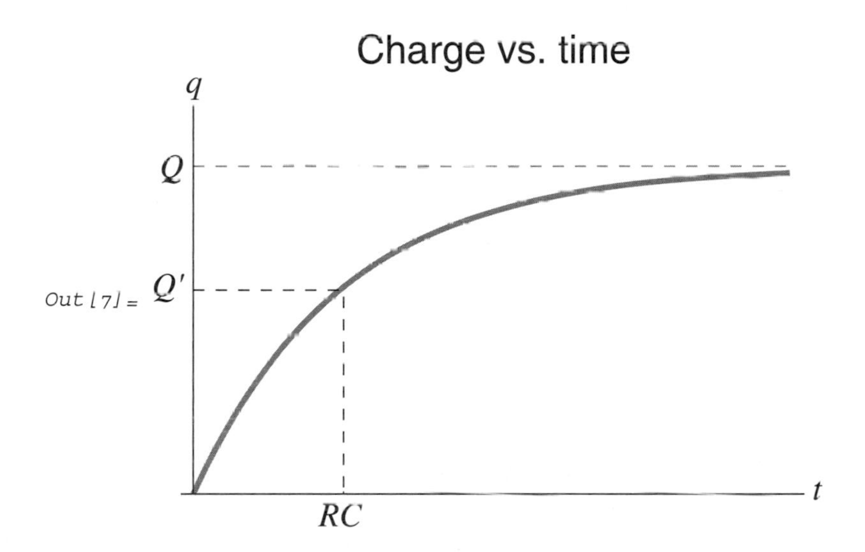

Note that style specifications in `Style` have higher precedence than those in `LabelStyle`.

With notebook front ends, we can also change the formats for texts in graphics by using the Font (or Show Fonts), Face, Size, Text Color, and Background Color submenus of the Format menu. Here is another way to produce the preceding plot:

```
In[8]:= Plot[{1, 1 - e^-t}, {t, 0, 4},
         PlotRange→{0, 1.2},
         PlotStyle→{Dashing[{0.015, 0.015}], Thickness[0.009]},
         AxesLabel→{"t", "q"}, PlotLabel→"Charge vs. time",
         Ticks→{{{1, "RC"}}, {{0.632, "Q'"}, {1, "Q"}}},
         Epilog→{Dashing[{0.015, 0.015}],
           Line[{{1, 0}, {1, 0.632}}], Line[{{0, 0.632}, {1, 0.632}}]}]
```

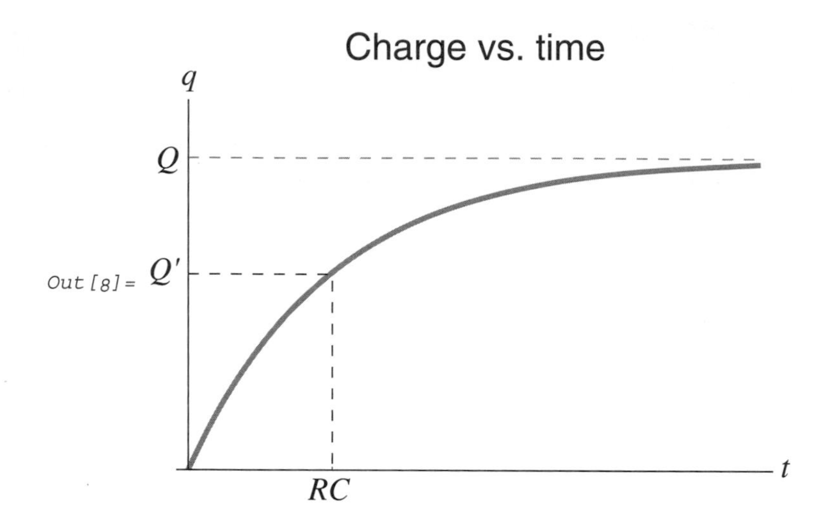

This section concludes with an example from geometric optics. Whereas Example 3.5.1 presents for a thin converging lens an unadorned ray diagram constructed solely with graphics primitives, this example displays for a concave mirror an enhanced ray diagram constructed with graphics directives and options as well as graphics primitives.

Example 3.5.3 For a spherical concave mirror, construct a ray diagram showing a virtual image.

The ray diagram comprises the following components:

- **mirror**

```
In[9]:= gr1 = {
          {RGBColor[0.250999, 0.878399, 0.815699],
           Tooltip[Disk[{-60, 0}, 64, {2Pi - Pi/6, 2Pi + Pi/6}], "mirror"]},
          {Lighter[Yellow, .7], Tooltip[Disk[{-60, 0}, 60, {2Pi - Pi/6 - .005,
              2Pi + Pi/6 + .005}], "None", ActionDelay -> 999]},
          {Thickness[0.004], Tooltip[Circle[{-60, 0}, 60,
             {2Pi - Pi/6 + 0.004, 2Pi + Pi/6 - 0.004}], "mirror"]}};
```

where `Tooltip[`*expr*, *label*`]` displays *label* as a tooltip while the mouse pointer is in the area where *expr* appears and the option setting `ActionDelay -> ` *t* specifies a delay of *t* seconds before the tooltip is displayed. Also, `Lighter[`*color*, *f*`]` represents a version of the specified color lightened by a fraction *f*.

- **principal axis**

```
In[10]:= gr2 = Tooltip[Line[{{-75, 0}, {75, 0}}], "principal axis"];
```

- object and image

```
In[11]:= gr3 = {Thickness[0.01],
          {Blue, Tooltip[Arrow[{{-20, 0}, {-20, 12}}], "object"]},
          {Dashed, Darker[RGBColor[1, 0.843104, 0], .25],
           Tooltip[Arrow[{{60, 0}, {60, 36}}], "image"]}};
```

where `Darker[`*color*,*f*`]` represents a version of the specified color darkened by a fraction *f*.

- incident and reflected rays

```
In[12]:= gr4 = {
           {Arrowheads[{{0.03, 0.75}}], Red, Tooltip[Arrow[
              {{-20, 12}, {-60 (1 - Cos[α]), 60 Sin[α]}}], "incident ray"]},
           {Red, Tooltip[Arrow[{{-20, 12}, {-72, -12 Tan[α]}}],
             "reflected ray"]}, {Arrowheads[{{0.03, 0.75}}], Red,
            Tooltip[Arrow[{{-20, 12}, {-δ, 12}}], "incident ray"]},
           {Red, Tooltip[Arrow[{{-δ, 12}, {-60, -30 Tan[β]}}],
             "reflected ray"]}} /.
          {α -> ArcTan[36/120], β -> ArcTan[36/90], δ -> 60 - √(60² - 12²)};
```

- apparent paths of the reflected rays

```
In[13]:= gr5 = {
           {Dashing[0.005],
            Tooltip[Line[{{-60 (1 - Cos[α]), 60 Sin[α]}, {60, 36}}],
             "apparent path of reflected ray"]},
           {Dashing[0.005], Tooltip[Line[{{-δ, 12}, {60, 36}}],
             "apparent path of reflected ray"]}} /.
          {α -> ArcTan[36/120], δ -> 60 - √(60² - 12²)};
```

- center of curvature, focal point, and their labels

```
In[14]:= gr6 = {
          {PointSize[0.015],
           Tooltip[Point[{-60, 0}], "center of curvature"],
           Tooltip[Point[{-30, 0}], "focal point"]},
          Tooltip[Text[C, {-60, -2}, {0, 1}], "center of curvature"],
          Tooltip[Text[F, {-30, -2}, {0, 1}], "focal point"]};
```

- mirror equation

```
In[15]:= gr7 = {Tooltip[Inset["1/p + 1/q = 1/f", {35, 12}], "mirror equation"]};
```

- **dimension lines**

```
In[16]:= gr8 = {
             Arrowheads[
              {{-.025, 0},
               {.015, 0, Graphics[Line[{{0, 1}, {0, -1}}]]}, {.05, .5,
                Graphics[Tooltip[Inset[Style[#3, Small], {0, .5}], #4]]},
               {.015, 1, Graphics[Line[{{0, 1}, {0, -1}}]]}, {.025, 1}}],
             Tooltip[Arrow[{#1, #2}], "dimension line"]} &@@#& /@
          {{{-20, -7}, {0, -7}, p, "object distance"},
           {{0, -7}, {60, -7}, q, "Image distance"},
           {{-30, -16}, {0, -16}, f, "focal length"}};
```

where each arrow is drawn with five arrowhead elements: left arrowhead, left vertical line, dimension text, right vertical line, and right arrowhead.

Graphics with the graphics elements and specified options generates the ray diagram:

```
In[17]:= Graphics[
          {gr1, gr2, gr3, gr4, gr5, gr6, gr7, gr8},
          BaseStyle -> FontFamily -> "Times", Frame -> True,
          FrameLabel -> "Ray Diagram for a Concave Mirror",
          FrameTicks -> {Table[i, {i, -60, 60, 20}], {-20, 0, 20}, None, None},
          PlotRangePadding -> 4, Background -> Lighter[Yellow, .7]]
```

Out[17]=

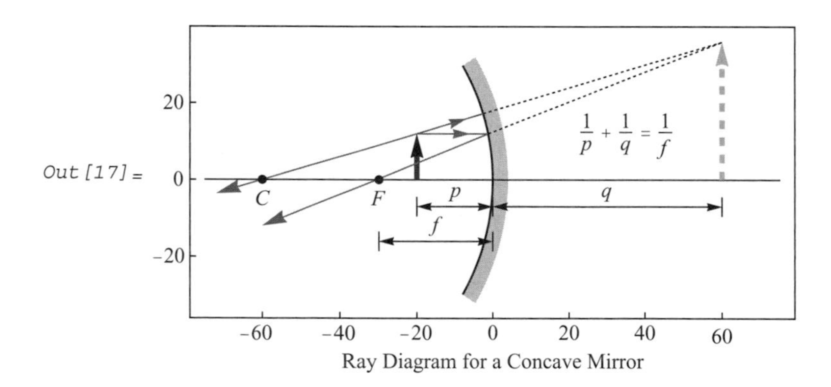

Moving the mouse pointer over an element of the graphic shows its name or description.

```
In[18]:= ClearAll[gr1, gr2, gr3, gr4, gr5, gr6, gr7, gr8]
```

■

3.5.2.4 Wave Motion

This section applies the two-dimensional graphics programming capabilities of *Mathematica* to an introductory physics problem.

The Problem A simple harmonic or sine wave travels in a long string stretched along the x direction. Animate the motion of a particle in the string, and demonstrate that the frequency and period of the simple harmonic motion of the particle equal those of the wave motion.

Physics of the Problem The equation of the wave is

$$y = A \sin 2\pi\left(\frac{x}{\lambda} - \frac{t}{T}\right) \tag{3.5.1}$$

where

y: the displacements of the particles in the y direction away from their equilibrium positions
x: the distance along the direction of propagation of the wave
t: time
A: amplitude, the magnitude of the maximum y displacement
λ: wavelength, the distance between two adjacent points having the same phase
T: period, the time for one complete vibration to pass a given point

The frequency f of the wave is defined as the number of vibrations per second passing a given point. It is related to the period by

$$f = \frac{1}{T} \tag{3.5.2}$$

Equation 3.5.1 implies that each particle of the string oscillates in the y direction with simple harmonic motion having the same period and, according to Equation 3.5.2, the same frequency as those of the wave motion.

If y, x, and t are in units of $A, \lambda/2\pi$, and $T/2\pi$, respectively, then Equation 3.5.1 can be written as

$$y = \sin(x - t) \tag{3.5.3}$$

Without loss of generality, we focus on the motion of the particle at $x = 0$. From Equation 3.5.3, the displacement of the particle as a function of time is

$$y = \sin(-t) \tag{3.5.4}$$

Solution with *Mathematica* To animate the motion of the particle as the wave passes by, we create a list of graphs of Equations 3.5.3 and 3.5.4 together for a number of instants and use **ListAnimate** to generate the animation:

```
In[1]:= ListAnimate[
          Table[
           Show[
            Plot[Sin[x - i (2 π/8)], {x, -4 π, 4 π}],
            Graphics[
             {{Blue, Dashed, Line[{{0, -1}, {0, 1}}]},
              PointSize[0.05], Red, Point[{0, Sin[-i (2 π/8)]}]}],
            Frame→True, FrameTicks→None, Background→GrayLevel[0.7],
            Axes→False, PlotRangePadding -> {None, 0.25}], {i, 0, 7}]]
```

Out[1]=

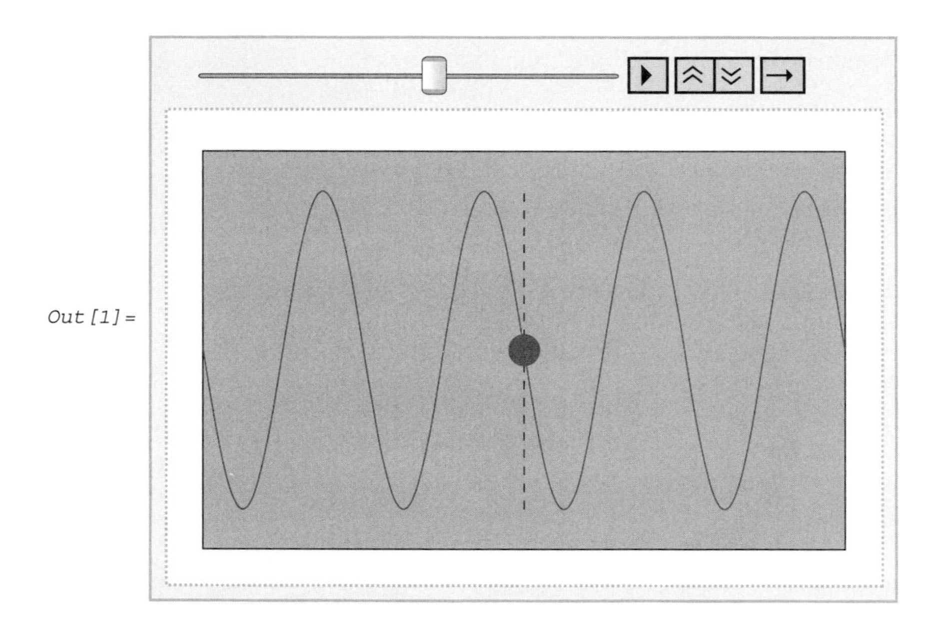

A vertical line is included in each graph to show that the direction of motion of the particle is transverse to the direction of wave propagation.

To start the animation if it is not already in progress, click the Play button; to stop the animation, click the Pause button. Click the Faster or Slower button to control the speed of the animation.

Note that the particle completes a cycle of motion as one wave vibration passes by it. In other words, the frequency and period of the motion of the particle equal those of the wave motion.

3.5.3 Three-Dimensional Graphics

3.5.3.1 Three-Dimensional Graphics Primitives

A number of three-dimensional graphics primitives are available in *Mathematica*:

Cuboid[{x_{min}, y_{min}, z_{min}}]	unit cube with edges oriented parallel to the axes; two opposite corners have coordinates $\{x_{min}, y_{min}, z_{min}\}$ and $\{x_{min}+1, y_{min}+1, z_{min}+1\}$
Cuboid[{x_{min}, y_{min}, z_{min}}, {x_{max}, y_{max}, z_{max}}]	cuboid (rectangular parallelepiped) with edges oriented parallel to the axes; two opposite corners have the specified coordinates
Cylinder[{{x_1, y_1, z_1}, {x_2, y_2, z_2}, r}]	cylinder of radius r around the line from $\{x_1, y_1, z_1\}$ to $\{x_2, y_2, z_2\}$; the default value for r is 1
GraphicsComplex[{pt_1, pt_2, ...}, *data*]	graphics complex in which coordinates given as integers i in graphics primitives in *data* are taken to be pt_i; *data* can be any nested list of graphics primitives and directives
Line[{{x_1, y_1, z_1}, {x_2, y_2, z_2}, ...}]	line joining the points $\{x_1, y_1, z_1\}$, $\{x_2, y_2, z_2\}$, ...
Line[{{{x_{11}, y_{11}, z_{11}}, {x_{12}, y_{12}, z_{12}}, ...}, ...}]	collection of lines
Point[{x, y, z}]	point with coordinates $\{x, y, z\}$
Point[{{x_1, y_1, z_1}, {x_2, y_2, z_2}, ...}]	collection of points
Polygon[{{x_1, y_1, z_1}, {x_2, y_2, z_2}, ...}]	filled polygon with the specified list of corners
Polygon[{{{x_{11}, y_{11}, z_{11}}, {x_{12}, y_{12}, z_{12}}, ...}, ...}]	collection of polygons
Sphere[{x, y, z}, r]	sphere of radius r centered at $\{x, y, z\}$; the default value for r is 1
Text[*expr*, {x, y, z}]	text corresponding to the textual form of *expr*, centered at the point $\{x, y, z\}$

Mathematica encloses each three-dimensional figure within a cuboidal box. The x-, y-, and z-axes are oriented parallel to the edges of the box and form a right-handed coordinate system. If the option setting **Boxed → False** is included in **Show** or **Graphics3D**, the box is not shown explicitly.

Coordinates for graphics primitives can be specified in either ordinary or scaled coordinates. Ordinary coordinates are given in the form {x, y, z}; scaled coordinates, running from 0 to 1 in each dimension, are given in the form **Scaled**[{sx, sy, sz}].

```
In[1]:= Graphics3D[
          {Cuboid[{0.5, 0.5, 0.5}, {0.95, 0.95, 0.95}],
           Cuboid[Scaled[{0.5, 0.5, 0.5}], Scaled[{0.95, 0.95, 0.95}]],
           Dashing[{0.02, 0.02}],
           Line[{{{0.5, 0.5, 0.5}, {0.5, 0.5, 0}},
             {{0.5, 0.5, 0}, {0.5, 0, 0}}, {{0.5, 0.5, 0}, {0, 0.5, 0}}}]},
          PlotRange→{{0, 2}, {0, 2}, {0, 2}}, Axes→True,
          AxesLabel→{"x", "y", "z"}]
```

Out[1]=

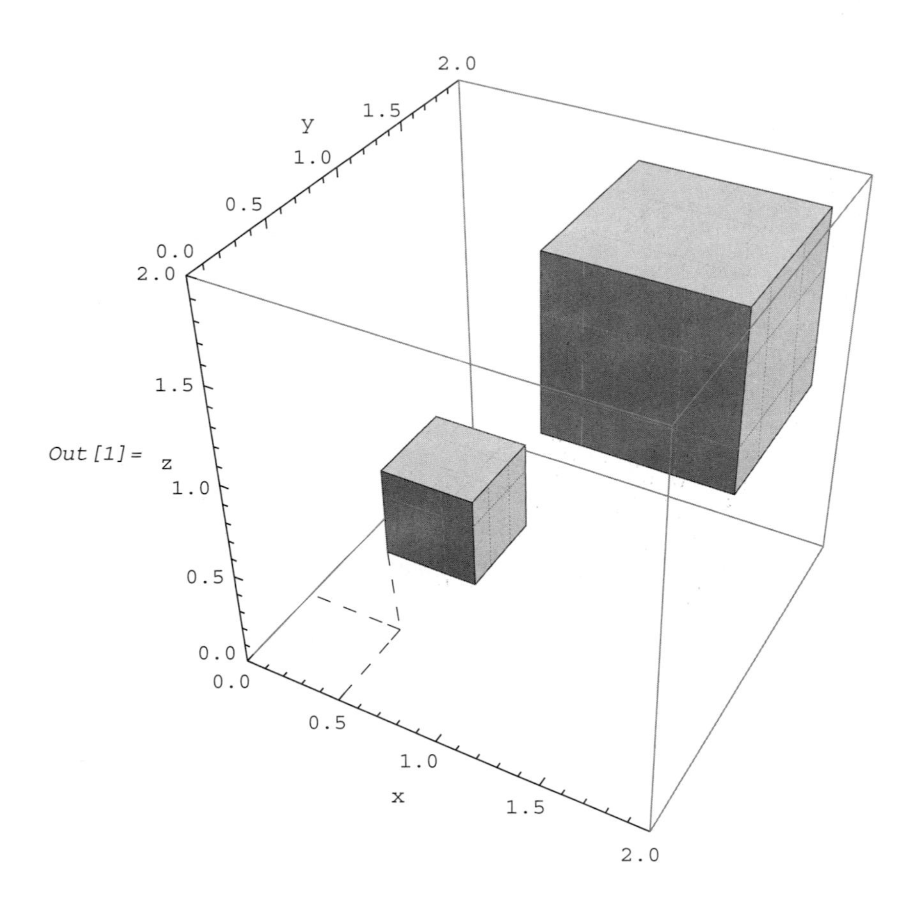

The opposite corners of the smaller cuboid are specified in ordinary coordinates, whereas those of the larger one are given in scaled coordinates. To facilitate visualization, three dashed lines are drawn for locating a corner of the smaller cuboid.

The corners of a polygon can form a nonconvex figure:

```
In[2]:= Graphics3D[Polygon[{{-1, -1, 0}, {-1, 1, 0},
          {1, 1, 0}, {1, -1, 0}, {0.5, 0, 0}, {-0.5, 0, 0}}],
         Axes→True, AxesLabel -> {"x", "y", "z"}]
```

Out[2]=

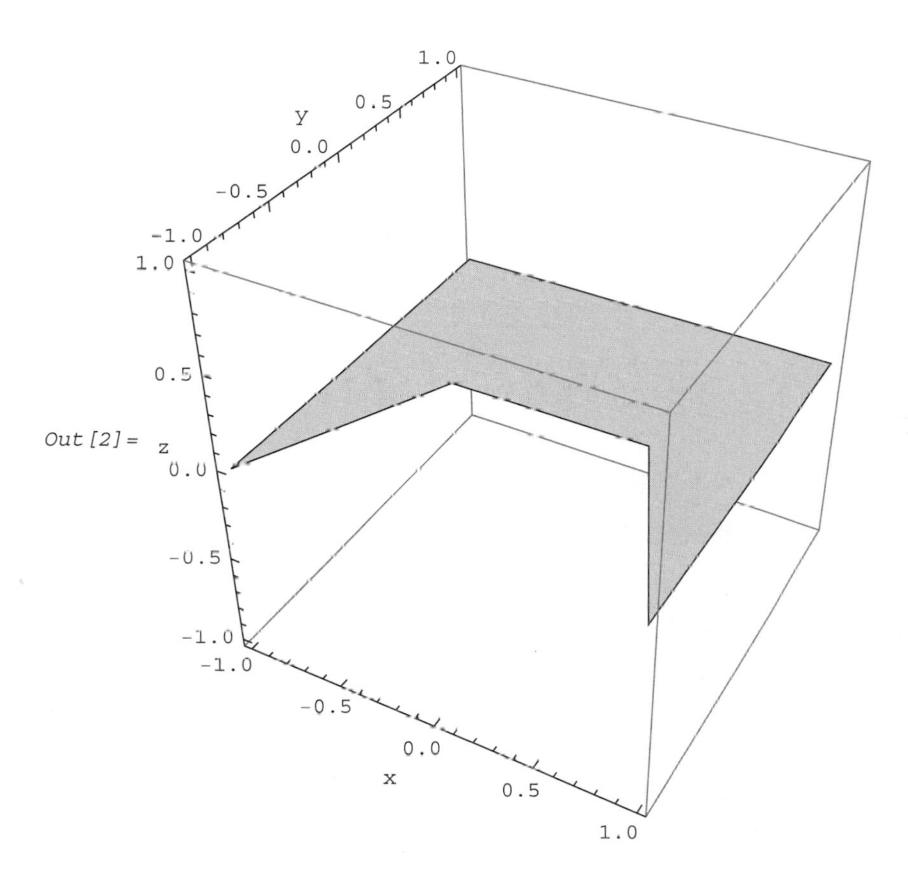

Polygons can be self-intersecting:

```
In[3]:= Graphics3D[
         Polygon[{
           {-1, -1, 0}, {-1, 1, 0}, {1, 1, 0},
           {1, -1, 0}, {0.5, 3, 0}, {0, -0.5, 0}, {-0.5, 3, 0}}],
         Axes→True, AxesLabel -> {"x", "y", "z"}]
```

Out[3]=

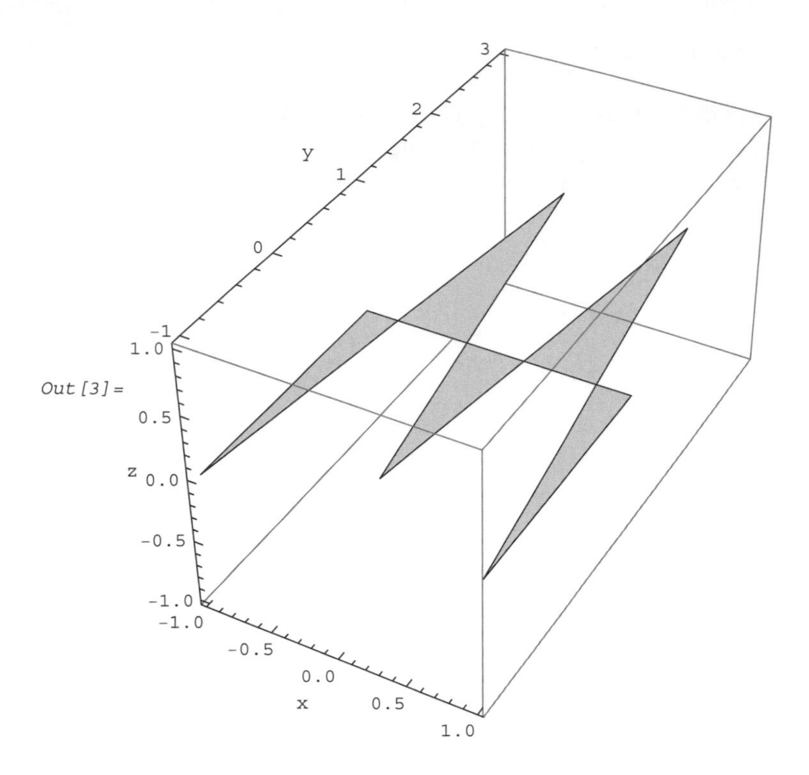

The corners of a polygon need not lie in a plane:

```
In[4]:= Graphics3D[Polygon[{{-1, -1, 0}, {-1, 1, 0},
           {1, 1, 0}, {1, -1, 0}, {0.5, 0, 1}, {-0.5, 0, 1}}],
         Axes→True, AxesLabel -> {"x", "y", "z"},
         ViewPoint→{3.17392, -1.06549, 0.49086}]
```

Out[4]=

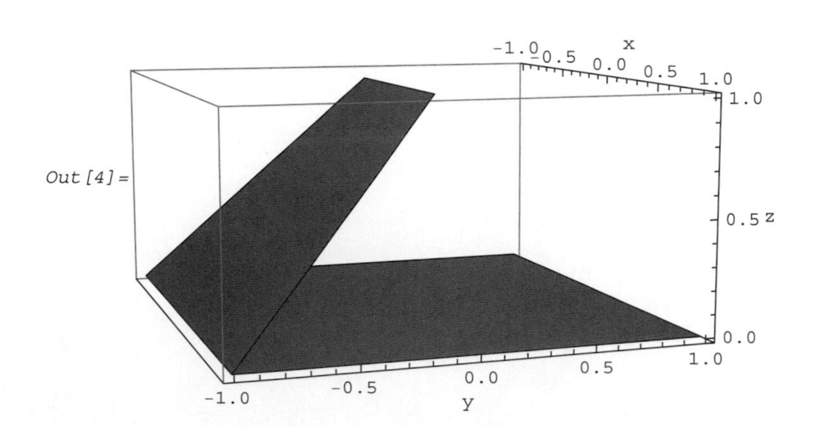

In the rendering of two-dimensional graphics, graphics objects whose expressions appear later as arguments of **Show** cover the earlier ones. Similarly, objects represented by graphics primitives that come later in the list for **Graphics** hide the earlier ones. In the displaying of three-dimensional graphics, the order of appearance of the expressions for graphics objects in **Show** or graphics primitives in **Graphics3D** is irrelevant. In this case, parts of the figure in front obscure those behind them.

```
In[5]:= Graphics3D[
          Polygon /@ {{{-1, -1, 0}, {-1, 1, 0}, {1, 1, 0}, {1, -1, 0}},
            {{-1, -1, -1.5}, {1, 1, -1.5}, {0, 0, 1}}}]
```

Out[5]=

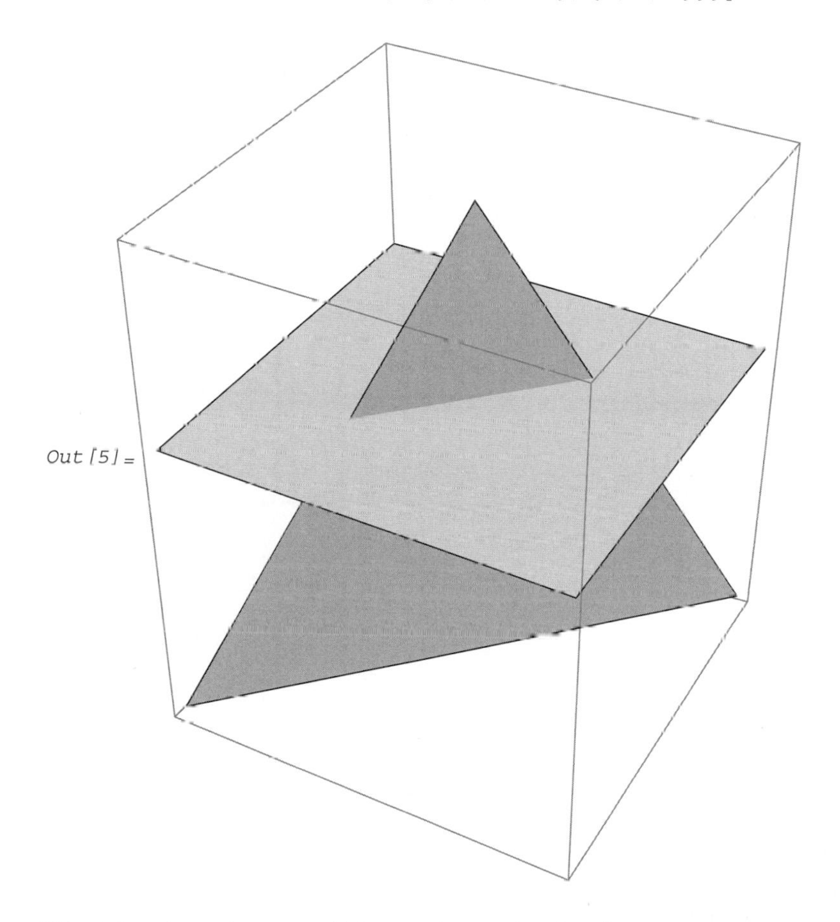

The triangle and square hide portions of each other.

To demonstrate the use of three-dimensional graphics primitives, we choose a problem from introductory electricity and magnetism.

Example 3.5.4 Construct a diagram depicting a snapshot of a linearly polarized, sinusoidal, and plane electromagnetic wave propagating in the positive y direction.

We begin with the plot of the electric and magnetic fields as a function of y at a particular instance of time:

```
In[6]:= gr1 = ParametricPlot3D[{{0, t, Sin[t]}, {Sin[t], t, 0}}, {t, 0, 4.2 π}];
```

ParametricPlot3D[{{f_x, f_y, f_z}, {g_x, g_y, g_z}, ...}, {t, t_{min}, t_{max}}] produces together several three-dimensional space curves parameterized by a variable t, which runs from $t_{\min}$ to $t_{\max}$. The following list contains the **Line** primitive representing the axes:

```
In[7]:= gr2 = {Line[{{{0, 0, 0}, {0, 4.5 π, 0}},
              {{0, 0, 0}, {0, 0, 1}}, {{0, 0, 0}, {1, 0, 0}}}]};
```

The **Line** and **Polygon** primitives in the following list represent the arrows for the electric and magnetic field vectors:

```
In[8]:= gr3 = Join[
          Line /@
           Flatten[
            Table[{{{0, i, 0}, {0, i, Sin[i]}},
              {{0, i, 0}, {Sin[i], i, 0}}}, {i, 0, 4 π, π/4}], 1],
          Polygon /@
           (Flatten[
              Table[{{{0, i, Sin[i]}, {0, i + Δ, Sin[i] - Sign[Sin[i]] Δ},
                 {0, i - Δ, Sin[i] - Sign[Sin[i]] Δ}}, {{Sin[i], i, 0},
                 {Sin[i] - Sign[Sin[i]] Δ, i + Δ, 0}, {Sin[i] - Sign[Sin[i]] Δ,
                  i - Δ, 0}}}, {i, 0, 4 π, π/4}], 1] /. Δ → 0.2)];
```

The **Text** primitives are for the labels:

```
In[9]:= gr4 =
         MapThread[
           Text,
           {MapThread[Style,
              {{"x", "y", "z", "E", "B"}, {opts1, opts1, opts1, opts2, opts2}}],
            {{1 + 0.2, 0.5, 0}, {0, 4.5 π + 0.5, 0}, {0, 0, 1 + 0.25},
              {0, 3 π/4 + 0.25, Sin[3 π/4] + 0.5},
              {Sin[3 π/4] + 0.5, 3 π/4 + 0.4, 0}}}] /.
          {opts1 → Sequence[FontFamily → "Times", Italic],
           opts2 → Sequence[FontFamily → "Times", Bold]};
```

We have taken the opportunity to illustrate the use of the function **MapThread**. **MapThread**[f, {{a_1, a_2, ...}, {b_1, b_2, ...}, ...}] gives {f[a_1, b_1, ...], f[a_2, b_2, ...], ...}.

Its use is elegant but optional. Perhaps, it may be clearer just to specify each piece of text explicitly in the form **Text[Style[***expr*, *options***], {***x*, *y*, *z***}]**. Finally, **Show** displays the graphics objects combined:

```
In[10]:= Show[gr1, Graphics3D[Join[gr2, gr3, gr4]], BoxRatios→{0.3, 1, 0.3},
          Boxed → False, Axes → False, PlotRange → {{-1, 1}, {0, 4.5π}, {-1, 1}},
          ViewPoint → {2.647, 1.137, 1.775}]
```

Out[10]=

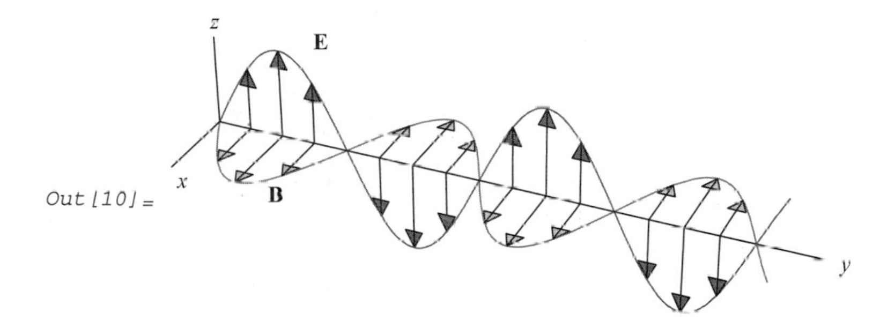

Note that **Show** can combine expressions created by built-in plotting functions with those constructed explicitly with graphics elements.

```
In[11]:= ClearAll[gr1, gr2, gr3, gr4]
```

■

3.5.3.2 Three-Dimensional Graphics Directives

Other than **Arrowheads**, the graphics directives introduced in Section 3.5.2.2 for two dimensions remain efficacious in three dimensions: **GrayLevel**, **RGBColor**, **Hue**, **Opacity**, **PointSize**, **AbsolutePointSize**, **Thickness**, **AbsoluteThickness**, **Dashing**, **AbsoluteDashing**, **EdgeForm**, **FaceForm**, and **Directive**. *Mathematica* provides two additional directives for three dimensions: **Glow** and **Specularity**.

Two- and three-dimensional graphics have the same graphics directives for the **Point** and **Line** primitives:

```
In[1]:= Graphics3D[
         {{PointSize[0.3], Opacity[0.35], Point[{0, 0, 0}]},
          {Hue[1], AbsoluteThickness[9], Line[{{0, 0, -1}, {0, 0, 1}}]}},
         PlotRange→{{-1, 1}, {-1, 1}, {-1, 1}}]
```

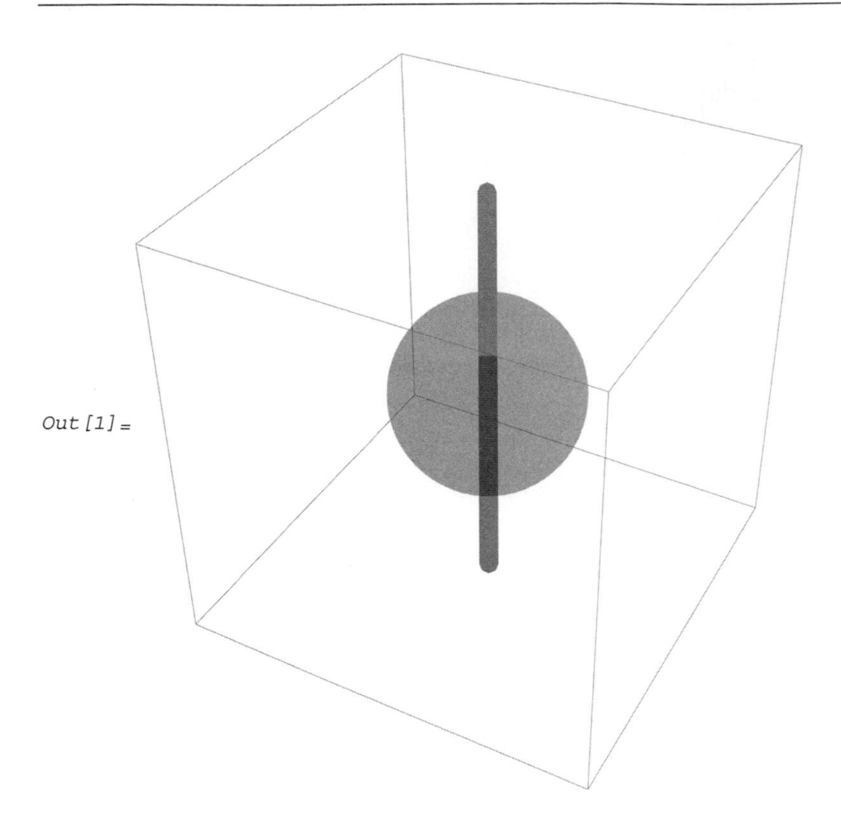

The diameter of the point is 30 percent of the total width of the graphic; the thickness of the line is approximately nine printer's points, or $9 \times 1/72$ of an inch.

EdgeForm determines how the edges of polygons, cuboids, and cylinders are drawn. With **EdgeForm[]**, no edges are drawn; with **EdgeForm[*g*]**, edges are drawn according to the specified graphics directive or list of graphics directives *g*. Whereas the default is to draw no edges for two-dimensional graphics, the default for three-dimensional graphics is to draw edges.

```
In[2]:= Graphics3D[{EdgeForm[{Red, Thickness[0.03]}],
          Polygon[{{-1, -1, -1.5}, {1, 1, -1.5}, {0, 0, 1}}],
          EdgeForm[], Cuboid[{0, -1, -1.5}, {0.75, 0, -0.5}]},
         Boxed→False]
```

Out[2]=

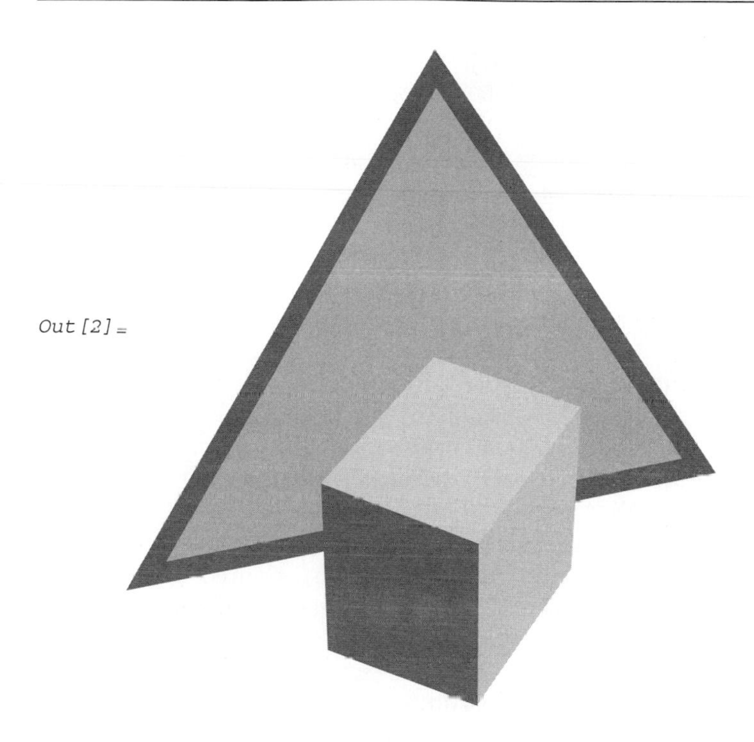

No edges are drawn for the cuboid; the edges of the triangle have a thickness of 3 percent of the total width of the graphic, and their color is red.

Four components determine the actual rendered colors of surfaces:

Component	Specify With
Glow	**`Glow`**
Diffuse reflection	**`GrayLevel`** or **`RGBColor`** (**`Black`** specifies no diffuse reflection, and **`White`** is the default)
Specular reflection	**`Specularity`**
Simulated lighting	**`Lighting`** (**`Lighting -> None`** specifies no simulated lighting, and **`Lighting -> Automatic`** is the default)

Section 3.5.3.3 will elaborate on simulated lighting; this section focuses on the other three components.

`Glow[`*color*`]`	surfaces of graphics objects that follow are to glow with *color*; `Glow[]`, the default, specifies that there is no glow
`GrayLevel[`*s*`]` or `RGBColor[`*r*`,`*g*`,`*b*`]`	surfaces of graphics objects that follow are to have the specified color for diffuse reflection; see Section 3.5.2.2 for `RGBColor` specifications for some common colors; in diffuse reflection, the surface appears equally bright when viewed from any direction (more precisely, in diffuse reflection, the radiance of the surface is independent of the viewing angle from the surface normal; for more information, see [PP93] and [Mey89]); the intensity of reflected light is $\cos\alpha$ times the intensity of the incident light, where α is the angle the incident light makes with the normal to the surface; with `GrayLevel[`*s*`]`, the surfaces are to reflect a fraction *s* of light that falls on them and the reflected light is to have the same color as the incident light; with `RGBColor[`*r*`,`*g*`,`*b*`]`, the red, green, and blue components of reflected light equal, respectively, *r*, *g*, and *b* times the corresponding components of the incident light
`Specularity[`*color*`]`	`GrayLevel[`*s*`]`, or simply *s*, surfaces of graphics objects that follow are to specularly reflect a fraction *s* of light that falls on them and the reflected light is to have the same color as the incident light; `RGBColor[`*r*`,`*g*`,`*b*`]`, or named color such as `Orange`, the red, green, and blue components of the reflected light are to be respectively *r*, *g*, and *b* times the corresponding components of the incident light; `Specularity[]`, `Specularity[0]`, or `Specularity[Black]`, the default, specifies no specular reflection
`Specularity[`*color*`,`*n*`]`	use specular exponent *n*; the default value is 1.5; shinier surfaces have higher values; intensity of the reflected light at an angle θ from the mirror-reflection direction is to be proportional to $\cos^n(\theta)$ for $\theta \leq 90°$ and 0 for $\theta > 90°$

With the specification `Lighting -> None`, diffuse and specular reflections become dormant and `Glow` alone determines the actual surface colors:

```
In[3]:= Graphics3D[
         {Glow[Red], Polygon[{{-1, -1, -1.5}, {1, 1, -1.5}, {0, 0, 1}}],
          Glow[GrayLevel[0.75]], Cuboid[{0, -1, -1.5}, {0.75, 0, -0.5}]},
         Boxed→False, Lighting -> None]
```

Out[3]=

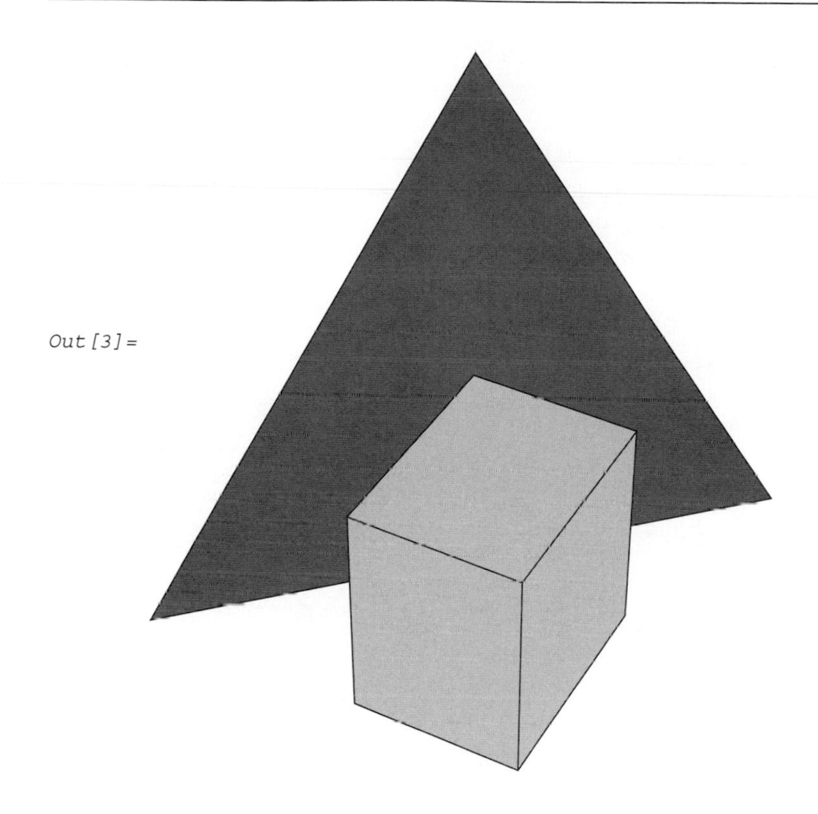

The color of the triangle is red and that of the cuboid is gray. The glow component is independent of the simulated lighting component. Even without the specification **Lighting -> None**, **Glow** alone still determines the actual surface colors as long as it is used in conjunction with the directives **Black** and **Specularity[]** that specify diffuse and specular reflections are to be dormant:

```
In[4]:= Graphics3D[
         {Black, Specularity[], Glow[Red],
          Polygon[{{-1, -1, -1.5}, {1, 1, -1.5}, {0, 0, 1}}],
          Glow[GrayLevel[0.75]], Cuboid[{0, -1, -1.5}, {0.75, 0, -0.5}]},
         Boxed→False]
```

Out[4]=

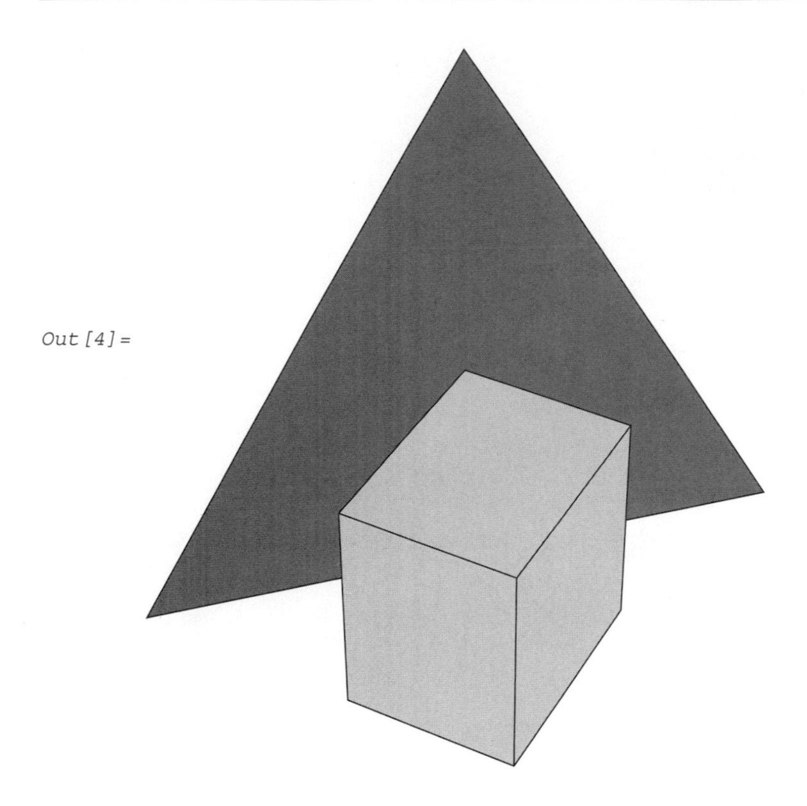

which is exactly the same graphic as that obtained previously. (**Specularity[]** can be omitted in the input because it is the default.)

With the default setting **Lighting -> Automatic,** the diffuse reflection component contributes to the actual rendered colors of surfaces:

```
In[5]:= GraphicsGrid[
          {{Graphics3D[{GrayLevel[0.75],
              Polygon[{{-1, -1, -1.5}, {1, 1, -1.5}, {0, 0, 1}}]}],
            Graphics3D[{Yellow, Polygon[{{-1, -1, -1.5},
                {1, 1, -1.5}, {0, 0, 1}}]}],
            Graphics3D[{Red, Polygon[{{-1, -1, -1.5},
                {1, 1, -1.5}, {0, 0, 1}}]}]}}]
```

Out[5]=

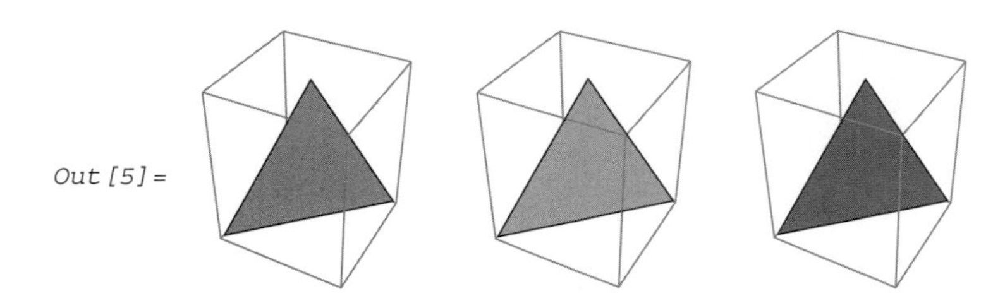

where **GraphicsGrid[{{**g_{11}**,**g_{12}**,...},...}]** generates a graphic in which the g_{ij} are laid out in a two-dimensional grid. Note that the colors of the triangles are not necessarily the specified ones because simulated lighting also affects the actual rendered colors.

FaceForm allows us to give different color specifications to the front and back faces of polygons. **FaceForm[**gf**,**gb**]** specifies the directive or list of directives gf for the front faces and the directive or list of directives gb for the back faces. What is a front face? When we look at the front face of a polygon, the consecutive points in **Polygon[{**$point_1$**,**$point_2$**,**$point_3$**,...}]**, which represents the polygon, proceed in a counterclockwise direction.

```
In[6]:= Graphics3D[
         {FaceForm[Glow[Red], {Glow[GrayLevel[0.75]], Opacity[0.75]}],
          Polygon[{{-1, -1, 0}, {1, -1, 0}, {1, 1, 0}, {-1, 1, 0}}],
          Polygon[{{-1, -1, -1.5}, {0, 0, 1}, {1, 1, -1.5}}]},
         Axes → True, AxesLabel → {"x", "y", "z"}, Lighting → None]
```

Out[6]=

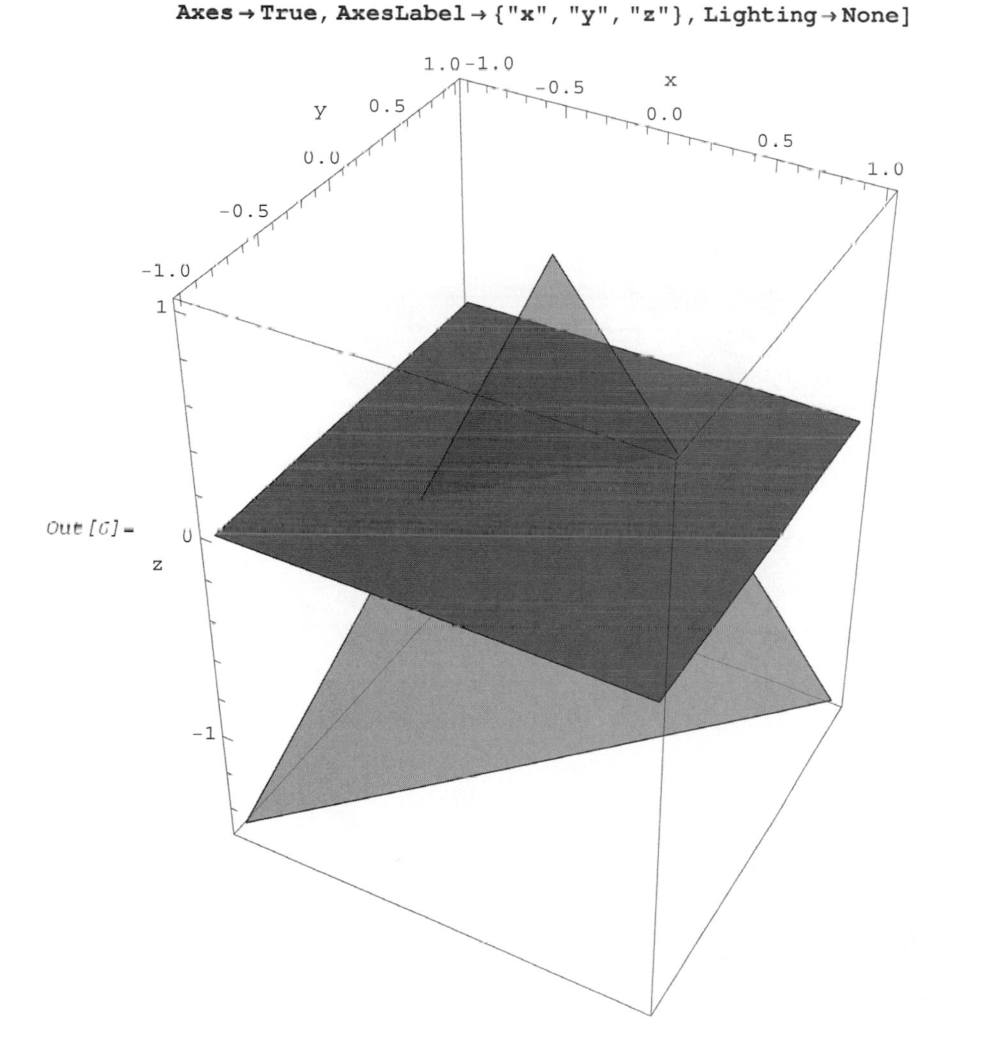

We are looking at the front of the square and the back of the triangle. With the default setting **Lighting -> Automatic**, adding the diffuse reflection component to the glow component changes the surface colors:

```
In[7]:= Graphics3D[
          {FaceForm[{Glow[Red], Blue},
            {Glow[GrayLevel[0.75]], Yellow, Opacity[0.75]}],
           Polygon[{{-1, -1, 0}, {1, -1, 0}, {1, 1, 0}, {-1, 1, 0}}],
           Polygon[{{-1, -1, -1.5}, {0, 0, 1}, {1, 1, -1.5}}]},
          Axes → True, AxesLabel → {"x", "y", "z"}]
```

Out[7]=

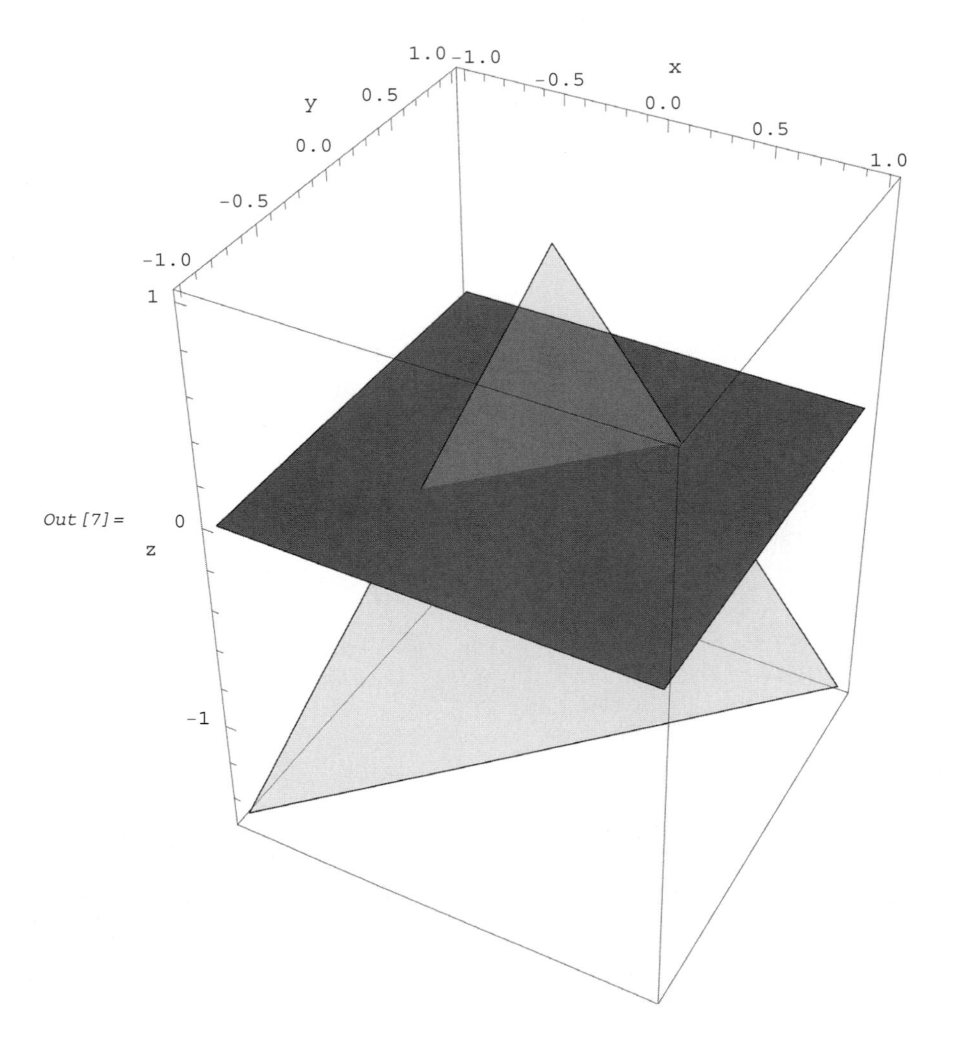

Using **PlotStyle**, we can include **FaceForm** in the three-dimensional plotting functions:

```
In[8]:= ParametricPlot3D[
          {Cos[ϕ]Sin[θ], Sin[ϕ]Sin[θ], Cos[θ]}, {ϕ, 0, 25Pi/16}, {θ, 0, Pi},
          PlotStyle→FaceForm[Glow[Yellow], Glow[GrayLevel[0.35]]],
          Boxed->False, Axes->False, Lighting->None]
```

Out[8]=

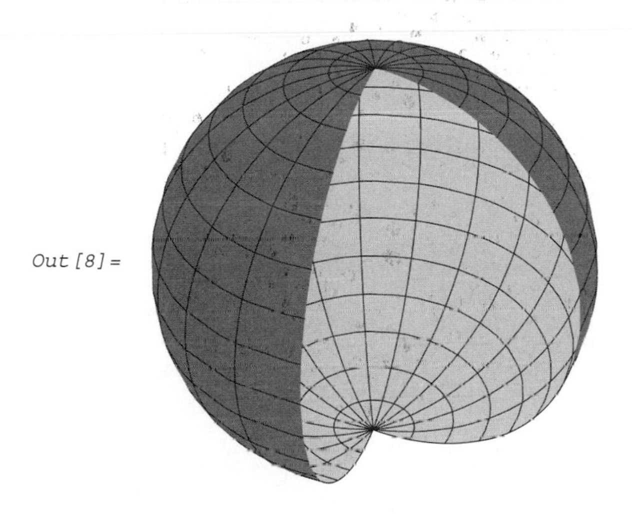

The front and back faces of the polygons have different colors; the inside of the sphere glows in bright yellow, and the outside is shrouded with dark gray.

Adding the specular reflection component to the glow, diffuse reflection, and simulated lighting components in coloring curved surfaces can produce interesting visual effects:

```
In[9]:= GraphicsGrid[
          {{Graphics3D[{Sphere[{0, 0, 0}]}],
            Graphics3D[{Black, Specularity[Yellow, 15], Sphere[{0, 0, 0}]}],
            Graphics3D[
             {Glow[Red], Yellow, Specularity[1, 10], Sphere[{0, 0, 0}]}],
            Graphics3D[{Glow[Red], Orange, Specularity[1, 20],
             Sphere[{0, 0, 0}]}]}}, ImageSize->7 × 72]
```

Out[9]=

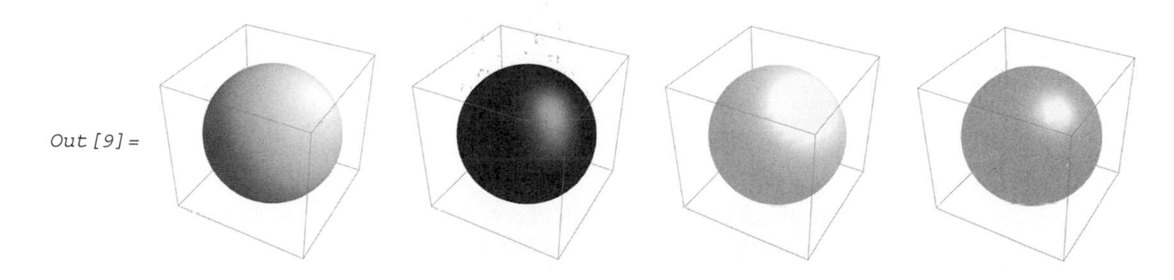

The three-dimensional plotting functions also support the four components for coloring surfaces. Let us revisit Example 2.3.21 that animates the $n = 2$ normal mode of vibration of the acoustic membrane:

```
In[10]:= Animate[
   ParametricPlot3D[{r Cos[θ], r Sin[θ],
     BesselJ[2, r] Sin[2 θ] Cos[(2 π/16) i]}, {r, 0, 5.13562}, {θ, 0, 2 π},
    PlotRange → {{5, -5}, {5, -5}, {-0.5, 0.5}}, PlotPoints → 25,
    PlotStyle -> {Glow[GrayLevel[0.1]], Orange, Specularity[1, 20]},
    BoxRatios → {1, 1, 0.4}, Viewpoint → {2.340, -1.795, 1.659},
    Boxed → False, Axes → False, Mesh -> None], {i, 0, 15, 1}]
```

Out[10]=

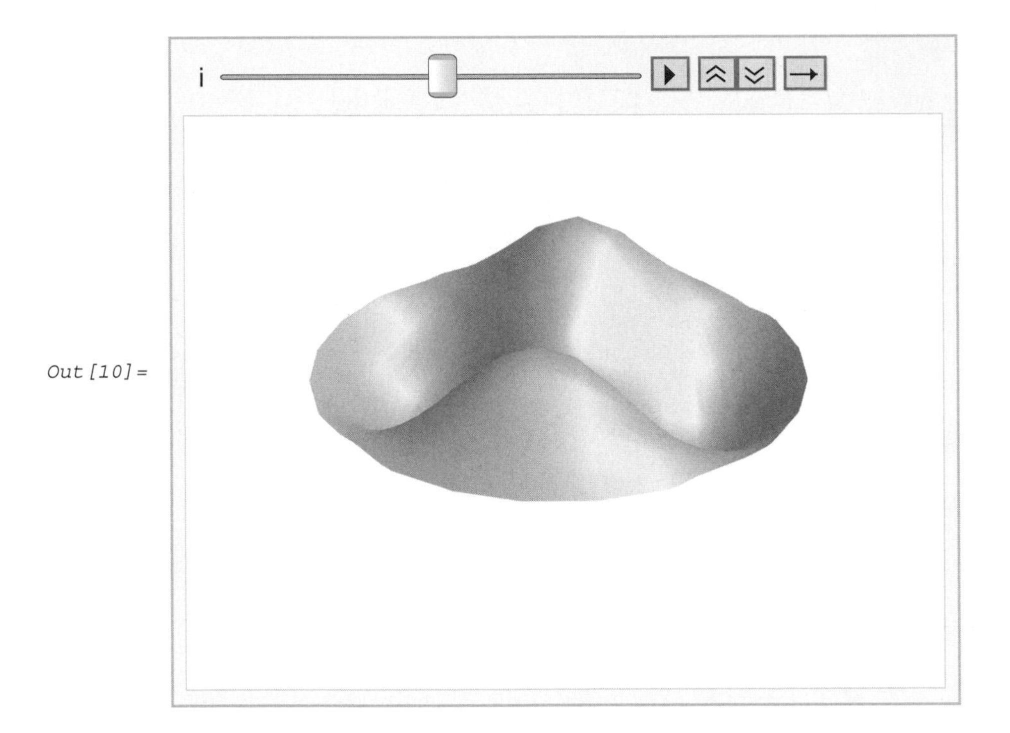

To illustrate the use of three-dimensional graphics directives, we select a problem from special relativity.

Example 3.5.5 An element of volume in the form of a small cube is at rest in the O' reference frame that is moving at a velocity $\mathbf{v}$ with respect to the O reference frame along the positive direction of the common y-axis. The two frames overlap when their origins coincide. Construct a figure depicting the situation.

Here are the graphics elements for the reference frame O:

```
In[11]:= gra1 = GraphicsComplex[
         {{0, 0, 0}, {2, 0, 0}, {2, 2, 0}, {0, 2, 0}, {2, 0, 2}, {0, 0, 2},
          {0, 2, 2}}, {EdgeForm[Thickness[0.004]], Glow[GrayLevel[0.85]],
          Polygon[{{1, 2, 3, 4}, {1, 2, 5, 6}, {1, 4, 7, 6}}]}];
```

Here are those for the reference frame O':

```
In[12]:= gra2 = GraphicsComplex[
         {{0.035, 1, 0.035}, {1.5, 1, 0.035}, {1.5, 2.5, 0.035}, {0.035, 2.5,
           0.035}, {1.5, 1, 1.5}, {0.035, 1, 1.5}, {0.035, 2.5, 1.5}},
         {EdgeForm[Thickness[0.004]], Glow[GrayLevel[1]],
          Polygon[{{1, 2, 3, 4}, {1, 2, 5, 6}, {1, 4, 7, 6}}]}];
```

The graphics elements for the cuboid are

```
In[13]:= gra3 = {Glow[GrayLevel[0.75]], Cuboid[{0.5, 1.75, 0.75}, {0.75, 2, 1}]};
```

The following **Line** primitive is for the lines indicating the position of the cuboid:

```
In[14]:= gra4 = GraphicsComplex[
         {{1.25/2, 3.75/2, 0.035}, {1.25/2, 3.75/2, 1.75/2},
          {1.25/2, 0.035, 0.035}, {0.035, 3.75/2, 0.035}},
         Line[{{1, 2}, {1, 3}, {1, 4}}]];
```

The graphics elements for the velocity vector are

```
In[15]:= gra5 = {{Thickness[0.004], Line[{{0.035, 1, 1}, {0.035, 1.7, 1}}]},
         Polygon[{{0.035, 1.75, 1}, {0.035, 1.75 - 0.1, 1 + 0.075},
           {0.035, 1.75 - 0.1, 1 - 0.075}}]};
```

The following **Text** primitives are for the labels:

```
In[16]:= gra6 = ({
           Text[Style["v", fn1], {0, 1.6, 1.2}],
           Text[Style["x", fn2], {2 - 0.2, 0.2, 0}],
           Text[Style["z", fn2], {0, 0.2, 2 - 0.2}],
           Text[Style["x'", fn2], {1.5 - 0.2, 1 + 0.2, 0}],
           Text[Style["y'", fn2], {0, 2.5 - 0.2, 0.2}],
           Text[Style["z'", fn2], {0, 1 + 0.2, 1.5 - 0.2}],
           Text[Style["O", fn2], {0, 0.125, 0.125}],
           Text[Style["O'", fn2], {0, 1 + 0.15, 0.15}]}/.
          {fn1 → Sequence[FontFamily → "Helvetica", Bold, 10],
           fn2 → Sequence[FontFamily → "Helvetica", Italic, 8]});
```

Finally, **Graphics3D** generates the graphic:

```
In[17]:= Graphics3D[{gra1, gra2, gra3, gra4, gra5, gra6},
           Viewpoint→{2.647, 1.137, 1.775}, Lighting→None, Boxed→False]
```

Out[17]=

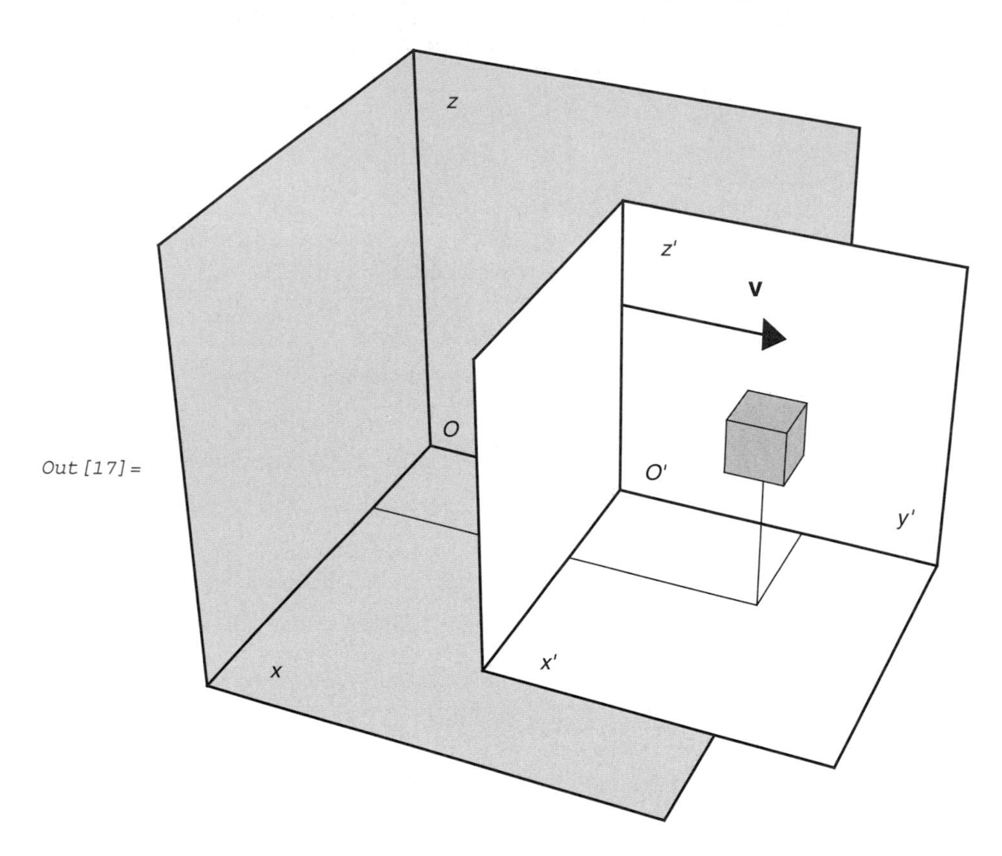


```
In[18]:= ClearAll[gra1, gra2, gra3, gra4, gra5, gra6]
```

■

3.5.3.3 Three-Dimensional Graphics Options

Whereas graphics directives control the rendering of particular graphics primitives in a graphics object, graphics options affect globally the display of the plot. *Mathematica* offers many graphics options for three-dimensional graphics.

Some of the graphics options, discussed in Section 3.5.2.3, work the same in three dimensions as in two dimensions: **Background**, **BaseStyle**, **DisplayFunction**, **FormatType**, **ImageMargins**, **ImagePadding**, **ImageSize**, **LabelStyle**, **Method**, and **PlotLabel**.

A number of the options, introduced in Section 3.5.2.3 for two dimensions, can be generalized to three dimensions:

Option Name	Default Value	Description
AlignmentPoint	**Center**	how three-dimensional objects should by default be aligned when they appear in the two-dimensional graphics primitive **Inset**; *opos* specifies that **Inset[***obj*,*pos***]** is equivalent to **Inset[***obj*,*pos*,*opos***]**—that is, align the inset so that position *opos* in the object lies at position *pos* in the enclosing graphic; *opos* is given in terms of the 0–1 coordinates relative to the whole image region; these coordinates, {*x*,*y*}, vary from 0 to 1 along the *x* and *y* directions, and the origin is located at the lower left corner of the image region; **Center** corresponds to the center of the image region; see the notes on **Inset** in Section 3.5.2.1
AspectRatio	**Automatic**	the ratio of height to width for the displayed two-dimensional image; **Automatic** preserves the natural three-dimensional projection and derives the ratio from **BoxRatios**—that is, the ratios of side lengths for the bounding box; if a numerical value is explicitly specified, the values of **BoxRatios** are forced to adjust so that the graphic may fill the image area
Axes	**False**	whether to include axes; **True** draws *x*-, *y*-, and *z*-axes on the edges of the cuboidal bounding box; **False** draws no axes
AxesLabel	**None**	labels to be put on axes; **None** gives no axis labels; *zlabel* specifies a label for the *z*-axis; {*xlabel*, *ylabel*, *zlabel*} puts labels on all three axes
AxesStyle	**{}**	how axes should be rendered; *style*, all axes are to be generated with the specified style; *style* can be a rule for an option such as **FontSize** or **FontFamily**; *style* can be a graphics directive such as **Thick**, **Red**, **Dashed**, **Thickness**, **Dashing**, or **Directive** (i.e., a combination of directives and options); {*xstyle*, *ystyle*, *zstyle*} specifies separate styles for each axis
BaselinePosition	**Automatic**	the baseline of a graphics object for purposes of alignment with surrounding text or other expressions; *pos*, position *pos* in an object

(Continued)

Option Name	Default Value	Description
		should align with the baseline of surrounding text or other expressions; *pos* can be **Automatic**, **Bottom**, **Top**, **Center**, and **Scaled**[*y*] (fraction *y* of the height of the object)
Epilog	{}	a list of two-dimensional graphics elements to be rendered after the main part of the graphics is rendered; coordinates of graphics primitives are given in terms of the 0−1 coordinates relative to the display area of the graphic; these coordinates $\{x, y\}$, vary from 0 to 1 along the *x* and *y* directions, and the origin is located at the lower left corner of the display area
PlotRange	**All**	the range of coordinates to include in a plot; **All** includes all points; **Automatic** drops outlying points; **Full** includes full range of original data; {*min*, *max*} specifies limits for *z*; $\{\{x_{min}, x_{max}\}, \{y_{min}, y_{max}\}, \{z_{min}, z_{max}\}\}$ gives limits for each coordinate; *s* is equivalent to $\{\{-s, s\}, \{-s, s\}, \{-s, s\}\}$
PlotRangePadding	**Automatic**	how much further axes etc. should extend beyond the range of coordinates specified by **PlotRange**; *p*, the same padding in all directions; $\{p_x, p_y, p_z\}$, different padding in x, y, and *z* directions; $\{\{p_{xL}, p_{xR}\}, \{p_{yF}, p_{yB}\}, \{p_{zB}, p_{zT}\}\}$, different padding on left and right, front and back, and bottom and top; *p* and p_i can be *s* (*s* coordinate units), **Scaled**[s] (a fraction *s* of the plot), **Automatic**, or **None**
PlotRegion	**Automatic**	the region that the plot should fill in the final display area; $\{\{sx_{min}, sx_{max}\}, \{sy_{min}, sy_{max}\}\}$ specifies the region in the 0−1 coordinates relative to the display area; these coordinates vary from 0 to 1 along the *x* and *y* directions, and the origin is located at the lower left corner of the display area; **Automatic** fills the final display area with the plot; when the plot does not fill the entire display area, the rest of the area is rendered according to the setting for the option **Background**

Option Name	Default Value	Description
Prolog	{}	a list of two-dimensional graphics elements to be rendered before the main part of the graphics is rendered; for specifying the coordinates of graphics primitives, see the notes for **Epilog** in this table
Ticks	**Automatic**	what tick marks (and labels) to draw if there are axes; **None** draws no tick marks; **Automatic** places tick marks automatically; {*xtick*, *ytick*, *ztick*} specifies tick mark options separately for each axis; for tick mark option specification along each axis, see the notes for **Ticks** in Section 2.5.2.3
TicksStyle	{}	how axis ticks (tick marks and tick labels) should be rendered; *style*, all ticks are to be rendered with the specified style; *style* can be a rule for an option such as **FontSize** or **FontFamily**; *style* can be a graphics directive such as **Thick**, **Red**, **Dashed**, **Thickness**, **RGBColor**, **Dashing**, or **Directive** (i.e., a combination of directives and options); *style* can also be a style name from the current stylesheet; {*xstyle*, *ystyle*, *zstyle*}, ticks on different axes should use different styles; explicit style specifications for **Ticks** override those for **TicksStyle**

Here are some of the graphics options that *Mathematica* provides specifically for three dimensions:

Option Name	Default Value	Description
AxesEdge	**Automatic**	on which edges of the bounding box should axes be drawn; **Automatic** uses an internal algorithm to decide on which exposed box edges axes should be drawn; {{dir_y, dir_z}, {dir_x, dir_z}, {dir_x, dir_y}} specifies on which three edges of the bounding box axes are drawn; the dir_i must be either +1 or −1 and specify whether axes are drawn on the edge of the box with a larger or smaller value of coordinate i, respectively; any pair {dir_i, dir_j} can be replaced by **Automatic** or **None**; **None** draws no axis
Boxed	**True**	whether to draw the edges of the cuboidal bounding box in a three-dimensional picture;

(Continued)

Option Name	Default Value	Description
		True draws the edges of the box; **False** omits the box
BoxRatios	**Automatic**	the ratios of side lengths for the bounding box; {s_x, s_y, s_z} specifies the side-length ratios; **Automatic** determines the ratios using the range of actual coordinate values in the graphic; if a numerical value is explicitly specified for the option **AspectRatio**, the values of **BoxRatios** are forced to adjust so that graphic may fill the image area
BoxStyle	**Automatic**	how the bounding box should be rendered; *style* can be a graphics directive or list of graphics directives such as **Dashing**, **Thickness**, **GrayLevel**, and **RGBColor**; **Automatic** uses a default style
FaceGrids	**None**	grid lines to draw on the faces of the bounding box; **All** draws grid lines on all faces; **None** draws no grid lines; {$face_1$, $face_2$, ...} draws grid lines on the specified faces; faces are specified as {dir_x, dir_y, dir_z}, where two of the dir_i must be 0 and the third one must be either +1 or −1; for example, **{1, 0, 0}** gives the $y-z$ face with largest x value; {{$face_1$, {$xgrid_1$, $ygrid_1$}}, ...} specifies an arrangement of grid lines for each face; {$xgrid_i$, $ygrid_i$} specifies the positions of grid lines in each direction; see the notes on **GridLines** in Section 3.5.2.3
FaceGridsStyle	**{}**	how face grids should be rendered; *style*, all face grid lines should be rendered by default with the specified style; *style* can be a graphics directive such as **Thick Red**, **Dashed**, **Thickness**, **RGBColor**, **Dashing**, or **Directive** (i.e., a combination of directives); {*xgrid*, *ygrid*, *zgrid*}, different grid directions should use different styles
Lighting	**Automatic**	what simulated lighting to use in coloring three-dimensional surfaces; **Automatic**, default colored light sources fixed relative to the display area; **None**, no simulated lighting; **"Neutral"**, white light sources in the default positions; {s_1, s_2, ...}, light sources $s_1, s_2, \ldots$

Option Name	Default Value	Description
RotationAction	**"Fit"**	how to render three-dimensional objects when they are interactively rotated; **"Fit"**, three-dimensional graphics are rescaled to fit in their image region at the end of every interactive rotation action; **"Clip"**, three-dimensional graphics are not rescaled at the end of a rotation action so that they may be clipped or leave extra space
SphericalRegion	**False**	whether the final image should be scaled so that a sphere drawn around the three-dimensional bounding box would fit in the display area specified; **True** scales images so that a sphere drawn around the bounding box always fits in the display area specified; the center of the sphere is at the center of the box; **False** scales images to be as large as possible, given the specified display area
ViewAngle	**Automatic**	the opening half-angle for a simulated camera used to view the three-dimensional scene; **All**, an opening half-angle sufficient to see everything; **Automatic**, a maximum half-angle of 35°; θ, an explicit half-angle in radians; changing the **ViewAngle** setting is like zooming a camera
ViewCenter	**Automatic**	the point that appears at the center of the final image; **Automatic** centers the whole bounding box in the final image; {x, y, z} specifies the point in scaled coordinates,which run from 0 to 1 across each dimension of the bounding box; **SphericalRegion→True** always centers the circumscribing sphere, regardless of the setting for **ViewCenter**
ViewPoint	**{1.3, -2.4, 2}**	the point in space from which three-dimensional objects are to be viewed; {x, y, z} gives the position of the viewpoint in a special scaled coordinate system in which the longest side of the bounding box has length 1 and the center of the box has coordinates **{0,0,0}**; for more information on **ViewPoint**, see Section 2.3.2.2
ViewRange	**All**	the range of distances from the view point to be included in displaying a three-dimensional

(Continued)

Option Name	Default Value	Description
		scene; **All**, a range sufficient to see everything; {*min*, *max*}, minimum and maximum distances
ViewVertical	{0,0,1}	what direction should be vertical in the final image; {*x*, *y*, *z*} specifies the direction in scaled coordinates, which run from 0 to 1 across each dimension of the bounding box; {*x*, *y*, *z*} and {*rx*, *ry*, *rz*} are equivalent specifications, where *r* is a real constant

To illustrate the use of some of these graphics options, let us create a figure that shows the relationship between rectangular and spherical coordinates:

```
In[1]:= Graphics3D[
    {{Opacity[0.3], Orange, Specularity[White, 5], Sphere[]},
     {Specularity[White, 40], Sphere[{x, y, z}, 0.025]},
     {Thickness[0.004], Line[{{0, 0, 0}, {x, y, z}}]},
     {Thick,
      Line[{{{0, 0, 0}, {1.5, 0, 0}},
         {{0, 0, 0}, {0, 0, 1.05}}, {{0, 0, 0}, {0, 1.15, 0}}}]},
     {Orange, Thickness[0.002], Dashing[{0.0075, 0.005}],
      Line[{{{x, y, 0}, {x, y, z}}, {{x, 0, 0}, {x, y, 0}},
         {{0, y, 0}, {x, y, 0}}, {{x, y, 0}, {0, 0, 0}}}]},
     Text[#1, #2] & @@ #& /@
      {{"x", {1.5, 0, 0.075}}, {"y", {0, 1.15, 0.08}},
       {"z", {0, 0.075, 1.075}}, {"r", 1/2 {x + 0.01, y + 0.15, z + 0.05}},
       {Style["θ", Plain], 1/3 {x, y - 0.09, z + 0.095}}, {Style["φ", Plain],
        {0.25, 0.125, 0}}, {"x", {x/2 + 0.1, y + 0.1, 0}},
       {"y", {x + 0.175, y/2 + 0.075, 0}}, {"z", {x, y + 0.05, z/2}}}},
    ViewPoint → {2.84775, 1.72478, 0.604528},
    Boxed -> False,
    BaseStyle -> Italic,
    PlotLabel -> "               Spherical Coordinates",
    LabelStyle -> {FontFamily -> "Helvetica", Plain, Bold}] /.
   {x -> Sqrt[3/2]/2, y -> Sqrt[3/2]/2, z -> 1/2}
```

Out[1]=

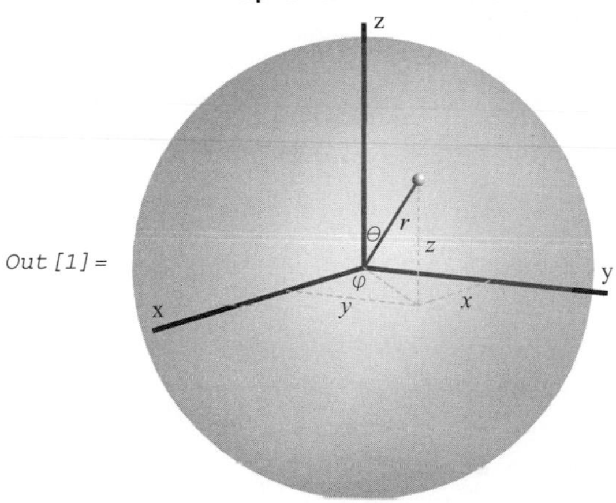

where **ViewPoint** selects a familiar viewpoint, **Boxed** removes the bounding box, **BaseStyle** italicizes the Roman letters, and **PlotLabel** together with **LabelStyle** adds a heading with the specified style. Note that **Style** for individual pieces of text and **LabelStyle** can be used together with **BaseStyle** and have higher priority. The figure shows the relationship between the rectangular and spherical coordinates:

$$x = r \sin\theta \cos\varphi$$

$$y = r \sin\theta \sin\varphi$$

$$z = r \cos\theta$$

Conversely,

$$r = \sqrt{x^2 + y^2 + z^2}$$

$$\theta = \tan^{-1} \frac{\sqrt{x^2 + y^2}}{z} \qquad (3.5.5)$$

$$\varphi = \tan^{-1} \frac{y}{x}$$

(Equation 3.5.5 will be useful later when we consider the rotation of an object about a vertical axis.)

With the setting **Axes → True**, the option **AxesEdge** specifies on which edges of the bounding box the axes should be drawn. For each axis, there are four choices. **AxesEdge → {{**$xdir_y$**,** $xdir_z$**}, {**$ydir_x$**,** $ydir_z$**}, {**$zdir_x$**,** $zdir_y$**}}** gives separate specifications for each of the x-, y-, and z-axes, where $idir_j$ must be either $+1$ or -1, depending on whether the i-axis is to be drawn on the box edge with a larger or smaller value of coordinate j, respectively. For example, consider nine spherical balls with equal radius tightly packed into a cubical box whose edge is 1 m long:

```
In[2]:= Graphics3D[
          {Lighter[Orange, 0.45], Specularity[White, 20],
           Append[
            Sphere[#, r] & /@
             Flatten[
              Table[{i coord, j coord, k coord}, {i, -1, 1, 2}, {j, -1, 1, 2},
               {k, -1, 1, 2}], 2], {Lighter[Blue, 0.5], Sphere[{0, 0, 0}, r]}],
           Opacity[0.2], Cuboid[{-0.5, -0.5, -0.5}]},
          ViewPoint → {3.03916, -1.23595, 0.828204},
          Boxed -> False,
          PlotLabel -> "Nine Spheres Packed in a Cube",
          LabelStyle -> Bold,
          Axes -> True,
          AxesLabel -> Evaluate[Style[#, Italic] & /@ {"x", "y", "z"}],
          AxesStyle -> Red,
          AxesEdge → {{-1, -1}, {1, -1}, {1, 1}}] //.
         {coord -> 1/2 - r, r -> (√3 - 1.5)}
```

Out[2]=

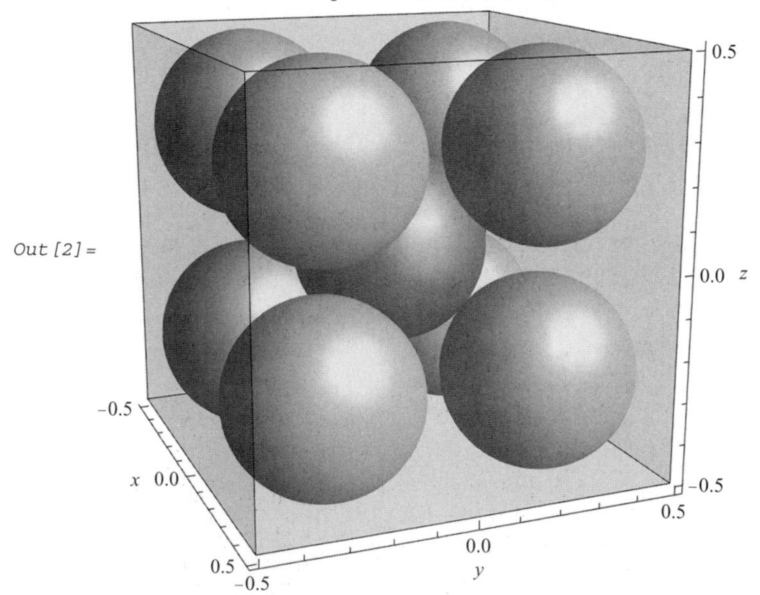

where **Axes** draws the axes, **AxesLabel** gives each axis a label, **AxesStyle** specifies a color for the axes, and **AxesEdge** positions the axes. The x-axis is near the box edge, with $y = -0.5$ and $z = -0.5$; the y-axis, with $x = 0.5$ and $z = -0.5$; and the z-axis, with $x = 0.5$ and $y = 0.5$. The

ReplaceRepeated operator `//.` repeatedly performs replacements until the result no longer changes.

As mentioned previously, four components determine the final rendered colors of three-dimensional surfaces: glow, diffuse reflection, specular reflection, and simulated lighting. Section 3.5.3.2 discussed glow, diffuse reflection, and specular reflection; this section considers simulated lighting. The option **Lighting** specifies the simulated lighting for coloring three-dimensional surfaces. It supports four settings:

Automatic	several default colored light sources fixed relative to the display area
None	no simulated lighting
"*name*"	named lighting configuration
{s_1, s_2, ...}	light sources $s_1, s_2, \ldots$

It allows several kinds of light sources:

{"**Ambient**", *col*}	uniform ambient light of color *col*
{"**Directional**", *col*, {pt_1, pt_2}}	directional light parallel to the vector from pt_1 to pt_2
{"**Point**", *col*, *pt*}	spherical point light source at position *pt*
{"**Spot**", *col*, {*pt*, *tar*}, α}	spotlight at *pt* aimed at the target position *tar* with half-angle α

Specifications of positions can take the forms:

{*x*, *y*, *z*}	ordinary coordinates
Scaled[{*x*, *y*, *z*}]	scaled coordinates
ImageScaled[{*x*, *y*, *z*}]	special coordinates fixed relative to the display area; used mainly with "Directional" light sources

Section 3.5.3.1 discussed ordinary and scaled coordinates. Because these coordinates are fixed relative to the cuboidal bounding box, rotating the bounding box—that is, altering the view point—rotates accordingly the light source and target positions specified in these coordinates:

```
In[3]:= GraphicsGrid[
         {{Graphics3D[Sphere[],
            Lighting -> {{"Spot", Yellow, {{0, 0, 2}, {0, 0, 0}}, {π/10, 0}}}],
           Graphics3D[Sphere[], Lighting ->
             {{"Spot", Yellow, {{0, 0, 2}, {0, 0, 0}}, {π/10, 0}}},
            ViewPoint → {1.58928, -0.258493, 2.97613}],
           Graphics3D[Sphere[], Lighting ->
             {{"Spot", Yellow, {{0, 0, 2}, {0, 0, 0}}, {π/10, 0}}},
            ViewPoint → {0.381771, -1.25199, 9.12037}]}},
         ImageSize -> Large, Spacings -> Scaled[0.3]]
```

Out[3]=

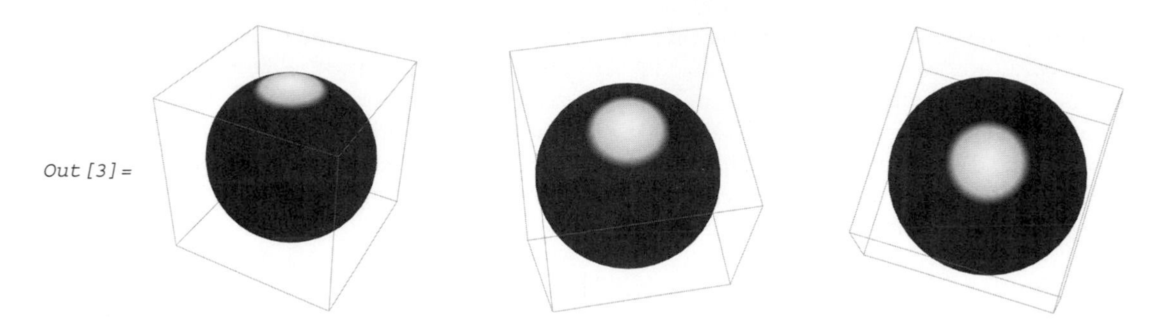

For each sphere, yellow spotlight with a half-angle of $\pi/10$ located at $\{0, 0, 2\}$ is aimed toward the target position at $\{0, 0, 0\}$. Because the light source and target positions are specified in ordinary coordinates fixed relative to the bounding box, the yellow spot moves with the box as the viewpoint changes from sphere to sphere. For the special image-scaled coordinates, x increases horizontally from left to right and y increases vertically from low to high across the display area. The z-axis is perpendicular to the display area and is directed toward the viewer. For use mainly with "Directional" light sources, the image-scaled coordinates are fixed relative to the display area. The light directions are therefore independent of the viewpoint:

```
In[4]:= GraphicsGrid[
         {{Graphics3D[Sphere[], Lighting -> {{"Directional", Yellow,
              {ImageScaled[{2, 2, 2}], ImageScaled[{0, 0, 0}]}}}],
           Graphics3D[Sphere[], Lighting -> {{"Directional", Yellow,
              {ImageScaled[{2, 2, 2}], ImageScaled[{0, 0, 0}]}}},
            ViewPoint → {1.58928, -0.258493, 2.97613}],
           Graphics3D[Sphere[], Lighting -> {{"Directional", Yellow,
              {ImageScaled[{2, 2, 2}], ImageScaled[{0, 0, 0}]}}},
            ViewPoint → {0.381771, -1.25199, 9.12037}]}},
         ImageSize -> Large, Spacings -> Scaled[0.3]]
```

Out[4]=

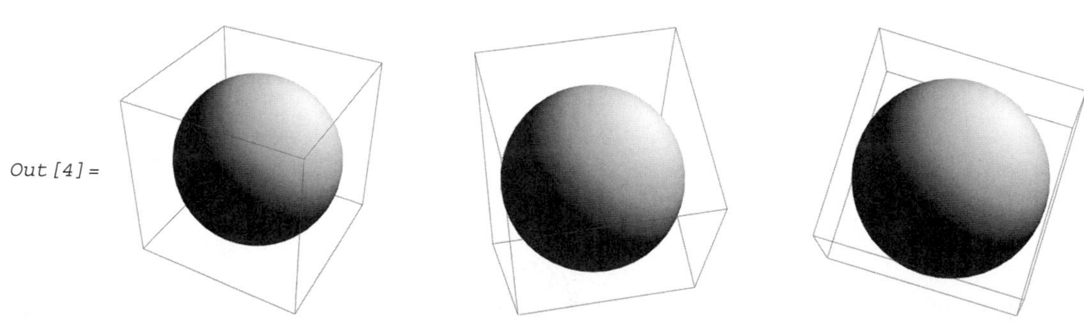

Parallel yellow light in the direction from **ImageScaled[{2, 2, 2}]** to **ImageScaled[{0, 0, 0}]** illuminates each sphere. Because the image-scaled coordinates are fixed relative to the display area, changing the orientation of the bounding box and altering the viewpoint have no effect on the light direction.

The graphics option **Lighting** can also be used as a graphics directive. For example, let us illuminate the center sphere differently in Nine Spheres Packed in a Cube discussed previously:

```
In[5]:= Graphics3D[
          {Lighter[Orange, 0.45], Specularity[White, 20],
           Append[
            Sphere[#, r] & /@
             Flatten[
              Table[{i coord, j coord, k coord}, {i, -1, 1, 2}, {j, -1, 1, 2},
               {k, -1, 1, 2}], 2], {Lighting -> {{"Ambient", GrayLevel[0.25]}},
             Lighter[Blue, 0.5], Sphere[{0, 0, 0}, r]}],
           Opacity[0.2], Cuboid[{-0.5, -0.5, -0.5}]},
          ViewPoint → {3.03916, -1.23595, 0.828204},
          Boxed -> False,
          PlotLabel -> "Nine Spheres Packed in a Cube",
          LabelStyle -> Bold,
          Axes -> True,
          AxesLabel -> Evaluate[Style[#, Italic] & /@ {"x", "y", "z"}],
          AxesStyle -> Red,
          AxesEdge → {{-1, -1}, {1, -1}, {1, 1}}] //.
         {coord -> 1/2 - r, r -> (√3 - 1.5)}
```

Out[5]=

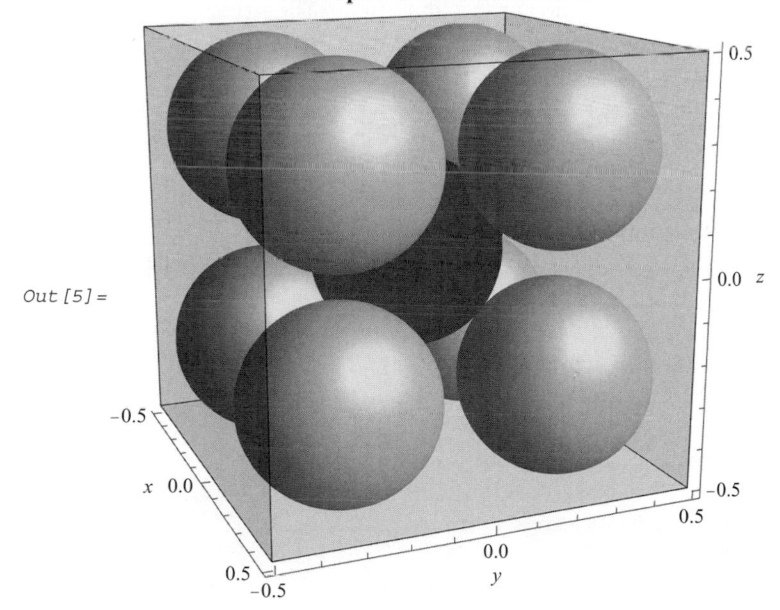

A subtlety of simulated lighting is that objects do not block light and consequently do not cast shadows:

```
In[6]:= Graphics3D[{Sphere[{0, 0, 0}], Sphere[{2.5, 0, 0}]},
          Axes -> True, AxesLabel -> {"x", "y", "z"},
          Lighting -> {{"Point", Yellow, {12, 0, 0}}},
          Ticks -> {Automatic, None, None}]
```

Out[6]=

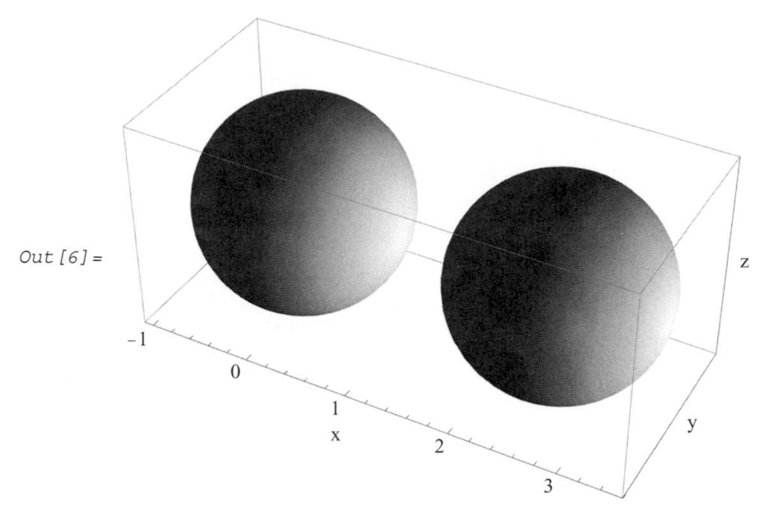

A yellow point source positioned at $\{12, 0, 0\}$ illuminates both spheres alike; the sphere closer to the light source does not block the light nor cast a shadow on the other sphere.

As noted previously, four components determine the actual rendered colors of surfaces: glow, diffuse reflection, specular reflection, and simulated lighting. **Manipulate** allows us to control these components interactively and observe the resultant colors:

```
In[7]:= Manipulate[
          Graphics3D[
           {Glow[a], b, Specularity[c, n], Sphere[{0, 0, 0}]},
           Lighting -> d, ImageSize -> Small], Style["Glow", Bold, Medium],
          {{a, Black, "Color"}, ColorSlider},
          Delimiter,
          Style["Diffuse Reflection", Bold, Medium],
          {{b, White, "Color"}, ColorSlider},
          Delimiter,
          Style["Specular Reflection", Bold, Medium],
          {{c, Black, "Color"}, ColorSetter},
          {{n, 1.5, "Exponent"}, 0, 80, Appearance -> "Labeled"},
          Delimiter,
          Style["Simulated Lighting", Bold, Medium],
          {{d, Automatic, "Pre-Defined"},
```

```
  {Automatic,
   None,
   "Neutral",
   {{"Ambient", GrayLevel[0.5]}},
   {{"Point", Cyan, Scaled[{1, 0.5, 1}]}},
   {{"Spot", White, {{0, 0, 2}, {0, 0, -2}}, π/2}}},
  ControlType → PopupMenu},
 {{d, InputField, "User-Defined"}},
 ControlPlacement -> Top]
```

Out[7]=

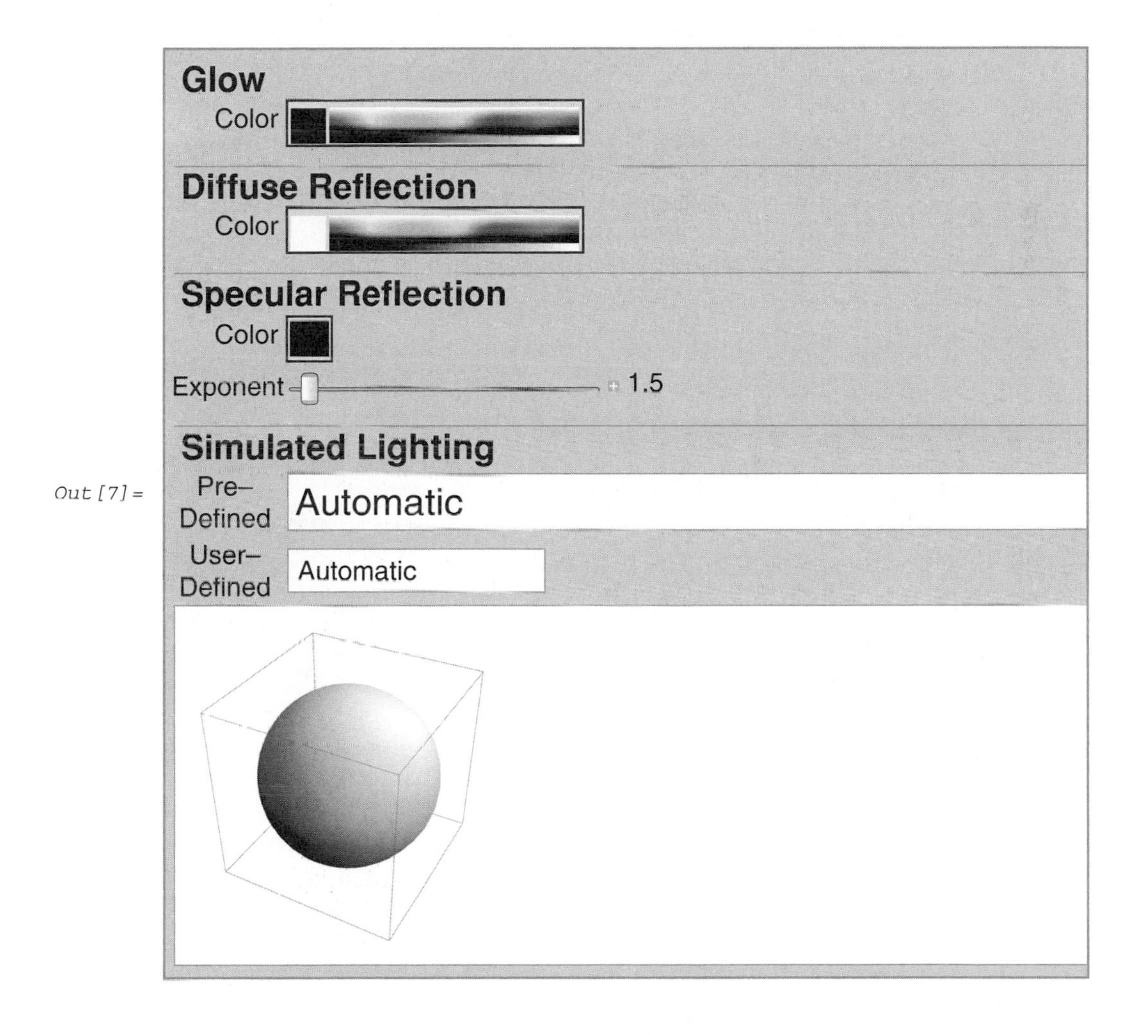

Section 2.3.3 introduced **Manipulate**. **Style["*string*",Bold,Medium]** includes *string* with the specified style as an annotation. **Delimiter** inserts a horizontal delimiter.

`{{u,` u_{init}`,` u_{lbl}`},ColorSlider}` yields a color slider with initial color u_{init} and label u_{lbl}; `{{u,` u_{init}`,` u_{lbl}`},ColorSetter}`, a color setter; `{{u,` u_{init}`,` u_{lbl}`},` u_{min}`,` u_{max}`}`, a manipulator with initial value u_{init}, label u_{lbl}, minimum value u_{min}, and maximum value u_{max}; and `{{u,` u_{init}`,` u_{lbl}`},` `{`u_1`,`u_2`, ...}`, a setter bar or popup menu with choices $u_1, u_2, \ldots$. The option setting **Appearance -> "Label"** displays the value of the variable as a label; **ControlType ->** *type* stipulates the use of control of the specified type; and **ControlPlacement ->** *pos* specifies the placement of controls at position *pos*. To select a color for glow and diffuse reflection, click the color on the corresponding color slider. To select a color for specular reflection, click the colored swatch to open a color picker dialog box. To select a value for the specular exponent, click in the slider area, drag the thumb to the desired position, or click the button ⊞ near the right end of the slider to show and use the controls. To specify a setting for the option **Lighting**, either select a predefined one from the popup menu or enter a user-defined one in the form $\{s_1, s_2, \ldots\}$, where s_i are the light sources, and then press *Enter* for Windows or *return* for Mac OSX.

With the default option setting **SphericalRegion → False**, *Mathematica* scales the objects in a three-dimensional picture so that it is as large as possible, given the final display area. Consequently, the sizes of the objects may depend on the viewpoint. This dependence of sizes of objects on their orientations can cause problems in an animation sequence involving different viewpoints. For example, consider the rotation of a cuboid. To begin, let us define the function **spin3D**, which generates rotational animations:

```
In[8]:= spin3D[obj_,opts___Rule] :=
          ListAnimate[
           Table[
            Show[obj,
              ViewPoint -> {r Sin[θ] Cos[φ + i (π/16)],
                r Sin[θ] Sin[φ + i (π/16)], r Cos[θ]},opts] /.
              {r -> 3.38378, θ -> 0.938431, φ -> -1.07437}, {i, 0, 31}]]
```

where $\{r, \theta, \varphi\} = \{3.38378, 0.938431, -1.07437\}$ are the spherical coordinates corresponding to the rectangular coordinates $\{x, y, z\} = \{1.3, -2.4, 2.0\}$ of the default viewpoint. (See Equation 3.5.5.) In the spinning of the object *obj* about the z-axis, the azimuthal angle φ varies with a range of 2π, while the radius r and the polar angle θ remain fixed. With the exception of **ViewPoint**, options for passing to **Show** may be included in **spin3D** after the first argument that specifies the three-dimensional object to be rotated. The function **spin3D** generates the animation showing the rotation of a cuboid:

```
In[9]:= spin3D[
         Graphics3D[Cuboid[{0, 0, 0}, {2, 0.5, 0.5}]],
         PlotRange→{{0, 2}, {0, 1}, {0, 1}}]
```

Out [9] =

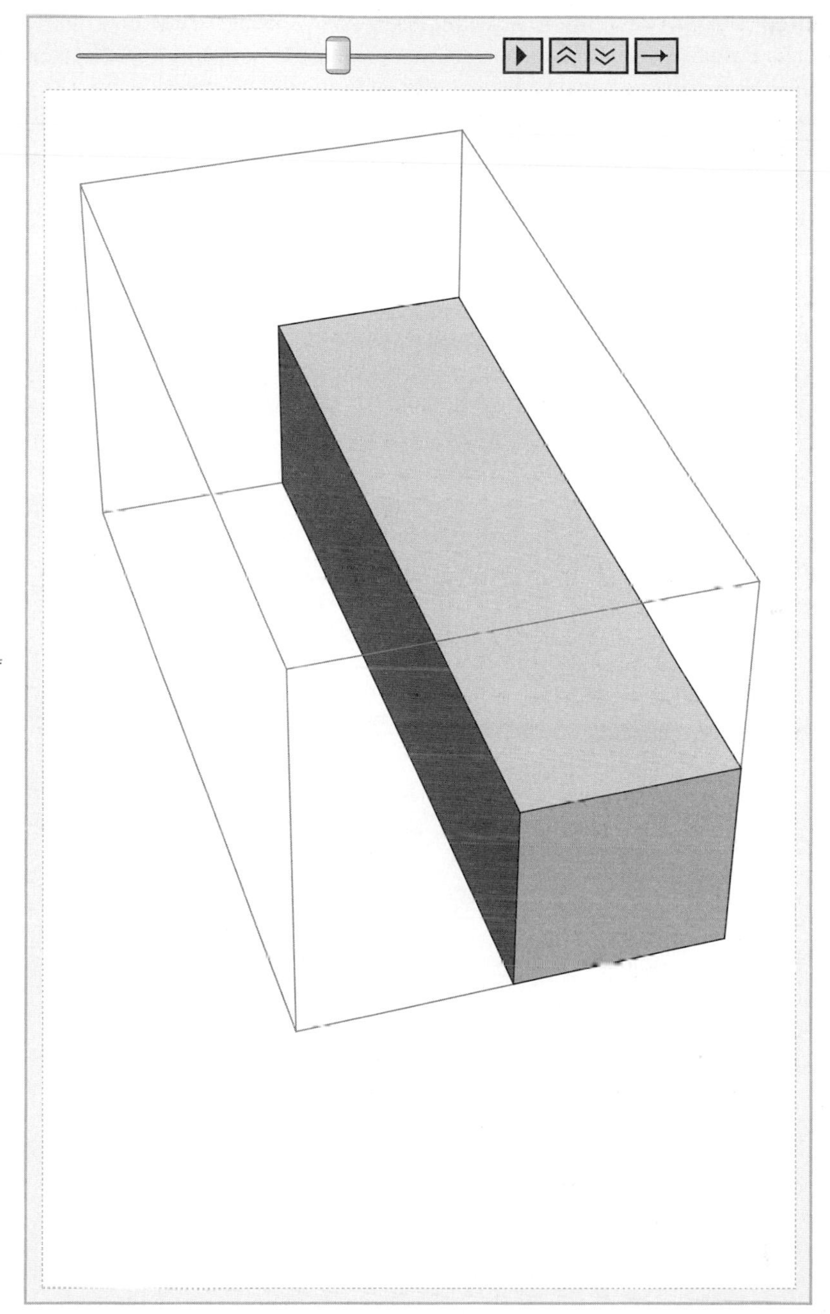

To start the animation if it is not already in progress, click the Play button; to stop the animation, click the Pause button. With the default option setting **SphericalRegion → False**, the cuboid wobbles as it rotates because its size depends on the orientation. Let us change the option setting to **SphericalRegion → True**:

```
In[10]:= spin3D[
          Graphics3D[Cuboid[{0, 0, 0}, {2, 0.5, 0.5}]],
          PlotRange → {{0, 2}, {0, 1}, {0, 1}}, SphericalRegion → True]
```

Out[10]=

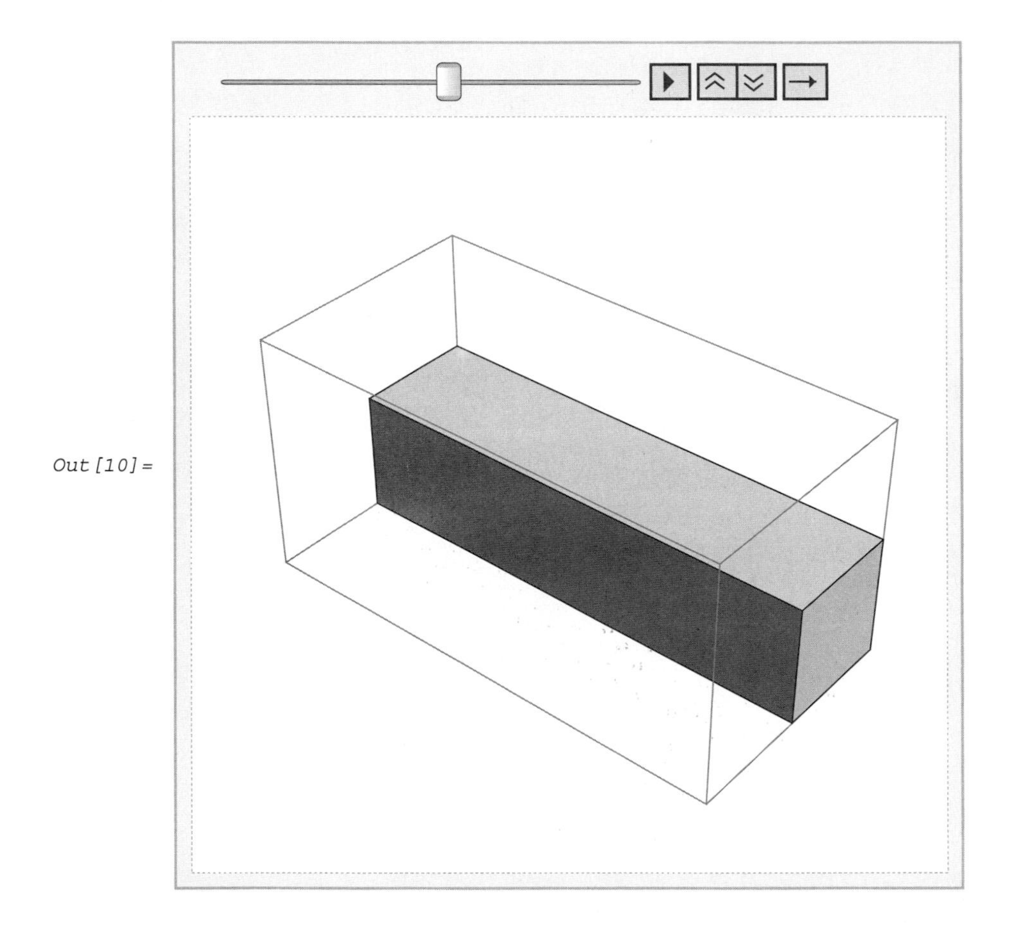

Again, to start the animation if it is not already in progress, click the Play button; to stop the animation, click the Pause button. With the option setting **SphericalRegion → True**, the cuboid rotates in a manner intended because *Mathematica* scales the cuboid so that a sphere drawn around the bounding box always fits the display area. In this case, the cuboid remains consistent in size, regardless of its orientation.

The options for **Graphics3D** form a subset of that for **Plot3D**. For the options supported by both functions, several default settings are different:

	Default for **Plot3D**	Default for **Graphics3D**
Axes	**True**	**False**
BoxRatios	**{1, 1, 0.4}**	**Automatic**
PlotRange	**{Full, Full, Automatic}**	**All**

As an example, **Plot3D** together with a number of options generates a three-dimensional plot:

```
In[11]:= Plot3D[Sin[Sqrt[x^2 + y^2]]/Sqrt[x^2 + y^2], {x, -8π, 8π}, {y, -8π, 8π},
  PlotPoints→50,
  Mesh -> None,
  PlotStyle -> {Lighter[Blue, .35], Specularity[White, 12]},
  PlotRange→{-0.5, 1.2},
  Boxed -> False,
  AxesLabel→{"x", "y", "z"},
  AxesStyle -> Directive[Thickness[0.002], Blue],
  BaseStyle→{12, Italic},
  Ticks→({spec, spec, {{0, Style["0", fs]}, {1, Style["1", fs]}}}//.
    {spec→({#, Style[#, fs]} & /@Table[(3i)π, {i, -2, 2}]),
     fs→{Plain, 9}}),
  Lighting→{{"Spot", White, {{0, 0, 2}, {0, 0, -2}}, Pi},
    {"Spot", White, {{20, - 20, .85}, {0, 0, 0}}, Pi/14}}]
```

Out[11]=

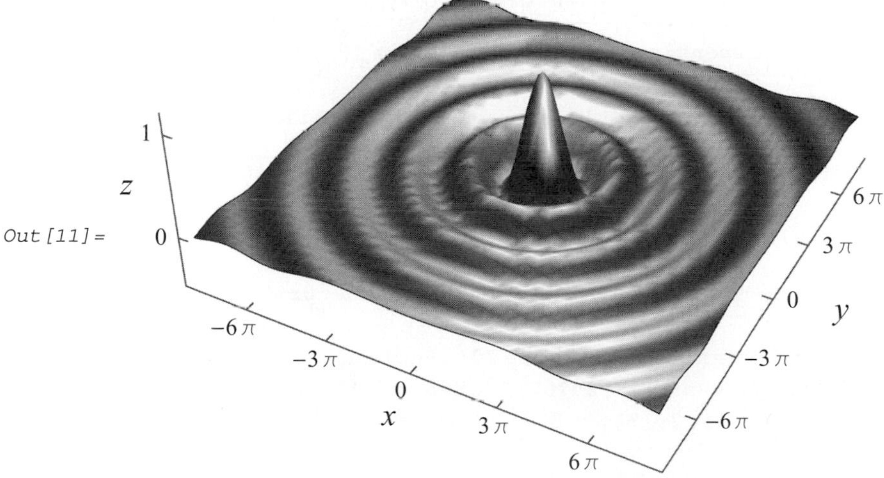

where **Ticks** draws tick marks and given labels at the specified positions. Note that **PlotPoints**, **Mesh**, and **PlotStyle** are not valid options for **Graphics3D**.

3.5.3.4 Crystal Structure

This section applies the three-dimensional graphics programming capabilities of *Mathematica* to an introductory solid state physics problem.

The Problem Construct the sodium chloride structure, showing a conventional cubic cell.

Physics of the Problem The sodium chloride structure consists of equal numbers of sodium and chlorine ions located at alternate sites of a simple cubic lattice. Since there are two kinds of ions, the structure no longer has the translational symmetry of the simple cubic Bravais lattice. It can be described as a face-centered cubic Bravais lattice with a basis of a sodium ion and a chlorine ion.

Solution with *Mathematica* Comments describing the program are embedded in the input.

```
In[1]:= (* a nested list of coordinates for some
         points of a simple cubic Bravais lattice *)
        coord = Table[{i, j, k}, {i, 0, 2}, {j, 0, 2}, {k, 0, 2}];

        (* lines joining the nearest neighbors *)
        lines = Line /@ Join[
            Flatten[coord, 1],
            Map[RotateRight, Flatten[coord, 1], {2}],
            Map[RotateLeft, Flatten[coord, 1], {2}]];

        (* spheres representing the sodium ions *)
        sodium =
          Sphere[#, 0.08] & /@ Select[Flatten[coord, 2], OddQ[Plus @@ #] &];

        (* spheres representing the chlorine ions *)
        chlorine =
          Sphere[#, 0.12] & /@ Select[Flatten[coord, 2], EvenQ[Plus @@ #] &];

        (* showing the sodium chloride crystal structure *)
        Graphics3D[
         {Thick, lines, Specularity[White, 20], Red, chlorine, Blue, sodium},
         Boxed → False, ViewPoint → {2.700, -1.825, 0.910},
         PlotRange → {{-0.15, 2.15}, {-0.15, 2.15}, {-0.15, 2.15}}]

        ClearAll[coord, lines, sodium, chlorine]
```

Out[5]=

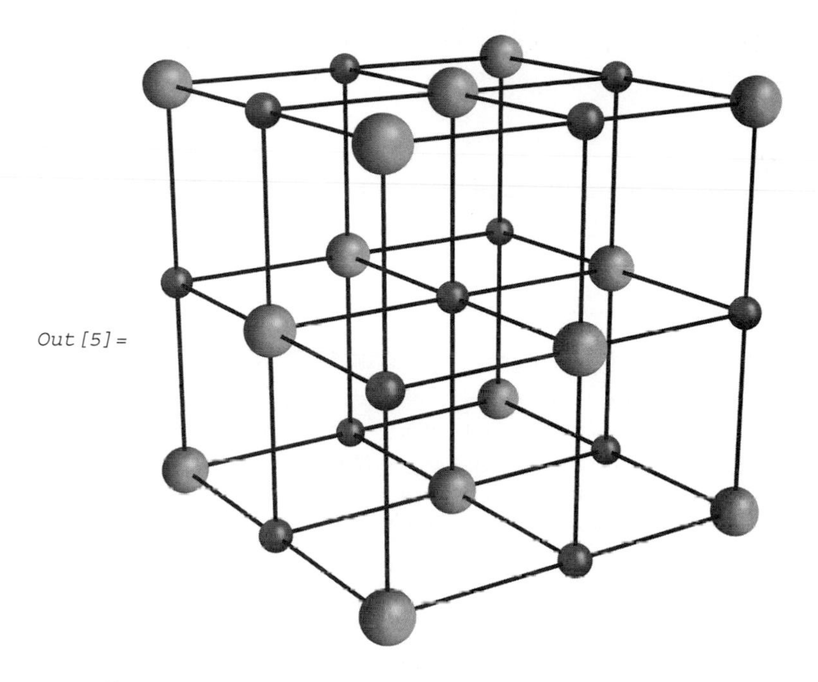

The smaller and larger spheres represent the sodium and chlorine ions, respectively.

3.5.4 Exercises

In this book, straightforward, intermediate-level, and challenging exercises are unmarked, marked with one asterisk, and marked with two asterisks, respectively. For solutions to most odd-numbered exercises, see Appendix C.

1. Construct a diagram, as shown, illustrating Kepler's second law stating that a line joining the Sun and any planet sweeps out equal areas in equal time intervals.

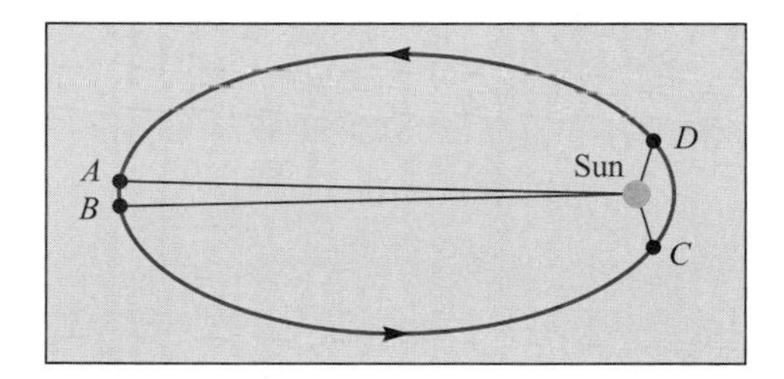

2. Draw an Atwood's machine.
3. Construct a six-pointed star by placing two triangles together with one inverted over the other, place a point with a diameter that is equal to 7.5% of the total width of the graphic at each corner, and give a random color to each point.

Answer:

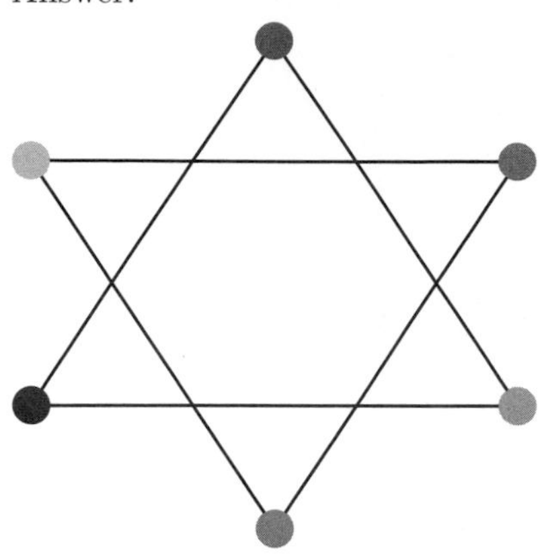

4. Draw a ray diagram for a spherical convex mirror.
5. Using graphics primitives and directives, produce the graphic

where the branches are green, the tree trunk is black, the base is red, and the letters are yellow.

*6. Animate the propagation of the electromagnetic wave shown in Example 3.5.3.

7. (a) Construct the face-centered cubic Bravais lattice, as shown.

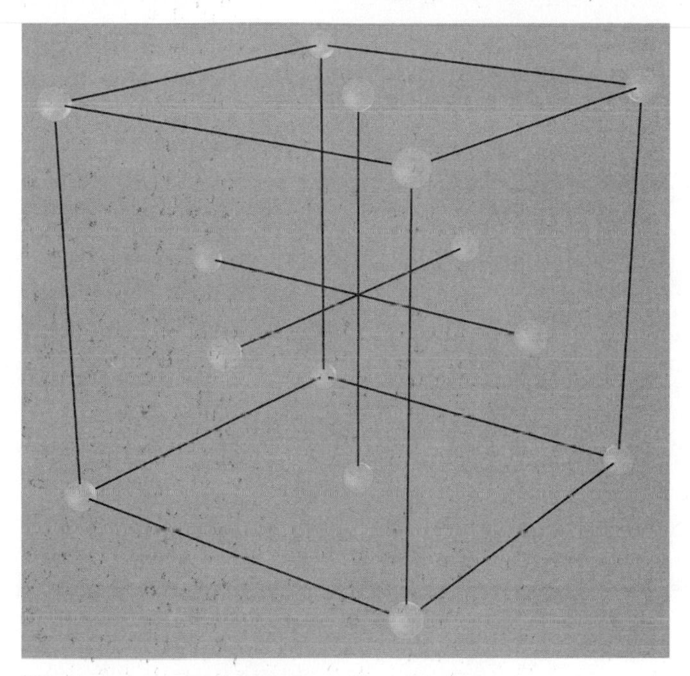

Hint: For **Graphics3D**, specify

```
ViewPoint → {1.901, -2.572, 1.104},
Lighting -> {{"Ambient", Red},
   {"Point", Orange, {0, -2, 0}}, {"Point", Darker[Yellow], {0, 2, 0}}}
```

(b) Generate an animation showing the face-centered cubic Bravais lattice spinning about the vertical axis through the center of the cube. Use the **ViewPoint** and **Lighting** specifications given in part (a). *Hint:* The vertical axis must remain stationary.

*8. The parametric equations for a space curve are

$$x = \frac{2t+1}{t-1}$$

$$y = \frac{t^2}{t-1}$$

$$z = t+2$$

(a) Find the equation of the plane on which the curve lies. *Hint:* The equation of a plane is

$$ax + by + cz + d = 0$$

Substitute the parametric equations into this equation, and then determine b, c, and d in terms of a. Use **Together**, **Numerator**, and **Collect**.

(b) Confirm and visualize the result of part (a) by plotting *together* the space curve and the plane on which it lies, as shown.

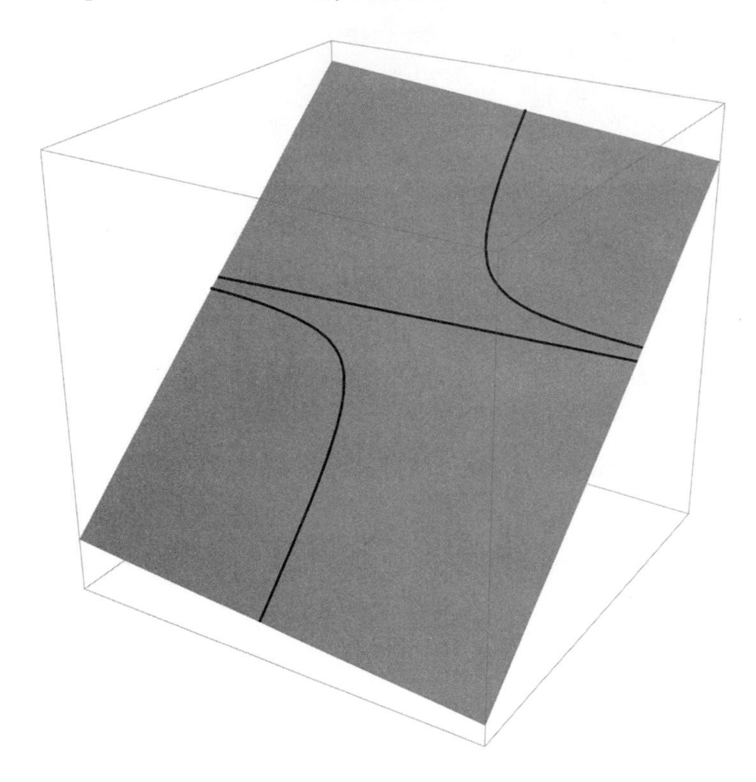

Hint: For **Show**, specify

```
Lighting -> {{"Directional", Red, {{2, -15, 10}, {-2, 15, -10}}}},
ViewPoint -> {1.61656, -2.60146, 1.43845}
```

(c) Animate the rotation of the space curve on the plane in part (b) about the vertical axis through the center of the bounding box.

3.6 PROGRAMMING STYLES

Before focusing on programming styles, let us expose a common pitfall en route to effective use of *Mathematica*. Dazzled by the power of *Mathematica*, we may be awed into relinquishing our mind to it. We must keep thinking for several important reasons.

First, *Mathematica* cannot save us from bad physics. Before solving a problem with *Mathematica*, clearly state the problem and carefully summarize the physics for the problem. As will be shown in Chapters 4–6, each notebook should have three sections: The Problem, Physics of the Problem, and Solution with *Mathematica*.

Second, *Mathematica* can make mistakes. Vigilant users can catch the mistakes! For instance, consider again Example 3.4.3 in Section 3.4.2. For the step potential

$$V(x) = \begin{cases} 0 & x < -3 \\ -2 & -3 \leq x < -1 \\ -1 & -1 \leq x < 1 \\ -2 & 1 \leq x < 3 \\ 0 & 3 \leq x \end{cases}$$

a single-clause definition is

```
In[1]:= V[x_] := Which[x< -3 || x≥3, 0, -3≤ x< -1 || 1≤ x<3, -2, -1≤ x<1, -1]
```

Mathematica gives

```
In[2]:= ∂x V[x] /. x→1
Out[2]= 0
```

The result is wrong because the potential V is in fact not differentiable at 1. To find the derivative, use the definition

```
In[3]:= myV[x_] := Piecewise[{{0, x < -3},
           {-2, -3 ≤ x < -1}, {-1, -1 ≤ x < 1}, {-2, 1 ≤ x < 3}, {0, 3 ≤ x}}]
```

Now *Mathematica* gives the correct result:

```
In[4]:= ∂x myV[x] /. x→1
Out[4]= Indeterminate
```

Third, built-in *Mathematica* functions may have subtleties that can be perilous to unwary users. For instance, as discussed in Section 2.4.11.2, **FindFit**[*data*, *model*, *pars*, *vars*] searches for a least-squares fit to a list of data according to the model containing the variables in the list *vars* and the parameters in the list *pars*. Yet, for some problems, specifications of starting values for the parameters are imperative; otherwise, *Mathematica* returns wrong results. In these cases, we must use **FindFit**[*data*, *model*, {{par_1, p_1}, {par_2, p_2}, ...}, *vars*] that starts the search for a fit with $\{par_1 \text{->} p_1, par_2 \text{->} p_2, \ldots\}$. For an example, see Exercise 38 in Section 2.4.12.

Finally, we must think in order to make efficient use of *Mathematica*. Improvement of seconds on the evaluation time may be immaterial. However, improvement of more than several minutes is worthwhile. For example, consider plotting the potential

$$V(x,y) = \frac{16}{\pi^2} \sum_{n=1}^{400} \sum_{m=1}^{400} \frac{\sin(2n-1)x \sin(2m-1)y}{(2n-1)(2m-1)}$$

which can be entered as

```
In[5]:= ClearAll[V]
```

In[6]:=

$$V[x\_, y\_] := \frac{16}{\pi^2} \sum_{n=1}^{400} \sum_{m=1}^{400} \frac{\text{Sin}[(2n-1)x]\ \text{Sin}[(2m-1)y]}{(2n-1)(2m-1)}$$

Plotting **V[x, y]** with the command

```
In[7]:= Plot3D[V[x, y], {x, 0, π}, {y, 0, π}, PlotRange→{0.9, 1.1},
          AxesLabel→{"x", "y", "V"}, Ticks -> None] // AbsoluteTiming

Out[7]:= $Aborted
```

takes hours on an iMac G4. (**AbsoluteTiming[***expr***]** evaluates *expr* and returns a list of the real time in seconds that have elapsed together with the result obtained.) By simply wrapping the function **Evaluate,** discussed in Section 2.3.1.1, around **V[x, y]** like

```
In[8]:= Plot3D[Evaluate[V[x, y]], {x, 0, π}, {y, 0, π}, PlotRange→{0.9, 1.1},
          AxesLabel→{"x", "y", "V"}, Ticks -> None] // AbsoluteTiming
```

Out[8]= {967.606758,

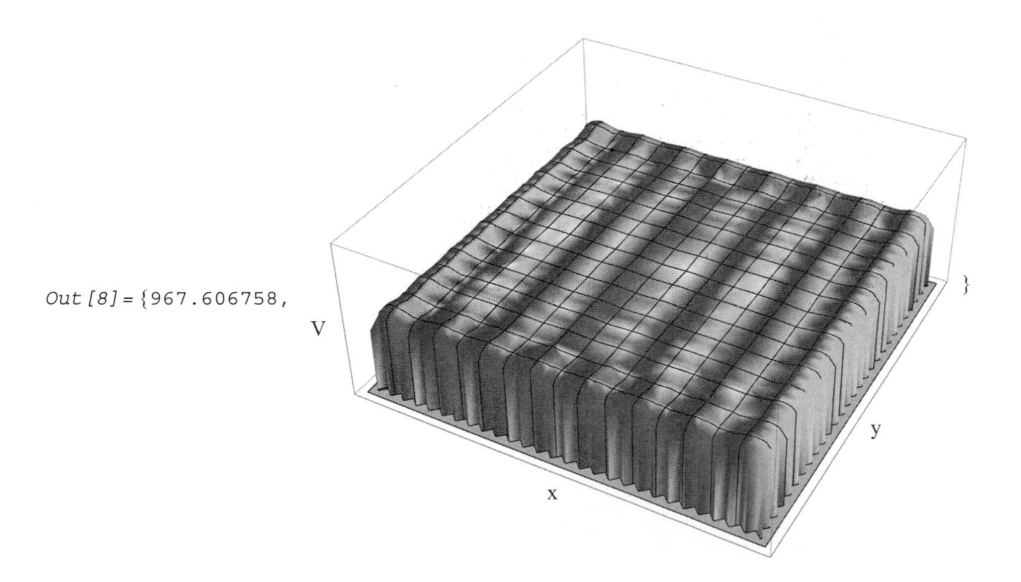

}

we can reduce the plotting time to approximately 16 minutes.

Although *Mathematica*'s interactive capabilities are sufficient for simple problems, sooner or later we must write programs in order to solve more advanced problems. Computer programs are sets of instructions for computers to solve particular problems. This book has already covered the essential programming concepts of *Mathematica*. There are many ways in which we can use them to solve a problem and to interpret and visualize the result. Our particular approach depends on the problem and on our computational background as well as temperament.

There are three popular programming styles: procedural, functional, and rule-based programming. Section 3.6.1 discusses procedural programming, which is the approach of traditional languages such as FORTRAN, Pascal, and C. If we are already familiar with one or more of these languages, we are naturally prone to write all *Mathematica* programs in this style. Be aware that for some problems, one of the other two programming styles may be more suitable. Section 3.6.2 examines functional programming and provides an example favoring this programming style. Similarly, Section 3.6.3 explicates rule-based programming.

A few words about clarity of programs are in order. The hallmark of computer programs is readability. Here is an opaque code for generating the crystal structure of sodium chloride:

```
Table[{i, j, k}, {i, 0, 2}, {j, 0, 2}, {k, 0, 2}];
Line/@Join[Flatten[%, 1], Map[RotateRight, Flatten[%, 1], {2}],
    Map[RotateLeft, Flatten[%, 1], {2}]];
Point/@Select[Flatten[%%, 2], OddQ[Plus @@ #] &];
Point/@Select[Flatten[%%%, 2], EvenQ[Plus @@ #] &];
Show[Graphics3D[
   {%%%, PointSize[0.06], %, PointSize[0.04], RGBColor[1, 0, 0], %%}],
  Boxed→False, ViewPoint→{2.700, -1.825, 0.910},
  PlotRange→{{-0.15, 2.15}, {-0.15, 2.15}, {-0.15, 2.15}}]
```

Such lack of clarity can be a major source of errors and reader frustration. Providing explanatory comments, following good formatting practices, and making assignments, if permitted, for temporary variables with descriptive names often add clarity to the program. Section 3.5.3.4 shows an improved version of this code.

```
In[9]:= ClearAll[V, myV]
```

3.6.1 Procedural Programming

A procedural program consists of a procedure discussed in Section 3.4. For altering the sequential flow of control within a procedure, the conditional functions **If**, **Which**, and **Switch** permit branching, and the iteration functions **Do**, **While**, and **For** allow looping. Procedural programs are often written as functions of the form

name[*arg*$_1$_, *arg*$_2$_, ...] := **Module**[{*name*$_1$, *name*$_2$ = *val*$_2$, ...}, *procedure*]

where the procedure is wrapped in a module. Within the procedure, there may be other modules or calls to other functions of this form. **Module** provides localization of symbols and thus sets up modules that are somewhat independent of each other.

The functions **electrostaticField** in Section 3.4.4 and **motion1DPlot** in Section 3.4.5 are procedural programs. This section presents an additional program. Consider again Example 2.1.9 on the motion of a planet under the gravitational attraction of the Sun. Whereas we used *Mathematica* interactively to plot the orbit of a planet before, we define here the function **planetMotion** to be a procedure that generates an animation of the motion of a planet.

A planet moves in a plane defined by its initial position vector $\mathbf{r}_0$ from the Sun located at the origin and its initial velocity $\mathbf{v}_0$. Its energy is

$$E = \frac{1}{2}mv^2 - \frac{GMm}{r}$$

where m, v, and r are the planet's mass, speed, and distance from the Sun, respectively. Also, G is the gravitational constant and M is the mass of the Sun. The angular momentum of the planet is

$$\mathbf{L} = m(\mathbf{r} \times \mathbf{v})$$

where $\mathbf{r}$ and $\mathbf{v}$ are, respectively, its position vector and velocity. If $L = 0$, the path of the planet is a straight line through the origin. If $L \neq 0$, its orbit is an ellipse, parabola, or hyperbola, depending on whether $E < 0$, $E = 0$, or $E > 0$, respectively. Since m is finite as well as positive and E together with L are constants of motion, we can express these conditions in terms of the parameters e and l defined by

$$e = \frac{E}{m} = \frac{v_{x0}^2 + v_{y0}^2}{2} - \frac{4\pi^2}{\sqrt{x_0^2 + y_0^2}}$$

and

$$l = \frac{L}{m} = \left| x_0 v_{y0} - y_0 v_{x0} \right|$$

where x_0, y_0, v_{x0}, and v_{y0} are the initial coordinates and components of the initial velocity and we have adopted a system of units in which length is in astronomical units and time is in years. In this system of units, $GM = 4\pi^2$. If $l = 0$, the path of the planet is a straight line. If $l \neq 0$, its orbit is an ellipse, parabola, or hyperbola, depending on whether $e < 0$, $e = 0$, or $e > 0$, respectively. For an elliptical orbit, the period of the motion is given by

$$\left(\left| \frac{e}{2\pi^2} \right| \right)^{-3/2}$$

(For a detailed discussion of the mechanics of planetary motion, see [Sym71].)

For generating an animation of the motion of a planet, the function **planetMotion** takes two arguments: a list of initial coordinates and a list of components of the initial velocity. Two-dimensional graphics options and the option **frameNumber** can be included as trailing arguments. The option **frameNumber**, with a default value of 30, specifies the number of frames in the animation sequence. Explanatory comments on **planetMotion** are embedded in the body of the function.

```
In[1]:= (* set up default value for the option frameNumber *)
        Options[planetMotion] = {frameNumber → 30};
```

```
In[2]:= planetMotion[{x0_, y0_}, {vx0_, vy0_}, opts___Rule] :=
  Module[
   {(* determine the energy parameter and
     assign it as initial value to a local variable *)
```

$$e = N\left[\frac{vx0^2 + vy0^2}{2} - \frac{4\pi^2}{\sqrt{x0^2 + y0^2}}\right],$$

```
   (* determine the angular momentum parameter
     and assign it as initial value to a local variable *)
   l = N[Abs[x0 vy0 - y0 vx0]],

   (* determine option value of frameNumber and
     assign it as initial value to a local variable *)
   n = frameNumber/.{opts}/.Options[planetMotion],

   (* select valid options for Show and assign
     the sequence as initial value to a local variable *)
   optShow = Sequence @@ FilterRules[{opts}, Options[Graphics]],

   (* declare other local variables *)
   orbit, period, x, y, t},

  Which[
   (* verify that initial conditions are real
    numbers and that frame number is a positive integer *)
   ((!(NumberQ[x0] && Im[x0] == 0)) || (!(NumberQ[y0] && Im[y0] == 0)) ||
     (!(NumberQ[vx0] && Im[vx0] == 0)) ||
     (!(NumberQ[vy0] && Im[vy0] == 0)) ||
     (!(IntegerQ[n] && Positive[n]))),
   Print["Initial conditions must be real numbers and
      frame number must be a positive integer."],

   (* check that the path is not a straight line *)
   l == 0,
   Print[
    "The initial conditions result in a straight line for the path
      whereas orbits of planets must be ellipses."],

   (* check that the orbit is not a hyperbola *)
   e > 0,
   Print[
    "The initial conditions result in a hyperbolic orbit whereas
      orbits of planets must be ellipses."],
```

```
(* check that the orbit is not a parabola *)
e == 0,
Print[
 "The initial conditions result in a parabolic orbit whereas
   orbits of planets must be ellipses."],

(* for the elliptical orbit *)
e < 0,

(* determine the period of the motion *)
period = N[Abs[e/(2 π^2)]^(-3/2)];

(* solve the equations of motion
 discussed in Example 2.1.9 of Section 2.1.18 *)

orbit = NDSolve[{x''[t] + (4 π^2 x[t])/(x[t]^2 + y[t]^2)^(3/2) == 0,

    y''[t] + (4 π^2 y[t])/(x[t]^2 + y[t]^2)^(3/2) == 0, x[0] == x0, y[0] == y0,

    x'[0] == vx0, y'[0] == vy0}, {x, y}, {t, 0, period}];

(* generate the animation *)
ListAnimate[

 Table[
  Show[
   (* the orbit *)
   ParametricPlot[Evaluate[{x[t], y[t]}/.orbit], {t, 0, period}],

   (* the Sun in yellow and the moving planet in red *)
   Graphics[{Yellow, PointSize[0.1], Point[{0, 0}], Red,
     PointSize[0.04], Point[{x[i], y[i]}/.orbit[[1]]]}],

   (* option for preserving the true shape of the orbit *}
   AspectRatio→Automatic,

   (* two-dimensional graphics options *)
   optShow],

  {i, 0, (1 - 1/n) period, period/n}]]]]
```

With this function, the sizes of the Sun and planet in the graphics are preset. For orbits of different sizes, we may wish that the function would allow us to adjust those of the Sun and planet. For adding this feature to **planetMotion**, see Exercise 1 in Section 3.6.4.

Let us call the function **planetMotion** with $x_0 = 1, y_0 = 0, v_{x0} = -2.5, v_{y0} = 5$, and several option specifications:

```
In[3]:= planetMotion[{1, 0}, {-2.5, 5},
          frameNumber → 40,
          PlotRange → {{-0.6, 1.21}, {-1.13, 0.6}},
          Axes → None, Frame → True,
          FrameTicks → None,
          Background → RGBColor[0.5, 0.85, 1]]
```

Out[3]=

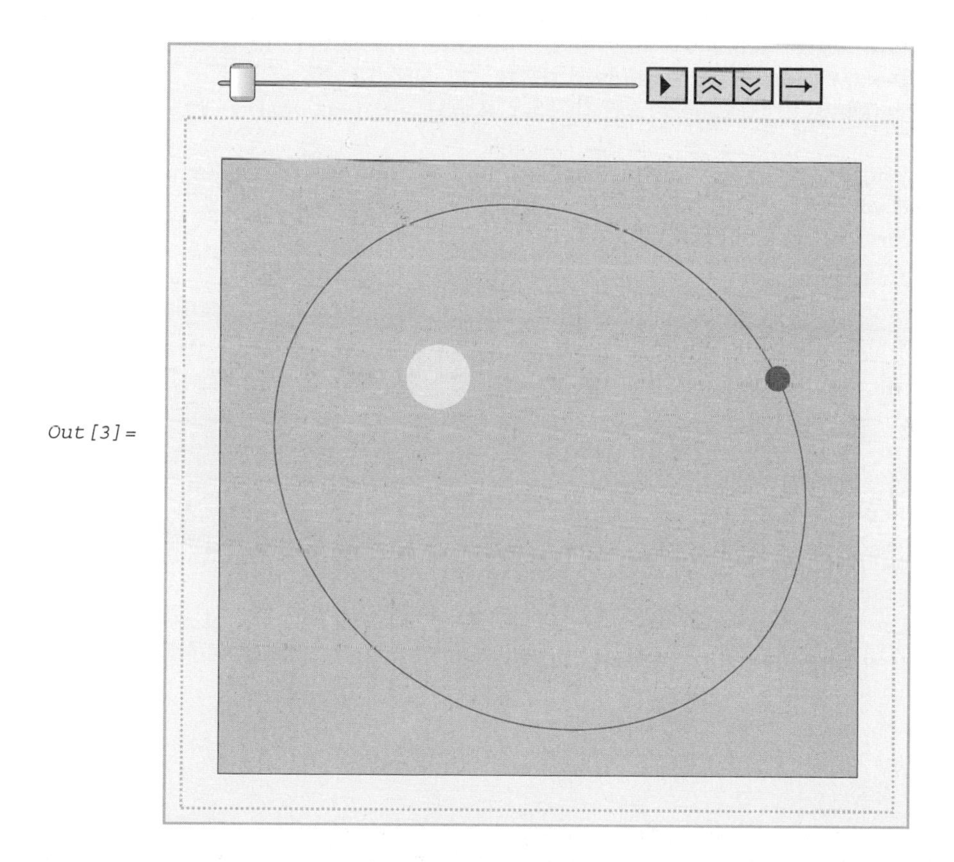

To animate, click the Play button—that is, the one with the right pointer. That the speed of the planet varies around the orbit is consistent with Kepler's second law, which states that the position vector of a planet from the Sun sweeps out equal areas in equal times.

```
In[4]:= ClearAll[planetMotion]
```

3.6.2 Functional Programming

A functional program, also known as a one-liner, consists of many nested functions—that is, a function acting on the results of applying other functions to the results of applying some other functions, and so forth, to some arguments. Functional programs are often written as functions of the form

name [arg_1_, arg_2_, ...] := *rhs*

where *rhs* is a single, albeit sometimes very long, expression of nested functions. When the function *name* is invoked, the values of the functions at the deepest level of the nest serve as arguments of the functions at the next level, and so on, terminating at the top-level function, the value of which becomes the value of the function *name*. For one-liners, *rhs* is made up solely of arg_i, pure functions, and built-in functions and constants. Auxiliary function definitions and temporary variable assignments are forbidden. For iterations, functional programs often employ the functions **Nest**, **NestList**, **FixedPoint**, and **FixedPointList**. For manipulating expressions, they frequently utilize the functions **Map**, **MapAll**, **MapAt**, **Apply**, **Flatten**, and **FlattenAt**.

To illustrate functional programming, we generate a bifurcation diagram for the logistic map. As introduced in Example 3.3.3 of Section 3.3.4, the logistic map is defined by the one-dimensional difference equation

$$x_{n+1} = f(x_n)$$

where $n = 0, 1, \ldots$ and

$$f(x) = \mu x(1 - x)$$

with $0 \leq x \leq 1$ and $0 \leq \mu \leq 4$.

A bifurcation diagram shows the long-term iterates x_n for each value of the parameter μ that varies over a range of values. Let us write a function for producing a bifurcation diagram. The function **bifurcation** takes five arguments: the initial value, minimum value of μ, maximum value of μ, increment of μ, and number of iterations. The function also accepts two-dimensional graphics options as trailing arguments.

```
In[1]:= bifurcation[x0_, μmin_, μmax_, Δ_, nmax_, opts___Rule] :=
          (ListPlot[
            Flatten[
             Table[
              Thread[
               List[
                μ,
                Drop[
                 NestList[(μ #(1 - #))&, x0, nmax],
                 200
```

```
                ]
              ]
                ],
        {μ, μmin, μmax, Δ}
              ],
    1
              ],
  opts
              ]
)
```

For clarity, we formatted the code to show the successive levels of the nest.

Let us dissect this function. For a given μ, **NestList[(μ#(1-#))&, x0, nmax]** generates the list

$$\{x_0, x_1, x_2, \ldots, x_{nmax}\}$$

Drop discards the first 200 elements of this list to ensure omission of transient behavior. Thus, **nmax** must be an integer greater than or equal to 200. **Thread** then produces the list

$$\{\{\mu, x_{200}\}, \{\mu, x_{201}\}, \{\mu, x_{202}\}, \ldots, \{\mu, x_{nmax}\}\}$$

because **Thread[**f**[**a **,** **{**b_1**,** b_2**,** **...,** b_n**}]]** returns the list

$$\{f[a, b_1], f[a, b_2], \ldots, f[a, b_n]\}$$

and, in our case, f is **List**, a is μ, and b_i is x_{199+i}. **Table** gives a nested list with μ taking on a sequence of values ranging from μ_{min} to μ_{max} in steps of Δ:

$$\{\{\{\mu_{min}, x_{200}\}, \{\mu_{min}, x_{201}\}, \{\mu_{min}, x_{202}\}, \ldots, \{\mu_{min}, x_{nmax}\}\},$$
$$\{\{\mu + \Delta, x_{200}\}, \{\mu + \Delta, x_{201}\}, \{\mu + \Delta, x_{202}\}, \ldots, \{\mu + \Delta, x_{nmax}\}\},$$
$$\{\{\mu + 2\Delta, x_{200}\}, \{\mu + 2\Delta, x_{201}\}, \{\mu + 2\Delta, x_{202}\}, \ldots, \{\mu + 2\Delta, x_{nmax}\}\},$$
$$\ldots,$$
$$\{\{\mu_{max}, x_{200}\}, \{\mu_{max}, x_{201}\}, \{\mu_{max}, x_{202}\}, \ldots, \{\mu_{max}, x_{nmax}\}\}\}$$

Flatten, with **1** as its second argument, flattens out the nested list to level 1 and yields a list of lists of coordinates for the points in the bifurcation diagram:

$$\{\{\mu_{min}, x_{200}\}, \{\mu_{min}, x_{201}\}, \{\mu_{min}, x_{202}\}, \ldots, \{\mu_{min}, x_{nmax}\},$$
$$\{\mu + \Delta, x_{200}\}, \{\mu + \Delta, x_{201}\}, \{\mu + \Delta, x_{202}\}, \ldots, \{\mu + \Delta, x_{nmax}\},$$
$$\{\mu + 2\Delta, x_{200}\}, \{\mu + 2\Delta, x_{201}\}, \{\mu + 2\Delta, x_{202}\}, \ldots, \{\mu + 2\Delta, x_{nmax}\},$$
$$\ldots,$$
$$\{\mu_{max}, x_{200}\}, \{\mu_{max}, x_{201}\}, \{\mu_{max}, x_{202}\}, \ldots, \{\mu_{max}, x_{nmax}\}\}$$

Finally, **ListPlot** plots these points.

Let us call the function **bifurcation** with $x_0 = 0.2, \mu_{min} = 2.9, \mu_{max} = 4.0, \Delta = 0.01, nmax = 500$, and several graphics option specifications:

```
In[2]:= bifurcation[0.2, 2.9, 4.0, 0.01, 500,
                    AxesOrigin -> {2.8, 0},
                    PlotRange -> {{2.8, 4}, {0, 1}},
                    AxesLabel -> {"μ", "xn"}]
```

Out[2]=

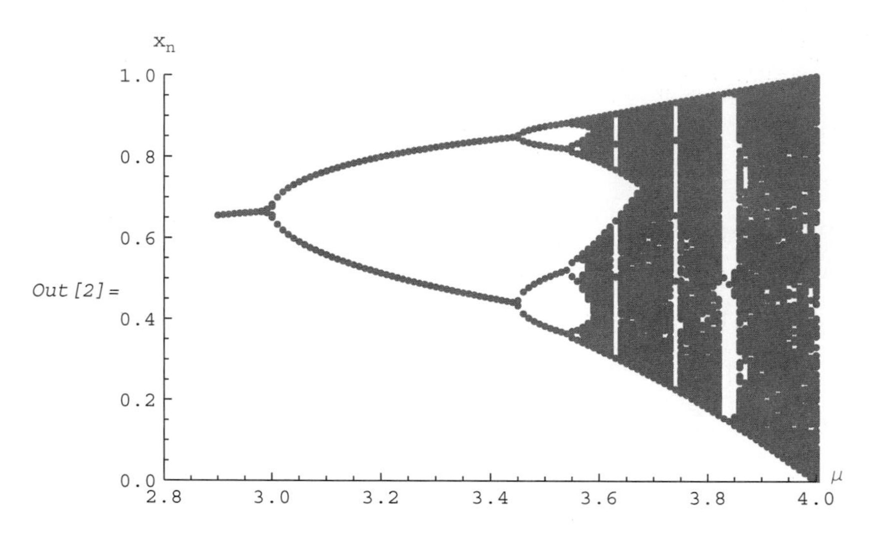

The bifurcation diagram shows period doublings, onset of chaotic behavior, and existence of windows of periodic behavior within the chaotic region. ([BG96] provides an excellent introduction to chaotic dynamics.)

Purists from the school of "strict" one-liners may be alarmed by the inclusion, on the *rhs* of the function definition for **bifurcation**, of the function **Table** because it contains a named iteration variable μ, whereas the only permissible named variables on the *rhs* are the argument names of **bifurcation**. This posture also excludes pure functions of the form **Function**[*var*, *body*] from strict one-liners. It even rules out some other built-in functions, such as **Sum**, **Integrate**, and **Solve**, unless their arguments are the argument names on the *lhs* of the function definition. Here is a strict one-liner for generating the bifurcation diagram for the logistic map:

```
In[3]:= bifurcationStrictOneLiner[
          x0_, μmin_, μmax_, Δ_, nmax_, opts___Rule] :=
         (ListPlot[
           Flatten[
            Map[
             Thread,
             Transpose[
```

```
          {NestList[
            (#+Δ)&,
            μmin,
            Round[(μmax-μmin)/Δ]
                   ],
           Map[
            Drop[#, 200]&,
            Map[
             NestList[#, x0, nmax]&,
             Map[
              Function,
              NestList[
                (#+Δ)&,
              μmin,
              Round[(μmax-μmin)/Δ]
                    ](#(1-#))
              ]
                ]
               ]
         }
                  ]
            ],
      1
            ],
    opts
            ]
   )
```

Choosing between **bifurcationStrictOneLiner** and **bifurcation** is a matter of aesthetics. The dissection of this function is left as an exercise (Exercise 5) in Section 3.6.4. (For further discussion of strict one-liners versus nonstrict or ordinary one-liners, see [Gra98].)

```
In[4]:= ClearAll[bifurcation, bifurcationStrictOneLiner]
```

3.6.3 Rule-Based Programming

Rule-based programming is *Mathematica*'s forte. A rule-based program consists of a collection of user-defined rewrite rules such as assignment statements and function definitions. We create rewrite rules with the functions **Set**, **UpSet**, **SetDelayed**, or **UpSetDelayed** in the forms

```
Set[lhs, rhs]
UpSet[lhs, rhs]
SetDelayed[lhs, rhs]
UpSetDelayed[lhs, rhs]
```

The corresponding special input forms are

lhs = *rhs*
lhs ^= *rhs*
lhs := *rhs*
lhs ^:= *rhs*

Often, *lhs* assumes the form

name[$patt_1$, $patt_2$, ...]

where $patt_i$ are patterns. We can also create rewrite rules with the syntax for dynamic programming so that *Mathematica* remembers the values it has found.

Rewrite rules reside in the global rule base, where more than one rule can be associated with a symbol. Whereas *Mathematica* stores the rules *lhs* = *rhs* and *lhs* ^= *rhs* with *rhs* evaluated, it stores the rules *lhs* := *rhs* and *lhs* ^:= *rhs* with *rhs* unevaluated. For example, let us create several rules:

```
In[1]:= ClearAll[a, b, f, g, p, q]

In[2]:= a = RandomInteger[10]
Out[2]= 10

In[3]:= f[n_]g[m_] ^= RandomInteger[{20, 30}]^(n m)
Out[3]= 21^(m n)

In[4]:= b:= RandomInteger[10]

In[5]:= p[n_]q[m_] ^:=RandomInteger[{20, 30}]^(nm)
```

The ? operator reveals how *Mathematica* stores these rules in the global rule base:

```
In[6]:= ?a

Global`a

a=10

In[7]:= ?f

Global`f

f/:f[n_] g[m_] = 21^(m n)
```

```
In[8]:= ?b
```

```
Global`b
```

```
b := RandomInteger[10]
```

```
In[9]:= ?p
```

```
Global`p
```

```
p/:p[n_] q[m_] := RandomInteger[{20, 30}]^(n m)
```

```
In[10]:= ClearAll[a, b, f, g, p, q]
```

When *Mathematica* encounters an expression that matches *lhs* of a rewrite rule in the global rule base composed of built-in functions and user-defined rewrite rules, it applies the rule. If the rule has the form *lhs* `=` *rhs* or *lhs* `^=` *rhs* and if *lhs* does not contain any patterns, it simply replaces the expression with *rhs*. If the rule has the form *lhs* `:=` *rhs* or *lhs* `^:=` *rhs* where *lhs*, as before, does not contain any patterns, it evaluates *rhs* and then replaces the expression with the result. If *lhs* contains patterns, it substitutes the actual values into the pattern names on *rhs* before applying the rule. If an expression matches *lhs* of several rules, *Mathematica* selects the rule according to the priority set forth in Section 3.3.2. That *Mathematica* uses user-defined rules before built-in ones bears emphasis. If the resulting expression or any of its parts again matches a rule, that rule is applied. The process continues until there are no more matching rules.

For implementing a set of mathematical relations, rule-based programming is usually the best choice. What are the formulas in mathematical tables? They are rewrite rules!

Let us illustrate rule-based programming with an example from quantum mechanics. With rules from the algebra of linear operators, we shall prove a number of commutator identities and simplify several commutators of functions of the one-dimensional position and momentum operators. ([Fea94] is the pioneering work on the application of *Mathematica* to quantum mechanics.)

The sum and product of two linear operators and the product of a linear operator with a number (more precisely, a scalar) are themselves linear operators. These operations of addition and multiplication obey all the rules of addition and multiplication of numbers, with one notable exception: Multiplication of linear operators is, in general, not commutative. If A, B, and C are linear operators and a is a number, then

$$AU = UA = A \tag{3.6.1}$$

$$A(B + C) = AB + AC \tag{3.6.2}$$

$$(A + B)C = AC + BC \tag{3.6.3}$$

$$A(BC) = (AB)C \tag{3.6.4}$$

$$A(aB) = aAB \tag{3.6.5}$$

$$(aA)B = aAB \tag{3.6.6}$$

where U is the unit or identity operator. (Pedants may object to our omission of the rule for the zero operator. Letting the symbol 0 denote here both the number zero and the zero operator makes the rule unnecessary and offers convenience without creating difficulties in computations.)

To implement these rules in *Mathematica*, we introduce the built-in function **NonCommutativeMultiply** because operator multiplication is often not commutative. **NonCommutativeMultiply[** *A* **,** *B* **]**, with the special input form *A* ** *B*, is a general associative, but noncommutative, form of multiplication. The associative law, Equation 3.6.4, is already built into **:

```
In[11]:= A** (B**C) == (A**B) **C
Out[11]= True
```

Equation 3.6.1 on the identity operator U can be entered as

```
In[12]:= Unprotect[NonCommutativeMultiply];
In[13]:= A_**U := A
In[14]:= U**A_ := A
```

where we have unprotected the built-in function **NonCommutativeMultiply** in order to specify rules for it. The distributive laws, Equations 3.6.2 and 3.6.3, can be entered as

```
In[15]:= A_** (B_+C_) := A**B + A**C
In[16]:= (A_+B_) **C_ := A**C + B**C
```

To enter Equations 3.6.5 and 3.6.6, we naturally think of the pattern **a_?NumberQ** for representing numbers. Yet a less restrictive pattern is needed because the number a, in the application of these rules, can actually be an algebraic expression representing a number and *Mathematica* does not recognize as numbers the expressions consisting of constants, numerals, and operation signs. For example, let a and b be constants—that is, symbols denoting particular numbers,

```
In[17]:= NumberQ[a] ^= True;
In[18]:= NumberQ[b] ^= True;
```

and evaluate

```
In[19]:= Cases[{a, b, (2/3)b, I Sqrt[a], b^3, a b, a/b, a^b}, a_? NumberQ]
Out[19]:= {a, b}
```

Mathematica recognizes only a and b as numbers. The less restrictive pattern **(x_. y_^n_. /;NumberQ[x] && NumberQ[y]&& NumberQ[n])** matches all the elements of the list in the previous example:

```
In[20]:= Cases[{a, b, (2/3)b, I Sqrt[a], b^3, ab, a/b, a^b},
           (x_. y_^n_./;NumberQ[x]&& NumberQ[y]&&NumberQ[n])]
```

$$Out[20]= \left\{a, b, \frac{2 b}{3}, i \sqrt{a}, b^3, a b, \frac{a}{b}, a^b\right\}$$

This pattern is sufficiently general in representing numbers for most problems in quantum mechanics. With the function

```
In[21]:= number3Q[x_, y_, n_] := NumberQ[x] && NumberQ[y] && NumberQ[n]
```

Equations 3.6.5 and 3.6.6 can be entered as

```
In[22]:= A_ ** (B_ (x_. y_^n_./;number3Q[x, y, n])) := ((xy^n) A ** B)
```

```
In[23]:= (A_ (x_. y_^n_./;number3Q[x, y, n])) ** B_ := ((xy^n)A ** B)
```

To avoid inadvertent modifications, we should restore the protection for **NonCommutative-Multiply**:

```
In[24]:= Protect[NonCommutativeMultiply];
```

Mathematica can now correctly simplify expressions consisting of linear operators and numbers. For example,

```
In[25]:= {(3A + B/5) **C, ((2a/Sqrt[5ab]) U) **A, ((a^2/b^3)U) ** (A+2B) **
           C, (Sqrt[2a] A) ** (B ** (3C/(5b^2)))} //PowerExpand//ExpandAll
```

$$Out[25]= \left\{3 A ** C + \frac{B**C}{5}, \frac{2 \sqrt{a} A}{\sqrt{5} \sqrt{b}}, \frac{a^2 A**C}{b^3} + \frac{2 a^2 B**C}{b^3}, \frac{3 \sqrt{2} \sqrt{a}\ A**B**C}{5 b^2}\right\}$$

With the rules (Equations 3.6.1 to 3.6.6) already in the global rule base, we can prove several commutator identities. The commutator of two linear operators is defined as

$$[A, B] = AB - BA$$

For *Mathematica*, we express this definition as

```
In[26]:= commutator[A_, B_] := A** B - B** A
```

Let us prove the identities

$$[A, B] = -[B, A] \tag{3.6.7}$$

$$[aA, bB] = ab[A, B] \tag{3.6.8}$$

$$[A, B + C] = [A, B] + [A, C] \tag{3.6.9}$$

$$[A + B, C] = [A, C] + [B, C] \tag{3.6.10}$$

$$[A, BC] = [A, B]C + B[A, C] \tag{3.6.11}$$

$$[AB, C] = A[B, C] + [A, C]B \tag{3.6.12}$$

$$[[A, B], C] + [[C, A], B] + [[B, C], A] = 0 \tag{3.6.13}$$

For Equation 3.6.7,

```
In[27]:= commutator[A, B] + commutator[B, A] == 0
Out[27]= True
```

For Equation 3.6.8,

```
In[28]:= commutator[a A, b B] - a b commutator[A, B] == 0 // ExpandAll
Out[28]= True
```

For Equation 3.6.9,

```
In[29]:= commutator[A, B + C] - commutator[A, B] - commutator[A, C] == 0
Out[29]= True
```

For Equation 3.6.10,

```
In[30]:= commutator[A + B, C] - commutator[A, C] - commutator[B, C] == 0
Out[30]= True
```

For Equation 3.6.11,

```
In[31]:= commutator[A, B ** C] - commutator[A, B] ** C - B ** commutator[A, C] == 0
Out[31]= True
```

For Equation 3.6.12,

```
In[32]:= commutator[A ** B, C] - A ** commutator[B, C] - commutator[A, C] ** B == 0
Out[32]= True
```

For Equation 3.6.13, the Jacobi identity,

```
In[33]:= commutator[commutator[A, B], C] +
           commutator[commutator[C, A], B] + commutator[commutator[B, C], A] == 0
Out[33]= True
```

The one-dimensional position and momentum operators x and p in quantum mechanics satisfy the commutator relation

$$[x,p] = i\hbar \tag{3.6.14}$$

Since the commutator of two operators is also an operator, the *rhs* of this commutation relation actually stands for

$$i\hbar U$$

where U is the identity operator. We can apply the commutation relation (Equation 3.6.14) to simplify a commutator with the function

```
In[34]:= xpCommutator[expr_] := ExpandAll[expr//.p**x:>x**p-iħU]
```

where ħ, Planck's constant divided by 2π, must be declared to be a number:

```
In[35]:= NumberQ[ħ] ^= True;
```

To simplify, for example, the commutators

$$[x,p^4]$$
$$[p,x^2p^2]$$
$$[xp^2,px^2]$$

we have

```
In[36]:= commutator[x, p**p**p**p] //xpCommutator
Out[36]= 4iħp**p**p
```

```
In[37]:= commutator[p, x**x**p**p] //xpCommutator
Out[37]= -2iħx**p**p
```

```
In[38]:= commutator[x**p**p, p**x**x] //xpCommutator
Out[38]= -6ħ^2 x**p-3iħx**x**p**p
```

Let us conclude this section with the evaluation of the commutator

$$[H,p]$$

for the one-dimensional harmonic oscillator whose Hamiltonian H is

$$H = \frac{p^2}{2m} + \frac{1}{2}m\omega^2x^2$$

where m and ω are the mass and frequency of the oscillator, respectively. After declaring m and ω to be constants,

```
In[39]:= NumberQ[m] ^= True;
In[40]:= NumberQ[ω] ^= True;
```

we have

```
In[41]:= commutator[p**p/(2m) + 1/2 m ω^2 x**x, p] //xpCommutator
Out[41]= i m x ω^2 ℏ
```

(For further discussions of operators, commutators, and their algebras, see [DK67], [Has91], [Mes00], and [Mor90].)

```
In[42]:= ClearAll[a, b, m, ω, ℏ, number3Q, xpCommutator]
```

3.6.4 Exercises

In this book, straightforward, intermediate-level, and challenging exercises are unmarked, marked with one asterisk, and marked with two asterisks, respectively. For solutions to most odd-numbered exercises, see Appendix C.

1. Modify the function **planetMotion** in Section 3.6.1 to allow option specifications for the sizes of the Sun and planet in the graphics.

*2. Write a function that generates an animation of the lecture demonstration discussed in Example 3.5.2. Each animation sequence should end when the projectile hits the target. Also, the function should have three arguments: the initial speed v_0 and elevation θ_0 of the projectile and the initial height h of the target.

*3. Using **Module**, write a function that accepts as the argument any list of integers each with a value in the interval (0, 100) and that returns a plot of the number of integers in each interval: (0, 0), (1, 5), (6, 10), ..., (91, 95), and (96, 100). The plot should not show the zero points for intervals with no integers; the function must not introduce unnecessary global variables with assigned values. For example,

```
scores = {52, 80, 59, 59, 44, 67, 53, 59, 50, 56, 69, 70, 58, 68, 79, 70,
   55, 66, 76, 73, 50, 68, 57, 59, 59, 52, 59, 45, 63, 56, 70, 47, 55,
   54, 64, 78, 48, 48, 56, 74, 45, 62, 58, 54, 57, 77, 45, 91, 61, 49,
   39, 46, 65, 52, 62, 50, 65, 58, 51, 59, 60, 62, 57, 46, 57, 57, 56,
   51, 48, 69, 67, 61, 66, 73, 61, 53, 44, 66, 62, 47, 58, 52, 69, 59,
   58, 57, 70, 81, 50, 74, 49, 56, 62, 78, 71, 56, 62, 53, 55, 51};
distribution[scores]
```

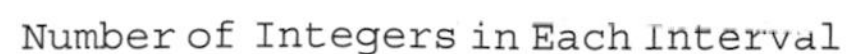

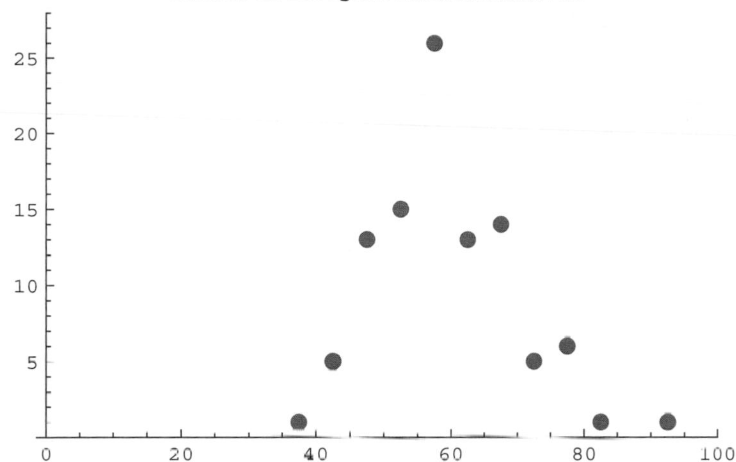

4. Modify the function **bifurcation** in Section 3.6.2 to generate a bifurcation diagram for the map

$$x_{n+1} = x_n e^{\mu(1-x_n)}$$

Produce a bifurcation diagram for $2.0 < \mu < 4.0$. *Hint:* Discard only the first 100 elements, and let $nmax = 300$.

5. In Section 3.6.2, the function **bifurcationStrictOneLiner** is defined as a nested function call—that is, the application of nested functions to some arguments. Following the dissection of the function **bifurcation** in the same section, take **bifurcationStrictOneLiner** apart and describe the result of each function application.

6. If A and B are linear operators and a is a scalar, show that

$$\frac{d^2 e^{-aB} e^{-aA} e^{a(A+B)}}{da^2} = [B, A]$$

as $a \to 0$. *Hint:* The symbol 1 denotes both the number 1 and the unit or identity operator U.

*7. The orbital angular momentum operator of a three-dimensional particle is

$$l = \mathbf{r} \times \mathbf{p}$$

where **r** and **p** are the position and linear momentum operators, the components of which satisfy the commutation relations

$$\begin{aligned} [r_i, r_j] &= 0 \\ [p_i, p_j] &= 0 \qquad i,j = 1,2,3 \\ [r_i, p_j] &= i\hbar\delta_{ij} \end{aligned} \tag{3.6.15}$$

where each subscript takes on the values 1, 2, and 3, corresponding to the x, y, and z components respectively; that is, $r_1 = x$, $p_1 = p_x$, and so forth. The Kronecker delta δ_{ij} is defined by

$$\delta_{ij} = \begin{cases} 0 & i \neq j \\ 1 & i = j \end{cases}$$

With the rules (Equations 3.6.1–3.6.6) and the commutation relations (Equation 3.6.15), verify

$$[l_x, l_y] = i\hbar l_z$$
$$[l_y, l_z] = i\hbar l_x$$
$$[l_z, l_x] = i\hbar l_y$$
$$\left[l_z, \boldsymbol{l}^2\right] = 0$$

where

$$\boldsymbol{l}^2 = l_x^2 + l_y^2 + l_z^2$$

*8. The three-dimensional isotropic harmonic oscillator is a particle of mass m moving in a central potential that is proportional to the square of its distance from the center. The Hamiltonian is

$$H = \frac{\mathbf{p}^2}{2m} + \frac{1}{2}k\mathbf{r}^2$$

where $\mathbf{r}$ and $\mathbf{p}$ are the position and linear momentum operators and k is the "spring constant." The orbital angular momentum operator is

$$\boldsymbol{l} = \mathbf{r} \times \mathbf{p}$$

In Cartesian coordinates, the Hamiltonian and the square of the orbital angular momentum operator are given as

$$H = \frac{p_x^2}{2m} + \frac{p_y^2}{2m} + \frac{p_z^2}{2m} + \frac{1}{2}k\left(x^2 + y^2 + z^2\right)$$
$$\boldsymbol{l}^2 = l_x^2 + l_y^2 + l_z^2$$

Show that H, $\boldsymbol{l}^2$, and l_z form a set of commuting operators. That is, show (a) H commutes with $\boldsymbol{l}^2$, (b) $\boldsymbol{l}^2$ commutes with l_z, and (c) l_z commutes with H. *Hint:* Use the rules (Equations 3.6.1–3.6.6) and the commutation relations (Equation 3.6.15).

9. Define a recursive function (i.e., a function that calls itself) that takes a list of random length of random integers as the argument and returns a list of the elements at odd positions (i.e., the first, third, fifth, ... integers). The recursive function must not contain the built-in function **Length**. For example,

```
f[{-60, -4, -64, 0, -85, -64, -20, -13, 49, -50, -16, -62, 66, -5}]
{-60, -64, -85, -20, 49, -16, 66}
```

3.7 PACKAGES

As discussed in Section 1.5, packages are files of definitions written in the *Mathematica* language, and a package must be loaded before its definitions are accessible. *Mathematica* comes with many standard packages, and we can write our own. What are the purposes of packages? Sections 2.2.7 and 2.2.16 warned that global symbols are potential pitfalls, and Section 3.4.1 showed how the function **Module** sets up local symbols. Using contexts, packages provide an encapsulation or localization mechanism for symbols in entire files so as to allow the use of the same symbol in a *Mathematica* session as well as in different packages. Packages also permit hiding implementations that are of no interest to users of the definitions or readers of the notebooks. Furthermore, they allow storing of definitions that are used often. Section 3.7.1 elaborates on the notion of contexts introduced in Section 1.5. Section 3.7.2 discusses the manipulation of contexts. Section 3.7.3 analyzes the construction of a sample package, and Section 3.7.4 provides a template for packages.

3.7.1 Contexts

Full names of *Mathematica* symbols have the form *context`short*, where *context`* is the context of a symbol and *short* is its short name. A context always ends with a backquote `, called a "context mark" in *Mathematica*. Symbols with the same short name but different contexts are different. For example, **physics`strangeness** and **psychology`strangeness** are different:

```
In[1]:= physics`strangeness === psychology`strangeness
Out[1]= False
```

SameQ[*lhs*, *rhs*], with the special input form *lhs* === *rhs*, yields **True** if the expression *lhs* is identical to *rhs*, and it yields **False** otherwise. Contexts are not symbols and must be enclosed by quotation marks when entered as arguments of commands. Symbols with common properties or functionalities or from one subject area are often put in the same context.

Contexts can be hierarchical. In other words, contexts can take the form *context*$_1$`*context*$_2$`...*context*$_n$`. For example, the context of the symbol **mechanics`kinematics`velocity** is **mechanics`kinematics`**:

```
In[2]:= Context[mechanics`kinematics`velocity]
Out[2]= mechanics`kinematics`
```

Context[*s*] returns the context of symbol *s*.

At any point in a *Mathematica* session, there is a current context, which is the value of the variable **$Context**. When a new symbol is introduced, it is put in the current context. For example, enter **charm**:

```
In[3]:= charm
Out[3]= charm
```

The default current context is **Global`**:

```
In[4]:= $Context
Out[4]= Global`
```

The context of **charm** is therefore **Global`**:

```
In[5]:= Context[charm]
Out[5]= Global`
```

We can refer to symbols in the current context by their short names:

```
In[6]:= Global`charm === charm
Out[6]= True
```

The symbol *`subcontext`* stands for *currentcontext`subcontext`*; that is, the full name of the context *`subcontext`* is *currentcontext`subcontext`*, where *currentcontext`* is the current context. For example, if the current context is **Global`**, the context of the symbol **`sound`wave** is **Global`sound`**:

```
In[7]:= {$Context, Context[`sound`wave]}
Out[7]= {Global`, Global`sound`}
```

Built-in *Mathematica* objects have the context **System`**. For example, **Plot3D** is in the context **System`**:

```
In[8]:= Context[Plot3D]
Out[8]= System`
```

When we invoke a short name associated with several contexts, to which symbol does it refer? *Mathematica* first searches the contexts in the context search path for a symbol with that short name. If it cannot find a symbol with that short name, *Mathematica* searches the current context for a symbol with the short name. Let us create, for example, the symbols **color**, **physics`color**, and **art`color**:

```
In[9]:= {color, physics`color, art`color}
Out[9]= {color, physics`color, art`color}
```

As mentioned previously, **$Context** returns the current context:

```
In[10]:= $Context
Out[10]= Global`
```

Therefore, the context of **color** is **`Global**:

```
In[11]:= Context[color]
Out[11]= Global`
```

The variable **$ContextPath** gives the context search path (i.e., the list of contexts, before the current context, that *Mathematica* searches for a symbol, in the order in which they appear in the list):

```
In[12]:= $ContextPath
Out[12]= {DocumentationSearch`,ResourceLocator`, JLink`,
          PacletManager`, WebServices`, System`, Global`}
```

(We have used *Mathematica* 6.0.2 to generate the preceding output—that is, Out[12]. Had we used *Mathematica* 6.0.0 or 6.0.1, the output would be slightly different: **{PacletManager`, WebServices`, System`, Global`}**. In both cases, only **System`** and **Global`** are germane to the discussion here.) Now prepend **physics`** and **art`** to the context search path:

```
In[13]:= $ContextPath = {"physics`","art`"} ~Join~ $ContextPath
Out[13]= {physics`,art`, DocumentationSearch`,ResourceLocator`,
          JLink`, PacletManager`, WebServices`, System`, Global`}
```

The short name **color** now refers to the symbol **physics`color** because **physics`** is the first context with a symbol having the short name **color** in the context search path:

```
In[14]:= Context[color]
Out[14]= physics`
```

The contexts of symbols exported by packages are the context names of the packages. There is an important point to remember: If two or more symbols have the same short name but different contexts, the definition of one symbol may shadow or be shadowed by those of the others. For example, let us define

```
In[15]:= Grad[f_]:= f^3
```

The symbol **Grad** is in the current context, which is **Global`**. Now load the package **VectorAnalysis`**. The function **Needs["*context*`"]** loads a package with the context name *context*`.

```
In[16]:= Needs["VectorAnalysis`"]
```

Grad::shdw:
Symbol Grad appears in multiple contexts {VectorAnalysis`, Global`};
definitions in context VectorAnalysis` may
shadow or be shadowed by other definitions.

Mathematica warns that the definition of **Grad** in the package may shadow or be shadowed by our previous definition. For this package, **CoordinateSystem** gives the name of the default coordinate system and **Coordinates[]** shows the default names of the coordinate variables in the default coordinate system:

```
In[17]:= {CoordinateSystem, Coordinates[]}
Out[17]= {Cartesian, {Xx, Yy, Zz}}
```

To illustrate the idea of shadowing, let us evaluate

```
In[18]:= Grad[Xx^2 Yy^2 Zz^2]
```

Out[18]= $\{2\,\mathrm{Xx}\,\mathrm{Yy}^2\,\mathrm{Zz}^2,\ 2\,\mathrm{Xx}^2\,\mathrm{Yy}\,\mathrm{Zz}^2,\ 2\,\mathrm{Xx}^2\,\mathrm{Yy}^2\,\mathrm{Zz}\}$

Mathematica returns **Grad[Xx^2 Yy^2 Zz^2]** in accordance with the definition of **Grad** in the package, which is consistent with vector analysis in mathematics and physics. Our previous definition of **Grad** is shadowed or ignored. To access our definition of **Grad**, we must first execute the command

```
In[19]:= Remove[Grad]
```

Remove::rmptc: Symbol Grad is Protected and cannot be removed. >>

where **Remove[**$symbol_1$**,**$symbol_2$**, ...]** removes symbols completely so that their names are no longer recognized by *Mathematica*. *Mathematica* informs us that the symbol **Grad** is protected. We could unprotect it with **Unprotect** before applying **Remove**:

```
Unprotect[Grad]
Remove[Grad]
```

Because Section 3.3.5 pointed out that using **Unprotect** can be perilous, let us avoid using it. Another way to access our definition is to alter the context search path:

```
In[20]:= $ContextPath = RotateRight[$ContextPath]
Out[20]= {Global`,VectorAnalysis`, physics`, art`, DocumentationSearch`,
          ResourceLocator`, JLink`, PacletManager`, WebServices`, System`}
```

Let us evaluate again

```
In[21]:= Grad[Xx^2 Yy^2 Zz ^2]
```

Out[21]= $\mathrm{Xx}^6\,\mathrm{Yy}^6\,\mathrm{Zz}^6$

Mathematica now returns the result according to our definition of **Grad**.

3.7.2 Context Manipulation

Mathematica provides several commands for manipulating contexts:

BeginPackage["*context*`"]	make *context*` and **System`** the only active contexts
BeginPackage["*context*`", {"$need_1$`", "$need_2$`", ...}]	call **Needs** on the $need_i$
EndPackage[]	restore **$Context** and **$ContextPath** to their values before the preceding **BeginPackage**, and prepend the current context to the list **$ContextPath**
Begin["*context*`"]	reset the current context
End[]	return the present context, and revert to the previous one

Let us illustrate the use of these commands. First, return the context search path to its default:

```
In[22]:= $ContextPath = {"DocumentationSearch`", "ResourceLocator`", "JLink`",
            "PacletManager`", "WebServices`", "System`", "Global`"}
Out[22]= {DocumentationSearch`, ResourceLocator`, JLink`,
          PacletManager`, WebServices`, System`, Global`}
```

Now enter

```
In[23]:= BeginPackage["test`"]
Out[23]= test`
```

BeginPackage["test`"] resets the values of both **$Context** and **$ContextPath**:

```
In[24]:= {$Context, $ContextPath}
Out[24]= {test`, {test`, System`}}
```

EndPackage[] restores the values of **$Context** and **$ContextPath** and prepends the current context to the list **$ContextPath**.

```
In[25]:= EndPackage[]
```

The current context is once again **Global`**:

```
In[26]:= $Context
Out[26]= Global`
```

The context search path is restored, but with context **test`** prepended:

```
In[27]:= $ContextPath
Out[27]= {test`,DocumentationSearch`,ResourceLocator`,
          JLink`,PacletManager`,WebServices`,System`,Global`}
```

Before proceeding, return the context search path to its default:

```
In[28]:= $ContextPath = {"DocumentationSearch`","ResourceLocator`","JLink`",
           "PacletManager`","WebServices`","System`","Global`"}
Out[28]= {DocumentationSearch`,ResourceLocator`,JLink`,
          PacletManager`,WebServices`,System`,Global`}
```

If **BeginPackage** has a second argument, the specified packages are loaded:

```
In[29]:= BeginPackage["test`",
          {"ResonanceAbsorptionLines`","VariationalMethods`"}]
Out[29]= test`
```

The current context becomes **test`**:

```
In[30]:= $Context
Out[30]= test`
```

The context search path now includes the additional contexts:

```
In[31]:= $ContextPath
Out[31]= {test`,ResonanceAbsorptionLines`,VariationalMethods`,System`}
```

Without altering the context search path, **Begin["mycontext`"]** resets the current context to **mycontext`**:

```
In[32]:= Begin["mycontext`"]
Out[32]= mycontext`
```

$Context shows the current context:

```
In[33]:= $Context
Out[33]= mycontext`
```

The context search path remains the same:

```
In[34]:= $ContextPath
Out[34]= {test`,ResonanceAbsorptionLines`,VariationalMethods`,System`}
```

Without changing the context search path, **End[]** returns the current context and reverts to the previous one:

```
In[35]:= End[]
Out[35]= mycontext`
```

The current context is once again **test`**:

```
In[36]:= $Context
Out[36]= test`
```

The context search path is unaffected:

```
In[37]:= $ContextPath
Out[37]= {test`, ResonanceAbsorptionLines`, VariationalMethods`, System`}
```

Before proceeding to the next section, let us return the current context and context search path to their defaults:

```
In[38]:= {$Context, $ContextPath} =
          {"Global`", {"DocumentationSearch`", "ResourceLocator`", "JLink`",
            "PacletManager`", "WebServices`", "System`", "Global`"}}
Out[38]= {Global`, {DocumentationSearch`, ResourceLocator`,
           JLink`, PacletManager`, WebServices`, System`, Global`}}
```

3.7.3 A Sample Package

With a sample package, this section shows how to use the commands introduced in Section 3.7.2 in setting up *Mathematica* packages.

3.7.3.1 The Problem

Write a package that provides the functions for the eigenenergies and normalized energy eigenfunctions of the hydrogen atom.

The normalized energy eigenfunctions of hydrogen can be written as

$$\psi_{nlm}(r,\theta,\phi) = R_{nl}(r)Y_{lm}(\theta,\phi)$$

where R_{nl} and Y_{lm} are the radial functions and the normalized spherical harmonics, respectively. In terms of the associated Laguerre polynomials L_p^q and the Bohr radius a, the radial functions are given by

$$R_{nl}(r) = a^{-3/2}\frac{2}{n^2}\sqrt{\frac{(n-l-1)!}{[(n+l)!]^3}}\,F_{nl}\left(\frac{2r}{na}\right)$$

with

$$F_{nl}(x) = x^l e^{-x/2} L_{n-l-1}^{2l+1}(x)$$

The associated Laguerre polynomials $L_p^q(x)$ are related to the generalized Laguerre polynomials **LaguerreL[p,q,x]** of *Mathematica* by

$$L_p^q(x) = (p+q)!\,\texttt{LaguerreL[p,q,x]}$$

The eigenenergies can be expressed as

$$E_n = -\frac{Ry}{n^2}$$

where Ry is the Rydberg constant.

The allowed quantum numbers n, l, and m are given by the rules

$$n = 1, 2, 3, \ldots$$
$$l = 0, 1, 2, \ldots, (n-1)$$
$$m = -l, -l+1, \ldots, 0, 1, 2, \ldots, +l$$

(For more information on the hydrogen atom, see [Lib03].)

3.7.3.2 The Package

What follows is a package that exports functions for the eigenenergies and eigenfunctions. The unit of eigenenergy can be the electron volt (eV), the joule (J), or the erg.

```
BeginPackage["HydrogenAtom`", "Units`"]

HydrogenAtom::usage =
 "HydrogenAtom` is a package that provides functions
   for the eigenenergies and normalized energy
   eigenfunctions of the hydrogen atom."

ψ::usage =
 "ψ[n, l, m, r, θ, ϕ] gives the normalized energy eigenfunctions
   in the spherical coordinates r, θ, and ϕ."
ℰ::usage =
 "ℰ[n] gives the eigenenergies of hydrogen in electron volts (eV).
   Joule or Erg may be entered as the value of the option
   Unit to express the energies in the specified unit."

a::usage = "The symbol a stands for the Bohr radius."
```

```
Unit::usage = "Unit is an option of ℰ[n]."

Options[ℰ] = {Unit → ElectronVolt}

Begin["`Private`"]

F[n_, l_, x_] :=
 ((x^l) Exp[-x/2] ((n+l)!)LaguerreL[n-l-1, 2l+1, x])

R[n_, l_, r_] := ((a^(-3/2)) (2/(n^2))
   Sqrt[(n-l-1)!/((n+l)!)^3] F[n, l, 2r/(na)])

HydrogenAtom::badarg =
 "You called `1` with `2` argument(s)! It must have `3` argument(s)."

ψ::quannum = "Your set of quantum numbers `1` is not allowed."

ℰ::badarg =
 "The first argument must be a positive integer. The second argument
   is optional and, if specified, must be one of the
   rules : Unit -> Joule or Unit -> Erg. You entered `1`."

ψ[n_Integer?Positive, l_Integer?NonNegative, m_Integer, r_, θ_, ϕ_] :=
 (R[n, l, r]SphericalHarmonicY[l, m, θ, ϕ]) /; (l ≤ (n-1) && -l ≤ m ≤ l)

ψ[n_, l_, m_, r_, θ_, ϕ_] := Message[ψ::quannum, {n, l, m}]

ψ[arg___/; Length[{arg}] ≠ 6] :=
 Message[HydrogenAtom :: badarg, ψ, Length[{arg}], 6]

ℰ[n_Integer?Positive, opts___Rule] :=
 (-Convert[Rydberg, Unit/.{opts}/.Options[ℰ]]/n^2)

ℰ[arg__/; Length[{arg}] < 3] := Message[ℰ::badarg, {arg}]

ℰ[arg___/; (Length[{arg}] == 0 || Length[{arg}] > 2)] :=
 Message[HydrogenAtom :: badarg, ℰ, Length[{arg}], "1 or 2"]

End[]

Protect[ψ, ℰ]

EndPackage[]
```

3.7.3.3 Analysis of the Package

This section analyzes the construction of the package in Section 3.7.3.2. In what follows, semicolons are put at the end of commands to suppress outputs that are normally absent in the actual loading of the package.

The package begins with the command

```
In[39]:= BeginPackage["HydrogenAtom`","Units`"];
```

Mathematica imports—that is, reads in—the package **Units`** if it is not already loaded, changes the current context to **HydrogenAtom`**, and sets the context search path to **{"HydrogenAtom`","Units`","System`"}**.

Next come the usage messages:

```
In[40]:= HydrogenAtom::usage =
          "HydrogenAtom` is a package that provides functions
            for the eigenenergies and normalized energy
            eigenfunctions of the hydrogen atom.";

         ψ::usage =
          "ψ[n,l,m,r,θ,ϕ] gives the normalized energy eigenfunctions
            in the spherical coordinates r,θ, and ϕ.";

         ℰ::usage =
          "ℰ[n] gives the eigenenergies of hydrogen in electron volts (eV).
            Joule or Erg may be entered as the value of the option
            Unit to express the energies in the specified unit.";

         a::usage = "The symbol a stands for the Bohr radius.";

         Unit::usage = "Unit is an option of ℰ[n].";
```

After loading the package, we can access these usage messages with the **?** operator just as we obtain those for the built-in objects. These messages serve another purpose of putting the symbols **HydrogenAtom**, **ψ**, **ℰ**, **a**, and **Unit** in the context **HydrogenAtom`**. Since the context **HydrogenAtom`** is included in the context search path after the package is loaded, these symbols, intended for export outside the package, are visible.

Options[*symbol***]** gives the list of default options assigned to *symbol*. The package specifies a default option for the function **ℰ** with the command

```
In[45]:= Options[ℰ] = {Unit→ElectronVolt};
```

Then comes the command

```
In[46]:= Begin["`Private`"];
```

The context search path remains **{"HydrogenAtom`", "Units`", "System`"}**, whereas the current context becomes **HydrogenAtom`Private`**.

At this point in the package, there are definitions for two auxiliary functions **F** and **R**:

```
In[47]:= F[n_,l_,x_]:=
           ((x^l)Exp[-x/2]((n+l)!)LaguerreL[n-l-1, 2l+1, x])

         R[n_, l_, r_]:= ((a^(-3/2))(2/(n^2))
             Sqrt[(n-l-1)!/((n+l)!)^3] F[n, l, 2r/(n a)])
```

These symbols are put in the context **HydrogenAtom`Private`**. Since this context is excluded from the context search path after the package is loaded, these symbols are invisible outside the package—they are intended solely for internal uses in the package.

Note that *Mathematica* does not support two-dimensional forms inside packages. For example, we must enter **x^l** rather than $\mathbf{x}^{l}$ in the definition for the function **F**.

Several error messages for the functions intended for export are then defined:

```
In[49]:= HydrogenAtom::badarg = "You called `1` with
             `2` argument(s)! It must have `3` argument(s).";

         ψ::quannum = "Your set of quantum numbers `1` is not allowed.";

         ℰ:: badarg =
           "The first argument must be a positive integer. The second
             argument is optional and, if specfied, must be one of
             the rules: Unit -> Joule or Unit -> Erg. You entered `1`.";
```

There are three definitions for the function ψ:

```
In[52]:= ψ[n_Integer?Positive, l_Integer?NonNegative, m_Integer, r_, θ_, ϕ_] :=
           (R[n, l, r] SphericalHarmonicY[l, m, θ, ϕ]) /; (l ≤ (n-1) && -l ≤ m ≤ l)

         ψ[n_, l_, m_, r_, θ_, ϕ_] := Message[ψ::quannum, {n, l, m}]

         ψ[arg___/; Length[{arg}] ≠ 6] :=
          Message[HydrogenAtom]::badarg, ψ, Length[{arg}], 6]
```

The last two definitions trigger error messages when incorrect arguments are entered. **Message**[*symbol*::*tag*, e_1, e_2, ...] prints the message *symbol*::*tag*. The values of the $e_1, e_2, \ldots$ replace `1`, `2`, ... in the message *symbol*::*tag* defined previously. The second definition returns an error message if unacceptable quantum numbers are specified:

```
In[55]:= ψ[-2, 1, 0, r, θ, ϕ]
        ψ::quannum : Your set of quantum numbers {-2, 1, 0} is not allowed.
```

The third definition gives an error message if there is a wrong number of arguments:

```
In[56]:= ψ[2, 1, 0, 2, r, θ, ϕ, z]
       HydrogenAtom::badarg :
        You called ψ with 8 argument(s)! It must have 6 argument(s).
```

There are also three definitions for the function $\mathcal{E}$:

```
In[57]:= ℰ[n_Integer?Positive, opts___Rule] :=
          (-Convert[Rydberg, Unit/.{opts}/.Options[ℰ]]/n^2)

         ℰ[arg__/; Length[{arg}] < 3] := Message[ℰ::badarg, {arg}]

         ℰ[arg___/; (Length[{arg}] == 0||Length[{arg}] > 2)] :=
          Message[HydrogenAtom::badarg, ℰ, Length[{arg}], "1 or 2"]
```

The second definition for $\mathcal{E}$ returns an error message if an unacceptable quantum number is entered or the option `Unit` is not specified in terms of a rule:

```
In[60]:= ℰ[-2]
       ℰ::badarg :
        The first argument must be a positive integer. The second
          argument is optional and, if specified, must be one of
          the rules : Unit -> Joule or Unit -> Erg. You entered {-2}.

In[61]:= ℰ[2, Unit]
       ℰ::badarg :
        The first argument must be a positive integer. The second
          argument is optional and, if specified, must be one of the
          rules: Unit -> Joule or Unit -> Erg. You entered {2, Unit}.
```

The third definition for $\mathcal{E}$ gives an error message if there is a wrong number of arguments:

```
In[62]:= ℰ[2, Unit → Erg, 3]
       HydrogenAtom::badarg : You called ℰ with
          3 argument(s)! It must have 1 or 2 argument(s).
```

Without changing the context search path, which is `{"HydrogenAtom`", "Units`", "System`"}`, the command `End` restores the current context to `HydrogenAtom``:

```
In[63]:= End[];

In[64]:= {$Context, $ContextPath}
```

```
Out[64]= {HydrogenAtom`, {HydrogenAtom`, Units`, System`}}
```

Protect protects the functions ψ and ℰ intended for export just as it does for built-in *Mathematica* objects:

```
In[65]:= Protect[ψ, ℰ];
```

Finally, **EndPackage[]** restores the current context to the one before the preceding **BeginPackage** was executed:

```
In[66]:= EndPackage[]
```

```
In[67]:= $Context
Out[67]= Global`
```

Also, the context **HydrogenAtom`** together with **Units`** is prepended to the context search path before the preceding **BeginPackage** was executed:

```
In[68]:= $ContextPath
Out[68]= {HydrogenAtom`, Units`, DocumentationSearch`, ResourceLocator`,
          JLink`, PacletManager`, WebServices`, System`, Global`}
```

As intended, the symbols in the context **HydrogenAtom`** are still visible, whereas those in the context **HydrogenAtom`Private`** are not.

```
In[69]:= F
Out[69]= F
```

```
In[70]:= R
Out[70]= R
```

We can obtain the eigenfunctions:

```
In[71]:= ψ[3, 2, -2, r, θ, φ]
```

$$\text{Out[71]=} \frac{e^{-\frac{r}{3a}-2i\phi}\, r^2\, \text{Sin}[\theta]^2}{162\, a^{7/2}\sqrt{\pi}}$$

We can determine the eigenenergies:

```
In[72]:= ℰ[1]
Out[72]= -13.6059 ElectronVolt
```

The energy can be expressed in terms of another unit:

```
In[73]:= ℰ[1, Unit → Joule]
Out[73]= -2.1799 × 10^-18 Joule
```

To use the package **HydrogenAtom`** like a standard *Mathematica* package, we must

1. Enter the commands of Section 3.7.3.2 into a cell of a new notebook.
2. Click the cell bracket and choose **Initialization Cell** in the Cell Properties submenu of the Cell menu.
3. For Windows, choose **Mathematica Package** in the "Save as type" pulldown menu of the Save As submenu of the File menu. For Mac OS X, choose **Mathematica Package** in the Format pulldown menu of the Save As submenu of the File menu.
4. Name the notebook **HydrogenAtom.m**.
5. For Windows, save the notebook in the directory C:\Program Files\Wolfram Research\Mathematica\6.0\AddOns\ExtraPackages. For Mac OS X, save the notebook in the Kernel folder of the Mathematica folder within the Library folder.
6. Close the new notebook, and click Don't Save in the dialog box.

3.7.4 Template for Packages

Here is a template for packages together with comments on its construction. It is an edited version of that suggested by Maeder [Mae91].

```
(* Skeleton` : a template for packages *)

(* set up the package context and import the necessary packages *)
BeginPackage["Skeleton`", {"Package1`", "Package2`"}]

(* usage messages for symbols intended for export *)
Skeleton::usage = "Skeleton` is a package that does nothing."
Function1::usage = "Function1[n] does nothing."
Function2::usage = "Function2[n, (m:17)] does even more nothing."

(* begin the private context *)
Begin["`Private`"]

(* definition of auxiliary functions *}
Aux[f_] := Do[something]

(* error messages for objects intended for export *)
Skeleton::badarg = "Sorry, you called `1` with argument `2`!"

(* definition of functions intended for export *)
```

```
Function1[n_]:= n
Function2[n_, m_:17]:=
 n m /; n < 5 || Message[Skeleton::badarg, Function2, n]

(* end the private context *)
End[]

(* protect symbols intended for export *)
Protect[Function1, Function2]

(* end the package context *)
EndPackage[]
```

3.7.5 Exercises

In this book, straightforward, intermediate-level, and challenging exercises are unmarked, marked with one asterisk, and marked with two asterisks, respectively. For solutions to most odd-numbered exercises, see Appendix C.

1. Write a package that exports a function for normalizing n-dimensional vectors, which are represented by n-element lists in *Mathematica*. The elements may be complex variables.

*2. Write a package that exports a function for vector addition in two dimensions. The function should take as its arguments the directions and magnitudes of an arbitrary number of coplanar vectors and an option for specifying whether the angles are in degrees or radians, and it should return the direction and magnitude of the resultant. Let the directions of vectors be specified by the angles they make counterclockwise from the positive x-axis.

*3. Write a package for exporting two functions: one for normalizing three-dimensional, single-particle wave functions in Cartesian coordinates and another for normalizing those in spherical coordinates.

4. Write a package that provides the functions for the eigenenergies and normalized energy eigenfunctions of a particle of mass m moving in a one-dimensional box with walls at $x = 0$ and $x = a$. *Hint:* See Exercise 96 in Section 2.2.20 for the equations for the eigenenergies and normalized energy eigenfunctions.

5. Write a package that provides the functions for the eigenenergies and normalized energy eigenfunctions of the simple harmonic oscillator. *Hint:* See Exercise 97 in Section 2.2.20 for the equations for the eigenenergies and normalized energy eigenfunctions.

6. Prepare the package **HydrogenAtom`** for use like a standard *Mathematica* package in accordance with the instructions given at the end of Section 3.7.3. Load the package with the **Needs** command. Then, load the package with the "**<<**" command. Finally, load the package again with the **Needs** command. What can be concluded about the advantages that **Needs** has over "**<<**"?

Part II: Physics with *Mathematica*

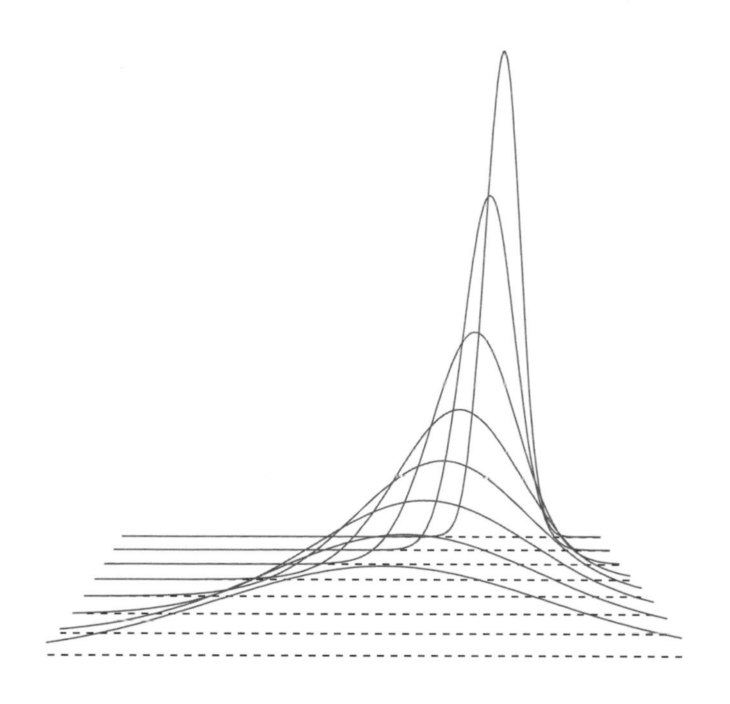

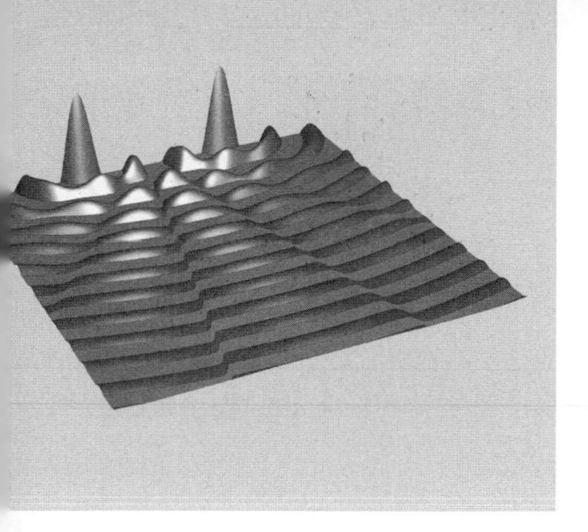

Chapter 4
Mechanics

4.1 FALLING BODIES

4.1.1 The Problem

A small body falls downward with an initial velocity v_0 from a height h near the surface of the earth, as in Figure 4.1.1. For low velocities (less than approximately 24 m/s), the effect of air resistance may be approximated by a frictional force proportional to the velocity. Find the displacement and velocity of the body, and determine the terminal velocity. Plot the speed as a function of time for several initial velocities.

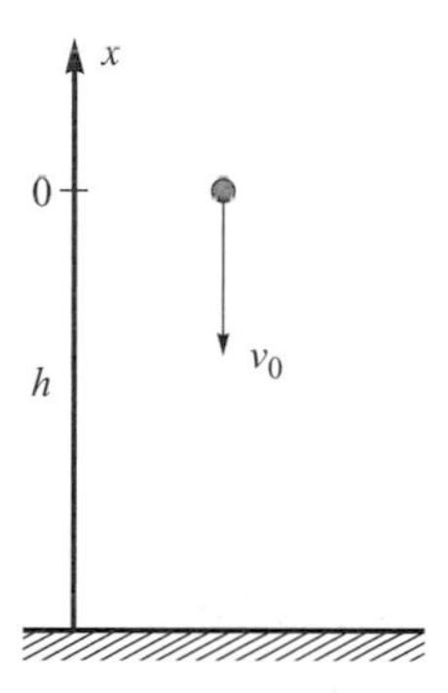

Figure 4.1.1. A small body falling downward with an initial velocity v_0 from a height h.

4.1.2 Physics of the Problem

Let the x-axis be directed upward, as in Figure 4.1.1. The net force on the object is

$$F = -mg - bv \tag{4.1.1}$$

where m is the mass, g is the magnitude of the acceleration due to gravity, and v is the velocity. The positive constant b depends on the size and shape of the object and on the viscosity of the air. From Newton's second law, the equation of motion is

$$m\frac{d^2x}{dt^2} = -mg - bv \tag{4.1.2}$$

with $x(0) = h$ and $v(0) = v_0$. We can rewrite Equation 4.1.2, a second-order equation, as two first-order equations:

$$\frac{dx}{dt} = v \tag{4.1.3}$$

$$\frac{dv}{dt} = -g - \frac{b}{m}v \tag{4.1.4}$$

(For further discussion of falling bodies, see [Sym71]; for more information on the effects of retarding forces on the motion of projectiles, refer to [TM04].)

4.1.3 Solution with *Mathematica*

```
In[1]:= ClearAll["Global`*"]
```

DSolve solves Equations 4.1.3 and 4.1.4 with the initial conditions $x(0) = h$ and $v(0) = v_0$:

```
In[2]:= sol =
          DSolve[{x'[t] == v[t], v'[t] == -g - (b/m)v[t], x[0] == h, v[0] == v0},
            {x[t], v[t]}, t] // ExpandAll
```

$$\text{Out[2]= } \left\{\left\{x[t] \to h + \frac{g m^2}{b^2} - \frac{e^{-\frac{b t}{m}} g m^2}{b^2} - \frac{g m t}{b} + \frac{m v0}{b} - \frac{e^{-\frac{b t}{m}} m v0}{b},\right.\right.$$
$$\left.\left. v[t] \to -\frac{g m}{b} + \frac{e^{-\frac{b t}{m}} g m}{b} + e^{-\frac{b t}{m}} v0\right\}\right\}$$

The solution for the displacement $x(t)$ and velocity $v(t)$ is given as a list of rules.

To determine the terminal velocity, let $bt/m >> 1$ in the rule for $v(t)$:

```
In[3]:= Last[Flatten[sol]] /. ((b t) / m → Infinity)
```

$$\text{Out[3]= } v[t] \to -\frac{g m}{b}$$

The terminal velocity v_t is $-gm/b$.

Let us determine the velocity as a function of time in a system of units where time, displacement, and, therefore, velocity are in units of m/b, gm^2/b^2, and gm/b, respectively:

```
In[4]:= v[t_] = (v[t]/. sol[[1]]) //. {gm/b→1, b/m→1}

Out[4]= -1 + e^-t + e^-t v0
```

We now plot the speed as a function of time for three illustrative initial velocities:

```
In[5]:= Plot[
         Evaluate[{Abs[v[t]] /. v0→ -3, Abs[v[t]] /. v0→-0.6,
           Abs[v[t]] /. v0→0, 1}], {t, 0, 3.5}, PlotRange→ All,
         Frame→ True, FrameLabel→ {"t (m/b)", "speed (gm/b)"},
         PlotStyle→ {{Red}, {}, {Blue}, {Dashing[{0.02}]}},
         Prolog→ {Text["v0 = 0", {0.95, 0.3}], Text["|v0| < |vt|", {2.8, 0.55}],
           Text["|v0| > |vt|", {1.5, 2.05}], Text["terminal speed, |vt|",
            {0.9, 1.2}], Line[{{0.95, 0.79}, {2.2, 0.6}}]}]
```

Out[5]=

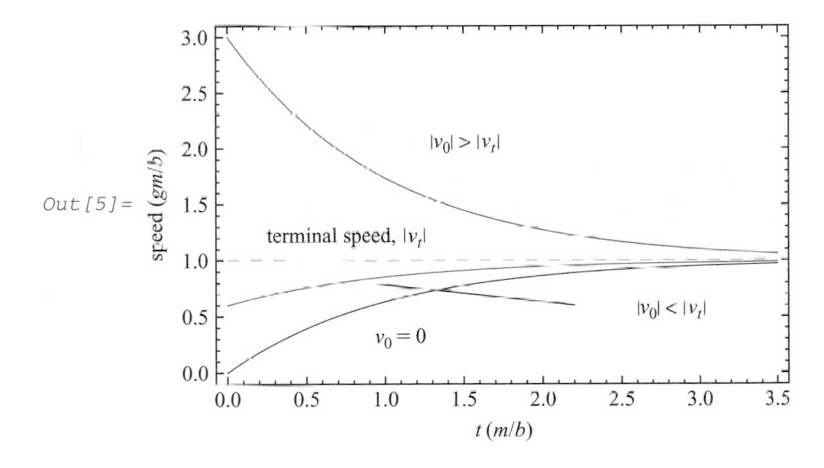

The bottom two curves indicate that the speed of the body increases, as expected, and eventually approaches the terminal speed during its descent. The top curve shows that if the initial speed exceeds the terminal speed, the falling object actually slows down and its speed approaches the terminal speed from above. (Note that although the velocity of the falling body is negative, its speed is positive.)

```
In[6]:= ClearAll[sol, v]
```

4.2 PROJECTILE MOTION

4.2.1 The Problem

A baseball of mass m leaves the bat with a speed v_0 at an angle θ_0 from the horizontal. The effect of air resistance may be approximated by a drag force $\mathbf{F}_D$ that depends on the square of the speed—that is,

$$\mathbf{F}_D = -mkv\mathbf{v}$$

where v and $\mathbf{v}$ are the speed and velocity of the ball, respectively, and the drag factor k is equal to $5.2 \times 10^{-3}\,\mathrm{m}^{-1}$. Define a function that generates an animation of the motion of the baseball. The function should have v_0 and θ_0 as arguments, and the animation should show the paths of the baseball with and without air resistance. Evaluate the function for $v_0 = 45\,\mathrm{m/s}$ and $\theta_0 = 60°$.

4.2.2 Physics of the Problem

The equations of motion for the baseball are

$$\frac{d^2x}{dt^2} = -k\sqrt{\left(\frac{dx}{dt}\right)^2 + \left(\frac{dy}{dt}\right)^2}\left(\frac{dx}{dt}\right) \tag{4.2.1}$$

$$\frac{d^2y}{dt^2} = -k\sqrt{\left(\frac{dx}{dt}\right)^2 + \left(\frac{dy}{dt}\right)^2}\left(\frac{dy}{dt}\right) - g \tag{4.2.2}$$

where the x and y directions are the horizontal and vertical directions, respectively, as in Figure 4.2.1, and g is the magnitude of the acceleration due to gravity.

If air resistance is negligible (i.e., $k = 0$), these equations can be solved analytically. The path of the baseball is given by the equation

$$y = \frac{v_{y0}}{v_{x0}}x - \frac{1}{2}\frac{g}{v_{x0}^2}x^2$$

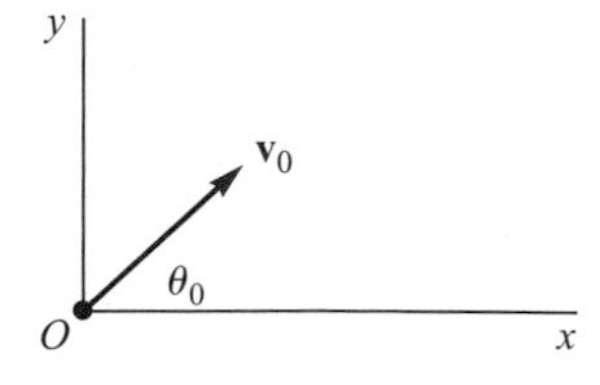

Figure 4.2.1. A baseball leaves the bat with an initial velocity $\mathbf{v}_0$ at an elevation angle θ_0.

where the initial position is chosen as the origin, and the initial x and y components of the velocity are

$$v_{x0} = v_0 \cos\theta_0$$

$$v_{y0} = v_0 \sin\theta_0$$

The maximum altitude occurs at

$$H = \frac{v_{y0}^2}{2g}$$

the range is

$$R = \frac{2\, v_{x0} v_{y0}}{g}$$

and the time of flight is

$$T = \frac{2\, v_{y0}}{g}$$

A realistic model for the flight of the baseball must include air resistance. Unfortunately, the equations of motion, Equations 4.2.1 and 4.2.2, can no longer be solved analytically; numerical methods must be employed. (For more information on the physics of baseball, consult [Ada95] and [Bra85].)

4.2.3 Solution with *Mathematica*

What follows is a function that generates an animation of the motion of the baseball. Comments giving the details about the function are embedded in the body of the function in the form (* *comment* *). These comments are ignored by the *Mathematica* kernel. The function takes two arguments: The first is the numerical value of the initial speed in SI units, and the second is that of the initial elevation angle in degrees. The initial speed is limited to 80 m/s because higher speeds are unlikely for baseballs, and the initial angle of elevation is restricted to the range 30° to 85° in order to accommodate the sizes of most monitor screens.

```
In[1]:= ClearAll["Global`*"]

In[2]:= projectile[v0_/; 0 < v0 ≤ 80, θ0_/; 30 ≤ θ0 ≤ 85] :=
         Module[
          {k = 5.2 10^-3, g = 9.81, vx0, vy0, x, y,
           R, H, tmax, T, pathWithoutAirResistance, sol},

          (* initial x and y components of velocity *)
```

```
vx0 = v0 Cos[θ0 °] // N;
vy0 = v0 Sin[θ0 °] // N;

(* range without air resistance *)

R = (2 vx0 vy0)/g;

(* maximum height without air resistance *)

H = vy0^2/(2 g);

(* time of flight without air resistance *)

T = (2 vy0)/g;

pathWithoutAirResistance =
 Plot[vy0/vx0 x - 1/2 g/vx0^2 x^2, {x, 0, R}, PlotRange → {0, H}];

(* numerical solution of the
 equations of motion with air resistance *)

sol = NDSolve[{x''[t] == -k Sqrt[x'[t]^2 + y'[t]^2] x'[t],
   y''[t] == -k Sqrt[x'[t]^2 + y'[t]^2] y'[t] - g, x[0] == 0,
   y[0] == 0, x'[0] == vx0, y'[0] == vy0}, {x, y}, {t, 0, T}];

(* x and y coordinates as a function of time *)

x[t_] = x[t] /. sol[[1]];
y[t_] = y[t] /. sol[[1]];

(* time of flight *)

tmax = t /. FindRoot[y[t] == 0, {t, T}];

(* animation *)

 ListAnimate[
  Table[
   Show[
    pathWithoutAirResistance,
```

```
Graphics[{AbsolutePointSize[7],
  Red, Point[{x[(tmax / 16)i], y[(tmax / 16)i]}]}],
ParametricPlot[{x[t], y[t]}, {t, 0, (tmax / 16) i + 0.0001},
 PlotStyle → {Blue, Dashing[{0.02, 0.02}]}], PlotRange →
 {{-0.01, R}, {-0.01, 1.02 H}}, AxesLabel → {"x(m)", "y(m)"},
AspectRatio → Automatic],
{i, 0, 16}]]
```

Let us generate the animation of the motion of the baseball with $v_0 = 45$ m/s and $\theta_0 = 60°$:

```
In[3]:= projectile[45, 60]
```

Out[3] =

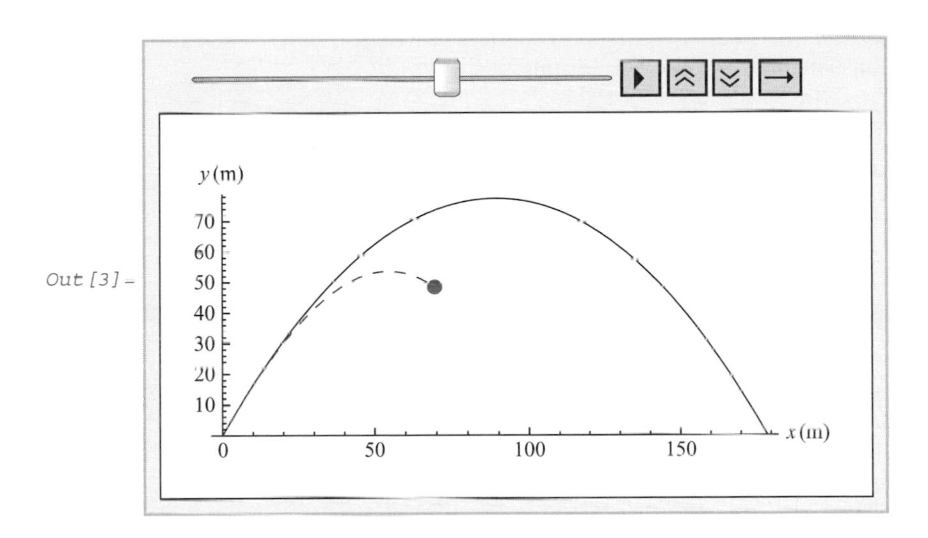

The solid line represents the path of the baseball without air resistance; the dashed line traces that with air resistance. To animate, click the Play button—that is, the one with the right pointer.

4.3 THE PENDULUM

4.3.1 The Problem

Solve the equations of motion numerically and plot the phase diagrams of the plane, the damped, and the damped, driven pendulums. For the damped, driven pendulum, also determine the Poincaré section and generate an animation of the pendulum.

4.3.2 Physics of the Problem

4.3.2.1 The Plane Pendulum

Consider a particle of mass m that is constrained by a rigid and massless rod to move in a vertical circle of radius L, as in Figure 4.3.1. The equation of motion is

$$\frac{d^2x}{dt^2} + \omega_0^2 \sin x = 0 \tag{4.3.1}$$

where x is the angle that the rod makes with the vertical,

$$\omega_0 = \sqrt{\frac{g}{L}} \tag{4.3.2}$$

and g is the magnitude of the acceleration due to gravity. We can rewrite Equation 4.3.1, a second-order equation, as two first-order equations:

$$\frac{dx}{dt} = v \tag{4.3.3}$$

$$\frac{dv}{dt} = -\omega_0^2 \sin x \tag{4.3.4}$$

If the unit of time t is taken to be $t_0 = 1/\omega_0$, Equation 4.3.4 assumes the form

$$\frac{dv}{dt} = -\sin x \tag{4.3.5}$$

where x is in radians and v is in radians/t_0.

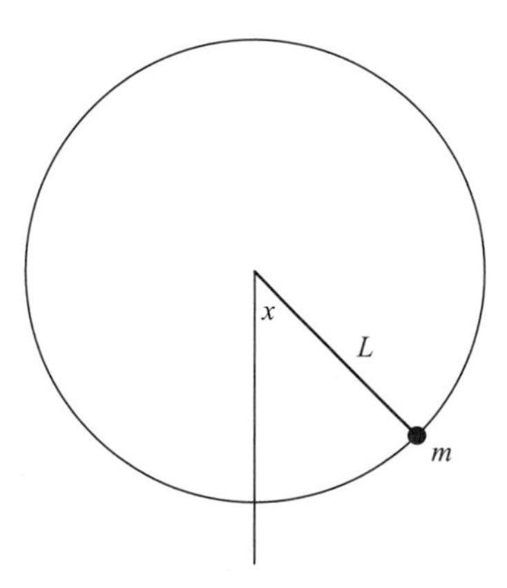

Figure 4.3.1. The plane pendulum.

4.3.2.2 The Damped Pendulum

We now include damping that is proportional to the velocity. Equation 4.3.5 becomes

$$\frac{dv}{dt} = -\sin x - av \tag{4.3.6}$$

where a is a parameter measuring the strength of the damping.

4.3.2.3 The Damped, Driven Pendulum

With a sinusoidal driving force, Equation 4.3.6 becomes

$$\frac{dv}{dt} = -\sin x - av + F\cos \omega t \tag{4.3.7}$$

where F and ω are the amplitude and frequency of the driving force, respectively.

The motion of the pendulum depends on the parameters a, F, and ω as well as on the initial conditions for x and v. Whether the motion is periodic or chaotic is determined solely by the parameters a, F, and ω. (For introductions to chaotic dynamics, consult [BG96] and [TM04]; for more advanced discussions, refer to [Moo87], [Ras90], and [Hil00].)

4.3.3 Solution with *Mathematica*

In this section, we solve the equations of motion and plot the phase diagrams of the plane, the damped, and the damped, driven pendulums. For the damped, driven pendulum, we also determine the Poincaré section and generate an animation of the pendulum.

```
In[1]:= ClearAll["Global`*"]
```

4.3.3.1 The Plane Pendulum

As an example, we choose the initial conditions $x = 0$ and $v = 2 - d$, where d is a small number. (What would happen if the initial conditions are such that $x = 0$ and $v \geq 2$? See Problem 5 in Section 4.5.) **NDSolve** solves Equations 4.3.3 and 4.3.5 together with the initial conditions:

```
In[2]:= points1 = NDSolve[{x'[t] == v[t], v'[t] == -Sin[x[t]],
            x[0] == 0, v[0] == 2 - 0.001}, {x, v}, {t, 0, 19.5}];
```

With the solution **points1**, we can plot the phase diagram:

```
In[3]:= ParametricPlot[Evaluate[{x[t], v[t]} /. points1],
          {t, 0, 19.5}, AxesLabel → {"x (rad)", "v (rad/t0)"}]
```

Out[3]=

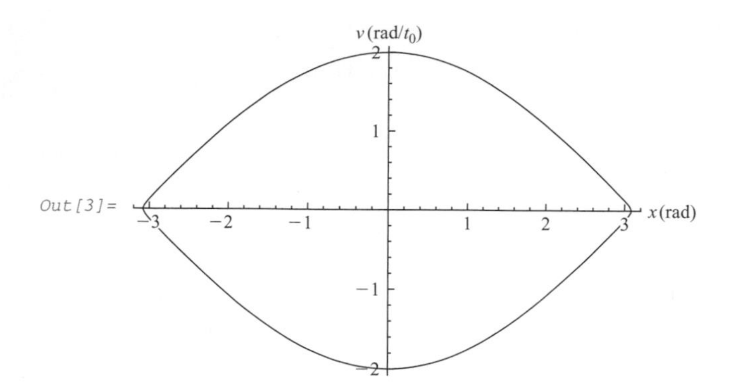

4.3.3.2 The Damped Pendulum

For an example, choose the initial conditions $x = 0$ and $v = 3$ and let $a = 0.2$. **NDSolve** solves Equations 4.3.3 and 4.3.6 together with the initial conditions:

```
In[4]:= points2 = NDSolve[{x'[t] == v[t], v'[t] == -Sin[x[t]] - 0.2 v[t],
          x[0] == 0, v[0] == 3}, {x, v}, {t, 0, 30}];
```

Let us plot the phase diagram:

```
In[5]:= ParametricPlot[Evaluate[{x[t], v[t]} /. points2],
          {t, 0, 30}, PlotRange → All, AxesLabel → {"x (rad)", "v (rad/t0)"}]
```

Out[5]=

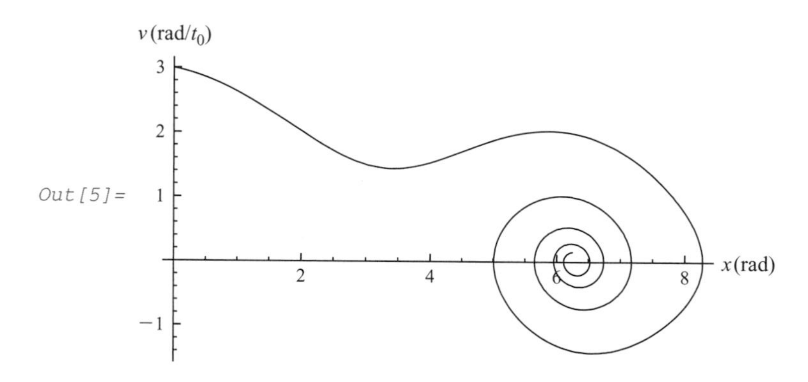

4.3.3.3 The Damped, Driven Pendulum

As an example, let $a = 0.2$, $F = 0.52$, and $\omega = 0.694$:

```
In[6]:= {a, F, ω} = {0.2, 0.52, 0.694};
```

Also, choose the initial conditions $x = 0.8$ and $v = 0.8$.

For t in the range 0 to $cycles(2\pi/\omega)$, **NDSolve** solves Equations 4.3.3 and 4.3.7 together with the initial conditions:

```
In[7]:= cycles = 50;
        sol3 = NDSolve[{x'[t] == v[t],
            v'[t] == -Sin[x[t]] - a v[t] + F Cos[ω t], x[0] == 0.8, v[0] == 0.8},
           {x, v}, {t, 0, cycles (2 π/ω)}, MaxSteps → 20000];
```

Mathematically, the angle x can take on values from $-\infty$ to $+\infty$; physically, x can only vary from $-\pi$ to $+\pi$. In the following, the phase diagram is constructed with all the points translated back to the interval $-\pi$ to $+\pi$ and with the lines connecting adjacent points omitted for clarity:

```
In[9]:= reduce[x_] := Mod[x, 2 π] /; Mod[x, 2 π] ≤ π;
        reduce[x_] := (Mod[x, 2 π] - 2 π) /; Mod[x, 2 π] > π
```

```
In[11]:= steps = 30;
         points3 = Flatten[Table[{x[t], v[t]} /. sol3,
             {t, 0, cycles (2 π/ω), (1/steps) (2 π/ω)}], 1];
         newpoints3 = {reduce[ # [[1]]], # [[2]]} & /@points3;
         ListPlot[newpoints3, PlotRange → {-3, 3},
          AxesLabel → {"x (rad)", "v (rad/t0)"}]
```

Out[14]=

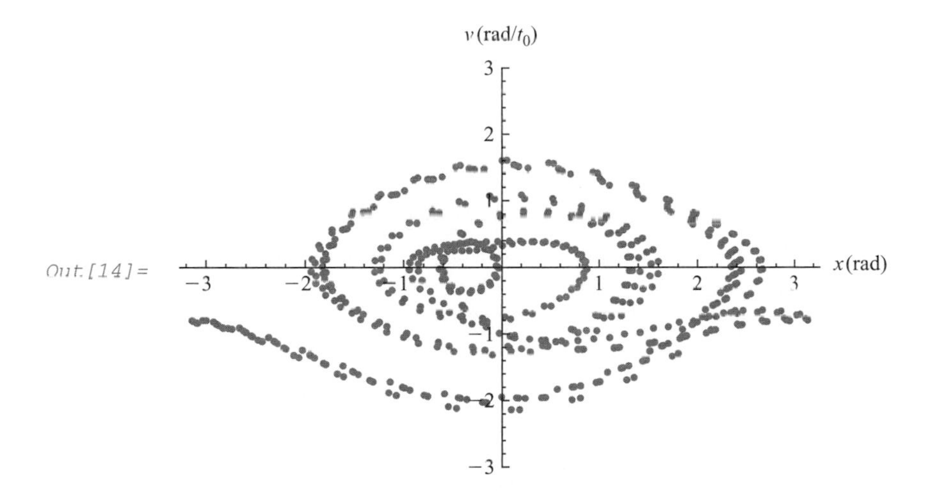

The Poincaré section provides a simplification of the phase diagram to reveal the essential features of the dynamics. For the phase diagram, the coordinates $(x(t), v(t))$ were determined for the values of time $t = 0, \Delta t, 2\Delta t, 3\Delta t, \ldots,$ where we have chosen $\Delta t = T/30$, in which T is the period of the driving force. For the Poincaré section, we select only those points for

$t = 0, T, 2T, 3T, \ldots$. To avoid transient effects, the points for the first 25 cycles are discarded in the following construction of the Poincaré section:

```
In[15]:= poincare3 = Table[newpoints3[[n]],
            {n, 1 + 25 steps, Length[newpoints3], steps}];

In[16]:= Length[poincare3]
Out[16]= 26

In[17]:= ListPlot[poincare3, PlotRange → {{-3, 3}, {-3, 3}},
         AxesLabel → {"x (rad)", "v (rad/t0)"}, PlotStyle → PointSize[0.015]]
```

Out[17]=

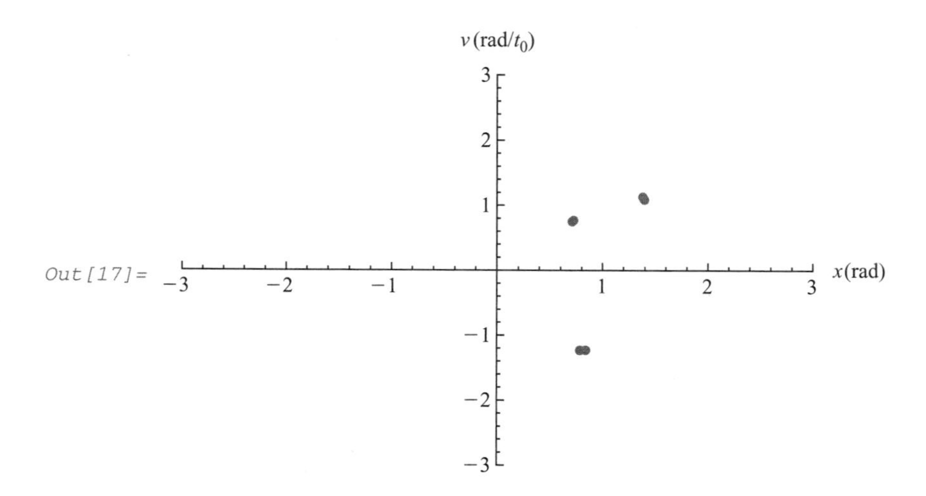

The Poincaré section has three dots and indicates periodic motion. (For an introduction to the Poincaré section, see [BG96] or [TM04].)

For another example, let $a = 1/2, F = 1.15$, and $\omega = 2/3$:

```
In[18]:= a = 1/2; F = 1.15; ω = 2/3;
```

Also, choose the initial conditions $x = 0.8$ and $y = 0.8$.

For t from 0 to $cycles(2\pi/\omega)$, **NDSolve** solves Equations 4.3.3 and 4.3.7 together with the initial conditions:

```
In[19]:= cycles = 180;
         sol4 = NDSolve[{x'[t] == v[t],
```

```
      v'[t] == -Sin[x[t]] - a v[t] + F Cos[ω t], x[0] == 0.8, v[0] == 0.8},
    {x, v}, {t, 0, cycles (2 π/ω)}, MaxSteps → 200000];
steps = 30;
points4 = Flatten[Table[{x[t], v[t]} /. sol4,
      {t, 0, cycles (2 π/ω), (1/steps) (2 π/ω)}], 1];
```

where we have generated a nested list of coordinates $\{x(t), v(t)\}$, with t ranging from 0 to $cycles(2\pi/\omega)$ in steps of size $\Delta t = (1/steps)(2\pi/\omega)$, for plotting the phase diagram and the Poincaré section. We must take computer solution of differential equations in chaotic dynamics with a grain of salt because sensitive dependence on initial conditions is a characteristic feature of chaos, and round-off error is an inherent limitation of numerical methods. (For further discussion, refer to [PJS92] and [Ras90].)

Let us plot the phase diagram with all the points translated back to the interval $x = -\pi$ to $+\pi$ and with the lines joining the points omitted for clarity:

```
In[23]:= newpoints4 = {reduce[#[[1]]], #[[2]]} & /@ points4;
         ListPlot[newpoints4, AxesLabel → {"x (rad)", "v (rad/t0)"}]
```

Out[24]=

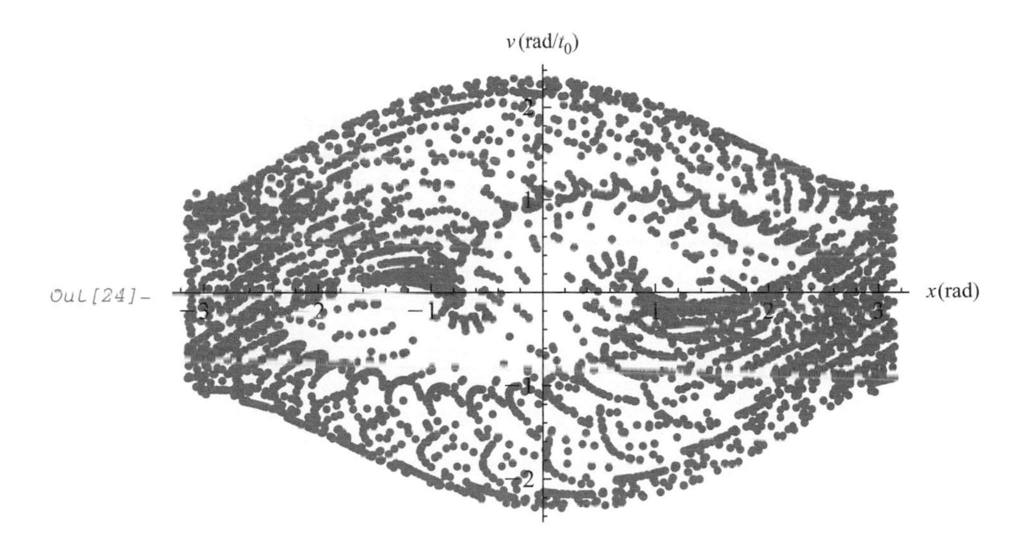

Next, we construct the Poincaré section:

```
In[25]:= poincare4 = Table[newpoints4[[n]],
             {n, 1 + 20 steps, Length[newpoints4], steps}];
         ListPlot[poincare4, PlotRange → {{-4, 4}, {-4, 4}},
          AxesLabel → {"x (rad)", "v (rad/t0)"}]
```

Out[26]=

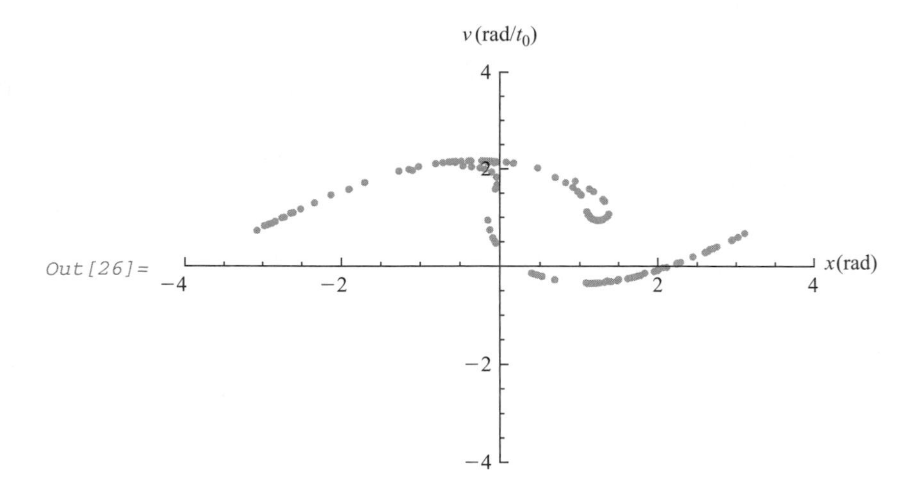

The Poincaré section has a complex geometry that is characteristic of chaotic states.

We end this notebook with an animation showing the chaotic motion of the pendulum. The package **ChaoticPendulum`** containing the function **Motion** is included in Section 4.3.3.4. To use this package like a standard *Mathematica* package, follow the instructions at the end of Section 3.7.3.3. **Motion**[*points*, *n*, *m*] generates an animation of the pendulum. The first argument *points* is for the nested list of coordinates $\{x(t_i), v(t_i)\}$ from the solution of the equation of motion, the second argument n is for the number of graphs of the pendulum to be plotted, and the third argument m is for the time increment. Only nonzero positive integers can be assumed by n and m. The time interval between two successive graphs is equal to m times the step size Δt previously specified. For given *points* and m, n has an upper limit

```
Ceiling[Length[points]/m]
```

Length[*points*] gives the number of elements in *points*, and **Ceiling**[*x*] is the least integer not smaller than x. Here is the number of elements in **points4**:

```
In[27]:= Length[points4]
Out[27]= 5401
```

Let us load the package **ChaoticPendulum`** and call the function **Motion**:

```
In[28]:= Needs["ChaoticPendulum`"]
         Motion[points4,1000,2]
```

Out [29] =

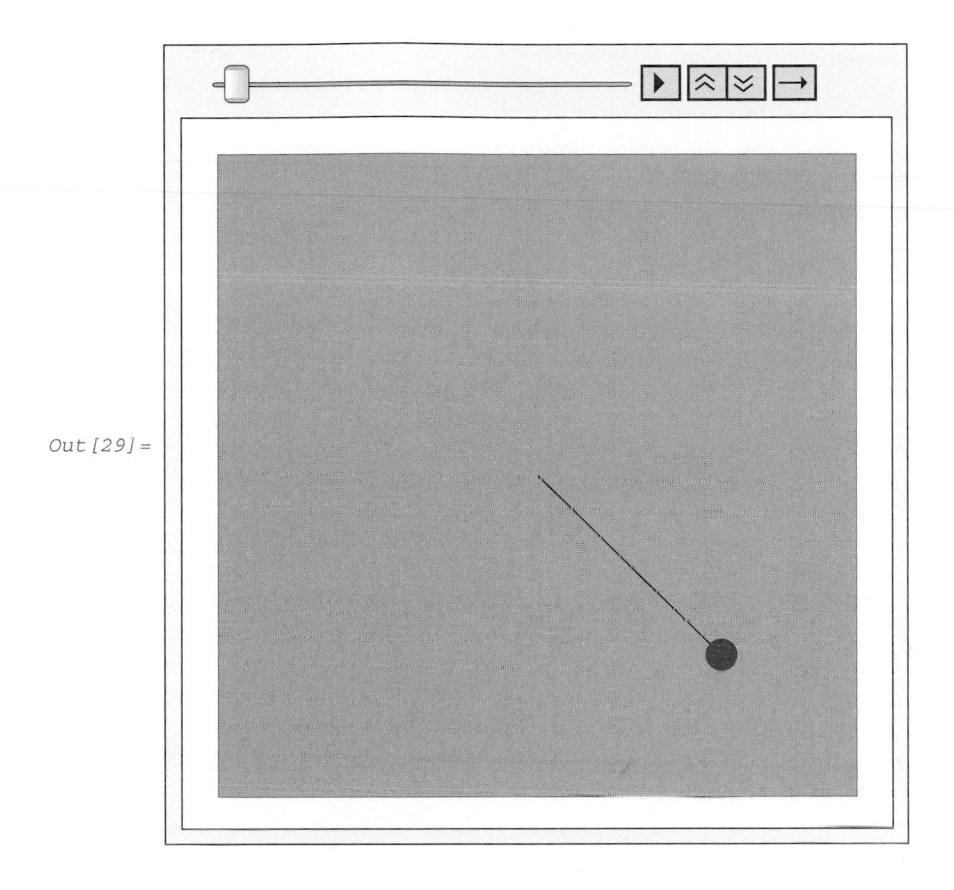

To animate, click the Play button that is, the one with the right pointer.

4.3.3.4 ChaoticPendulum` : A *Mathematica* Package

```
(* To use this package like a standard Mathematica package,
follow the instructions at the end of Section 3.7.3.3. *)

BeginPackage["ChaoticPendulum`"]

ChaoticPendulum::usage =
 "ChaoticPendulum` is a package for generating
   an animation of the pendulum."

Motion::usage =
 "Motion[points, n, m] generates an animation of the pendulum."
```

```
Begin["`Private`"]

Motion::badarg =
 "For the given nested list of coordinates and your choice
   of time increment, the maximum number of graphs that
   can be plotted is `1`. You requested `2` graphs."

Motion[points_List, n_Integer?Positive, m_Integer?Positive] := (
   list = Table[points[[i, 1]], {i, 1, Length[points], m}];
   x = Sin[list];
   y = -Cos[list];
   ListAnimate[
    Table[
     Show[
      Graphics[{Line[{{0, 0}, {x[[k]], y[[k]]}}],
        PointSize[0.05], Red, Point[{x[[k]], y[[k]]}]}],
      PlotRange→ {{-1.25, 1.25}, {-1.25, 1.25}},
      AspectRatio→ Automatic, Frame→ True, FrameTicks→ None,
      Background→ Lighter[Blue, 0.5]], {k, n}]]
 ) /; n≤Ceiling[Length[points]/m]

Motion[points_List,n_Integer?Positive, m_Integer?Positive] :=
 (Message[Motion::badarg, Ceiling[Length[points]/m], n])/;
  n>Ceiling[Length[points]/m]

End[]

Protect[Motion]

EndPackage[]
```

4.4 THE SPHERICAL PENDULUM

4.4.1 The Problem

The spherical pendulum consists of a particle of mass m constrained by a rigid and massless rod to move on a sphere of radius R. With several choices of initial conditions for the pendulum, explore its motion.

4.4.2 Physics of the Problem

We adopt a system of units in which $m = 1$, $R = 1$, and $g = 1$, where g is the magnitude of the acceleration due to gravity. Let θ and φ be the spherical coordinates of the particle, and let the fixed end of the rod be the origin, as in Figure 4.4.1. In terms of the spherical coordinates,

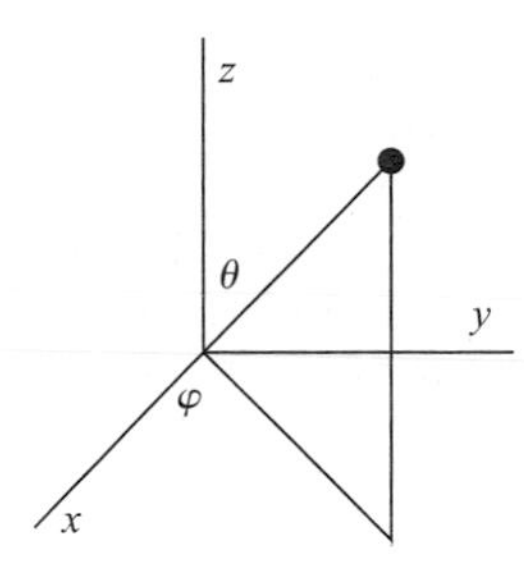

Figure 4.4.1. The spherical pendulum.

the kinetic energy is

$$T = \frac{1}{2}\left(\dot{\theta}^2 + \sin^2\theta\,\dot{\varphi}^2\right) \tag{4.4.1}$$

The gravitational potential energy relative to the horizontal plane is

$$V = \cos\theta \tag{4.4.2}$$

Hence, the Lagrangian, $T - V$, can be expressed as

$$L = \frac{1}{2}\left(\dot{\theta}^2 + \sin^2\theta\,\dot{\varphi}^2\right) - \cos\theta \tag{4.4.3}$$

The Lagrange equations of motion give

$$\frac{d\dot{\theta}}{dt} - \sin\theta\cos\theta\,\dot{\varphi}^2 - \sin\theta = 0 \tag{4.4.4}$$

$$\frac{d}{dt}\left(\sin^2\theta\,\dot{\varphi}\right) = 0 \tag{4.4.5}$$

Equations 4.4.4 and 4.4.5 together with the initial conditions θ_0, θ_0, φ_0, and $\dot{\varphi}_0$ completely specify the motion of the system. (For an introduction to Lagrangian mechanics, see [Sym71] or [TM04].)

We gain insight into the motion of the pendulum by recognizing that there are two constants of the motion: the angular momentum about the z-axis and the total energy. Since the coordinate φ is ignorable (i.e., it does not appear explicitly in the Lagrangian), the corresponding generalized momentum (also known as canonical momentum or conjugate momentum)

$$p_\varphi = \frac{\partial L}{\partial\dot{\varphi}} = \sin^2\theta\,\dot{\varphi} \tag{4.4.6}$$

is a constant of the motion, as can be seen from Equation 4.4.5. The total energy

$$E = T + V = \frac{1}{2}\left(\dot{\theta}^2 + \sin^2\theta\,\dot{\varphi}^2\right) + \cos\theta \tag{4.4.7}$$

can also be shown to be a constant of the motion because the spherical pendulum is a natural system (i.e., the Lagrangian has the form $T_2 - V$, where the kinetic energy T_2 contains only terms quadratic in the generalized velocities and the potential energy V is velocity independent), the Lagrangian of which does not depend explicitly on time. (For a detailed discussion of the energy integral of the motion, see [CS60].) In terms of the angular momentum p_φ, Equations 4.4.4 and 4.4.5 can be written as

$$\ddot{\theta} - \frac{\cos\theta}{\sin^3\theta} p_\varphi^2 - \sin\theta = 0 \tag{4.4.8}$$

$$\dot{\varphi} = \frac{p_\varphi}{\sin^2\theta} \tag{4.4.9}$$

and Equation 4.4.7 can be put in the form

$$E = \frac{1}{2}\left(\dot{\theta}^2 + \frac{p_\varphi^2}{\sin^2\theta}\right) + \cos\theta \tag{4.4.10}$$

Let us define an effective potential

$$U(\theta) = \frac{1}{2}\frac{p_\varphi^2}{\sin^2\theta} + \cos\theta \tag{4.4.11}$$

Equation 4.4.10 becomes

$$\frac{1}{2}\dot{\theta}^2 = E - U(\theta) \tag{4.4.12}$$

If $p_\varphi = 0$, Equations 4.4.8 and 4.4.9 are the equations of motion of a simple pendulum oscillating in a plane perpendicular to the xy plane. If $p_\varphi \neq 0$, Equation 4.4.9 indicates that, except for the two singularities at θ equals 0° and 180°, the sign of $\dot{\varphi}$ remains constant. Thus, the particle rotates about the z-axis. We obtain further insight into the motion of the pendulum by examining the effective potential curve in Figure 4.4.2. The effective potential rises to infinity at $\theta = 0°$ and $\theta = 180°$. Therefore, these angles are inaccessible to the system unless the total energy is infinite because $\dot{\theta}^2$ in Equation 4.4.12 cannot be negative. In general, the angle θ oscillates between two turning points θ_1 and θ_2, which are the roots of the equation

$$E - U(\theta) = 0 \tag{4.4.13}$$

If $E = U(\theta_m)$, where θ_m is the angle for the minimum value of $U(\theta)$, these roots coincide and the particle rotates about the z-axis at the fixed angle θ_m and, from Equation 4.4.9, with constant $\dot{\varphi}$.

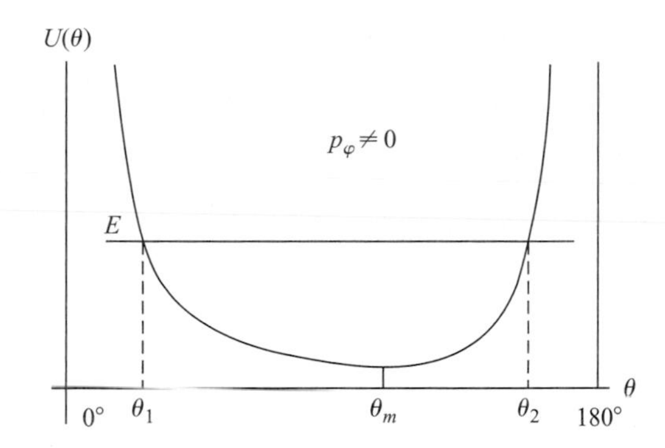

Figure 4.4.2. Effective potential for the spherical pendulum.

4.4.3 Solution with *Mathematica*

```
In[1]:= ClearAll["Global`*"]
```

We begin with the definition of a function for plotting the effective potential together with the total energy. The first argument of the function is θ_0 in degrees from 0° to 180°; the second argument, $\dot{\theta}_0$; and the third argument, $\dot{\varphi}_0$. Options, except **AxesLabel**, **PlotStyle**, and **Ticks**, for passing to the **Plot** function may be specified after the third argument.

```
In[2]:= effectivePotential[
          θ0_ /; 0 ≤ θ0 ≤ 180, θdot0_, φdot0_, opts___Rule] :=
        Module[{pφ},
         pφ = Sin[θ0 °]^2 φdot0;
         Plot[{1/2 pφ^2/Sin[θ °]^2 + Cos[θ °], 1/2 (θdot0^2 + pφ^2/Sin[θ0 °]^2) + Cos[θ0 °]},
          {θ, 0 + 10^-3, 180 - 10^-3},
          Ticks → {{0, 90, 180}, Automatic}, AxesLabel → {"θ(°)", "U(mgR)"},
          PlotStyle → {{Thickness[0.0075]}, {Dashing[{0.05, 0.05}]}}, opts]]
```

We now define the function that generates an animation of the spherical pendulum. The first argument of the function is θ_0 in degrees from 0° to 180°; the second argument, $\dot{\theta}_0$ limited between -10 and 10; the third argument, φ_0 in degrees from 0° to 360°; the fourth argument, $\dot{\varphi}_0$ confined between -10 and 10; the fifth argument, the interval of time $tmax$ restricted from 1 to 60 for the animation sequence; and the sixth argument, the number of frames nv generated with a minimum value of 2 and a computed default value of the integer closest to $3tmax$. Options, with the exception of **PlotRange** and **BoxRatios**, for passing to the **Show** function may be specified after these arguments. The construct of the function is explained in the body of the function with statements in the form (* *comment* *). The maximum possible values for *tmax* and *nv* depend on the available memory. With *nv* = 2, the function generates two frames: one for time $t = 0$ and another for $t = tmax$.

```
In[3]:= sphericalPendulum[θ0_ /; 0 ≤ θ0 ≤ 180, θdot0_ /; Abs[θdot0] < 10,
         φ0_ /; 0 ≤ φ0 ≤ 360, φdot0_ /; Abs[φdot0] < 10,
         tmax_ /; 1 ≤ tmax ≤ 60, nv_Integer:999, opts___Rule] :=

        Module[{n = nv, pφ, sol, θ, φ, x, y, z, sphere},

         (* number of frames must be an integer greater than one *)

         If[n > 1,

          (* computed default for number of frames *)

          If[n == 999, n = Round[3 tmax]];

          (* angular momentum *)

          pφ = (Sin[θ0 °]^2 φdot0 // N);

          (* numerical solution of equations of motion *)

          sol =
           NDSolve[{θ''[t] - Cos[θ[t]]/Sin[θ[t]]^3 pφ^2 - Sin[θ[t]] == 0,
             φ'[t] == pφ/Sin[θ[t]]^2, θ[0] == θ0 °, θ'[0] == θdot0, φ[0] == φ0 °},
            {θ, φ}, {t, 0, tmax + 0.01}, MaxSteps → 6000];

          (* spherical coordinates *)

          θ = θ /. First[sol];
          φ = φ /. First[sol];

          (* rectangular coordinates *)

          x[t_] := Sin[θ[t]] Cos[φ[t]];
          y[t_] := Sin[θ[t]] Sin[φ[t]];
          z[t_] := Cos[θ[t]];

          (* animation *)
```

```
sphere =
 ParametricPlot3D[
  {{0, Sin[t], Cos[t]}, {Sin[t], 0, Cos[t]}}, {t, 0, 2 π}];

ListAnimate[
 Table[
  Show[

   (* background sphere *)

   sphere, Graphics3D[Line[{{0, 0, -1}, {0, 0, 1}}]],

   (* trace *)

   ParametricPlot3D[{x[t], y[t], z[t]}, {t, 0, (tmax / (n - 1)) i +
      0.0001}, PlotPoints → (25 + Round[12 tmax / (n - 1) i])],

   (* pendulum *)

   Graphics3D[
    {Thickness[0.0125], Red, Line[{{0, 0, 0}, {x[(tmax / (n - 1)) i],
        y[(tmax / (n - 1)) i], z[(tmax / (n - 1)) i]}}],
     PointSize[0.04], Red,
     Point[{x[(tmax / (n - 1)) i],
       y[(tmax / (n - 1)) i], z[(tmax / (n - 1)) i]}]}],

   (* options *)

   PlotRange → {{-1.25, 1.25}, {-1.25, 1.25}, {-1.25, 1.25}},
   BoxRatios → {1, 1, 1}, opts,
   AxesLabel → {"x (R)", "y (R)", "z (R)"}],

  {i, 0, n - 1}]],

(* error message *)

sphericalPendulum::badarg =
 "Number of frames must be an integer greater than one!
You requested `1` frame(s).";
Message[sphericalPendulum::badarg, nv]]]
```

With the functions **effectivePotential** and **sphericalPendulum**, we are ready to explore the motion of the spherical pendulum with several choices of initial conditions.

4.4.3.1 $\theta_0 = 120$, $\dot{\theta}_0 = 0$, $\varphi_0 = 45$, $\dot{\varphi}_0 = 0$

Since $p_\varphi = 0$, the spherical pendulum behaves as a simple pendulum. The function **effective-Potential** plots the effective potential together with the total energy:

```
In[4]:= effectivePotential[120, 0, 0]
```

Out[4]=

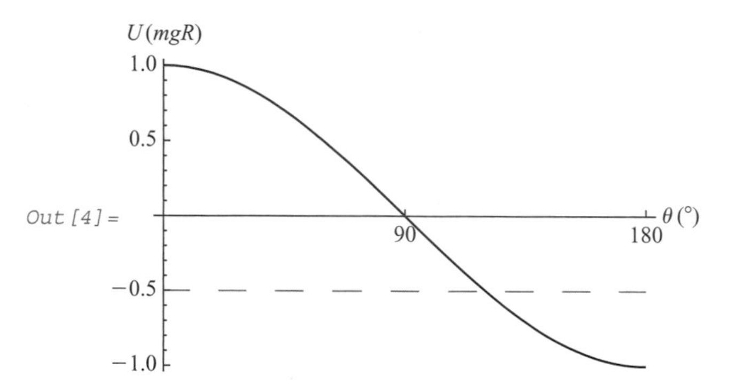

In the graph, the solid curve and the dashed line represent the effective potential and the total energy, respectively. The effective potential has a minimum at $\theta = 180°$. The pendulum oscillates about this angle, and the turning point $\theta = 120°$ occurs at the intersection of the effective potential curve and the total energy line. The function **sphericalPendulum** generates the animation:

```
In[5]:= sphericalPendulum[120, 0, 45, 0, 27,
         Background→RGBColor[0.640004, 0.580004, 0.5]]
```

Out[5] =

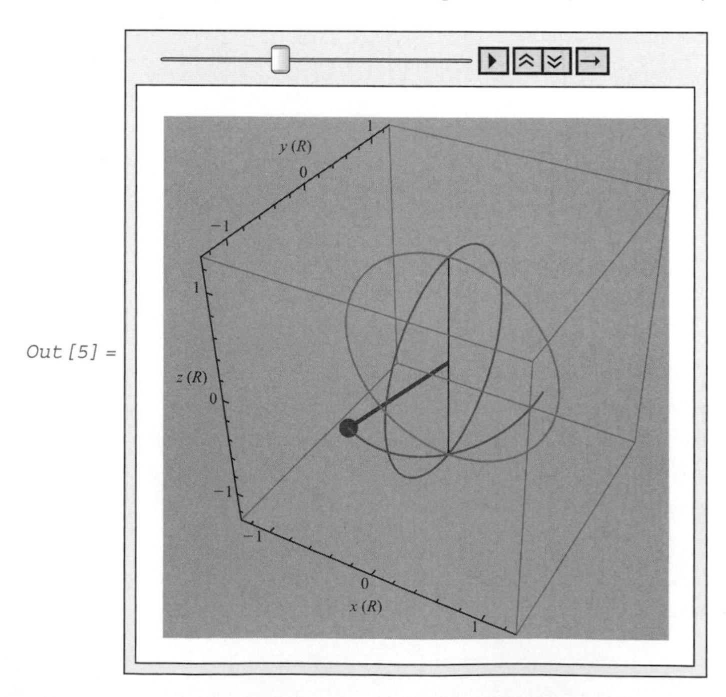

To animate, click the Play button—that is, the one with the right pointer. The curve traces the path of the particle from its initial position. In this case, the spherical pendulum behaves like a simple pendulum, as expected.

4.4.3.2 $\theta_0 = 135$, $\dot{\theta}_0 = 0$, $\varphi_0 = 90$, $\dot{\varphi}_0 = 2^{1/4}$

Let us plot the effective potential:

```
In[6]:= effectivePotential[135, 0, 2^(1/4), PlotRange→{-0.5, 2.0}]
```

Out[6] =

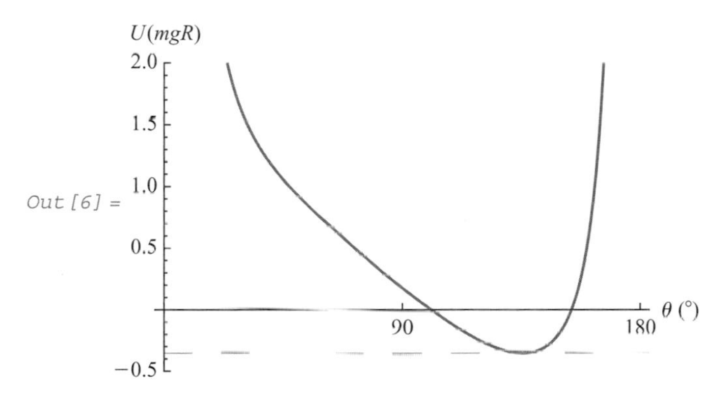

Since $p_\varphi \neq 0$ and $E = U(\theta_m)$, the particle rotates uniformly in a horizontal circle about the z-axis. We now generate the animation:

```
In[7]:= sphericalPendulum[135, 0, 90, 2^(1/4), 31,
          Background→RGBColor[0.640004, 0.580004, 0.5]]
```

Out[7] =

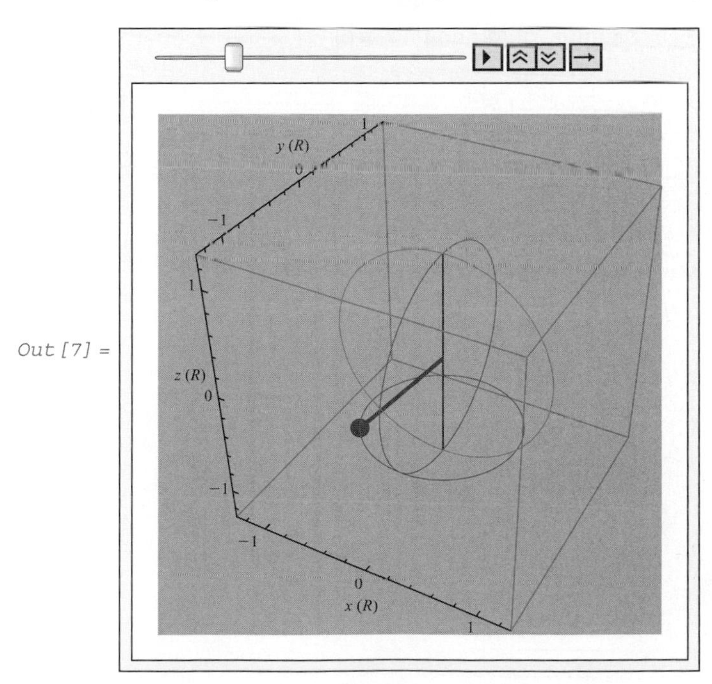

To animate, click the Play button. The pendulum indeed rotates uniformly about the z-axis.

4.4.3.3 $\theta_0 = 135$, $\dot{\theta}_0 = 2.5$, $\varphi_0 = 90$, $\dot{\varphi}_0 = 1.5 \times 2^{1/4}$

Again, we plot the effective potential:

```
In[8]:= effectivePotential[135, 2.5, 1.5×2^(1/4), PlotRange→{-0.25, 4.0}]
```

Out[8] =

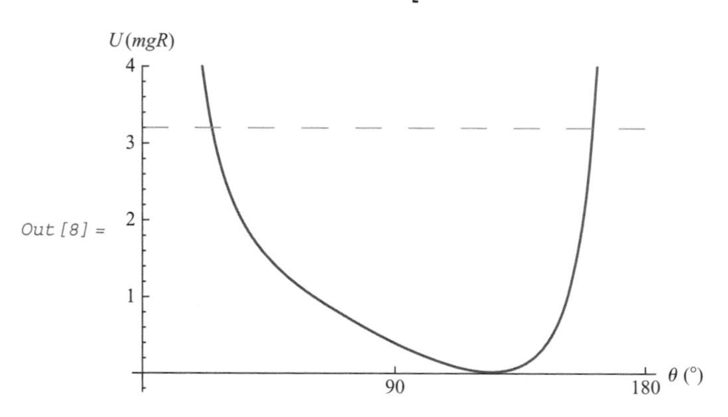

The particle rotates about the z-axis while the angle θ oscillates between two turning points, $\theta \approx 24°$ and $\theta \approx 162°$, which occur at the intersections of the effective potential curve and the energy line. Let us generate the animation:

```
In[9]:= sphericalPendulum[135, 2.5, 90, 1.5×2^(1/4), 40,
          Background→RGBColor[0.640004, 0.580004, 0.5]]
```

Out[9]=

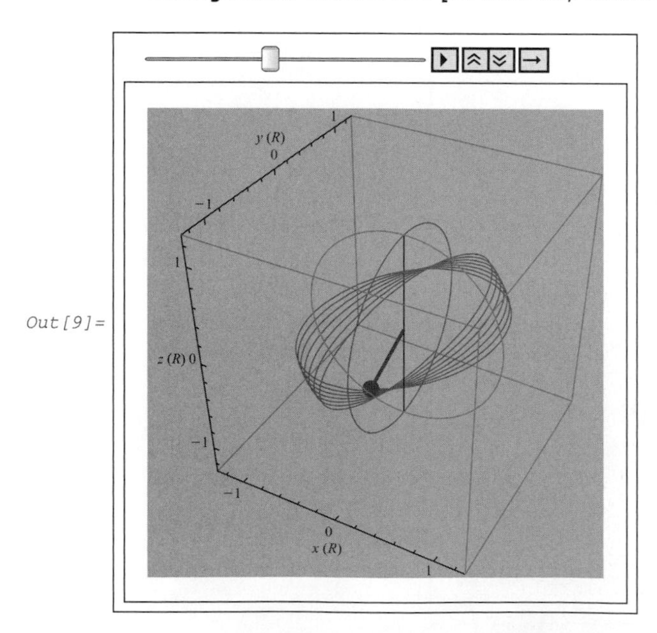

To animate, click the Play button. We see that the orbit of the pendulum rotates about an axis.

4.4.3.4 $\theta_0 = 120$, $\dot{\theta}_0 = 0.75$, $\varphi_0 = 90$, $\dot{\varphi}_0 = 2.0 \times 2^{1/4}$

Once more, we plot the effective potential:

```
In[10]:= effectivePotential[120, 0.75, 2.0 × 2^(1/4), PlotRange → {0, 4}]
```

Out[10]=

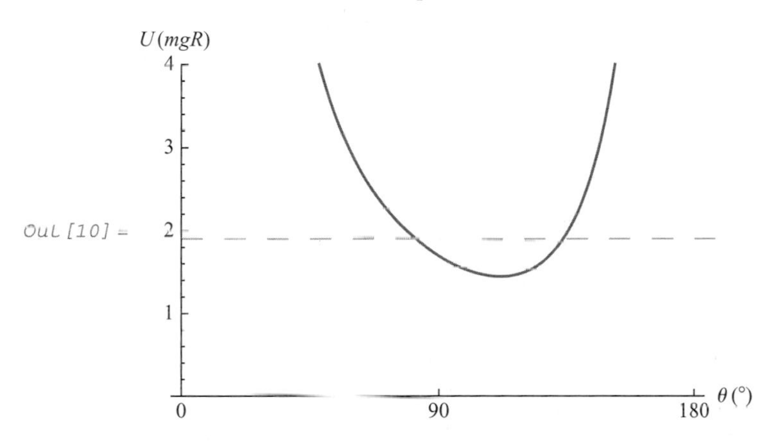

Here, the turning points for the angle θ are approximately 77° and 127°. We now generate the animation:

```
In[11]:= sphericalPendulum[120, 0.75, 90, 2.0 × 2^(1/4),
          40, Background → RGBColor[0.640004, 0.580004, 0.5]]
```

Out[11]=

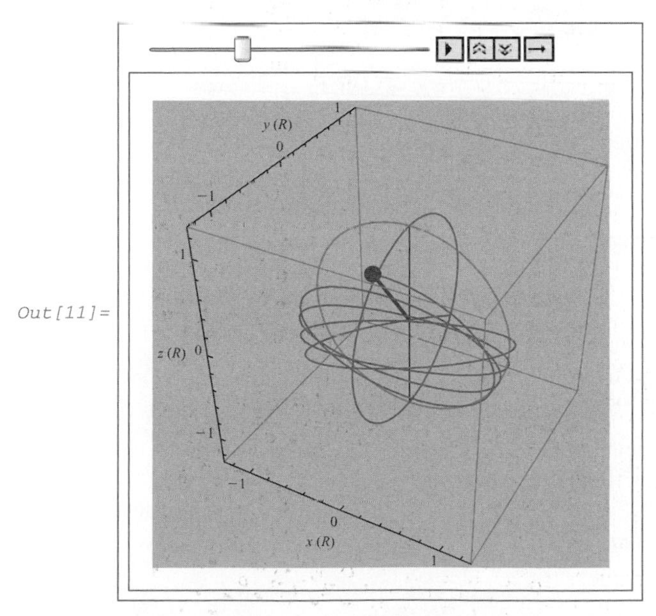

To animate, click the Play button. We see that the motion of the pendulum is more complex than those of the preceding cases. As discussed previously, the angular momentum about the z-axis must remain constant during the motion.

It is instructive to observe the motion of the pendulum from another viewpoint.

```
In[12]:= sphericalPendulum[120, 0.75, 90, 2.0 × 2^(1/4),
          40, Background→RGBColor[0.640004, 0.580004, 0.5],
          ViewPoint→{3.225, 1.018, 0.109}]
```

Out[12]=

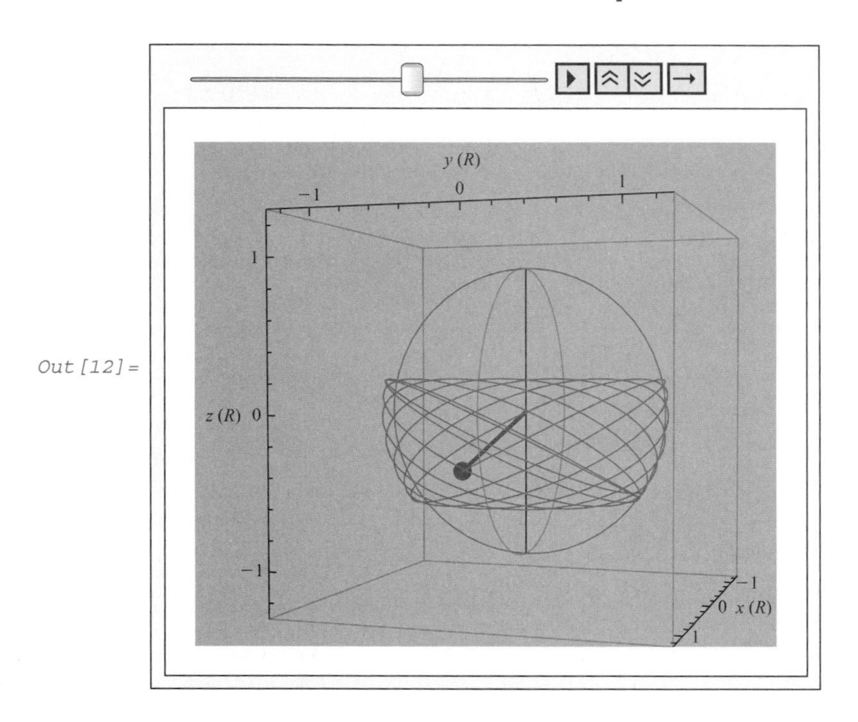

The path of the particle is clearly confined between two planes parallel to the xy plane; that is, the angle θ oscillates between two turning points.

```
In[13]:= ClearAll[effectivePotential, sphericalPendulum]
```

4.5 PROBLEMS

In this book, straightforward, intermediate-level, and challenging problems are unmarked, marked with one asterisk, and marked with two asterisks, respectively. For solutions to selected problems, see Appendix D.

1. A physicist of mass m_p (including paint) is about to paint her house. She places a uniform ladder of length L and mass m_L against the house at an angle θ with respect to the

ground. Being cautious, she decides to do a quick calculation to see how far up the ladder she can safely climb. Suppose there are rollers at the top of the ladder—that is, let the coefficient of (static) friction between the ladder and the house be zero, let the coefficient of (static) friction between the ladder and the ground be μ, and express the position of the physicist measured up the ladder from its base as fL. What fraction of the total length of the ladder does she calculate she can climb before the ladder slips, and what are the forces that the ground and the house exert on the ladder? Obtain numerical answers for the values $m_L = 9.50\,\text{kg}$, $m_p = 88.5\,\text{kg}$, $\mu = 0.55$, $\theta = 50°$, and $g = 9.80\,\text{m/s}^2$. (For solutions of this problem with MACSYMA, Maple, Mathcad, *Mathematica*, and Theorist, see [CDSTD92a].)

2. In Section 4.1.3, can we replace **v0** by the two-dimensional form $\mathbf{v_0}$ in **DSolve**? If not, why not?
3. Consider a body falling from rest near the surface of the earth. For a small heavy body with large terminal velocity, the effect of air resistance may be approximated by a frictional force proportional to the square of the speed. Determine the motion of the body.

*4. A projectile of mass m leaves the origin with a speed v_0 at an angle θ_0 from the horizontal. Assume that the effect of air resistance may be approximated by a frictional force $-b\mathbf{v}$, where $\mathbf{v}$ is the velocity. (a) Obtain an equation for the trajectory—that is, $y = f(x)$, where the x- and y-axes are along the horizontal and vertical directions, respectively. *Hint:* Use **DSolve** and **Solve**. (b) Then let $m = 0.14\,\text{kg}$, $v_0 = 45\,\text{m/s}$, $\theta_0 = 60°$, and $b = 0.033\,\text{kg/s}$ in the equation obtained in part (a). On the same graph, plot the path of the projectile with and without air resistance.

5. Section 4.3.3.1 specified, for the plane pendulum, the initial conditions $x = 0$ and $v = 2 - d$, where d is a small number. (a) What is the significance of this choice of initial conditions? (b) Solve Equations 4.3.3 and 4.3.5 and plot the phase diagram for the initial conditions $x = 0$ and $v = 2 + d$, where d is again a small number. Explain the resulting phase diagram. (c) Repeat part (b) for the initial conditions $x = 0$ and $v = 0$. Is the resulting phase diagram what you expected? If not, why not?
6. Generate an animation of the periodic motion of a damped, driven pendulum with $a = 0.2$, $F = 0.52$, and $\omega = 0.694$. *Hint:* See Section 4.3.3; discard points for the early cycles to avoid transient effects.

*7. Consider the Duffing–Holmes equation for the forced vibrations of a buckled elastic beam or the forced oscillations of a particle in a two-well potential:

$$\frac{d^2x}{dt^2} + \gamma\frac{dx}{dt} - \frac{1}{2}x\left(1 - x^2\right) = F\cos\omega t$$

Solve the equation numerically, plot the phase diagram, and determine the Poincaré section for the following parameters: (a) $\gamma = 0.15$, $\omega = 0.8$, $F = 0.32$; (b) $\gamma = 0.15$, $\omega = 0.8$, $F = 0.2$; and (c) $\gamma = 0.15$, $\omega = 0.8$, $F = 0.15$.

8. In Section 4.4.3, the first five arguments of the function **sphericalPendulum** are restricted. Are these restrictions necessary? Justify your answer.

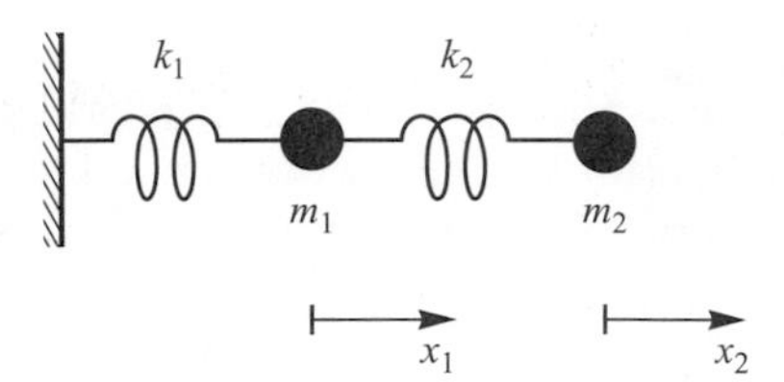

Figure 4.5.1. Two attached one-dimensional harmonic oscillators.

*9. Consider again Example 2.1.9 on the motion of a planet. (a) Plot the trajectory for the initial conditions $x(0) = 1$, $y(0) = 0$, $x'(0) = 0$, and $y'(0) = 8$. Since $x'(0) = 0$ and $y'(0) = 8$, should the orbit be a circle? Explain. (b) Plot the trajectory for $x(0) = 1$, $y(0) = 0$, $x'(0) = -2.5$, and $y'(0) = 8$. Should the foci of the ellipse be located on the x- or y-axis? Explain. (c) Find a set of initial conditions for which the orbit is a parabola. (d) Find a set of initial conditions for which the orbit is a hyperbola. *Hint:* See the discussion in Section 3.6.1.

*10. Two one-dimensional harmonic oscillators are attached as in Figure 4.5.1, where x_1 and x_2 are the displacements of m_1 and m_2, respectively. Each displacement is measured from the equilibrium position, and the direction to the right is positive. While m_2 is held in place, m_1 is moved to the left a distance a. At time $t = 0$, both masses are released. (a) Determine their subsequent motions. (b) What are the normal or characteristic frequencies of the system? (c) Plot $x_1(t)$ and $x_2(t) + 2$ on the same graph. (d) Is the motion of the system periodic? *Hint:* Use `Rationalize`. (e) Generate an animation of the motion of the coupled oscillators. For this problem, consider only the case of equal mass and identical spring constant—that is, $m_1 = m_2 = m$ and $k_1 = k_2 = k$.

**11. Write a function that generates an animation of the motion, as viewed from the laboratory frame, of a system of two identical masses under their mutual gravitational attraction.

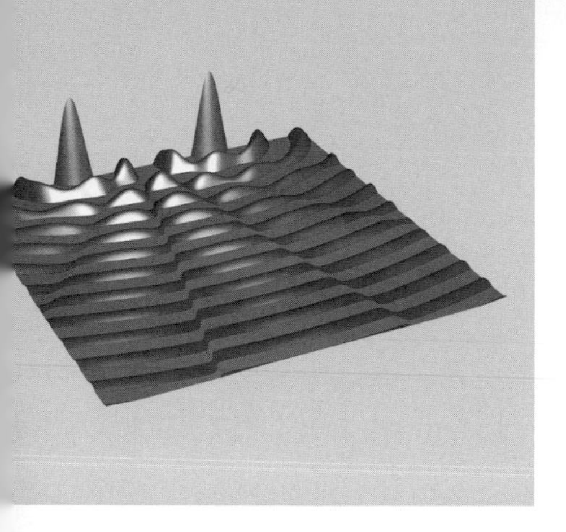

Chapter 5
Electricity and Magnetism

5.1 ELECTRIC FIELD LINES AND EQUIPOTENTIALS

5.1.1 The Problem

Plot the electric field lines and equipotentials for two point charges q_1 and q_2 when (a) $q_1 = q_2 = +q$ and (b) $q_1 = +2q$ and $q_2 = -q$.

5.1.2 Physics of the Problem

5.1.2.1 Electric Field Lines

At position $\mathbf{r}$, the electric field of a charge distribution consisting of a discrete distribution of N point charges $q_1, q_2, \ldots, q_N$ at positions $\mathbf{r}_1, \mathbf{r}_2, \ldots, \mathbf{r}_N$, respectively, and a continuous distribution of volume charge density $\rho(\mathbf{r}')$ in the volume V and surface charge density $\sigma(\mathbf{r}')$ on the surface S that bounds V is

$$\begin{aligned}\mathbf{E}(\mathbf{r}) = \frac{1}{4\pi\varepsilon_0}\sum_{i=1}^{N} q_i \frac{\mathbf{r}-\mathbf{r}_i}{|\mathbf{r}-\mathbf{r}_i|^3} + \\ \frac{1}{4\pi\varepsilon_0}\int_V \frac{\mathbf{r}-\mathbf{r}'}{|\mathbf{r}-\mathbf{r}'|^3}\rho(\mathbf{r}')dv' + \frac{1}{4\pi\varepsilon_0}\int_S \frac{\mathbf{r}-\mathbf{r}'}{|\mathbf{r}-\mathbf{r}'|^3}\sigma(\mathbf{r}')da'\end{aligned} \tag{5.1.1}$$

where ε_0 is the permittivity of free space. The electric field is a vector point function or a vector field.

To represent the electric field graphically, we can draw vectors at points on a grid such that at each point the direction of the vector is in the direction of the field and the length of the vector is proportional to the magnitude of the field. **VectorFieldPlots`**, a standard *Mathematica* package, contains functions for generating such field maps. When the lengths of vectors vary over wide ranges, these plots do not provide clear representations of the fields. As an example, let us map on the xy plane the electric field of two equal positive charges:

```
In[1]:= Needs["VectorFieldPlots`"]

        GradientFieldPlot[
          -1/Sqrt[(x + 1)^2 + y^2] - 1/Sqrt[(x - 1)^2 + y^2], {x, -2, 2}, {y, -2, 2}]
```

$$-\frac{1}{\sqrt{(x+1)^2+y^2}}-\frac{1}{\sqrt{(x-1)^2+y^2}},\ \{x,-2,2\},\ \{y,-2,2\}$$

Out[2] =

For most vectors in this picture, only the heads are visible.

Electric field lines offer a better means for visualization. These directed curves are drawn such that at each point the tangent to the curve is in the direction of the electric field and the number of lines per unit area perpendicular to the lines (the density of the lines) is proportional to the magnitude of the electric field. Electric field lines have the following properties:

1. The lines must begin on positive charges and end on negative charges or at infinity. It is possible for lines to come from infinity and terminate on negative charges.

2. The number of lines leaving or entering a charge is proportional to the magnitude of the charge. If n lines are at charge q, then $n' = |q'/q|n$ lines must be at charge q'. The value of the proportionality constant is chosen so that the graph shows the essential features of the electric field.
3. Electric field lines cannot cross each other except at points of equilibrium where the electric field vanishes.
4. The field lines near a charge are symmetrical about the charge and radially directed.

(For an introduction to electric field lines, see [FGT05]; for an advanced treatment, refer to [Jea66].)

5.1.2.2 Equipotentials

The electric field in Equation 5.1.1 can be written as

$$\mathbf{E}(\mathbf{r}) = -\nabla\varphi(\mathbf{r}) \tag{5.1.2}$$

where the scalar function $\varphi(\mathbf{r})$, called the electric potential, is

$$\varphi(\mathbf{r}) = \frac{1}{4\pi\varepsilon_0}\sum_{i=1}^{N}\frac{q_i}{|\mathbf{r}-\mathbf{r}_i|} + \frac{1}{4\pi\varepsilon_0}\int_V \frac{\rho(\mathbf{r}')}{|\mathbf{r}-\mathbf{r}'|}dv' + \frac{1}{4\pi\varepsilon_0}\int_S \frac{\sigma(\mathbf{r}')}{|\mathbf{r}-\mathbf{r}'|}da' \tag{5.1.3}$$

Equipotentials are regions where the electric potential has constant values. For a system of point charges, the equipotentials are surfaces (lines in two dimensions). Except at points of equilibrium where the electric field vanishes, the equipotential surfaces do not intersect and the electric field lines are perpendicular to these surfaces everywhere.

5.1.2.3 Electric Field Lines and Equipotentials for Two Point Charges

Let the midpoint between the two point charges be the origin of the coordinate system and the x-axis be directed from q_1 to q_2, as in Figure 5.1.1. For these charges, it is necessary to plot only the electric field lines and equipotentials on the upper xy plane because of the rotational symmetry about the x-axis. However, for clarity, we show the lines on the entire xy plane. To obtain a three-dimensional picture, rotate the plot about the x-axis.

For these charges, Equation 5.1.3 gives

$$\varphi(x, y) = \frac{1}{4\pi\varepsilon_0}\left(\frac{q_1}{\sqrt{(x+d)^2+y^2}} + \frac{q_2}{\sqrt{(x-d)^2+y^2}}\right) \tag{5.1.4}$$

If q_1 and q_2, x and y, and $\varphi(x, y)$ are expressed in units of q, d, and $q/4\pi\varepsilon_0 d$, respectively, Equation 5.1.4 becomes

$$\varphi(x, y) = \frac{q_1}{\sqrt{(x+1)^2+y^2}} + \frac{q_2}{\sqrt{(x-1)^2+y^2}} \tag{5.1.5}$$

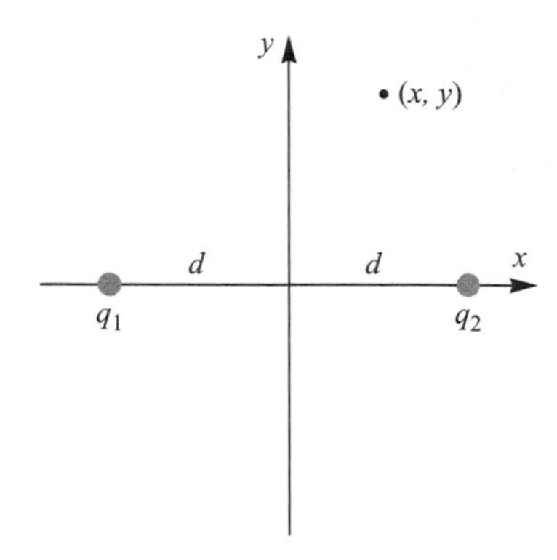

Figure 5.1.1. A coordinate system for two point charges.

Equation 5.1.2 for the electric field implies

$$\mathbf{E}(x, y) = -\frac{\partial\varphi(x, y)}{\partial x}\mathbf{i} - \frac{\partial\varphi(x, y)}{\partial y}\mathbf{j} \tag{5.1.6}$$

where **i** and **j** are unit vectors in the x and y directions, respectively.

5.1.3 Solution with *Mathematica*

There is a simple method for generating the electric field lines of collinear charges (see Problem 2 in Section 5.4). Here, we adopt another algorithm in order to illustrate list manipulation and offer easy generalization for noncollinear charges.

5.1.3.1 $q_1 = q_2 = +q$

```
In[3]:= ClearAll["Global`*", Subscript]
```

For the electric field lines, let us define a set of starting points symmetrically distributed around and near q_1:

```
In[4]:= x0[n_] := N[-1 + 0.2 Cos[nπ/12]]
        y0[n_] := N[0.2 Sin[nπ/12]]
```

For each point with coordinates (x, y) on an electric field line, the coordinates of the adjacent point a distance ds away down the line are

$$\left(x + \frac{E_x}{E}ds,\ y + \frac{E_y}{E}ds\right)$$

A function for such mapping from one point to another is

```
In[6]:= f[{x_, y_}] := {x+0.02 Ex[x, y]/ℰ[x, y], y+0.02 Ey[x, y]/ℰ[x, y]}
```

where ds is set equal to 0.02, and **Ex**, **Ey**, and ℰ are the x component, y component, and magnitude of the electric field, respectively. The Greek letter **E** used here is different from the similar-looking keyboard letter **E**, which represents the exponential constant in *Mathematica*; the subscripts of **E** are the script letters *x* and *y* rather than the ordinary letters **x** and **y**; ℰ is a capital script letter.

The components of the electric field are given by Equation 5.1.6 in terms of the electric potential φ in Equation 5.1.5:

```
In[7]:= φ[x_, y_] := 1/Sqrt[(x+1)^2 + y^2] + 1/Sqrt[(x-1)^2 + y^2]

        Ex[x_, y_] = -∂x φ[x, y];

        Ey[x_, y_] = -∂y φ[x, y];
```

The magnitude of the electric field can be determined from its components:

```
In[10]:= ℰ[x_, y_] = Sqrt[Ex[x, y]^2 + Ey[x, y]^2];
```

Starting with a point that has coordinates (x, y), a function for generating a list of coordinates of consecutive points down an electric field line in the plotting region is

```
In[11]:= g[{x_, y_}]:=FixedPointList[f, {x, y},
           SameTest → ((Sqrt[(#2[[1]])^2 + (#2[[2]])^2] > 3.0) || (#2[[1]] > 0) &)]
```

FixedPointList[*f*, *x*, **SameTest** -> *test*] generates the list $\{x, f[x], f[f[x]], f[f[f[x]]], \ldots\}$, stopping when the function *test* yields **True**. The function *test* is normally used to compare, in the form **(Abs[#1- #2]** < *condition* **&)**, two consecutive elements of the list; we use it here to test whether the second element of this pair reaches outside the plotting region, which is a quarter-circle located in the second quadrant and centered at the origin.

Now apply **g** to each element in a list of starting points to obtain a nested list of coordinates for the electric field lines in the second quadrant:

```
In[12]:= g/@ Table[{x0[n], y0[n]}, {n, 1, 11, 2}];
```

The coordinates in the other quadrants can easily be determined from the symmetries across the x- and y-axes:

```
In[13]:= coordinates1 = Join[%, %/.{x_, y_} → {x, -y},
            %/.{x_, y_} → {-x, y}, %/.{x_, y_} → {-x, -y}];
```

ListLinePlot generates the electric field lines, and **ContourPlot** maps the equipotentials on the same graph:

```
In[14]:= SetOptions[ListLinePlot, PlotStyle → Red];
         Show[
          ListLinePlot/@ coordinates1,
          ContourPlot[φ[x, y], {x, -2, 2}, {y, -2, 2}, ContourShading → False,
           PlotRange → {1.1, 6.0}, Contours → 16, PlotPoints → 50,
           Frame → False, ContourStyle → {{Blue, Dashing[{0.01, 0.01}]}}],
          Graphics[{Text["+q", {-1, 0}], Text["+q", {1, 0}]}],
          AspectRatio → Automatic,
          PlotRange → {{-2, 2}, {-2, 2}}, Axes → False]
```

Out[15]=

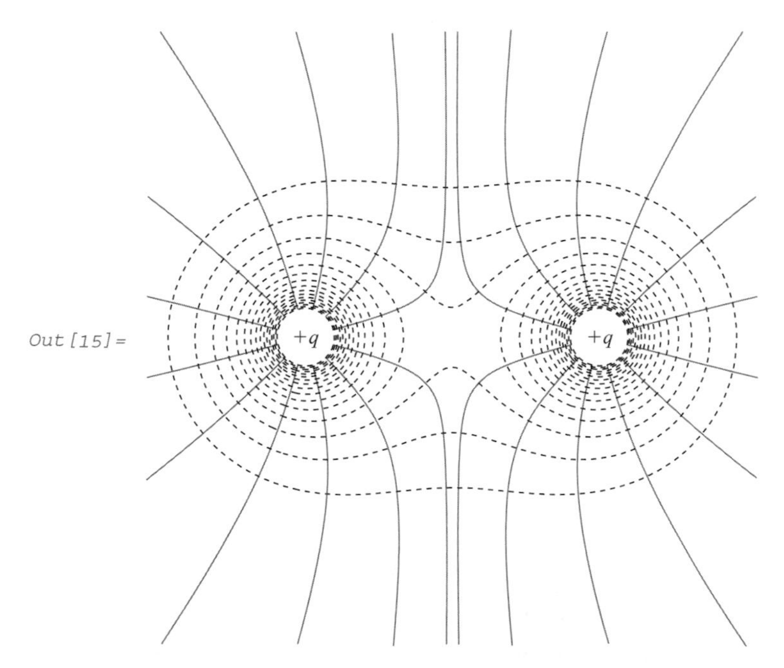

The solid lines directed from the charges to infinity are the electric field lines, and the dashed closed contours are the equipotentials. The electric field lines and equipotentials are perpendicular to each other everywhere. Since the equipotentials are plotted with a sequence of equally spaced values for the potential, the electric field is strong in regions where the equipotentials are close together, in accordance with Equation 5.1.6. The origin is a point of equilibrium where the electric field vanishes. The uniqueness of this point is not obvious from the graph. Let us plot the electric field lines and the equipotential through this point:

```
In[16]:= (* generate coordinates for electric field lines *)
         coordinates2 = g/@ {{-1+0.2, 0}, {0, 0.01}};

         (* enter coordinates of vertices of arrowheads
          for labeling the senses of electric field lines *)
         arrowheads = {{{0.5, 0}, {0.6, 0.1}, {0.6, -0.1}},
             {{0, 1}, {0.1, 0.9}, {-0.1, 0.9}}};

        Show[
         ListLinePlot/@
          Join[coordinates2, coordinates2/.{x_, y_}→{x, -y},
           coordinates2/.{x_, y_}→{-x, y},
           coordinates2/. {x_, y_}→{-x, -y}],
         ContourPlot[φ[x, y], {x, -2, 2}, {y, -2, 2}, ContourShading→False,
          Contours→{2}, PlotPoints→100, Frame→False,
          ContourStyle→{{Blue, Dashing[{0.01, 0.01}]}}], Graphics[
          Polygon/@arrowheads~Join~(arrowheads/.{x_, y_}→{-x, -y})],
         Graphics[{Text["+q", {-1, 0}], Text["+q", {1, 0}]}],
         AspectRatio→Automatic,
         PlotRange→{{-2, 2}, {-2, 2}}, Axes→False]
```

Out[18]=

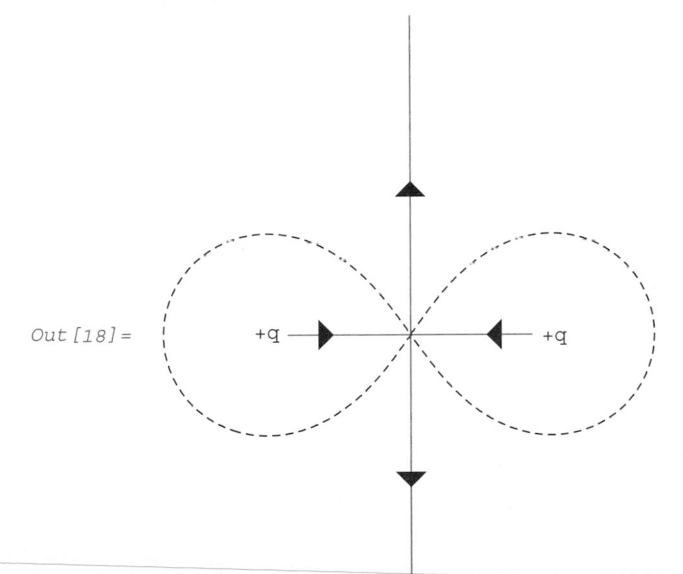

At the origin, an equipotential intersects itself and the two electric field lines, one from each charge, coalesce before splitting in two lines, one going to $+\infty$ and the other to $-\infty$. Also, the electric field lines are not perpendicular to the equipotential there.

5.1.3.2 $q_1 = +2q$ and $q_2 = -q$

```
In[19]:= ClearAll["Global`*", Subscript]
```

For these charges, it is not sufficient to calculate the coordinates for the electric field lines in only one quadrant because the reflection symmetry across the y-axis no longer exists. Here, we determine the coordinates within a semicircle centered at the origin, less two small semicircles centered at the charges. The procedure outlined in the preceding section for mapping the electric field can easily be modified:

```
In[20]:= x0[n_] := N[-1 + 0.2 Cos[nπ/16]]
         y0[n_] := N[0.2 Sin[nπ/16]]
         f[{x_, y_}] := {x + 0.02 Ex[x, y]/ℰ[x, y], y + 0.02 Ey[x, y]/ℰ[x, y]}
         φ[x_, y_] := 2/Sqrt[(x + 1)^2 + y^2] - 1/Sqrt[(x - 1)^2 + y^2]
         Ex[x_, y_] = -∂x φ[x, y];
         Ey[x_, y_] = -∂y φ[x, y];
         ℰ[x_, y_] = Sqrt[Ex[x, y]^2 + Ey[x, y]^2];
         g[{x_, y_}] := FixedPointList[f, {x, y}, SameTest →
             (Sqrt[(#2[[1]])^2 + (#2[[2]])^2] > 3.5 || Sqrt[((#2[[1]]) - 1)^2 + (#2[[2]])^2] < 0.2 &)]
         coordinates3 = g/@ Table[{x0[n], y0[n]}, {n, 1, 15, 2}];
         SetOptions[ListLinePlot, PlotStyle → Red];
         Show[
          ListLinePlot/@
           Join[coordinates3, coordinates3 /. {x_, y_} → {x, -y}],
          ContourPlot[φ[x, y], {x, -3, 3}, {y, -3, 3}, ContourShading → False,
           PlotRange → {-5.0, 10}, Contours → 24, PlotPoints → 50,
           Frame → False, ContourStyle → {{Blue, Dashing[{0.01, 0.01}]}}],
          Graphics[{Text["+2q", {-1, 0}], Text["-q", {1, 0}]}],
          AspectRatio → Automatic,
          PlotRange → {{-3, 3}, {-3, 3}}, Axes → False]
```

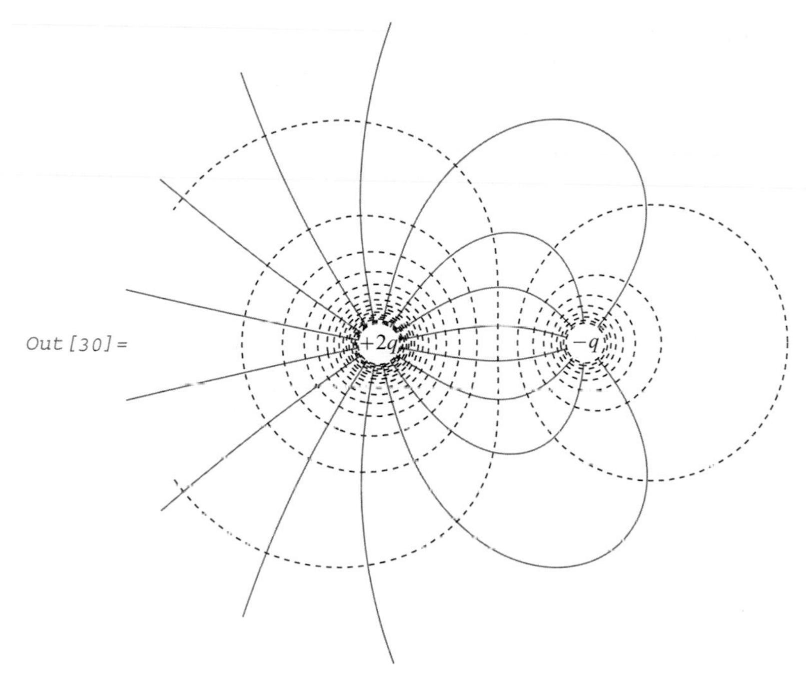

The solid lines are the electric field lines and the dashed lines are the equipotentials. Of the 16 lines emanating from $q_1 = +2q$, half of them terminate on $q_2 = -q$, as they should because $|q_1/q_2| = 2$, and the other half go to infinity. The point of equilibrium at $\{5.83, 0\}$ is not shown because it lies outside the plotting region.

5.2 LAPLACE'S EQUATION

5.2.1 The Problem

Figure 5.2.1 shows two grounded, semi-infinite, parallel electrodes separated by a distance b. A third electrode located at $x = 0$ is maintained at potential V_0. For the points between the plates, solve Laplace's equation analytically and numerically and plot the electric potential.

5.2.2 Physics of the Problem

The electric potential V between the plates satisfies Laplace's equation

$$\nabla^2 V = 0 \tag{5.2.1}$$

with the boundary conditions

$$V = 0 \quad y = 0 \tag{5.2.2}$$

$$V = 0 \quad y = b \tag{5.2.3}$$

$$V = V_0 \quad x = 0 \tag{5.2.4}$$

$$V \to 0 \quad x \to \infty \tag{5.2.5}$$

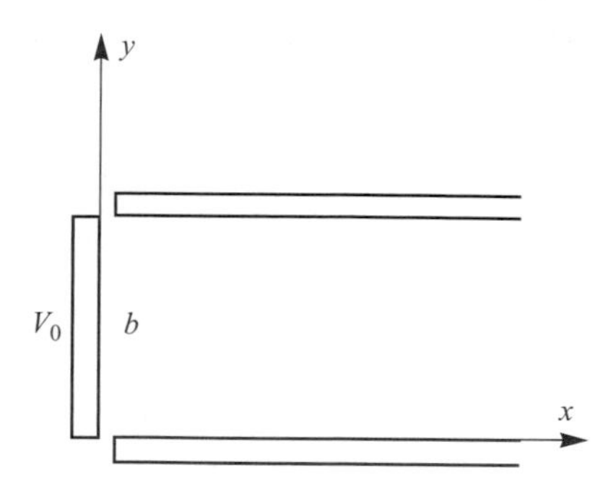

Figure 5.2.1. Two grounded, plane electrodes, one at $y = 0$ and the other at $y = b$, lie parallel to the xz plane and extend along the positive x direction from zero to infinity. At $x = 0$, a third electrode, in the form of a plane strip, is maintained at potential V_0. These electrodes are infinite along the z direction perpendicular to the page.

Because of the translational symmetry along the z direction, the potential must be independent of the z coordinate, and Equation 5.2.1 becomes

$$\frac{\partial^2 V}{\partial x^2} + \frac{\partial^2 V}{\partial y^2} = 0 \tag{5.2.6}$$

In what follows, let x as well as y be expressed in units of b and V in units of V_0.

5.2.2.1 Analytical Solution

With the method of separation of variables and Fourier analysis, we can determine the solution to the two-dimensional Laplace's equation with the boundary conditions. The potential V at any point (x, y, z) between the plates is

$$V(x, y) = \frac{4}{\pi} \sum_{m}^{\infty} \frac{1}{(2m - 1)} \sin[(2m - 1)\,\pi y]\, e^{-(2m-1)\pi x} \tag{5.2.7}$$

where m varies over all positive integers, and V depends only on x and y. (For a derivation of Equation 5.2.7, see [LC70] or [Gri99].)

5.2.2.2 Numerical Solution

For a numerical solution of the two-dimensional Laplace's equation, we approximate the boundary condition of Equation 5.2.5 by

$$V \to 0 \quad x \to x_f \tag{5.2.8}$$

where x_f is finite but sufficiently large so that the solution has the desired accuracy. Thus, we are to determine the potential at points inside a rectangular region bounded by the lines $x = 0$, $x = x_f$, $y = 0$, and $y = 1$.

Let us solve Laplace's equation with the finite difference method that imposes a grid of points (x_i, y_i) on the rectangular region such that

$$x_i = i\Delta \quad i = 0, 1, 2, \ldots, imax \tag{5.2.9}$$

$$y_j = j\Delta \quad j = 0, 1, 2, \ldots, jmax \tag{5.2.10}$$

and approximates the second derivatives by

$$\frac{\partial^2 V}{\partial x^2} \to \frac{V_{i+1,j} - 2V_{i,j} + V_{i-1,j}}{\Delta^2} \tag{5.2.11}$$

$$\frac{\partial^2 V}{\partial y^2} \to \frac{V_{i,j+1} - 2V_{i,j} + V_{i,j-1}}{\Delta^2} \tag{5.2.12}$$

where Δ is the separation between adjacent points and we have introduced the notation

$$V_{i,j} = V(x_i, y_j)$$

(For more information on the finite difference method, see [DeV94] and [KM90].) With Equations 5.2.11 and 5.2.12, Equation 5.2.6 becomes a finite difference equation

$$\frac{V_{i+1,j} - 2V_{i,j} + V_{i-1,j}}{\Delta^2} + \frac{V_{i,j+1} - 2V_{i,j} + V_{i,j-1}}{\Delta^2} = 0 \tag{5.2.13}$$

Solving for $V_{i,j}$, we find

$$V_{i,j} = \frac{1}{4}\left[V_{i+1,j} + V_{i-1,j} + V_{i,j+1} + V_{i,j-1}\right] \tag{5.2.14}$$

which states that the potential at any interior grid point inside the boundary is the average of the four nearest neighbors. This equation is a discrete approximation of the two-dimensional Laplace's equation (Equation 5.2.6) and approaches the latter as Δ goes to zero. The potential at the boundary grid points is given by Equations 5.2.2–5.2.4 and 5.2.8, which imply

$$\begin{aligned} V_{i,0} &= 0 \\ V_{i,jmax} &= 0 \\ V_{0,j} &= 1 \\ V_{imax},j &= 0 \end{aligned} \tag{5.2.15}$$

for $i = 0, 1, 2, \ldots, imax$ and $j = 1, \ldots, jmax - 1$.

Equation 5.2.14 yields a system of $(imax - 1) \times (jmax - 1)$ linear equations in the unknowns $V_{i,j}$ with $i = 1, 2, \ldots, imax - 1$ and $j = 1, 2, \ldots, jmax - 1$. These unknowns can, in principle, be determined exactly with the methods for solving systems of linear equations, such as the Gaussian elimination or factorization methods (see [Pat94]). When the number of equations is large, these direct methods require such an enormous amount of computation that iterative methods, called relaxation methods, are often used. In one relaxation method, the Jacobi method, Equation 5.2.14 is written as

$$V_{i,j}^{n+1} = \frac{1}{4}\left[V_{i+1,j}^{n} + V_{i-1,j}^{n} + V_{i,j+1}^{n} + V_{i,j-1}^{n}\right] \tag{5.2.16}$$

where the superscripts denote the iteration numbers. We start with an initial guess $V_{i,j}^{0}$ for the potential at the interior grid points and use the set of equations given by Equation 5.2.16 to compute the first iterate $V_{i,j}^{1}$, which is then substituted back into the equations to determine the second iterate $V_{i,j}^{2}$, and so on. In this manner, the initial guess eventually converges to the correct solution of Equation 5.2.14. In practice, the iteration terminates when the condition

$$\left|V_{i,j}^{n+1} - V_{i,j}^{n}\right| < tol \tag{5.2.17}$$

is satisfied for all interior grid points. The parameter *tol* is chosen to be small enough to ensure that the solution attains a specified level of accuracy.

The rate of convergence of the Jacobi method depends on the initial guess. The closer the initial guess is to the correct solution, the faster is the convergence. The rate of convergence can be substantially improved by using the successive overrelaxation (SOR) method, in which Equation 5.2.16 is replaced by

$$V_{i,j}^{n+1} = (1-w)V_{i,j}^{n} + \frac{w}{4}\left[V_{i+1,j}^{n} + V_{i-1,j}^{n+1} + V_{i,j+1}^{n} + V_{i,j-1}^{n+1}\right] \tag{5.2.18}$$

with $1 < w < 2$. The relaxation parameter w affects the rate of convergence, and the optimal value for w depends on the size of the rectangle, the grid spacing Δ, and the iteration parameter *tol*. (For further discussion of this method, see [Pat94] and [Won92].)

Aside from the usual round-off error, there are three kinds of errors in the numerical solution: boundary error in replacing Equation 5.2.5 with Equation 5.2.8, discretization error in approximating Equation 5.2.6 by Equation 5.2.14, and iteration error in using relaxation methods to determine the correct solution to Equation 5.2.14. We can reduce these errors by increasing x_f and decreasing Δ as well as *tol*. Of course, the sizes of these parameters are limited by the required computational resources.

• 5.2.3 Solution with *Mathematica*

```
In[1]:= ClearAll["Global`*"]
```

5.2.3.1 Analytical Solution

The solution given in Equation 5.2.7 is exact. Before plotting V as a function of x and y, we must examine the convergence of the series. The series converges very rapidly because of the factor $1/(2m-1)$ and the exponential, except for $x \approx 0$. At $x = 0$, the exponential equals 1. Therefore, we must determine how many terms are needed to approximate the boundary condition $V(0, y) = 1$. The electric potential is

```
In[2]:= v[x_, y_, mmax_] := 4/π Sum[1/(2m-1) Sin[(2m-1)π y] Exp[-(2m-1)π x], {m, 1, mmax}]
```

$$v[x\_, y\_, mmax\_] := \frac{4}{\pi}\sum_{m=1}^{mmax}\frac{1}{(2m-1)}\mathrm{Sin}[(2m-1)\pi y]\,\mathrm{Exp}[-(2m-1)\pi x]$$

where **mmax** is the number of terms in the series and **v** is the alias of the potential V. Here are the plots of $V(0, y)$ with 10, 200, and 400 terms:

```
In[3]:= Plot[Evaluate[v[0, y, 10]], {y, 0, 1},
         PlotRange→{0, 1.3}, AxesLabel→{"y(b)","V(V0)"}]
```

Out[3] =

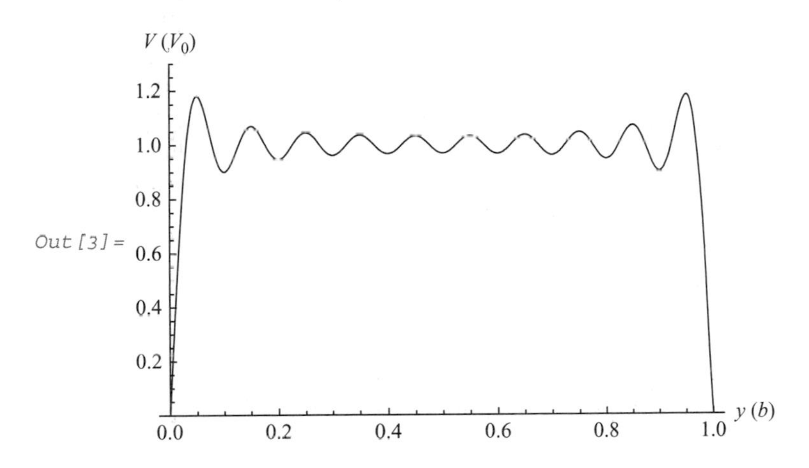


```
In[4]:= Plot[Evaluate[v[0, y, 200]], {y, 0, 1},
         PlotRange→{0, 1.3}, AxesLabel→{"y(b)", "V(V0)"}]
```

Out[4] =

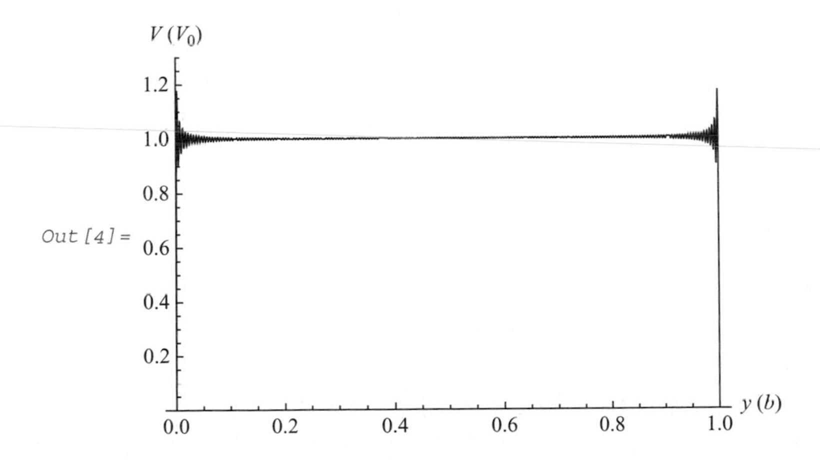


```
In[5]:= Plot[Evaluate[v[0, y, 400]], {y, 0, 1},
         PlotRange→{0, 1.3}, AxesLabel→{"y(b)", "V(V0)"}]
```

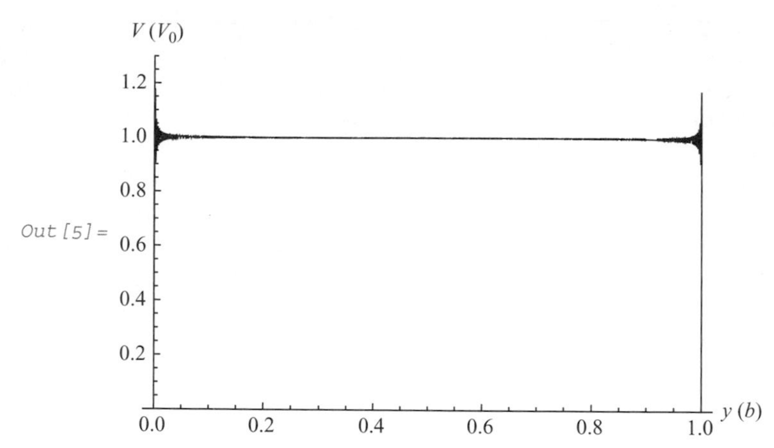

With 400 terms, the series gives an excellent approximation of the boundary condition $V(0, y) = 1$.

Let us plot the potential $V(x, y)$ with 400 terms:

```
In[6]:= Plot3D[Evaluate[v[x, y, 400]], {x, 0, 1.3},
          {y, 0, 1}, PlotRange→{0, 1}, Boxed→False,
          ViewPoint→{1.741, -2.565, 1.356}, PlotPoints→27,
          BoxRatios→{1, 1, 1}, AxesLabel→{"x(b)", "y(b)", "V(V0)"}]
```

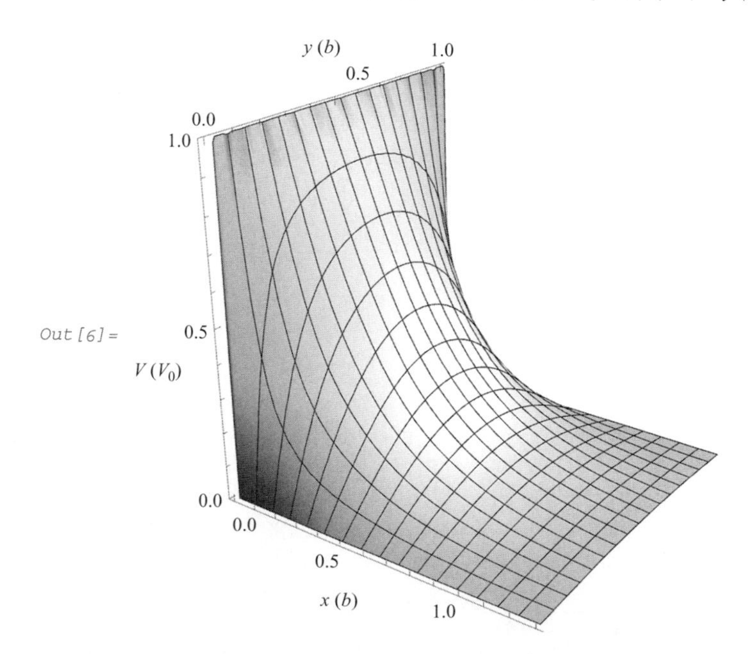

5.2.3.2 Numerical Solution

As discussed in Section 5.2.2.2, we must make an initial guess for the potential at the interior grid points and choose the values for the parameters w, x_f, Δ, and *tol*. While the initial guess and all the parameters affect the rate of convergence, the boundary limit x_f, the grid spacing Δ, and the tolerance *tol* also influence the error in the solution.

Although any initial guess for the potential at the interior grid points converges toward the correct solution of Equation 5.2.14, a good guess substantially improves the rate of convergence—that is, the number of necessary iterations. (See Problem 8 of Section 5.4.) From the boundary conditions (Equation 5.2.15), we surmise that the initial potential is a product of an exponentially decreasing function of the coordinate x and a sinusoidal function of the coordinate y such that the potential vanishes at $y = 0$ and $y = 1$. Let us just take the first term of the analytical series solution in Equation 5.2.7. With Equations 5.2.9 and 5.2.10 for $i = 1, 2, \ldots, imax - 1$ and $j = 1, 2, \ldots, jmax - 1$, we have

```
In[7]:= V[0][i_Integer?Positive/; i < imax,
          j_Integer?Positive/; j < jmax] :=
        V[0][i, j] = N[4/π Sin[π j Δ] Exp[-π i Δ]]
```

Note that dynamic programming, introduced in Section 3.3.3, is used in the definition in order to reduce the amount of computation.

We set $w = 1.0$ for the moment and discuss its appropriate value later. This choice of w reduces the successive overrelaxation method to what is known as the Gauss–Seidel method.

```
In[8]:= w = 1.0;
```

The smaller the Δ and *tol*, the smaller the error. However, their sizes are limited by the necessary computation effort. We choose the reasonable values

```
In[9]:= {Δ, tol} = {1/20, 0.00025};
```

Equation 5.2.10 implies $1 = jmax\Delta$. Thus,

```
In[10]:= jmax = 20;
```

The choice of x_f is somewhat arbitrary. We may start with a reasonable guess and determine the solution to Equation 5.2.14. Then, we increase x_f by a fixed amount and determine another solution, and so on, until the solutions of the last two iterations differ by at most a specified small constant, resulting in an acceptable accuracy in the final solution. We may also seek guidance from the available analytical solution. It can be shown that the rate of convergence and the error in the solution are rather insensitive to choices for x_f as long as they are reasonable. (See Problem 9 of Section 5.4.) Let $x_f = 5/2$. Equation 5.2.9 implies $x_f = imax\Delta$. Hence,

```
In[11]:= imax = 50;
```

The iteration equation (Equation 5.2.18) becomes

```
In[12]:= V[n_Integer?Positive][i_Integer?Positive/; i < imax,
           j_Integer?Positive/; j < jmax] :=
          V[n][i, j] = ((1 - w)V[n - 1][i, j] + w/4 (V[n - 1][i + 1, j] +
               V[n][i - 1, j] + V[n - 1][i, j + 1] + V[n][i, j - 1]))
```

Again, it is important to use dynamic programming in the definition.

Equation 5.2.15 for the boundary conditions requires

```
In[13]:= V[n_Integer?NonNegative]
           [i_Integer?NonNegative/; i ≤ imax, 0] := V[n][i, 0] = 0
         V[n_Integer?NonNegative][i_Integer?NonNegative/; i ≤ imax, jmax] :=
          V[n][i, jmax] = 0
         V[n_Integer?NonNegative][0, j_Integer?Positive/; j < jmax] :=
          V[n][0, j] = 1
         V[n_Integer?NonNegative][imax, j_Integer?Positive/; j < jmax] :=
          V[n][imax, j] = 0
```

To implement the condition in Equation 5.2.17, we use the **For** loop discussed in Section 3.4.3.2:

```
In[17]:= For[
           m = 1,
           Max[Table[Abs[V[m][i, j] - V[m - 1][i, j]],
              {i, 1, imax - 1}, {j, 1, jmax - 1}]] ≥ tol, m++,
           If[m > 40, Print["Iteration limit of 40 exceeded."]; Break[]];
           Print["Iteration Number = " <> ToString[m]]];
       Iteration Number = 1
       Iteration Number = 2
       Iteration Number = 3
       Iteration Number = 4
       Iteration Number = 5
       Iteration Number = 6
       Iteration Number = 7
       Iteration Number = 8
       Iteration Number = 9
       Iteration Number = 10
       Iteration Number = 11
```

```
Iteration Number = 12
Iteration Number = 13
Iteration Number = 14
Iteration Number = 15
Iteration Number = 16
Iteration Number = 17
Iteration Number = 18
Iteration Number = 19
Iteration Number = 20
Iteration Number = 21
Iteration Number = 22
Iteration Number = 23
Iteration Number = 24
Iteration Number = 25
Iteration Number = 26
```

For each iteration, the command **Print** prints the iteration number. The operator **<>** with the full name **StringJoin** concatenates two strings and **ToString[*expr*]** gives a string corresponding to the printed form of *expr*. The loop terminates when Equation 5.2.17 or the first argument of the **If** function is true. The function **If** provides a graceful exit and prints a message when the iteration number exceeds the specified limit.

An important result of the **For** loop is the final iteration number at which the loop terminates:

```
In[18]:= Print["Final Iteration Number = " <> ToString[m]]
         Final Iteration Number = 27
```

With this number already assigned to **m**, we can plot the electric potential:

```
ListPlot3D[
 Table[V[m][i, j], {j, 0, jmax}, {i, 0, 26}],
 DataRange→{{0, 26Δ}, {0, jmaxΔ}},
 BoxRatios→{1, 1, 1}, PlotRange→{0, 1}, Boxed→False,
 Axes→True, AxesLabel→{"x(b)", "y(b)", "V(V0)"},
 ViewPoint→{1.741, -2.565, 1.356}]
```

ListPlot3D[*array*] generates a three-dimensional plot of a surface representing an array of height values. We have suppressed the plot of the electric potential because it looks identical to that in Section 5.2.3.1 for the analytical solution.

For comparison, let us tabulate the values of the electric potential from the analytical and numerical solutions:

```
In[19]:= Table[Evaluate[{v[iΔ, jΔ, 400]// N, V[m][i, j]}],
           {i, 0, 50, 5}, {j, 2, 10, 2}];
```

```
In[20]:= PaddedForm[
          TableForm[
           Prepend[Flatten[#, 1]&/@ Transpose @
              {Table["x = " <> ToString[iΔ//N], {i, 0, 50, 5}], %}, Prepend[
             Table[" y = " <> ToString[jΔ//N], {j, 2, 10, 2}], {}]]], {4, 3}]
```

```
Out[20]//PaddedForm=
             y = 0.1     y = 0.2     y = 0.3     y = 0.4     y = 0.5
             0.997       0.999       0.999       0.999       0.999
 x = 0.
             1.000       1.000       1.000       1.000       1.000
             0.218       0.379       0.477       0.529       0.545
 x = 0.25
             0.219       0.380       0.478       0.530       0.545
             0.085       0.159       0.215       0.249       0.261
 x = 0.5
             0.085       0.160       0.216       0.250       0.261
             0.038       0.071       0.098       0.115       0.120
 x = 0.75
             0.038       0.071       0.098       0.115       0.121
             0.017       0.032       0.045       0.052       0.055
 x = 1.
             0.017       0.032       0.045       0.052       0.055
             0.008       0.015       0.020       0.024       0.025
 x = 1.25
             0.008       0.015       0.020       0.024       0.025
             0.004       0.007       0.009       0.011       0.011
 x = 1.5
             0.004       0.007       0.009       0.011       0.011
             0.002       0.003       0.004       0.005       0.005
 x = 1.75
             0.002       0.003       0.004       0.005       0.005
             0.001       0.001       0.002       0.002       0.002
 x = 2.
             0.001       0.001       0.002       0.002       0.002
             0.000       0.001       0.001       0.001       0.001
 x = 2.25
             0.000       0.001       0.001       0.001       0.001
             0.000       0.000       0.000       0.000       0.000
 x = 2.5
             0.000       0.000       0.000       0.000       0.000
```

The upper number in each entry is from the analytical solution, and the lower number is from the numerical solution. The agreement between the two solutions is excellent to three digits after the decimal point. Including the values for $y > 0.5$ in the table is unnecessary because the potential is symmetric about $y = 0.5$.

In our computations, we have set $w = 1.0$. What is the optimal value for the relaxation parameter w? The following table shows the final iteration numbers for $jmax = 20$, $tol = 0.00025$, and w ranging from 1.0 to 1.9 in steps of 0.1. (See Problem 7 of Section 5.4.)

Relaxation parameter w	1.0	1.1	1.2	1.3	1.4	1.5	1.6	1.7	1.8	1.9
Final iteration number	27	23	21	19	17	16	15	21	28	> 40

The relaxation parameter w indeed controls the rate of convergence, and its optimal value is 1.6.

In this section, we have used *Mathematica* as an interactive system rather than as a programming language in order to explain and justify our solution of Laplace's equation. For symbol localization, clarity, convenience, and elegance, we should assemble the commands into procedures in function definitions if we are to use them for more than one set of parameter values or if they compose only a subset of the commands in a *Mathematica* session. For example, we can write a procedural program in the form of a function that prints the final iteration number and plots the electric potential. The function takes four arguments: The first must be an even integer and is for *jmax*, which equals $1/\Delta$; the second is for the relaxation parameter w, which must be a real number between 1 and 2; the third is for the tolerance *tol*, which must be a positive real number; and the fourth is for the graphics options.

```
laplace2D[jmax_Integer?EvenQ, w_Real/; 1 < w < 2, tol_Real?Positive,
  opts___] := Module[{imax = (5/2) jmax, Δ = 1/jmax, m, V},

  (* initial guess *)
  V[0][i_Integer?Positive/; i < imax,
    j_Integer?Positive/; j < jmax] :=
   V[0][i, j] = N[4/π Sin[π j Δ] Exp[-π i Δ]];

  (* boundary conditions *)
  V[n_Integer?NonNegative][i_Integer?NonNegative/; i ≤ imax, 0] :=
   V[n][i, 0] = 0; V[n_Integer?NonNegative][
    i_Integer?NonNegative/; i ≤ imax, jmax] := V[n][i, jmax] = 0;
  V[n_Integer?NonNegative][0, j_Integer?Positive/; j < jmax] :=
   V[n][0, j] = 1; V[n_Integer?NonNegative][imax,
    j_Integer?Positive/; j < jmax] := V[n][imax, j] = 0;

  (* iteration equation *)
  V[n_Integer?Positive][
    i_Integer?Positive/; i < imax, j_Integer?Positive/; j < jmax] :=
   V[n][i, j] = ((1 - w) V[n - 1][i, j] + w/4 (V[n - 1][i + 1, j] +
        V[n][i - 1, j] + V[n - 1][i, j + 1] + V[n][i, j - 1]));

  (* final iteration number *)
  For[
   m = 1,
   Max[Table[Abs[V[m][i, j] - V[m - 1][i, j]],
      {i, 1, imax - 1}, {j, 1, jmax - 1}]] ≥ tol, m++,
```

```
  If[m > 40, Print["Iteration limit of 40 exceeded."]; Break[]]];
 Print["Final Iteration Number = " <> ToString[m]];

 (* graphics *)
 ListPlot3D[
  Table[V[m][i, j], {j, 0, jmax}, {i, 0, 26}],
  DataRange→{{0, 26Δ}, {0, jmax Δ}}, opts,
  BoxRatios→{1, 1, 1}, PlotRange→{0, 1},
  Boxed→False, AxesLabel→{"x(b)", "y(b)", "V(V0)"},
  ViewPoint→{1.741, -2.565, 1.356}]]
```

5.3 CHARGED PARTICLE IN CROSSED ELECTRIC AND MAGNETIC FIELDS

5.3.1 The Problem

For a charge moving in uniform electric and magnetic fields that are perpendicular to each other, solve the equation of motion, determine the nature of the motion, and plot the trajectories for several choices of field strengths and initial conditions. (For a comparison of solutions of this problem with MACSYMA, Maple, Mathcad, *Mathematica*, and Theorist, see [CDSTD92b].)

5.3.2 Physics of the Problem

On a particle of charge q and mass m moving in electric and magnetic fields $\mathbf{E}$ and $\mathbf{B}$, the Lorentz force is

$$\mathbf{F} = q\,(\mathbf{E} + \mathbf{v} \times \mathbf{B}) \tag{5.3.1}$$

where $\mathbf{v}$ is the particle velocity. Thus, the equation of motion can be written as

$$m\frac{d^2\mathbf{r}}{dt^2} = q\,\mathbf{E} + q\frac{dr}{dt} \times \mathbf{B} \tag{5.3.2}$$

in which $\mathbf{r}$ is the position vector of the particle. We orient the coordinate axes so that the uniform crossed fields are given by

$$\mathbf{E} = E_0\,\hat{\mathbf{y}} \quad \mathbf{B} = B_0\,\hat{\mathbf{z}} \tag{5.3.3}$$

and locate the origin so that the initial position of the particle is specified by

$$\mathbf{r}\,(0) = 0 \tag{5.3.4}$$

If $\mathbf{v}_0$ is the initial velocity,

$$\frac{d\,\mathbf{r}\,(0)}{dt} = \mathbf{v}_0 \tag{5.3.5}$$

5.3.3 Solution with *Mathematica*

```
In[1]:= ClearAll["Global`*", Subscript]
```

Mathematica represents vectors by lists. In Cartesian coordinates, the position vector **r** is defined as

```
In[2]:= r[t_]:={x[t], y[t], z[t]}
```

With **EF** and **BF** as, respectively, aliases of **E** and **B**, Equation 5.3.3 takes the form

```
In[3]:= EF:={0, Ε0, 0}
        BF:={0, 0, B0}
```

where the symbol **Ε** in $\mathbf{Ε_0}$ is a capital Greek letter rather than an ordinary keyboard letter. To solve Equation 5.3.2 subject to the initial conditions (Equations 5.3.4 and 5.3.5), first obtain a list of their Cartesian components:

```
In[5]:= Thread/@{m r''[t] == q EF + q r'[t] × BF,
           r[0] == {0, 0, 0}, r'[0] == {vx0, vy0, vz0}}//Flatten

Out[5]= {m x''[t] == q B0 y'[t], m y''[t] == q Ε0 - q B0 x'[t], m z''[t] == 0,
           x[0] == 0, y[0] == 0, z[0] == 0, x'[0] == vx0, y'[0] == vy0, z'[0] == vz0}
```

Then **DSolve** solves this system of equations:

```
In[6]:= sol = DSolve[%, {x[t], y[t], z[t]}, t]//.{B0 -> m ω/q, Ε0 -> vd B0}//
           Simplify//ExpandAll
```

$$\text{Out[6]:=}\left\{\left\{x[t]\to t\,v_d-\frac{\text{Sin}[t\,\omega]\,v_d}{\omega}+\frac{\text{Sin}[t\,\omega]\,v_{x0}}{\omega}+\frac{v_{y0}}{\omega}-\frac{\text{Cos}[t\,\omega]\,v_{y0}}{\omega},\right.\right.$$
$$\left.\left.y[t]\to\frac{v_d}{\omega}-\frac{\text{Cos}[t\,\omega]\,v_d}{\omega}-\frac{v_{x0}}{\omega}+\frac{\text{Cos}[t\,\omega]\,v_{x0}}{\omega}+\frac{\text{Sin}[t\,\omega]\,v_{y0}}{\omega},\ z[t]\to t\,v_{z0}\right\}\right\}$$

where we have introduced the definitions

$$\omega = \frac{q}{m}B_0 \tag{5.3.6}$$

$$v_d = \frac{E_0}{B_0} \tag{5.3.7}$$

To determine the nature of the particle's motion, convert the rules returned by **DSolve** into equations:

```
In[7]:= eqn = sol[[1]]/. Rule → Equal
```

$$Out[7]=\ \left\{x[t]==t\,v_d-\frac{\text{Sin}[t\,\omega]\,v_d}{\omega}+\frac{\text{Sin}[t\,\omega]\,v_{x0}}{\omega}+\frac{v_{y0}}{\omega}-\frac{\text{Cos}[t\,\omega]\,v_{y0}}{\omega},\ y[t]==\frac{v_d}{\omega}-\frac{\text{Cos}[t\,\omega]\,v_d}{\omega}-\frac{v_{x0}}{\omega}+\frac{\text{Cos}[t\,\omega]\,v_{x0}}{\omega}+\frac{\text{Sin}[t\,\omega]\,v_{y0}}{\omega},\ z[t]==t\,v_{z0}\right\}$$

Subtract the term **(t v_d + v_{y0}/ω)** from both sides of the first equation and then square both sides of the resultant equation:

```
In[8]:= Map[((#- (t vd + vy0/ω))^2)&, eqn[[1]]]
```

$$Out[8]=\ \left(-t\,v_d-\frac{v_{y0}}{\omega}+x[t]\right)^2==\left(-\frac{\text{Sin}[t\,\omega]\,v_d}{\omega}+\frac{\text{Sin}[t\,\omega]\,v_{x0}}{\omega}-\frac{\text{Cos}[t\,\omega]\,v_{y0}}{\omega}\right)^2$$

Repeat with the second equation but this time subtract the term **(v_d/ω − v_{x0}/ω)** from both sides of the equation:

```
In[9]:= Map[((#- (vd/ω - vx0/ω))^2)&, eqn[[2]]]
```

$$Out[9]=\ \left(-\frac{v_d}{\omega}+\frac{v_{x0}}{\omega}+y[t]\right)^2==\left(-\frac{\text{Cos}[t\,\omega]\,v_d}{\omega}+\frac{\text{Cos}[t\,\omega]\,v_{x0}}{\omega}+\frac{\text{Sin}[t\,\omega]\,v_{y0}}{\omega}\right)^2$$

Add the two equations:

```
In[10]:= Thread[Plus[%%, %], Equal]
```

$$Out[10]=\ \left(-t\,v_d-\frac{v_{y0}}{\omega}+x[t]\right)^2+\left(-\frac{v_d}{\omega}+\frac{v_{x0}}{\omega}+y[t]\right)^2==\left(-\frac{\text{Sin}[t\,\omega]\,v_d}{\omega}+\frac{\text{Sin}[t\,\omega]\,v_{x0}}{\omega}-\frac{\text{Cos}[t\,\omega]\,v_{y0}}{\omega}\right)^2+\left(-\frac{\text{Cos}[t\,\omega]\,v_d}{\omega}+\frac{\text{Cos}[t\,\omega]\,v_{x0}}{\omega}+\frac{\text{Sin}[t\,\omega]\,v_{y0}}{\omega}\right)^2$$

Finally, simplify the *rhs* of the equation:

```
In[11]:= MapAt[Simplify,%,2]
```

$$Out[11]=\ \left(-t\,v_d-\frac{v_{y0}}{\omega}+x[t]\right)^2+\left(-\frac{v_d}{\omega}+\frac{v_{x0}}{\omega}+y[t]\right)^2==\frac{v_d^2-2v_dv_{x0}+v_{x0}^2+v_{y0}^2}{\omega^2}$$

This is the equation of a circle on the *xy* plane, with radius

$$R=\left(\frac{v_d^2-2v_dv_{x0}+v_{x0}^2+v_{y0}^2}{\omega^2}\right)^{1/2} \tag{5.3.8}$$

and center at coordinates

$$x_c = \frac{v_{y0}}{\omega} + v_d t \tag{5.3.9}$$

$$y_c = \frac{v_d}{\omega} - \frac{v_{x0}}{\omega} \tag{5.3.10}$$

The motion along the z direction is described by

```
In[12]:= eqn[[3]]
Out[12]= z[t] == t vz0
```

Hence, the general motion of the particle consists of three components: (a) circular motion perpendicular to **B** with angular frequency ω, called the cyclotron frequency; (b) uniform translational motion parallel to **B**; and (c) constant drifting motion in the direction of $\mathbf{E} \times \mathbf{B}$ with velocity v_d, called the drift velocity.

To plot the trajectories of the particle for several choices of numerical values of field strengths and initial conditions, we first obtain the Cartesian coordinates of the particle as a function of time from the solution returned by `DSolve`:

```
In[13]:= {x[t_], y[t_], z[t_]} = {x[t], y[t], z[t]}/.sol[[1]];
```

For a given charge, specifying ω and v_d is equivalent to specifying E_0 and B_0. Let us express time in units of $1/\omega$ and velocity in units of v_d—that is, $\omega = 1$ and $v_d = 1$. We now plot the paths for $v_{z0} = 0.75$, $v_{y0} = 0$, and $v_{x0} = -1, 0, \ldots, 4$:

```
In[14]:= g[i_] :=
          (vx0 = i;
           Show[
            ParametricPlot3D[{x[t], y[t], 0}, {t, 0, 22}],
            ParametricPlot3D[r[t], {t, 0, 22}],
            Graphics3D[
             {Thickness[0.008],
              Line[{{0, 0, 0}, {0, 0, 5}}],
              Line[{{0, 0, 0}, {0, 5, 0}}],
              Line[{{0, 0, 0}, {22, 0, 0}}],
              Text[
               Style["E", FontFamily→ "Times", Bold, 12], {1.3, 5.5, 0}],
              Text[Style["B", FontFamily→ "Times", Bold, 12], {0, 0, 7}],
              Text[Style["vx0 = " <> ToString[vx0], FontFamily→ "Times", 12],
               Scaled[{0.4, 0.9, 0.85}]]}],
            Axes→False, PlotRange→ {{-2, 24}, {-7, 7}, {0, 22}},
            ImageSize -> 140])

In[15]:= {ω, vd, vz0, vy0} = {1, 1, 0.75, 0};
         Print["\nω = ", ω, " vd = ",
```

```
        vd, " vz0 = ", vz0, " vy0, = ", vy0, "\n"];
    Grid[Partition[Table[g[i], {i, -1, 4}], 3], Spacings -> {1.5, 2.0}]
```

$\omega = 1 \quad v_d = 1 \quad v_{z0} = 0.75 \quad v_{y0} = 0$

Out[17]=

$v_{x0} = -1$ $v_{x0} = 0$ $v_{x0} = 1$

$v_{x0} = 2$ $v_{x0} = 3$ $v_{x0} = 4$

The trajectories of the particle with $v_{y0} = 0$ are shown together with their projections on the xy plane. If $v_{y0} \neq 0$, the paths are essentially the same except that the x_c coordinate of the center at $t = 0$ is shifted and the radius R is enlarged in accordance with Equations 5.3.8 and 5.3.9. (**Grid**[{{$expr_{11}$, $expr_{12}$, ...}, {$expr_{21}$, $expr_{22}$, ...}, ...}] is an object that formats with the $expr_{ij}$ arranged in a two-dimensional grid. **Spacings** -> {$spec_x$, $spec_y$} specifies the horizontal and vertical spacings between successive elements.)

Let us generate an animation of a representative motion of the particle:

```
In[18]:= {ω, vd, vx0, vy0, vz0} = {1, 1, 4, 0, 0.75};
         ListAnimate[
          Table[
           Show[
            ParametricPlot3D[r[t], {t, 0, i + 0.0001}],
            Graphics3D[
             {{Thickness[0.008],
               Line[{{0, 0, 0}, {0, 0, 5}}],
               Line[{{0, 0, 0}, {0, 5, 0}}],
               Line[{{0, 0, 0}, {24, 0, 0}}]},
              Text[
               Style["E", FontFamily→"Times", Bold, 10], {1.3, 5.5, 0}],
```

```
    Text[Style["B", FontFamily→"Times", Bold, 10], {0, 0, 7}],
    {Dashing[{0.01, 0.01}], Line[{r[i], {x[i], y[i], 0}}]},
    {PointSize[0.03], Point[r[i]], GrayLevel[0.55],
     Point[{x[i], y[i], 0}]}}], Axes→False,
  PlotRange→{{-2, 25}, {-7, 7}, {0, 22}}], {i, 0, 25}]]
```

Out[19]=

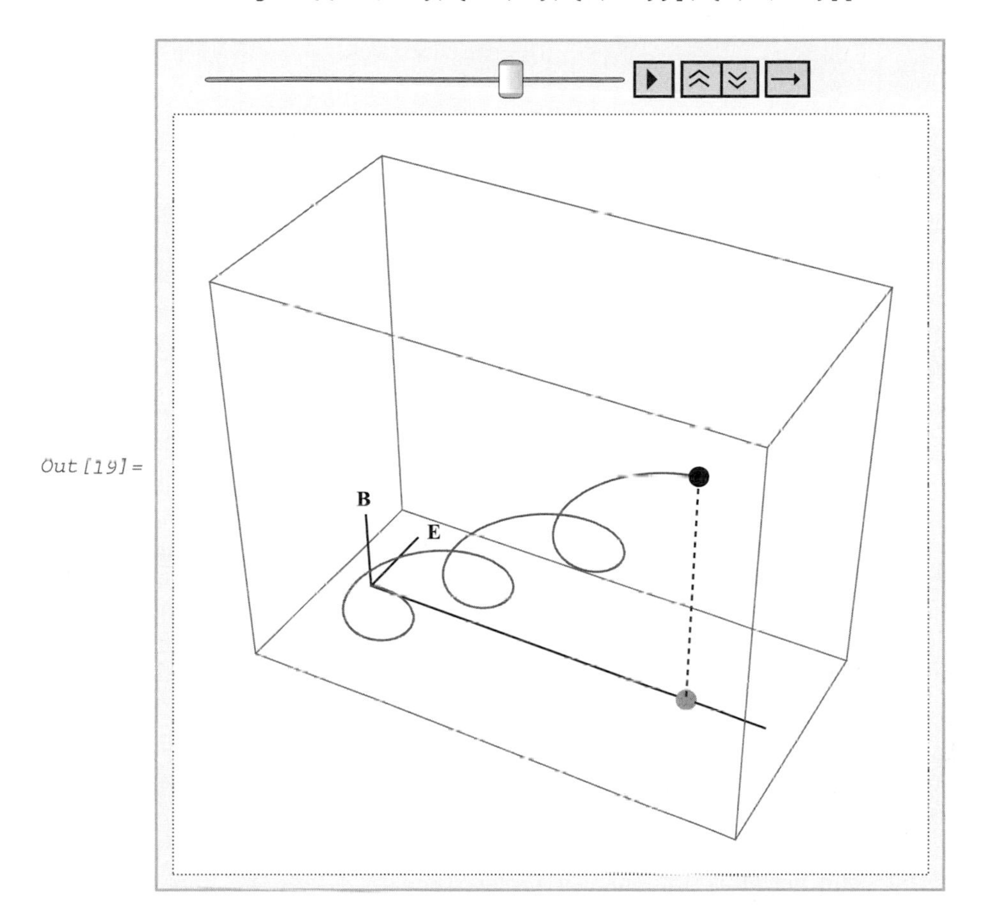

The vertical dashed line connects the particle with its projection on the *xy* plane. To animate, click the Play button—that is, the button with the right pointer.

```
In[20]:= ClearAll[r, EF, BF, sol, eqn, x, y, z, ω, g, Subscript]
```

5.4 PROBLEMS

In this book, straightforward, intermediate-level, and challenging problems are unmarked, marked with one asterisk, and marked with two asterisks, respectively. For solutions to selected problems, see Appendix D.

1. Determine the electric field for points on the axis of a uniformly charged disk of radius a and charge density b. Verify that at large distances from the disk, the electric field approaches that produced by a point charge.
2. Consider a distribution of collinear point charges $q_1, q_2, q_3, \ldots, q_N$, at positions $x_1, x_2, x_3, \ldots, x_N$, respectively. Show that the differential equation for the electric field lines on the xy plane is

$$\frac{dy}{dx} = \frac{E_y}{E_x}$$

Verify that the solution to this equation is

$$\sum_{i=1}^{N} \frac{q_i(x - x_i)}{\sqrt{(x - x_i)^2 + y^2}} = constant$$

With **ContourPlot**, generate the electric field lines when there are (a) 2 equal charges; (b) 2 equal but opposite charges; (c) 2 charges, $+2q$ and $-q$; and (d) 21 equal and evenly spaced charges. What can be inferred from the plot in (d)?
3. In Section 5.1.3, can the keyboard letters **x** and **y** rather than the script letters x and y be the subscripts of the Greek letter **E**? Explain.
*4. Three equal point charges q are at the corners of an equilateral triangle of sides b. For the plane of the triangle, locate the points of equilibrium where the electric field vanishes. *Hints:* Generalize the equations in Section 5.1.2.3. Also, `FindRoot[{eqn1, eqn2, ...}, {{x, x0}, {y, y0}, ...}]` searches for a numerical solution to the simultaneous equations eqn_i. To determine the starting points, use **Find Selected Function** in the Help menu to obtain information on the function **ContourPlot**.
**5. Plot the electric field lines and equipotentials for (a) two equal but opposite point charges and (b) three equal positive point charges at the corners of an equilateral triangle.
*6. For a distribution of two point charges, $+2q$ and $-q$, plot the electric field lines and equipotential that pass through the point of equilibrium.
*7. Modify the function **laplace2D** in Section 5.2.3.2 to omit plotting of the electric potential. The new function, named **newLaplace2D**, prints the final iteration number and the table for comparing values of the electric potential from the analytical and numerical solutions. The function **newLaplace2D** takes only three arguments because the fourth argument for the graphics options is no longer necessary. Evaluate `newLaplace2D[jmax, w, tol]` for

$$jmax = 20, 40$$
$$w = 1.2, 1.3, 1.4, 1.5, 1.6, 1.7$$
$$tol = 0.00025, 0.0025.$$

Does the optimal value for w depend on $jmax$ and tol? How do the rate of convergence and the error in the numerical solution depend on $jmax$, w, and tol? Note that $jmax$ equals $1/\Delta$, where Δ is the grid spacing, and that x_f equals 5/2. (The legitimate values $w = 1.1, 1.8$, and 1.9 have been dropped to make this problem slightly less time-consuming.)

*8. Replace the initial guess for the electric potential of the interior grid points in the function **newLaplace2D** of Problem 7 with

$$V^0_{i,j} = \frac{4}{\pi}\sum_{m=1}^{30}\frac{1}{(2m-1)}\sin[(2m-1)\pi y]e^{-(2m-1)\pi x}$$

Call the new function **newLaplace2Dg1**. Repeat with the initial guess

$$V^0_{i,j} = f(y)e^{-\pi x}$$

where

$$f(y) = \begin{cases} +2y & 0 \le y \le 1/2 \\ -2(y-1) & 1/2 < y \le 1 \end{cases}$$

Call this function **newlaplace2Dg2**. Evaluate **newLaplace2D**, **newLaplace2Dg1**, and **newlaplace2Dg2** for $jmax = 20$, $tol = 0.00025$, and $w = 1.2, 1.3, 1.4, 1.5, 1.6, 1.7$. What can be concluded about the dependence of the optimal value for w, the rate of convergence, and the error in the numerical solution on the initial guess?

*9. Modify the function **newLaplace2D** in Problem 7 to include a fourth argument for the boundary limit x_f. Name this function **xfnewLaplace**. Evaluate this function for $jmax = 20$, $tol = 0.00025$, $w = 1.6$, and $x_f = 5/2, 3$, and $7/2$. Are the rate of convergence and the error in the numerical solution sensitive to the choice for x_f?

**10. The figure shows two sets of parallel conducting plates of finite width but infinite length along the z direction perpendicular to the page. The plates separated by a distance b are grounded, and those at a distance a apart are maintained at a potential V_0. Consider the case $a = 2b$. For the points enclosed by the plates, solve analytically and numerically Laplace's equation and plot the electric potential. *Hint:* For the analytical solution, see [LC70] or [Gri99].

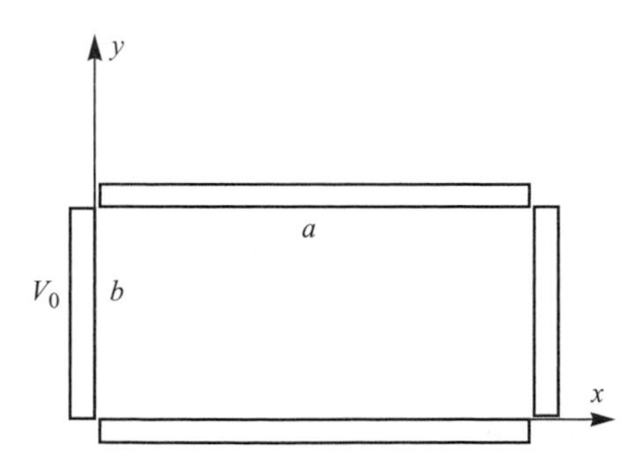

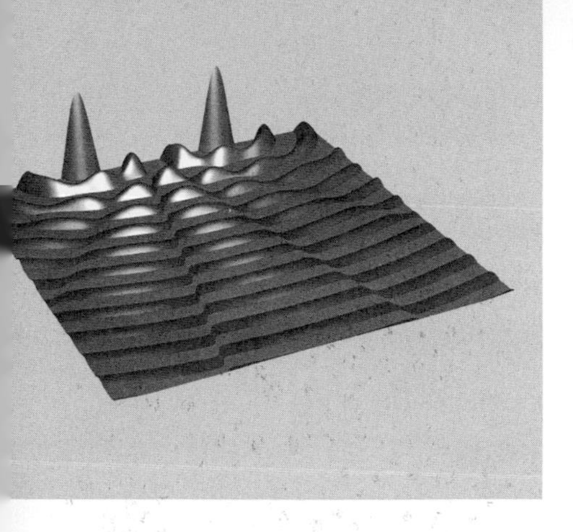

Chapter 6
Quantum Physics

6.1 BLACKBODY RADIATION

6.1.1 The Problem

The Planck formula for blackbody radiation is

$$u(\nu, T) = \frac{8\pi h}{c^3} \frac{\nu^3}{e^{h\nu/kT} - 1} \tag{6.1.1}$$

where $u(\nu, T)$ is the energy density as a function of frequency ν and temperature T, k is Boltzmann's constant, c is the speed of light, and h is Planck's constant. Obtain an expression for the energy density as a function of wavelength λ and temperature T. Also, plot this energy density for several temperatures. Then, derive Wien's displacement law relating T and λ_{max}, the wavelength for which the energy density is maximal. Finally, use 5700 K as an estimate of the Sun's surface temperature to determine λ_{max} for solar radiation.

6.1.2 Physics of the Problem

The frequency ν, the wavelength λ, and the speed of light c are related by

$$\nu = \frac{c}{\lambda} \tag{6.1.2}$$

Thus,

$$d\nu = -\frac{c}{\lambda^2} d\lambda \tag{6.1.3}$$

The energy density as a function of wavelength λ, $u(\lambda, T)$, is given in terms of the energy density as a function of frequency ν, $u(\nu, T)$, by

$$u(\lambda, T)d\lambda = -u(\nu, T)d\nu \tag{6.1.4}$$

(For an elaboration of this equation, see [Eis61].) With Equation 6.1.3, Equation 6.1.4 takes the form

$$u(\lambda, T) = u(\nu, T)\frac{c}{\lambda^2} \tag{6.1.5}$$

From Equations 6.1.1, 6.1.2, and 6.1.5, we have

$$u(\lambda, T) = \frac{8\pi hc}{\lambda^5}\frac{1}{e^{hc/\lambda kT} - 1} \tag{6.1.6}$$

Wien's displacement law can be written as

$$\lambda_{\max} = \frac{0.201405\, hc}{k}\frac{1}{T} \tag{6.1.7}$$

Therefore, $\lambda_{\max}$ is displaced toward shorter wavelengths as the temperature increases. (For further discussions of blackbody radiation, see [ER85], [Gas96], and [Eis61].)

6.1.3 Solution with *Mathematica*

```
In[1]:= ClearAll["Global`*"]
```

6.1.3.1 $u(\lambda, T)$ at Several Temperatures

Equation 6.1.6 for the energy density can be entered as

```
In[2]:= u[λ_, T_] := (8 π h c)/λ^5 1/(ⅇ^((h c)/(λ k T)) - 1)
```

To plot $u(\lambda, T)$ at several temperatures, we need the values for c, h and k, which can be obtained from the package **PhysicalConstants`**. Let us load the package:

```
In[3]:= Needs["PhysicalConstants`"]
```

In plotting graphs, we do not want to carry the units along. To discard the units of c, h, and k and to express u in units of 10^6 joule/m^3-m and λ in units of 10^4 Å (i.e., 10^{-6} m), define a function ρ:

```
In[4]:= ρ[λ_, T_] := u[λ 10^-6, T]/10^6 /. {c → SpeedOfLight (Second/Meter),
            h → PlanckConstant/(Joule Second),
            k → BoltzmannConstant (Kelvin/Joule)}
```

Let us plot the energy density at four selected temperatures:

```
In[5]:= style1[label_] := Thread[Style[label, FontFamily -> "Times"]]
In[6]:= Plot[
         Evaluate[Table[ρ[λ, i 1000], {i, 4, 7}]],
         {λ, 0.1, 1.5}, PlotRange → {0, 3.15}, Frame → True,
         PlotLabel -> style1["Energy Density vs. Wavelength"],
         FrameTicks → {{0, 0.5, 1.0, 1.5}, {0, 1, 2, 3}, None, None},
         FrameLabel -> style1[{"λ(10^4 Å)", "u(10^6 joule/m^3 - m)"}]]
```

Out[6]=

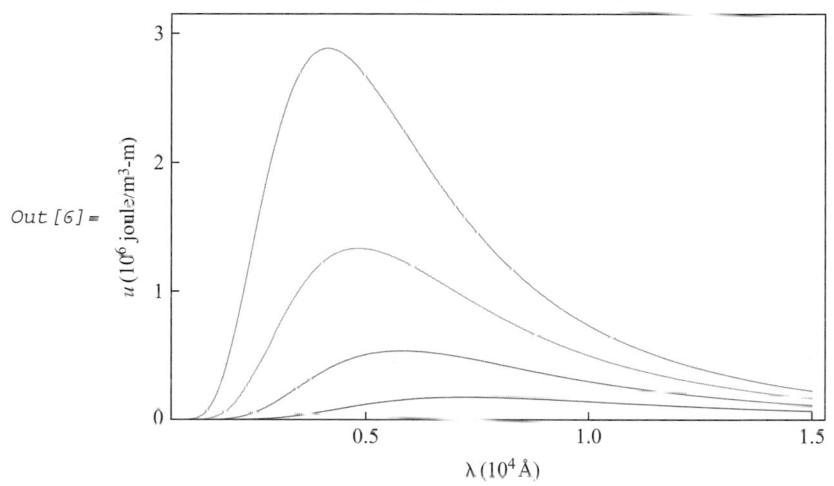

The curve at the top is for $T = 7000$ K, the next one down is for $T = 6000$ K, the one below that is for $T = 5000$ K, and the bottom curve is for $T = 4000$ K. Consistent with Wien's displacement law, $\lambda_{\max}$ decreases as T increases.

6.1.3.2 Wien's Displacement Law

To determine the wavelength $\lambda_{\max}$ at which $u(\lambda, T)$ is a maximum for a given T, we set the derivative of $u(\lambda, T)$ with respect to λ to zero and solve for λ.

```
In[7]:= dudλ = ∂λ u[λ, T] // Together
```

$$\text{Out[7]=} -\frac{8\,c\,h\,\pi\left(-c\,e^{\frac{c\,h}{k\,T\,\lambda}}\,h - 5\,k\,T\,\lambda + 5\,e^{\frac{c\,h}{k\,T\,\lambda}}\,k\,T\,\lambda\right)}{\left(-1 + e^{\frac{c\,h}{k\,T\,\lambda}}\right)^2 k\,T\,\lambda^7}$$

This derivative is zero if its numerator is zero unless, of course, the denominator is also zero. The function **Numerator** picks out the numerator:

```
In[8]:= Numerator[dudλ]
```

```
Out[8]= -8 c h π (-c e^(c h/(k T λ)) h - 5 k T λ + 5 e^(c h/(k T λ)) k T λ)
```

Let us make a change of variable from λ to x such that

$$\lambda = \frac{ch/kx}{T} \tag{6.1.8}$$

```
In[9]:= %/. λ -> ((c h)/(k x))/T // Simplify

Out[9]= (8 c^2 h^2 π (5 + e^x (-5 + x)))/x
```

In terms of x, the condition $du(\lambda, T)/d\lambda = 0$ becomes

$$5 + e^x(-5 + x) = 0 \tag{6.1.9}$$

Dividing Equation 6.1.9 by e^x, we have the equation for x:

```
In[10]:= %[[5]]/e^x == 0 // Simplify

Out[10]= 5 e^-x + x == 5
```

Observe that the solution is $x \approx 5$ since $5e^{-x} \ll 5$ for $x = 5$. **FindRoot** yields a more accurate solution:

```
In[11]:= FindRoot[%, {x, 5}]

Out[11]= {x → 4.96511}
```

(We can also determine graphically the starting value for **FindRoot** with the procedure outlined in Section 2.3.1.4.) Wien's displacement law follows immediately from Equation 6.1.8 together with the value of x just calculated:

```
In[12]:= λmax = ((c h)/(k x))/T /. %

Out[12]= (0.201405 c h)/(k T)
```

What remains is the verification of our assumption that the denominator of the derivative $du(\lambda, T)/d\lambda$ is not zero at $\lambda = \lambda_{\text{max}}$:

```
In[13]:= Denominator[dudλ] /. λ → λmax

Out[13]= (0.272306 c^7 h^7)/(k^6 T^6)
```

It is indeed not zero.

6.1.3.3 λ_{max} for Solar Radiation

With 5700 K as the estimated surface temperature of the Sun, Wien's displacement law gives λ_{max} for solar radiation:

```
In[14]:= λmax /. {T → 5700 Kelvin, c → SpeedOfLight,
             h → PlanckConstant, k → BoltzmannConstant}
Out[14]= 5.0838 × 10^-7 Meter
```

Measured in angstroms, λ_{max} for solar radiation is

```
In[15]:= % 10^10 Å / Meter
Out[15]= 5083.8 Å
```

FindMaximum provides another method for determining λ_{max} (see Example 2.3.11):

1. Evaluate

```
In[16]:= ClickPane[
          Plot[
           Evaluate[ρ[λ, 5700]],
           {λ, 0.1, 1.5}, Frame → True, PlotRange → {0, 1.1},
           PlotLabel -> style1["Energy Density vs. Wavelength"],
           FrameTicks → {{0, 0.5, 1.0, 1.5}, {0, 0.5, 1}, None, None},
           FrameLabel -> style1[{"λ(10^4 Å)", "u(10^6 joule/m^3 - m)"}]],
          (λρcoord = #) &]
```

Out[16]=

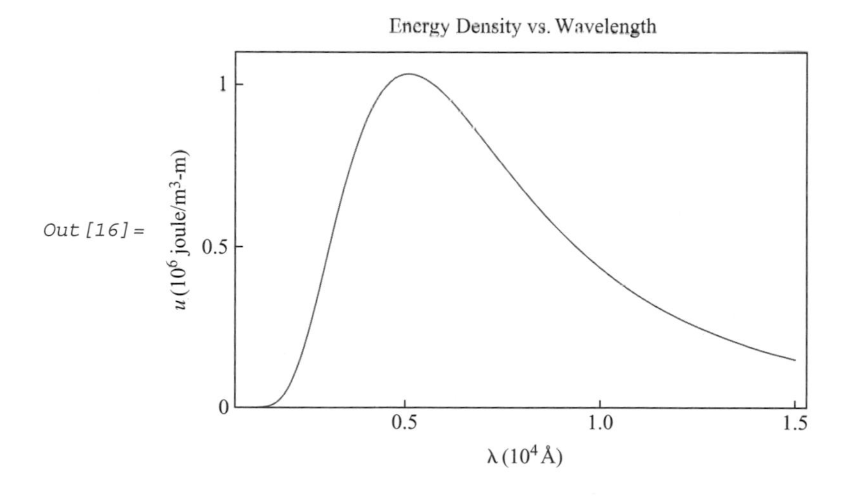

2. With the mouse pointer, click the point of the maximum of u.
3. Evaluate

```
In[17]:= FindMaximum[ρ[λ, 5700], {λ, λρcoord[[1]]}]
Out[17]= {1.03298, {λ→0.50838}}
```

We have found $\lambda_{max} = 0.50838\,10^4$ Å (i.e., 5083.8 Å), which is the same as that determined previously with Wien's displacement law.

```
In[18]:= ClearAll[u, ρ, dudλ, style1, λρcoord, Subscript]
```

6.2 WAVE PACKETS

6.2.1 The Problem

Demonstrate graphically the spreading of the one-dimensional, free-particle, Gaussian wave packet.

6.2.2 Physics of the Problem

The one-dimensional, free-particle, Gaussian wave packet can be written as

$$\Psi(x,t) = \left[\frac{1}{2\pi L^2\left(1 + i\frac{\hbar}{2mL^2}t\right)^2}\right]^{1/4} e^{i(k_0 x - \omega_0 t)} \exp\left[-\frac{1}{4L^2}\frac{(x - v_{gr}t)^2}{\left(1 + i\frac{\hbar}{2mL^2}t\right)}\right]$$

where m is the mass of the particle, and L is the position uncertainty at time $t = 0$. Also,

$$v_{gr} = \frac{\hbar k_0}{m}$$

$$\omega_0 = \frac{\hbar k_0^2}{2m}$$

and k_0 is the average wave number.

The position probability density, $\Psi^*(x,t)\Psi(x,t)$, can be expressed as

$$P(x,t) = \left[\frac{1}{2\pi L^2\left(1 + \frac{\hbar^2}{4m^2L^4}t^2\right)}\right]^{1/2} \exp\left[-\frac{1}{2L^2}\frac{(x - v_{gr}t)^2}{\left(1 + \frac{\hbar^2}{4m^2L^4}t^2\right)}\right] \tag{6.2.1}$$

Let x, t, and $P(x,t)$ be measured, respectively, in units of L, T, and A, where

$$T = \frac{2mL^2}{\hbar}$$

$$A = \frac{1}{\sqrt{2\pi L^2}}$$

Equation 6.2.1 becomes

$$P(x,t) = \left[\frac{1}{1+t^2}\right]^{1/2} \exp\left[-\frac{(x-vt)^2}{2\,(1+t^2)}\right] \tag{6.2.2}$$

in which

$$v = \frac{2mL}{\hbar} v_{\text{gr}}$$

In units of L, the position uncertainty or the width of the wave packet at time t is

$$\Delta x(t) = \sqrt{1+t^2} \tag{6.2.3}$$

(For a discussion that "plumbs the depths of a Gaussian wave packet," see [Mor90].)

6.2.3 Solution with *Mathematica*

```
In[1]:= ClearAll["Global`*"]
```

There are several ways to demonstrate the spreading of the wave packet. The simplest is to plot $\Delta x(t)$ given in Equation 6.2.3:

```
In[2]:= style1[label_] := Thread[Style[label, FontFamily -> "Times"]]
```

```
In[3]:= Plot[{√(1+t²), t}, {t, 0, 6},
          PlotRange → {0, 6}, PlotStyle → {{}, {Dashing[{0.02, 0.02}]}},
          AxesLabel → style1[{"t(T)", "Δx(L)"}]]
```

Out[3]=

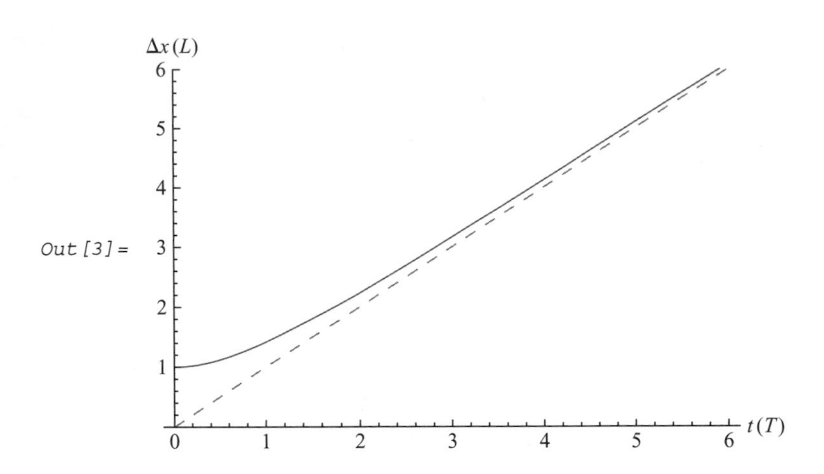

A dashed straight line is added to the graph to show that $\Delta x(t) \approx t$ for $t \gg 1$. The function **style1** defines a style for the axes labels.

Another way to show the spreading of the wave packet is to plot $P(x, t)$ in Equation 6.2.2 for several values of t. In what follows, we set $v = 1$.

In[4]:= **P[x_, t_] :=** $\frac{1}{\sqrt{1+t^2}}$ **Exp**$\left[-\frac{(x-t)^2}{2(1+t^2)}\right]$

```
In[5]:= style2[label_] := Thread[Style[label, FontFamily -> "Times", Italic]]
```

```
In[6]:= Plot[
  Evaluate[Table[P[x, i], {i, 0, 3}]], {x, -4, 12},
  PlotStyle→{{Thickness[0.01], Orange}, {Red},
    {Dashing[{0.05, 0.05}], Blue}, {Dashing[{0.01, 0.015}]}},
  PlotRange→{0, 1.05}, AxesLabel→style2[{"x(L)", "P(x, t)(A)"}],
  BaseStyle→{FontFamily->"Times"},
  Epilog→{Text["Spreading of a\nGaussian Wave Packet", {8.5, 0.8}],
    Text[HoldForm[t = 0], {2, 0.9}], Text[HoldForm[t = T], {3.5, 0.6}],
    Text[HoldForm[t = 2 T], {5.2, 0.4}],
    Text[HoldForm[t = 3 T], {8, 0.24}]}]
```

Out[6]=

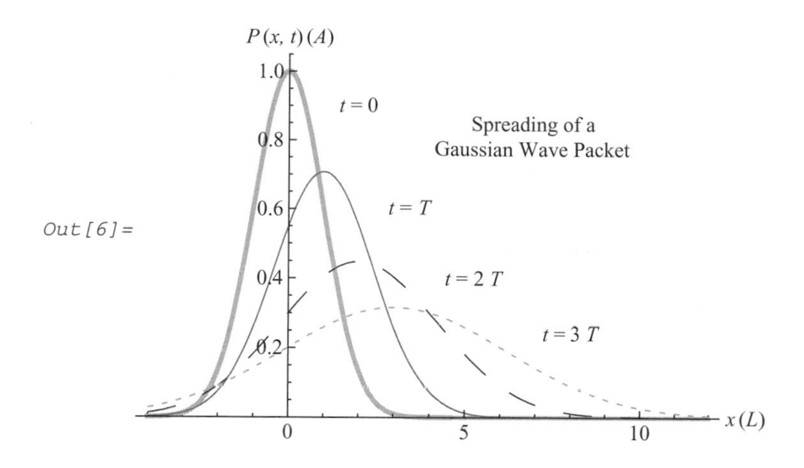

There is another way to plot $P(x, t)$ for several values of t:

```
In[7]:= g[n_] :=
          ParametricPlot3D[{x, n, P[x, n]}, {x, -5.6, 21}, PlotPoints→100]

In[8]:= Show[
         Table[
          {g[i], Graphics3D[{Dashing[{0.01, 0.01}],
             Line[{{-5.6, i, 0}, {21, i, 0}}]}]}, {i, 0, 7}],
         PlotRange→{{-5.6, 21}, {0, 7}, {0, 1}}, ViewPoint→
          {2.230, 2.323, 1.040}, BoxRatios→{1, 2, 0.5}, Boxed→False,
         AxesLabel→style2[{"      x (L)", "t (T) ", "P(x, t) (A)"}],
         Ticks→{Automatic, Automatic, {0, 1}}]
```

Out[8]=

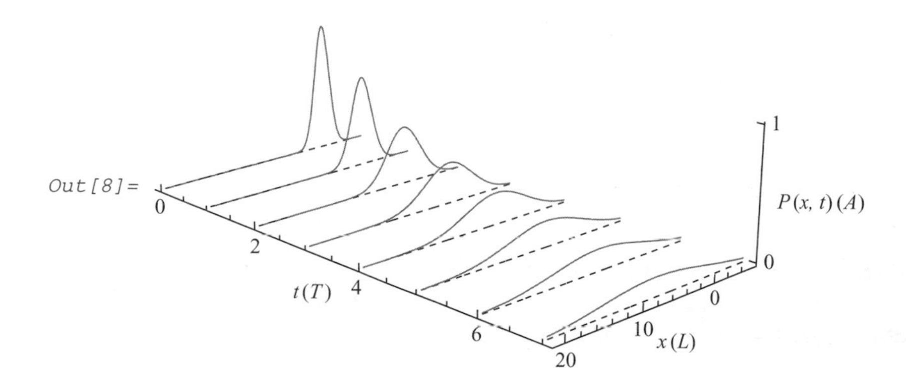

Here, the added dimension is time. Let us observe the spreading of the wave packet from another viewpoint:

```
In[9]:= Show[%, BoxRatios→{1, 1, 1},
         Boxed→True, ViewPoint→{0.017, 3.375, 0.245}]
```

Out[9]=

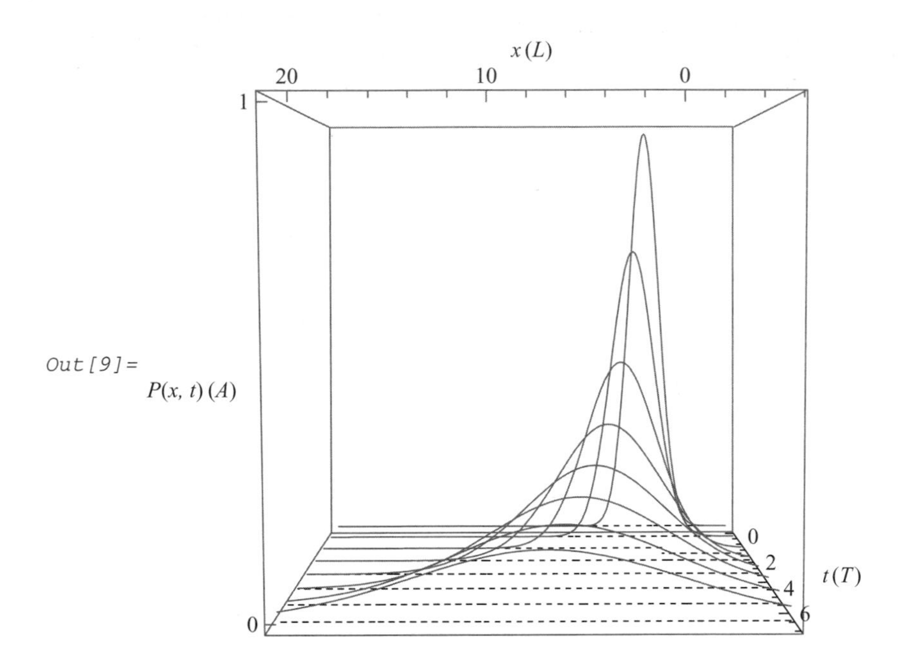

We can generate an animation of the spreading of the wave packet:

```
In[10]:= ListAnimate[
          Table[
           Show[
            {g[i],
             Graphics3D[
              {Dashing[{0.01, 0.01}], Line[{{-5.6, i, 0}, {21, i, 0}}]}]},
            PlotRange→{{-5.6, 21}, {0, 7}, {0, 1}}, BoxRatios→{1, 1, 1},
            ViewPoint→{0.017, 3.375, 0.245},
            AxesLabel→style2[{"x (L)", "t (T)", "P(x, t) (A)"}],
            Ticks→None], {i, 0, 7, 1/8}]]
```

Out[10]=

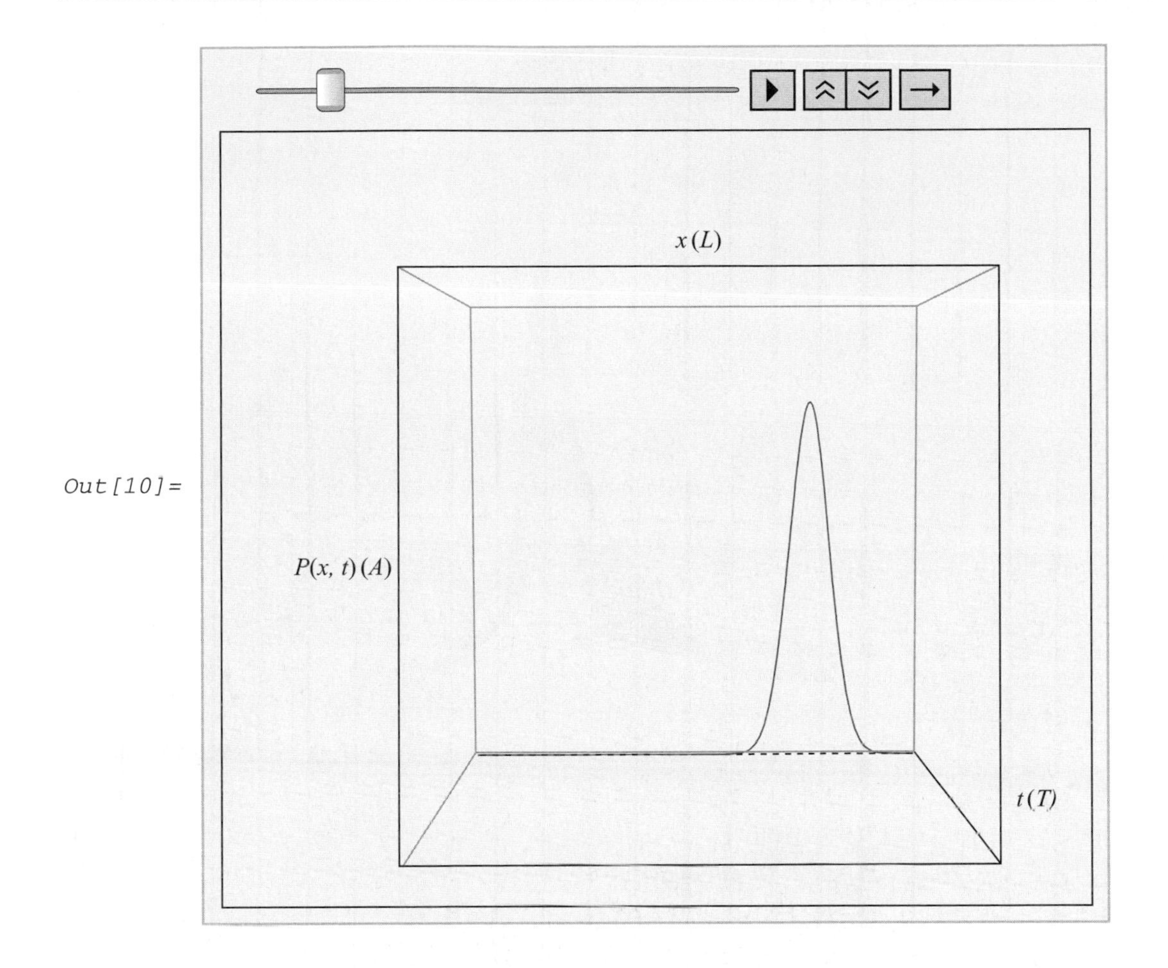

To animate, click the Play button.

Finally, we generate an animation of the spreading of the wave packet that leaves a trace:

```
In[11]:= ListAnimate[
          Table[
           Plot3D[P[x, t], {x, -5, 14}, {t, 0, i / 8 + 0.05},
            PlotRange → {{-5, 14}, {0, 5}, {0, 1}},
            ViewPoint → {0.017, 3.375, 0.245}, BoxRatios → {1, 1, 1},
            PlotPoints → 45, Ticks → None, AxesLabel →
             style2[{"x  (L)", "t  (T)", "P(x, t)  (A)"}]], {i, 0, 40}]]
```

Out[11]=

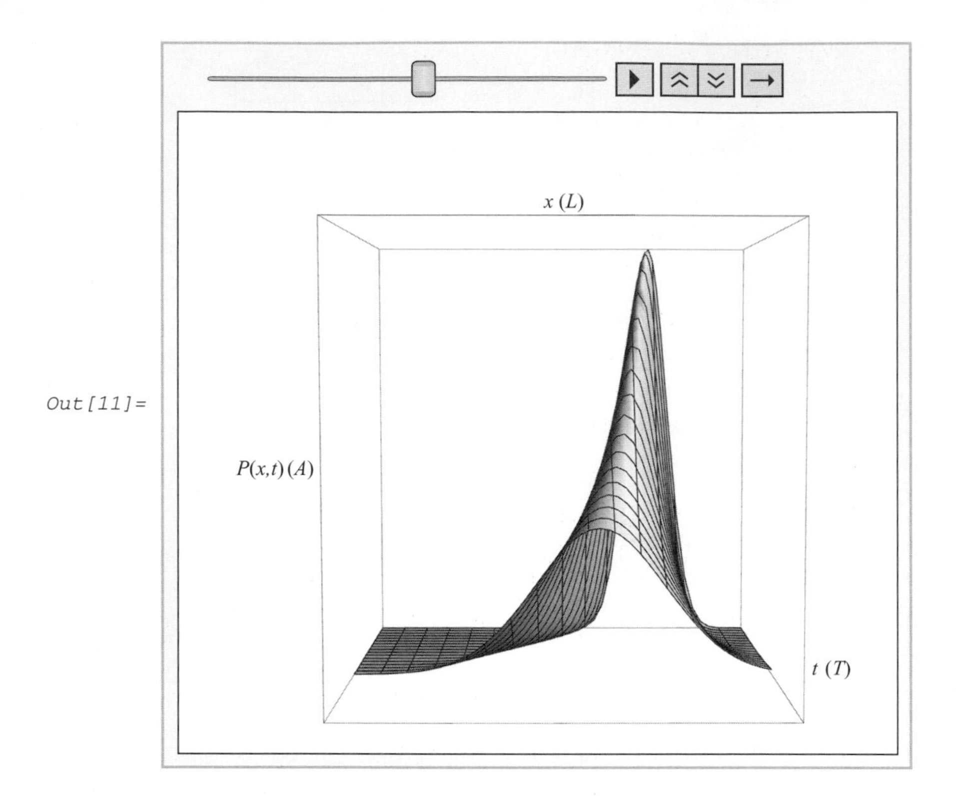

To animate, click the Play button.

```
In[12]:= ClearAll[P,g]
```

6.3 PARTICLE IN A ONE-DIMENSIONAL BOX

6.3.1 The Problem

Animate the time evolution of a wave packet in an infinite one-dimensional potential well.

6.3.2 Physics of the Problem

Consider a particle of mass m moving in a one-dimensional potential

$$V(x) = \begin{cases} 0 & |x| < a \\ \infty & \text{elsewhere} \end{cases} \tag{6.3.1}$$

In a system of units with $m = 1, a = 1$, and $\hbar = 1$, the solutions to the time-independent Schrödinger equation are

$$\phi_n^{(-)}(x) = \sin n\pi x \tag{6.3.2}$$

$$\phi_n^{(+)}(x) = \cos\left[n - \frac{1}{2}\right]\pi x \tag{6.3.3}$$

and the corresponding energy eigenvalues are

$$E_n^{(-)} = n^2 \pi^2 / 2 \tag{6.3.4}$$

$$E_n^{(+)} = \left[n - \frac{1}{2}\right]^2 \pi^2 / 2 \tag{6.3.5}$$

for all positive integers n.

The wave function $\Psi(x,t)$ can be expressed as an infinite superposition

$$\Psi(x,t) = \sum_{n=1}^{\infty} \left[c_n^{(+)} \phi_n^{(+)}(x) e^{-iE_n^{(+)}t} + c_n^{(-)} \phi_n^{(-)}(x) e^{-iE_n^{(-)}t}\right] \tag{6.3.6}$$

with

$$c_n^{(\pm)} = \int_{-1}^{1} \phi_n^{(\pm)*}(x) \Psi(x,0) dx \tag{6.3.7}$$

Let us consider the case in which $\Psi(x,0)$ is a Gaussian function—that is,

$$\Psi(x,0) = \left(\frac{1}{2\pi L^2}\right)^{1/4} e^{ik_0 x} e^{-x^2/4L^2} \tag{6.3.8}$$

where the width L and the average wave number k_0 of the wave packet are in units of a and $1/a$, respectively. The Gaussian function cannot actually be an initial wave function since it does not satisfy the boundary conditions that all wave functions must vanish at $x = \pm 1$. In what follows, we compute the coefficients $c_n^{(\pm)}$ that give an initial wave function $\Psi(x,0)$ that approximates a Gaussian function and satisfies the boundary conditions. The approximation is reasonable if the width of the Gaussian function is a fraction of that of the potential well. Equation 6.3.8 can be written as

$$\Psi(x,0) = \Psi^{(+)}(x,0) + i\Psi^{(-)}(x,0) \tag{6.3.9}$$

with

$$\Psi^{(+)}(x,0) = \left(\frac{1}{2\pi L^2}\right)^{1/4} \cos k_0 x \; e^{-x^2/4L^2} \tag{6.3.10}$$

$$\Psi^{(-)}(x,0) = \left(\frac{1}{2\pi L^2}\right)^{1/4} \sin k_0 x \; e^{-x^2/4L^2} \tag{6.3.11}$$

Note that $\phi_n^{(+)}(x)$ and $\Psi^{(+)}(x,0)$ are even functions of x, whereas $\phi_n^{(-)}(x)$ and $\Psi^{(-)}(x,0)$ are odd functions. Equations 6.3.7 and 6.3.9 imply

$$c_n^{(+)} = 2\int_0^1 \phi_n^{(+)*}(x)\Psi^{(+)}(x,0)dx \tag{6.3.12}$$

$$c_n^{(-)} = 2i\int_0^1 \phi_n^{(-)*}(x)\Psi^{(-)}(x,0)dx \tag{6.3.13}$$

The probability density is

$$P(x,t) = \Psi^*(x,t)\Psi(x,t) \tag{6.3.14}$$

Our task is to explore the time evolution of the probability density.

6.3.3 Solution with *Mathematica*

```
In[1]:= ClearAll["Global`*"]
```

6.3.3.1 Function Definitions

In this section, we define a function that generates an animation of the time evolution of the probability density. The first argument of the function is the Gaussian width L restricted from 1/8 to 1/4; the second argument, the average wave number k_0 limited from 15 to 30; and the third argument, the number m of graphs generated, which must be a positive integer. (Why should L and k_0 be restricted? See Problem 7 in Section 6.7.) Options for passing to the **Plot** function may be specified after the third argument. Statements explaining the construct of the function are embedded in the body of the function in the form (∗ *comment* ∗) ignored by the kernel.

```
In[2]:= particle[L_ /; 1/8 ≤ L ≤ 1/4, k0_ /; 15 ≤ k0 ≤ 30, m_Integer ? Positive,
      opts___] := Module[{SuperPlus, SuperMinus, Ψ, P, Δt = 16/(m π)},
        (* evaluate the coefficients with Equations 6.3.12 and 6.3.13
         and set those less than 10^-2 identically to zero, i.e.,
        determine a set of coefficients that gives a reasonable
         approximation of the Gaussian function for the initial
         wave function that satisfies the boundary conditions *)
        c+n_ := c+n = Chop[2 NIntegrate[Evaluate[ϕ+n [x] Ψ+ [x] // N], {x, 0, 1}],
          10^-2];
        c-n_ := c-n = Chop[2 ⅈ NIntegrate[Evaluate[ϕ-n [x] Ψ- [x] // N], {x, 0, 1}],
          10^-2];
```

```
Ψ⁺[x_] := (1/(2 π L²))^(1/4) Cos[k0 x] Exp[-x²/(4 L²)];

Ψ⁻[x_] := (1/(2 π L²))^(1/4) Sin[k0 x] Exp[-x²/(4 L²)];

ϕ⁺n_[x_] := Cos[(n - 1/2) π x];

ϕ⁻n_[x_] := Sin[n π x];

(* define the wave function and its complex conjugate *)

Ψ[x_, t_] := Sum[(c⁺n ϕ⁺n[x] Exp[-ⅈ e⁺n t] + c⁻n ϕ⁻n[x] Exp[-ⅈ e⁻n t]), {n, 1, 14}];

e⁺n_ := (n - 1/2)² π² / 2;

e⁻n_ := n² π² / 2;

Ψ*[x_, t_] := (Ψ[x, t] /. Complex[p_, q_] → Complex[p, -q]);

(* define the probability density *)

P[x_, t_] := Ψ[x, t] Ψ*[x, t] // N // Expand // Chop;

(* generate the animation *)

ListAnimate[
 Table[
  Plot[
   Evaluate[P[x, t]], {x, -1, 1}, opts,
   Ticks → {{-1, 0, 1}, None}, AxesLabel → {"x (a)", "P(x, t)"},
   PlotRange → {0, 6}, Prolog → {Line[{{1, 0}, {1, 6}}],
     Line[{{-1, 0}, {-1, 6}}]}], {t, 0, 16/π - Δt, Δt}]]]
```

6.3.3.2 Animation

As an example, consider the case with $L = 1/4$, $k_0 = 20$, and $m = 336$. Let us generate the animation:

```
In[3]:= particle[1/4, 20, 336, PlotStyle -> Red,
          Background → Lighter[Blue, 0.85]]
```

Out[3]=

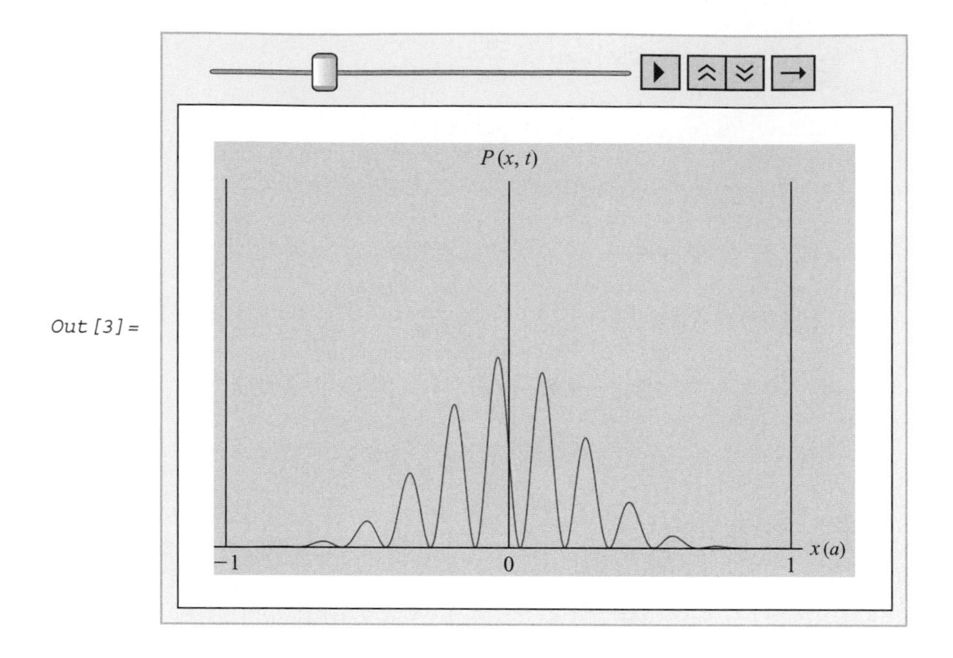

To animate, click the Play button. (For the rationale behind choosing $m = 336$, see Problem 8 in Section 6.7.) We observe the interference among the component waves as the wave packet sloshes between the two walls. It is a mistake to assume that the probability density will eventually become uniform within the well. On the contrary, the probability density restores its original shape in time $T = 2\pi/E_1^{(+)} = 16/\pi$, in our system of units. The animation generated spans the entire time interval T.

```
In[4]:= ClearAll[particle]
```

6.4 THE SQUARE WELL POTENTIAL

6.4.1 The Problem

Determine analytically and numerically the bound states of a particle in a one-dimensional square potential well.

6.4.2 Physics of the Problem

Consider a particle of mass m moving in a one-dimensional potential

$$V(x) = \begin{cases} 0 & |x| > a \\ -V_0 & |x| < a \end{cases} \tag{6.4.1}$$

with $V_0 > 0$. The time-independent Schrödinger equation is

$$-\frac{\hbar^2}{2m}\frac{d^2u(x)}{dx^2} + V(x)u(x) = E\,u(x) \tag{6.4.2}$$

where $u(x)$ is the eigenfunction corresponding to the energy eigenvalue E. For bound states, $E < 0$.

6.4.2.1 Analytical Solution

Let

$$k = \frac{\sqrt{2m|E|}}{\hbar} \tag{6.4.3}$$

$$q = \frac{\sqrt{2m\,(V_0 - |E|)}}{\hbar} \tag{6.4.4}$$

$$\lambda = \frac{2mV_0a^2}{\hbar^2} \tag{6.4.5}$$

$$y = qa \tag{6.4.6}$$

Equations 6.4.4–6.4.6 imply that the energy eigenvalue E can be expressed as

$$E = -\left(1 - \frac{y^2}{\lambda}\right) V_0 \tag{6.4.7}$$

and Equations 6.4.3, 6.4.5, and 6.4.7 suggest that k can be written as

$$k = \frac{\sqrt{\lambda - y^2}}{a} \tag{6.4.8}$$

As a consequence of the symmetry of the potential, the time-independent Schrödinger equation (Equation 6.4.2) has two classes of solutions that are bounded at infinity. The even solutions are

$$u(x) = \begin{cases} Be^{kx} & x < -a \\ A\cos qx & -a < x < a \\ Be^{-kx} & x > a \end{cases} \tag{6.4.9}$$

The continuity of the eigenfunctions and their derivatives at $x = +a$ (or $x = -a$) imposes the conditions

$$B = A\cos qa\, e^{ka} \tag{6.4.10}$$

$$k = q\tan qa \tag{6.4.11}$$

In terms of λ and y defined in Equations 6.4.5 and 6.4.6, Equation 6.4.11 can be written as

$$\frac{\sqrt{\lambda - y^2}}{y} = \tan y \tag{6.4.12}$$

The corresponding equations for the odd solutions are

$$u(x) = \begin{cases} De^{kx} & x < -a \\ C \sin qx & -a < x < a \\ -De^{-kx} & x > a \end{cases} \tag{6.4.13}$$

$$D = -C \sin qa \, e^{ka} \tag{6.4.14}$$

$$k = -q \cot qa \tag{6.4.15}$$

$$\frac{\sqrt{\lambda - y^2}}{y} = -\cot y \tag{6.4.16}$$

(For more information on the bound states in a square potential well, see [Gas96].)

6.4.2.2 Numerical Solution

Although specifying values for the potential at $x = \pm a$ is unnecessary for the analytical solution of the time-independent Schrödinger equation (Equation 6.4.2), it is prudent, for the numerical solution, to modify Equation 6.4.1 to include the reasonable value $-V_0$ for $V(x)$ at these points. Thus,

$$V(x) = \begin{cases} 0 & |x| > a \\ -V_0 & |x| \leq a \end{cases} \tag{6.4.17}$$

with $V_0 > 0$. Equation 6.4.2 can be written as

$$\frac{d^2u(x)}{dx^2} = F(x)u(x) \tag{6.4.18}$$

where lengths are measured in units of a—that is, $a = 1$—and

$$F(x) = \begin{cases} -\lambda e & |x| > 1 \\ -\lambda(e+1) & |x| \leq 1 \end{cases} \tag{6.4.19}$$

with the energy parameter

$$e = \frac{E}{V_0} \tag{6.4.20}$$

The parameter λ measuring the strength of the potential well is defined in Equation 6.4.5. For bound states, $0 > e > -1$.

As mentioned in Section 6.4.2.1, the symmetry of the potential implies that there are two classes of solutions: even solutions and odd solutions (see [Sch55]). For even solutions, $u'(0) = 0$. Without loss of generality, we let $u(0) = 1$ and determine its appropriate value later with the normalization condition

$$\int_{-\infty}^{\infty} u^2(x)dx = 1 \tag{6.4.21}$$

For the odd solutions, $u(0) = 0$. We may set $u'(0) = 1$ and impose the normalization condition on the eigenfunctions later. For both even and odd solutions, it is only necessary to numerically solve Equation 6.4.18 for $x \geq 0$.

6.4.3 Solution with *Mathematica*

```
In[1]:= ClearAll["Global`*"]
```

As an example, we take, rather arbitrarily, $\lambda = 16$:

```
In[2]:= λ = 16;
```

We consider first the analytical solution and then the numerical solution for the bound states. (Section 6.4.3 is based on [HT94].)

6.4.3.1 Analytical Solution

Equation 6.4.12 is a transcendental equation, and its roots can be determined only by graphical or numerical methods. To determine its roots, we can follow the procedure outlined in Section 2.3.1.4:

1. Evaluate

```
In[3]:= ClickPane[Plot[{Sqrt[λ - y^2]/y, Tan[y]},
            {y, 0, 4}, PlotRange→{0, 5}], {yzcoord = #} &]
```

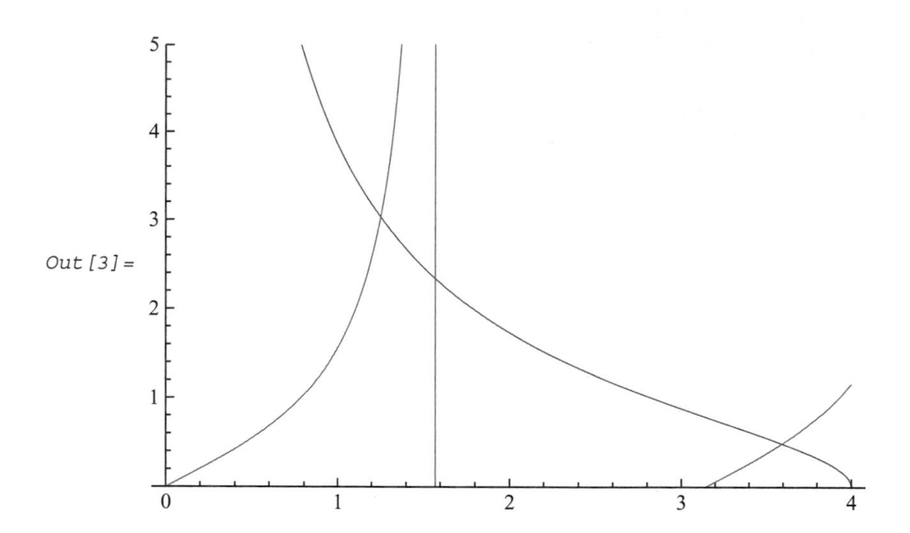

(The vertical line represents an asymptote of the tangent function.)

2. Evaluate

```
In[4]:= Dynamic[yzcoord]
Out[4]= yzcoord
```

3. With the mouse pointer, click an intersection. Then copy and paste the first element of the list now displayed in step 2 (i.e., *Out[4]*) to a list named **startvalue**.

```
startvalue = {1.2721070114013822`}
```

4. With the mouse pointer, click the other intersection. Then copy and append the first element of the list displayed now in step 2 to **startvalue**.

```
startvalue = {1.2721070114013822`, 3.5915582873092213`}
```

5. Evaluate

```
In[5]:= startvalue = {1.2721070114013822`, 3.5915582873092213`};
```

6. Evaluate

$$In[6] := \texttt{Table}\left[\texttt{FindRoot}\left[\frac{\sqrt{\lambda - y^2}}{y} == \texttt{Tan[y]},\ \{y,\ \texttt{startvalue[[i]]}\}\right],\ \{i,\ \texttt{Length[startvalue]}\}\right]$$

```
Out[6]= {{y→1.25235}, {y→3.5953}}
```

7. The roots are

```
In[7]:= {y[1], y[3]} = y /. %
Out[7]= {1.25235, 3.5953}
```

We can also determine the roots of Equation 6.4.16:

1. Evaluate

In[8]:= `ClickPane`$\left[\texttt{Plot}\left[\left\{\frac{\sqrt{\lambda - y^2}}{y}, -\texttt{Cot}[y]\right\}, \{y, 0, 4\}, \texttt{PlotRange} \rightarrow \{0, 5\}\right], (\texttt{yzcoord} = \#)\ \&\right]$

Out[8]=

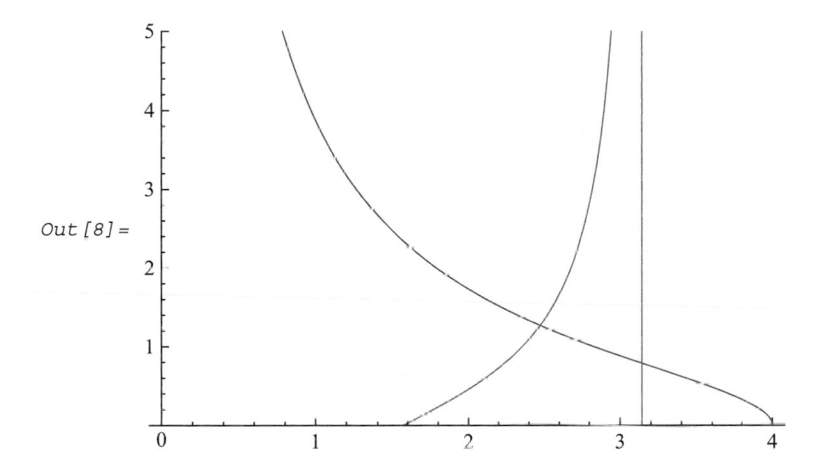

(Again, the vertical line represents an asymptote of the tangent function.)

2. With the mouse pointer, click the intersection.
3. Evaluate

In[9]:= `FindRoot`$\left[\frac{\sqrt{\lambda - y^2}}{y} == -\texttt{Cot}[y], \{y, \texttt{yzcoord}[[1]]\}\right]$

```
Out[9]= {y→2.47458}
```

4. The root is

```
In[10]:= y[2] = y /. %
Out[10]= 2.47458
```

When a root of Equation 6.4.12 or 6.4.16 is known, Equation 6.4.7 gives the corresponding energy eigenvalue in terms of V_0:

```
In[11]:= E[n_]:= -(1 - y[n]^2/λ) V0
```

where the capital Greek letter **Ε** is the alias of E. The potential and the energy eigenvalues are illustrated to scale in the following:

```
In[12]:= V[x_ /;Abs[x] > 1]:= 0
         V[x_ /;Abs[x] < 1]:= -1
         Plot[V[x], {x, -2, 2},
          PlotStyle→Thickness[0.0125], PlotRange→{0.05, -1.05},
          Axes→False, Ticks→{None, Automatic},
          BaseStyle→{FontFamily->"Times", FontSize→12},
          Epilog→({
             Line[{{1, E[1]}, {-1, E[1]}}], Text["E1", {1.25, E[1]}],
             Text[ToString[E[1] /. V0→1] <> " V0", {0, E[1] + 0.055}],
             Line[{{1, E[2]}, {-1, E[2]}}], Text["E2", {1.25, E[2]}],
             Text[ToString[E[2] /. V0→1] <> " V0", {0, E[2] + 0.055}],
             Line[{{1, E[3]}, {-1, E[3]}}], Text["E3", {1.25, E[3]}],
             Text[ToString[E[3] /. V0→1] <> " V0", {0, E[3] + 0.055}],
             Text[" -V0", {-1.25, -1}], Text["0", {-0.85, 0}]} /. V0→1)]
```

Out[14]=

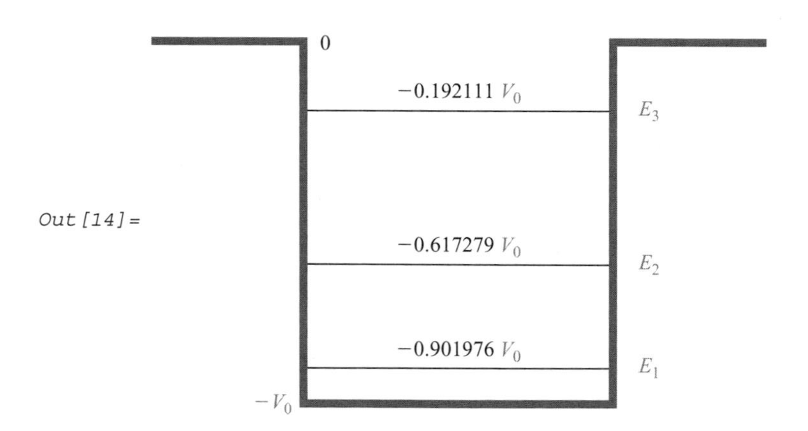

We have used the function **ToString** and the operator **<>** with the full name **StringJoin** in several **Text** primitives; **ToString**[*expr*] gives a string corresponding to the printed form of *expr*; "s_1" **<>** "s_2" **<>** ... or **StringJoin**["s_1", "s_2", ...] yields a string consisting of a concatenation of the s_i.

If y is known, Equations 6.4.6 and 6.4.8 yield q and k, respectively. Equations 6.4.9 and 6.4.10 or Equations 6.4.13 and 6.4.14 then give the yet to be normalized $u(x)$. To plot the eigenfunctions, we define a plotting function in order to streamline the input. The first argument of the function accepts a root of Equation 6.4.12 or 6.4.16, namely, **y[1]**, **y[2]**, or **y[3]**. The second argument takes on either the unnormalized (i.e., $A = 1$) even eigenfunction given in Equations 6.4.9 and 6.4.10 or the odd one specified in Equations 6.4.13 and 6.4.14. Explanatory comments are embedded into the body of the function, and lengths are measured in units of a—that is, $a = 1$.

```
In[15]:= eigenfunctionPlot[y_, u_] :=
  Module[{normalizationConstant, normalizedu, energy},

   (* Equation 6.4.6 *)
   q = y;

   (* Equation 6.4.8 *)
   k = √(λ - y²);

   (* normalize the eigenfunction *)
   normalizationConstant = 1/√(NIntegrate[u[x]², {x, -∞, ∞}]);
   normalizedu[x_] = normalizationConstant u[x];

   (* Equation 6.4.7 *)
   energy = -(1 - y²/λ);

   (* plot the eigenfunction *)
   Plot[normalizedu[x], {x, -3, 3},
    Epilog → {Line[{{1, -1}, {1, 1}}], Line[{{-1, -1}, {-1, 1}}]},
    Frame → True, FrameLabel → {"x (a)", "u(x) (1/√a)"},
    ImageSize → {4 × 72, 3 × 72},
    PlotLabel → "E = " <> ToString[energy] <> "V0"]]
```

Let **u[1] [x]**, **u[2] [x]**, and **u[3] [x]** represent the unnormalized eigenfunctions corresponding to the energy eigenvalues E_1, E_2, and E_3, respectively. Note that *Mathematica* expressions can have complicated heads.

```
In[16]:= uEven[x_] := Piecewise[{{(Cos[q]Exp[k])Exp[k x], x < -1},
     {Cos[q x], Abs[x] ≤ 1}, {(Cos[q]Exp[k])Exp[-k x], x > 1}}]
   uOdd[x_] := Piecewise[{{(-Sin[q]Exp[k])Exp[k x], x < -1},
     {Sin[q x], Abs[x] ≤ 1}, {(Sin[q]Exp[k])Exp[-k x], x > 1}}]

In[18]:= u[1] = u[3] = uEven;
   u[2] = uOdd;
```

We now plot the eigenfunctions:

```
In[20]:= Column[Table[eigenfunctionPlot[y[n], u[n]], {n, 1, 3}]]
```

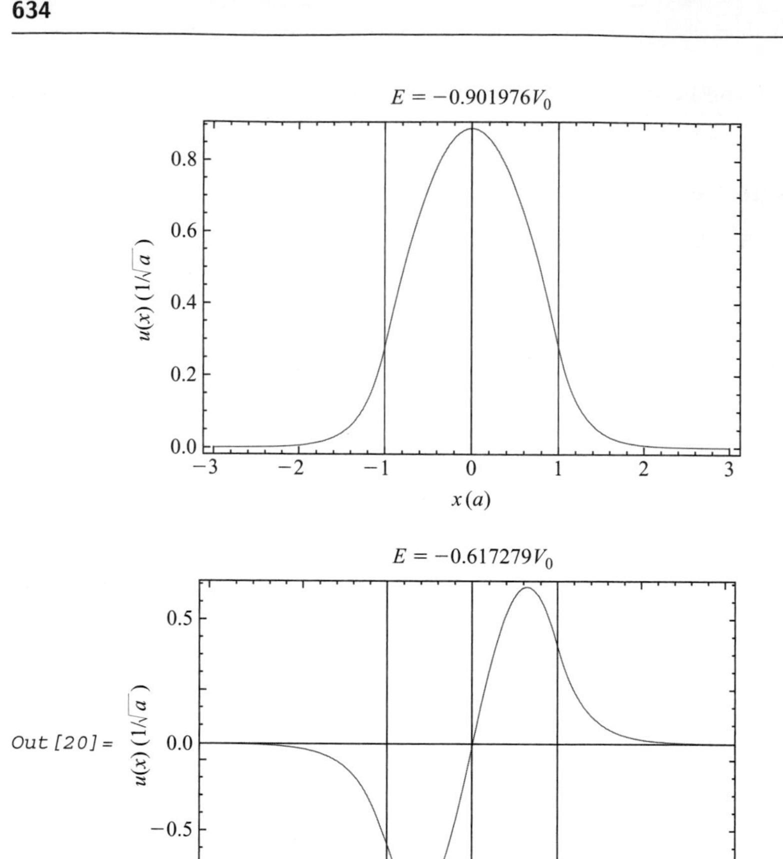

E = −0.901976V0
0.8
0.6
0.4
0.2
0.0
u(x) (1/√a)
−3
−2
−1
0
1
2
3
x (a)
E = −0.617279V0
Out [20] =
0.5
0.0
−0.5
u(x) (1/√a)
−3
−2
−1
0
1
2
3
x (a)

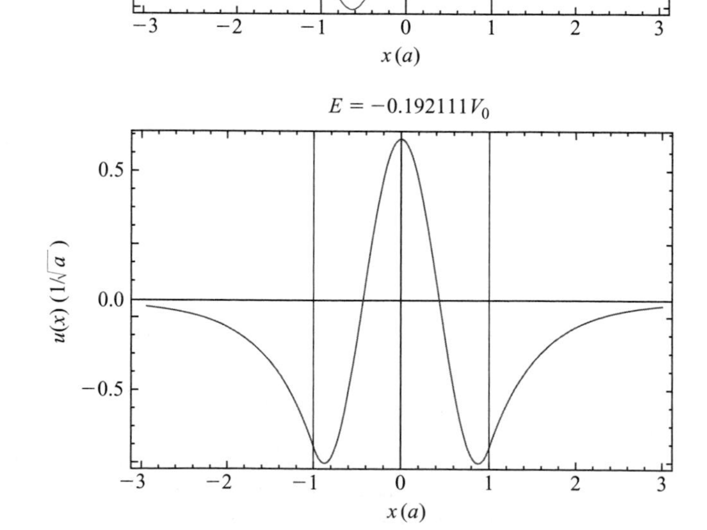

E = −0.192111V0
0.5
0.0
−0.5
u(x) (1/√a)
−3
−2
−1
0
1
2
3
x (a)

The vertical lines depict the boundaries of the potential well.

```
In[21]:= ClearAll[yzcoord, V, q, k, y,
           uEven, uOdd, u, E, eigenfunctionPlot]
```

6.4.3.2 Numerical Solution

Following Equation 6.4.19, we define

```
In[22]:= F[x_] := Piecewise[{{-λ e, Abs[x] > 1}, {-λ(e + 1), Abs[x] ≤ 1}}]
```

Given a value for the energy parameter e, we can solve Equation 6.4.18 for the corresponding eigenfunction $u(x)$. However, only for certain discrete values of the energy parameter are the eigenfunctions bounded—that is,

$$u(\infty) = 0 \tag{6.4.22}$$

For numerical analysis, we choose a point $x = x_f$ at a finite distance from the potential well but sufficiently far away from it so that Equation 6.4.22 can be replaced by the condition of $u(x)$ approaching zero at $x = x_f$. The choice of x_f depends on λ. For $\lambda = 16$, we take, somewhat arbitrarily, $x_f = 4$. (To ascertain the validity of the choice for x_f, see Problem 9 in Section 6.7.)

To find the values of e for which the eigenfunctions approach zero at $x = 4$, we define the function

```
In[23]:= eqnEven[r_?NumberQ] := (
            e = r;
            sol = NDSolve[
             {u''[x] == F[x]u[x], u'[0] == 0, u[0] == 1}, u, {x, 0, 4}];
            u[4] /. sol[[1]])
```

for the even solutions, and the function

```
In[24]:= eqnOdd[r_?NumberQ] := (
            e = r;
            sol = NDSolve[
             {u''[x] == F[x]u[x], u'[0] == 1, u[0] == 0}, u, {x, 0, 4}];
            u[4] /. sol[[1]])
```

for the odd solutions. With a value for e as their argument, these functions return the value of $u(4)$ and name **sol** as the solution to Equation 6.4.18. Thus, finding the values of e for which the eigenfunctions equal zero at $x = 4$ is the same as finding the roots of the equation *eqnEven*$(r) = 0$ for the even solutions and of the equation *eqnOdd*$(r) = 0$ for the odd solutions. Furthermore, the solutions **sol** for these values of e yield the corresponding bound state eigenfunctions. (Of course, we should verify graphically that these functions indeed approach zero at $x = 4$ to eliminate the remote possibility that they happen to cross the x-axis at that point.) **FindRoot[***eqn***[***r***] == 0, {***r***,** r_0**}]** introduced in Section 2.3.1.4 searches for a root using the Newton–Raphson method. Using this method requires that the function *eqn*(r) is differentiable. (See Problem 18 of Section 3.3.7.) Well, *eqnEven*(r) and *eqnOdd*(r) are not:

```
In[25]:= D[eqnEven[r], r]
Out[25]= eqnEven'[r]
```

Mathematica cannot differentiate the function and leaves the derivative in a symbolic form. Fortunately, **FindRoot**[*eqn*[*r*] == 0, {*r*, r_0, r_1}] uses a variant of the secant method that does not require *eqn*(r) being differentiable. Yet it requires two starting values r_0 and r_1. Although it is not necessary that *eqn*(r_0) and *eqn*(r_1) have opposite signs, it is important that r_0 and r_1 are sufficiently close to a root.

Consider the even solutions. To determine the appropriate values for r_0 and r_1, we generate a list of $\{e_i, eqnEven(e_i)\}$ with e_i ranging from -1 to 0 in steps of 0.1 as $-1 < e < 0$ for bound states:

```
In[26]:= Table[{i, eqnEven[i]}, {i, -1, 0, 0.1}]
Out[26]= {{-1., 81377.4}, {-0.9, -735.146},
          {-0.8, -16150.5}, {-0.7, -12767.1}, {-0.6, -7007.68},
          {-0.5, -3049.7}, {-0.4, -1040.}, {-0.3, -238.913},
          {-0.2, -6.5943}, {-0.1, 22.8457}, {0., 8.42799}}
```

We look for the pairs (e_i, e_{i+1})—that is, two consecutive values of e—for which $(eqnEven(e_i)eqnEven(e_{i+1})) \leq 0$ because each of these pairs can serve as a set of starting values (r_0, r_1). As mentioned previously, although it is not necessary that *eqn*(r_0) and *eqn*(r_1) have opposite signs, the fact that they do ensures the existence of at least one root between r_0 and r_1. Then, if r_0 and r_1 are also close together, they must be close to a root. To automate the selection of starting values, partition the list above into nonoverlapping sublists of two pairs with an offset of 1:

```
In[27]:= Partition[%, 2, 1]
Out[27]= {{{-1., 81377.4}, {-0.9, -735.146}},
          {{-0.9, -735.146}, {-0.8, -16150.5}},
          {{-0.8, -16150.5}, {-0.7, -12767.1}},
          {{-0.7, -12767.1}, {-0.6, -7007.68}},
          {{-0.6, -7007.68}, {-0.5, -3049.7}},
          {{-0.5, -3049.7}, {-0.4, -1040.}},
          {{-0.4, -1040.}, {-0.3, -238.913}},
          {{-0.3, -238.913}, {-0.2, -6.5943}},
          {{-0.2, -6.5943}, {-0.1, 22.8457}},
          {{-0.1, 22.8457}, {0., 8.42799}}}
```

Then choose those sublists in which the second elements of the pairs have opposite signs:

```
In[28]:= Select[%, (Sign[#[[1, 2]]]Sign[#[[2, 2]]] ≤ 0)&] //N
Out[28]= {{{-1., 81377.4}, {-0.9, -735.146}},
          {{-0.2, -6.5943}, {-0.1, 22.8457}}}
```

Finally, pick out the pairs (e_i, e_{i+1}) from the nested list:

```
In[29]:= {#[[1, 1]], #[[2, 1]]}& /@%
Out[29]= {{-1., -0.9}, {-0.2, -0.1}}
```

The two sets of starting values for **FindRoot**[*eqn*[*r*] == 0, {*r*, r_0, r_1}] are $\{-1, -0.9\}$ and $\{-0.2, -0.1\}$. We can define a function to carry out all the foregoing operations for locating the starting values. This function has four arguments. The first argument is for the minimum value of e; the second argument is for the maximum value of e. The third argument plus 1 equals the number of sampling values for e, and the last argument accepts either *eqnEven* or *eqnOdd*.

```
In[30]:= initialValues[emin_, emax_, n_, eqn_] := (
           {#[[1, 1]], #[[2, 1]]}& /@ (Select[Partition[
               Table[{i, eqn[i]}, {i, emin, emax, (emax - emin) /n}], 2, 1],
             (Sign[#[[1, 2]]]Sign[#[[2, 2]]] ≤ 0)&] //N))
```

For the even solutions, we have

```
In[31]:= evenvalues = initialValues[-1, 0, 10, eqnEven]
Out[31]= {{-1., -0.9}, {-0.2, -0.1}}
```

For the odd solutions, we have

```
In[32]:= oddvalues = initialValues[-1, 0, 10, eqnOdd]
Out[32]= {{-0.7, -0.6}}
```

For each pair of starting values, **FindRoot**[*eqn*[*r*] == 0, {*r*, r_0, r_1}] returns the energy parameter e and names **sol**, as a result of the call to *eqn*(e), the corresponding bound state solution to Equation 6.4.18. For the starting values $\{-1., -0.9\}$, we have

```
In[33]:= FindRoot[eqnEven[r] == 0, {r, Sequence @@ evenvalues[[1]]}]
Out[33]= {r -> 0.901976}
```

where **Sequence @@** *list* gives a sequence of the elements of *list*. With Equation 6.4.20, the energy eigenvalue is

```
In[34]:= E[1] = V0 r /. %
Out[34]= -0.901976 V0
```

where the capital Greek letter **E** is the alias of E, and the associated eigenfunction is

```
In[35]:= u[1][x_/;0 ≤ x ≤ 4] = u[x] /. sol[[1]];
         u[1][x_/; -4 ≤ x < 0] := u[1][-x]
```

Similarly, we have, for the starting values $\{-0.7, -0.6\}$,

```
In[37]:= FindRoot[eqnOdd[r] == 0, {r, Sequence @@ oddvalues[[1]]}]
Out[37]= {r→-0.617279}

In[38]:= E[2] = V₀ r /. %
Out[38]= -0.617279 V₀

In[39]:= u[2][x_/;0 ≤ x ≤ 4] = u[x] /. sol[[1]];
         u[2][x_/; -4 ≤ x < 0] := -u[2][-x]
```

and, for the starting values $\{-0.2, -0.1\}$,

```
In[41]:= FindRoot[eqnEven[r] == 0, {r, Sequence @@ evenvalues[[2]]}]
Out[41]= {r→-0.192105}

In[42]:= E[3] = V₀ r /. %
Out[42]= -0.192105 V₀

In[43]:= u[3][x_/;0 ≤ x ≤ 4] = u[x] /. sol[[1]];
         u[3][x_/; -4 ≤ x< 0] := -u[3][-x]
```

The energy eigenvalues agree to at least four significant figures with those determined analytically in Section 6.4.3.1. We can produce the diagram, drawn to scale, of the potential and the energy eigenvalues:

```
V[x_/;Abs[x] > 1] := 0
V[x_/;Abs[x] < 1] := -1
Plot[V[x], {x, -2, 2},
 PlotStyle→Thickness[0.0125], PlotRange→{0.05, -1.05},
 Axes→False, Ticks→{None, Automatic},
 BaseStyle→{FontFamily→"Times", FontSize→12},
 Epilog→({
     Line[{{1, E[1]}, {-1, E[1]}}], Text["E₁", {1.25, E[1]}],
     Text[ToString[E[1] /. V₀→1] <> " V₀", {0, E[1] + 0.055}],
     Line[{{1, E[2]}, {-1, E[2]}}], Text["E₂", {1.25, E[2]}],
     Text[ToString[E[2] /. V₀→1] <> " V₀", {0, E[2] + 0.055}],
     Line[{{1,E[3]}, {-1, E[3]}}], Text["E₃", {1.25, E[3]}],
     Text[ToString[E[3] /. V₀→1] <> " V₀", {0, E[3] + 0.055}],
     Text["-V₀", {-1.25, -1}], Text["0", {-0.85, 0}]} /. V₀→1)]
```

The diagram is not shown because it is essentially the same as that in Section 6.4.3.1.

We can plot the normalized eigenfunctions:

```
In[45]:= plotFunc[u_] :=
          Module[{normalizationConstant, normalizedu},
```

```
normalizationConstant = 1/Sqrt[2 NIntegrate[u[x]^2, {x, 0, 4}]];
normalizedu[x_] = normalizationConstant u[x];
Plot[normalizedu[x], {x, -3, 3},
  Epilog→{Line[{{1, -1}, {1, 1}}], Line[{{-1, -1}, {-1, 1}}]},
  Frame→True, FrameLabel→{"x(a)", "u(x)(1/√a)"},
  ImageSize->{4×72, 3×72},
  PlotLabel->"E = "<>ToString[E[n]/.V0→1]<>"V0"]]
Column[Table[plotFunc[u[n]], {n, 1, 3}]]
```

We have also omitted the graphs of the eigenfunctions because they appear identical to those from the analytical solution in Section 6.4.3.1.

Our numerical method, a form of what is known as the shooting method, can be adapted for other one-dimensional potentials, and it can still be used when the time-independent Schrödinger equation cannot be solved by even the most powerful analytical tools. However, a few words of caution are in order. With this numerical method, it is quite possible to miss some energy eigenvalues, but this pitfall can be avoided. A general property of one-dimensional bound states can serve as a safeguard. If the bound states are arranged in the order of increasing energies $E_1, E_2, \ldots, E_n, \ldots$, the nth eigenfunction has $n-1$ nodes (see [Mes00]). Thus, if, for example, the eigenfunctions corresponding to two consecutive energies in our spectrum have, respectively, n and $n+2$ nodes, we must have missed an energy eigenvalue. In determining the appropriate starting values r_0 and r_1 with **initialValues[**e_i, e_f, n, eqn**]**, consider making n bigger or reducing $e_f - e_i$ to scan regions that are likely to have roots. (For a discussion of the shooting method, see [Pat94].)

```
In[46]:= ClearAll[λ, F, e, sol, eqnEven, eqnOdd,
           initialValues, evenvalues, oddvalues, E, u, V, plotFunc]
```

6.5 ANGULAR MOMENTUM

6.5.1 The Problem

The Hamiltonian of a force-free rigid rotator is

$$H = \frac{1}{2I_1}L_x^2 + \frac{1}{2I_2}L_y^2 + \frac{1}{2I_3}L_z^2 \tag{6.5.1}$$

where I_1, I_2, and I_3 are the principal moments of inertia, and L_x, L_y, and L_z are the components of the total angular momentum operator along the principal axes. Find the eigenvalues of H when the total angular momentum quantum number equals 1, 2, and 3.

6.5.2 Physics of the Problem

6.5.2.1 Angular Momentum in Quantum Mechanics

In classical mechanics, angular momentum of a particle is a vector defined in the coordinate space in terms of the position and momentum vectors of the particle. Corresponding to this classical dynamical variable is a quantum mechanical observable, called the orbital angular momentum. It is a vector operator obtained according to the general correspondence rule. The components of the orbital angular momentum, as we will show in the next section, obey a set of commutation relations. A particle can also have an intrinsic angular momentum or spin, which has no classical analog and cannot be defined in the coordinate space. The commutation relations for the orbital angular momentum are so fundamental that spin is defined in terms of them together with a supplementary condition on the spectrum of the square of the spin. These commutation relations provide the general definition of all angular momenta even including those of a system of particles.

In general, angular momentum is a vector operator, the components of which are observables satisfying the commutation relations

$$\begin{aligned} [J_x, J_y] &= i\hbar J_z \\ [J_y, J_z] &= i\hbar J_x \\ [J_z, J_x] &= i\hbar J_y \end{aligned} \tag{6.5.2}$$

An observable is a Hermitian operator possessing a complete, orthonormal set of eigenvectors. Not all Hermitian operators possess such a set of eigenvectors; those capable of representing physical quantities do.

It follows from Equation 6.5.2 that

$$[\mathbf{J}^2, J_z] = 0 \tag{6.5.3}$$

where the square of the angular momentum is

$$\mathbf{J}^2 = J_x^2 + J_y^2 + J_z^2 \tag{6.5.4}$$

Since $\mathbf{J}^2$ and J_z commute with each other, they possess a common set of eigenvectors. The eigenvalue equations can be written as

$$\mathbf{J}^2 |jm\rangle = \hbar^2 j(j+1) |jm\rangle \tag{6.5.5}$$

$$J_z |jm\rangle = \hbar m | jm\rangle \tag{6.5.6}$$

From the general definition of angular momentum, we can show the following properties for the spectrum of $\mathbf{J}^2$ and J_z, and for the matrix elements of J_+, J_-, and J_z:

(A) The only possible values of j are

$$j = 0, \frac{1}{2}, 1, \frac{3}{2}, 2, \frac{5}{2}, \ldots, \infty \tag{6.5.7}$$

(B) For a given j, the only possible values of m are

$$m = -j, -j+1, \ldots, +j \tag{6.5.8}$$

(C) The matrix elements of J_+, J_-, and J_z are

$$\langle jm | J_z | j'm' \rangle = m\hbar \delta_{jj'} \delta_{mm'} \tag{6.5.9}$$

$$\langle jm | J_\pm | j'm' \rangle = \hbar \sqrt{j(j+1) - mm'}\, \delta_{jj'} \delta_{mm' \pm 1} \tag{6.5.10}$$

where we have introduced two Hermitian conjugate operators

$$J_\pm = J_x \pm iJ_y \tag{6.5.11}$$

(For an introduction to angular momentum in quantum mechanics, see [Lib03]; for an advanced treatment, see [Mes00].)

6.5.2.2 Orbital Angular Momentum

In classical mechanics, the angular momentum $\mathbf{l}$ of a particle about a point O is defined as

$$\mathbf{l} = \mathbf{r} \times \mathbf{p} \tag{6.5.12}$$

where $\mathbf{r}$ and $\mathbf{p}$ are respectively the position vector from O and the linear momentum of the particle. Corresponding to this classical dynamical variable is a quantum mechanical observable called the orbital angular momentum $\mathbf{l}$. This vector operator is obtained by replacing the vectors $\mathbf{r}$ and $\mathbf{p}$ with their corresponding observables $\mathbf{r}$ and $\mathbf{p}$ in Equation 6.5.12. The orbital angular momentum $\mathbf{l}$ is given by

$$\mathbf{l} = \mathbf{r} \times \mathbf{p} \tag{6.5.13}$$

The components of $\mathbf{r}$ and $\mathbf{p}$ satisfy the commutation relations

$$\begin{aligned} [r_i, r_j] &= 0 \\ [p_i, p_j] &= 0 \\ [r_i, p_j] &= i\hbar\delta_{ij} \end{aligned} \tag{6.5.14}$$

where i and j can be x, y, or z. From Equations 6.5.13 and 6.5.14, we can show

$$\begin{aligned} [l_x, l_y] &= i\hbar l_z \\ [l_y, l_z] &= i\hbar l_x \\ [l_z, l_x] &= i\hbar l_y \end{aligned} \tag{6.5.15}$$

In wave mechanics, Equation 6.5.13 takes the form

$$\mathbf{l} = -i\hbar \mathbf{r} \times \nabla \tag{6.5.16}$$

Consider a system of N particles. The orbital angular momentum of the nth particle about a point O is

$$\mathbf{l}^{(n)} = \mathbf{r}^{(n)} \times \mathbf{p}^{(n)} \tag{6.5.17}$$

The total orbital angular momentum $\mathbf{L}$ is defined as

$$\mathbf{L} = \sum_{n=1}^{N} \mathbf{l}^{(n)} \tag{6.5.18}$$

From Equation 6.5.15 for each $\mathbf{l}^{(n)}$, Equations 6.5.17 and 6.5.18, and the fact that the components of the orbital angular momenta for different particles commute with each other, we can show

$$\begin{aligned} [L_x, L_y] &= i\hbar L_z \\ [L_y, L_z] &= i\hbar L_x \\ [L_z, L_x] &= i\hbar L_y \end{aligned} \tag{6.5.19}$$

Thus, we have verified that the total orbital angular momentum $\mathbf{L}$ is an angular momentum according to the general definition in Section 6.5.2.1. Therefore, $\mathbf{L}$ assumes all the properties of $\mathbf{J}$. However, there is one important difference. The fact that $\mathbf{L}$ is defined in the coordinate space puts a further restriction on the spectrum of $\mathbf{L}^2$ and L_z. The eigenvalue equations can be written as

$$\mathbf{L}^2 \left| lm \right\rangle = \hbar^2 l(l+1) \left| lm \right\rangle \tag{6.5.20}$$

$$L_z \left| lm \right\rangle = \hbar m \left| lm \right\rangle \tag{6.5.21}$$

Whereas j in Equation 6.5.7 can take on integral and half-odd integral values, l must be an integer. To show l is an integer, we first show that the quantum number l_i for each particle must be an integer. This can be done by considering the eigenvalue equation of l_z for each particle in wave mechanics where the orbital angular momentum of a particle is defined in Equation 6.5.16. The single-valuedness of the wave function in coordinate space requires m_i and, consequently, l_i to be integers. (For discussions of how the single-valuedness of the wave function leads to integral values for m_i, see [Lib03], [Mes00], and [Tow92]; for dissenting opinions, see [Gas96] and [Oha90].) Since l_i are integers for all particles, the theorem for addition of angular momenta requires l must be an integer. (For an introduction to the addition of angular momenta, see Sections 9.4 and 9.5 of [Lib03].) Properties (A), (B), and (C) in Section 6.5.2.1 now, for $\mathbf{L}$, become

(A′) The only possible values of l are

$$l = 0, 1, 2, \ldots, \infty \tag{6.5.22}$$

(B′) For a given l, the only possible values of m are

$$m = -l, -l+1, \ldots, +l \tag{6.5.23}$$

(C′) The matrix elements of L_+, L_-, and L_z are

$$\langle lm\,|L_z|\,l'm'\rangle = m\hbar\delta_{ll'}\delta_{mm'} \tag{6.5.24}$$

$$\langle lm\,|L_\pm|\,l'm'\rangle = \hbar\sqrt{l(l+1) - mm'}\;\delta_{ll'}\delta_{mm'\pm 1} \tag{6.5.25}$$

where

$$L_\pm = L_x \pm iL_y \tag{6.5.26}$$

Equation 6.5.26 implies

$$L_x = \frac{1}{2}(L_+ + L_-) \tag{6.5.27}$$

$$L_y = \frac{1}{2i}(L_+ - L_-) \tag{6.5.28}$$

6.5.2.3 The Eigenvalue Problem

If $I_1 = I_2 = I_3$, the problem of finding the eigenvalues of H in Equation 6.5.1 is trivial because, in this case,

$$H = \frac{\mathbf{L}^2}{2I} \tag{6.5.29}$$

The eigenvalues are simply

$$\frac{\hbar^2 l(l+1)}{2I} \tag{6.5.30}$$

where the values of l are given in Equation 6.5.22.

If $I_1 = I_2 \neq I_3$, the Hamiltonian becomes

$$H = \frac{\mathbf{L}^2}{2I_1} + \left(\frac{1}{2I_3} - \frac{1}{2I_1}\right)L_z^2 \tag{6.5.31}$$

The eigenvalues are

$$\frac{\hbar^2 l(l+1)}{2I_1} + \left(\frac{1}{2I_3} - \frac{1}{2I_1}\right)\hbar^2 m^2 \tag{6.5.32}$$

The values for l and m are given in Equations 6.5.22 and 6.5.23.

If $I_1 \neq I_2 \neq I_3$, we diagonalize the matrix of H in the $\{|lm>\}$ representation to find the eigenvalues. Let

$$a = \frac{1}{2I_1} \quad b = \frac{1}{2I_2} \quad c = \frac{1}{2I_3} \tag{6.5.33}$$

The Hamiltonian can be written as

$$H = aL_x^2 + bL_y^2 + cL_z^2 \tag{6.5.34}$$

The matrix of H can be determined with Equations 6.5.22–6.5.25, 6.5.27, 6.5.28, and 6.5.34. Since there are an infinite number of eigenvectors $|lm>$, we are faced with a horrendous task of diagonalizing an infinite dimensional matrix. Fortunately, from Equations 6.5.24, 6.5.25, 6.5.27, 6.5.28, and 6.5.34, we observe that $<lm|H|l'm'>$ vanishes unless $l = l'$. In other words, we can diagonalize the submatrix for each l one at a time. Since the order of the submatrix is $2l + 1$, it is still not a simple matter unless l is small.

6.5.3 Solution with *Mathematica*

```
In[1]:= ClearAll["Global`*"]
```

Equation 6.5.24 gives us the matrix elements of L_z:

```
In[2]:= Lz[n_, m_] := m ℏ KroneckerDelta[n, m]
```

Equation 6.5.25 gives us the matrix elements of L_+:

```
In[3]:= L+[l_, n_, m_] := ℏ √(l (l + 1) - n m) KroneckerDelta[n, m + 1]
```

Since L_- is the Hermitian conjugate of L_+ and the matrix elements of L_+ are real, the matrix elements of L_- are

```
In[4]:= L-[l_, n_, m_] := L+[l, m, n]
```

Equations 6.5.27 and 6.5.28 give us the matrix elements of L_x and L_y:

```
In[5]:= Lx[l_, n_, m_] := (L+[l, n, m] + L-[l, n, m])/2
        Ly[l_, n_, m_] := (L+[l, n, m] - L-[l, n, m])/(2ⅈ)
```

For a given l, the matrices of L_x, L_y, L_z, and H are

```
In[7]:= Lx[l_] := Table[Lx[l, n, m], {n, l, -l, -1}, {m, l, -l, -1}]
        Ly[l_] := Table[Ly[l, n, m], {n, l, -l, -1}, {m, l, -l, -1}]
```

```
Lz[l_] := Table[Lz[n, m], {n, l, -l, -1}, {m, l, -l, -1}]
H[l_] := (a Lx[l].Lx[l] + b Ly[l].Ly[l] + c Lz[l].Lz[l])
```

The eigenvalues of H for a given l are given by

```
In[11]:= eigenvaluesH[l_] := Eigenvalues[H[l]]
```

We can now determine the eigenvalues of H when the total angular momentum quantum number equals 1, 2, and 3:

```
In[12]:= one = eigenvaluesH[1] // Factor
```

Out[12]= $\{(a+b)\hbar^2, (a+c)\hbar^2, (b+c)\hbar^2\}$

```
In[13]:= two = Simplify[eigenvaluesH[2], ħ > 0]
```

Out[13]= $\{(4a+b+c)\hbar^2, (a+4b+c)\hbar^2, (a+b+4c)\hbar^2,$
$2\left(a+b+c-\sqrt{a^2+b^2-bc+c^2-a(b+c)}\right)\hbar^2,$
$2\left(a+b+c+\sqrt{a^2+b^2-bc+c^2-a(b+c)}\right)\hbar^2\}$

```
In[14]:= three = Simplify[eigenvaluesH[3], ħ > 0]
```

Out[14]= $\{4(a+b+c)\hbar^2, \left(5a+5b+2c-2\sqrt{4a^2+4b^2-bc+c^2-a(7b+c)}\right)\hbar^2,$
$\left(5a+5b+2\left(c+\sqrt{4a^2+4b^2-bc+c^2-a(7b+c)}\right)\right)\hbar^2,$
$\left(2a+5b+5c-2\sqrt{a^2+4b^2-7bc+4c^2-a(b+c)}\right)\hbar^2,$
$\left(2a+5b+5c+2\sqrt{a^2+4b^2-7bc+4c^2-a(b+c)}\right)\hbar^2,$
$\left(5a+2b+5c-2\sqrt{4a^2+b^2-bc+4c^2-a(b+7c)}\right)\hbar^2,$
$\left(5a+2b+5c+2\sqrt{4a^2+b^2-bc+4c^2-a(b+7c)}\right)\hbar^2\}$

Let us diagram the eigenvalues of H (i.e., the energy eigenvalues) for the arbitrarily chosen parameters $a = 1, b = 2$, and $c = 5$:

```
In[15]:= Show[
  Graphics[
   {{Line[{{1, #}, {2, #}}] & /@one, Line[{{3, #}, {4, #}}] & /@two,
     Line[{{5, #}, {6, #}}] & /@three} /. {a→1, b→2, c→5, ħ→1},
    Text[Style["l = 1", ts], {1.5, 12}],
    Text[Style["l = 2", ts], {3.5, 28}],
    Text[Style["l = 3", ts], {5.5, 12}]}] /.
  ts→Sequence[FontFamily-> "Times", 9],
 Frame→True, AspectRatio->1/GoldenRatio,
 FrameTicks→{None, Automatic, None, None},
```

```
FrameLabel → {None, "Energy (ℏ²/(2I₁))"},
PlotLabel → "The Energy Spectrum", Background → Lighter[Red, 0.75]]
```

Out[15]=

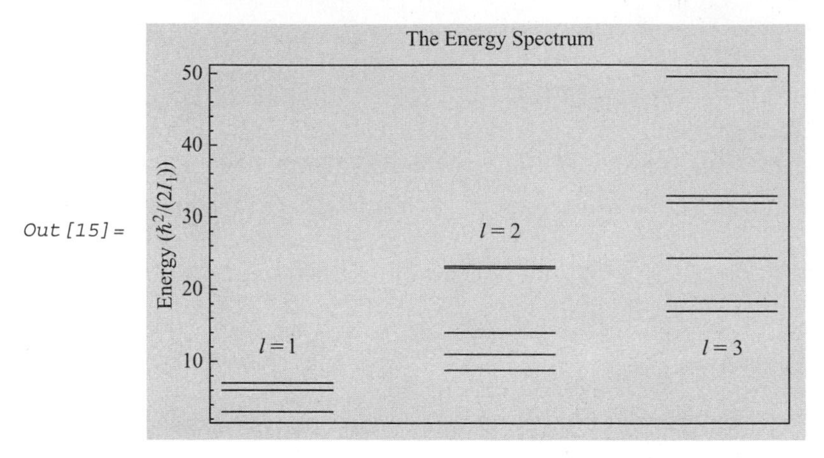

For $l = 2$, there are actually five energy levels; for $l = 3$, there are seven levels. In both cases, the lines for the two highest levels are too close together to be resolved.

For $l \geq 4$, the function **eigenvaluesH** returns some or all of the eigenvalues of H in terms of expressions of the form **Root**[*f*, *k*] that represents the kth root of the polynomial equation $f(x) = 0$. To express the eigenvalues in terms of explicit radicals, apply **ToRadicals** to the result returned by **eigenvaluesH**. **ToRadicals**[*expr*] attempts to transform all **Root** objects in *expr* to explicit radicals. If the degree of the polynomial in a **Root** object is higher than 4, **ToRadicals** cannot always transform the object to radicals because the transformation may not mathematically be done. Also, if the **Root** objects in *expr* contain parameters, **ToRadicals**[*expr*] may not yield a result that is equivalent to *expr* for all values of the parameters. (For a validation of **ToRadicals**, see Problem 11 in Section 6.7.)

If we only wish to diagonalize numerically the Hamiltonian matrix with specified values for the principal moments of inertia I_1, I_2, and I_3 (i.e., a, b, and c), we can define the functions

```
In[16]:= NH[l_, a_, b_, c_] := (a Lx[l].Lx[l] + b Ly[l].Ly[l] + c Lz[l].Lz[l])/ℏ²
         NeigenvaluesH[l_Real, a_Real, b_Real, c_Real] :=
          Eigenvalues[NH[l, a, b, c]]ℏ²
```

for which the arguments *must* be entered as approximate real numbers. As an example, let us determine the energy eigenvalues for $l = 12$, $a = 1$, $b = 5$, and $c = 2$:

```
In[18]:= NeigenvaluesH[12.0, 1.0, 5.0, 2.0] //Chop
Out[18]= {738.45ℏ², 738.45ℏ², 658.862ℏ², 658.862ℏ², 586.315ℏ²,
          586.315ℏ², 520.852ℏ², 520.852ℏ², 462.555ℏ², 462.554ℏ²,
          411.584ℏ², 411.555ℏ², 368.453ℏ², 367.989ℏ², 335.258ℏ²,
          331.241ℏ², 313.359ℏ², 298.202ℏ², 292.985ℏ², 263.765ℏ²,
          263.146ℏ², 224.49ℏ², 224.458ℏ², 179.724ℏ², 179.723ℏ²}
```

Molecules and nuclei often have rotational energy levels characteristic of a rigid rotator. For more information, see [McM94] and [Tho94].

```
In[19]:= ClearAll[L, H, eigenvaluesH, one, two, three, NH, NeigenvaluesH]
```

6.6 THE KRONIG–PENNEY MODEL

6.6.1 The Problem

Consider the Kronig–Penney model. Plot the energy of the electron as a function of the wave vector.

6.6.2 Physics of the Problem

Consider an electron in a one dimensional crystal. The time-independent Schrödinger equation is

$$\left(-\frac{\hbar^2}{2m}\frac{d^2}{dx^2} + V(x)\right)\psi(x) = E\psi(x) \tag{6.6.1}$$

where the potential $V(x)$ has the periodicity of the crystal lattice so that

$$V(x + na) = V(x) \tag{6.6.2}$$

for all integers n.

There is an important theorem, called Bloch's theorem, in solid state physics. For an electron in a one-dimensional periodic potential, it states that the solutions of the time-independent Schrödinger equation can be chosen so that associated with each ψ is a wave vector q such that

$$\psi(x + na) = e^{iqna}\psi(x) \tag{6.6.3}$$

for all integers n.

In the Kronig–Penney model, the potential is taken as a series of repulsive delta function potentials:

$$V(x) = \frac{\hbar^2\lambda}{2ma}\sum_{n=-\infty}^{\infty}\delta(x - na) \tag{6.6.4}$$

Away from the points $x = na$, the potential V is zero, and Equation 6.6.1 becomes a free-particle equation. In the region $(n-1)a \leq x \leq na$, the solution can be written as

$$\psi(x) = A_n \sin k(x - na) + B_n \cos k(x - na) \tag{6.6.5}$$

and in the region $na \leq x \leq (n+1)a$, it can take the form

$$\psi(x) = A_{n+1} \sin k[x-(n+1)a] + B_{n+1} \cos k[x-(n+1)a] \tag{6.6.6}$$

We have for the energy

$$E = \frac{\hbar^2 k^2}{2m} \tag{6.6.7}$$

It can be shown with the time-independent Schrödinger equation that if the potential contains a term like $\kappa\delta(x-na)$, the derivative of ψ is not continuous at $x = na$. Rather, it obeys the relation

$$\left[\frac{d\psi}{dx}\right]_{na+\varepsilon} - \left[\frac{d\psi}{dx}\right]_{na-\varepsilon} = \frac{2m}{\hbar^2}\kappa\psi(na) \tag{6.6.8}$$

with ε arbitrarily small and positive.

The wave functions in Equations 6.6.5 and 6.6.6 must satisfy the following conditions: the continuity of the wave function at $x = na$, Equation 6.6.8 for the derivative of the wave function, and Equation 6.6.3 with $n = 1$. Applying these conditions to the wave functions, we have, after some algebra, the eigenvalue condition

$$\cos qa = \cos ka + \frac{1}{2}\lambda\frac{\sin ka}{ka} \tag{6.6.9}$$

The energy E can now be determined as a function of the wave vector q. Given a wave vector q, we can use Equation 6.6.9 to find k, and Equation 6.6.7 gives us E. (For details of the Kronig–Penney model, see [Gas96], [Lib03], [Gri05], and [Par92].)

6.6.3 Solution with *Mathematica*

```
In[1]:= ClearAll["Global`*"]
```

We would like to plot the energy E as a function of the wave vector q. To do so, we must solve Equation 6.6.9. However, this equation cannot be solved easily because, for every q, the number of solutions for k is infinite. Fortunately, the programming capability of *Mathematica* can circumvent this difficulty. For plotting the energy as a function of the wave vector, we define the function **kpmPlot** that takes three arguments: The first is for the dimensionless constant λ that characterizes the strength of the potential in Equation 6.6.4, the second is for the number of energy bands to be plotted, and the third is for the number of points to be used for plotting the right branch of each energy band. The first argument must be a positive number, the second argument must be a positive integer, and the third argument must be an integer larger than 10. Comments elucidating the construction of the function are included in its body in the form `(* comment *)`.

```
In[2]:= kpmPlot[λ_?Positive, nb_Integer?Positive, mp_Integer/; mp>10] :=
          Module[{K, Q, ε, r},
```

```
(* For the right branches, find k for each q,
using Equation 6.6.9. Let K = k a and Q = q a. *)
K[n_, 1] = n π;
Q[n_, m_] := n π - (m - 1) π / (mp - 1);
K[n_, m_] := K[n, m] = FindRoot[
    Cos[Q[n, m]] == Cos[x] + (λ/2) (Sin[x]/x), {x, K[n, m - 1]}] [[1, 2]];

(* For the right branches, calculate E for each k,
using Equation 6.6.7. Let ℰ = E 2ma²/ℏ². *)
ℰ[n_, m_] := K[n, m]^2;

(* Determine the coordinates of the right branches. *)
r[n_] := r[n] = Table[{Q[n, m], ℰ[n, m]}, {m, 1, mp}];

(* Plot E as a function of q. *)
Show[
 Graphics[Flatten[Table[
    {Line[r[n]], Line[{-#[[1]], #[[2]]} & /@ r[n]]}, {n, 1, nb}]]],
 Ticks → {Table[{i π, i "π"}, {i, -nb, nb}], Automatic},
 Axes → True, AxesLabel → {"q (1/a)", "E (ℏ²/2ma²)"},
 AspectRatio -> 1/GoldenRatio]]
```

As an example, let us call the function with $\lambda = 3\pi$, $nb = 4$, and $mp = 35$:

```
In[3]:= kpmPlot[3 π, 4, 35]
```

Out[3]=

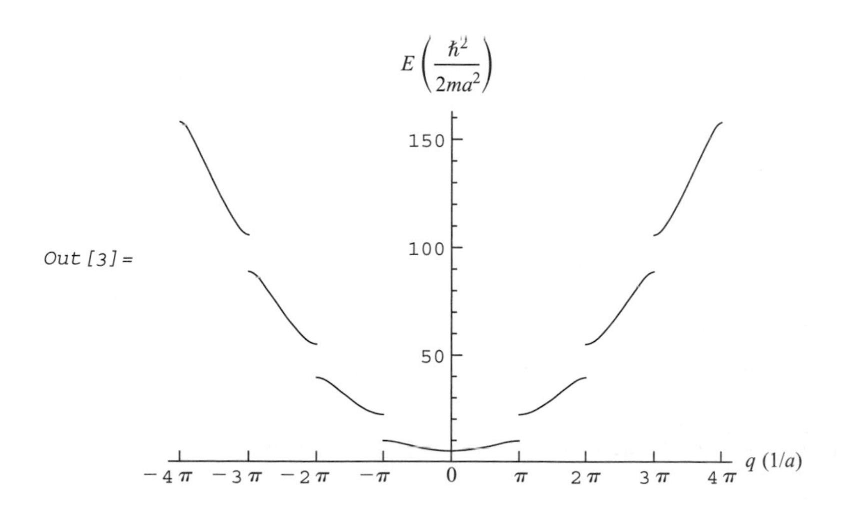

The graph shows an energy band structure. There are allowed energy bands separated by energy gaps, regions that are forbidden. The existence of energy gaps has important consequences in the transport properties of electrons. (For an in-depth discussion of energy bands, see, for example, [AM76]. For an earlier version of this notebook, see [Tam91].)

6.7 PROBLEMS

In this book, straightforward, intermediate-level, and challenging problems are unmarked, marked with one asterisk, and marked with two asterisks, respectively. For solutions to selected problems, see Appendix D.

1. Consider a particle in a one-dimensional box with ends at $x = 0$ and $x = 2$. Suppose the probability density is given by

$$P(x) = \begin{cases} \frac{15}{16}\left(x^2 - \frac{1}{4}x^4\right) & 0 \leq x \leq 2 \\ 0 & \text{elsewhere} \end{cases}$$

   (a) Plot the probability density $P(x)$.
   (b) Show that $P(x)$ is correctly normalized.
   (c) Calculate the expectation value of x.
   (d) Determine the root-mean-square deviation of x from the mean.

2. Consider a nonstationary state that is a superposition of the first two eigenstates of a one-dimensional harmonic oscillator:

$$\psi(x,t) = \frac{1}{\sqrt{2}}\left[\exp(-iE_0 t/\hbar)u_0(x) + \exp\left(-iE_1 t/\hbar\right)u_1(x)\right]$$

   where u_n and E_n are the normalized energy eigenfunctions and energy eigenvalues, respectively. Animate the time evolution of the probability density

$$P(x,t) = \left|\psi(x,t)\right|^2$$

   Indicate on the frames the classically forbidden regions.

3. Consider a one-dimensional harmonic oscillator with mass m and spring constant k. Let the initial wave function $\psi(x,0)$ be a Gaussian wave packet:

$$\psi(x,0) = \left(\frac{1}{2\pi\sigma^2}\right)^{1/4} \exp\left[-\frac{(x-x_0)^2}{4\sigma^2}\right]$$

   It can be shown (see [BD95]) that the probability density at time t is

$$P(x,t) = \frac{1}{\sqrt{2\pi}} \frac{2\sigma}{\sqrt{\sigma_0^4 s^2 + 4\sigma^4 c^2}} \exp\left[-\frac{2\sigma^2}{\sigma_0^4 s^2 + 4\sigma^4 c^2}(x - cx_0)^2\right]$$

where c and s are cos ωt and sin ωt, respectively. Also,

$$\omega = \sqrt{k/m}$$

$$\sigma_0 = \sqrt{\hbar/m\omega}$$

Let $x_0 = 3\sigma_0$. On a three-dimensional graph and with $\sigma = \sigma_0/(2\sqrt{2})$, plot the probability density for times $t_n = (2\pi/\omega)(n/16)$, where $n = 0, 1, 2, \ldots, 31$. Repeat with $\sigma = \sigma_0/\sqrt{2}$ and $\sigma = 3\sigma_0/(2\sqrt{2})$.

*4. A deuteron has spin 1. Consider two deuterons, with spin operators $\mathbf{S}_1$ and $\mathbf{S}_2$. Denote the one-particle spin eigenstates by $\kappa_i(m_s)$ so that

$$\mathbf{S}_i^2\kappa_i(m_s) = 1(1+1)\hbar^2\kappa_i(m_s)$$

$$S_{iz}\kappa_i(m_s) = m_s\hbar\kappa_i(m_s)$$

where $m_s = +1, 0, -1$, and $i = 1, 2$. The total spin $\mathbf{S}$ is defined as

$$\mathbf{S} = \mathbf{S}_1 + \mathbf{S}_2$$

If the orbital angular momentum of the two-deuteron system in their center-of-mass system is $\mathbf{L}$, the total angular momentum is defined as

$$\mathbf{J} = \mathbf{L} + \mathbf{S}$$

(a) Determine the eigenvalues and eigenstates of $\mathbf{S}^2$ and S_z. Express the eigenstates in terms of the single-particle spin eigenstates $\kappa_i(m_s)$. (b) Taking cognizance of the symmetrization requirement on the total state under a two-particle exchange, list the possible total angular momentum states in the spectroscopic notation

$$^{2S+1}L_J$$

for $L = 0, 1, 2$, and 3. (c) Express the $^5D_4$ states in terms $\kappa_L(M_L)$ and $\kappa_S(M_S)$ that are the eigenstates of $\mathbf{L}^2$ and L_z and those of $\mathbf{S}^2$ and S_z, respectively.

*5. Determine analytically and numerically the energy eigenvalue and the associated eigenfunction for the ground state of a particle of mass m moving in the potential

$$V(x) = \begin{cases} \infty & |x| > a \\ 0 & a/2 < |x| < a \\ V_0 & |x| < a/2 \end{cases}$$

with $V_0 > 0$. Let

$$V_0 = \frac{\pi^2\hbar^2}{32ma^2}$$

*6. Find the energy eigenvalues and the corresponding eigenfunctions for the ground state and the first two excited states of an anharmonic oscillator with potential energy of the form

$$V(x) = \frac{A}{2}x^2 + \frac{B}{2}x^4$$

First, convert the time-independent Schrödinger equation to the dimensionless form

$$\frac{d^2\psi}{du^2} = -\left(\varepsilon - u^2 - \delta u^4\right)\psi$$

Then, consider the case $\delta = 0.25$.

7. In Section 6.3.3, the arguments L and k_0 of the function **`particle`** are restricted. Why? Explain.

8. Enter into *Mathematica* the function definition, in Section 6.3.3, for **`particle`** with the symbol **`P`** deleted from the local variable list and **`//N`** removed from the definition of **`P[x, t]`**. Evaluate **`particle[1/4, 20, 1]`** and **`Chop[P[x, t], 10^-2]`**. (a) Determine the frequency f_{max} of the fastest time-oscillating term of **`Chop[P[x, t], 10^-2]`**. (b) Calculate its period T_{min}. (c) If $\Delta t = 16/m\pi = T_{min}/2$, what is m?

9. In Section 6.4.3.2, we take $x_f = 4$. Determine the energy eigenvalues for the choices $x_f = 2, 3, 5, 6, 7$, and 8. What can be concluded from the results?

10. Using the numerical solutions in Section 6.4.3.2 for the bound states of a particle in a one-dimensional square potential well, determine the probability of finding the particle outside the potential well when it is in the (a) ground state, (b) first excited state, and (c) second excited state.

11. Consider the functions **`ToRadicals`**, **`eigenvaluesH`**, and **`NeigenvaluesH`**, which were introduced in Section 6.5.3. (a) Evaluate **`eigenvaluesH[6]`** and then **`ToRadicals[eigenvaluesH[6]]`**, and compare the results. (b) Set $\hbar = 1$, and then show that **`Sort[Chop[N[ToRadicals[eigenvaluesH[6]] /. {a→1, b→7, c→4}]]]`** is equivalent to **`Sort[NeigenvaluesH[6.0, 1.0, 7.0, 4.0]]`**.

12. In Section 6.6.3, the function **`kpmPlot` plots the energy bands in the extended zone scheme. Modify the function so that it plots the energy bands in the reduced zone scheme. (For an introduction to energy bands, see [Kit86].)

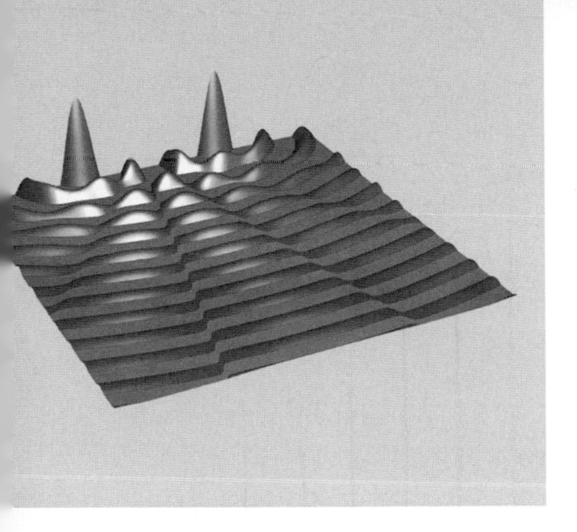

Appendix A
The Last Ten Minutes

What follows is a last-minute addendum to Section 2.2.2, as this book goes to press. **Trig** is an option for the algebraic transformation functions (introduced in Section 2.2.1) except **ComplexExpand** and **PowerExpand**. With **Trig -> True**, these functions attempt to transform trigonometric functions in algebraic manipulations; with **Trig -> False**, they leave the trigonometric functions unchanged. **Trig -> False** is the default for these functions except **Simplify** and **FullSimplify** whose default is **Trig -> True**. For example, **Apart** leaves $2(\cos(\alpha) - \cos(\beta))\csc(\alpha)\csc(\beta)$ unaltered:

```
In[1]:= Apart[2 (Cos[α] - Cos[β]) Csc[α] Csc[β]]
Out[1]= 2 (Cos[α] - Cos[β]) Csc[α] Csc[β]
```

With **Trig -> True**, **Apart** rewrites the trigonometric expression:

```
In[2]:= Apart[2 (Cos[α] - Cos[β])Csc[α]Csc[β], Trig -> True]
```

$$\text{Out[2]=}\ \frac{1}{2}\mathrm{Csc}\left[\frac{\beta}{2}\right]\mathrm{Sec}\left[\frac{\alpha}{2}\right]\left(-\mathrm{Sin}\left[\frac{\alpha}{2}-\frac{\beta}{2}\right]-\mathrm{Sin}\left[\frac{\alpha}{2}+\frac{\beta}{2}\right]\right)+ \frac{1}{2}\mathrm{Csc}\left[\frac{\alpha}{2}\right]\mathrm{Sec}\left[\frac{\beta}{2}\right]\left(-\mathrm{Sin}\left[\frac{\alpha}{2}-\frac{\beta}{2}\right]+\mathrm{Sin}\left[\frac{\alpha}{2}+\frac{\beta}{2}\right]\right)$$

For another example, **Simplify** transforms $1 + \tan^2(\theta)$ into $\sec^2(\theta)$:

```
In[3]:= Simplify[1 + Tan[θ]^2]
Out[3]= Sec[θ]^2
```

With **Trig -> False**, **Simplify** no longer effects the transformation:

```
In[4]:= Simplify[1 + Tan[θ]^2, Trig -> False]
Out[4]= 1 + Tan[θ]^2
```

If you have difficulties with any inputs or questions about any outputs in this book, examine as well as execute, with *Mathematica* 6, a copy of the relevant code on the accompanying compact disc, and then check the result. Error reports will be appreciated. Please donate them to pttl@humboldt.edu.

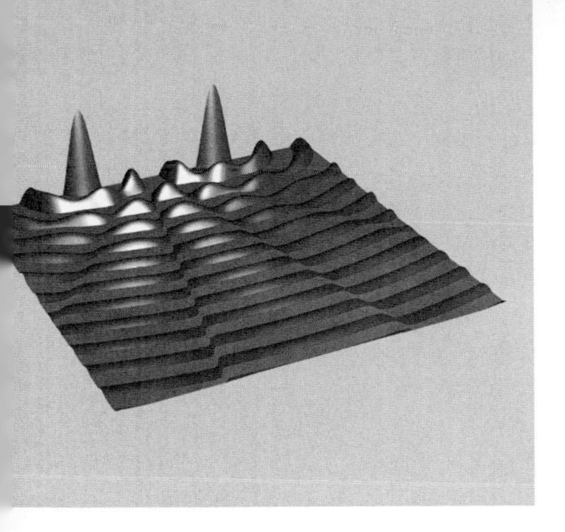

Appendix B
Operator Input Forms

This appendix presents a table of some common operator input forms together with the corresponding full forms and examples. To obtain more information on these operators, use the "front end help" (discussed in Section 1.6.3) on the corresponding full-form function names. For example, to obtain information on the operator "<<", highlight its function name "**Get**" in a notebook and choose **Help ▸ Find Selected Function**. For many of these operators, we can also invoke online help with the **?** operator, as described at the end of Section 1.3. For example, to obtain information on "<<", enter "**?<<**".

Operator Input Form	Full Form	Example
symbol::*tag*	**MessageName**[*symbol*, "*tag*"]	**Sqrt::argx**
<<*name*	**Get**["*name*"]	**<<VectorAnalysis`**
p **?** *test*	**PatternTest**[*p*, *test*]	**n_Integer?NonNegative**
expr1[*expr2*, ...]	*expr1*[*expr2*, ...]	**Exp[I k x]**
expr1[[*expr2*, ...]]	**Part**[*expr1*, *expr2*, ...]	**mylist[[{3, 5}]]**
expr1⟦*expr2*, ...⟧	**Part**[*expr1*, *expr2*, ...]	**ourlist⟦1, 2⟧**
x++	**Increment**[*x*]	**i = 1; While[i < = 3, (Print[i^2]; i++)]**
x--	**Decrement**[*x*]	**For[i = 7; t = x, i^2 > 10, i--, t = t^2 + i; Print[Expand[t]]]**

++*x*	**PreIncrement**[*x*]	**For[i = 1, i <= 3, ++ i, Print[i^2]]**
--*x*	**PreDecrement**[*x*]	**Table[{n -- , -- n}, {n, 1, 5}]**
expr1 @ *expr2*	*expr1* [*expr2*]	**H[V[x]]@ψ[x]**
expr1 ~ *expr2* ~ *expr3*	*expr2* [*expr1*, *expr3*]	**{a, b} ~ Join ~ {c, d, e, f}**
f /@ *expr*	**Map**[*f*, *expr*]	**f/@{a, b, c}**
f //@ *expr*	**MapAll**[*f*, *expr*]	**f//@{{a, b}, {c, d}}**
f @@ *expr*	**Apply**[*f*, *expr*]	**Plus @@ {a, b, c, d}**
n!	**Factorial**[*n*]	**100!**
f '	**Derivative[1]**[*f*]	**x′[t]**
f ''...' (*n* times)	**Derivative**[*n*][*f*]	**x′′[t]**
"*s1*" <> "*s2*" <> ...	**StringJoin**["*s1*", "*s2*", ...]	**"Happy" <> "New" <> "Year!"**
x^*y*	**Power**[*x*, *y*]	**(a + b)^2**
x^y	**Power**[*x*, *y*]	$(\mathbf{a}+\mathbf{b})^2$
$\surd z$	**Sqrt**[*z*]	$\surd(\mathbf{a}^2+\mathbf{b}^2)$
$\sqrt{z}$	**Sqrt**[*z*]	$\sqrt{\mathbf{a}^2+\mathbf{b}^2}$
$\int f\, dx$	**Integrate**[*f*, *x*]	$\int \frac{3}{1+\mathbf{x}^2}\, d\mathbf{x}$
$\int_{xmin}^{xmax} f\, dx$	**Integrate** [*f*, **List**[*x*, *xmin*, *xmax*]]	$\int_1^4 \sqrt{1+\mathbf{x}^2}\, d\mathbf{x}$
$\partial_x f$	**D**[*f*, *x*]	$\partial_{\mathbf{x}}$**Cos[x]**
a ** *b* ** *c*	**NonCommutativeMultiply**[*a*, *b*, *c*]	**p ** x ** x**
a × *b*	**Cross**[*a*, *b*]	$\{a_1, b_1, c_1\} \times \{a_2, b_2, c_2\}$
a.*b*.*c*	**Dot**[*a*, *b*, *c*]	**{x, y}.{{a, b}, {c, d}}.{x, y}**
-*x*	**Times**[-1, *x*]	$-(\mathbf{y}+2)^2$
+*x*	**Plus**[*x*] or *x*	**+3**
x / *y*	**Times**[*x*, **Power**[*y*, -1]] or **Divide**[*x*, *y*]	$(\mathbf{a}^2+\mathbf{b}^2)/(\mathbf{a}+\mathbf{b})$
x ÷ *y*	**Times**[*x*, **Power**[*y*, -1]] or **Divide**[*x*, *y*]	$(\mathbf{a}^2+\mathbf{b}^2) \div (\mathbf{a}+\mathbf{b})$
x y z	**Times**[*x*, *y*, *z*]	**m v r**
x * *y* * *z*	**Times**[*x*, *y*, *z*]	**m * v * r**
x × *y* × *z*	**Times**[*x*, *y*, *z*]	**m × v × r**
x + *y* + *z*	**Plus**[*x*, *y*, *z*]	**1 + 2a + b**
x - *y*	**Plus**[*x*, **Times**[-1, *y*]]	$\mathbf{a}^2 - \mathbf{b}^2$
list1 ⋂ *list2*	**Intersection**[*list1*, *list2*]	**{1, 2, 3} ⋂ {2, 3, 4, 5}**
list1 ⋃ *list2*	**Union**[*list1*, *list2*]	**{1, 2, 3} ⋃ {2, 3, 4, 5}**
lhs == *rhs*	**Equal**[*lhs*, *rhs*]	$\mathbf{x}^2 + 1 == 0$
lhs ⩵ *rhs*	**Equal**[*lhs*, *rhs*]	$\mathbf{x}^2 + 1$ ⩵ 0
lhs != *rhs*	**Unequal**[*lhs*, *rhs*]	**f[n_Integer/;n != 1]** := $\frac{1}{(n-1)^2}$

<table>
<tr><td>lhs ≠ rhs</td><td>Unequal[lhs, rhs]</td><td>f[n_Integer/;n≠1] := $\frac{1}{(n-1)^2}$</td></tr>
<tr><td>x > y</td><td>Greater[x, y]</td><td>Cases[%, x_Integer/; (80 > x > 10)]</td></tr>
<tr><td>x >= y</td><td>GreaterEqual[x, y]</td><td>V[x_/;Abs[x] >= 1] := 1</td></tr>
<tr><td>x ≥ y</td><td>GreaterEqual[x, y]</td><td>V[x_/;Abs[x] ≥1] := 1</td></tr>
<tr><td>x < y</td><td>Less[x, y]</td><td>V[x_/;x <- 3||x ≥ 3] := 0</td></tr>
<tr><td>x <= y</td><td>LessEqual[x, y]</td><td>V[x_] := Which[
x < -3||x >= 3, 0,
-3 <= x < -1||1 <= x < 3, -2,
-1 <= x < 1, -1]</td></tr>
<tr><td>x ≤ y</td><td>LessEqual[x, y]</td><td>V[x_] := Which[
x < -3||x ≥ 3, 0,
-3 ≤ x < -1||1 < x < 3, -2,
-1 ≤ x < 1, -1]</td></tr>
<tr><td>lhs === rhs</td><td>SameQ[lhs, rhs]</td><td>Head[a + b] === Plus</td></tr>
<tr><td>lhs =!= rhs</td><td>UnsameQ[lhs, rhs]</td><td>x = 1.0;
While[1 + $\frac{1}{x}$ =!= x, x = 1 + $\frac{1}{x}$];x</td></tr>
<tr><td>x ∈ dom</td><td>Element[x, dom]</td><td>r ∈ Reals</td></tr>
<tr><td>!expr</td><td>Not[expr]</td><td>!(NumberQ[x_0]&&Im[x_0] == 0)</td></tr>
<tr><td>¬ expr</td><td>Not[expr]</td><td>¬(NumberQ[x_0]&&Im[x_0] == 0)</td></tr>
<tr><td>expr1 && expr2 && expr3</td><td>And[expr1, expr2, expr3]</td><td>IntegerQ[η]&&
Positive[η]&&η ≤ 50</td></tr>
<tr><td>expr1 ∧ expr2 ∧ expr3</td><td>And[expr1, expr2, expr3]</td><td>IntegerQ[η] ∧
Positive[η] ∧ η ≤ 50</td></tr>
<tr><td>expr1 || expr2 || expr3</td><td>Or[expr1, expr2, expr3]</td><td>x < -3||x >= 3</td></tr>
<tr><td>expr1 ∨ expr2 ∨ expr3</td><td>Or[expr1, expr2, expr3]</td><td>x < -3 ∨ x ≥ 3</td></tr>
<tr><td>p..</td><td>Repeated[p]</td><td>q:{{_, _, _}..}</td></tr>
<tr><td>p...</td><td>RepeatedNull[p]</td><td>{{___}...}</td></tr>
<tr><td>p1 | p2</td><td>Alternatives[p1, p2]</td><td>f[x_Rational|x_Integer]</td></tr>
<tr><td>s:obj</td><td>Pattern[s, obj]</td><td>h:Sin[x_]</td></tr>
<tr><td>p:v</td><td>Optional[p, v]</td><td>y_:0</td></tr>
<tr><td>patt/;test</td><td>Condition[patt, test]</td><td>x_/;Abs[x] ≤ 1</td></tr>
<tr><td>lhs -> rhs</td><td>Rule[lhs, rhs]</td><td>{m -> 1, a -> 1, ℏ -> 1}</td></tr>
<tr><td>lhs → rhs</td><td>Rule[lhs, rhs]</td><td>{m → 1, a → 1, ℏ → 1}</td></tr>
<tr><td>lhs :> rhs</td><td>RuleDelayed[lhs, rhs]</td><td>f[x_] :> x</td></tr>
<tr><td>lhs :→ rhs</td><td>RuleDelayed[lhs, rhs]</td><td>f[x_] :→ x</td></tr>
<tr><td>expr/.rules</td><td>ReplaceAll[expr, rules]</td><td>2x + 10/.x → 10</td></tr>
<tr><td>expr//.rules</td><td>ReplaceRepeated[expr, rules]</td><td>x^2 + y^2//.{x → b + 5, b → 1}</td></tr>
<tr><td>x += dx</td><td>AddTo[x, dx]</td><td>x += Δ</td></tr>
<tr><td>x -= dx</td><td>SubtractFrom[x, dx]</td><td>x -= 0.5</td></tr>
<tr><td>x *= c</td><td>TimesBy[x, c]</td><td>x *= 4b^2</td></tr>
<tr><td>x /= c</td><td>DivideBy[x, c]</td><td>x /= 2</td></tr>
<tr><td>body &</td><td>Function[body]</td><td>$\frac{1}{1+\#}$&</td></tr>
</table>

expr//*f*	*f*[*expr*]	(x^2 + 1)//FullForm
lhs = *rhs*	Set[*lhs*, *rhs*]	a = 1
lhs := *rhs*	SetDelayed[*lhs*, *rhs*]	f[x_] := x^2 - 1
lhs ^= *rhs*	UpSet[*lhs*, *rhs*]	Im[a]^= 0
lhs ^:= *rhs*	UpSetDelayed[*lhs*, *rhs*]	f[g[x_]]^ := x^2
f/:*lhs* = *rhs*	TagSet[*f*, *lhs*, *rhs*]	a/:Im[a] = 0
f/:*lhs*:= *rhs*	TagSetDelayed[*f*, *lhs*, *rhs*]	g/:f[g[x_]] := x^2
lhs = .	Unset[*lhs*]	t = .
f/:*lhs* =.	TagUnset[*f*, *lhs*]	g/:f[g[x_]] =.
expr1; *expr2*; *expr3*	CompoundExpression[*expr1*, *expr2*, *expr3*]	r = 3; s = √7 + r; r^2 + s^2
expr1; *expr2*;	CompoundExpression[*expr1*, *expr2*, Null]	func[y_] := (Print[Plot[Sin[x], {x, 0, 2π}]]; x = y;)
#	Slot[1]	3 # &
#*n*	Slot[*n*]	(#1^2 + #2^2) &
##	SlotSequence[1]	{##}&
##*n*	SlotSequence[*n*]	Plus[##4]&
%	Out[]	N[%,2]
%%	Out[-2]	Show[%, %%]
%%...%(*n times*)	Out[-*n*]	%%%^2 + 1
%*n*	Out[*n*]	x[t_] = x[t]/.%14
_	Blank[]	_^_
_*h*	Blank[*h*]	_Symbol
__	BlankSequence[]	x : {_, __}
__*h*	BlankSequence[*h*]	{__List}
___	BlankNullSequence[]	{___, x_, x_, x_, ___}
___*h*	BlankNullSequence[*h*]	___Integer
_.	Optional[Blank[]]	Default[f] = Null; f[x_, _.] := {x}
*s*_	Pattern[*s*, Blank[]]	x_
*s*_*h*	Pattern[*s*, Blank[*h*]]	n_Integer
*s*__	Pattern[*s*, BlankSequence[]]	f[x__]
*s*__*h*	Pattern[*s*, BlankSequence[*h*]]	x__List
*s*___	Pattern[*s*, BlankNullSequence[]]	f[a_, b_, c___]
*s*___*h*	Pattern[*s*, BlankNullSequence[*h*]]	opts___Rule
*s*_.	Optional[Pattern[*s*, Blank[]]]	x_ + y_.

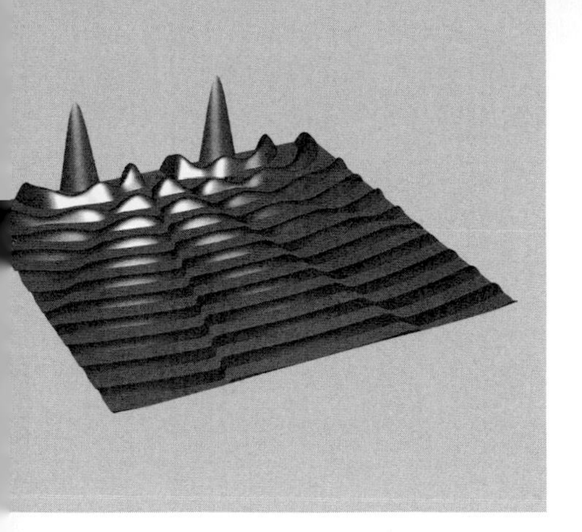

Appendix C
Solutions to Exercises

This appendix provides solutions to selected exercises in Chapters 2 and 3.

SECTION 2.1.20

1.
```
(10 Sqrt[(10.8 x 10^3)/300] + 4)^(1/3)
```

3.
```
ScientificForm[((2.54^(3/5) Sqrt[1.15 x 10^-2]) + (5.11^(2/5)))/
   Sqrt[2.32 x 10^-5], 3]
```

5.
```
ScientificForm[
  (((3 x 10^2)^3) ((2 x 10^-5)^(1/3)))/Sqrt[3.63 x 10^-8], 3]
```

7.
```
ScientificForm[N[Sqrt[(-2 -5)^2 + (5 + 3)^2]], 3]
```

9.
```
ScientificForm[400 Cos[35 Degree] //N, 3]
```

11.
```
θ = N[ArcCos[2/3]/Degree]
Tan[θ Degree]^2
```

13.
```
N[EulerGamma, 25]
```

15.
```
A = {{1, 0, -1}, {2, 4, 7}, {5, 3, 0}}
B = {{6, 1}, {0, 4}, {-2, 3}}
A.B // MatrixForm
Clear[A, B]
```

17.
```
mymatrix = {{-2, 1, 3}, {0, -1, 1}, {1, 2, 0}}
Inverse[mymatrix] // MatrixForm
Clear[mymatrix]
```

19.
```
data = {{0.608, 0.05}, {0.430, 0.10}, {0.304, 0.20},
  {0.248, 0.30}, {0.215, 0.40}, {0.192, 0.50}}
Fit[data, {1, 1/V, 1/(V^2)}, V]
Clear[data]
```

21.
```
ourdata = {{0.5, 4.2}, {1.0, 16.1}, {1.5, 35.9}, {2.0, 64.2}}
Fit[ourdata, {t^2}, t]
2 × 16.0328 ft/s^2
Clear[ourdata]
```

23.
```
data = {{0.1, 1.1}, {0.3, 3.2}, {0.5, 4.9}, {0.7, 6.8}, {0.9, 9.1}}
Fit[data, {x}, x]
9.98182 lb/in
```

25.
```
N[(3 Pi + 7 I) Cos[37 Degree] + (2 + 8 I) Exp[-3 I + 2]]
```

27.
```
RandomInteger[{1, 100}, 20]
Max[%]
```

29. (a)
```
RandomInteger[{0, 1}, 10]
Count[%, 0]
N[%/10]
```

(b)
```
RandomInteger[{0, 1}, 100]
Count[%, 0]
N[%/100]
```

(c)
```
N[Count[RandomInteger[{0, 1}, 10 000], 0]/10 000]
```

(d) The fractional outcome approaches the probability as N becomes large.

31.
```
NSolve[4 β^7 - 16 β^4 + 17 β^3 + 6 β^2 - 21 β + 10 == 0, β]
```

33.
```
NIntegrate[(x^3)/(Exp[x] -1), {x, 0, 1}]
```

35.
```
NIntegrate[(x^2)/Sqrt[(x^7) + 1], {x, 2, ∞}]
```

37.
```
NIntegrate[Log[x]/Sqrt[1 - x^2], {x, 0, 1}] // Chop
```

Note that the integrand approaches 0 as x approaches 1.

39.
```
NIntegrate[1/((x - 1)^2), {x, 0, 1}]
```

41.
```
(a^3) NIntegrate[Sqrt[x^2 + y^2], {x, -1, 0}, {y, -Sqrt[1 - x^2], 0}]
```

43.
```
(4 k a^6) NIntegrate[(x^2 + z^2)^(3/2), {y, 0, 4},
  {x, 0, Sqrt[1 + y^2]}, {z, 0, Sqrt[1 + y^2 - x^2]}] // Chop
```

45.
```
NDSolve[{y''[x] + (Sin[x]^2) y'[x] + 3 y[x]^2 == Exp[-x^2],
  y[0] == 1, y'[0] == 0}, y, {x, 0, 3}]
Table[y[x] /. %[[1]], {x, 3}]
Plot[Evaluate[y[x] /. %%], {x, 0, 3}, AxesLabel → {"x", "y"}]
```

47.
```
sol = NDSolve[
  {y''[t] == Sin[t] y[t] + t, y[0] == 0, y'[0] == 1}, y, {t, 0, 6.4}]
Plot[Evaluate[y[t] /. sol], {t, 0, 6.4}, AxesLabel → {"t", "y"}]
Clear[sol]
```

49.
```
(1/π)NIntegrate[
  r/Sqrt[((2.5 - r Cos[θ])^2) + ((1.2 - r Sin[θ])^2) + 3.7^2],
  {r, 0, 1}, {θ, 0, 2π}]
```
in units of $q/(4\pi\varepsilon_0 a)$, where ε_0 is the permittivity constant.

SECTION 2.2.20

1. `Apart[(x^2 + 2x - 1)/(2x^3 + 3x^2 - 2x)]`
3. `Factor[x^4 + 2x^3 - 3x - 6]`
5. `Factor[a x^2 + a y + b x^2 + b y]`
7. `Together[(1/((x^2) -16)) - ((x + 4)/((x^2) - 3x - 4))]`
9. `Together[((1/x^2) - (1/y^2))/((1/x) + (1/y))]`
11. `Together[(3/((x^3) - x)) + (4/((x^2) + 4x + 4))] //ExpandDenominator`
13. `Apart[(x^4 - 2x^2 + 4x + 1)/(x^3 - x^2 - x + 1)]`
15. `PowerExpand[(54(x^3)(y^6)z)^(1/3)]`
17. `PowerExpand[Log[Sqrt[x]/x] + Log[(E x^2)^(1/4)]]`
19. `Simplify[(Sec[α] - 2Sin[α])(Csc[α] + 2Cos[α])Sin[α]Cos[α]]`
21. `TrigExpand[Sin[7θ]]`
23. `TrigFactor[(Sin[x]^3)Sin[3x] + (Cos[x]^3)Cos[3x]] //Simplify`
25. `TrigFactor[-Sin[α - β - γ] + Sin[α + β - γ] + Sin[α - β + γ] - Sin[α + β + γ]]`
27. `FullSimplify[Sec[x]Tan[x](Sec[x] + Tan[x]) + (Sec[x] - Tan[x])]`
29.
```
Simplify[(Sec[x] - Csc[x])/(Tan[x] + Cot[x]) ==
   (Tan[x] - Cot[x])/(Sec[x] + Csc[x])]
```
31. `FunctionExpand[Gamma[z]Gamma[-z]]`
33. `FunctionExpand[D[HermiteH[n, x], x]]`
35. `FullSimplify[D[Gamma[z + 1], z] == Gamma[z] + z D[Gamma[z], z]]`
37. `FunctionExpand[Gamma[z]Gamma[z + 1/2]]`
39.
```
Simplify[Sin[n π - θ], ((n - 1/2)/2) ∈ Integers]
Table [2k + 1/2, {k, -10, 10}]
```
41. `Simplify[ArcSin[Sin[u]], -π/2 ≤ u ≤ π/2]`
43. `Coefficient[((4a + x)^2)((7 + b x^2)^4), x, 0]`
45. `Numerator[(x^2 + 3x + 2)/(x^2 - 1)]`
47.
```
Needs["Units`"]
Needs["PhysicalConstants`"]
h = PlanckConstant
me = ElectronMass
mp = ProtonMass
```
(a)
```
ℰk = 10 ElectronVolt
Convert[h/Sqrt[2 me ℰk], Meter] //PowerExpand
```

(b)
```
ℰk = 10 Mega ElectronVolt
Convert[h/Sqrt[2 mp ℰk], Meter] // PowerExpand
Clear[h, me, mp, ℰk]
```

49.
```
Solve[{4 x + 5 y == 5, 6 x + 7 y == 7}, {x, y}]
```

51.
```
Solve[Sqrt[x + 2] + 4 == x, x]
```

53.
```
Solve[Sqrt[2 x - 3] - Sqrt[x + 7] == 2, x]
```

55.

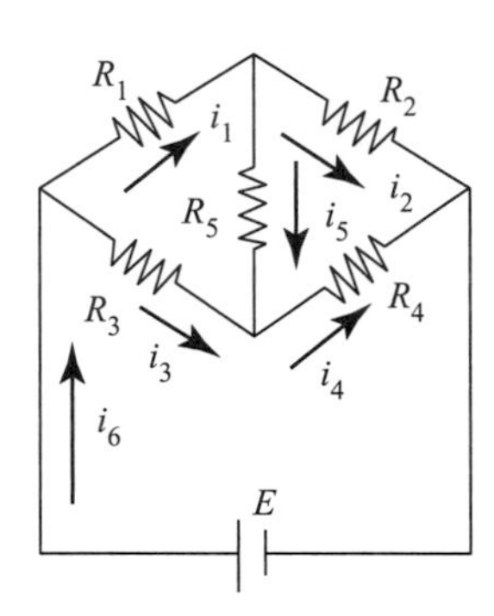


```
Solve[{i1 + i3 == i6, i2 + i5 == i1, i2 + i4 == i6, emf - i1 R1 - i2 R2 == 0,
    - i1 R1 - i5 R5 + i3 R3 == 0, -i2 R2 + i4 R4 + i5 R5 == 0},
  {i1, i2, i3, i4, i5, i6}] // Simplify
% /. {emf → 12 V, R1 → 2.0 Ω, R2 → 4.0 Ω,
  R3 → 6.0 Ω, R4 → 2.0 Ω, R5 → 3.0 Ω} /. V/Ω → A
```

57.
```
D[Sin[Exp[x^2]], x]
```

59.
```
D[x ArcSinh[x/3] - Sqrt[9 + x^2], x] // Simplify
```

61.
```
D[Sin[x y]/Cos[x + y], x]
```

63.
```
Dt[x^2 + 3 x y - y^2] // Simplify
% /. {x → 2, y → 3, Dt[x] → 0.05, Dt[y] → -0.04}
```

65.
```
P = n R T/V
Dt[P, t, Constants → {n, R}]
% /. {Dt[T, t, Constants → {n, R}] → α,
   Dt[V, t, Constants → {n, R}] → β} // Simplify
Clear[P]
```

67.
```
Integrate[5 Log[t] - Exp[-3 t], t]
```

69.
```
Integrate[4/(1 + x^2), {x, 0, 1}]
```

71.
```
Integrate[(2 x^2 - x + 4)/(x^3 + 4 x), x]
D[%, x]
% // Simplify
```

73.
```
Integrate[(x^2) Exp[-2 a x^2], {x, -∞, ∞}, Assumptions → a > 0]
```

75.
```
Integrate[(Sin[x]^p) x, {x, 0, π}, Assumptions → {p > -1}]
```

77. (a)
```
sol = Solve[Integrate[(A/(b^2 + x^2))^2,
        {x, -∞, ∞}, Assumptions → b > 0] == 1, A][[2]]
```

(b)
```
Integrate[(x(A/(b^2+x^2))^2/.sol),
  {x, -∞, ∞}, Assumptions→b>0]
```
We do not need *Mathematica* to determine this. Because the integrand is odd, the integral must vanish.

(c)
```
Integrate[((x^2)(A/(b^2+x^2))^2/.sol),
  {x, -∞, ∞}, Assumptions→b>0]
Clear[sol]
```

79.
```
Sum[(Cos[x]^n)/(2^n), {n, 0, ∞}]
```

81.
```
Sum[i^3, {i, n}]
```

83.
```
Series[Sec[x], {x, π/3, 3}]
```

85.
```
Limit[
   Log[10, x-a]/((a-b)(a-c)) + Log[10, 2(x-b)]/((b-c)(b-a)) +
    Log[10, x-c]/((c-a)(c-b)), x→∞] //FullSimplify
```

87.
```
Limit[1/(Exp[x] - Exp[x-x^(-2)]), x→∞]
```

89.
```
Limit[(1+Sin[4x])^Cot[x], x→0, Direction→ -1]
```

91.
```
Limit[(Sqrt[(t^2)+9]-3)/(t^2), t→0]
```

93.
```
DSolve[y''[x] - 2y'[x] == Sin[4x], y[x], x]
```

95. (a)
```
sol = DSolve[ϕ''[x] + (k^2)ϕ[x] == 0, ϕ[x], x]
```
(b)
```
((ϕ[x] /. sol[[1]]) /. x→0) == 0
```
(c)
```
Simplify[(Sin[k a] /. k→ (n π/a)) == 0, n ∈ Integers]
```
(d)
```
Reduce[{Sin[k a] == 0, a ≠ 0}, k, GeneratedParameters→n]
```
(e)
```
Solve[Integrate[(C2 Sin[n π x/a])^2, {x, 0, a},
     Assumptions→{n ∈ Integers, n>0}] == 1, C2][[2, 1]]
En == ((ℏ^2)(k^2)/(2m)) /. k→ (n π/a)
ϕn == C2 Sin[k x] /. {k→ (n π/a), C2→Sqrt[2/a]}
```
(f)
```
Integrate[(2/a)Sin[n π x/a]Sin[m π x/a], {x, 0, a},
  Assumptions→{m ∈ Integers, n ∈ Integers, m ≠ n}]
Clear[sol]
```

97. (a)
```
E[n_] := (ℏ ω (n+1/2))
```
where **E** is the capital Greek letter epsilon.
```
Table[E[n], {n, 0, 3}]
```
(b)
```
ϕ[n_, x_] := ((2^(-n/2))(n!^(-1/2))(((m ω)/(ℏ π))^(1/4))
    HermiteH[n, Sqrt[m ω/ℏ]x]Exp[-((m ω)/(2ℏ))x^2])
Table[ϕ[n, x], {n, 0, 3}]
Clear[E, ϕ]
```

99.
```
FourierTransform[Piecewise[{{-A, -b<x<0}, {A, 0<x<b}}],
  x, -k, Assumptions→b>0] //FullSimplify
```

101.
```
InverseFourierTransform[1/(1+I(k^3)), k, -x]
FourierTransform[%, x, -k]
Simplify[% == 1/(1+I(k^3))]
```

103.
```
RSolve[{A[n] == A[n-1] + r A[n-1], A[0] == P}, A[n], n]
A[n] /. %[[1, 1]] /. {P→1000, r→.08, n→30}
```

105. Apply `Simplify[■]` three times:

```
            1 - Cos[2 x]
--------------------------------------------------
(Csc[x] + Cot[x]) (1 - Cos[x]) (Tan[2 x] + Cot[2 x])

            1 - Cos[2 x]
-------------------------------------------
(Csc[x] + Cot[x]) (1 - Cos[x]) (2 Csc[4 x])

   1 - Cos[2 x]
------------------
Sin[x] (2 Csc[4 x])

   1 - Cos[2 x]
------------------
(2 Csc[4 x] Sin[x])
```

SECTION 2.3.5

1.
```
Plot[x/(x^2+1), {x, -10, 10}]
```

3.
```
sol =
 NDSolve[{y''[t] == -(t^3)y[t] + 1, y[0] == 1, y'[0] == 0}, y, {t, 0, 8}]
Plot[Evaluate[y[t]/.sol], {t, 0, 8}, AxesLabel→{"t", "y"}]
Clear[sol]
```

5.
```
n = 5;
sol = NDSolve[{ψ''[ξ] + (-ξ^2 + 2 n + 1) ψ[ξ] == 0,
    ψ[0] == 0, ψ'[0] == 1}, ψ, {ξ, -5, 5}]
{ψ[-5], ψ[5]}/.sol[[1]]
Plot[Evaluate[ψ[ξ]/.sol], {ξ, -5, 5}, AxesLabel -> {"ξ", "ψ"}]
Clear[n, sol]
```

7.
```
sol = NDSolve[{x''[t] - (12/(x[t]^13)) + (6/(x[t]^7)) == 0,
    x[0] == 1.02, x'[0] == 0}, x, {t, 0, 10}]
Plot[Evaluate[{x[t], x'[t]}/.sol], {t, 0, 10},
 PlotStyle→{{}, {Dashing[{0.01, 0.02}]}},
 PlotLabel -> "x(t) and x'(t)"]
Clear[sol]
```

9.
```
Plot[x^2 + Cos[32 x], {x, -5, 5}, AxesLabel→{"x", "f"}]
```

The plot cannot be right since the function is an even function.

```
Plot[x^2 + Cos[32 x], {x, -5, 5},
 AxesLabel→{"x", "f"}, PlotPoints→120]
```

This is more like it.

11. (a)
```
Z = 1 + Exp[-ϵ/(k T)]
```

The average energy E is

```
E = k(T^2)D[Log[Z], T] // Simplify
```

where we have assigned the average energy to the capital Greek letter epsilon **Ε**. (The ordinary keyboard **E** has built-in meaning in *Mathematica*; it is the exponential constant e.) Let

$$\varepsilon = E/\epsilon$$

$$t = kT/\epsilon$$

Thus,

$$\varepsilon = \frac{1}{1 + e^{1/t}}$$

```
Plot[1/(1+Exp[1/t]), {t, 0, 4}, AxesLabel→{"T (ϵ/k)", "E (ϵ)"}]
```

(b) The heat capacity C is

```
D[Ε, T]
```

Let

$$c = C/k$$

$$t = kT/\epsilon$$

Thus,

$$c = \frac{e^{1/t}}{\left(1 + e^{1/t}\right)^2 t^2}$$

```
Plot[(Exp[1/t])/(((1+Exp[1/t])^2)(t^2)),
  {t, 0, 4}, AxesLabel→{"T (ϵ/k)", "C (k)"}]
```

(c) The entropy S is

$$S = k \ln Z + \frac{E}{T}$$

```
k Log[Z] +Ε/T
```

Let

$$s = S/k$$

$$t = kT/\epsilon$$

Thus,

$$s = \frac{1}{\left(1 + e^{1/t}\right) t} + \ln\left(1 + e^{-1/t}\right)$$

```
Plot[(1/((1+Exp[1/t])t))+Log[1+Exp[-1/t]], {t, 0, 4},
  PlotRange->All, AxesLabel→{"T (ϵ/k)", "S (k)"}]
Clear[Z, E]
```

13.
```
Series[Sin[x]Cos[x], {x, 0, 7}] //Normal
Plot[{%, Sin[x]Cos[x]}, {x, -2, 2}]
```
The series is a good approximation of $f(x)$ for $|x| \lesssim 1.3$.

15.
```
S[n_, x_]:=Sum[(((-1)^k)/((k!)^2))((x/2)^(2k)), {k, 0, n}]
Plot[Evaluate[{BesselJ[0, x], S[1, x], S[2, x], S[3, x], S[4, x]}],
  {x, -5, 5}, PlotRange→{-0.75, 1.25},
  PlotStyle→{{Black, Thickness[0.0125], Dashing[Medium]},
    {Green, Thick}, {Magenta, Thick}, {Blue, Thick}, {Red, Thick}},
  Background→Lighter[Blue, 0.9],
  PlotLabel→"Partial sums of the Bessel function"]
Clear[S]
```

17.
```
Plot[1/(Exp[x-12]+1), {x, 0, 20},
  PlotRange→{0, 1.1}, PlotStyle→Thickness[0.008],
  AxesLabel→{"ε(kT)", "n(ε)"}, Ticks→{{{12, "μ"}}, {0.5, 1}}]
Plot[1/(Exp[x]+1), {x, -10, 6}, Frame→True,
  GridLines→{Table[i, {i, -10, 6, 2}], Table[i, {i, 0, 1, 0.2}]},
  PlotStyle→Thickness[0.008],
  FrameLabel→{"(ε-μ)(kT)", "n(ε)"},
  PlotLabel->"Fermi-Dirac Distribution",
  FrameTicks->{Table[i, {i, -10, 6, 2}], Automatic, None, None},
  FrameStyle→Thickness[0.005]]
```

19.
```
ClickPane[Plot[{6(Sech[x]^2), Sqrt[5-x^2]},
   {x, -Sqrt[5], Sqrt[5]}], (xycoord=#)&]
Dynamic[xycoord]
startvalue={-2.2190927280183064`, -1.1662385139950224`,
   1.166238513995022`, 2.219092728018306`};
Table[FindRoot[6(Sech[x]^2) == Sqrt[5-x^2],
   {x, startvalue[[i]]}], {i, Length[startvalue]}]
Clear[xycoord, startvalue]
```

21.
```
ClickPane[Plot[BesselJ[0, x], {x, 0, 16}], (xycoord=#)&]
Dynamic[xycoord]
startvalue={2.4017801178387854`, 5.53453679328068`,
   8.728720070201827`, 11.922903347122976`, 14.871380218127111`};
Table[FindRoot[BesselJ[0, x] == 0, {x, startvalue[[i]]}],
  {i, Length[startvalue]}]
ourstartvalue={3.9374451548201064`, 10.202958505703895`};
Table[FindMinimum[BesselJ[0, x], {x, ourstartvalue[[i]]}],
  {i, Length[ourstartvalue]}]
Clear[xycoord, startvalue, ourstartvalue]
```

23.

$$c = \left(\frac{1}{t}\right)^2 \frac{e^{1/t}}{\left(e^{1/t} - 1\right)^2}$$

```
ClickPane[Plot[{((1/t)^2)(Exp[1/t]/(Exp[1/t] - 1)^2), 0.3552},
   {t, 0.1, 0.3}], (tccoord = #) &]
FindRoot[((1/t)^2)(Exp[1/t]/(Exp[1/t] - 1)^2) == 0.3552,
  {t, tccoord[[1]]}]
358.5K/t/.%
Clear[tccoord]
```

25.

```
ClickPane[Plot[{Abs[(Sin[x] - x)/Sin[x]], 0.01}, {x, 0, 0.3}],
   (xycoord = #) &]
FindRoot[Abs[(Sin[x] - x)/Sin[x]] == 0.01, {x, xycoord[[1]]}]
%[[1, 2]]/Degree
Clear[xycoord]
ClickPane[Plot[{Abs[(Sin[x] - x)/Sin[x]], 0.005}, {x, 0, 0.2}],
  (xycoord = #) &]
FindRoot[Abs[(Sin[x] - x)/Sin[x]] == 0.005, {x, xycoord[[1]]}]
%[[1, 2]]/Degree
Clear[xycoord]
```

27.

```
mydata = {{4, 5}, {6, 8}, {8, 10}, {9, 12}};
Fit[mydata, {1, x}, x]
plot1 = Plot[%, {x, 3, 10}, AxesLabel -> {"x", "y"}]
plot2 = ListPlot[mydata,
   AxesLabel -> {"x", "y"}, PlotStyle -> PointSize[0.02]]
Show[plot1, plot2]
Clear[mydata, plot1, plot2]
```

29.

```
xlist = Table[i, {i, 0, 2π, 2π/9}];
Length[%]
ylist = Sin[xlist];
Transpose[{xlist, ylist}];
ListPlot[%, PlotStyle -> PointSize[0.025],
   AxesLabel -> {"xlisti", "ylisti"}]
Clear[xlist, ylist]
```

31.

```
data = {{0, 1}, {0.1, 1.05409}, {0.2, 1.11803}, {0.3, 1.19523},
     {0.4, 1.29099}, {0.5, 1.41421}, {0.6, 1.58114},
     {0.7, 1.82574}, {0.8, 2.23607}, {0.9, 3.16228}};
Fit[data, {1, x, x^2, x^3}, x]
Plot[%, {x, 0, 1}, AxesLabel → {"x", "f(x)"}]
ListPlot[data, PlotStyle → PointSize[0.02],
   AxesLabel → {"x", "f(x)"}]
Show[%, %%]
Clear[data]
```

33.
```
ParametricPlot[{θ-0.5Sin[θ],1-0.5Cos[θ]},
  {θ, 0, 7π}, PlotRange->{{0, 7π}, {0, 2}},
  Ticks→{Range[0, 7π, π], Range[0, 2]},
  AxesLabel→{"x (r)", "y (r)"},
  ImageSize->72 * 7.5, PlotStyle->{Red, Thick}]
```

```
ParametricPlot[{θ-1.5Sin[θ],1-1.5Cos[θ]},
  {θ, 0, 7π}, PlotRange->{{-π/6, 7π}, {-1, 3}},
  Ticks→{Range[π, 7π, 2π], Range[-1, 3]},
  AxesLabel→{"x (r)", "y (r)"},
  ImageSize->72 * 7.5, PlotStyle->{Red, Thick}]
```

35.
```
ParametricPlot[{√3 Sin[t], √5 Cos[t]},
  {t, 0, 2π}, AxesLabel->{"x", "y"}]
```

37.
```
Do[vdP=NDSolve[{x'[t]==v[t], v'[t]==0.5(1-x[t]^2)v[t]-x[t],
     v[0]==0, x[0]==i}, {x, v}, {t, 0, 50}];
  Print[ParametricPlot[Evaluate[{x[t], v[t]}/.vdP],
    {t, 0, 7π}, AxesLabel→{"x", "v"}]], {i, 1, 3, 0.5}]
Clear[vdP]
```

39.
```
Plot3D[x^2+y^2, {x, -2, 2}, {y, -2, 2},
  BoxRatios->{1, 1, 1}, AxesLabel->{"x", "y", "z"}]
```

41.
```
Plot3D[Sin[x+y], {x, 0, 2π},
  {y, 0, 2π}, AxesLabel→{"x", "y", "f(x, y)"}]
```

43.
```
Plot3D[x(y^2)-x^3, {x, -2, 2}, {y, -2, 2}, BoxRatios→{1, 1, 1},
  Boxed→False, Axes→False, ViewPoint→{-0.060, 3.354, 0.447}]
```

45.
```
Plot3D[Exp[-(x+y)^2], {x, -2, 2}, {y, -2, 2},
  AxesLabel→{"x", "y", "f"}, Ticks→{{-2, 0, 2}, {-2, 0, 2}, None}]
```

47.
```
ListAnimate[Table[Plot3D[(t+x)((x^2)-3(y^2)), {x, -2, 2},
    {y, -2, 2}, BoxRatios→{1, 1, 1}, PlotRange→{-150, 150},
    Boxed→False, Axes→False], {t, -12, 12, 1}]]
```

49.
```
ListAnimate[Table[Plot[Sin[x-i(2π/16)]+0.75Sin[x+i(2π/16)],
    {x, 0, 8π}, PlotRange→{-2, 2},
    AxesLabel->{"x (1/k)", "y (A)"}], {i, 0, 15, 1}]]
```

There are antinodes but no nodes.

There is another way to generate the animation:

```
Do[Print[Plot[Sin[x-i(2π/16)]+0.75Sin[x+i(2π/16)],
   {x, 0, 8π}, PlotRange→{-2, 2},
   AxesLabel->{"x (1/k)", "y (A)"}]], {i, 0, 15}]
```

To animate, select all the Print cells and choose **Graphics ▸ Rendering ▸ Animate Selected Graphics**. Again, there are antinodes but no nodes.

51.
```
ParametricPlot3D[{{u Cos[v], u Sin[v], u^2},
  {u Cos[v], u Sin[v], (1/2)(u Cos[v] + 3 u Sin[v] + 3)}},
 {u, 0, 2}, {v, 0, 2 π}, Boxed→False, Axes→False]
```

53.
```
g[i_]:=
 Plot[
  {Exp[-16 (x - i)^2] + 1.5 Exp[-(x + i)^2],
   1.5 Exp[-(x + i)^2] + 3, Exp[-16 (x - i)^2] + 5}, {x, -3.0, 3.0},
  PlotStyle→{{Thickness[0.01], Black},
    {Thickness[0.005], Red}, {Thickness[0.005], Blue}},
  PlotRange→{-0.1, 6.25}, Axes→False,
  Frame -> True, FrameTicks -> None]
Grid[Partition[Table[g[i], {i, -2.0, 3.5, 0.5}], 3]]
Clear[g]
```

55.
```
Plot[{y Tan[y], Sqrt[16 - y^2]},
 {y, 0, 8}, PlotRange→{{0, 5}, {0, 5}}]
startvalue = {{1.254, 3.802}, {3.606, 1.739}};
Table[FindRoot[y Tan[y] == Sqrt[16 - y^2], {y, startvalue[[i, 1]]}],
 {i, Length[startvalue]}]
```

SECTION 2.4.12

1.
```
RandomInteger[20, {5, 5}]
% // MatrixForm
```

3. (a) For $n = 10^2$, we have for the outcomes

```
RandomInteger[{1, 2}, 10^2]
```

where "1" represents heads up and "2" represents tails up.
We can generate the distribution of getting heads and tails:

```
distribution = Table[Count[%, i], {i, 1, 2}]
```

To produce a bar chart of this distribution, let us use the function **BarChart** in the package **BarCharts`**:

```
Needs["BarCharts`"]
BarChart[distribution, BarLabels→{"heads", "tails"}]
```

(b) For $n = 10^3, 10^4, 10^5$, and 10^6, we have

```
Do[Print[
  BarChart[Table[Count[RandomInteger[{1, 2}, 10^j], i], {i, 1, 2}],
   BarLabels→{"heads", "tails"},
   PlotLabel -> "n = " <> ToString[10^j]]], {j, 3, 6}]
```

(c) Probability indicates the relative frequency that a particular event would occur in the long run. Here, after a very great number of tosses, the distribution became fairly uniform and the fractional number of times getting, for example, heads approached the value 1/2, the probability of a coin landing heads up with a single toss.

5.
```
Table[(x^n) - 1, {n, 1, 11, 2}]
Drop[%, {2, 3}]
```

7.
```
mymatrix = Array[a, {4, 4}]
% // MatrixForm
mymatrix[[All, 3]]
% // MatrixForm
mymatrix[[Range[2, 4], {3, 4}]]
% // MatrixForm
Clear[mymatrix]
```

9.
```
Select[{14, 29, 30, 35, 53, 86, 42,
  76, 16, 98, 87, 54, 100, 69, 20, 101, 3}, OddQ]
```

11.
```
mylistA = RandomInteger[{301, 600}, 20]
mylistB = RandomInteger[{301, 600}, 20]
ReplacePart[mylistA, mylistB[[12]], 2]
Clear[mylistA, mylistB]
```

13.
```
{a, {b, c}, a, {d, {e, {f, {a, c, g}}}}}
Flatten[%]
Position[%%, {a, c, g}]
FlattenAt[%%%, {{2}, {4, 2, 2, 2}}]
```

15.
```
mylist = {a, {d, {e, {f, {g, h}}}}, {{i, {j, k}}, {p, q, {r, s}}}}
Position[mylist, {g, h}]
Position[mylist, {r, s}]
FlattenAt[mylist, {{2, 2, 2, 2}, {3, 2, 3}}]
Clear[mylist]
```

17.
```
bigger[x_] := x > 50
someRandoms = RandomInteger[{1, 200}, 30]
Select[someRandoms, bigger]
Clear[bigger, someRandoms]
```

19.
```
yourlist = Flatten[Table[{n, l, ml, ms}, {n, 2, 4},
   {l, 0, n - 1}, {ml, -l, l}, {ms, -1/2, 1/2}], 3]
yourtest[x_] := x[[3]] == -2
Select[yourlist, yourtest]
Clear[yourlist, yourtest]
```

21.
```
noduplicates[list_] := Union[list] === Sort[list]
noduplicates[{d, c, a, b}]
noduplicates[{d, d, d, a}]
noduplicates[{a, a, a, a}]
Clear[noduplicates]
```

23.
```
onlyduplicates[list_] := Length[Union[list]] == 1
onlyduplicates[{d, c, a, b}]
```

```
onlyduplicates[{a, a, a, d}]
onlyduplicates[{a, a, a, a}]
onlyduplicates[{x^2 - 1, x^2 - 1, x^2 - 1, x^2 - 1, (x - 1) (x + 1)}]
onlyduplicates[{{-1, -1, 2}, {-1, -1, 2}, {-1, -1, 2}, {-1, -1, 2}}]
Clear[onlyduplicates]
```

25.
```
Range[2, 16, 2]^2
Apply[Plus, %]/Length[%]
```

27.
```
ourlist =
 {84, 79, 30, 45, 51, 86, 42, 57, 6, 98, 3, 87, 14, 100, 69, 20}
```

(a) `Apply[Plus, ourlist]/Length[ourlist] //N`

(b) `Total[ourlist]/Length[ourlist] //N`

(c) `Mean[ourlist] //N`

(d) `(Plus @@ Take[Sort[ourlist], {8, 9}])/2`

(e) `Median[ourlist]`

(f)
```
crit[x_] := x > 50
Select[ourlist, crit]
```

(g)
```
newlist = Partition[ourlist, 4]
% // MatrixForm
```

(h) `newlist[[3]]`

(i) `Transpose[newlist][[3]]`

or

`newlist[[All, 3]]`

(j)
```
Flatten[newlist]
Clear[ourlist, crit, newlist]
```

29.
```
Range[100]
Partition[%, Length[%]/2]
MapAt[Reverse, %, 2]
%[[1]] + %[[2]]
101 Length[%]
%/100
```

31.
```
T = {15, 20, 25, 30, 40, 50, 60, 70, 80, 90,
   100, 110, 120, 130, 140, 150, 160, 170, 180, 190, 200,
   210, 220, 230, 240, 250, 260, 270, 280, 290, 298.1};
Cp = {0.311, 0.605, 0.858, 1.075, 1.452, 1.772, 2.084,
   2.352, 2.604, 2.838, 3.060, 3.254, 3.445, 3.624, 3.795,
   3.964, 4.123, 4.269, 4.404, 4.526, 4.639, 4.743, 4.841,
   4.927, 5.010, 5.083, 5.154, 5.220, 5.286, 5.350, 5.401};
Integrate[Interpolation[Transpose[{T, Cp/T}]][x], {x, 15, 298.1}]
Clear[T, Cp]
```

33.
```
Needs["VectorAnalysis`"]
SetCoordinates[Cartesian[x, y, z]];
a = {ax[x, y, z], ay[x, y, z], az[x, y, z]};
```

```
f = func[x, y, z];
Simplify[Curl[f a] - f Curl[a] + CrossProduct[a, Grad[f]]]
Clear[f, a]
```

35.

```
data = {{0.46, 0.19}, {0.69, 0.27}, {0.71, 0.28}, {1.04, 0.62},
    {1.11, 0.68}, {1.14, 0.70}, {1.14, 0.74}, {1.20, 0.81},
    {1.31, 0.93}, {2.03, 2.49}, {2.14, 2.73}, {2.52, 3.57},
    {3.24, 3.90}, {3.46, 3.55}, {3.81, 2.87}, {4.06, 2.24},
    {4.93, 0.65}, {5.11, 0.39}, {5.26, 0.33}, {5.38, 0.26}};
FindFit[data, a1 Exp[-(1/2) (((x - a2) /a3)^2)], {a1, a2, a3}, x]
```

Thus, $a_1 = 3.97924, a_2 = 2.99725$, and $a_3 = 1.00152$.

```
Show[ListPlot[data, PlotStyle -> PointSize[0.025]],
 Plot[Evaluate[a1 Exp[-(1/2) (((x - a2) /a3)^2)] /.%], {x, 0, 6}],
 AxesLabel → {"x", "y"}]
Clear[data]
```

37.

```
tlist = Table[15 i, {i, 58}];
countlist = {775, 479, 380, 302, 185, 157, 137, 119, 110, 89, 74, 61,
    66, 68, 48, 54, 51, 46, 55, 29, 28, 37, 49, 26, 35, 29, 31,
    24, 25, 35, 24, 30, 26, 28, 21, 18, 20, 27, 17, 17, 14, 17,
    24, 11, 22, 17, 12, 10, 13, 16, 9, 9, 14, 21, 17, 13, 12, 18};
data = Transpose[{tlist, countlist}];
FindFit[data, a1 + a2 Exp[-t/a3] + a4 Exp[-t/a5],
 {a1, a2, a3, a4, a5}, t, PrecisionGoal → 6]
decay[t_] = (a1 + a2 Exp[-t/a3] + a4 Exp[-t/a5]) /.%
Show[ListPlot[data, PlotStyle → PointSize[0.015]],
 Plot[decay[t], {t, 15, 870}, PlotRange -> All],
 AxesLabel -> {"time (sec)", "counts"}]
Clear[tlist, countlist, data, decay]
```

39.

```
data = {{100, -160}, {200, -35},
    {300, -4.2}, {400, 9.0}, {500, 16.9}, {600, 21.3}};
B = Interpolation[data]
Show[Plot[B[T], {T, 100, 600}],
 ListPlot[data, PlotStyle → PointSize[0.02]], Frame → True,
 FrameLabel → {"T (K)", "B (cm^3/ mol)"}, Axes → None]
B[450] (cm^3) / mol
Clear[data, B]
```

41. Let us begin with the assignment:

```
distances = {1023.56, 1023.47, 1023.51, 1023.49,
   1023.51, 1023.48, 1023.50, 1023.53, 1023.48, 1023.52};
```

The sample mean (in ft) is

```
NumberForm[Mean[distances], 7]
```

The sample standard deviation (in ft) is

```
NumberForm[StandardDeviation[distances], 2]
```

The standard deviation of the mean (in ft) is

```
NumberForm[%/Sqrt[Length[distances]], 1]
```

Thus, the best estimates for the distance between the two points and its uncertainty are 1023.505 ft and 0.009 ft, respectively. The experimental result is usually stated as 1023.505 ± 0.009 ft.

```
Clear[distances]
```

43. (a) Let us begin with the assignment for the list of numbers of decays observed in 1 minute:

```
dpm = {9, 11, 15, 16, 19, 22, 24}
```

For each number of decays observed in 1 minute, there is a number of times observed. The list of numbers of times observed is

```
freq = {2, 2, 2, 1, 1, 1, 1}
```

and the list of fractional numbers of times observed is

```
frac = freq/Total[freq]
```

The best estimate of the mean number of decays in 1 minute is the sample mean:

```
μ = Total[dpm frac] //N
```

The best estimate of the standard deviation is

```
NumberForm[Sqrt[μ], 2]
```

The best estimate of the standard deviation of the mean σ_μ is given by

```
NumberForm[Sqrt[μ/Total[freq]], 2]
```

(b) To plot the bar histogram, we use the function **GeneralizedBarChart** in the package **BarCharts`**. Here is the bar histogram of the fractional number of "times observed" versus the "number of decays" together with the plot of the Poisson distribution with $\mu = 15.1$:

```
Needs["BarCharts`"]
Show[GeneralizedBarChart[Transpose[{dpm, frac, Table[0.2, {7}]}]],
  Plot[(μ^x Exp[-μ])/((x)!), {x, 9, 24},
    PlotStyle→Thickness[0.0125]], Frame→True,
FrameLabel→{"Decay Number", "frac times observed"},
FrameTicks→{Automatic, Automatic, None, None}, Axes→False]
Clear[dpm, freq, frac, μ]
```

45.
```
myaccumulate[list_]:=
  Table[Sum[list[[n]], {n, 1, i}], {i, 1, Length[list]}]
myaccumulate[{a, b, c, d}]
```

```
myaccumulate[{a, b, c, d, e}]
myaccumulate[{{a, b}, {c, d}, {e, f}}]
myaccumulate[{{a, b}, {c, d}, {e, f}, {g, h}}]
Clear[myaccumulate]
```

47.
```
MyData = {{"t(s)", "d(ft)"},
   {0.5, 4.2}, {1.0, 16.1}, {1.5, 35.9}, {2.0, 64.2}};
Export["C:\APGTM\MyFolder\Grav.txt", MyData, "Table"]
Export["C:\APGTM\MyFolder\Grav1.xls", MyData]
Clear[MyData]
```

49.
```
data =
  Import["/2nd Edition 2008/APGTM2ND/OurFolder/ACDC.txt", "Table"]
mydata = Drop[data, 9]
func[x_] := Take[x, {2, 13}]
ourdata = func/@mydata
newdata = ourdata // Flatten
concentration = Interpolation[newdata]
Plot[concentration[x], {x, 1, 480},
 Axes → False, Frame → True, FrameTicks →
  {{Automatic, Automatic}, {Transpose[{Table[i, {i, 1, 433, 72}],
      Table[1964 + i, {i, 1, 37, 6}]}], Automatic}},
 FrameLabel → {"Year", "Concentration (ppm)"}]
Clear[data, mydata, func, ourdata, newdata, concentration]
```

SECTION 2.5.4

1. (i) $\texttt{Inverse@}\begin{pmatrix} 16 & 0 & 0 \\ 0 & 14 & -6 \\ 0 & -6 & -2 \end{pmatrix} \texttt{ // MatrixForm}$

$\begin{pmatrix} 16 & 0 & 0 \\ 0 & 14 & -6 \\ 0 & -6 & -2 \end{pmatrix} \texttt{. Inverse@}\begin{pmatrix} 16 & 0 & 0 \\ 0 & 14 & -6 \\ 0 & -6 & -2 \end{pmatrix} \texttt{ // MatrixForm}$

(ii) $\texttt{NIntegrate}\left[\frac{r}{\sqrt{(1 - r\,\texttt{Cos}[\varphi])^2 + (2 - r\,\texttt{Sin}[\varphi])^2 + 5^2}},\ \{r, 0, 1\}, \{\varphi, 0, 2\pi\}\right]$

(iii) $\texttt{Assuming}\left[\frac{c\,h}{k\,T} > 0,\ \int_0^{\infty} \frac{2\pi h c^2}{\lambda^5 \left(e^{\frac{h c}{\lambda k T}} - 1\right)}\, d\lambda\right] /.\ \frac{2\pi^5 k^4}{15 c^2 h^3} \to \sigma$

(iv) $\texttt{H}[V\_]\texttt{@}\psi\_ := \left(-\frac{\hbar^2}{2\,m}\partial_{x,x}\psi + V\psi\right)$

$\texttt{H}[\texttt{V}[x]]\texttt{@}\psi[x] == E\,\psi[x]$

$\psi[x] = e^{i\,k\,x}$

$\texttt{H}[0]\texttt{@}\psi[x] == E\,\psi[x]$

```
Solve[%, E]
Clear[H, ψ]
```

(v)

```
Plot3D[1/Sqrt[(x + 1)^2 + y^2] - 1/Sqrt[(x - 1)^2 + y^2], {x, -15, 15},
  {y, -15, 15}, Ticks → {Automatic, Automatic, {-0.04, 0, 0.04}},
  AxesLabel → {"x (b)", "y (b)", "V (kQ/b)"}]
Plot3D[1/Sqrt[(x + 1)^2 + y^2] - 1/Sqrt[(x - 1)^2 + y^2], {x, -15, 15},
  {y, -15, 15}, PlotRange → {-0.125, 0.125}, PlotPoints → 40,
  BoxRatios → {1, 1, 1}, Boxed → False, Axes → False]
```

(vi)

```
T = {15, 20, 25, 30, 40, 50, 60, 70, 80, 90,
    100, 110, 120, 130, 140, 150, 160, 170, 180, 190, 200,
    210, 220, 230, 240, 250, 260, 270, 280, 290, 298.1};
Cp = {0.311, 0.605, 0.858, 1.075, 1.452, 1.772, 2.084,
    2.352, 2.604, 2.838, 3.060, 3.254, 3.445, 3.624, 3.795,
    3.964, 4.123, 4.269, 4.404, 4.526, 4.639, 4.743, 4.841,
    4.927, 5.010, 5.083, 5.154, 5.220, 5.286, 5.350, 5.401};
```

where $\mathcal{C}$ is the capital script c.

```
CpOverT = Cp/T;
Sum[1/2 (CpOverT[[i + 1]] + CpOverT[[i]]) (T[[i + 1]] - T[[i]]), {i, 1, Length[T] - 1}]
```

with the units cal/K.

```
Transpose[{T, Cp}];
ListPlot[%, PlotStyle → PointSize[0.015],
  AxesLabel → {"T (K)", "Cp (cal/K)"}]
Clear[T, Subscript]
```

(vii)

```
Needs["VectorAnalysis`"]
SetCoordinates[Spherical];
Coordinates[]
SetCoordinates[Spherical[r, θ, ϕ]]
V = -E0 (1 - (a/r)^3) r Cos[θ];
```

where **E** is the capital Greek letter epsilon.

```
Laplacian[V] // Simplify
V/.r → a
V/.((a/r)^3 → 0)
-Grad[V] // Simplify
ε0 %[[1]]/.r → a
```

$$2\pi a^2 \int_0^{\pi} \% \,\mathrm{Sin}[\theta]\, d\theta$$

```
Clear[V]
```

3. (a) $\sum_{n=0}^{\infty} n \frac{1}{n!} \left(\frac{t}{\tau_0}\right)^n e^{-t/\tau_0}$

$$\sum_{n=0}^{\infty} \frac{n \left(\frac{t}{\tau_0}\right)^n e^{-\frac{t}{\tau_0}}}{n!}$$

(b) $\frac{\sum_{n=0}^{\infty} n\, e^{-\frac{n \epsilon}{k T}}}{\sum_{n=0}^{\infty} e^{-\frac{n \epsilon}{k T}}}$

$$\frac{\sum_{n=0}^{\infty} n\, e^{-\frac{n \epsilon}{k T}}}{\sum_{n=0}^{\infty} e^{-\frac{n \epsilon}{k T}}}$$

5. $-\frac{1}{\omega_1} \int_{\omega_1 t}^{0} e^{-\gamma t} e^{\frac{\gamma - \beta}{\omega_1} z}\, \mathrm{Sin}[z]\, dz$ `// TraditionalForm`

7. (a) $r^2 \frac{1}{24 a^3} \frac{r^2}{a^2} e^{-r/a}$;

(b) $\mathrm{Solve}[\partial_r \% == 0, r][[4]]$

9. $\frac{1}{e^{(\varepsilon - \mu)/(k T)} + 1}$ /.

$\{\varepsilon \to 7.20\,\mathrm{eV}, \mu \to 7.11\,\mathrm{eV}, k \to 8.617 \times 10^{-5}\,\mathrm{eV/K}, T \to 300\,\mathrm{K}\}$

11. `Needs["Units`"]`

(a) $\mathcal{C} = 5.10 \times 10^{-6}$ Farad

$f_0 = 1.30 \times 10^3$ Hertz

where we have used the capital script $\mathcal{C}$ to denote the capacitance because the keyboard capital C has built-in meaning in *Mathematica*. The inductance is given by

$$\mathrm{ScientificForm}\left[L = \mathrm{Convert}\left[\mathrm{Solve}\left[f_0 == \frac{1}{2\pi\sqrt{L\,\mathcal{C}}}, L\right][[1,1,2]], \mathrm{Henry}\right]\right]$$

For a discussion on obtaining parts of expressions, see Section 3.1.3.1.

(b) $V_{rms} = 11.0$ Volt

$\bar{P}_0 = 25.0$ Watt

At resonance, we have

$$I_{rms} = \frac{V_{rms}}{R}$$

and $\cos\phi = 1$. Thus, the average power delivered by the source at resonance can be written as

$$\overline{P}_0 = \frac{V_{\text{rms}}^{\;2}}{R}$$

The resistance is given by

```
R = Convert[V_rms^2/P̄_0, Ohm]
```

(c) The power factor is given by

```
Convert[R/Sqrt[R^2 + (X_L - X_C)^2] /. {X_L → 2 π f L, X_C -> 1/(2 π f C)} /.
  f → 2.31 × 10^3 Hertz, 1]
```

We have avoided making assignments for X_L and X_C because L and C already have assigned values:

```
{X_L, X_C}
```

Using transformation rules allows us to bypass this difficulty.

```
Clear[C, L, R, Subscript, OverBar]
```

SECTION 3.1.4

1. (a) `x - y // FullForm`

   (b) `x^2/y^2 // FullForm`

   (c) `{a, {b, {c, d}}, {e, f}} // FullForm`

   (d) `(a + I b)/(c - I d) // FullForm`

   (e) `Integrate[x^2, x] // Hold // FullForm`

   (f) `DSolve[y''[x] - 3 y'[x] - 18 y[x] == x Exp[4 x], y[x], x] // Hold // FullForm`

   (g) `f[a] + g[c, d] /. f[x_] → x^3 // Hold // FullForm`

   (h) `Sin /@ (a + b + c) // Hold // FullForm`

   (i)
   ```
   CrossProduct[a, CrossProduct[b, c]] +
     CrossProduct[b, CrossProduct[c, a]] +
     CrossProduct[c, CrossProduct[a, b]] // FullForm
   ```

3. `complexConjugate[expr_] := (expr /. Complex[x_, y_] → Complex[x, -y])`

   For example,

   ```
   complexConjugate[(a + ⅈ b)/(c - ⅈ d)^2]
   Clear[complexConjugate]
   ```

5.
   ```
   Level[√3/2 + 1/2 (-π/3 + x) - 1/4 √3 (-π/3 + x)^2 - 1/12 (-π/3 + x)^3 + (-π/3 + x)^4/(16 √3),
     {3, 4}, Heads → True]
   ```

7. `Level[x^3 + (1 + z)^2, {2}]`

9. `Rest[Part[a + b/c + (d + e)/(1 + f/(1 - g/h)), 3, 2, 1, 2, 2, 1, 2]]`

11.
```
myexpr = a + b/c + (d + e/g)/(r + s/(1 + t));
Position[myexpr, 1 + t]
Part[myexpr, 3, 2, 1, 2, 2, 1]
Clear[myexpr]
```

13.
```
Select[{π, ϕ, 2 π, ⅇ^x, f₃, 4 + √7 ⅈ,
  4 + 2 ⅈ, 7.12, P̄, "Chem/Phyx 340", a + b}, AtomQ]
```

15.
```
Select[{x, Pi, ⅇ, ∞, °, "It is a wonderful world.",
  44/100, {a, {b, c}}, 4 + 5 ⅈ}, NumericQ]
```

17.
```
mytest[expr_] := FreeQ[expr, y]
Select[
 1 + 3 x + 3 x^2 + x^3 + 3 y + 6 x y + 3 x^2 y + 3 y^2 + 3 x y^2 + y^3, mytest]
Clear[mytest]
```

19.
```
Table[{i}, {i, 2, 8}]
Insert[h[e₁, e₂, e₃, e₄, e₅, e₆, e₇], b, Table[{i}, {i, 2, 8}]]
```

21.
```
yourlist = {{a, {b, c}}, {d, {e, {f, {g, h}}}}, {{i, j}, k}};
```
(a)
```
FlattenAt[yourlist, Position[yourlist, {g, h}]]
```
(b)
```
Flatten[yourlist, 2]
Clear[yourlist]
```

23.
```
t = {15, 20, 25, 30, 40, 50, 60, 70, 80, 90,
   100, 110, 120, 130, 140, 150, 160, 170, 180, 190, 200,
   210, 220, 230, 240, 250, 260, 270, 280, 290, 298.1};
cp = {0.311, 0.605, 0.858, 1.075, 1.452, 1.772, 2.084,
    2.352, 2.604, 2.838, 3.060, 3.254, 3.445, 3.624, 3.795,
    3.964, 4.123, 4.269, 4.404, 4.526, 4.639, 4.743, 4.841,
    4.927, 5.010, 5.083, 5.154, 5.220, 5.286, 5.350, 5.401};
```
(a)
```
cp/t
```
(b)
```
f[x_, y_] := x/y
Thread[f[cp, t]]
Clear[cp, t, f]
```

25.
```
mult[expr_] := expr (x - 1)
Map[mult, x^2/(x - 1) == 1/(x - 1)]
Clear[mult]
```

27.
```
ourexpr = a + b/c + (d + e/f)/(r + s/t)
FullForm[ourexpr]
Position[ourexpr, Power]
```

For the positions of all the numerators, replace the last two elements of each sublist by a single element `1`.

```
num[x_] := Append[Drop[x, -2], 1]
MapAt[func, ourexpr, num/@%%]
Clear[ourexpr, num]
```

29. `Apply[Equal, x → v0x t + 1/2 ax t^2]`

31. `Apply[List, Expand[5 a + 3 b + (2 c + 5 a)/7]]`

Here is another way to do this:

```
List@@Expand@(5 a + 3 b + (2 c + 5 a)/7)
```

SECTION 3.2.9

1.
```
λ[n_Integer/;n > 2] := 364.5 n^2/(n^2 - 4) nm
{λ[-1], λ[2], λ[3, 4], λ[3.5], λ[3]}
ClearAll[λ]
```

3.
```
sgn[x_?Positive] := 1
sgn[x_?Negative] := -1
Plot[sgn[x], {x, -2, 2}, AxesLabel → {"x", "sgn(x)"}]
ClearAll[sgn]
```

5.
```
vectorClassify[{x_, y_}/; (x > 0 && y > 0)] := 1
vectorClassify[{x_, y_}/; (x < 0 && y > 0)] := 2
vectorClassify[{x_, y_}/; (x < 0 && y < 0)] := 3
vectorClassify[{x_, y_}/; (x > 0 && y < 0)] := 4
vectorClassify[{x_, 0}/;x > 0] := 5
vectorClassify[{x_, 0}/;x < 0] := 6
vectorClassify[{0, y_}/;y > 0] := 7
vectorClassify[{0, y_}/;y < 0] := 8
```

Another definition:

```
ClearAll[vectorClassify]

vectorClassify[{_?Positive, _?Positive}] := 1
vectorClassify[{_?Negative, _?Positive}] := 2
vectorClassify[{_?Negative, _?Negative}] := 3
vectorClassify[{_?Positive, _?Negative}] := 4
vectorClassify[{_?Positive, 0}] := 5
vectorClassify[{_?Negative, 0}] := 6
vectorClassify[{0, _?Positive}] := 7
vectorClassify[{0, _?Negative}] := 8
```

```
{vectorClassify[{-10, 0}], vectorClassify[{2, 0}],
 vectorClassify[{0, -5.2}], vectorClassify[{-2.6, 4}],
 vectorClassify[{3, -3}], vectorClassify[{3π/2, -2.1}]}

{vectorlassify[{1, 1, 2}],
 vectorClassify[{0, 0}], vectorClassify[1, 0]}

ClearAll[vectorClassify]
```

7.
```
Cases[{a, {a}, {a, a}, {b, c}, {d, {e, f}}, {Sin[a x], Sin[a x]},
   {a, a, a}, {Sin[b x], Tan[c x]}}, {x_, y_}/; x=!=y, Infinity]
```

9. $\left\{\left(1-x+\text{Sin}\left[(a-b)^2\right]\right)^2, \sqrt{(a-b)^3\,\text{Cos}\left[(4-x)^3\right]},\ (a-3)\,\text{ArcTan}\left[\text{Sin}\left[\pi\left((x-1)(x+1)-x^2\right)\right]\right]\right\}$ /.

```
Sin[z_] :→ Sin[Expand[z]]
```

11.
```
SetAttributes[f, Orderless]
f[x_Symbol, y_Integer, z_Complex] := {x, y, z}
f[2 + I, g, 5]
ClearAll[f]
```

13.
```
func[x_Symbol, y_Integer:7] := y x
{func[x, 10], func[x], func[2, 5], func[x, y, 4], func[{x, 6}]}
ClearAll[func]
```

15.
```
myRange[n_Integer:1, m_Integer] := Table[i, {i, n, m}] /; m ≥ n
{myRange[5], myRange[1], myRange[0], myRange[-7]}
{myRange[-3, 2], myRange[2, 4], myRange[5, 2], myRange[3.5, 5.5]}
ClearAll[myRange]
```

17.
```
ourPattern =
   a_Symbol_0 | (a_Symbol_n_ x_Symbol^n_. /; IntegerQ[n] && a =!= x);
```

Cases$\left[\left\{\frac{b_{-3}}{x^3}, \frac{b_{-2}}{x^2}, \frac{b_{-1}}{x}, \frac{x_{-3}}{x^3}, \frac{x_{-2}}{x^2}, \frac{x_{-1}}{x}, b_0, x\,b_1, x^2\,b_2, x^3\,b_3, x_2\,x^2, 3\,y^4, a+b, \sqrt{1+x}, \text{Sin}[x]\,x, x_{11}\,y^{11}, \frac{1}{z}\right\}, \text{ourPattern}\right]$

```
ClearAll[ourPattern]
```

19.
```
ourfunc[r_Rational | r_Integer] := Abs[r]

{ourfunc[-3/5], ourfunc[-5], ourfunc[7],
 ourfunc[x], ourfunc[-3/5, -5], ourfunc[1.2]}

ClearAll[ourfunc]
```

21.
```
characterList = Characters[ToString[N[Pi, 770]]];
findNines[{"3", ".", a___, "9", "9", "9", "9", "9", "9", ___}] :=
 Length[{a}]
findNines[characterList]
ClearAll[characterList, findNines]
```

23.
```
ourcharacters = Characters[ToString[N[√5, 6200]]];
Print["Number of 7's    Number of Digits"]
ourcharacters/.{"2", ".", a___, "7", "7", b : "7".., ___}:>
   "Ha!"/; Print[Length[{b}] + 2, "          ", Length[{a}]];
ClearAll[ourcharacters]
```

25. Here are several acceptable definitions:
```
mymean[{x__}] := (Plus[x])/Length[{x}]/;NumberQ[Plus[x]]
yourmean[x : {__}] := (Plus @@ x)/Length[x]/;NumberQ[Plus @@ x]
ourmean[x_List/; Length[x]≥1] :=
  (Plus @@ x)/Length[x]/;NumberQ[Plus @@ x]
mylist = Table[RandomInteger[1000], {RandomInteger[{1, 100}]}];
{mymean[mylist], yourmean[mylist], ourmean[mylist]}
yourlist = {1, a, x, y, b}
{mymean[yourlist], yourmean[yourlist], ourmean[yourlist]}
ClearAll[mymean, yourmean, ourmean, mylist, yourlist]
```

27.
```
myfunc[x__List] := Length/@{x}
{myfunc[{1}, {2, 3, 4, 5}, {6, 7}, {8, 9}],
 myfunc[{a, b}, {c, 5}], myfunc[x, 3, {1}]}
ClearAll[myfunc]
```

SECTION 3.3.7

1. (a)
```
f[x_] := 1 + x^3
ClearAll[f]
```

(b)
```
func[x_] := {x, x^2}
ClearAll[func]
```

(c)
```
myf[z_] := z/.x→y
ClearAll[myf]
```

(d)
```
g[x_, y_] := 1/x^y
ClearAll[g]
```

(e)
```
h[x_, y_, z_] := (x, y^z)
ClearAll[h]
```

(f)
```
myg[y_] := -ⅈ D[y, x]
ClearAll[myg]
```

(g)
```
g[y_] := y^3
f[x_] := g/@x
ClearAll[g, f]
```

(h)
```
hisf[x_] := Apply[And, Map[OddQ, x]]
ClearAll[hisf]
```

(i)
```
herf[x_] := Delete[x, RandomInteger[{1, Length[x]}]]
ClearAll[herf]
```

3.
```
Cases[{1, a, 2.0, 5.0, 4, x^2, Sin[x]}, _Integer?(#> 3 &)]
```

5.
```
Select[{{4.294, 3.757, 7.222}, {9.240, 3.008, 1.001},
  {0.696, 3.826, 0.375}, {1.931, 4.814, 5.422},
  {7.161, 3.665, 0.212}, {1.809, 4.298, 7.333},
  {5.745, 3.287, 1.215}, {1.901, 3.335, 0.022},
  {8.769, 3.246, 0.137}}, (Times @@ #) > 20&]
```

7.
```
Table[{RandomInteger[10], RandomInteger[10], RandomInteger[10]},
 {RandomInteger[10]}]
Select[%, EvenQ[Plus @@ #] &]
Select[%%, OddQ[Plus @@ #] &]
```

9.
```
Needs["VectorAnalysis`"]
CoordinateSystem
Coordinates[]
SetCoordinates[Cartesian[x, y, z]]
A = {Ax[x, y, z], Ay[x, y, z], Az[x, y, z]};
B = {Bx[x, y, z], By[x, y, z], Bz[x, y, z]};

Simplify[Curl[CrossProduct[A, B]] -
  (DotProduct[B, {D[#, x], D[#, y], D[#, z]}] &) /@ A +
  (DotProduct[A, {D[#, x], D[#, y], D[#, z]}] &) /@ B -
 A Div[B] + B Div[A]]
ClearAll[A, B]
```

11. (a)
```
{v₀, θ₀} // FullForm
```

(b)
```
{v₀, θ₀} = {55 m/s, 30 °};
{v₀, θ₀}
?? Subscript
ClearAll[Subscript]
{v₀, θ₀}
```

13. (a)
```
hermite[n_Integer /; n > 1, z_] :=
 Expand@(2 z hermite[n - 1, z] - 2 (n - 1) hermite[n - 2, z])
hermite[0, z_] := 1
hermite[1, z_] := 2 z
totalCalls[n_Integer /; n > 1, z_] :=
  Trace[hermite[n, z], hermite[_Integer, z]] // Flatten // Length
totalCalls[20, z]
```

(b)
```
partialCalls[n_Integer /; n > 1, k_Integer ? NonNegative, z_] :=
  (Trace[hermite[n, z], hermite[k, z]] // Flatten // Length) /; n ≥ k

partialCalls[20, 3, z]
```

(c)
```
newTotalCalls[n_Integer /; n > 1, z_] := (ClearAll[H];
  H[m_Integer /; m > 1, x_] :=
    H[m, x] = Expand@(2 x H[m - 1, x] - 2 (m - 1) H[m - 2, x]);
  H[0, x_] := 1;
  H[1, x_] := 2 x;
```

```
  Trace[H[n, z], H[_Integer, z]]//Flatten//Length)
newTotalCalls[20, z]
newPartialCalls[n_Integer/; n>1,
   k_Integer?NonNegative, z_]:= (ClearAll[H];
    H[m_Integer/; m>1, x_]:=
     H[m, x] = Expand@(2 x H[m-1, x] - 2(m-1)H[m-2, x]);
    H[0, x_]:=1;
    H[1, x_]:=2 x;
    Trace[H[n, z], H[k, z]] //Flatten//Length)/; n≥k
newPartialCalls[20, 3, z]
ClearAll[hermite, totalCalls,
 partialCalls, H, newTotalCalls, newPartialCalls]
```

15. (a)

```
m = 3.00 10^-3;
b = 0.0300;
g = 9.8;
vT = -g m/b
```

The terminal speed is 0.98 m/s. Ninety-nine percent of the terminal speed is 0.9702 m/s:

```
Abs[%]0.99
```

(b)

```
t[0] = 0;
x[0] = 0;
v[0] = 0;
a[n_]:=a[n] = -g- (b/m)v[n]
Δt = 0.005;
t[n_]:=t[n] =t[n-1] + Δt
v[n_]:=v[n] =v[n-1] +a[n-1] Δt
x[n_]:=x[n] =x[n-1] +v[n-1]Δt
velocity = Table[v[n], {n, 0, 90}];
time = Table[t[n], {n, 0, 90}];
ListPlot[Transpose[{time, velocity}],
 AxesLabel→{"time(s)", "velocity(m/s)"}]
position = Table[x[n], {n, 0, 90}];
ListPlot[Transpose[{time, position}],
 AxesLabel→{"time(s)", "position(m)"}]
TableForm[Transpose[{time, position, velocity}], TableHeadings→
  {None, {"Time(s)", "Position(m)", "Velocity(m/s)"}}]
ClearAll[m, b, g, vT, t, x, v, a, Δt, velocity, time, position]
```

(c)

```
m = 3.00 10^-3;
b = 0.0300;
g = 9.8;
sol = NDSolve[{x'[t] == v[t], v'[t] == -g - (b v[t])/m, x[0] == 0, v[0] == 0},
  {x, v}, {t, 0, 0.465}]
```

```
TableForm[Table[{t, x[t], v[t]}/.sol[[1]], {t, 0, 0.465, 0.005}],
 TableHeadings→
  {None, {"Time(s)", "Position(m)", "Velocity(m/s)"}}]
ClearAll[m, b, g, sol]
```

Note that there are discrepancies between the results of Euler's method and **NDSolve**. Decreasing the step size would improve the accuracy of the Euler method solution.

17. (a) `FoldList[Plus, 0, {a, b, c, d}]`

(b) `FoldList[Times, 1, {a, b, c, d}]`

(c) `Fold[10 #1 + #2&, 0, {1, 0, 3, 5, 7, 3}]`

19. `f[x_] := μ(1 - 2 Abs[x - 1/2])`

(a) The point $x = 0$ is a stable fixed point when the map parameter μ is less than $1/2$:

```
μ = RandomReal[1/2]
0 == f[0]
Table[FixedPoint[f, RandomReal[],
   SameTest→(Abs[#1 - #2] < 10^-11 &)]//Chop, {10}]
```

Here is a graph of $f(x)$ together with the straight line x:

```
Plot[{f[x], x}, {x, 0, 1}, PlotRange→{0, 1}]
```

(b) The point $x = 0$ is a fixed point when the map parameter μ is greater than $1/2$:

```
μ = RandomReal[{1/2, 1}]
0 == f[0]
```

The point is not stable or attracting:

```
ListLinePlot[Transpose[{Range[0, 100], NestList[f, 0.0001, 100]}],
 PlotRange→{0, 1}]
```

Another point $x = 2\mu/(1 + 2\mu)$ is also a fixed point when the map parameter μ is greater than $1/2$:

```
2μ/(1 + 2μ) == f[2μ/(1 + 2μ)]
```

The point is also not stable or attracting:

```
ListLinePlot[
 Transpose[{Range[0, 100], NestList[f, 2μ/(1 + 2μ) + 0.0001, 100]}],
 PlotRange→{0, 1}]
```

Here is a graph of $f(x)$ together with the straight line x:

```
Plot[{f[x], x}, {x, 0, 1}, PlotRange→{0, 1}]
ClearAll[f, μ]
```

21. (a) `√a² // FullForm`

The head of $\sqrt{a^2}$ is **Power** rather than **Sqrt**. Thus, the additional rules associated with **Sqrt** do not apply to $\sqrt{a^2}$.

(b)
```
Im[a]^=0;
Positive[a]^=True;
Im[b]^=0;
Negative[b]^=True;
Unprotect[Power];
√(x_^2/; (Im[x] == 0 && Positive[x])) := x
√(x_^2/; (Im[x] == 0 && Negative[x])) := -x
Protect[Power];
{Sqrt[a²], Sqrt[b²]}
{√a², √b²}
```

SECTION 3.4.6

1. Single-clause definition:

```
scf[x_]:=If[x ≥ 0, x, -x]
```

Another single-clause definition:

```
anotherscf[x_]:=Piecewise[{{x, x ≥ 0}, {-x, x < 0}}]
```

Multiclause definition:

```
mcf[x_/; x ≥ 0]:=x
mcf[x_/; x < 0]:= -x
```

Derivative of f at 0:

```
∂x{scf[x], anotherscf[x], mcf[x]}/.x→0
```

The answer **1** for **scf'[0]** is wrong because f is not differentiable at 0, and we cannot find $f'(x)$ with the multiclause definition.

```
ClearAll[scf, anotherscf, mcf]
```

3. (a)
```
scV[x_]:=
 Which[x < -2|| -1 ≤ x < 1, 0, -2 ≤ x < -1, -2, 1≤ x < 3, -1, x ≥ 3, 1]
anotherscV[x_]:=Piecewise[{{0, x < -2|| -1 ≤ x < 1},
   {-2, -2 ≤ x < -1}, {-1, 1 ≤ x < 3}, {1, x ≥ 3}}]
mcV[x_/; x < -2|| -1 ≤ x < 1]:= 0
mcV[x_/; -2 ≤ x< -1]:= -2
mcV[x_/;1 ≤ x< 3]:= -1
mcV[x_/;x ≥ 3]:= 1
```

(b)
```
Plot[scV[x], {x, -4, 4}, AxesLabel→{"x(a)", "V(b)"}]
Plot[anotherscV[x], {x, -4, 4}, AxesLabel→{"x(a)", "V(b)"}]
Plot[mcV[x], {x, -4, 4}, AxesLabel→{"x(a)", "V(b)"}]
ClearAll[scV, anotherscV, mcV]
```

5.
```
func[x_/;VectorQ[x, NumberQ]]:= If[Count[x, 0] == 0,
  (1/#)&/@x, Print[Count[x, 0], " zeroes in list"]]
func[{1, 2, 3, 4, a}]
func[{2, 3, 4, 10}]
func[{0, 3, 0, 0}]
ClearAll[func]
```

7. (a)
```
mySum[n_Integer?Positive]:=
 Module[{i = 1, total = 0},
  While[i ≤ n, total = total + Prime[i]; i++]; total]
mySum[10^6]//Timing
```

(b)
```
yourSum[n_Integer?Positive]:= Plus@@Table[Prime[i], {i, n}]
yourSum[10^6]//Timing
```

(c)
```
ourSum[n_Integer?Positive]:= Sum[Prime[i], {i, 1, n}]
ourSum[10^6]//Timing
ClearAll[mySum, yourSum, ourSum]
```

9. (a)
```
Options[binomialExpansion] = {caption→False, exponent→2};

binomialExpansion[x_Symbol, y_Symbol, opts___Rule] :=
 Module[{caption, exponent},
  Print[{opts}];
  Print[Options[binomialExpansion]];
  Print[caption];
  Print[exponent];
  caption = caption/.{opts}/.Options[binomialExpansion];
  exponent = exponent/.{opts}/.Options[binomialExpansion];
  Print[caption];
  Print[exponent];
```

```
  If[caption === True, Print["Expansion of ", (x+y)^exponent]];
  Expand[(x + y)^exponent]]
```

```
binomialExpansion[a, b]
```

```
binomialExpansion[a, b, caption → True, exponent → 10]
```

Within **Module**, **caption** and **exponent** are declared local variables and, thus, become **caption$***n* and **exponent$***n*. In the evaluation of **caption = caption/. {opts}/.Options[binomialExpansion]** and **exponent = exponent/.{opts}/. Options[binomialExpansion]**, **caption$***n* and **exponent$***n* do not match the (unaltered) option names **caption** and **exponent** in **{opts}** and **Options[binomial Expansion]** and, therefore, remain unchanged because the option rules do not apply.

```
ClearAll[binomialExpansion]
```

(b)
```
ClearAll["Global`*"]
Options[binomialExpansion] = {caption → False, exponent → 2};
binomialExpansion[x_Symbol, y_Symbol, opts___Rule] :=
 Module[{cap, exp},
  cap = caption/.{opts}/.Options[binomialExpansion];
  exp = exponent/.{opts}/.Options[binomialExpansion];
  If[cap === True, Print["Expansion of ", (x + y)^exp]];
  Expand[(x + y)^exp]]
```

```
binomialExpansion[a, b]
```

```
binomialExpansion[a, b, exponent → 10, caption → True]
```

Here, the local variables are **cap** and **exp** rather than **caption** and **exponent**.

```
ClearAll[binomialExpansion]
```

11. (a) Let V, x, and m be measured in units of V_0, a, and m, respectively.

```
V[x_] := - (1 + x^2)/(8 + x^4)
Plot[V[x], {x, -10, 10}, AxesLabel → {"x(a)", "V(V₀)"}]
F = -∂ₓ V[x]//Together
Plot[F, {x, -10, 10}, AxesLabel → {"x(a)", "F(V₀/a)"}]
```

(b) **Important**: Enter the definition of **motion1DPlot** in section 3.4.5 here.

```
(* name the options and specify their default values *)
Options[motion1DPlot] =
  {positionPlot → True,
```

```
    velocityPlot → True,
    accelerationPlot → True,
    combinationPlot → True,
    positionAFLabel → {"t (s)", "x (m)"},
    velocityAFLabel → {"t (s)", "v (m/s)"},
    accelerationAFLabel → {"t (s)", "a (m/s^2)"},
    combinationAFLabel → {"t (s)", None}};

motion1DPlot[a_, x0_, v0_, tmax_, opts___Rule] :=
 Module[
  {(* declare local variables *)
    sol, curves = {}, plotx, plotv, plota,

    (* determine option values and assign
     them as initial values to local variables *)
    position = positionPlot/.{opts}/.Options[motion1DPlot],
    velocity = velocityPlot/.{opts}/.Options[motion1DPlot],
    acceleration =
     accelerationPlot/.{opts}/.Options[motion1DPlot],
    combination = combinationPlot/.{opts}/.Options[motion1DPlot],
    positionLabel = positionAFLabel/.{opts}/.
      Options[motion1DPlot], velocityLabel =
     velocityAFLabel/.{opts}/.Options[motion1DPlot],
    accelerationLabel = accelerationAFLabel/.{opts}/.
      Options[motion1DPlot], combinationLabel =
     combinationAFLabel/.{opts}/.Options[motion1DPlot],

    (* select valid options for Plot and Show and
     assign them as initial values to local variables *)
    optPlot = Sequence @@ FilterRules[{opts}, Options[Plot]],
    optShow = Sequence @@ FilterRules[{opts}, Options[Graphics]]},

    (* set text of a warning message *)
    motion1DPlot::argopt =
     "Each of the values for the options positionPlot,
       velocityPlot, accelerationPlot, and
       combinationPlot must be either True or False.";

   (* verify option specifications *)
   If[Count[{position, velocity,
       acceleration, combination}, True | False] =!= 4,
    Message[motion1DPlot::argopt]; Return[$Failed]];

   (* solve the equation of motion numerically *)sol =
   NDSolve[{x''[t] == a, x[0] == x0, x'[0] == v0}, x, {t, 0, tmax}];
```

```
(* plot position vs. time *)
If[position,
 plotx = Plot[Evaluate[x[t] /.sol], {t, 0, tmax},
  PlotLabel→"position vs. time", AxesLabel→positionLabel,
  Ticks→Automatic, FrameLabel→positionLabel,
  FrameTicks→Automatic, Evaluate[optPlot],
  PlotRange→All, Axes→False, Frame→True];
 Print[plotx];
 AppendTo[curves, plotx]];

 (* plot velocity vs. time *)
 If[velocity,
  plotv = Plot[Evaluate[x'[t] /.sol],
     {t, 0, tmax}, PlotLabel→"velocity vs.time",
     AxesLabel→velocityLabel, Ticks→Automatic,
     FrameLabel→velocityLabel, FrameTicks→Automatic,
     Evaluate[optPlot], PlotStyle→Dashing[{0.03, 0.03}],
     PlotRange→All, Axes→False, Frame→True];
  Print[plotv];
  AppendTo[curves, plotv]];

  (* plot acceleration vs. time *)
  If[acceleration,
   plota = Plot[Evaluate[a/.sol], {t, 0, tmax},
     PlotLabel→"acceleration vs.time",
     AxesLabel→accelerationLabel, Ticks→Automatic,
     FrameLabel→accelerationLabel, FrameTicks→Automatic,
     Evaluate[optPlot], PlotStyle→RGBColor[1, 0, 0],
     PlotRange→All, Axes→False, Frame→True];
   Print[plota];
   AppendTo[curves, plota]];

  (* combine the plots *)
  If[(combination)&&(Length[curves] > 1),
   Show[curves,
    PlotLabel→"combination", AxesLabel→combinationLabel,
    Ticks→{Automatic, None}, FrameLabel→combinationLabel,
    FrameTicks→{Automatic, None}, optShow]]]
```

E greater than zero:

```
motion1DPlot[- 2(-8 x[t] + 2 x[t]^3 + x[t]^5) / (8 + x[t]^4)^2,
  -10, 1, 19.1, combinationPlot→False,
```

```
positionAFLabel → {"t (a √(m/V0))", "x (a)"},
velocityAFLabel → {"t (a √(m/V0))", "v (√(V0/m))"},
accelerationAFLabel → {"t (a √(m/V0))", "a (V0/ma)"}]
```

E between $-V_0/8$ and zero:

```
motion1DPlot[ - 2 (-8 x[t] + 2 x[t]^3 + x[t]^5) / (8 + x[t]^4)^2,
  3.5, 0, 47, combinationPlot → False,
  positionAFLabel → {"t (a √(m/V0))", "x (a)"},
  velocityAFLabel → {"t (a √(m/V0))", "v (√(V0/m))"},
  accelerationAFLabel → {"t (a √(m/V0))", "a (V0/ma)"}]
```

E between $-V_0/4$ and $-V_0/8$:

```
motion1DPlot[ - 2 (-8 x[t] + 2 x[t]^3 + x[t]^5) / (8 + x[t]^4)^2,
  2.25, 0, 14.5, combinationPlot → False,
  positionAFLabel → {"t (a √(m/V0))", "x (a)"},
  velocityAFLabel → {"t (a √(m/V0))", "v (√(V0/m))"},
  accelerationAFLabel → {"t (a √(m/V0))", "a (V0/ma)"}]
ClearAll[V, F, motion1DPlot]
```

13. (a)

```
(* name the options and specify their default values *)
Options[ourmotion1DPlot] =
  {positionPlot → True,
   velocityPlot → True,
   accelerationPlot → True,
   combinationPlot → True,
   positionAFLabel → {"t (s)", "x (m)"},
   velocityAFLabel → {"t (s)", "v (m/s)"},
   accelerationAFLabel → {"t (s)", "a (m/s^2)"},
   combinationAFLabel → {"t (s)", None}};

ourmotion1DPlot[eqn_, x0_, v0_, tmax_, opts___Rule] :=
```

```
Module[
 {(* declare local variables *)
  sol, curves = {}, plotx, plotv, plota,

  (* determine option values and assign
    them as initial values to local variables *)
  position = positionPlot/.{opts}/.Options[ourmotion1DPlot],
  velocity = velocityPlot/.{opts}/.Options[ourmotion1DPlot],
  acceleration = accelerationPlot/.{opts}/.
    Options[ourmotion1DPlot], combination =
   combinationPlot/.{opts}/.Options[ourmotion1DPlot],
  positionLabel = positionAFLabel/.{opts}/.
    Options[ourmotion1DPlot], velocityLabel =
   velocityAFLabel/.{opts}/.Options[ourmotion1DPlot],
  accelerationLabel = accelerationAFLabel/.{opts}/.
    Options[ourmotion1DPlot], combinationLabel =
   combinationAFLabel/.{opts}/.Option[ourmotion1DPlot],

 (* select valid options for Plot and Show and
  assign them as initial values to local variables *)
 optPlot = Sequence @@ FilterRules[{opts}, Options[Plot]],
 optShow = Sequence @@ FilterRules[{opts}, Options[Graphics]]},

(* set texts of warning messages *)
ourmotion1DPlot::argtype =
 "One or more arguments entered are of the wrong type.";
ourmotion1DPlot::argopt = "Each of the values for the options
   positionPlot, velocityPlot, accelerationPlot, and
   combinationPlot must be either True or False.";

(* check argument types *)
If[!((Head[eqn] === Equal) && (NumberQ[x0]) &&
     (Im[x0] == 0) && (NumberQ[v0]) && (Im[v0] == 0) &&
     (NumberQ[tmax]) && (Im[tmax] == 0) && (Positive[tmax])),
 Message[ourmotion1DPlot::argtype]; Return[$Failed]];

(* verify option specifications *)
If[Count[{position, velocity,
    acceleration, combination}, True | False] = ! = 4,
 Message[ourmotion1DPlot::argopt]; Return[$Failed]];

(* solve the equation of motion numerically *)
sol = NDSolve[{eqn, x[0] == x0, x'[0] == v0}, x, {t, 0, tmax}];

(* plot position vs. time *)
```

```
If[position,
 plotx = Plot[Evaluate[x[t] /. sol], {t, 0, tmax},
   PlotLabel → "position vs. time", AxesLabel → positionLabel,
   Ticks → Automatic, FrameLabel → positionLabel,
   FrameTicks → Automatic, Evaluate[optPlot],
   PlotRange → All, Axes → False, Frame → True];
 Print[plotx];
 AppendTo[curves, plotx]];

(* plot velocity vs. time *)
If[velocity,
 plotv = Plot[Evaluate[x'[t] /. sol],
   {t, 0, tmax}, PlotLabel → "velocity vs. time",
   AxesLabel → velocityLabel, Ticks → Automatic,
   FrameLabel → velocityLabel, FrameTicks → Automatic,
   Evaluate[optPlot], PlotStyle → Dashing[{0.03, 0.03}],
   PlotRange → All, Axes → False, Frame → True];
 Print[plotv];
 AppendTo[curves, plotv]];

(* plot acceleration vs. time *)
If[acceleration,
 plota = Plot[Evaluate[(eqn[[2]]) /. sol], {t, 0, tmax},
   PlotLabel → "acceleration vs.time",
   AxesLabel → accelerationLabel, Ticks → Automatic,
   FrameLabel → accelerationLabel, FrameTicks → Automatic,
   Evaluate[optPlot], PlotStyle → RGBColor[1, 0, 0],
   PlotRange → All, Axes → False, Frame → True];
 Print[plota];
 AppendTo[curves, plota]];

(* combine the plots *)
If[(combination) && (Length[curves] > 1),
 Show[curves, PlotLabel → "combination",
  AxesLabel → combinationLabel, Ticks → {Automatic, None},
  FrameLabel → combinationLabel,
  FrameTicks → {Automatic, None}, optShow]]
     ]
ClearAll[ourmotion1DPlot]
```

(b)

```
(* name the options and specify their default values *)
Options[newmotion1DPlot] =
  {PositionPlot → True,
   velocityPlot → True,
   accelerationPlot → True,
   combinationPlot → True,
```

```
    positionAFLabel → {"t(s)", "x(m)"},
    velocityAFLabel → {"t(s)",  "v(m/s)"},
    accelerationAFLabel → {"t(s)", "a(m/s²)"},
    combinationAFLabel → {"t(s)", None}};

newmotion1DPlot[
 eqn_ /; Head[eqn] === Equal,
 x0_ /; (NumberQ[x0] && Im[x0] == 0),
 v0_ /; (NumberQ[v0] && Im[v0] == 0),
 tmax_ /; (NumberQ[tmax] && Im[tmax] == 0 && Positive[tmax]),
 opts___Rule] :=
Module[
 {(* declare local variables *)
  sol, curves = {}, plotx, plotv, plota,

  (* determine option values and assign
    them as initial values to local variables *)
  position = positionPlot /. {opts} /. Options[newmotion1DPlot],
  velocity = velocityPlot /. {opts} /. Options[newmotion1DPlot],
  acceleration = accelerationPlot /. {opts} /.
    Options[newmotion1DPlot], combination =
   combinationPlot /. {opts} /. Options[newmotion1DPlot],
  positionLabel = positionAFLabel /. {opts} /.
    Options[newmotion1DPlot], velocityLabel =
   velocityAFLabel /. {opts} /. Options[newmotion1DPlot],
  accelerationLabel = accelerationAFLabel /. {opts} /.
    Options[newmotion1DPlot], combinationLabel =
   combinationAFLabel /. {opts} /. Options[newmotion1DPlot],

  (* select valid options for Plot and Show and
   assign them as initial values to local variables *)
  optPlot = Sequence @@ FilterRules[{opts}, Options[Plot]],
  optShow = Sequence @@ FilterRules[{opts}, Options[Graphics]]},

 (* set text of a warning message *)
 newmotion1DPlot::argopt =
 "Each of the values for the options positionPlot,
   velocityPlot, accelerationPlot, and
    combinationPlot must be either True or False.";

(* verify option specifications *)
If[Count[{position, velocity,
    acceleration, combination}, True | False] =!= 4,
 Message[newmotion1DPlot::argopt]; Return[$Failed]];

(* solve the equation of motion numerically *)
sol = NDSolve[{eqn, x[0] == x0, x'[0] == v0}, x, {t, 0, tmax}];
```

```
(* plot position vs. time *)
If[position,
 plotx = Plot[Evaluate[x[t]/.sol], {t, 0, tmax},
   PlotLabel → "position vs. time", AxesLabel → positionLabel,
   Ticks → Automatic, FrameLabel → positionLabel,
   FrameTicks → Automatic, Evaluate[optPlot],
   PlotRange → All, Axes → False, Frame → True];
 Print[plotx];
 AppendTo[curves, plotx]];

(* plot velocity vs. time *)
If[velocity,
 plotv = Plot[Evaluate[x'[t]/.sol],
   {t, 0, tmax}, PlotLabel → "velocity vs. time",
   AxesLabel → velocityLabel, Ticks → Automatic,
   FrameLabel → velocityLabel, FrameTicks → Automatic,
   Evaluate[optPlot], PlotStyle → Dashing[{0.03, 0.03}],
   PlotRange → All, Axes → False, Frame → True];
 Print[plotv];
 AppendTo[curves, plotv]];

(* plot acceleration vs. time *)
If[acceleration,
 plota = Plot[Evaluate[(eqn[[2]])/.sol], {t, 0, tmax},
   PlotLabel → "acceleration vs. time",
   AxesLabel → accelerationLabel, Ticks → Automatic,
   FrameLabel → accelerationLabel, FrameTicks → Automatic,
   Evaluate[optPlot], PlotStyle → RGBColor[1, 0, 0],
   PlotRange → All, Axes → False, Frame → True];
 Print[plota];
 AppendTo[curves, plota]];

(*combine the plots *)
If[(combination) && (Length[curves] > 1),
 Show[curves,
  PlotLabel → "combination", AxesLabel → combinationLabel,
  Ticks → {Automatic, None}, FrameLabel → combinationLabel,
  FrameTicks → {Automatic, None}, optShow]]
     ]

newmotion1DPlot[eqn_, x0_, v0_, tmax_, opts___Rule]:=
 (newmotion1DPlot::argtype =
   "One or more arguments entered are of the wrong type.";
  Message[newmotion1DPlot::argtype]; Return[$Failed])

ClearAll[newmotion1DPlot]
```

SECTION 3.5.4

1.
```
ClearAll["Global`*"]
```
$$y[x\_]:=\frac{1}{2}\sqrt{4-x^2}$$
```
g1 = {{Thickness[0.0045], Darker[Red], Circle[{0, 0}, {2, 1}]}};
g2 = {Arrow[{{0.100, 1.00}, {-0.100, 1.00}}],
    Arrow[{{-0.100, -1.00},{0.100, -1.00}}]};
g3 = Line/@{
     {{1.73, 0}, {1.85, y[1.85]}},
     {{1.73, 0}, {1.85, -y[1.85]}},
     {{1.73, 0}, {-1.992, y[1.992]}},
     {{1.73, 0}, {-1.992, -y[1.992]}}};
g4 = {{RGBColor[1, 0.5, 0], Disk[{1.73, 0}, 0.10]}};
g5 = {
    {PointSize[0.022], Point/@{{1.85, y[1.85]}, {1.85, -y[1.85]},
       {-1.992, y[1.992]}, {-1.992, -y[1.992]}}}};
g6 = {
    Text["A", {-2.12, y[1.97]}, {1, 0}],
    Text["B", {-2.12, -y[1.97]}, {1, 0}],
    Text["C", {1.97, -y[1.725]}, {-1, 0}],
    Text["D", {1.97, y[1.80]}, {-1, 0}],
    Text[Style["Sun", Plain, Bold], {1.45, 0.09}, {0, -1}]};
Graphics[Join[g1, g2, g3, g4, g5, g6],
 PlotRange→{{-2.5, 2.5}, {-1.2, 1.2}},
 Frame -> True, FrameTicks -> None,
 ImagePadding -> 1, Background -> Lighter[Blue, 0.85],
 BaseStyle ->{FontFamily ->"Times", 14, Italic, Bold}]
```

3.
```
ClearAll["Global`*"]
twoTriangles = {
    Thickness[0.008],
    Line[{{-4, 0}, {4, 0}, {0, 6}, {-4, 0}}],
    Line[{{-4, 4}, {4, 4}, {0, -2}, {-4, 4}}]};
sixVertices = {
    PointSize[0.075],
    Transpose[{
      Table[RGBColor[Random[], Random[], Random[]], {6}],
      Point/@{{-4, 0}, {4, 0}, {0, 6}, {-4, 4}, {4, 4}, {0, -2}}}]};
Graphics[{twoTriangles, sixVertices}]
```

5.
```
Graphics[
  {{Green, Polygon[{{0, 10}, {5, 16}, {10, 10}}],
    Polygon[{{1, 15}, {5, 21}, {9, 15}}],
```

```
    Polygon[{{2, 20}, {5, 25}, {8, 20}}]},
   Rectangle[{4.5, 5}, {5.5, 10}],
   {Red, Rectangle[{2, 2}, {8, 5}]},
   Text[Style["Merry X'mas",
     FontFamily→"Times", 14, Yellow], {5, 3.35}]}]
```

7. (a)

```
coord1 = Flatten[Table[{i, j, k}, {i, 0, 1}, {j, 0, 1}, {k, 0, 1}], 2];
```

$$\text{coord2} = \left\{\left\{0, \frac{1}{2}, \frac{1}{2}\right\}, \left\{1, \frac{1}{2}, \frac{1}{2}\right\}, \left\{\frac{1}{2}, 0, \frac{1}{2}\right\}, \left\{\frac{1}{2}, 1, \frac{1}{2}\right\}, \left\{\frac{1}{2}, \frac{1}{2}, 0\right\}, \left\{\frac{1}{2}, \frac{1}{2}, 1\right\}\right\};$$

```
fcc = Graphics3D[{
    {Red, Specularity[White, 4],
     Sphere[#, 0.04]&/@Join[coord1, coord2]},
    Thickness[0.003], Line/@Partition[coord2, 2],
    Line/@Select[
      Flatten[
       Table[{coord1[[i]], coord1[[j]]}, {i, 1, 8}, {j, i+1, 8}], 1],
      ((#[[1, 1]] - #[[2, 1]])^2 + (#[[1, 2]] - #[[2, 2]])^2 +
          (#[[1, 3]] - #[[2, 3]])^2) === 1 &]},
   PlotRange→{{-0.05, 1.05}, {-0.05, 1.05}, {-0.05, 1.05}},
   ViewPoint→{1.901, -2.572, 1.104},
   Boxed->False, Background->Lighter[Blue, 0.25],
   Lighting->{{"Ambient", Red}, {"Point", Orange, {0, -2, 0}},
     {"Point", Darker[Yellow], {0, 2, 0}}}]
```

(b)

```
spin3D[obj_, opts___Rule] :=
 ListAnimate[
  Table[
   Show[obj,
     ViewPoint->{r Sin[θ] Cos[φ + i (π/16)],
       r Sin[θ] Sin[φ + i (π/16)], r Cos[θ]}, opts]/.
    {r->3.38346, θ->1.23842, φ->-0.934299}, {i, 0, 31}]]
spin3D[fcc, SphericalRegion->True]
```

SECTION 3.6.4

1.

```
(* set up default values for the options frameNumber,
sunSize,and earthSize *)Options[newplanetMotion] =
  {frameNumber→30, sunSize→0.1, earthSize→0.04};
```

```
newplanetMotion[{x0_, y0_}, {vx0_, vy0_}, opts___Rule] :=
 Module[
  {(* determine the energy parameter and
    assign it as initial value to a local variable *)
   e = N[(vx0^2 + vy0^2)/2 - 4 π^2/Sqrt[x0^2 + y0^2]],

   (* determine the angular momentum parameter and
    assign it as initial value to a local variable *)
   l = N[Abs[x0 vy0 - y0 vx0]],

   (* determine option values of frameNumber,
   sunSize, and earthSize,
   and assign them as initial values to local variables *)
   n = frameNumber/.{opts}/.Options[newplanetMotion],
   ss = sunSize/.{opts}/.Options[newplanetMotion],
   es = earthSize/.{opts}/.Options[newplanetMotion],

   (* select valid options for Show and assign the
    sequence as initial value to a local variable *)
   optShow = Sequence@@FilterRules[{opts}, Options[Graphics]],

   (* declare other local variables *)
   orbit, period, x, y, t},

  Which[
   (* verify that initial conditions are real numbers
    and that frame number is a positive integer *)
   ((! (NumberQ[x0] && Im[x0] == 0)) ||
     (! (NumberQ[y0]&&Im[y0] == 0)) ||
     (! (NumberQ[vx0]&&Im[vx0] == 0)) ||
     (! (NumberQ[vy0]&&Im[vy0] == 0)) ||
     (! (IntegerQ[n]&&Positive[n]))),
   Print["Initial conditions must be real numbers and
      frame number must be a positive integer."],

   (* check that the path is not a straight line *)
   l == 0,
   Print[
    "The initial conditions result in a straight line for the
      path whereas orbits of planets must be ellipses."],
```

```
(* check that the orbit is not a hyperbola *)
e>0,
Print[
 "The initial conditions result in a hyperbolic orbit whereas
   orbits of planets must be ellipses."],

(* check that the orbit is not a parabola *)
e == 0,
Print[
 "The initial conditions result in a parabolic orbit whereas
   orbits of planets must be ellipses."],

(* for the elliptical orbit *)
e<0,

(* determine the period of the motion *)
period = N[Abs[e/(2 π^2)]^(-3/2)];

(* solve the equations of motion
 discussed in Example 2.1.9 of Section 2.1.18 *)

orbit = NDSolve[{x''[t] + (4 π^2 x[t])/(x[t]^2 + y[t]^2)^(3/2) == 0,
   y''[t] + (4 π^2 y[t])/(x[t]^2 + y[t]^2)^(3/2) == 0, x[0] == x0, y[0] == y0,
   x'[0] == vx0, y'[0] == vy0}, {x, y}, {t, 0, period}];

(* generate the animation *)
ListAnimate[
 Table[
  Show[
   (* the orbit *)
   ParametricPlot[
    Evaluate[{x[t], y[t]}/.orbit], {t, 0, period}],

   (* the Sun in yellow and the moving planet in red *)
   Graphics[{Yellow, PointSize[ss], Point[{0, 0}], Red,
     PointSize[es], Point[{x[i], y[i]}/.orbit[[1]]]}],

   (* option for preserving the true shape of the orbit *)
   AspectRatio→Automatic,
```

```
        (* two-dimensional graphics options *)
        optShow],
```

$$\left\{i,\ 0,\ \left(1-\frac{1}{n}\right)\text{period},\ \frac{\text{period}}{n}\right\}]]]]]$$

```
newplanetMotion[{1, 0}, {-2.5, 5},
  frameNumber → 40,
  PlotRange → {{-0.6, 1.21}, {-1.13, 0.6}},
  Axes → None, Frame → True,
  FrameTicks → None,
  Background → RGBColor[0.5, 0.85, 1],
  sunSize → 0.2]

ClearAll[newplanetMotion]
```

3.
```
distribution[x_ /; VectorQ[x, IntegerQ]] :=
  Module[{whichBin, binlabels, numberinBin, binpositions},
    whichBin = Ceiling[x/5];
    binlabels = Range[Min[whichBin], Max[whichBin]];
    numberinBin = Count[whichBin, #] &/@binlabels;
    binpositions = ((5 binlabels) -2.5) /. (elem_ /; elem < 0) → 0;
    ListPlot[
      DeleteCases[Transpose[{binpositions, numberinBin}], {_, 0}],
      PlotStyle → PointSize[0.025],
      PlotRange → {{-1.5, 100}, {0, Max[numberinBin] + 2}},
      PlotLabel → "Number of Integers in Each Interval"]]

scores = {52, 80, 59, 59, 44, 67, 53, 59, 50, 56, 69, 70, 58, 68, 79,
    70, 55, 66, 76, 73, 50, 68, 57, 59, 59, 52, 59, 45, 63, 56, 70, 47,
    55, 54, 64, 78, 48, 48, 56, 74, 45, 62, 58, 54, 57, 77, 45, 91, 61,
    49, 39, 46, 65, 52, 62, 50, 65, 58, 51, 59, 60, 62, 57, 46, 57, 57,
    56, 51, 48, 69, 67, 61, 66, 73, 61, 53, 44, 66, 62, 47, 58, 52, 69,
    59, 58, 57, 70, 81, 50, 74, 49, 56, 62, 78, 71, 56, 62, 53, 55, 51};

distribution[scores]

ClearAll[distribution, scores]
```

5. For an illustration, drop only five elements and use the values:

```
x0 = 0.2; μmin = 2.9; μmax = 3.0; Δ = 0.01; nmax = 10;
```

Here is the dissection:

```
NestList[
  (#+Δ)&,
 μmin,
 Round[(μmax - μmin)/Δ]
        ]

 %(#(1 - #))
 Map[Function, %]
 Map[NestList[#, x0, nmax]&, %]
 Map[Drop[#, 5]&, %]
 NestList[
   (#+Δ)&,
  μmin,
  Round[(μmax - μmin)/Δ]
         ]

  Transpose[{%, %%}]
  Map[Thread, %]
  Flatten[%, 1]
  ListPlot[%]
  ClearAll[x0, μmin, μmax, Δ, nmax]
```

7.
```
Unprotect[NonCommutativeMultiply];
A_ ** U := A
U ** A_ := A
A_ ** (B_ + C_) := A ** B + A ** C
(A_ + B_) ** C := A ** C + B ** C
number3Q[x_, y_, n_]:= NumberQ[x] && NumberQ[y] && NumberQ[n]
A_ ** (B_(x_. y_^n_./; number3Q[x, y, n])):= ((xy^n)A ** B)
(A_(x_. y_^n_./; number3Q[x, y, n])) ** B_ := ((xy^n)A ** B)
Protect[NonCommutativeMultiply];
commutator[A_, B_]:= A ** B - B ** A
NumberQ[ℏ]^= True;
xp3DCommutator[expr_]:=
 ExpandAll[expr//. {pₓ ** x :→ x ** pₓ - ⅈ ℏ u, p_y ** y :→ y ** p_y - ⅈ ℏ u,
   p_z ** z :→ z ** p_z - ⅈ ℏ u, x ** y :→ y ** x, x ** z :→ z ** x,
   y ** z :→ z ** y, p_y ** x :→ x ** p_y, p_z ** x :→ x ** p_z, p_x ** y :→ y ** p_x,
   p_z ** y :→ y ** p_z, p_x ** z :→ z ** p_x, p_y ** z :→ z ** p_y,
   p_x ** p_y :→ p_y ** p_x, p_x ** p_z :→ p_z ** p_x, p_y ** p_z :→ p_z ** p_y}]
l_x = y ** p_z - z ** p_y;
l_y = z ** p_x - x ** p_z;
l_z = x ** p_y - y ** p_x;
(commutator[l_x, l_y] - ⅈ ℏ l_z == 0) // xp3DCommutator
(commutator[l_y, l_z] - ⅈ ℏ l_x == 0) // xp3DCommutator
(commutator[l_z, l_x] - ⅈ ℏ l_y == 0) // xp3DCommutator
```

```
(commutator[lz, lx ** lx + ly ** ly + lz ** lz] == 0) // xp3DCommutator
ClearAll[number3Q, commutator, ℏ, xp3DCommutator, Subscript]
```

9.
```
f[{x_, y_, z___}] := {x} ~Join~ f[{z}]
f[{x_}]:= {x}
f[{}] = {};
Table[RandomInteger[{-100, 100}], {RandomInteger[20]}]
f[%]
ClearAll[f]
```

SECTION 3.7.5

5.
```
BeginPackage["HarmonicOscillator`"]

HarmonicOscillator::usage =
 "HarmonicOscillator` is a package that provides
   functions for the eigenenergies and normalized energy
   eigenfunctions of the simple harmonic oscillator."

ϕ::usage = "ϕ[n, x] gives the normalized energy
    eigenfunctions of the simple harmonic oscillator."

ℰ::usage = "ℰ[n] gives the eigenenergies of
    the simple harmonic oscillator in terms of ℏω."

ω::usage = "The symbol ω stands for the oscillator frequency."

m::usage = "The symbol m stands for the mass."

ℏ::usage =
 "The symbol ℏ stands for h/2π, with h being Planck's constant."

Begin["`Private`"]

HarmonicOscillator::badarg =
 "You called `1` with `2` argument(s)!
   It must have `3` argument(s)."

ϕ::quannum =
 "The quantum number must be a nonnegative integer. Your
   quantum number `1` is not allowed."

ℰ::badarg =
 "The argument must be a nonnegative integer. You entered `1`."

ϕ[n_Integer?NonNegative, x_]:=
 ((2^(-n/2))(n!^(-1/2))(((mω)/(ℏπ))^(1/4))
   HermiteH[n, Sqrt[mω/ℏ]x]Exp[-((mω)/(2ℏ))x^2])

ϕ[n_, x_]:= Message[ϕ::quannum, {n}]
```

```
ϕ[arg___ /; Length[{arg}] ≠ 2] :=
 Message[HarmonicOscillator::badarg, ϕ, Length[{arg}], 2]

ℰ[n_Integer?NonNegative] := (ℏ ω (n + 1/2))

ℰ[arg_]:= Message[ℰ::badarg, {arg}]

ℰ[arg___/; (Length[{arg}] ≠ 1)] :=
 Message[HarmonicOscillator::badarg, ℰ, Length[{arg}], 1]

End[]

Protect[ϕ, ℰ]

EndPackage[]
```

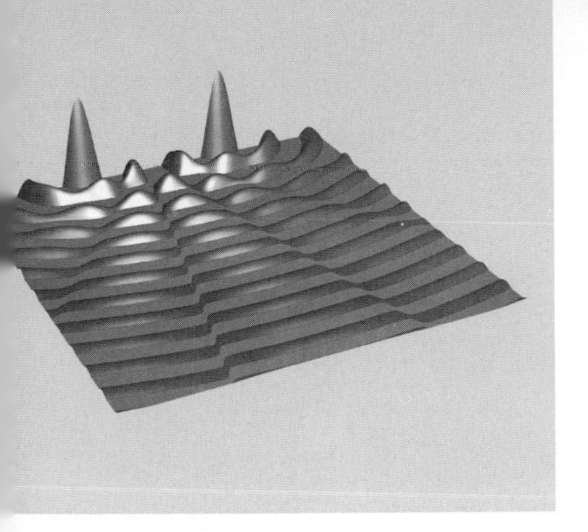

Appendix D
Solutions to Problems

This appendix provides solutions to selected problems in Chapters 1, 2, 4, 5, and 6.

SECTION 1.7

3. `Factor[45 + 63 x + 32 x^2 + 16 x^3 + 3 x^4 + x^5]`

7. The letter **J** (red) in **BesselJ[2,r,J]** is an excess argument; both it and the comma to its left should be removed. The symbol **pi** (blue) is a global symbol that has no value assigned; it should be **Pi**, which is the *Mathematica* name for the constant π. **BoxRatio** (red) is an unrecognized option name; it should be **BoxRatios**. The first left curly bracket (purple) to the right of **ViewPoint** and the last right square bracket (purple) of the input are syntax errors; they should be removed. The correct input is

```
ParametricPlot3D[
 {r Cos[theta], r Sin[theta], BesselJ[2, r]Sin[2 theta]Cos[Pi]},
 {r, 0, 5.13562}, {theta, 0, 2 Pi},
 PlotPoints → 25, BoxRatios → {1, 1, 0.4},
 ViewPoint → {2.340, -1.795,1.659}]
```

SECTION 2.6

1. $\texttt{ScientificForm}\left[\texttt{Abs}\left[\frac{1.735 \times 10^{-7}\,(\texttt{Sin}[23°])\,e^{35.7}}{\sqrt{3 + 7\,i\,(2 + 9\,i)^2}}\right], 3\right]$

3. $\sum_{k=1}^{n} k^4$

5.
```
Solve[{2x+y == 3z, 6x+24+4y == 0, 20-5x+2z == 0}, {x, y, z}]
{2x+y == 3z, 6x+24+4y == 0, 20-5x+2z == 0} /. %
```

7.
```
Solve[{5i1+7i2 == 4, 5i1-2i3 == 32, i1+i3 == i2}, {i1, i2, i3}]
ScientificForm[N[%], 3]
```

9.
```
NSolve[{x^2+2xy+y^2 == 1, x^3+x^2 y+y^2+y^3 == 4}, {x, y}]
```

11. $\int \frac{x^3}{\sqrt{a^2 - x^2}} dx$
```
∂x % // Simplify
```

13.
```
NDSolve[{i'[t]+2i[t] == Sin[t], i[0] == 0}, i, {t, 0, 1.1}]

numericali = (i /. %)[[1]]

DSolve[{i'[t]+2i[t] == Sin[t], i[0] == 0}, i[t], t]

analyticali[t_] = (i[t] /. %)[[1]]

TableForm[
 Table[{t, numericali[t], analyticali[t]}, {t, 0, 1, 0.1}] // Chop,
 TableHeadings → {None, {"t", "numerical", "analytical"}},
  TableSpacing -> {5, 5}]

Clear[numericali,analyticali]
```

15.
```
DSolve[y''[t] - 2y[t] == 3e^(-t^2), y[t], t]
```

17.
```
NDSolve[{ψ''[x] + 6Sech[x]^2 ψ[x] == 4ψ[x], ψ[0] == 1, ψ'[0] == 0},
 ψ, {x, -5, 5}]

Plot[Evaluate[ψ[x] /. %], {x, -5, 5}, AxesLabel → {"x", "ψ"}]
```

19.
```
data = Transpose[{{1.00, 2.00, 4.00, 6.00, 8.00, 10.0, 12.0},
    {3100, 2450, 1480, 910, 545, 330, 200}}];

FindFit[data, R0 e^(-λt), {R0, λ}, t]
{%[[1, 2]] "decays/min",
 (%[[2, 2]]/60) "min^-1 ", Log[2]/(%[[2, 2]]/(60 "min"))}

ClearAll[data]
```

21.
```
ClickPane[
 Plot[{(Sin[x])^2, Tanh[x/2] + 1/5}, {x, -1, 3}], (xycoord = #) &]

Dynamic[xycoord]
```

```
startvalue =
  {-0.2749188687992582`, 0.942049142327307`, 1.8547751506722305`};

Table[FindRoot[(Sin[x])^2 == Tanh[x/2] + 1/5, {x, startvalue[[i]]}],
  {i, Length[startvalue]}]

ClearAll[xycoord, startvalue]
```

23.
```
Plot3D[x^2 + y^2, {x, -2, 2}, {y, -2, 2},
  BoxRatios -> {1, 1, 1}, AxesLabel -> {"x", "y", "z"}]
Show[%, ViewPoint -> {-0.079, -3.353, 0.447}]
Show[%, ViewPoint -> {-0.000, -0.000, 3.384}]
Show[%, ViewPoint -> {-2.171, 0.515, 2.544}]
```

25.
```
Show[Plot3D[0.25 x + 0.25 y + 2.5, {x, -3, 3}, {y, -3, 3}],
  Plot3D[x^2 + y^2, {x, -2, 2}, {y, -2, 2}], BoxRatios -> {1, 1, 1},
  Boxed -> False, Axes -> False, PlotRange -> All]
```

27.
```
FourierTransform[Piecewise[{{Sin[ω0 t], Abs[t] < T}}],
  t, -ω, Assumptions -> T > 0]

Simplify[% == - i/Sqrt[2 π] (Sin[(ω - ω0) T]/(ω - ω0) - Sin[(ω + ω0) T]/(ω + ω0))]
```

29.
```
Simplify[DSolve[{D[v[t]/Sqrt[1 - v[t]^2/c^2], t] == F/m, v[0] == 0}, v[t], t],
  F > 0 && m > 0 && c > 0 && t ≥ 0]

DSolve[{v'[t] == F/m, v[0] == 0}, v[t], t]

Simplify[DSolve[{x'[t] == %%[[2, 1, 2]], x[0] == 0}, x[t], t],
  F > 0 && m > 0 && c > 0 && t ≥ 0]

DSolve[{x''[t] == F/m, x'[0] == 0, x[0] == 0}, x[t], t]
```

Let the units of x, v, and t be mc^2/F, c, and mc/F, respectively.

```
Plot[{t/Sqrt[1 + t^2], t}, {t, 0, 3},
  PlotRange -> {0, 1.2}, AxesLabel -> {"t (mc/F)", "v (c)"},
  PlotStyle -> {{Thickness[0.012], Red}, Blue}]

Plot[{Sqrt[1 + t^2] - 1, 1/2 t^2}, {t, 0, 2},
  PlotRange -> {0, 1.5}, AxesLabel -> {"t (mc/F)", "x (mc^2/F)"},
  PlotStyle -> {{Thickness[0.015], Red}, Blue}]
```

SECTION 4.5

9.
```
l[x0_, y0_, vx0_, vy0_] := Abs[x0 vy0 - y0 vx0]

e[x0_, y0_, vx0_, vy0_] := ((vx0^2 + vy0^2)/2 - 4 π^2/Sqrt[x0^2 + y0^2]) //N

τ[x0_, y0_, vx0_, vy0_] := Abs[e[x0, y0, vx0, vy0]/(2 π^2)]^(-3/2)

plotOrbit[x0_, y0_, vx0_, vy0_, tmax_, opts___Rule] :=
 Module[{orbit},
  orbit =
   NDSolve[{x''[t] + (4 π^2 x[t])/(x[t]^2 + y[t]^2)^(3/2) == 0,
     y''[t] + (4 π^2 y[t])/(x[t]^2 + y[t]^2)^(3/2) == 0, x[0] == x0, y[0] == y0,
     x'[0] == vx0, y'[0] == vy0}, {x, y}, {t, 0, tmax}];
  ParametricPlot[Evaluate[{x[t], y[t]} /. orbit], {t, 0, tmax},
   AspectRatio -> Automatic, AxesLabel → {"x(AU)", "y(AU)"}, opts]]
```

(a)
```
l[1, 0, 0, 8]
e[1, 0, 0, 8]
τ[1, 0, 0, 8]
plotOrbit[1, 0, 0, 8, 4.28825]
```

(b)
```
l[1, 0, -2.5, 8]
e[1, 0, -2.5, 8]
τ[1, 0, -2.5, 8]
plotOrbit[1, 0, -2.5, 8, 9.65493]
```

(c)
```
l[1, 0, -2 π, 2 π]
e[1, 0, -2 π, 2 π]
plotOrbit[1, 0, -2 π, 2 π, 0.35, Ticks → {Automatic, {-0.5, 0, 0.5}}]
```

(d)
```
l[0, -2, π, 3 π]
e[0, -2, π, 3 π]
plotOrbit[0, -2, π, 3 π, 0.35, PlotRange → {{-1.12, 0.75}, {-2, 1.5}},
 Ticks → {{-1, -0.5, 0.5, 1}, Automatic}]
ClearAll[l, e, τ, plotOrbit]
```

SECTION 5.4

3. The keyboard letters **x** and **y** cannot be the subscripts of the Greek letter **E** because the full form, for example, of **Ex[x, y]** is **Subscript[E,x][x,y]**, where the two **x**'s are identical

and, therefore, the same symbol **x** is used to label the x component of the electric field and also to denote the x coordinate of a point. The definition for **f**, for example, is no longer valid.

7.
```
newLaplace2D[jmax_Integer?EvenQ,
  w_Real/;1 < w < 2, tol_Real?Positive]:=
 Module[{imax = (5/2) jmax, Δ = 1/jmax, m, V, v, com},

  Print["\nw = " <> ToString[w]];

  (* initial guess *)
  V[0][i_Integer?Positive/; i < imax,
    j_Integer?Positive /; j < jmax]:=

   V[0][i, j] = N[4/π Sin[π j Δ] Exp[-π i Δ]];

   (* boundary conditions *)
   V[n_Integer?NonNegative][i_Integer?NonNegative /; i ≤ imax,
     0]:= V[n][i, 0] = 0; V[n_Integer?NonNegative][
     i_Integer?NonNegative /; i ≤ imax, jmax]:= V[n][i, jmax] = 0;
   V[n_Integer?NonNegative][0, j_Integer?Positive /; j < jmax]:=
    V[n][0, j] = 1; V[n_Integer?NonNegative][imax,
     j_Integer?Positive /; j < jmax]:= V[n][imax, j] = 0;

   (* iteration equation *)
   V[n_Integer?Positive][i_Integer?Positive/; i < imax,
     j_Integer?Positive/; j < jmax]:=
    V[n][i, j] = ((1 - w) V[n - 1][i, j] + w/4 (V[n - 1][i + 1, j] +
          V[n][i - 1, j] + V[n - 1][i, j + 1] + V[n][i, j - 1]));

   (* final iteration number *)
   For[
    m = 1,
    Max[Table[Abs[V[m][i, j] - V[m - 1][i, j]],
        {i, 1, imax - 1}, {j, 1, jmax - 1}]] ≥ tol, m++,
    If[m > 60, Print["Iteration limit of 60 exceeded."]; Break[]]];
   Print["Final Iteration Number = " <> ToString[m]];

   (* comparison *)
   v[x_, y_, kmax_]:=
    4/π Sum[1/(2 k - 1) Sin[(2 k - 1) π y] Exp[-(2 k - 1) π x], {k, 1, kmax}];
   com = Table[Evaluate[{v[i Δ, j Δ, 400] // N, V[m][i, j]}],
     {i, 0, imax, imax/10}, {j, jmax/10, jmax/2, jmax/10}];
```

```
Print[
 PaddedForm[
  TableForm[
   Prepend[
    Flatten[#, 1]&/@Transpose@{Table["x = " <>
         ToString[iΔ//N], {i, 0, imax, imax/10}], com},
    Prepend[
     Table["y = " <> ToString[jΔ//N],
      {j, jmax/10, jmax/2, jmax/10}], {}]]], {4, 3}]]
```

SECTION 6.7

11.

```
Lz[n_, m_] := mℏKroneckerDelta[n, m]
L+[l_, n_, m_] := ℏ√(l (l + 1) - nm) KroneckerDelta[n, m + 1]
L-[l_, n_, m_] := L+[l, m, n]
Lx[l_, n_, m_] := (L+[l, n, m] + L-[l, n, m])/2
Ly[l_, n_, m_] := (L+[l, n, m] - L-[l, n, m])/(2ⅈ)
Lx[l_] := Table[Lx[l, n, m], {n, l, -l, -1}, {m, l, -l, -1}]
Ly[l_] := Table[Ly[l, n, m], {n, l, -l, -1}, {m, l, -l, -1}]
Lz[l_] := Table[Lz[n, m], {n, l, -l, -1}, {m, l, -l, -1}]
H[l_] := (a Lx[l].Lx[l] + b Ly[l].Ly[l] + c Lz[l].Lz[l])

eigenvaluesH[l_] := Eigenvalues[H[l]]

NH[l_, a_, b_, c_] := (a Lx[l].Lx[l] + b Ly[l].Ly[l] + c Lz[l].Lz[l])/ℏ^2
NeigenvaluesH[l_Real, a_Real, b_Real, c_Real] :=
 Eigenvalues[NH[l, a, b, c]] ℏ^2
```

(a)

```
eigenvaluesH[6]
ToRadicals[eigenvaluesH[6]]
```

(b)

```
ℏ = 1;
Sort[Chop[N[ToRadicals[eigenvaluesH[6]] /. {a→1, b→7, c→4}]]] ==
 Sort[NeigenvaluesH[6.0, 1.0, 7.0, 4.0]]

ClearAll[Subscript, H, eigenvaluesH, NH, NeigenvaluesH, ℏ]
```

References

[Ada95] R. K. Adair. The Physics of Baseball. **Physics Today**, 48(5):26–31, 1995.

[AM76] N. W. Ashcroft and N. D. Mermin. **Solid State Physics**. Holt, Rinehart & Winston, New York, 1976.

[Bai99] R. Baierlein. **Thermal Physics**. Cambridge University Press, Cambridge, UK, 1999.

[BG96] G. L. Baker and J. P. Gollub. **Chaotic Dynamics: An Introduction**, Second Edition. Cambridge University Press, Cambridge, UK, 1996.

[BR03] P. R. Bevington and D. K. Robinson. **Data Reduction and Error Analysis**, Third Edition. McGraw-Hill, New York, 2003.

[Bra85] P. J. Brancazio. Trajectory of a Fly Ball. **The Physics Teacher**, 20–23, January 1985.

[BD95] S. Brandt and H. D. Dahmen. **The Picture Book of Quantum Mechanics**, Second Edition. Springer-Verlag, New York, 1995.

[CDSTD92a] D. Cook, R. Dubisch, G. Sowell, P. Tam, and D. Donnelly. A Comparison of Several Symbol-Manipulating Programs: Part I. **Computers in Physics**, 6(4):411–419, 1992.

[CDSTD92b] D. Cook, R. Dubisch, G. Sowell, P. Tam, and D. Donnelly. A Comparison of Several Symbol-Manipulating Programs: Part II. **Computers in Physics**, 6(5):530–540, 1992.

[Cop91] V. T. Coppola. Working Directly with Subexpressions. **The *Mathematica* Journal**, 1(4): 41–44, 1991.

[CS60] H. C. Corben and P. Stehle. **Classical Mechanics**, Second Edition. Wiley, New York, 1960.

[Cra91] R. E. Crandall. ***Mathematica* for the Sciences**. Addison-Wesley, Redwood City, CA, 1991.

[DK67] P. Dennery and A. Krzywicki. **Mathematics for Physicists**. Harper & Row, New York, 1967.

[DeV94] P. L. DeVries. **A First Course in Computational Physics**. Wiley, New York, 1994.

[Eis61] R. M. Eisberg. **Fundamentals of Modern Physics**. Wiley, New York, 1961.

[ER85] R. Eisberg and R. Resnick. **Quantum Physics of Atoms, Molecules, Solids, Nuclei, and Particles**, Second Edition. Wiley, New York, 1985.

[Fea94] J. M. Feagin. **Quantum Methods with *Mathematica***. Springer-Verlag, New York, 1994.

[FGT05] P. M. Fishbane, S. G. Gasiorowicz, and S. T. Thornton. **Physics for Scientists and Engineers**, Third Edition. Prentice Hall, Upper Saddle River, NJ, 2005.

[Gar94] A. L. Garcia. **Numerical Methods for Physics**. On Preface to the First Edition. Prentice Hall, Englewood Cliffs, NJ, 1994.

[Gas96] S. Gasiorowicz. **Quantum Physics**, Second Edition. Wiley, New York, 1996.

[Gos03] A. Goswami. **Quantum Mechanics**, Second Edition. Waveland Press, Long Grove, IL, 2003.

[GTC07] H. Gould, J. Tobochnik, and W. Christian. **An Introduction to Computer Simulation Methods: Applications to Physical Systems**, Third Edition. Addison-Wesley, San Francisco, 2007.

[Gra98] J. W. Gray. **Mastering *Mathematica*: Programming Methods and Applications**, Second Edition. Academic Press, San Diego, 1998.

[Gri99] D. J. Griffiths. **Introduction to Electrodynamics**, Third Edition. Prentice Hall, Upper Saddle River, NJ, 1999.

[Gri05] D. J. Griffiths. **Introduction to Quantum Mechanics**, Second Edition. Prentice Hall, Upper Saddle River, NJ, 2005.

[HRK02] D. Halliday, R. Resnick, and K. S. Krane. **Physics**, Fifth Edition. Wiley, New York, 2002.

[Hal58] P. R. Halmos. **Finite-Dimensional Vector Spaces**, Second Edition. Van Nostrand, Princeton, NJ, 1958.

[HT94] M. Hanson and P. Tam. **The Square Well Potential**. Unpublished, 1994.

[Has91] S. Hassani. **Foundations of Mathematical Physics**. Allyn & Bacon, Boston, 1991.

[Hil00] R. C. Hilborn. **Chaos and Nonlinear Dynamics**, Second Edition. Oxford University Press, New York, 2000.

[Jea66] J. H. Jeans. **The Mathematical Theory of Electricity and Magnetism**, Fifth Edition. Cambridge University Press, Cambridge, UK, 1966.

[KW04] C. D. Keeling and T. P. Whorf. **Atmospheric CO_2 Records from Sites in the SIO Air Sampling Network**. http://cdiac.esd.ornl.gov/trends/co2/sio-mlo.htm, 2004.

[Kit86] C. Kittel. **Introduction to Solid State Physics**, Sixth Edition. Wiley, New York, 1986.

[KM90] S. E. Koonin and D. C. Meredith. **Computational Physics**, Fortran Version. Addison-Wesley, Redwood City, CA, 1990.

[Lea04] S. M. Lea. **Mathematics for Physicists**. Brooks/Cole, Belmont, CA, 2004.

[Lib03] R. L. Liboff. **Introductory Quantum Mechanics**, Fourth Edition. Addison-Wesley, San Francisco, 2003.

[LC70] P. Lorrain and D. R. Corson. **Electromagnetic Fields and Waves**, Second Edition. Freeman, San Francisco, 1970.

[Mae91] R. E. Maeder. **Programming in *Mathematica***, Second Edition. Addison-Wesley, Redwood City, CA, 1991.

[McM94] S. M. McMurry. **Quantum Mechanics**. Addison-Wesley, Wokingham, UK, 1994.

[Mes00] A. Messiah. **Quantum Mechanics**, Translated from French by G. M. Temmer. Dover, Mineola, NY, 2000.

[Mey89] J. R. Meyer-Arendt. **Introduction to Classical and Modern Optics**, Third Edition. Prentice Hall, Englewood Cliffs, NJ, 1989.

[Moo87] F. C. Moon. **Chaotic Vibrations**. Wiley, New York, 1987.

[Mor90] M. A. Morrison. **Understanding Quantum Physics**. Prentice Hall, Englewood Cliffs, NJ, 1990.

[MF53] P. M. Morse and H. Feshbach. **Methods of Theoretical Physics**. McGraw-Hill, New York, 1953.

[Oha90] H. C. Ohanian. **Principles of Quantum Mechanics**. Prentice Hall, Englewood Cliffs, NJ, 1990.

[Par92] D. Park. **Introduction to the Quantum Theory**, Third Edition. McGraw-Hill, New York, 1992.

[Pat94] V. A. Patel. **Numerical Analysis**. Saunders, Fort Worth, TX, 1994.

[PP93] F. L. Pedrotti and L. S. Pedrotti. **Introduction to Optics**, Second Edition. Prentice Hall, Englewood Cliffs, NJ, 1993.

[PJS92] H. Peitgen, H. Jürgens, and D. Saupe. **Chaos and Fractals: New Frontiers of Science**. Springer-Verlag, New York, 1992.

[Ras90] S. N. Rasband. **Chaotic Dynamics of Nonlinear Systems**. Wiley, New York, 1990.

[Sch55] L. I. Schiff. **Quantum Mechanics**, Second Edition. McGraw-Hill, New York, 1955.

[Sch00] D. V. Schroeder. **An Introduction to Thermal Physics**. Addison-Wesley, San Francisco, 2000.

[SMM97] R. A. Serway, C. Moses, and C. A. Moyer. **Modern Physics**, Second Edition. Brooks/Cole, Belmont, CA, 1997.

[SJ04] R. A. Serway and J. W. Jewett. **Physics for Scientists and Engineers with Modern Physics**, Sixth Edition. Brooks/Cole, Belmont, CA, 2004.

[Ste03] J. Stewart. **Calculus**, Fifth Edition. Brooks/Cole, Belmont, CA, 2003.

[Sym71] K. R. Symon. **Mechanics**, Third Edition. Addison-Wesley, Reading, MA, 1971.

[Tam91] P. Tam. Physics and *Mathematica*. **Computers in Physics**, 5(3):342–348, 1991.

[Tay82] J. R. Taylor. **An Introduction to Error Analysis: The Study of Uncertainties in Physical Measurements**. University Science, Mill Valley, CA, 1982.

[Tho92] W. J. Thompson. **Computing for Scientists and Engineers**. Wiley, New York, 1992.

[Tho94] W. J. Thompson. **Angular Momentum**. Wiley, New York, 1994.

[TM04] S. T. Thornton and J. B. Marion. **Classical Dynamics of Particles and Systems**, Fifth Edition. Brooks/Cole, Belmont, CA, 2004.

[Tow92] J. S. Townsend. **A Modern Approach to Quantum Mechanics**. McGraw-Hill, New York, 1992.

[Vve93] D. D. Vvedensky. **Partial Differential Equations with *Mathematica***. Addison-Wesley, Wokingham, UK, 1993.

[WP05] S. Wagon and R. Portmann. How Quickly Does Water Cool? ***Mathematica* in Education and Research**, 10(3):1–9, 2005.

[WGK05] P. R. Wellin, R. J. Gaylord, and S. N. Kamin. **An Introduction to Programming with *Mathematica***, Third Edition. Cambridge University Press, Cambridge, UK, 2005.

[Wol91] S. Wolfram. ***Mathematica*: A System for Doing Mathematics by Computer**, Second Edition. Addison-Wesley, Redwood City, CA, 1991.

[Wol96] S. Wolfram. **The *Mathematica* Book**, Third Edition. Wolfram Media, Champaign, IL, 1996.

[Wol03] S. Wolfram. **The *Mathematica* Book**, Fifth Edition. Wolfram Media, Champaign, IL, 2003.

[Won92] S. S. M. Wong. **Computational Methods in Physics and Engineering**. Prentice Hall, Englewood Cliffs, NJ, 1992.

Index

B

Q

R